GEOCHEMICAL EXPLORATION 1972

GEOCHEMICAL EXPLORATION 1972

Proceedings of the
Fourth International Geochemical Exploration Symposium,
held in London from 17 to 20 April, 1972,
and organized by the Institution of Mining and Metallurgy
for the Association of Exploration Geochemists

Edited by M. J. Jones

THE INSTITUTION OF MINING AND METALLURGY

Library of Congress Catalog Card Number: 72-97896

ISBN 0 900488 16 6

Contents

Organizing Committee

Professor J. S. Webb (Chairman)

Dr. J. W. Aucott
Dr. S. H. U. Bowie
Dr. F. W. D. Cornwall
Professor G. R. Davis
Dr. M. J. Gallagher
Dr. P. L. Lowenstein
W. T. Meyer
Dr. O. W. Nicolls
Jane Plant
R. Rice
Dr. J. S. Tooms

Foreword

The Fourth International Geochemical Exploration Symposium, held in London from 17 to 20 April, 1972, was organized by the Institution of Mining and Metallurgy for the Association of Exploration Geochemists. The occasion was notable as being the first on which this series of symposia had been held outside North America.

The proceedings were opened by Mr. A. Grant, M.P., Under Secretary of State for Industrial Development, his speech being followed by introductory remarks by Dr. J. A. Hansuld, President of the Association of Exploration Geochemists, and the Ninth Sir Julius Wernher Memorial Lecture entitled 'Basic and applied geochemists in search of ore' by Professor K. C. Dunham (now Sir Kingsley Dunham), F.R.S., Director of the Institute of Geological Sciences.*

In soliciting contributions to the Symposium, the Organizing Committee decided to concentrate on the outstanding problems of prospecting in thick transported overburden and exploration for 'blind' ore deposits at depth. Other prime topics included new techniques and instrumentation, marine mineral exploration, and the development of geostatistical and other aids to the interpretation of geochemical data. Case histories as such were excluded, except insofar as these served to evaluate a new technique or approach to a particular exploration problem.

The normal Institution practice of distributing preprints of papers in advance of a symposium was varied on this occasion: extended synopses of 41 papers intended for presentation (together with 11 synopses of papers which could not be fitted into the Symposium programme) were published in the *Handbook*. The late withdrawal of three papers necessitated a rearrangement of the technical sessions, and 38 papers were presented and discussed as indicated below.

Session 1

Chairman, Professor J. S. Webb

Papers by G. B. Gott and J. M. Botbol; and D. M. Hausen, J. W. Ahlrichs and J. R. Odekirk

Sessions 2 and 3

Chairmen, Dr. H. E. Hawkes and Professor G. H. Friedrich

Papers by A. S. Dass, R. W. Boyle and W. M. Tupper; L. V. Tauson and V. D. Kozlov; P. H. Davenport and I. Nichol; R. J. Watling, G. R. Davis and W. T. Meyer; E. C. Smith and G. R. Webber; and G. J. S. Govett

Sessions 4 and 5

Chairmen, Dr. R. W. Boyle and Dr. A. Kvalheim

Papers by R. E. Learned and R. Boissen; M. B.

* Professor Dunham's lecture was published in *Bull. Instn Min. Metall.* no. 789, Aug. 1972, 13–8.

Mehrtens, J. S. Tooms and A. G. Troup; U. McL. Michie, M. J. Gallagher and A. Simpson; R. J. Allan, E. M. Cameron and C. C. Durham; and I. R. Jonasson and R. J. Allan

Sessions 6 and 7
Chairmen, Professor B. M. Shmakin and Professor I. Nichol

Papers by L. N. Ovchinnikov *et al.*; P. L. Lowenstein and R. J. Howarth; A. N. Eremeev *et al.*; R. Holmes and J. S. Tooms; R. G. Garrett; and W. R. Hesp and D. Rigby

Sessions 8 and 9
Chairmen, Professor F. C. Jaffé and Dr. R. L. Erickson

Papers by R. C. Obial and C. H. James; I. Nichol and J. O. Larsson;* R. J. Howarth; R. I. Dubov; B. Bølviken and R. Sinding-Larsen; and M. Dall'Aglio and F. Tonani

Sessions 10 and 11
Chairmen, Dr. W. R. Hesp and Professor M. Dall'Aglio

Papers by J. C. Robbins; W. T. Meyer and Y. C. Y. Lam Shang Leen; C. L. Sainsbury; H. A. Wollenberg; S. H. U. Bowie, P. R. Simpson and C. M. Rice; Jane Plant and R. F. Coleman; and A. E. Hubert and H. W. Lakin

Sessions 12 and 13
Chairmen, Dr. S. H. U. Bowie and H. W. Lakin

Papers by R. C. Leake and J. W. Aucott; H. Kunzendorf; B. Bølviken *et al.*; W. L. Plüger and G. H. Friedrich; S. E. Kesler and J. C. Van Loon; and G. H. Friedrich *et al.*

Grateful thanks are due to the Chairmen of the technical sessions for their skilful handling of a crowded schedule.

The content of the 37 papers in this volume testifies to a truly remarkable surge in the rate of advance—both in method and concept—as a result of increasing research and development during recent years. The base for this growth in interest and activity can be found in the world-wide acceptance now achieved by geochemistry in mineral exploration, and can confidently be expected to provide a firm foundation for future progress. A measure of the extent of this interest is given by the registered attendance at the Symposium, which totalled some 450, and included representatives from industry, government agencies and universities in 37 countries.

The outcome of any symposium ultimately depends on the concentrated efforts and enthusiasm of the many people responsible for each of the different facets that go to make up the programme as a whole. On behalf of the Institution and my colleagues on the Organizing Committee I would like to thank the authors (including those whose work could not be presented at the technical sessions), those who contributed to the discussions, and the reviewers of the papers submitted. We are also grateful to the Institution of Electrical Engineers for the excellent rooms and facilities which were put at our disposal, and for the hospitality of Her Majesty's Government at the Lancaster House reception. Special thanks are due to those who organized the field excursions, which followed the Symposium, and to MINA, the ladies' social organization of the Institution, for the ladies' programme, which was greatly appreciated. Acknowledgment of the efforts of the Institution's editorial staff—the Editor, Michael Jones, and his colleagues Mrs. M. Orde and Mrs. G. M. Lawrencepulle—to ensure the prompt publication of *Geochemical exploration 1972* is also appropriate.

In conclusion, to all those who attended the Fourth International Geochemical Exploration Symposium, we sincerely hope that you found the meeting both enjoyable and stimulating, and look forward to seeing you again at the next Symposium in 1974.

John S. Webb
Chairman, Organizing Committee
October, 1972

* Regrettably, this paper ('Comparison of multivariate techniques in the interpretation of geochemical data') was not received in its final form before the Organizing Committee's deadline for the submission of manuscripts.

Opening Speeches

Professor M. G. Fleming
President of the Institution of Mining and Metallurgy for the session 1971–72, said:

Ladies and Gentlemen, it is my great pleasure to welcome you to the Fourth International Geochemical Exploration Symposium. It is an especial pleasure to have so many overseas visitors, and for you, particularly, we hope that this will be a rewarding and thoroughly enjoyable week. Certainly, the omens are good: the programme which Professor Webb and his Organizing Committee have put together looks varied and imaginative, and we have experts from many countries to provide what, I hope, will be a lively and fruitful discussion. The weather is promising and London is at its best in spring sunshine. So, on this optimistic note, I invite Mr. Anthony Grant to officially open the Symposium.

Mr. A. Grant
M.P., Parliamentary Under Secretary of State for Industrial Development, said:

It gives me great pleasure to welcome you to London, especially since your visit coincides with an important step for industrial development in this country. Last month the Secretary of State for Trade and Industry announced the setting up of a new Industrial Development Executive. This will operate under the direction of a new Minister for Industrial Development, who was appointed just before the Executive started work a week ago. The Executive will have a growing part to play on the industrial scene in this country. One of its functions will be to administer incentives for investment, including investment in the mining industry.

It is particularly appropriate that this Symposium should be held here in London. Much of the pioneer work on the application of geochemical techniques to search for mineral deposits was carried out by Professor Webb and his colleagues at Imperial College. Furthermore, the British Isles have a long history of mining and there has been a considerable revival of interest here, particularly in metal mining, in the course of the last decade. As you probably know, at one time the United Kingdom was the world's leading supplier of copper and tin, but as our more easily accessible deposits were worked out, our domestic metal mining industry declined and, apart from a few tin mines in Cornwall, production of metal ores virtually ceased.

That situation is, however, changing. Major discoveries of gas and oil have been made offshore, and many important mining companies are taking a keen interest in our potential as a source of non-ferrous metal ores and industrial minerals, such as potash, which have not previously been mined here to any extent. Last year Wheal Jane tin mine in Cornwall went into production. This is the first major new metal mine in the United Kingdom for 50 years. A major new potash mine is also under construction in north Yorkshire.

With the aid of geochemistry, supplemented by matching geophysics, deposits can now be detected well below the surface. In north Aberdeenshire, for example, mining companies are already exploring for nickel in an area with no previous history of mining following a general soil investigation by the Macaulay Institute in Aberdeen. Successive Governments in this country have been concerned to encourage companies to use these techniques.

I should like to add a few remarks about the steps taken by the Government to assist mineral development. The Industrial Development Executive is part of the Department of Trade and Industry, which also includes the Warren Spring Laboratory with its high reputation for the quality of its work on mineral processing.

Outside Government, but supported by it, and working closely with the Industrial Development Executive, is the Institute of Geological Sciences, whose Director, Dr. Kingsley Dunham, will be addressing you this afternoon. This, too, as you know, enjoys a fine reputation and its work is known throughout the world.

One of its major tasks has been a Uranium Reconnaissance Programme, which is now entering its fifth year. As a result, a new uranium province has come to light in northern Scotland and is being further investigated by mining companies to ascertain whether grade and tonnages of ore are sufficient to sustain production.

All over the world there is growing concern about the *environment* and the effects on it of mining. In a small country such as Britain there are many competing uses for land. We have, however, established through our planning legislation a system which allows all the environmental issues affecting a particular project to be examined together. Before an exploration programme involving any interference with land is started or a new mine is constructed it is necessary for planning permission to be obtained from the appropriate authority, normally the local Council. If planning permission is refused, there is a right of appeal to the appropriate Minister. Once planning permission has been granted, a mining company can go ahead with its approved plans confident that the work will not be held up in the later stages. Moreover, I do not need to tell you that there are advantages in working in a developed country such as Britain—a readily available market, good infrastructure and political stability.

The main role of the new Industrial Development Executive, of which I am a Minister, in relation to mining is to do all it can to help the industry to explore for and to establish new mines in this country. We recognize the problems of exploration for the more valuable minerals and the exceptional and unavoidable risks that it involves. Contributions are available under the Mineral Exploration Act, which was passed earlier this year, towards approved expenditure incurred in exploring for and evaluating mineral deposits in Great Britain. For the ores of non-ferrous metals and for fluorspar, barium minerals and potash up to 35% of qualifying costs, including geochemical and geophysical surveys, may be paid. The qualifying costs relate to work on both exploration *and* the evaluation needed to enable a decision to be taken on the viability of an orebody. We have had an encouraging response to the scheme and have received some 100 applications to date from more than 20 companies, more than half of which are based overseas. Most projects are in the early stages, but the companies have already committed in excess of £2 500 000.

Here, as in most other countries, we have regional problems, and last month new Government proposals were announced to assist certain parts of the country where there is a high level of unemployment. Substantial grants and other assistance will be available for investment in these areas and will be available towards capital expenditure on mining works. This will apply to all minerals, even though they are outside the scheme I have just mentioned.

Other measures recently taken to encourage investment in the assisted areas allow foreign-owned United Kingdom companies to borrow sterling in this country without limit for new direct investment in the assisted areas.

I have mentioned only a few points about the arrangements for mineral exploration and development in this country. But, when the Symposium is over, I should like you to take back with you the clear message that the United Kingdom has many advantages for companies, including overseas companies, who are considering possibilities for new exploration programmes.

Finally, I should like to pay tribute to the work of the Institution of Mining and Metallurgy generally and to the interest it has shown in mineral development in this country. The Institution has a proud record in its studies of all aspects of mineral development and has made a valuable contribution to the application of geochemistry, in particular, through its meetings and publications. It is very appropriate that the Institution should be your hosts at the Association's first symposium outside North America.

Mr. Chairman, Ladies and Gentlemen, I hope that you will find the Symposium and its associated activities enjoyable and stimulating and that the meeting will make a useful contribution in extending knowledge of geochemistry.

Professor Fleming

Thank you Mr. Grant; your speech is evidence of the lively interest in geological exploration and mining which is now being taken by the Government. It is welcomed by this gathering, and the attitude which it expresses is also much appreciated by the Institution of Mining and Metallurgy.

It is now my pleasure to call on Dr. J. A. Hansuld.

Dr. J. A. Hansuld

President of the Association of Exploration Geochemists for the session 1971–72, said:

Mr. Chairman, Ladies and Gentlemen, on behalf of the Association of Exploration Geochemists I wish to extend a warm welcome to this the Fourth International Geochemical Exploration Symposium. Although known as the *Fourth* Symposium, it is, in many ways, also the first. It is the first to schedule social events, the first to include a ladies' programme and the first to plan field trips. In my view, however, the two most significant 'firsts' are that it is the first symposium to be held under the auspices of the Association of Exploration Geochemists and the first to be held outside North America. The second of these is a function and out-growth of the first, as the holdings of these symposia and founding of the Association are very closely related. Since I have been directly or indirectly involved with all four symposia, it is to this relationship and the evolution of the symposia and Association that I wish to address my remarks.

The first three symposia were organized in a rather informal way by small groups of dedicated exploration geochemists in Canada and the United States. The first was held in Ottawa in 1966, under the sponsorship of the Canadian National Advisory Committee on Research in the Geological Sciences. With little publicity, what started out to be a small gathering to examine the status of geochemical prospecting in Canada grew into a Symposium attended by more than 250 participants from six countries. It marked a real milestone in bringing together for the first time under the name of exploration geochemistry members of industry, government and university from widely separated areas. The eagerness and enthusiasm with which the meeting was received and attended clearly demonstrated the growing interest and need to hold such a meeting. The three days of technical sessions and informal discussions provided many opportunities for establishing the background and set the pattern for future symposia.

The second Symposium was held in Colorado in 1968, under the co-sponsorship of the Colorado School of Mines and the United States Geological Survey. It was declared international in scope and attracted nearly 500 participants from 16 countries. In view of the rapidly expanding interest in geochemical exploration, the Symposium presented a timely opportunity to obtain an expression of interest on matters of concern to the profession. A questionnaire was circulated during the meeting to canvass registrants' views on two matters: *(a)* the scheduling of future symposia and *(b)* the formation of a professional organization. The registrants unanimously approved the holding of meetings at two-year intervals and many indicated that some sort of professional organization devoted to the field of exploration geochemistry was needed. Accordingly, two important steps were taken: the holding of the next symposium was agreed upon, and, at the suggestion of Allan Coope, a group of geochemists met to discuss the establishment of a professional organization. As a result of this meeting, a volunteer steering committee, headed by Dr. Coope, was formed to draft a constitution.

The third Symposium, held in Toronto in 1970, was sponsored by the Canadian Institute of Mining and Metallurgy and the Society of Economic Geologists. It attracted more than 700 delegates from 26 countries. In the opening session Dr. Coope reviewed the need for founding an organization to advance the recognition of exploration geochemistry as a profession and provide a medium for future communication among interested parties. I quote: 'The overwhelming response to the 1970 International Geochemical Exploration Symposium is an acknowledgment of the importance of geochemistry in mineral exploration today and also an expression of confidence in the future usefulness of geochemical techniques in the search for ore'.*

During the Symposium a special meeting was held and the Association of Exploration Geochemists was officially founded with the adoption of a constitution and the election of officers and a council.

Many of the founding members were also those who had organized the three previous symposia, and it was agreed that the Association would assume the responsibility of scheduling future

* Coope, J. A. Opening address. In *Geochemical exploration* (Montreal: CIM, 1971), 5. *(CIM Spec. vol. 11)*

meetings. Also, with the founding of the Association, the administrative vehicle was established to consider sponsoring meetings outside North America. One of the first orders of business of the newly formed Association was to choose a time and location for the next Symposium. It was only fitting that the Council selected the proposal of Professor Webb and the IMM to host the 1972 Symposium. Imperial College, through the work of Professor Webb, his colleagues and students, has been recognized throughout the world as a centre of applied geochemical research and education for nearly a quarter of a century. The Association is very fortunate and honoured to be able to pay tribute to Professor Webb's accomplishments by holding its first official symposium in London.

The Association, immediately upon its founding, embarked on a very ambitious programme, mainly through the work of such committees as Analyses, Bibliography, Case History, Computer Applications, Constitution, Publication, Research and Education, and Admissions. These committees, comprising members from many parts of the world, already have made a number of solid accomplishments, many of which are firsts in the field of exploration geochemistry. In addition to sponsoring this Symposium, the Association has published its first special volume in the form of an up-dated bibliography.* An agreement has been signed with Elsevier Publishing Company for the printing of the Association's official publication, *The Journal of Geochemical Exploration*, the first scientific publication in the western world devoted exclusively to the field of exploration geochemistry. The journal will be quarterly: the first issue is scheduled for distribution in June of this year and it will mark a historical milestone for the Association and the profession. Another major achievement has been the preparation of much needed geochemical bulk standards. These standards will provide the basis for tying in and comparing analytical methods and data on a worldwide basis, the importance and implication of which with respect to establishing background values and its application to ecology and pollution are limitless. Surveys have been conducted on computer applications and research and education in an effort to improve techniques and services to the industry. During the past year quarterly *Newsletters* were instituted to keep the membership informed of the various committees' progress and other Association activities.

Before closing, I would like to make a few remarks about the administration of the Association. The Association was founded in North America and, because the bulk of the membership resides in Canada and the United States, the officers and councillors have been North American residents. As a new organization this localization of the administration has been, in my view, a real asset in contributing to the functioning and progress of the Association. As the membership is expanding, however, especially outside North America, and the Association is becoming more firmly established, the geographical coverage of the administration is and should be broadened. About 25% of our nearly 400 members reside outside North America. Regional Correspondents have been appointed to represent groups of members in the United Kingdom and the Republic of Ireland, Europe and Australia; and the first Regional Councillor was recently elected by our Australian members. It is important to the growth and success of the Association that as many members as possible, representing all regions and interests, contribute to and participate in the activities and administration of the Association.

In closing, I wish to take this opportunity to thank Professor Webb, his Organizing Committee, the Institution of Mining and Metallurgy, and the Department of Trade and Industry of Her Majesty's Government for hosting this Symposium. The meeting provides an opportunity for exploration geochemists from many parts of the world to get together, compare notes on matters of mutual interest and contribute to the affairs of the Association. I particularly wish to express my appreciation to all delegates, especially those who have travelled from afar, for their interest and support. As another *first* in the international development of exploration geochemistry, I am sure we can all look forward to a very enjoyable, informative and rewarding symposium.

* *Exploration geochemistry*, compiled by H. E. Hawkes (Toronto: Association of Exploration Geochemists, 1972), 118 p.

550.42:546:553.412.4(796.311)

Zoning of major and minor metals in the Coeur d'Alene mining district, Idaho, U.S.A.

Garland B. Gott, B.A.

Joseph M. Botbol, PH.D.

Both of the U.S. Geological Survey, Denver, Colorado, U.S.A.

Synopsis

Lead–zinc–silver replacement veins in Precambrian rocks of the Belt Supergroup are the principal ore deposits in the Coeur d'Alene district, the most important silver producer in the United States. The main ore minerals are galena, sphalerite and tetrahedrite. The host rocks are mainly quartzites and argillites. Cretaceous quartz-monzonites locally intruded the Belt rocks.

During the recent geochemical study of the district by the U.S. Geological Survey approximately 8000 soil samples and 4000 rock samples were collected along roads and ridges with a sample spacing of about one sample per 500 ft of traverse line, the area sampled covering about 200 square miles. A review of computer-generated geochemical maps shows that the surface dispersion patterns of selected metals in soils are areally equal to or greater than the dispersion patterns of these metals in rocks.

The metal distribution found in the district shows the effects of different mobilities of the metals during primary mineralization. Many of the metals are zoned both vertically and laterally with respect to local centres of mineralization as well as to major geologic features of the entire district. Patterns of trace silver clearly define the various mineral belts in which the silver-bearing deposits are shallow or of only moderate depth. The presence of tetrahedrite is indicated by anomalous silver and antimony in the surficial materials, whereas argentiferous galena is indicated by anomalous silver and lead.

In contrast, tellurium seems to have been driven beyond the ore metals at the time the ore deposits were emplaced, and it forms haloes far out from the centres of mineralization. It would, therefore, seem to be a useful pathfinder element for deeply buried deposits. Manganese, like tellurium, forms haloes far beyond the ore metals.

Cadmium is more temperature-sensitive than zinc, and the two metals tend to be fractionated from each other around monzonite intrusives. Sphalerite in or near the intrusives is relatively depleted in cadmium. Because of the high temperature of the intrusion, the cadmium has moved down the temperature gradient (away from the centre of the intrusion) and formed a cadmium-rich halo around the stocks. The data indicate that cadmium, relative to zinc is depleted in vein material and enriched in the wallrock of the ore deposits.

One of the major faults within the district is a post-mineralization right-lateral fault that has shifted the northern part of the district 16 miles eastward relative to the southern part. On the basis of district geology and geochemical fabric this part of the district can be restored to approximately its original position. Several potentially large exploration targets are then apparent. Additional targets are suggested by the distribution patterns of silver, antimony, tellurium and manganese.

Computer-generated perspectives indicate that a trend of high silver and copper extends southeastward from the established Wallace–Kellogg part of the district. The geochemical data pertaining to this trend have not been fully evaluated, but the perspectives suggest that the trend is of geochemical, and possibly of economic, significance.

This paper presents some preliminary geochemical data pertaining to the zoning relations of several metals in the Coeur d'Alene district of northern Idaho, U.S.A. Lead–zinc–silver replacement veins in strongly folded quartzites and argillites of the Precambrian Belt Supergroup are the principal ore deposits. In the Coeur d'Alene district the exposed Belt Supergroup is composed of fine-grained conformable clastic rocks 25 000–30 000 ft thick. An unknown thickness of the lower part of the supergroup is not exposed. Locally, these rocks have been intruded by Cretaceous quartz-monzonites (Fig. 1). Galena, sphalerite and tetrahedrite are the most important ore minerals; small amounts of chalcopyrite are present in most of the ore deposits. Siderite is an abundant gangue mineral, and pyrite is ubiquitous. Magnetite, pyrrhotite and arsenopyrite are gangue minerals in low concentrations. The district is the most important silver producer in the United States.

be collected throughout the district with a sample density sufficient to give adequate control. Orientation studies indicated that soil samples collected from a depth greater than 6 in were generally free from contamination; therefore, all soil samples were collected at or below this depth. Rock samples also were collected from outcrop and float in order to evaluate the validity of the soil sampling.

Samples were collected at 0·1-mile intervals on jeep traverses and generally at 300- to 500-ft intervals on walking traverses (Fig. 2). To guide the sampling programme most of the samples were analysed for several elements in mobile laboratories in the field. The remainder of the analytical work was done in the Denver laboratories of the Geological Survey. Silver, tellurium, cadmium, zinc and copper were determined by wet-chemical methods by use of atomic absorption spectrophotometry. Antimony was determined by another wet-chemical method.[6] All other elements were determined by a semiquantitative spectrographic method.[4, 6]

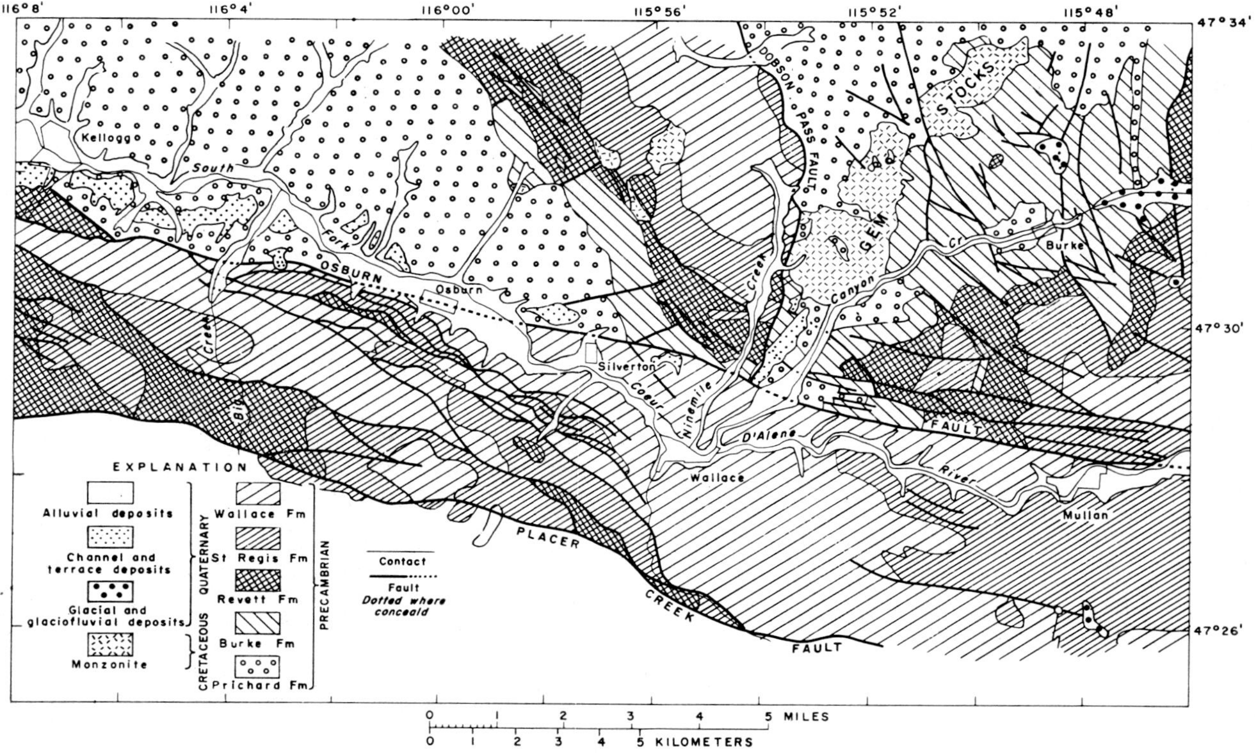

Fig. 1 Generalized geologic map of Coeur d'Alene district, Shoshone County, Idaho. Modified from Hobbs et al.[5] (plates 1–5)

The large mass of geochemical data was computerized for ease of manipulation into forms from which it could be evaluated. Many of the illustrations in this paper, therefore, are either completely or largely computer-generated.

Geochemical investigations

Materials sampled and analytical procedures

About 8000 soil samples and 4000 rock samples were collected within the district. The scarcity of outcrops made it obvious at the outset that soils were the only sample medium which could

Graphics

Two computer-implemented graphics techniques have been used to illustrate the geochemical relationships in the Coeur d'Alene district. Prior

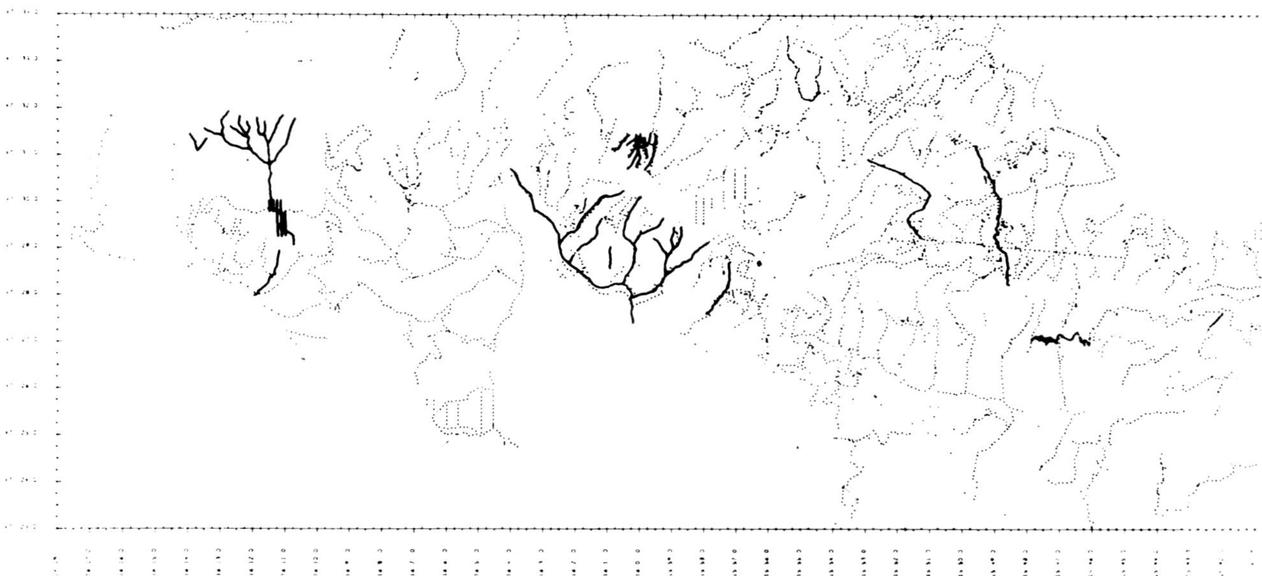

Fig. 2 Soil sample locality map. Dots indicate soil sample localities; lines show traverses along which samples were collected at 100-ft intervals

to graphical implementation, the original analytical data were expanded to include ratios and reciprocals of the ratios of the more common sulphide elements. All data were then gridded to a rectangular coordinate system with mesh points 1000 ground feet apart.

The original data points are transposed to grid coordinates or mesh points by drawing a circle of arbitrary size around each mesh point, all data points within each circle then being mathematically shifted to the coordinates of the mesh point. Next, each point is weighted according to its distance from the mesh point; as a result, close-lying data points have more influence than outlying data points on the final value to be used at the mesh point. After data points have been weighted and projected to a mesh point, multiplicity of values is created at the mesh point. Multiplicity is then removed by averaging of ratio variables and selection of the highest value of elemental variables. The extent of interpolation between values is determined by the size of the circle drawn around each mesh point. The larger the circumscribed circle, the larger is the number of data points represented by a single value. Thus, larger circle sizes have a smoothing effect on the resulting geochemical surface. For most of the illustrations given here, a circle of radius 1000 ground feet was used.

The computed geochemical values are plotted at coordinate intersections on an x–y flat-bed plotter equipped with a 50- by 60-in table. All contouring of iso-concentration was done manually to permit interpretation of stratigraphic and structural discontinuities, and, in areas of low sample density, extrapolation beyond the bounds set in the gridding program.

An interactive cathode ray tube (CRT) was used to display the many geochemical variables computed for the Coeur d'Alene data set. The method of representation on the CRT was by perspective diagram. In order to visualize a perspective diagram, imagine an elastic net draped over a solid, but invisible, model of the geochemical surface. The irregular geochemical surface is reflected in the lines of the net, and it is these lines that are displayed on the television screen of the interactive terminal. By simple command from the keyboard of the CRT, any of the ordered gridded arrays that are stored in the computer can be displayed on the CRT within a few seconds. A large number of line segments were required to correctly represent the geochemical surfaces of the Coeur d'Alene data set, and this number of lines resulted in overflows of the storage buffers in the CRT. The only practical solution to this problem was to reduce the size of the coordinate system. Consequently, the original 1000-ft grid was reduced to a grid where the mesh points are 4000 ft apart. Although some loss of detail resulted, the salient geochemical features were preserved, and the perspectives served a very useful purpose—the rapid review of a multitude of variables.

The interactive terminal permits viewing any geochemical model from any point in space. In addition, one can manipulate and, in most cases, enhance the model by exaggeration of the vertical scale. After a suitable model has been developed on the interactive terminal, a single command directs the computer to produce a 35-mm microfilm of the image on the CRT.

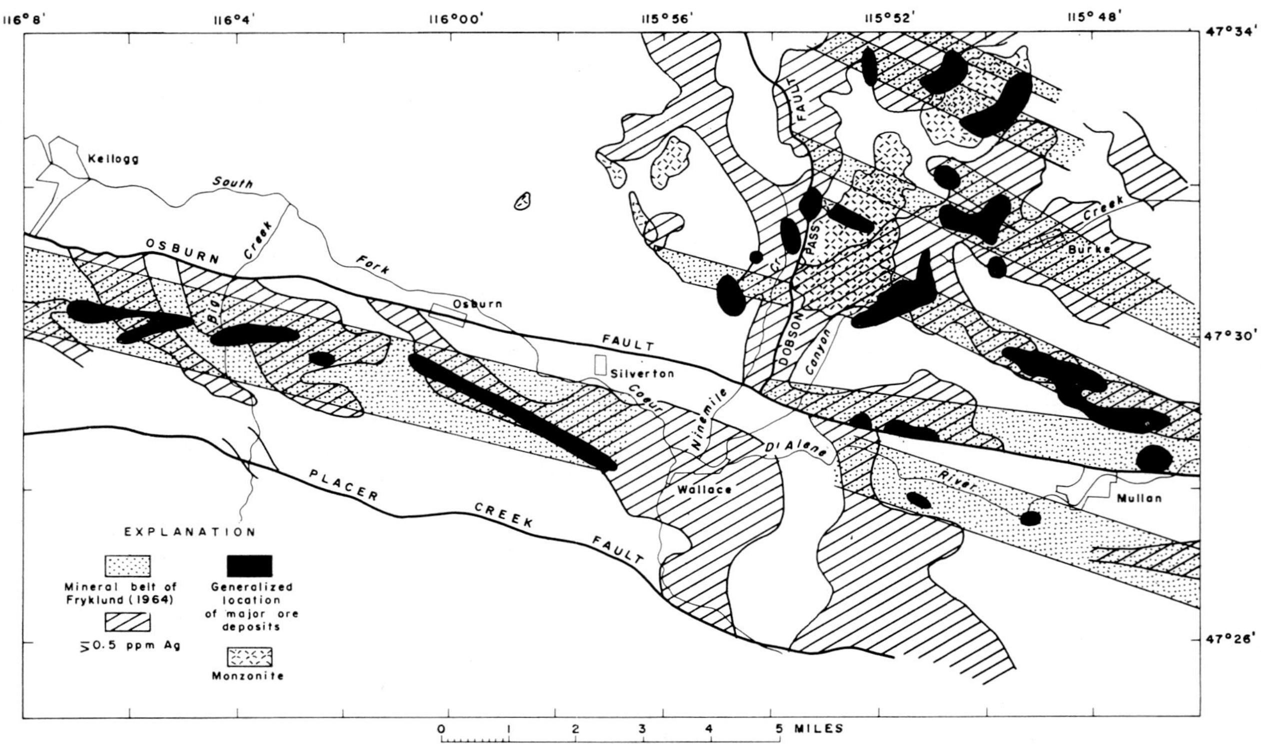

Fig. 3 Mineral belts, distribution of silver in rocks and location of major orebodies, Coeur d'Alene district

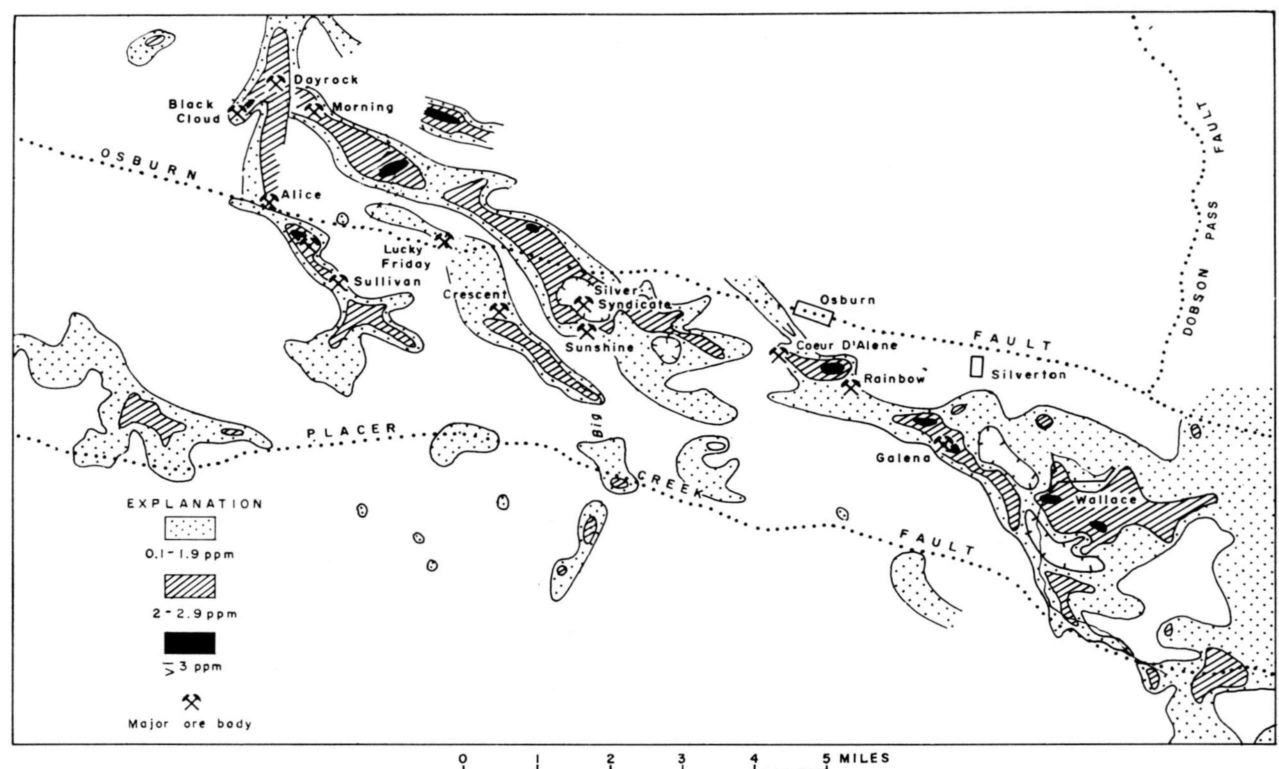

Fig. 4 Map showing restored location of major orebodies and silver dispersion patterns as they were prior to movement along Osburn, Dobson Pass and Placer Creek faults

Mineral belts

The ore deposits can be grouped into distinct mineral belts. Fryklund[2] (p. 6 and plate 2) showed the location of twelve mineral belts and sub-belts within the district, and Crosby[1] (p. 171, Fig. 2) showed the location of the mineral belts that are north of the Osburn fault. The remarkably good correspondence between these mineral belts, the distribution of silver in the rocks and the major ore deposits is shown in Fig. 3. Other metals also define parts of these belts.

Evidence presented by Hobbs *et al.*[5] indicated that right-lateral movement along the Osburn fault has drastically altered the geometry of the district. Their evidence indicates that the northern part of the district has been shifted a total of 16 miles eastward relative to the southern part. Initially, right-lateral movement amounted to about 13 miles; then movement began along the Dobson Pass fault, and the lower plate of this block was faulted an additional 3 miles eastward.

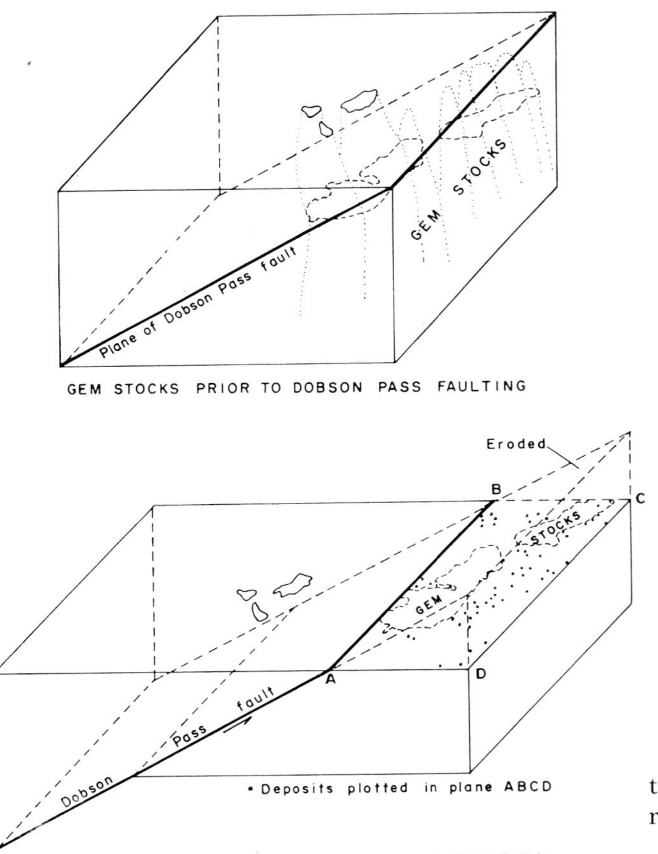

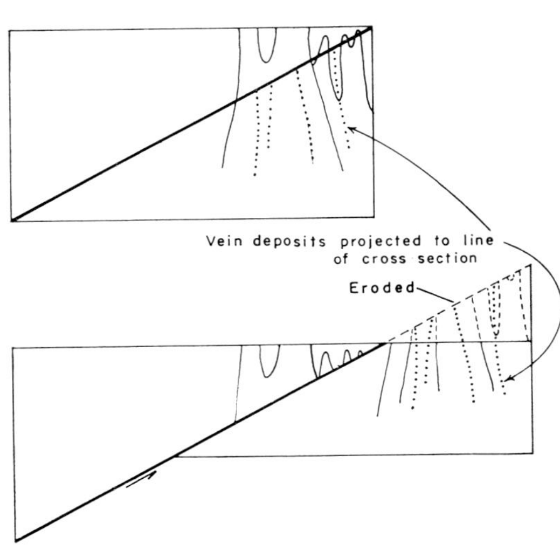

Fig. 5 Gem stocks before and after Dobson Pass faulting

An attempt has been made to show the district as it might have been before movement on the Osburn and Dobson Pass faults (Fig. 4). The restoration, made on the basis of the best fit for both the geology and geochemistry, indicates that, before right-lateral movement occurred, several of the mineral belts were continuous across the present trace of the Osburn fault. According to this restoration, the district was originally much more compact than it is today; ignoring the western and southwestern parts, the district was perhaps of the order of 12 miles long and 3 miles wide.

Movement along the Dobson Pass fault resulted in the eastward offset of the Gem stocks, leaving their isolated tops as illustrated in Fig. 5. As a result of this movement many vein deposits were exposed at the surface east of the fault and have been the locus of extensive mining activity in the past. An interval of the Belt rocks containing the middle part of the veins has been eroded. If this interpretation is correct, similar vein deposits may exist in the upper plate west of the Dobson Pass fault and above the original position of the Gem stocks. As is shown below, this interpretation is supported by extensive haloes of tellurium and by soils enriched in cadmium relative to zinc at the present surface.

Zoning of the metals

The preliminary data show that many of the metals are zoned both laterally and vertically with respect to local centres of mineralization and to major geologic features of the entire district. This zoning is best indicated by the dispersion patterns of silver, antimony, tellurium and manganese and the ratio Zn/Cd. Lead, zinc, copper and arsenic are constituents of the ore minerals, but, by themselves, they do not define the ore deposits and the mineral belts so well as do the ratios of these elements to silver and copper. Mercury, in small concentrations, is also present, apparently in the crystal lattice of some of the sulphides, but because of its erratic distribution in the oxidized zone it is not particularly useful as a pathfinder element. In general, the shallow and intermediate-depth ore deposits are best indicated by dispersion haloes of silver and/or antimony, whereas the deep deposits are best indicated by dispersion haloes of tellurium, manganese and enriched cadmium relative to zinc.

silver is most useful as an indicator of deposits of shallow or intermediate depth.

Antimony, when associated with silver, shows promise of being a useful indicator of tetrahedrite deposits. Throughout the Wallace–Kellogg mineral belt, whose ore deposits are shallow to intermediate in depth, the surface dispersion pattern of antimony seems to reflect all the major deposits where tetrahedrite is the principal ore mineral (Fig. 7). High antimony content near Burke in the northeastern part of the district may indicate the presence of stibnite, which has been reported to have been mined in Stanley mine (Fryklund,[2] p. 15). Distribution of antimony does not appear to correlate with the mineral belts north of the Osburn fault.

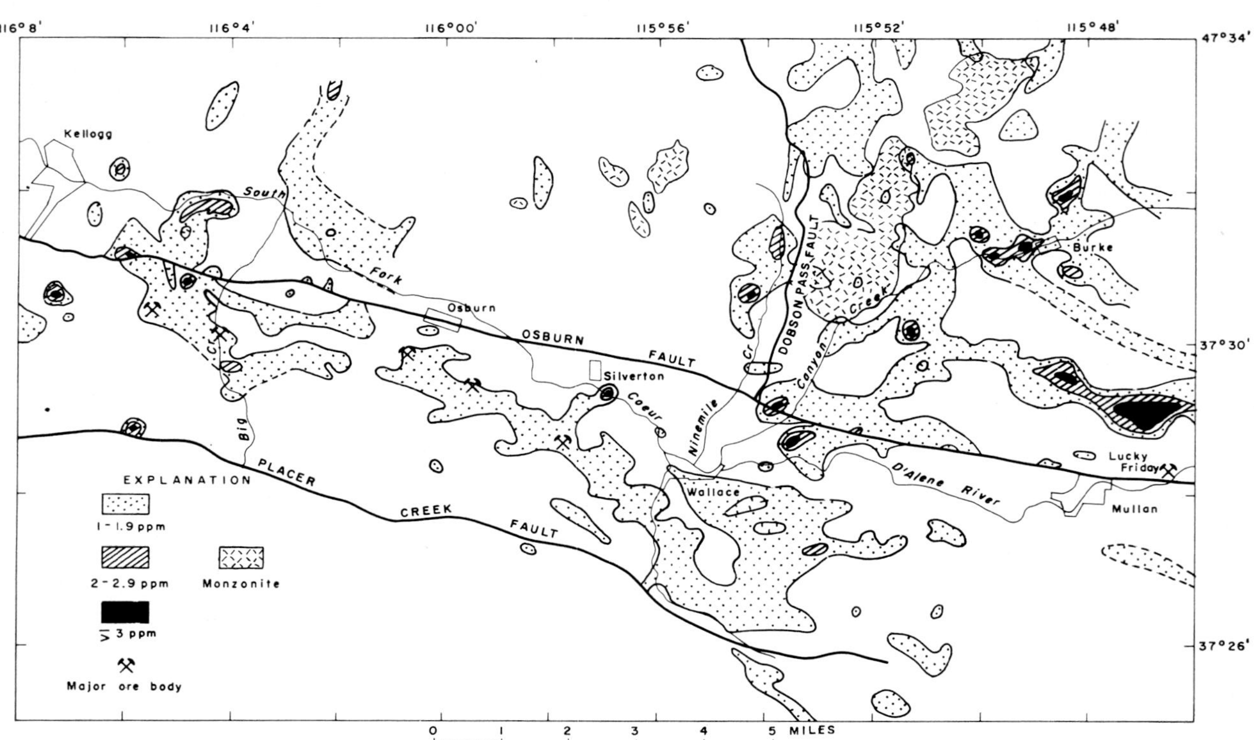

Fig. 6 Distribution of silver in soils of Coeur d'Alene district

Silver (Fig. 6) particularly well defines the mineral belts in the northern part of the district where the ore deposits are relatively shallow or were exposed at the time of their discovery. The dispersion pattern for silver in the area between Wallace and Kellogg, south of the Osburn fault, where the ore deposits are somewhat deeper, is less striking, but, on the whole, still definitive. Silver apparently has no usefulness as an indicator metal in the area around the Lucky Friday mine, about one mile east of Mullan, where the main part of the ore deposit is more than 2000 ft deep. These relations seemingly indicate that

The lack of correlation of silver and antimony distribution over the mineral belts in the northern part of the district suggests that tetrahedrite is relatively scarce, and that the silver probably occurs with the galena in that area. Perhaps tetrahedrite deposits can be distinguished from argentiferous galena deposits by determination of the dispersion patterns of antimony and silver.

Tellurium in the soils does not define the shallow mineral belts in the northern part of the district, but it does correlate with some of the intermediate-depth deposits in the Wallace–Kellogg area (Fig. 8). A tellurium halo overlies the Lucky Friday

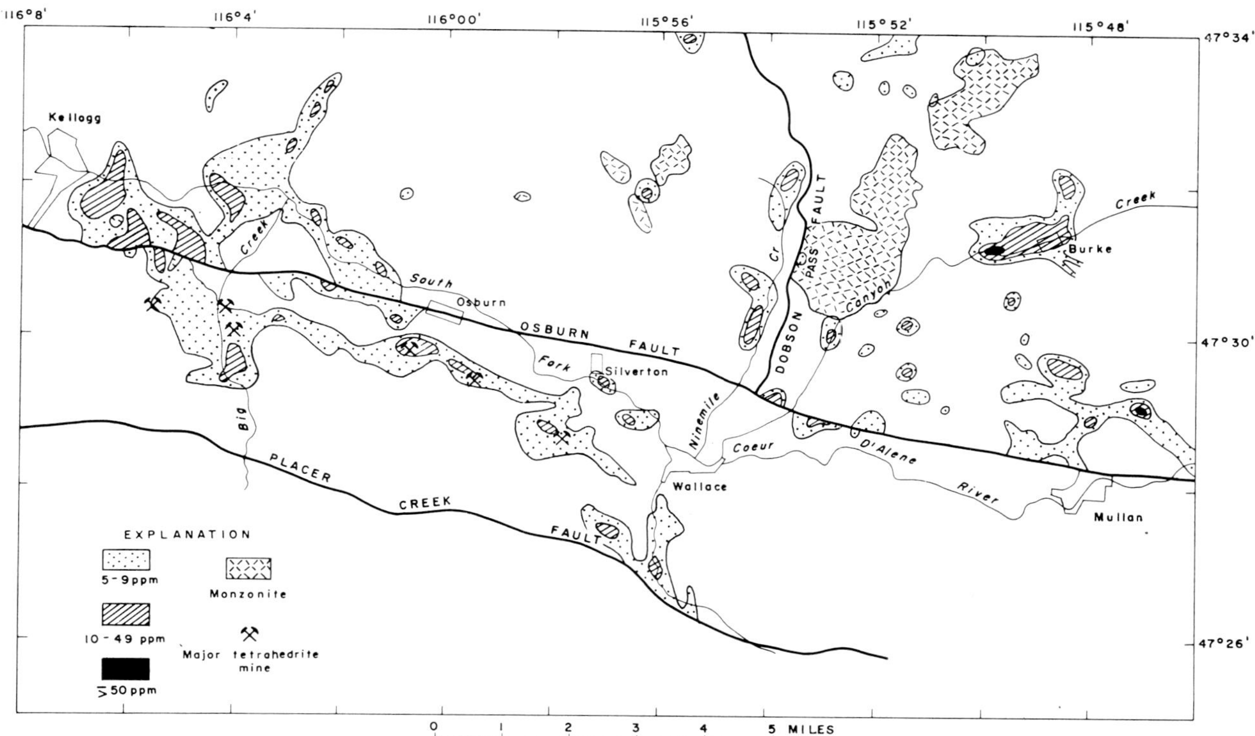

Fig. 7 Distribution of antimony in soils of Coeur d'Alene district

deposit 1 mile east of Mullan, where the main orebody is more than 2000 ft below the surface. This implies that, during ore deposition, tellurium was driven far from the centre of mineralization. Erosion removed the tellurium halo over the shallow deposits, but has not cut deeply enough to remove the halo over the deep deposits.

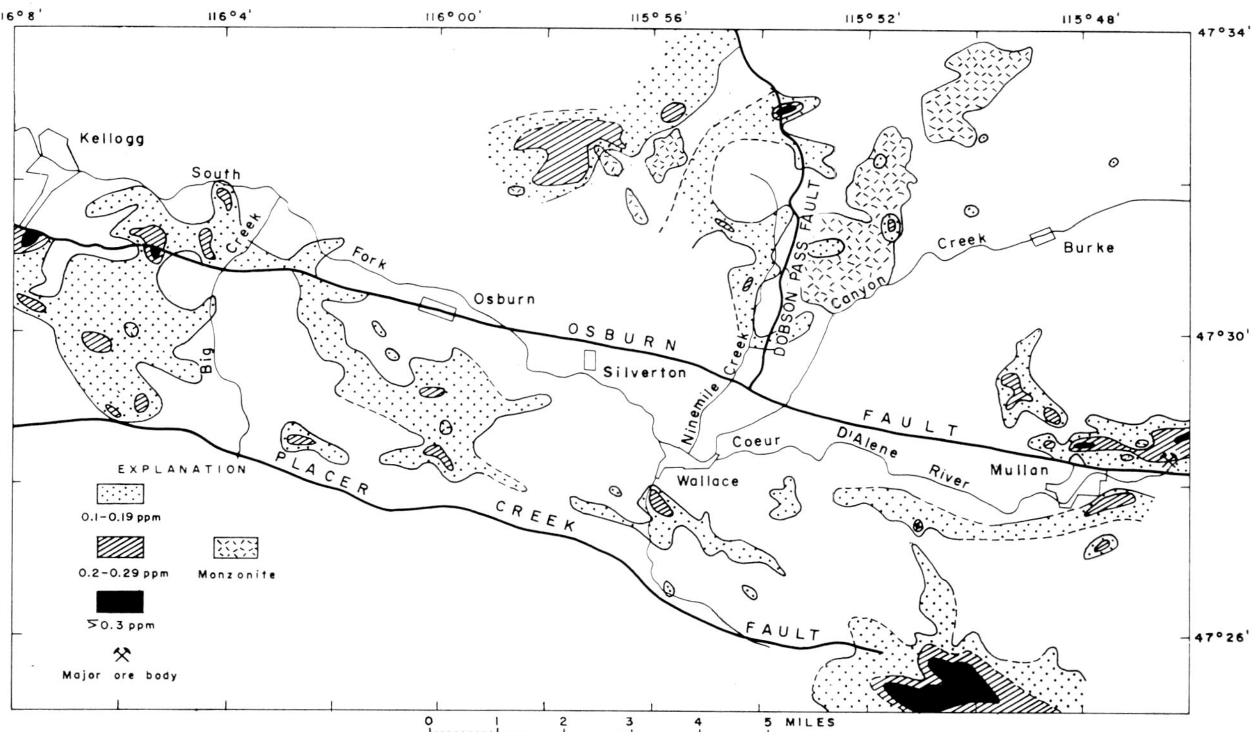

Fig. 8 Distribution of tellurium in soils of Coeur d'Alene district

There are several productive low-silver lead–zinc deposits peripheral to the Gem stocks. Tellurium does not occur near these deposits, probably because it was driven upward and outward beyond the sulphide ores, and was subsequently removed by erosion. As described before, at the time that the Gem stocks were emplaced they were about 3 miles farther west than they now are, the small isolated patches of monzonite now present west of the Dobson Pass fault being their decapitated tops in the hanging-wall. The tellurium associated with the isolated monzonite patches is, therefore, a tellurium halo directly above the original position of the Gem stocks. The Dobson Pass fault, along which the stocks were displaced, dips 29° to the west. If that angle of dip persists at depth, the fault plane would be 5000–7000 ft deep in the vicinity of the monzonite outliers. Thus, the vertical migration of tellurium beyond the lead, zinc and silver ore deposits may have been of the order of thousands of feet.

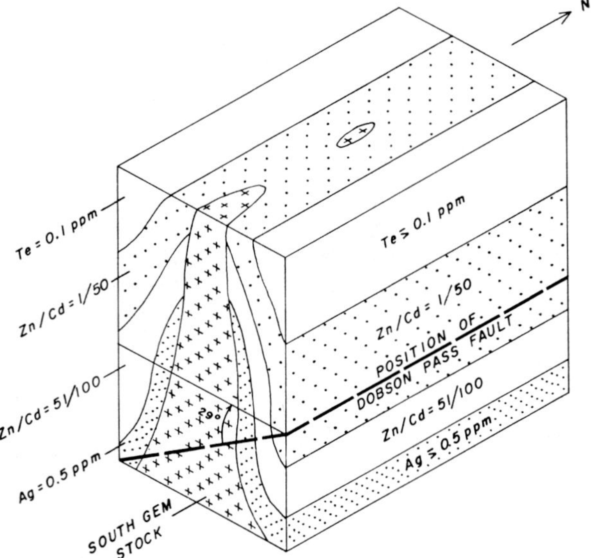

Fig. 9 Zoning of silver, enriched cadmium relative to zinc, and tellurium around South Gem stock, Coeur d'Alene district

The inferred zoning of silver, cadmium relative to zinc, and tellurium (Fig. 9) is based on geochemical data gathered from the area of the present position of the Gem stocks and the area surrounding the outlying monzonite patches west of the Gem stocks. These data were used to restore the zoning pattern as it is thought to have existed prior to movement on the Dobson Pass fault.

Manganese also seems to be an indicator of the more deeply buried deposits, suggesting that, like tellurium, it too was driven far from the centre of mineralization. Many of the relatively shallow deposits in the northeastern part of the district occur in areas where the soils are not enriched in manganese. The manganese concentration increases over intermediate-depth deposits between

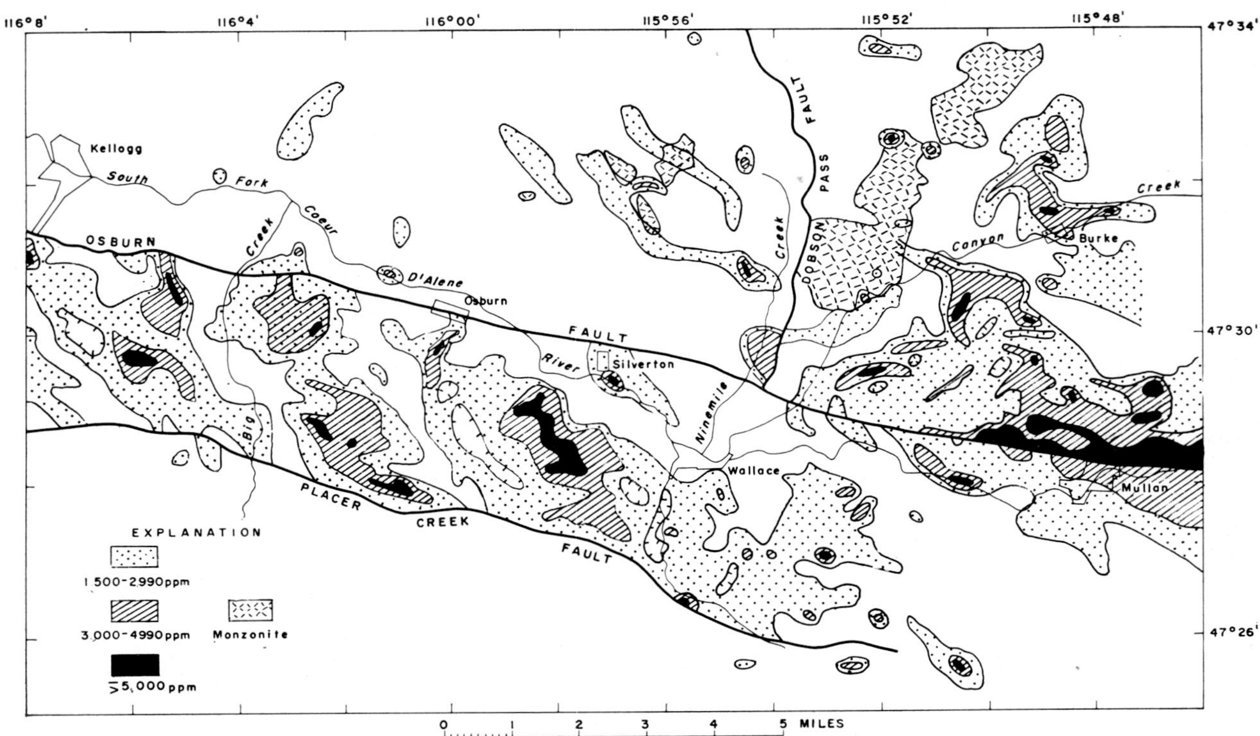

Fig. 10 Distribution of manganese in soils of Coeur d'Alene district

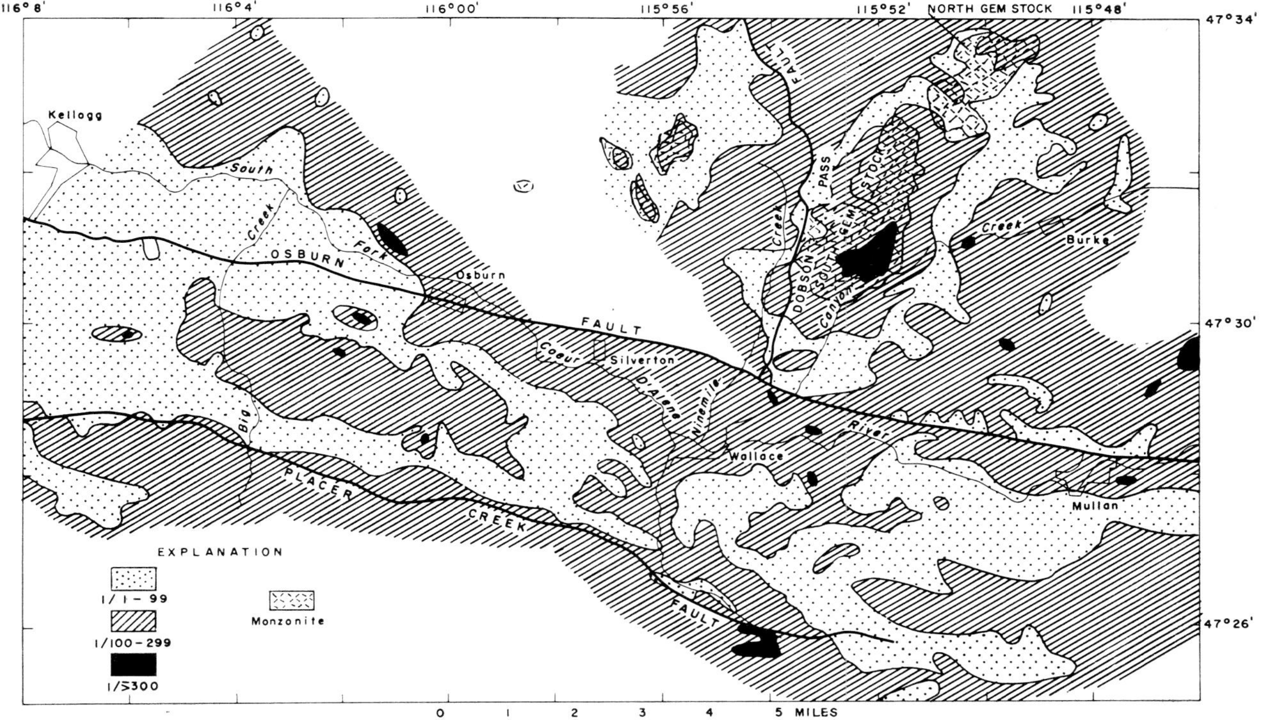

Fig. 11 Ratio Zn/Cd in soils of Coeur d'Alene district

Wallace and Kellogg and further increases over the deeper Lucky Friday mine east of Mullan (Fig. 10).

West of Wallace, in the area between the Osburn and Placer Creek faults, the manganese forms systematic ladder-shaped dispersion patterns that conform, in part, to the dispersion patterns of silver in rocks (Figs. 3 and 10). These dispersion patterns are oblique to the faults shown on the geologic map (Fig. 1) and oblique to the trend of the ore deposits in the Wallace–Kellogg area. To our knowledge there has been no fault or fracture system mapped that conforms to these patterns, and their origin is unexplained. It can be speculated, however, that the manganese and, in the northern end of the 'ladder rungs', the silver were originally emplaced when an early fracture system was open. Later, new lines of tension were exerted between the Osburn and Placer Creek faults, resulting in new structures that apparently exerted some control on the emplacement of later ore deposits throughout the Wallace–Kellogg area. In this regard it should be pointed out that the known ore deposits within the Wallace–Kellogg area (Figs. 3 and 10) are restricted to the manganese highs and are absent from the manganese gaps.

The ratio Zn/Cd in soils ranges from 15 to 1150 within the Coeur d'Alene district. Cadmium, a strongly chalcophile element, follows zinc in sulphide deposits much more closely than it does

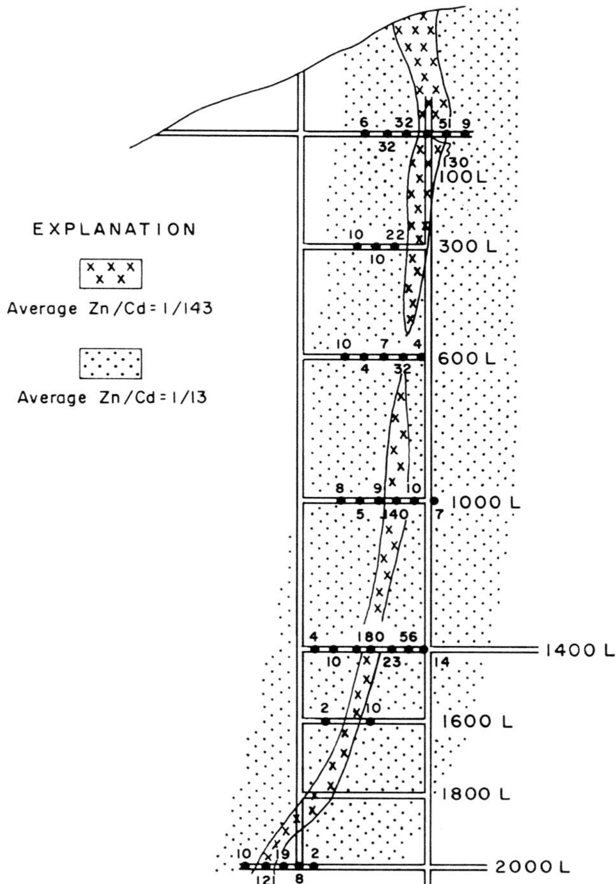

Fig. 12 Ratio Zn/Cd in Lucky Friday mine

lead and copper, and it is contained chiefly in sphalerite. According to Goldschmidt[3] (p. 271), sphalerite which formed at high temperatures will accept less cadmium in its crystal lattice than sphalerite formed at lower temperatures. The cadmium-rich halo around the Gem stock (Fig. 11) probably resulted from the elevated temperature of the intrusive, which caused the cadmium

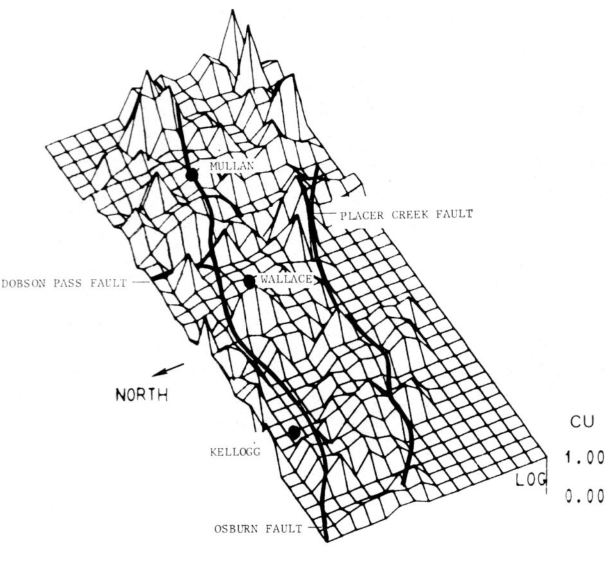

Fig. 13 Distribution of copper

to fractionate from the zinc. The halo completely encircles the south stock and passes through the north stock—a seeming indication that the north stock already existed when the south stock was emplaced. This interpretation is supported by age determinations of 94 m.y. for the south stock and 116 m.y. for the north stock (Hobbs *et al.*,[5] p. 51).

As was discussed earlier, field evidence indicates that the Gem stocks have been faulted

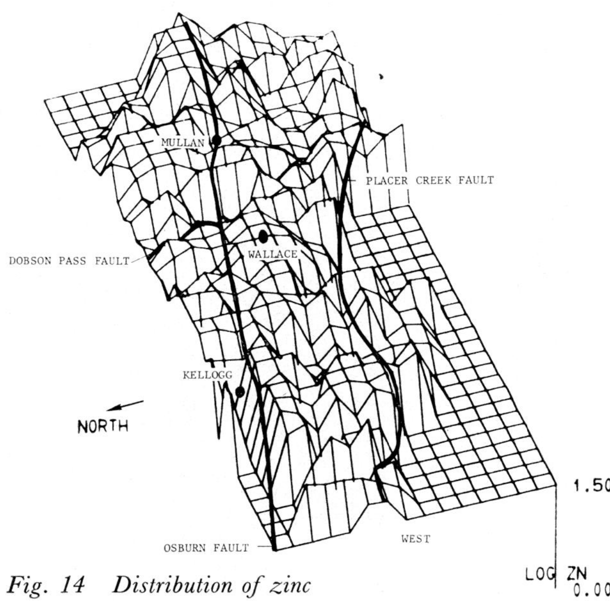

Fig. 14 Distribution of zinc

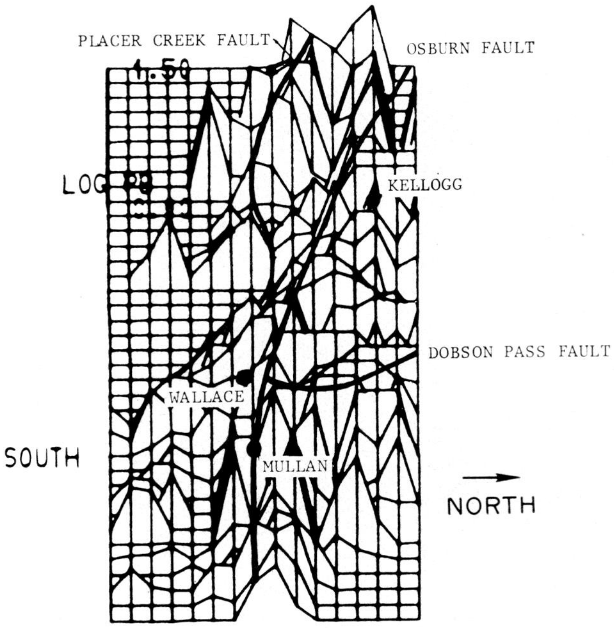

Fig. 15 Distribution of lead

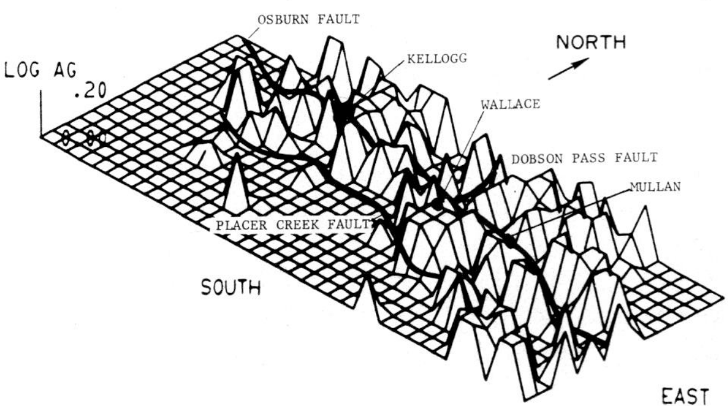

Fig. 16 Distribution of silver

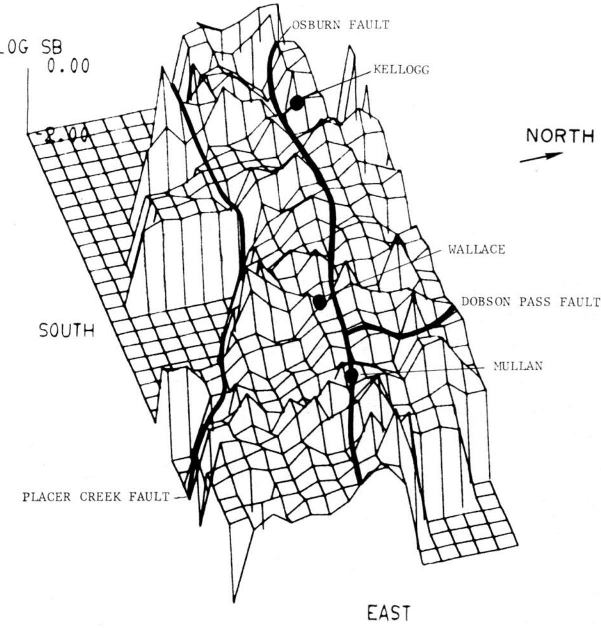

Fig. 17 Distribution of antimony

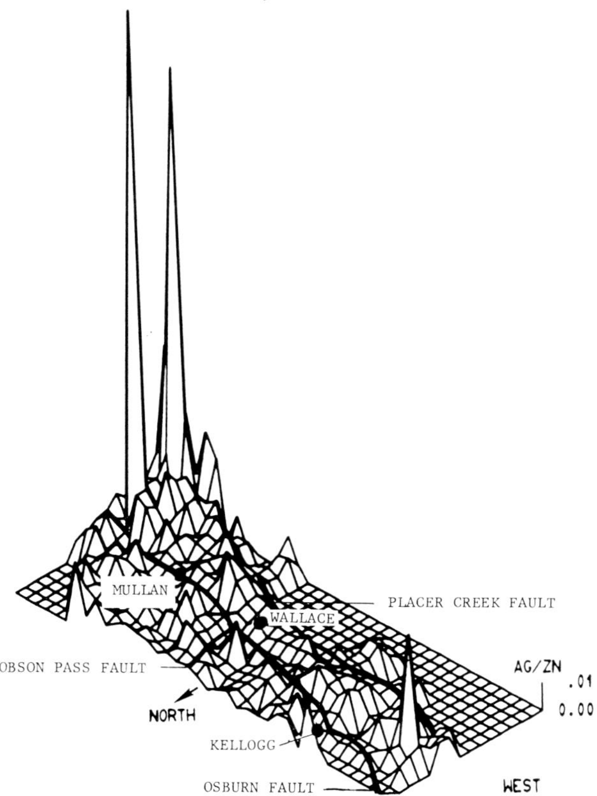

Fig. 18 Ratio Ag/Zn in perspective

eastward about 3 miles from their original position. The cadmium-rich halo around the small patches of monzonite to the west is similar to the halo around the south Gem stock and may represent a cadmium halo above the original position of the Gem stocks. The present surface in the vicinity of the small patches of monzonite in the upper plate is 5000–7000 ft above the Dobson Pass fault. Presumably, zinc and cadmium were fractionated through this distance as a result of the subjacent heat.

Another example of zinc–cadmium fractionation is to be found in Lucky Friday mine. Samples supplied by Hecla Mining Company of vein material and wallrock down to the 3650-ft level show that vein material is impoverished in cadmium, whereas wallrocks are greatly enriched in cadmium. Data to the 2000-ft level are shown in Fig. 12. Where possible, the wallrock was sampled at distances of 50, 100 and 150 ft on both sides of the vein. Within that distance the cadmium in the wallrock is increased tenfold over cadmium in the vein material. The target presented by this halo is many times larger than the target presented by the actual vein.

After Figs. 6, 7, 8, 10 and 11 were constructed, soil sampling was extended into the area west of the Dobson Pass fault and north of the Osburn fault and in an extensive area to the southeast. These additional data are included in the perspectives shown in Figs. 13–22. Except for Figs. 15, 16 and 17, which are viewed from the southeast, all the perspectives are viewed from the northwest.

Perspectives of the log distribution of copper, zinc, lead, silver and antimony (Figs. 13–17) show that a high trend of copper and silver extends from the known Wallace–Kellogg mineral belt into the relatively unexplored southeast

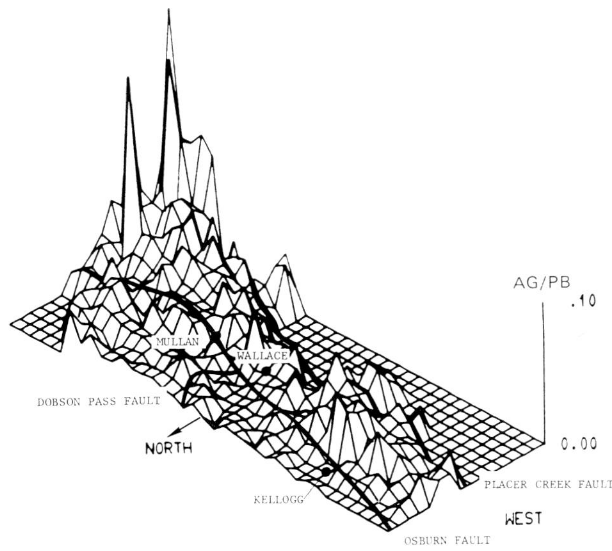

Fig. 19 Ratio Ag/Pb in perspective

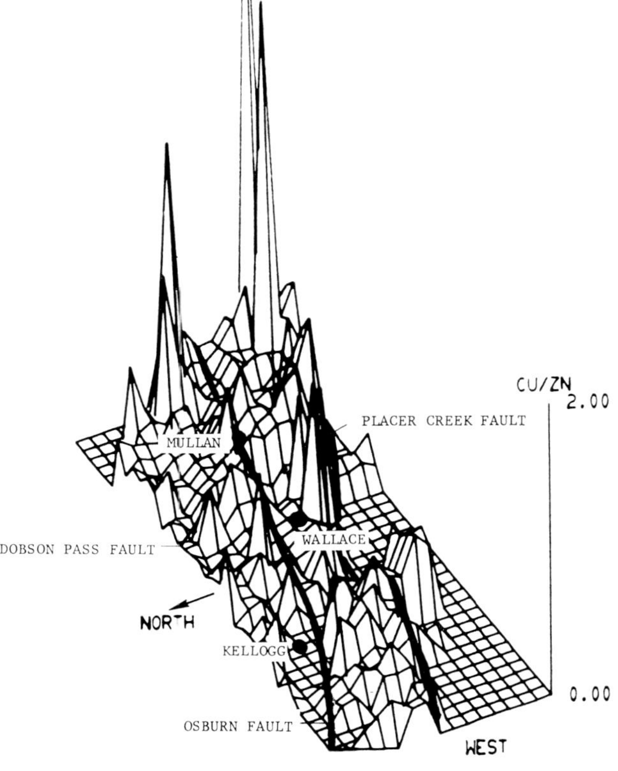

Fig. 20 Ratio Cu/Zn in perspective

corner of the mapped area, whereas lead, zinc and antimony tend to decrease in that direction.

This trend is greatly enhanced when silver and copper are examined in ratio form with lead and zinc (Figs. 18–21). The Ag/Zn ratios (Fig. 18) and Ag/Pb ratios (Fig. 19) show a particularly well-defined trend starting from the west side of the perspectives through the established Wallace–Kellogg part of the mining district and continuing

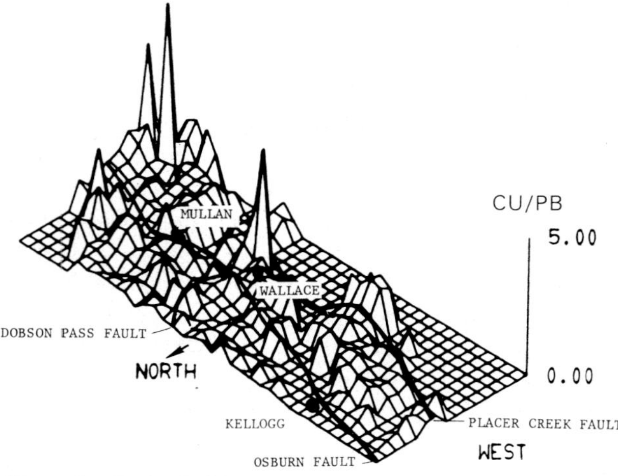

Fig. 21 Ratio Cu/Pb in perspective

with increasing silver, relative to lead and zinc, to the southeast. The ratios Cu/Zn (Fig. 20) and Cu/Pb (Fig. 21) show a similar trend, and high tellurium rims high silver in the southeast (Fig. 22).

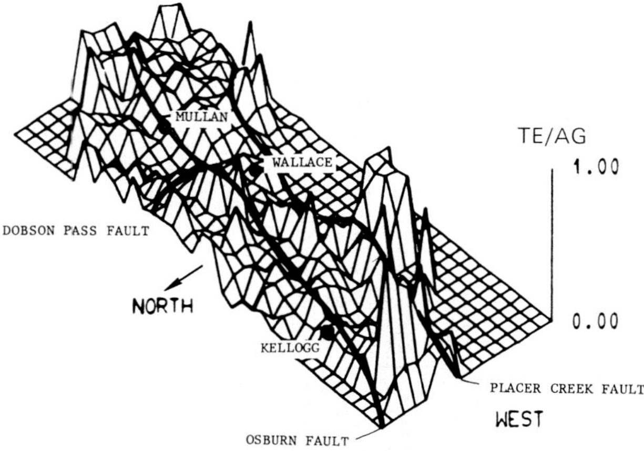

Fig. 22 Ratio Te/Ag in perspective

Summary

These geochemical studies show that the dispersion patterns of selected metals in soils are areally extensive and spatially related to the mineral belts. A vertical and lateral zoning is also evident among the metals. Depending on the depth to the ore deposits, the mineral belts are outlined by anomalous silver, tellurium or manganese and, to a lesser degree, by other metals. A qualitative measure of the temperature of formation of sphalerite is suggested by the ratio Zn/Cd. Favourable areas for the occurrence of tetrahedrite are indicated by the dispersion patterns for antimony and silver. The depth to a deposit can be roughly estimated by studying the relative concentration and distribution of silver, tellurium and manganese. Computer-generated perspectives can economically and rapidly portray a large mass of geochemical data.

Acknowledgment

The authors gratefully acknowledge the generous permission of the National Center for Atmospheric Research (Boulder, Colorado), supported by the National Science Foundation, for use of their facilities for interactive graphics. The assistance of John B. Cathrall, Theodore M. Billings and Jack B. Fife of the Geological Survey was invaluable. The authors greatly appreciate the samples of vein material and wallrock from Lucky Friday mine supplied by the Hecla Mining Company.

The analytical support of the following U.S. Geological Survey chemists and technicians is appreciated: Richard N. Babcock, Craig A. Curtis, James G. Frisken, David J. Grimes, Jerry R. Hassemer, James D. Hoffman, Roy T. Hopkins, Reinhard W. Leinz, Elwin L. Mosier, Jerry M. Motooka, John C. Negri, Richard M. O'Leary, Zelia C. Stephenson, Susan H. Truesdell, James H. Turner, Robert L. Turner, John Watterson and Eric P. Welsch.

The paper is published by permission of the Director, U.S. Geological Survey.

References

1. Crosby, G. M. A preliminary examination of trace mercury in rocks, Coeur d'Alene district, Idaho. *Colo. Sch. Mines Q.*, **64**, Jan. 1969, 169–94.
2. Fryklund, V. C. Jr. Ore deposits of the Coeur d'Alene district, Shoshone county, Idaho. *Prof. Pap. U.S. geol. Surv.* 445, 1964, 103 p.
3. Goldschmidt, V. M. *Geochemistry* (Oxford: Clarendon Press, 1954), 730 p.
4. Grimes, D. J. and Marranzino, A. P. Direct-current arc and alternating-current spark emission spectrographic field methods for the semi-quantitative analysis of geologic materials. *Circ. U.S. geol. Surv.* 591, 1968, 6 p.
5. Hobbs, S. W. *et al.* Geology of the Coeur d'Alene district, Shoshone county, Idaho. *Prof. Pap. U.S. geol. Surv.* 478, 1965, 139 p.
6. Ward, F. N. *et al.* Analytical methods used in geochemical exploration by the U.S. Geological Survey. *Bull. U.S. geol. Surv.* 1152, 1963, 100 p.

550.84:543.062:546.74:546.22

Application of sulphur and nickel analyses to geochemical prospecting

D. M. Hausen, PH.D.

J. W. Ahlrichs, B.S.

J. R. Odekirk, M.S.

All of Newmont Exploration, Ltd., Danbury, Connecticut, U.S.A.

Synopsis

Specimens of basic and ultrabasic rocks from mineralized prospects in Canada and Australia have been analysed for sulphur and nickel, and compared for different rock types. Log–log plots of nickel against sulphur values suggest a method for evaluating the mineralization potential of specific rock types. Assay data group together characteristically for different rock types, indicating a geochemical method for correlating lithology and possibly identifying rock types from a given area.

A comparison of serpentinites from mineralized and unmineralized localities suggests that concentrations and ratios of sulphur and nickel (near 1 to 1) may be used as a guide to more favourably mineralized districts. Sulphur–nickel plots are presented for serpentinites from several known producers and prospects, including the Thompson Belt, Texmont mine, Mount Keith and Kambalda.

A method is proposed for screening favourable prospects based on a comparison of nickel and sulphur analyses with rock types.

In the past several years exploration for low-grade nickel deposits in ultrabasic and basic rocks has been especially active in Canada and Australia. The discovery and development of two low-grade, large-tonnage disseminated nickel deposits at Mount Keith in Western Australia and Dumont in northwest Quebec were announced in 1970. Disseminated deposits, averaging 0·2–0·7% nickel, have been prospected by various geological, geophysical and geochemical techniques.

Geochemists have been pursuing the subject, as evidenced by the large number of publications. A variety of analyses has been used to interpret compositional rock types and their origin in an attempt to delineate mineralization trends. Analyses of interest often include silica, alumina, ferric and ferrous iron, chromium, magnesia, lime, soda, potash, sulphur, nickel, copper and other base metals. In most cases only a few select elements, e.g. nickel, sulphur, copper, chromium and magnesium, are determined at Newmont.

In this study methods and applications of sulphur and nickel are described. These analyses, coupled with mineralogical and petrological studies of suites of ultrabasic and basic rocks from greenstone belts of Australia and Canada, have provided an opportunity to compare rock geochemistry in the two shield areas. Concentrations of sulphur and nickel have been compared with specific rock types from several localities as a guide to mineralization.

Methods of plotting and correlating nickel and sulphur analyses on a log–log scale for each rock type from several properties are described. Sulphur–nickel ratios are also compared between highly mineralized and less mineralized serpentinites from different localities.

General background

Several nickel properties have been explored by Newmont in Australia and Canada: three of these are referred to as *A*, *B* and *C*. Geographic locations of properties have been omitted to avoid the release of information of proprietary importance to the company. These properties are compared geochemically on a regional basis with published data from several well-known mineralized properties, which include the Katiniq Sill, Ungava and the McWatters deposit.

In addition to mineralogical studies, all samples from prospects *A*, *B* and *C* were analysed for nickel and sulphur. Initially, sulphur analyses were obtained (1) to evaluate the availability of sulphur and (2) to obtain rough estimates for silicate nickel in addition to sulphide nickel. During the course of these studies sulphur and nickel values were noted to be segregated according to rock types. Consequently, sulphur–nickel assays were plotted to evaluate their significance for rock-type classification.

Sulphur–nickel ratios for serpentinites from Newmont prospects are also compared with those from disseminated nickel deposits from Mount Keith, Dumont, the Thompson Belt and Texmont. Sulphur–nickel and pyrrhotite–pentlandite ratios are characteristically low for most of these properties. Favourable sulphur–nickel ratios for nickel prospecting of serpentinites are suggested by these studies.

No attempt was made to describe the mineralogy and petrology of the rocks in detail. Rock-type classifications were based on X-ray diffraction analysis aided by thin-section studies of select samples for each rock type. Particular emphasis is placed on analytical techniques and methods of plotting data by rock type for comparison of different mineralized prospects. Log–log plots of total sulphur and nickel appear to offer a preliminary means of evaluating types and favourable environments for nickel mineralization.

Analytical methods

Total nickel analyses

Several techniques were investigated to obtain rapid and inexpensive total nickel assays for large numbers of samples, the techniques comprising a combination of atomic absorption and X-ray fluorescence. Use of atomic absorption only is relatively expensive and time-consuming because the samples under consideration require an extra fusion step for complete dissolution of nickel.

Reliable nickel values were obtained by computerized X-ray fluorescence techniques coupled with occasional checks by use of atomic absorption, several hundred basic and ultrabasic rock samples from Australia and Canada having been analysed in this way. The essential requirement for accurate nickel values is that standard samples of known composition contain matrices similar to the samples under analysis. Matrices differed between samples from different localities, but they did not affect accuracy when sample suites were analysed separately. Five to ten samples from each suite were analysed by atomic absorption for use as standards for X-ray fluorescent assays.

Nickel analyses by X-ray fluorescence involve five counts of 10 sec each per sample on the nickel peak at 48·70°, 2θ and background counts at 47·70°, 2θ. The data are punched on computer tape and then fed to an IBM 1130 computer, which calculates and reports nickel percentages in a final typed form. Approximately every tenth or twentieth sample is monitored by atomic absorption.

Sulphide and silicate nickel analyses

Several methods of partial extraction for the evaluation of sulphide and silicate nickel were evaluated at the Newmont laboratory, three of these being noted below with the times required for complete dissolution of sulphide minerals.

Method	*Time required to dissolve sulphides*
(1) Ascorbic acid–H_2O_2[1]	18 h
(2) Water–bromine[2]	2 h
(3) Alcohol–bromine[3]	18 h

Solutions which contain sulphide nickel are analysed by atomic absorption. The difference between dissolved nickel and total nickel equals *silicate* nickel. Accuracy of each method was evaluated microscopically in polished mounts under incident light. The time required to dissolve sulphides was determined by several tests run at various time-intervals: the water–bromine technique required only 2 h, whereas 18 h was required by the ascorbic acid–H_2O_2 technique. Miscroscopically, however, silicates in residues from both techniques showed varying degrees of attack, and some of the silicate nickel may also have been removed. Additional work is required to determine optimum contact time for the ascorbic acid–H_2O_2 and the water–bromine methods by which all sulphides are removed with minimal attack on silicates.

Table 1 Nickel and sulphur analyses of rock types from prospect A

Location (depth, ft)		Rock type	% Nickel	% Sulphur
GS prospect				
GP-10	(604)	Peridotite	0·15	0·991
GP-10	(754)	Peridotite	0·80	0·337
GP-10	(763)	Peridotite	0·13	1·03
GP-10	(779)	Peridotite	0·10	1·12
GP-10	(785)	Peridotite	0·08	0·82
GP-10	(787)*	Massive sulphides in peridotite	0·25	5·35
GP-10	(788)	Peridotite	0·08	1·25
DD/D1	(680)*	Massive sulphides in peridotite	0·31	2·38
DD/D1	(681)	Peridotite	0·24	1·97
MP prospect				
MPD1	(100)	Garnet amphibolite	0·003	0·012
MPD1	(235)	Garnet amphibolite	0·005	0·205
MPD1	(240)	Garnet amphibolite	0·008	0·011
MPD1	(288)	Garnet amphibolite	0·010	1·59
MPD1	(295)	Garnet amphibolite	0·011	0·016
MPD1	(448)	Quartz–plagioclase–garnet gneiss	0·002	0·023
MPD3	(321)	Garnet amphibolite	0·010	0·019
MPD3	(338)	Garnet amphibolite	0·005	0·408
MPD3	(434)	Pyroxenite–plagioclase gneiss	0·010	0·010
G2 prospect				
G2H1	(243)	Pyroxenite	0·033	0·025
G2H3	(171)	Pyroxenite	0·036	0·059
G2H3	(172)	Pyroxenite	0·13	0·109
G2H14	(221)	Amphibolite	0·037	0·019
G2H15	(208)	Amphibolite	0·022	0·093
G2H16	(222)	Quartz–plagioclase–biotite gneiss	0·003	0·014
NA prospect				
NAH3	(140)	Olivinite	0·099	0·080
NAH4	(172)	Amphibolite	0·017	0·037
NAH9	(181)	Pyroxene granulite	0·009	0·052
NAH13	(164)	Pyroxene granulite	0·007	0·014
NAH14	(66)	Pyroxene granulite	0·007	0·104
TBD prospect				
TBD 4	(308)*	Massive sulphides in pyroxenite	0·38	3·32
TBD 4	(339)	Olivinite	0·20	0·18
TBD 5	(442)	Olivinite	0·18	0·19
TWD prospect				
TWD 1	(378)	Pyroxenite	0·021	0·078
TWD 1	(388)	Serpentinite	0·38	0·052
TWD 1	(413)	Serpentinite	0·35	0·111
TWD 1	(431)	Anthophyllite–serpentine rock	0·045	0·016
TWD 2	(728)*	Garnet amphibolite containing massive sulphides	0·05	4·84

Continued overleaf

Table 1—continued

Location (depth, ft)		Rock type	% Nickel	% Sulphur
Y1 prospect				
Y1H1	(202)	Muscovite–biotite–quartz–plagioclase gneiss	0·005	0·026
Y1H2	(182)	Muscovite–biotite–quartz–plagioclase gneiss	0·004	0·201
Y1H2	(183)	Muscovite–biotite–quartz–plagioclase gneiss	0·003	0·008
Y1H4	(170)	Garnet amphibolite	0·018	0·012
Y1H12	(125)	Pyroxene granulite	0·005	0·063
Y1H15	(172)	Garnet amphibolite	0·020	0·012
Y1D1	(286)	Serpentinite	0·39	0·044
Y1D1	(380)	Serpentinite	0·35	0·052
Y1D1	(397)	Serpentinite	0·45	0·037
Y1D1	(434)	Serpentinite	0·41	0·054
Y1D1	(439)	Garnet amphibolite	0·08	0·005
Y1D1	(444)	Garnet amphibolite	0·01	0·041
GR0 group				
GR0-101		Anorthosite	0·001	0·030
GR0-102		Diorite gneiss	0·002	0·046
GR0-104		Diorite gneiss	0·004	0·034
GR0-105		Amphibolite	0·009	0·041
GR0-107		Pyroxene granulites	0·005	0·077
GR0-108		Pyroxene granulites	0·004	0·045
GR0-109		Pyroxene granulites	0·004	0·049
GR0-110		Pyroxene granulites	0·004	0·051
GR0-111		Pyroxene granulites	0·005	0·13
GR0-112		Meta-norite	0·026	0·049
GR0-113		Meta-norite	0·049	0·044
GR0-114		Pyroxenite	0·065	0·037

*Specimens containing massive sulphides; not used in calculations.

Total sulphur analyses

In the writers' opinion, analyses for total sulphur and nickel are highly desirable, being preferred in many cases to partial extraction techniques. Their observations for serpentinites suggest that if sufficient sulphur is present, silicate nickel is at a minimum. In magmatic liquids large concentrations of sulphur have an affinity for nickel. The nickel then tends to form pentlandite, lesser amounts crystallizing with silicates. If sulphur is in excess, pyrrhotite and possibly pyrite will occur. If sufficient sulphur is introduced hydrothermally, nickel may be removed from earlier silicates and may be precipitated as sulphides.

A statistical treatment described by Cameron and co-workers[4] for ore elements in samples collected from 61 localities in Canada indicated that sulphur, in addition to copper, is an extremely useful element for gaining information regarding ore potential in ultramafic rocks.

For total sulphur values in the samples reported here analyses were performed by conventional gravimetric techniques. Presently, sulphur is determined by a Leco combustion furnace, which provides more rapid results. Assays for total nickel and sulphur for properties *A*, *B* and *C* are compared in Tables 1, 2 and 3.

Mineralogical analyses

All samples were analysed by X-ray diffraction supplemented by examination of polished thin sections. Semiquantitative mineralogical data were compared with sulphur and nickel values.

Regional abundance of sulphur and nickel

Attempts were made to provide a regional view of sulphur and nickel analyses without regard to lithology. Abundance of sulphur and nickel on a regional basis is a helpful indicator of ore potential when compared with other known mineralized localities.

Frequency distributions of nickel and sulphur

Table 2 Nickel and sulphur assays of rock types from prospect B

Location	(depth, ft)	Rock type	% Nickel	% Sulphur
JH-4	(143)	Tremolite–serpentine–chlorite rock	0·19	0·022
JH-4	(150)	Tremolite–serpentine–chlorite rock	0·17	0·242
JH-4	(175)	Tremolite–serpentine–chlorite rock	0·14	0·151
JH-4	(188·5)	Tremolite–serpentine–chlorite rock	0·31	0·510
JH-4	(214)	Tremolite–serpentine–chlorite rock	0·19	0·346
JH-4	(215)	Tremolite–serpentine–chlorite rock	0·22	0·467
JH-4	(217)	Tremolite–serpentine–chlorite rock	0·11	1·22
JH-4	(219·5)	Tremolite–serpentine–chlorite rock	0·12	0·179
JH-4	(220·6)	Anthophyllite–serpentine rock	0·48	1·65
JH-4	(221·5)*	Massive sulphides in anthophyllite	3·24	7·08
JH-4	(222)*	Massive sulphides in anthophyllite	4·26	23·73
JH-4	(222·5)	Massive sulphides in anthophyllite	1·89	7·78
JH-4	(235)	Amphibolite	0·018	7·78
JH-8	(139)	Tremolite–serpentine–chlorite rock	0·13	0·23
JH-8	(165·5)	Actinolite amphibolite	0·22	0·49
JH-8	(174)	Tremolite–serpentine–chlorite rock	0·17	0·07
JH-8	(194)	Tremolite–serpentine–chlorite rock	0·13	0·09
JH-8	(195)	Tremolite–serpentine–chlorite rock	0·26	0·25
JH-8	(214)	Tremolite–serpentine–chlorite rock	0·046	0·02
JH-8	(216)	Tremolite–serpentine–chlorite rock	0·072	0·70
JH-8	(219·9)	Tremolite–serpentine–chlorite rock	0·20	1·35
JH-8	(222)	Anthophyllite–serpentine rock	1·04	1·11
JH-8	(223·5)	Anthophyllite–serpentine rock	0·26	0·16
JH-8	(228·5)*	Massive sulphides in anthophyllite rock	2·44	12·86
JH-8	(231·2)*	Massive sulphides in anthophyllite rock	9·33	37·21
JH-8	(231·5)*	Massive sulphides in anthophyllite rock	7·96	36·22
JH-8	(233)	Anthophyllite–serpentine rock	1·12	8·79
JH-8	(233·4)	Anthophyllite–serpentine rock	1·79	10·00
JH-8	(235·5)	Amphibolite	1·30	3·54
JH-8	(237·5)*	Massive sulphides in anthophyllite rock	6·85	19·86
JH-8	(238)	Amphibolite	0·23	1·61
JH-8	(243·5)	Amphibolite	0·015	0·15
JH-8	(246)	Amphibolite	0·043	0·28
JH-8	(250·9)	Amphibolite	0·021	0·19

*Specimens containing massive sulphides; not used in calculations.

Table 3 Nickel and sulphur assays of rock types from prospect C

Area	% Nickel	% Sulphur	Area	% Nickel	% Sulphur
Amphibolites			*Amphibolites*		
1	0·02	0·04	5	0·01	0·09
1	0·02	0·07	5	0·01	0·14
1	0·04	0·03	5	0·04	0·01
4	0·03	0·03	5	0·01	0·03
5	0·06	0·04	5	0·01	0·07
5	0·01	0·77	5	0·007	0·20
5	0·02	0·15	5	0·006	1·15
5	0·01	0·07	5	0·07	0·05
5	0·01	0·05	5	0·09	0·13

Continued overleaf

Table 3—continued

Area	% Nickel	% Sulphur
Amphibolites—continued		
5	0·07	0·13
5	0·06	0·03
5	0·04	0·11
5	0·01	0·11
5	0·05	0·81
5	0·04	0·45
0	0·05	0·05
0	0·01	0·97
0	0·09	0·07
0	0·01	0·01
0	0·10	0·03
0	0·01	0·02
6	0·04	0·13
6	0·05	0·05
6	0·05	0·04
Tremolite–chlorite rocks		
1	0·13	0·05
1	0·17	0·24
1	0·15	0·12
1	0·11	0·11
1	0·20	0·03
1	0·15	0·05
4	0·10	0·32
4	0·03	0·03
4	0·11	0·01
4	0·13	0·09
4	0·10	0·14
4	0·11	0·07
4	0·13	0·03
5	0·10	0·02
5	0·10	0·05
5	0·09	0·02
5	0·12	0·01
0	0·20	0·27
0	0·10	0·01
0	0·15	0·05
0	0·09	0·06
6	0·14	0·06
6	0·16	0·06
7	0·11	0·02
7	0·15	0·06
7	0·15	0·06
7	0·13	0·07
Serpentinites		
2	0·24	0·05
2	0·28	0·02
2	0·22	0·11
2	0·24	0·07
2	0·21	0·05
3	0·21	0·21
Serpentinites—continued		
4	0·28	0·21
5	0·34	0·02
5	0·30	0·01
5	0·23	0·14
Anthophyllite–fremolite–chlorite rocks		
5	0·09	0·17
5	0·15	0·15
5	0·16	0·15
5	0·16	0·12
5	0·16	0·13
5	0·11	0·11
5	0·11	0·05
5	0·13	0·14
5	0·12	0·06
Tremolite–chlorite–serpentine rocks		
1	0·17	0·08
1	0·14	0·09
1	0·21	0·02
1	0·21	0·11
1	0·19	0·04
1	0·17	0·06
1	0·17	0·01
0	0·18	0·27
0	0·19	0·06
0	0·16	0·16
0	0·17	0·04
0	0·23	0·05
0	0·21	0·13
0	0·15	0·14
0	0·17	0·06
0	0·14	0·16
0	0·15	0·03
0	0·14	0·11
6	0·21	0·05
7	0·17	0·14
7	0·18	0·06
7	0·18	0·02
7	0·14	0·10
7	0·14	0·06
7	0·15	0·03

analyses have been plotted as histograms in Fig. 1 for the Katiniq Sill, Ungava, Quebec, the Thompson Belt, Manitoba, and the McWatters deposit, Ontario.[5] Fig. 2 represents histograms for Newmont prospects *A*, *B* and *C*. Comparison of Figs. 1 and 2 indicates (1) that the average nickel and sulphur contents of properties *A* and *C* are much lower than samples from the Thompson Belt, Ungava, the McWatters deposit and

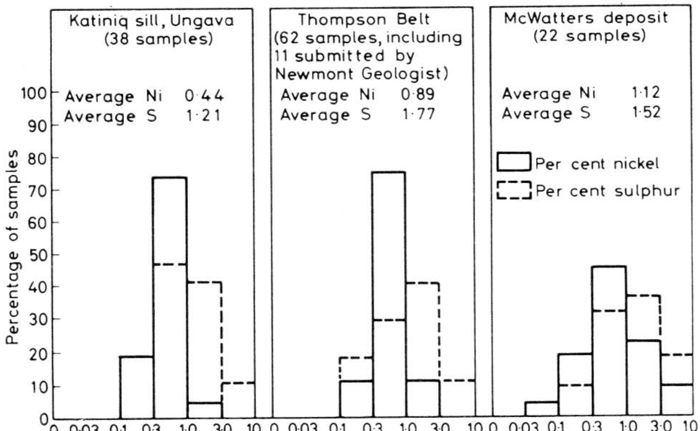

Fig. 1 Histograms of sulphur and nickel assays for known mineralized localities

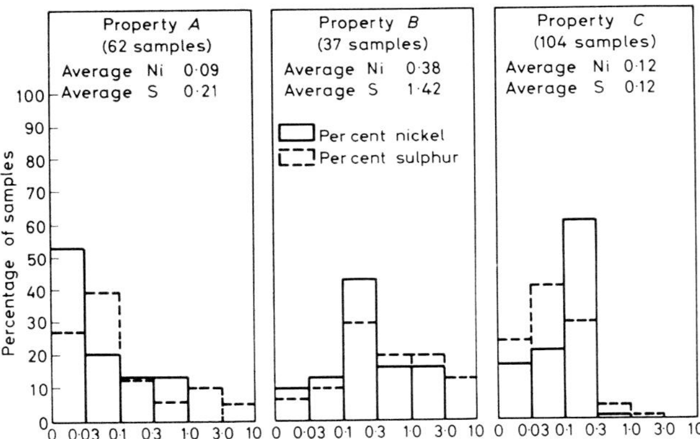

Fig. 2 Histograms of sulphur and nickel assays for specimens from properties A, B and C

property B; (2) that nickel in most of the samples from the three known mineralized properties is in the range 0·3–1·0%: at properties B and C the largest number of samples is in the 0·1–0·3% range (less than 0·03% at property A); and (3) that sulphur assays have a slightly greater range, reaching maximum frequencies in the 0·3–3·0% range for Ungava, Thompson and McWatters. Maximum frequencies for properties B and C are in the 0·1–0·3% range, whereas most sulphur values at property A are less than 0·03%.

Lower concentrations of nickel and sulphur occur at property A than at B and C. Of the three properties, B appears to be the most favourable, being similar to Ungava. Regionally, properties A and C are sulphur-deficient, suggesting that they are less favourable areas for nickel sulphide mineralization.

From the regionally evaluated data the less favourable prospects can be screened. Also, samples with more favourable sulphur–nickel values can be further analysed by X-ray diffraction to gain information regarding the composition of rock types in which sulphur and nickel are concentrated. Hence, less favourable areas can be rejected and the geologist can concentrate effort on areas more favourable for nickel mineralization.

Correlation of sulphur and nickel assays with rock types

Nickel and sulphur analyses become more meaningful when they are compared with specific rock types for any given prospect. Plots of nickel against sulphur on log–log paper for the various rock types have been especially revealing, since data tend to group characteristically for each rock type. Sulphur–nickel plots are presented in Figs. 3, 4 and 5 for properties A, B and C, respectively, and sulphur and nickel assays are shown for each rock type in Tables 1, 2 and 3, respectively.

Property *A*

Serpentinites

Serpentinites contain the highest nickel values in property A, ranging from 0·35 to 0·45% Ni, but they have relatively low sulphur values (between 0·03 and 0·11% S). Values tend to group closely together on nickel–sulphur plots (Fig. 3), suggesting a genetic influence and an alternative

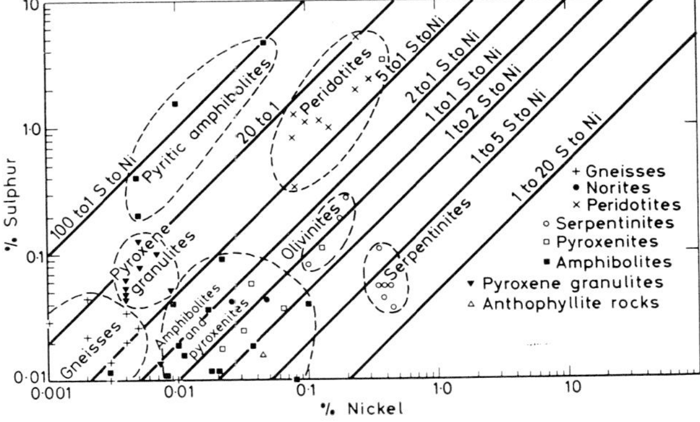

Fig. 3 Sulphur–nickel plots showing grouping of rock types from property A

means of characterizing serpentinites from other rock types. Specimens were obtained from the *TWD* and *Y1* prospects, and are typically characterized by low sulphur–nickel ratios, ranging from about 1:5 to 1:20. Most of the nickel is contained in silicate fractions as nickeliferous serpentine, with insufficient sulphur to produce economic concentrations of pentlandite. Only bare traces of pyrrhotite and pentlandite have been identified in most serpentinites from property A.

Olivinites

Olivinites contain slightly less nickel and more sulphur than the serpentinites, and show a similar tendency to group together near serpentinites in Ni–S plots. Values range from 0·1 to 0·2% Ni and from 0·08 to nearly 0·3% S, falling mostly near a 1:1 ratio.

Specimens of olivinite are mostly from the *NA* and *TBD* prospects. A specimen of pyroxenite from *G2* prospect also falls into this group on the basis of nickel and sulphur assays.

Olivinites consist mostly of olivine, with minor amounts of pyroxenes and serpentine. In most cases they appear to represent recrystallized or sheared dunites, which have probably been affected by two or more cycles of metamorphism and retrograde serpentinization. Traces of pentlandite, chalcopyrite and pyrrhotite have been identified in most olivinite specimens.

Peridotites

Specimens of peridotite (harzburgite) from the *GS* sampling area contain the highest sulphur of any rock type from property *A*, ranging from 0·3 to 5% sulphur and from 0·08 to 0·3% nickel and occupying a unique location on the Ni–S plot (Fig. 3). Sulphur–nickel ratios range from about 5 to 20, indicating a significant excess of sulphur over nickel.

The *GS* prospect appears favourable from a standpoint of excess sulphur, but nickel values are low in most samples. Select specimens show structural control of sulphides along fractures and contacts of late aplitic intrusions in the peridotite. Moderate pyrite and pyrrhotite, and minor pentlandite and chalcopyrite, have been identified.

Pyroxene granulites

Property *A* specimens of pyroxene granulite from sampling areas designated *NA*, *GR0* and *Y1* are largely grouped together in Fig. 3, ranging from 0·004 to 0·009% nickel and 0·04 to 0·13% sulphur. Although nickel values are very low, an excess of sulphur is indicated by sulphur–nickel ratios ranging from 5 to slightly above 20.

Pyroxene granulites do not appear promising either as host or source rocks for nickel.

Gneissic rocks

Lowest nickel values (0·001–0·005% Ni) are contained in gneissic rocks, represented by select specimens from the *MPD*, *Y1*, *G2* and *GR0* sampling areas. Sulphur, too, is low, ranging from 0·01 to nearly 0·05% S. On the basis of nine gneissic specimens, they appear to be unfavourable for nickel exploration in property *A*.

Amphibolites and pyroxenites

Amphibolites and pyroxenites have been grouped together in Fig. 3 because they tend to cluster together on the nickel–sulphur plot. They also tend to grade from one to the other in mineral composition. Specimens from various sampling areas, including *G2*, *NA*, *MPD*, *TWD*, *Y1* and *GR0*, range mostly from about 0·01 to 0·1% Ni and from 0·01 to nearly 0·1% S. They range from about 1/5 to 5/1, grouping about a 1 to 1 S–Ni ratio.

Two specimens of apparent noritic origin in this group have been partially amphibolitized, but contain appreciable amounts of relic pyroxene, and appear to represent intermediate stages of metamorphism between pyroxenites and amphibolites.

Pyritic amphibolites

An anomalous group of amphibolites, rich in sulphides and represented by several specimens from the *MPD* and *TWD* prospects, are termed 'pyritic amphibolites'. These sulphide-rich amphibolites are relatively low in nickel (0·005–0·05%), but contain anomalously high sulphur (0·2–4%) and occupy an unusual position in the nickel–sulphur plot (Fig. 3). Sulphur–nickel ratios are highest in property *A*, ranging from 50 to more than 100.

Pyritic amphibolites may be of potential interest where they occur near nickel-rich serpentinites, and may serve as guides to sulphide-rich areas.

Property *B*

Rock types at property *B* (Fig. 4) were classified as (1) tremolite–chlorite–serpentine rocks, (2) amphibolites and (3) anthophyllite–serpentine rocks. A comparison was made between the *GS* peridotite rom property *A* and the tremolite–chlorite–serpentinite rocks of property *B*. The latter are believed to represent altered ultrabasics near peridotite in primary composition. Averages of nickel and sulphur for specimens (exclusive of those containing massive sulphides) are 0·12% Ni and 1·07% S for the *GS* prospect—compared with 0·38% Ni and 1·42% S for property *B*. Most mineralized specimens from property *B* fall near 5 to 1, sulphur to nickel, and slightly above 5 to 1 for the *GS* prospect (Fig. 3).

Property *B* specimens from drill-holes show a broad correlation between sulphur and nickel

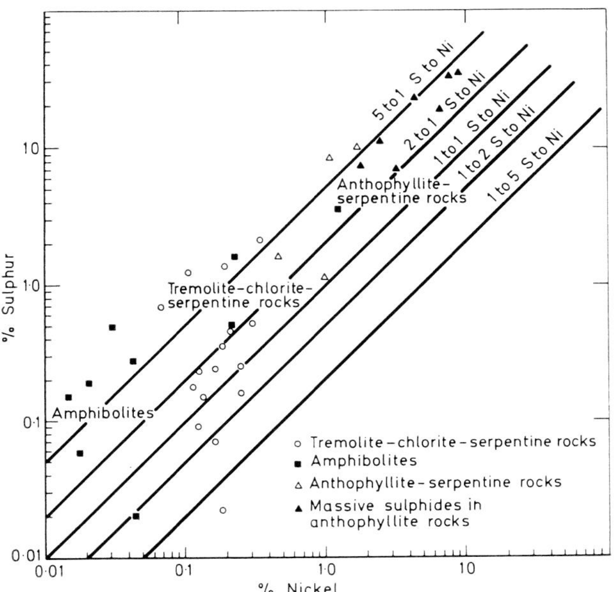

Fig. 4 Correlation of sulphur and nickel assays with rock types at property B

values and an excellent example of rock type dependency, where highest assays (>0·5% Ni, >1·0% S) occur in rocks containing significant amounts of anthophyllite. Intermediate values (0·1–0·4% Ni, 0·1–2% S) are found mostly in tremolite–chlorite–serpentine rocks, and lowest values (<0·05% Ni and <0·05% S) occur in amphibolites. Sulphides at property *B* are largely controlled by structure, occurring along fractures near the footwall contact, where anthophyllite is a predominant alteration feature of tremolite.

Property *C*

Three general rock types are indicated in Fig. 5—amphibolites, tremolite–chlorite rocks and serpentinites. Tremolite–chlorite rocks may be subgrouped into transitional anthophyllite–tremolite–chlorite rocks and tremolite–chlorite–serpentine rocks.

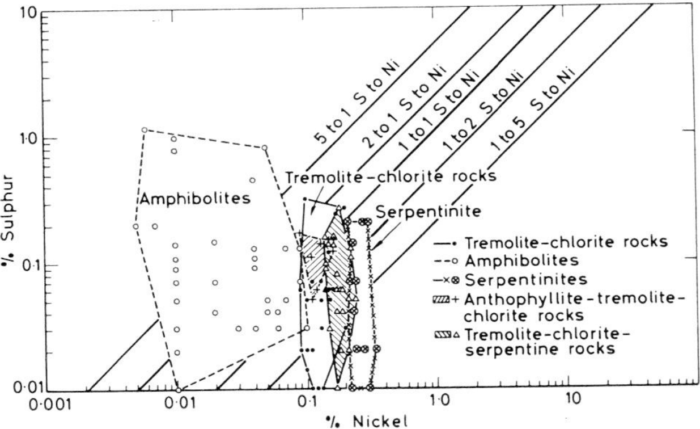

Fig. 5 Sulphur–nickel plots for rock types at property C

In the three general rock types serpentinites are completely isolated from higher metamorphic grades of tremolite–chlorite rocks and amphibolites, although there is slight overlap of the latter two. The tremolite–chlorite–serpentine rocks grade from serpentinites into tremolite–chlorite rocks and overlap slightly with anthophyllite–tremolite–chlorite rocks, the other transitional type.

Nickel values increase from left to right, amphibolites containing the least (0·005–0·1%) and serpentinites the largest amounts (0·2–0·34%). Decreasing nickel appears to correlate with decreasing amounts of serpentine, and, generally, with increasing grades of metamorphism. Most of the nickel is contained in the lattices of the serpentine rather than in sulphides. With the exception of several amphibolites, most rocks from property *C* are sulphur-poor, ranging from 0·01 to slightly greater than 0·3%.

The gradational geochemical plots in Fig. 5 between serpentinites and tremolite–chlorite rocks are supported mineralogically by microscopic observations. Serpentine is partially replaced by tremolite during later, higher-grade metamorphism. Serpentine is absent in tremolite–chlorite rocks, suggesting that complete amphibolitization took place. Later stages of higher-grade metamorphism prevailed in anthophyllite–tremolite–chlorite rocks, resulting in partial replacement of tremolite by anthophyllite.

Originally, there were eight broad areas of interest at prospect *C*. Based on sulphur and nickel values, all but two rock types can be eliminated from further consideration. Anthophyllite–tremolite–chlorite rocks and/or serpentinites appear to be the most favourable for nickel prospecting. Both rock types contain favourable sulphur–nickel ratios (between 2:1 and 1:2). Also, the serpentinites with the highest sulphur values (0·21%) are located near an inferred domal structure.

Anthophyllite-bearing rocks are located approximately one-half mile from a past producer of nickel. The importance of anthophyllite rocks as a guide for nickel prospecting is unknown, but they have been reported to be associated closely with massive sulphides at prospect *B* and also described at Ungava and the Manibridge deposit.[5]

Correlation of sulphur and nickel assays in serpentinites from properties *A* and *C* with other localities

The term 'serpentinite' is used for rocks which contain greater than 60% serpentine. Percentages of total nickel and sulphur given in Table 4 are

Table 4 Nickel and sulphur assays of serpentinites

Location		% Ni	% S	Estimated % silicate Ni
Moak Lake	*Thompson Belt*	1·29	7·40	0·10
Mystery Lake	*Thompson Belt*	0·71	3·03	0·36
Pipe Lake	*Thompson Belt*	1·07	1·38	0·45
Pipe Lake	*Thompson Belt*	0·42	0·25	0·18
Manibridge	*Thompson Belt*	0·50	0·26	0·17
Bucko Lake[5]	*Thompson Belt*	0·80	1·17	n.d.
Bucko Lake[5]	*Thompson Belt*	0·70	0·95	n.d.
Bowden Lake[5]	*Thompson Belt*	0·96	1·40	n.d.
Composite	*Mount Keith*	0·76	1·06	0·20
27-1855	*Mount Keith*	0·54	0·49	n.d.
29-1259	*Mount Keith*	0·84	1·45	n.d.
31-174	*Mount Keith*	0·89	1·09	n.d.
32-168	*Mount Keith*	0·78	0·80	n.d.
32-1235	*Mount Keith*	0·75	1·19	n.d.
Ore	*Texmont*	3·60	3·91	n.d.
30 ft into footwall	*Texmont*	0·31	0·30	n.d.
60 ft into footwall	*Texmont*	0·24	0·22	n.d.
150 ft into footwall	*Texmont*	0·52	0·44	n.d.
10 ft into footwall	*Durkin shoot*	0·29	0·16	n.d.
100 ft into footwall	*Durkin shoot*	0·29	0·20	n.d.
150 ft into footwall	*Durkin shoot*	0·22	0·14	n.d.
Composite (outcrop samples)	*Dumont*	0·30	0·32	0·20
E-14 515 ft	*Dumont*	0·81	0·23	0·17
E-14 1145 ft	*Dumont*	0·55	0·24	0·29
TWD 1 388 ft	*Property A*	0·38	0·05	n.d.
TWD 1 413 ft	*Property A*	0·35	0·11	n.d.
Y1D1 286 ft	*Property A*	0·39	0·04	n.d.
Y1D1 380 ft	*Property A*	0·35	0·05	n.d.
Y1D1 397 ft	*Property A*	0·45	0·04	n.d.
Y1D1 434 ft	*Property A*	0·41	0·05	n.d.
Area 2 801	*Property C*	0·24	0·05	0·21
802	*Property C*	0·28	0·02	0·27
803	*Property C*	0·22	0·11	0·18
804	*Property C*	0·24	0·07	0·20
805	*Property C*	0·21	0·05	0·19
Area 3 901	*Property C*	0·21	0·21	0·06
	Property C	0·28	0·21	n.d.
Area 5 11013	*Property C*	0·34	0·02	0·33
11014	*Property C*	0·30	0·01	0·30
11015	*Property C*	0·23	0·14	0·13
Area 6 10022	*Property C*	0·22	0·07	0·19
1107	*Property C*	0·23	0·04	0·21
Area 7 11031	*Property C*	0·24	0·02	0·22
11033	*Property C*	0·22	0·01	0·21

n.d., not determined.

plotted on a log–log scale in Fig. 6. Silicate nickel analyses are also shown in Table 4 for selected samples of serpentinite. Serpentinites from properties *A* and *C* are compared with known mineralized localities in Australia and Canada, including several areas along the Thompson Belt, Dumont, Texmont, Mount Keith and the Durkin shoot at Kambalda.

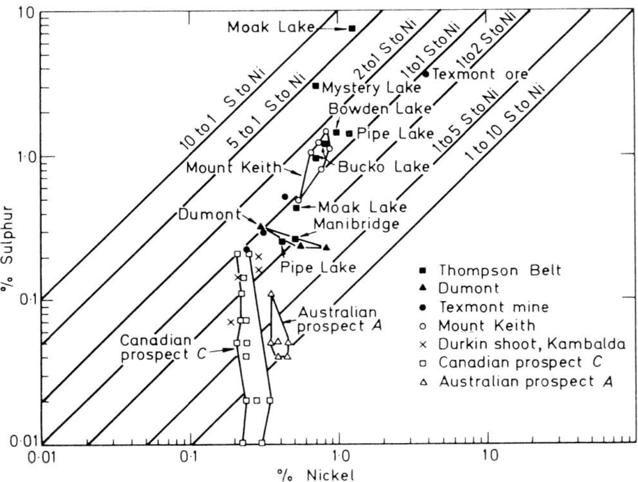

Fig. 6 Sulphur–nickel plots of serpentinites

Property *B* was not compared in the sulphur–nickel plots (Fig. 6) because rock types correspond more closely to meta-peridotites than serpentinites. Only a few serpentinites were sampled from the known mineralized localities, but the resulting sulphur and nickel assays served to provide information regarding favourable sulphur–nickel ratios for serpentine prospects with mineralizing potential.

Fig. 6 shows random sulphur–nickel plots for serpentinites from the Thompson Belt, including Moak Lake, Mystery Lake, Pipe Lake, Manibridge, Bucko Lake[6] and Bowden Lake.[6] Poorest correlations resulted for those samples which have sulphur–nickel ratios ranging from about 1:2 at Manibridge to 5:1 at Moak Lake. Most Thompson Belt serpentinites, however, including Pipe Lake, Moak Lake, Bucko Lake and Bowden Lake, contain sulphur–nickel ratios which range from about 1:1 to 2:1, these plots being similar to those shown for Mount Keith. Plots for serpentinites from Mount Keith, the Durkin shoot, Texmont and Dumont are grouped closely together, yet each appears to be individually clustered, depending on characteristic sulphur and nickel values.

Serpentinites from Texmont were sampled at intervals of 30, 60 and 150 ft into the footwall from the ore. The ore assayed 3·60 and 3·91% nickel and sulphur, respectively. Sulphur–nickel ratios decrease progressively from the ore (1·09) to 150 ft into the footwall (0·85). This variation also occurs at the Durkin shoot. Sulphur–nickel ratios decrease from 0·68 at 10 ft into the hangingwall to a low of 0·37 at 150 ft into the hangingwall. Wallrock serpentinites at Durkin shoot are sulphur-poor and locally contain heazlewoodite (Ni_3S_2) and millerite (NiS), in addition to pentlandite. Sulphur–nickel ratios for selected specimens from Durkin shoot correspond closely with the richest sulphur-bearing serpentinites from property *C*.

Two of the three serpentinites from Dumont contain sulphur–nickel ratios which are less than 1:2, and heazlewoodite and awaruite (Ni_3Fe) were identified in these two samples.

Serpentinites from property *A*, grouped separately from all others in Fig. 6, contain moderate amounts of nickel ($>0·3\%$) but minimal sulphur (0·11% or less). No metallics and only trace pentlandite were identified—suggesting that most of the nickel was never removed from the serpentine lattices. This may be due to the apparent deficiency of sulphur in the environment. Conditions for nickel mineralization are less favourable at property *A*. Little, if any, nickel appears to have been released from the serpentinites, and none of the rock types in the vicinity of the sulphur-poor serpentinites is a likely host for economic nickel mineralization.

At property *C* most serpentinites are sulphur-poor (range, 0·01–0·21%). Nickel values are consistently within the range 0·2–0·34%. Sulphur-poor serpentinites represent unfavourable environments and should be eliminated from nickel prospecting.

Two serpentinites from property *C* near an inferred domal structure contain sulphur and nickel in nearly equivalent amounts of about 0·2%, with minor silicate nickel. Such areas of sulphur-rich serpentinites may favour additional exploration.

Most of the S/Ni ratios for favourable serpentinites approach 1:1, with a broad range of 1:2 to 2:1. If the ratios are significantly less, an unfavourable sulphur-poor environment is indicated, suggesting that nickel has not combined with sulphur but has remained in serpentine lattices. If larger ratios occur, a sulphur-rich environment is indicated—especially favourable if nickel values are high.

In summary, log–log plots in Fig. 6 show an overall trend, the slope of which transgresses the iso-lines of S/Ni ratios. Lowest sulphur–nickel plots are represented by property *C* and property *A*. Intermediate S/Ni ratios pass through Dumont, Manibridge, Pipe Lake, Moak Lake, Mount Keith and Texmont. Highest ratios are represented by specimens from Mystery Lake and Moak Lake. There appears to be a general tendency for S/Ni ratios to increase with increasing nickel in the vicinity of orebodies. The tendency for clustering of data is a characteristic of each serpentinite locality, and may possibly be used as an index to favourability.

Conclusions

Rock specimens collected on a regional scale from several mineralized properties in Australia and Canada contain low concentrations of nickel and sulphur. Plots of nickel–sulphur values on a log–log scale for each rock type were found to be useful in correlating geochemistry with lithology and in evaluating the ore-bearing potential of rock types, for the following reasons. (1) Plotted data usually show a tendency to group characteristically for each rock type, providing an alternative means of correlating rocks of different lithology by means of geochemical assays. (2) Serpentinites contain concentrations and ratios of sulphur and nickel that are often characteristic of mining districts or favourably mineralized localities, and offer a means of distinguishing serpentinites having potential for nickel mineralization. (3) The tendency for clustering of data is a characteristic of each serpentinite locality, and may be used as an index to favourability in some instances. Mineralized serpentinites from the Thompson Belt and from Mount Keith have S/Ni ratios near 1·0.

Acknowledgment

This investigation was made possible through the approval of the Newmont Mining Corporation, and under the guidance and assistance of many geologists from Newmont Proprietary, Ltd., and Newmont Mining Corporation of Canada, Ltd. The authors are especially grateful to R. J. Searls, B. P. Webb, J. A. Coope and G. W. H. Norman for their enthusiastic support and sponsorship of this type of geochemical study on Newmont exploration prospects.

References

1. Lynch, J. J. The determination of copper, nickel and cobalt in rocks by atomic absorption spectrometry using a cold leach. In *Geochemical exploration* (Montreal: CIM, 1971), 313–4. *(CIM Spec. vol. 11)*
2. Czamanske, G. K. and Ingamells, C. O. Selective chemical dissolution of sulfide minerals: a method of mineral separation. *Am. Miner.*, **55**, 1970, 2131–4.
3. Spira, P. and Themelis, N. J. The solubility of copper in slags. *J. Metals, N.Y.*, **21**, April 1969, 35–42.
4. Cameron, E. M. Siddeley, G. and Durham, C. C. Distribution of ore elements in rocks for evaluating ore potential: nickel, copper, cobalt and sulphur in ultramafic rocks of the Canadian Shield. In *Geochemical exploration* (Montreal: CIM, 1971), 298–313. *(CIM Spec. vol. 11)*
5. Coats, C. J. A. and Brummer, J. J. Geology of the Manibridge nickel deposit, Wabowden, Manitoba. *Spec. Pap. geol. Ass. Can.* no. 9, 1971, 155–65.
6. Wilson, H. D. B. *et al.* Geochemistry of some Canadian nickeliferous ultrabasic intrusions. In *Magmatic ore deposits: a symposium* Wilson, H. D. B. ed. (Lancaster, Penna: The Economic Geology Publishing Co., 1969), 294–309. (*Econ. Geol. Monogr.* 4)
7. Woodall, R. and Travis, G. A. The Kambalda nickel deposits, Western Australia. In *Mining and petroleum geology* (London: IMM, 1970), 517–33. (*Proc. 9th Commonw. Min. Metall. Congr. 1969, vol. 2*)

550.42:546:553.412 (713.8)

Endogenic haloes of the native silver deposits, Cobalt, Ontario, Canada

A. S. Dass, PH.D.

Geological Survey of India, Calcutta, India

R. W. Boyle, PH.D.

Geological Survey of Canada, Ottawa, Ontario, Canada

W. M. Tupper, PH.D.

Carleton University, Ottawa, Ontario, Canada

Synopsis

The economic mineral deposits of the Cobalt area are extremely rich silver-bearing veins cutting Cobalt Series sediments, Nipissing diabase, Keewatin greenstone and associated rocks. The mineral assemblage of the veins is complex, consisting of carbonates, native silver and a great variety of nickel–cobalt arsenides.

Wallrock alteration haloes enclosing the native silver–nickel–cobalt arsenide veins are well developed in the Nipissing diabase, Keewatin greenstone, rhyolite porphyry and lamprophyre, but they are much less pronounced in the Cobalt Series sediments. Chloritization, sericitization, carbonatization, albitization, pyritization and arsenopyritization took place in the wallrocks adjacent to the veins. H_2O, CO_2 and Fe_2O_3 increase and SiO_2 decreases in the wallrocks towards the veins. The iron content in chlorite increases, the sodium content of plagioclase increases and the calcium content of plagioclase decreases towards the veins.

The endogenic haloes of trace elements are coincident with the zones of vein clusters. The dispersion of Ag, As, Ni, Co, Sb, Mn and Hg outward from the veins in the Cobalt Series sediments is broad, and veins may be indicated as far as 80–100 ft away by these elements. The dispersion pattern in the Keewatin greenstone can be traced 50–60 ft outward from the veins. Haloes are very much restricted in the Nipissing diabase. The dispersion patterns of Cu, Pb and Zn are also broad and coincident with the veins.

A close association of silver–nickel–cobalt ores with the Keewatin sedimentary interflows and certain volcanic flows has been recognized. Furthermore, the Keewatin greenstone and interflow sediments contain a higher than average background amount of Ag, As, Sb, Ni, Co, Hg, Cu, Pb and Zn. These rock units may be the source of the ore constituents, the younger Nipissing diabase serving as an energy source for their mobilization. Thus, the most favourable prospecting localities may be restricted to areas where the Keewatin interflows adjoin the Nipissing diabase, or where they underlie the Cobalt Series sediments adjacent to the diabase. The pronounced trace-element dispersion patterns in the wallrocks can be used to locate the narrow silver veins in the above rock units.

This paper summarizes the results of research on endogenic haloes which occur around the veins of native silver and nickel–cobalt arsenides at Cobalt, Ontario. The chemical analyses for both major and minor elements in the altered and unaltered wallrocks, chlorites and feldspars in the altered and unaltered rocks, and the carbonate gangue minerals in the veins were carried out by spectrographic, X-ray fluorescent and wet-chemical methods.

The silver deposits of the Cobalt area occur in the Timiskaming district of northern Ontario, about 250 miles northwest of Ottawa and some 75 miles NNW of North Bay (Fig. 1 (*a*)).

Previous studies of the Cobalt area include those by Miller,[15, 16] Hore,[10] Knight,[12] Hriskevich[11] and Halls and Stumpfl.[8] The mineralogy of the silver–nickel–cobalt ores of Cobalt was examined by Bastin,[1] Halls and co-workers[7] and Petruk.[18] Thomson[21] revised the geological maps

80°50'
48°
79°20'
48°
Matachewan
Englehart
Elk Lake
Montreal
River
Gowganda
Notre-Dame
du-Nord
New Liskeard
Haileybury
Lady Evelyn Lake
Cobalt
Latchford
Ville Marie
Lake
Timiskaming
Timagami
47°
80°50'
47°
79°20'
10 0 10
Miles

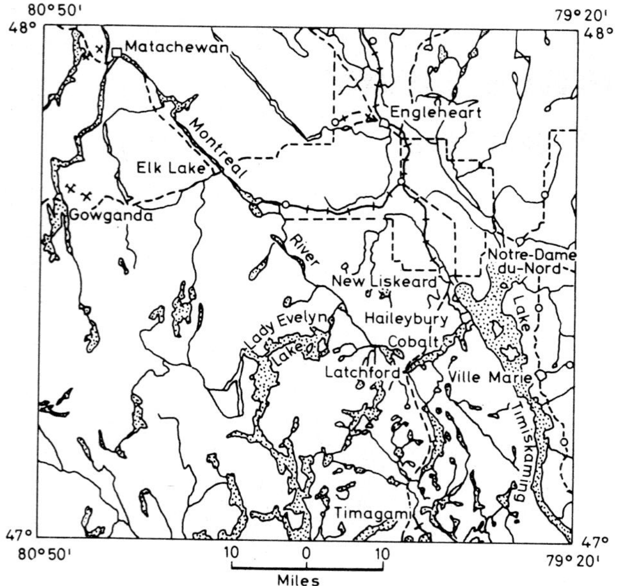

Fig. 1(a) Key map showing location of Cobalt area, Ontario

of the area. Symons[19] examined the palaeomagnetism of the Nipissing diabase at Cobalt. A volume on the geology, mineralogy and geochemistry of the Cobalt–Gowganda silver deposits has recently appeared.[2]

General geology and structure

The Cobalt area lies in the Superior Province of the Precambrian Shield. The oldest (Archaean) rocks of the area are steeply dipping Keewatin lavas (greenstones) and interflow sedimentary rocks, overlain in places by steeply dipping greywacke, quartzite and conglomerate of the Timiskaming Series (Fig. 1(*b*)). Both the Keewatin and Timiskaming rocks are intruded by Lorrain granites and hornblende syenite. The Keewatin metavolcanic rocks (greenstone), for the most part, correspond to basalts and andesites. They are fine-grained, with pillow structures in the upper part, and massive in the lower part.

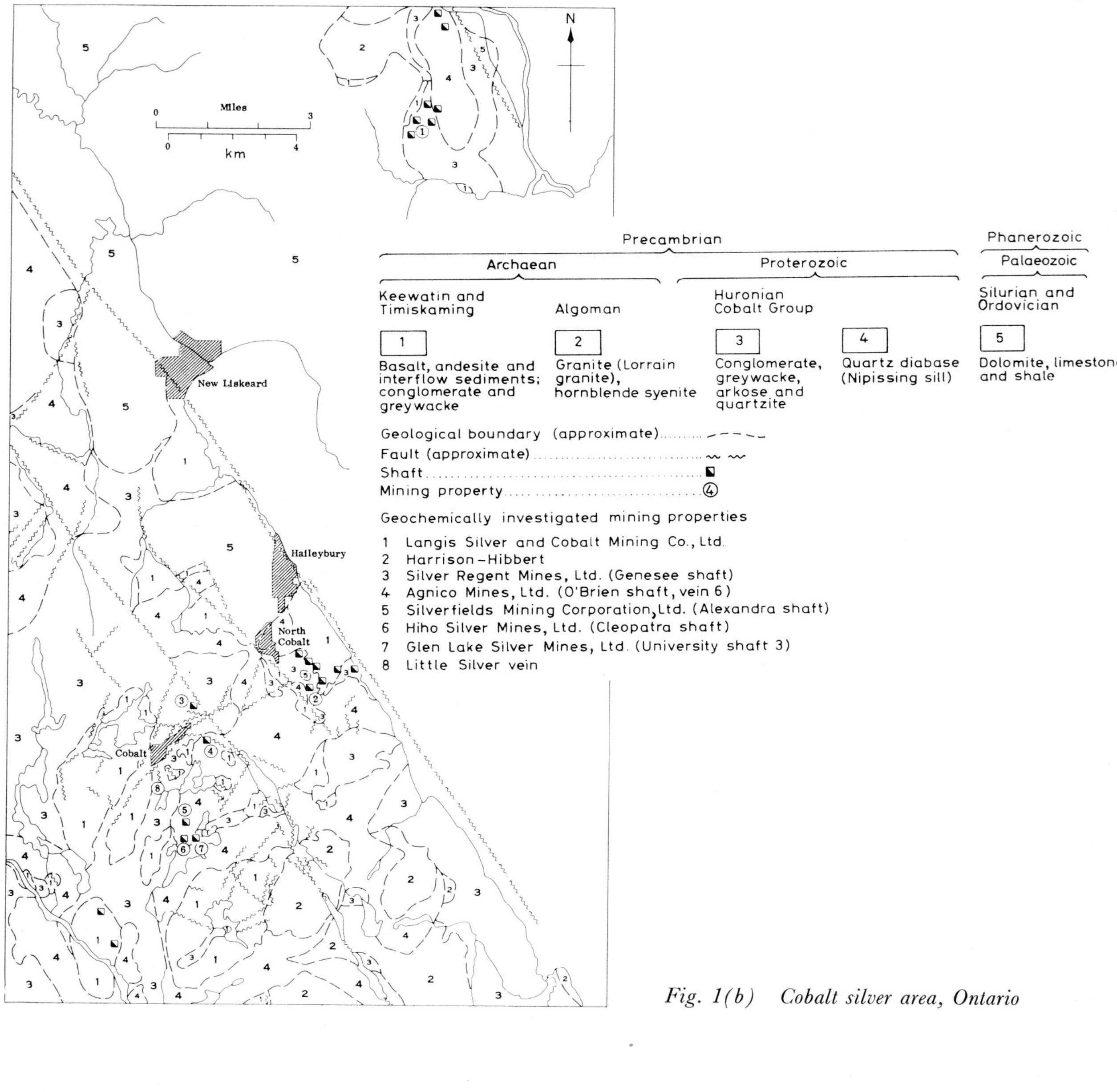

N
Miles
0 3
0 4
km
New Liskeard
Haileybury
North Cobalt
Cobalt
Precambrian
Archaean
Proterozoic
Phanerozoic
Palaeozoic
Keewatin and Timiskaming
Algoman
Huronian Cobalt Group
Silurian and Ordovician
1
Basalt, andesite and interflow sediments; conglomerate and greywacke
2
Granite (Lorrain granite), hornblende syenite
3
Conglomerate, greywacke, arkose and quartzite
4
Quartz diabase (Nipissing sill)
5
Dolomite, limestone and shale
Geological boundary (approximate)
Fault (approximate)
Shaft
Mining property
Geochemically investigated mining properties
1 Langis Silver and Cobalt Mining Co., Ltd.
2 Harrison–Hibbert
3 Silver Regent Mines, Ltd. (Genesee shaft)
4 Agnico Mines, Ltd. (O'Brien shaft, vein 6)
5 Silverfields Mining Corporation, Ltd. (Alexandra shaft)
6 Hiho Silver Mines, Ltd. (Cleopatra shaft)
7 Glen Lake Silver Mines, Ltd. (University shaft 3)
8 Little Silver vein

Fig. 1(b) Cobalt silver area, Ontario

Locally, they are intruded by felsite, feldspar porphyry and quartz porphyry. Metasedimentary rocks include iron formation, tuff, mica and graphitic schists, slate, chert, greywacke and quartzite. Well-developed chlorite spots, about 0·1–0·3 in in diameter, are locally present in the Keewatin greenstones.

The Proterozoic Cobalt Series, which consists of conglomerate and quartzite with a flat to gentle dip, lies with profound unconformity on the Archaean rocks. According to Moore,[17] the Cobalt Series at the Silverfields deposit is divided into four groups (from top to bottom): well-bedded greywacke, conglomerate (0–10% pebbles); grey quartzite, conglomerate (40–70% pebbles); well-bedded greywacke, conglomerate (70–90% pebbles); and bedded greywacke, conglomerate (10–40% pebbles).

Many of the rock types of the Cobalt Series contain dark chlorite spots, from 0·1 to 0·4 in in diameter, which usually follow bedding.

Both the Archaean and the Proterozoic rocks are intruded by a gently dipping Nipissing diabase sill (or composite sills). The age of the Nipissing diabase, determined by the potassium–argon method, is 2095 m.y.,[14] and by the rubidium–strontium method 2180 ± 40 m.y.[22] The main body is in the form of a large undulating sheet with gentle dip, but the dips steepen on the flanks of arches and basins. The total thickness is more than 1000 ft. The intrusion took place near the contact between the nearly horizontal beds of the Cobalt Series and the steeply dipping Keewatin rocks. No flow structures, primary lineation or foliation were noted in the diabase. It is composed (from top to bottom)[11] of the following rocks: fine-grained diabase at top contact; quartz diabase (in many places with coarse texture); hypersthene diabase ± olivine; quartz diabase; and fine-grained diabase at bottom contact.

Ordovician and Silurian limestones, dolomites, shales and sandstones lie unconformably on the Precambrian rocks, and these, in turn, are overlain unconformably by Pleistocene glacial and post-glacial deposits.

The Keewatin and Timiskaming rocks are highly folded into vertical attitudes. The axes of the folds trend northwest. The beds of the Cobalt Series have a low dip. The Nipissing diabase sill has gentle dips.

The major sets of faults, one striking northeast–southwest and the other northwest–southeast, are marked by linear, topographic depressions. The Cobalt Lake fault represents the first set, and the Cross Lake fault, Lake Timiskaming fault and Mackenzie fault represent the second. Unusual joint structures, which have a peculiar cylindroidal pattern, similar to that described by Eakins,[5] are developed in the Nipissing diabase.

Vein fissures, fractures and shear zones are well developed in the rocks of the area. Both single and multiple fractures intersect and ramify at various angles. They commonly dip steeply or vertically and vary in width from a few inches to several inches, and extend several hundreds of feet both along strike and down dip.

Mineral deposits

The economic mineral deposits of the Cobalt area are extremely rich silver-bearing veins cutting the Cobalt Series sediments, Nipissing diabase, Keewatin greenstones and associated rocks. The silver veins contain, essentially, native silver, carbonates, nickel–cobalt arsenides, a few sulphides and a variety of other minerals. The veins may be single or multiple; they ramify and intersect and pinch and swell throughout their extent. Most are short and narrow; strike lengths and dip extents are generally measured in a few hundreds of feet, and widths in inches or fractions of inches. Most veins are steep to vertically dipping, but some have dips as low as 30°. The silver veins cut all rocks in the area, and are younger than the sulphide pods and layers in the Keewatin interflow sedimentary beds.

The most important ore mineral is native silver. It is massive or forms specks in rosettes and bodies of niccolite, safflorite, skutterudite, gersdorffite, rammelsbergite and arsenopyrite. It also forms sheet or leaf at the vein–wallrock contact or in tiny fractures in the wallrocks. Proustite, associated with argentite, occurs locally. Most of the arsenides have a concretionary or tubercular texture and occur as individual rosettes, clusters of rosettes and veinlets. Cylindrical tubercles of these arsenides locally have a dendritic pattern. The arsenide minerals locally have a zoning in the veins from top to bottom: Ni–As, Ni–Co–As, Co–As, Co–Fe–As and Fe–As.[18] The high-grade silver ore is commonly confined to the Ni–Co–As and Co–As zones.

Carbonates with minor quartz are the principal gangue minerals at Cobalt. The carbonates include ankerite, dolomite, calcite and minor siderite. Along a vein the proportion of carbonates to arsenides varies within a short distance.

The other known mineral deposits of the area comprise disseminated base-metal sulphides in the interflow sedimentary beds in the Keewatin greenstones.[3] The minerals of interest in these beds include pyrite, pyrrhotite, chalcopyrite, sphalerite, galena, arsenopyrite and marcasite. The beds vary in width from a few inches to several tens of feet or more.

In addition to the above two types of mineral

deposit, the lower part of the Cobalt Series sediments locally contains disseminated base-metal sulphides, mainly pyrite, galena, sphalerite and chalcopyrite.

Some of the silver veins in the district are partly oxidized to a clay-like mass consisting of limonite, wad and secondary carbonates associated with coatings of annabergite, erythrite, malachite and azurite. This type of alteration is, however, limited at Cobalt.

Basins and arches in the diabase sheet locally control the fissures and fractures. Contacts between the Keewatin and Cobalt Series, between the top of the Nipissing diabase and the overlying Keewatin, and along the bottom of the diabase sill with an undulating formation are important in localizing the deposits.[20]

Sulphide mineralization in the interflow beds is controlled essentially by stratigraphic features, later modified by contortion and dragging during folding.

Disseminated base-metal sulphides are localized in the lower part of the Cobalt Series, particularly in the more porous quartzite and the gritty conglomerate. The sulphides in these zones are pyrite, galena, sphalerite and chalcopyrite. They form aggregates and tiny discontinuous stringers in the matrix of the quartzite and conglomerate.

Wallrock alteration

Wallrock alteration haloes enclosing the native silver–nickel–cobalt arsenide veins have developed in the Nipissing diabase, Keewatin greenstone, rhyolite porphyry and lamprophyre. They are much less pronounced in the Cobalt Series sediments. Fresh rock grades transitionally veinward through a light grey bleached band to a characteristic, narrow, dark chloritic band adjacent to the vein. These bands have bilateral symmetry with respect to the vein. Alteration zones are normally 6–8 in wide, rarely exceeding 12 in. Since the physical, mineralogical and chemical changes are similar in most rock types of this area, the alteration is described for one vein system from each rock group.

Wallrock alteration in the Nipissing diabase

Alteration in the Nipissing diabase is marked by a narrow dark green band, adjacent to the vein, which grades into a light, greyish white bleached zone through a zone of chlorite spots into the grey unaltered diabase. The minerals affected during alteration are augite, hypersthene, hornblende, biotite, plagioclase and magnetite; quartz and pyrite are relatively unaltered. Weak propylitization pervades most of the diabase, resulting in a slight alteration of augite and plagioclase. Augite and hypersthene have been partly altered to hornblende. Hornblende was partly biotized and chloritized, with the formation of a minor amount of iron oxides, and plagioclase has been sericitized, with the development of some epidote.

Four zones have been recognized in the altered wallrock adjacent to the vein.[4] Zone 1 corresponds to the spotted chlorite alteration zone, which grades into the greyish white bleached zone 2 and outwards to the unaltered diabase. Zone 2 grades into the greyish white to dark green zone 3. Zone 3 is a transitional zone between zone 2 and the dark green zone 4, which occurs adjacent to the vein. Contacts between the various zones are gradational. On the basis of alteration mineral assemblages, the zonal arrangements may be grouped as follows: (1) unaltered diabase; (2) sericite–epidote–chlorite–carbonate (zones 1 and 2); and (3) chlorite–carbonate–albite (zones 3 and 4).

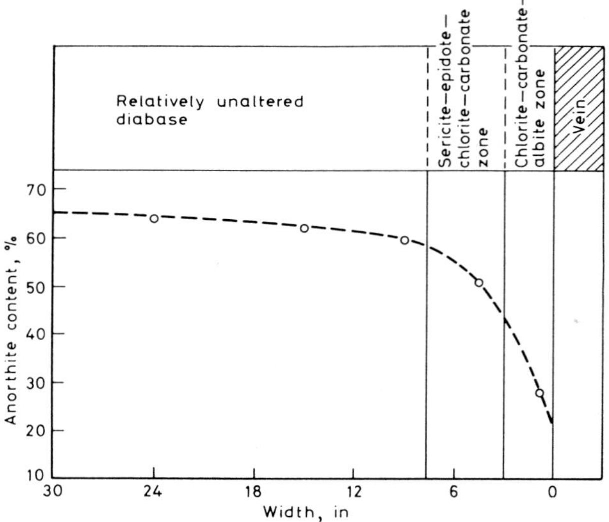

Fig. 2 Variation in composition of plagioclase in diabase wallrock, 6th level, Keeley deposit, South Lorrain

Some epidotization, carbonatization and chloritization of augite and hornblende have taken place in zones 1 and 2. Augite, hornblende, biotite and epidote have been completely altered to chlorite and some carbonate in zones 3 and 4. Chlorite forms sheaves and rosettes, and is an iron-rich variety. It belongs to the ripidolite–pycnochlorite group.[9] Feldspars have been altered to epidote, with some chlorite and carbonate, during chloritization of the ferromagnesian minerals. Within zone 3 dolomite grades to ankerite towards the vein; secondary albite begins to develop in zone 3. Titaniferous magnetite has been altered to ilmenite. Leucoxene was formed from alteration of ilmenite. Zone 3 grades into zone 4 with an increase of ankerite, albite and iron-rich chlorite. Fresh secondary albite is well developed in zone 4. Plagioclase becomes more sodic in the alteration zones towards the vein (Fig. 2). Hematite flakes are disseminated in the

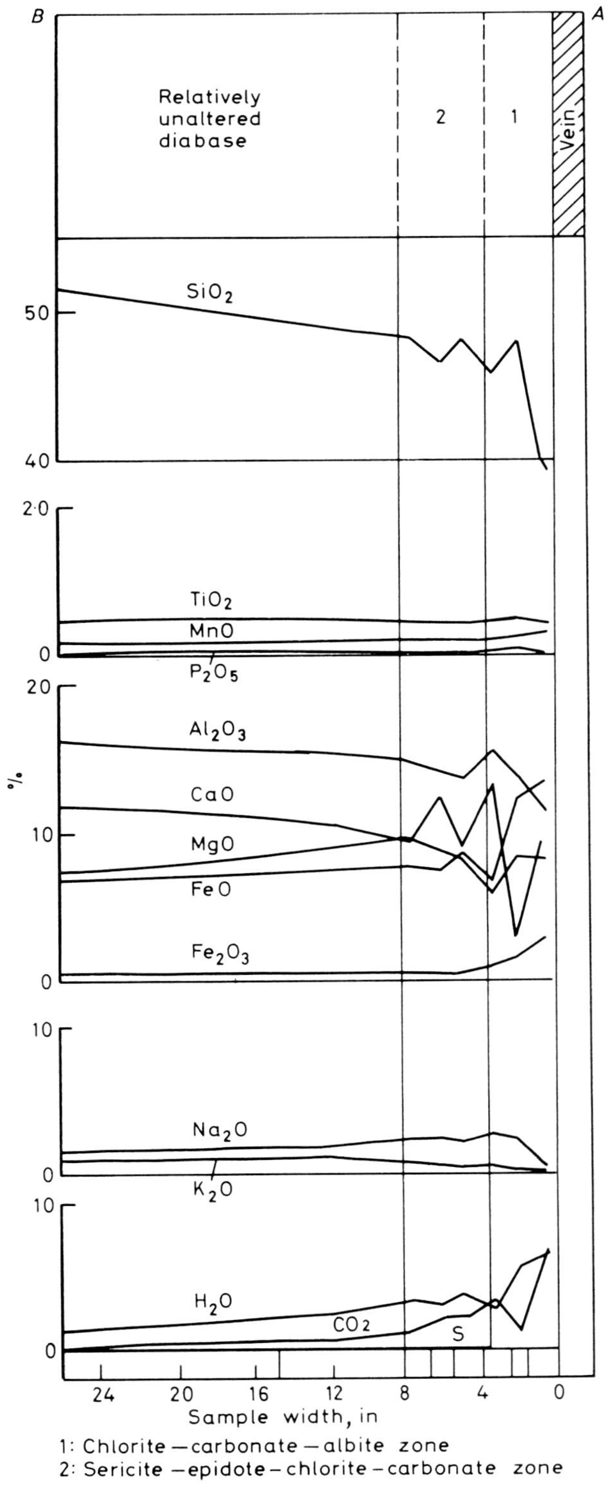

Fig. 3 Chemical variation in alteration haloes of diabase, Glen Lake adit, Glen Lake Silver Mines, Ltd.

albite, where red alteration occurs adjacent to veins. Apatite has been nearly destroyed in this zone.

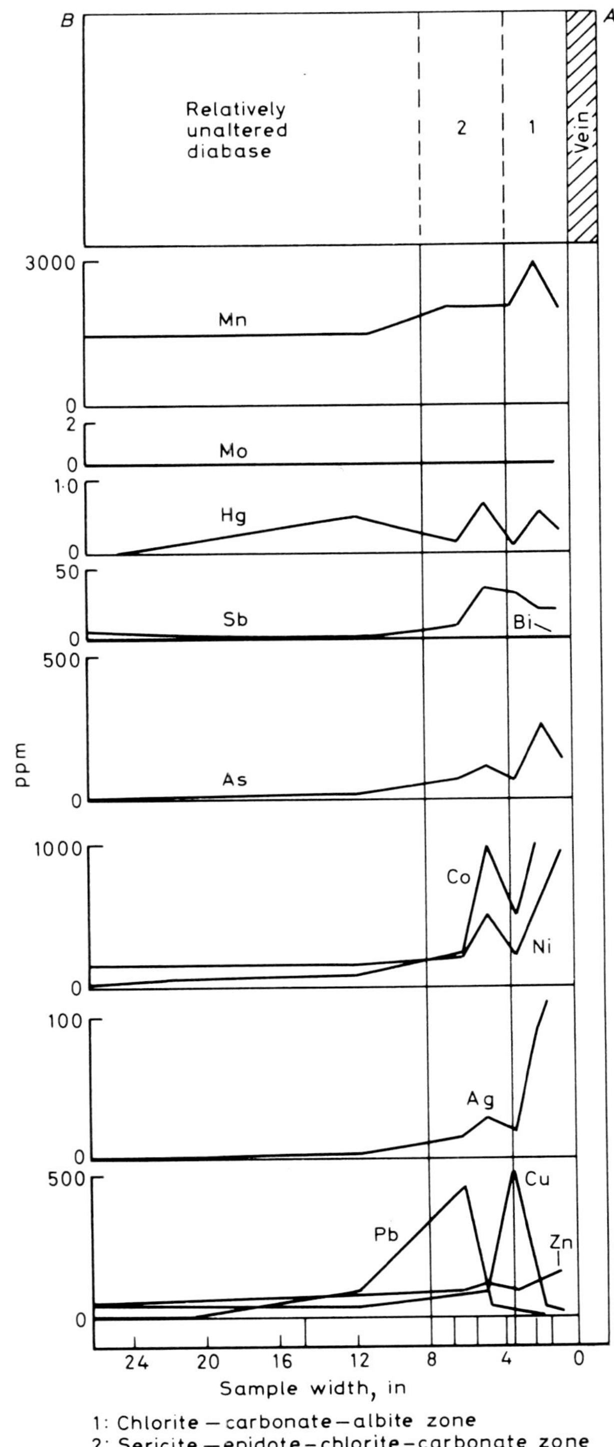

Fig. 4 Distribution of minor elements in alteration haloes of diabase, Glen Lake adit, Glen Lake Silver Mines, Ltd.

The gains and losses of elements were determined from composition–volume relationships.[6] SiO_2, Al_2O_3 and K_2O decrease in both the sericite–epidote–chlorite–carbonate and chlorite–carbonate–albite zones in comparison with the relatively unaltered diabase (Fig. 3). This decrease is paralleled by an increase in CO_2, H_2O and Fe_2O_3. Trends are not consistent for the other elements.

Distribution of cations in the chlorite of the wallrock shows that Fe^{+2}, Fe^{+3}, Al and Mg increase in the sericite–epidote–chlorite–carbonate and chlorite–carbonate–albite zones towards the veins, whereas Si, Ca, K, Mn and Ti decrease in these zones.

Distribution of the minor elements is shown in Fig. 4. The endogenic haloes of elements about the vein are restricted to the alteration zone. Ni, Co, As and Sb increase moderately in the sericite–epidote–chlorite–carbonate and chlorite–carbonate–albite zones, Ag does not change much, except in the chlorite–carbonate–albite zone, where it increases adjacent to the vein. Trends of Cu, Pb, Hg and Mn are not consistent in either of these zones. Zn, Bi and Mo do not change appreciably from background values.

Wallrock alteration in the Cobalt Series sediments

Mineral assemblages of wallrock alteration in the greywacke, conglomerate and quartzite of the Cobalt Series are grouped into (1) a quartz–sericite–carbonate zone and (2) a chlorite–carbonate–sericite zone adjacent to the vein.[4] The quartz–sericite–carbonate zone consists, essentially, of a matted aggregate of chlorite, grains of quartz, highly altered sericitized feldspar, abundant carbonate, some chlorite, and iron oxides. The ferromagnesian minerals of the mafic rock fragments have been altered to fine-grained epidote, carbonate, chlorite and iron oxides. In the chlorite–carbonate–sericite zone the ferromagnesian minerals have been completely altered to chlorite and carbonate. Sericitized feldspars have been altered to albite and fine-grained sericite. Chlorite belongs to the ripidolite–pycnochlorite group.[9] The fine-grained matrix is completely chloritized and ankeritized.

The gains and losses of elements were calculated from composition–volume relationships.[6] The SiO_2 and Al_2O_3 contents are lower in the quartz–sericite–carbonate and chlorite–carbonate–sericite zones compared with the relatively unaltered rock (Fig. 5). This decrease is paralleled by an increase of CO_2, H_2O, CaO, MgO and Fe_2O_3 (total). S, K_2O and MnO increase slightly and Na_2O and TiO_2 decrease slightly in the chlorite–carbonate–sericite zone.

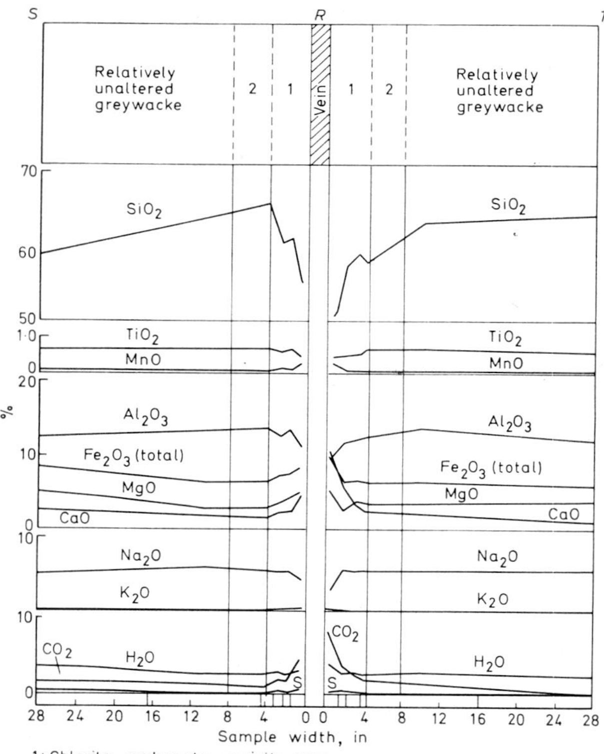

Fig. 5 Chemical variation in alteration haloes of greywacke, vein 11B, 400 W level, Silverfields Mining Corporation, Ltd.

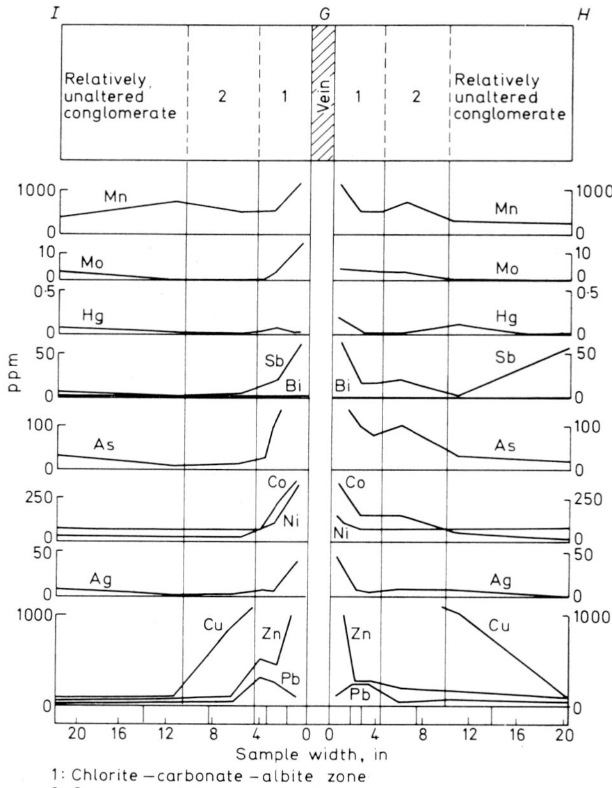

Fig. 6 Distribution of minor elements in alteration haloes of conglomerate, vein 8, 300 W level, Silverfields Mining Corporation, Ltd.

The distribution of minor elements is presented in Fig. 6. The wallrocks enclosing the vein are enriched in Cu, Zn, Ag, Ni, Co, As, Sb, Mo and Mn compared with the background values, and the endogenic haloes have a bilateral

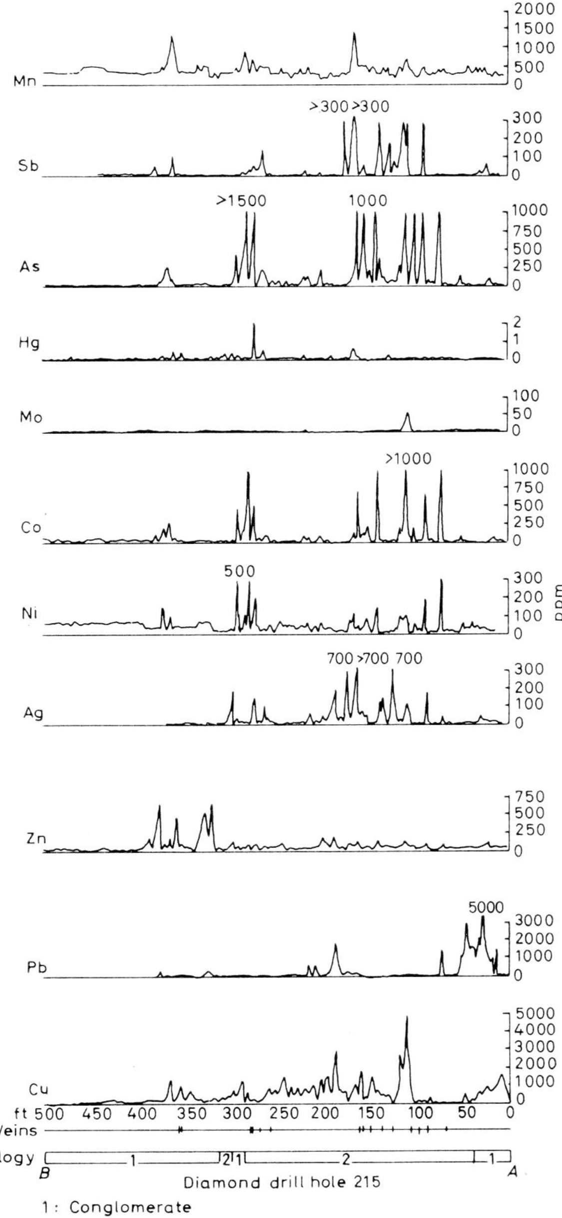

Fig. 7 Distribution of minor elements in Cobalt sediments, 300 level, Silverfields Mining Corporation, Ltd.

symmetry about the vein. Pb increases slightly towards the vein, but, locally, the distribution of Pb is erratic. Hg does not change appreciably. The concentration of Bi does not change from the background value.

Dispersion patterns in the Silverfields and Little Silver deposits indicate that the endogenic haloes of Ag, Ni, Co, As and Sb adjacent to

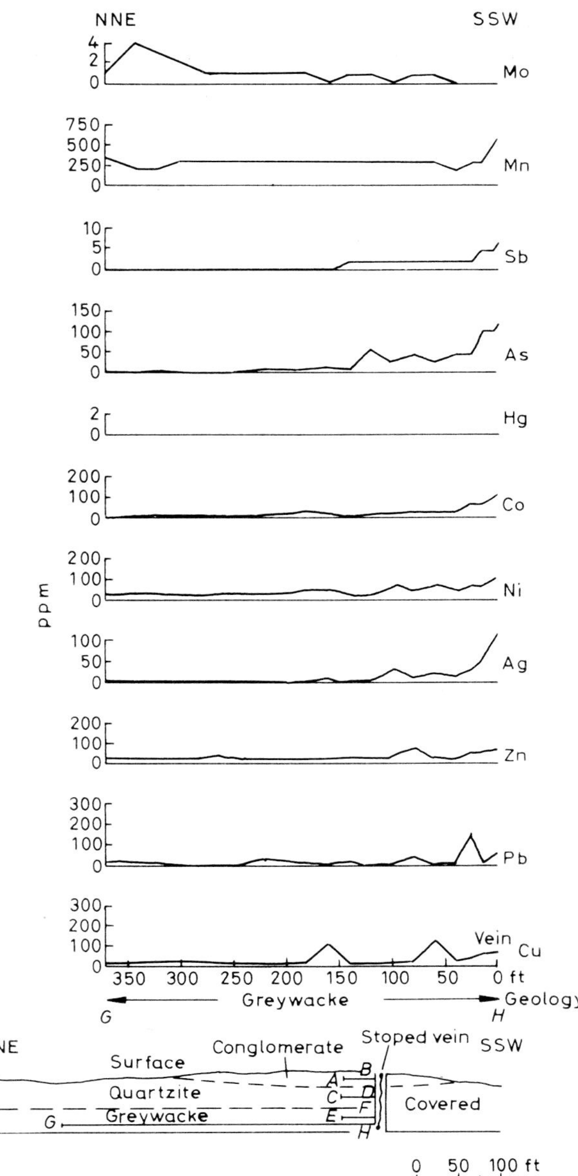

Fig. 8 Distribution of minor elements in Cobalt sediments (greywacke), Little Silver vein, and geological section in vicinity of vein

veins extend outwards up to 90 ft or more (Figs. 7 and 8).

Wallrock alteration in the Keewatin rocks

Mineral assemblages of wallrock alteration in the Keewatin greenstones are grouped into (1) a chlorite–epidote–carbonate–sericite zone and (2) a sericite–carbonate–chlorite zone adjacent to the vein.[4] The chlorite–epidote–carbonate–sericite zone corresponds to the greenish-grey to greyish-white bleached band. Plagioclase has been sericitized and epidotized. The chlorite is mostly iron-poor. Titaniferous magnetite has been altered to ilmenite. The sericite–carbonate–chlorite zone corresponds to the dark green band at the vein–

wallrock contact. Carbonatization and sericitization are intense and widespread, the carbonates and sericite replacing most of the feldspar and epidote and part of the chlorite. The proportion of sericite and carbonate increases and chlorite decreases towards the vein. The carbonate is mainly ankerite. The chlorite is mostly iron-rich and belongs to the ripidolite–pycnochlorite group.[9]

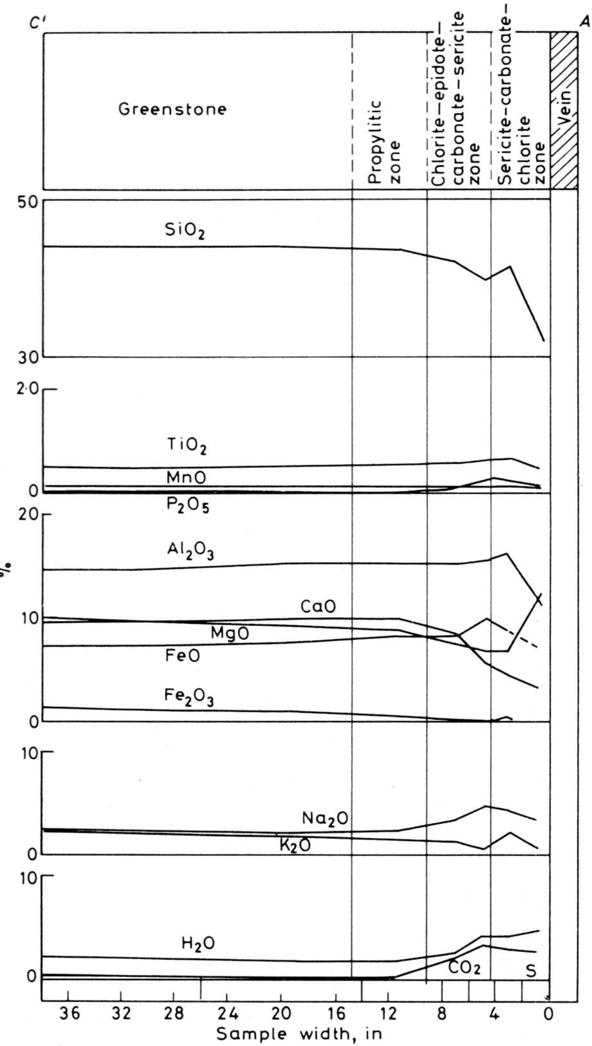

Fig. 9 Chemical variation in alteration haloes of greenstone, 291 level, Hiho Silver Mines, Ltd.

The distribution of the major elements in the alteration zones is presented in Fig. 9. The SiO_2, CaO and Fe_2O_3 contents decrease in the chlorite–epidote–carbonate–sericite and sericite–carbonate–chlorite zones compared with the relatively unaltered rock. This decrease is paralleled by an increase of H_2O and CO_2. Na_2O, Al_2O_3, FeO, TiO_2 and P_2O_5 increase in the chlorite–epidote–carbonate–sericite zone and decrease in the sericite–carbonate–chlorite zone, whereas the reverse is true for MgO. S and MnO do not vary.

The distribution of the cations in the chlorite of the Keewatin wallrock indicates that Si, Ca, K and Ti decrease in the chlorite–epidote–carbonate–sericite and sericite–carbonate–albite zones, whereas the reverse is true for Al, Mg and Fe^{+2}. Mn and Fe^{+3} increase slightly in the sericite–carbonate–chlorite zone.

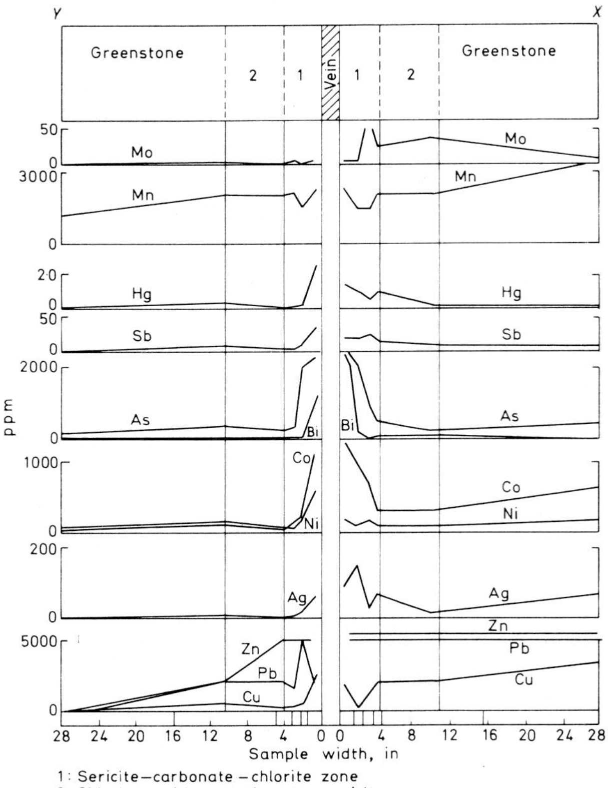

Fig. 10 Distribution of minor elements in alteration haloes of greenstone, vein 8, 500 level, Silverfields Mining Corporation, Ltd.

The distribution of minor elements in the Keewatin wallrock is given in Fig. 10. Co, As and Hg are enriched in the sericite–carbonate–chlorite zone compared with the background values. Ni, Ag, Bi and Sb increase slightly in the sericite–carbonate–chlorite zone, and Zn and Pb increase in both zones. Cu, Mn and Mo have an erratic distribution.

Discussion

Wallrock alteration haloes enclosing the Cobalt deposits demonstrate convergent alteration in that intense alteration has changed andesite, basalt, lamprophyre, rhyolite porphyry, diabase, greywacke, conglomerate and quartzite to a uniform rock type, which consists of chlorite, sericite and ankerite. The lateral zonation adjacent to the veins appears to have developed contemporaneously, or nearly so, with the deposition of the vein minerals. The evidence in favour of a contemporaneous origin is (1) that certain constituents, such as carbonates, chlorite, various sulphides and arsenides, present in the altered zones, are the same as those in the veins, and (2) that the same type of alteration zones are present in both the ore-bearing and gangue-bearing veins without ore.

Two generations of chlorite are present: chlorite spots in the Cobalt Series sediments and Keewatin rocks, developed at an early stage by low-grade contact metamorphism, associated with intrusion of the Nipissing diabase, and younger chlorite, developed as a result of wallrock alteration adjacent to the vein during vein formation. Hematitization associated with albitization involves the development of red alteration haloes in places. The red haloes were formed by wallrock alteration processes, and the red colour is due to a large concentration of minutely disseminated hematite flakes in albite.

The combined effects of chloritization, carbonatization, albitization, sericitization, pyritization and arsenopyritization involve an introduction of CO_2, H_2O, S, As and minor Na and Mn, and a removal of Si and Al. Some Ca, Mg, Fe, K, P and Ti were redistributed within the alteration zones, and some migrated into the vein.

Wallrock alteration resulted mainly from the transfer of H_2O, CO_2 and S into the walls from the vein fluid. Migration of constituents took place where a chemical potential gradient developed between the vein system and the wallrock.[4] Migration of elements from the vein to the wall took place by two possible processes: (1) mass transport in solution and (2) diffusion of material through the solution. Mass transport in solution took place along fractures, pores, grain boundaries and other discontinuities in rocks, under a pressure gradient. Where mass transport in the solution was slow, migration took place by diffusion in an aqueous medium. According to Korzhinskii,[13] in diffusion processes the chemical potentials of components in two adjacent zones tend to equalization, whereas the concentrations of components in pore solutions of two adjacent zones may be different. In flow of solution these conditions are reversed.

Guides for exploration

Three features may be used in the Cobalt area as a guide to the discovery of silver veins: wallrock alteration, vein gangue materials and association of certain rock types.

Wallrock alteration

Wallrock alteration products can be used under certain circumstances for the exploration of silver veins. The alteration zones are only a few inches wide, and their use as a guide to ore is, therefore, limited. The primary dispersion of minor elements in the wallrocks is extensive, however, and this parameter has considerable potential as an exploration tool.

Physical

The well-developed dark chlorite zone which adjoins the vein grades outward into a light greyish-white bleached zone, which exhibits good bilateral symmetry with respect to the vein. These coloured zones are only a few inches wide, and although these are very diagnostic, they can be used only in a detailed exploration programme and not on a regional scale.

Mineralogical

The mineralogical regularity in the wallrock alteration is evidenced by the tendency of dissimilar host rocks, such as the Nipissing diabase, Cobalt Series sediments and Keewatin rocks, to converge to a uniform type of alteration closely related to the silver veins. Although these mineralogical zones are distinct, they are narrow, and can only be used in a detailed exploration programme and not on a regional scale.

Spotted chloritic alteration is widespread in the mineralized zones of the Cobalt Series sediments and Keewatin rocks adjacent to the upper and lower contacts of the Nipissing diabase. No significant correlation could be made between the spotted chloritic alteration and an individual vein, but the distribution of silver veins is confined to this zone of spotted alteration. The spotted alteration is, therefore, useful in a general way as a guide in the search for silver deposits.

Chemical

The distribution of minor elements in various rocks may serve as a valuable guide to ore. The primary dispersion of minor elements in the wallrocks depends on the nature of the host rock. The primary halo is widespread in the Cobalt Series sediments, less extensive in the Keewatin rocks, and relatively local in the Nipissing diabase.

The dispersion of Ag, As, Ni, Co, Sb, Mn and Hg outward from the veins in the Cobalt Series sediments is broad, and veins may be indicated as far as 80–100 ft away by these elements. The dispersion patterns of these elements in the Keewatin rocks have a similar trend and can be traced 50–60 ft outward from the vein. The distribution in diabase outward from the vein is much restricted and can be traced only a few feet. Thus, the pronounced trace-element dispersion patterns in certain rock types can be used on a broad scale to locate the narrow silver veins.

Vein gangue materials

Various types of carbonate gangue minerals are associated with different grades of silver veins, and these may be used in a detailed exploration programme.

Grey to pinkish-grey ankerite and dolomite with minor calcite are dominant in the high-grade silver veins in the Cobalt Series sediments, buff to pinkish-white calcite with minor ankerite and dolomite dominate in the intermediate grade, and calcite with recrystallized quartz is dominant in the barren veins.

The silver ores in the Nipissing diabase are associated with grey to dark grey and pink rhombohedral and sugary calcite with minor ankerite and dolomite, whereas the fine-grained pure white to salmon-pink calcite with quartz is devoid of silver.

In the Keewatin rocks grey to dark grey ankerite and dolomite, with some pink calcite, dominate in rich to intermediate-grade silver veins.

Association of certain rock types

A close association of the silver–nickel–cobalt ores with the Keewatin interflow sedimentary rocks and certain volcanic flows has been recognized. Furthermore, the Keewatin greenstones and interflow sedimentary rocks contain higher than average background amounts of Ag, As, Sb and Hg. These rocks are postulated as the source of the vein metals, the younger Nipissing diabase serving as an energy source for their mobilization. Thus, the most favourable prospecting localities may be restricted to areas where the Keewatin interflows adjoin the Nipissing diabase, or where they underlie the Cobalt Series sediments adjacent to the diabase.

Such conditions are found in the Cobalt, South Lorrain, Casey Township and Gowganda areas, and it is suggested that prospecting and exploration should be concentrated on those areas.

Summary and conclusions

The economic mineral deposits of the Cobalt area are predominantly rich silver-bearing veins cutting the Cobalt Series sediments, Nipissing diabase, Keewatin greenstones and associated rocks. The silver veins contain, essentially, native silver, carbonates, nickel–cobalt arsenides, a few sulphides and a variety of other minerals.

Wallrock alteration haloes which enclose the veins have developed in all the host rocks, and the physical, mineralogical and chemical changes are similar in most. Alteration is marked in the Nipissing diabase, Keewatin greenstones, rhyolite porphyry and lamprophyre, but is much less pronounced in the Cobalt Series sediments. Wallrock alteration demonstrates convergent alteration in that intense alteration has changed andesite, basalt, lamprophyre, rhyolite porphyry, diabase, greywacke, conglomerate and quartzite to a uniform rock type—mainly chlorite, sericite and ankerite. The combined effects of chloritization, carbonatization, albitization, sericitization, pyritization and arsenopyritization involved an introduction of CO_2, H_2O, S, As and minor Na and Mn, and a removal of Si and Al. Some Ca, Mg, Fe, K, P and Ti were redistributed within the alteration zones, and some migrated into the vein.

The dispersion of Ag, As, Sb, Ni, Co, Mn and Hg in the Cobalt Series sediments outwards from the vein is broad, and the endogenic haloes extend outward up to 90 ft in places. A somewhat similar pattern is present in the Keewatin rocks, but the endogenic haloes are generally more restricted. The dispersion is narrow in the Nipissing diabase.

Wallrock alteration resulted mainly from the transfer of materials into the walls from the vein fluid, and migration of constituents took place where a chemical potential gradient developed between the vein system and the wallrock. Migration of elements in the wallrock from the vein took place by mass transport in solution and diffusion of material through solution. Wallrock alteration products can be used under certain circumstances for the exploration of silver veins. The endogenic haloes of minor elements in the wallrocks are extensive, and this parameter has considerable potential as an exploration tool. Mineralogical zones can be used in a detailed exploration programme. Spotted chloritic alteration is widespread in the mineralized zones in the Cobalt Series sediments and Keewatin rocks adjacent to the upper and lower contacts of the Nipissing diabase. This type of alteration is useful in a general way as a guide in the search for silver deposits. Various types of carbonate gangue minerals may be used in a detailed exploration programme.

A close association of silver–nickel–cobalt ores with the Keewatin sedimentary interflows and certain volcanic flows has been recognized, and the most favourable prospecting localities should be restricted to areas where the Keewatin interflows adjoin the Nipissing diabase, or where they underlie the Cobalt Series sediments adjacent to the diabase.

Acknowledgment

This paper is based, in part, on work for a Ph.D. thesis (by A. S. D.) undertaken at Carleton University under the supervision of R.W.B. and W.M.T. Carleton University, the National Research Council (Grant No. B-176) and the Ontario Department of University Affairs (Grant No. 2058–38) provided financial support, and the Geological Survey of Canada sponsored two seasons of field work and one summer of laboratory work.

References

1. Bastin, E. S. Significant replacement textures at Cobalt and South Lorraine, Ontario. *Econ. Geol.*, **45**, 1950, 808–17.
2. Berry, L. G. ed. The silver–arsenide deposits of the Cobalt–Gowganda region, Ontario. *Can. Mineralogist*, **11**, pt 1, 1971, 1–429.
3. Boyle, R. W. *et al.* Research in geochemical prospecting methods for native silver deposits, Cobalt area, Ontario, 1966. *Pap. geol. Surv. Can.* 67–35, 1969, 91 p.
4. Dass, A. S. Wall rock alteration enclosing the silver deposits, Cobalt, Ontario. Ph.D. thesis, Carleton University, Ottawa, 1970.
5. Eakins, P. R. Cylindroidal jointing in diabase at Gowganda, Ontario. *Proc. Geol. Ass. Can.*, **13**, 1962, 85–93.
6. Gresens, R. L. Composition–volume relationships of metasomatism. *Chem. Geol.*, **2**, 1967, 47–65.
7. Halls, C. Clark, A. M. and Stumpfl, E. F. Some observations on silver–antimony phases from Silverfields mine, Ontario, Canada. *Trans. Instn Min. Metall. (Sect. B: Appl. earth sci.)*, **76**, 1967, B19–24.
8. Halls, C. and Stumpfl, E. F. Geology and ore deposition, Western Kerr Lake Arch, Cobalt, Ontario. In *Mining and petroleum geology* (London: IMM, 1970), 241–84. *(Proc. 9th Commonw. Min. Metall. Congr. 1969, vol. 2)*
9. Hey, M. H. A new review of the chlorites. *Mineralog. Mag.*, **30**, 1954, 277–92.
10. Hore, R. E. Differentiation products in quartz diabase masses of the silver fields of Nipissing, Ontario. *Econ. Geol.*, **6**, 1911, 51–9.
11. Hriskevich, M. E. Petrology of the Nipissing diabase sheet of the Cobalt area of Ontario. Ph.D. thesis, Princeton University, New Jersey, 1952.
12. Knight, C. W. Geology of the mine workings of Cobalt and South Lorrain silver areas. *Rep. Ont. Dep. Mines*, **31**, pt 2 1922, 374 p.
13. Korzhinskii, D. S. The theory of metasomatic zoning. *Mineralium Depos.*, **3**, 1968, 222–31.
14. Lowdon, J. A. *et al.* Age determinations and geological studies. *Pap. geol. Surv. Can.* 62–17, 1963, 140 p.
15. Miller, W. G. The cobalt–nickel arsenides and silver deposits of Timiskaming, 3rd edn. *Rep. Ont. Dep. Mines*, **16**, pt 2 1908, 212 p.
16. Miller, W. G. The cobalt–nickel arsenides and silver deposits of Timiskaming, 4th edn. *Rep. Ont. Dep. Mines*, **19**, pt 2 1913, 279 p.
17. Moore, H. A. Silverfields Mining Corporation Limited. In *Centennial field excursion, northwestern Quebec and northern Ontario* (Montreal: CIM, 1967), 146–9.
18. Petruk, W. Mineralogy and origin of the Silverfields silver deposit in the Cobalt area, Ontario. *Econ. Geol.*, **63**, 1968, 512–31.
19. Symons, D. T. A. Paleomagnetism of the Nipissing diabase, Cobalt area, Ontario. *Can. J. Earth Sci.*, **7**, 1970, 86–90.
20. Thomson, R. Cobalt camp. In *Structural geology of Canadian ore deposits, volume 2* (Montreal: CIM, 1957), 377–88.
21. Thomson, R. *Cobalt silver area, maps 2050, 2051 and 2052; 1:12,000* (Toronto: Ontario Department of Mines, 1964).
22. Van Schmus, W. R. The geochronology of the Blind River–Bruce Mines area, Ontario, Canada. *J. Geol.*, **73**, 1965, 755–80.

550.42:552.3:553.2

Distribution functions and ratios of trace-element concentrations as estimators of the ore-bearing potential of granites

L. V. Tauson, PROF.

V. D. Kozlov, M.SC.

Both of the Institute of Geochemistry, Siberian Branch of the Academy of Sciences of the U.S.S.R., Irkutsk, U.S.S.R.

Synopsis

There are five main types of granitoids, which differ geochemically: plagiogranites, ultra-metamorphic granites, palingenic granites, plumasitic acid granites and agpaitic leucogranites. Decisive with respect to the ore-bearing potential of the granite magma is the emanation concentration of elements as a result of their migration together with the volatile components of the magma and concentration in dome projections of intrusive bodies and in the residual melts of magmatic chambers. There, the potential ore-bearing granites are characterized by an irregular distribution of rare elements, and this assists in their identification by a sharp increase in the concentration dispersion.

This is accompanied by the formation of wide emanation haloes of diverse elements. Identification of geochemical types of granitoids and, in particular, of potential ore-bearing intrusives is made possible by certain element concentration ratios, the most informative being the Ba/Rb ratio; in high-fluorine granites the Li·1000/K ratio should also be employed. Alternative ratios—e.g. F/Li and Li/Zn—are discussed.

The question as to the source of the ore-forming materials involved in endogenic ore formation is considered to be one of the basic problems of geology. Most investigators are now convinced that one of the main sources of ore-forming materials is igneous rock. In some cases a direct genetic link between rock and ore can be established, but in others the two show a paragenetic relationship to a general source. The difficulty of deciding on the magmatic sources of endogenic ores is increased by the fact that ore formation usually involves only one-thousandth part of the ore-forming materials dispersed and hidden in the igneous rocks. Even if it is agreed that endogenic ore formation produces ten times more ore dispersed in haloes in the rocks enclosing the orebodies than in the orebodies themselves, the amount of material separating as ore from the magmatic source will not exceed one-thousandth of the total amount of ore-forming materials dispersed in the rocks.

Exceptions are the volatile components of magma (water, fluorine, chlorine, carbon dioxide, sulphur, etc.), the bulk of which leave the magmatic chamber during crystallization of the magma. The problem of the source of ore-forming materials involved in endogenic ore formation might be positively solved if two basic questions could be answered: (1) What are the geological and geochemical peculiarities in the origin and development of potential ore-bearing magma chambers? and (2) Under what geological conditions are potentially ore-bearing intrusions able to produce concentrations of ore?

In trying to answer the first question the proposition naturally arises that the primary reason for higher concentrations of trace (ore-forming) elements in emanative magma chambers and in ore-forming hydrothermal solutions is that the parent magma is initially enriched in these elements. It would also be right to say that the enrichment of ore-forming elements in post-magmatic emanations is controlled by peculiarities that occur during magma crystallization.

The enrichment of hydrothermal solutions with ore-forming elements may, therefore, be explained by both primary enrichment of the magma and by peculiarities in crystallization and differentiation. It is very important to estimate the relative importance of these two factors.

It is probably fair to say that the concept of 'metallogenic specialization of magma', which states that there need only be an enrichment of the primary magma in the ore-forming elements found in its genetically related deposits, is now generally rejected. Read's famous phrase states that 'there are granites and granites': granites differ not only in their mode of formation but geochemically.

On this basis they fall into five main categories: (1) plagiogranites, which are terminal acid differentiates of gabbroic magma; (2) ultra-metamorphic granites, which form by partial melting of highly metamorphosed rocks; (3) palingenic granites, which form as a result of complete remelting of different metamorphic rocks in the earth's crust: these can apparently be further subdivided into normal and subalkaline palingenic granites; (4) plumasitic leucogranites, often termed 'tin-bearing': these are genetically the latest acid differentiates of large chambers of normal palingenic granitic magma or acid differentiates of abyssal magma chambers of alkali basalts with a high content of potassium and volatiles; and (5) agpaitic leucogranites, 'rare-metal' or 'columbite-bearing' according to some workers: these may be considered genetically as terminal differentiates of large chambers of subalkaline palingenic granite magma or as acid differentiates of abyssal chambers of alkali basalt magma with a lower volatile content.

Preliminary data on the trace-element composition of different geochemical types of granite are presented in Table 1. It should be noted that the plagiogranite examples are from the plagiogranite complexes of eastern Sayan, the ultra-metamorphic granites are from Aldan,[6] the agpaitic leucogranites are from eastern Mongolia,[4] and the examples of palingenic granite and plumasitic leucogranite are from eastern and central Transbaikalia. It is evident that the trace-element content of the different granitic rock types varies considerably. For certain elements the contents differ by from 50 to 100 times (e.g. lithium,

*Table 1 Content of trace elements in granitoids of different geochemical types**

Elements	Plagiogranites	Ultra-metamorphic leucocratic granites	Palingenic granites	Plumasitic leucocratic granites	Agpaitic leucocratic granites	Vinogradov,[1] ppm
K	0·5	5·4	3·3	4·0	3·8	3·3
Na	3·2	2·5	2·9	2·8	3·2	2·8
F	0·015	0·014	0·06	0·30	0·1	0·08
Li	2	8	36	97	34	40
Rb	4	140	140	400	130	200
Be	0·6	0·6	3·5	6·8	4·9	5·5
Sr	190	420	300	100	7·4	300
Ba	180	1600	750	200	45	830
Sn	2·7	2·6	5·3	6·3	7·4	3·0
W	0·7†	0·7†	2·0	4·1	2	1·5
Mo	1·3†	1·4	1·6	1·4	—	1·0
Zn	70	43	45	57	69	60
Pb	4·4	14	25	30	22·5	20
Zr	50	238·8†	200†	260‡	1500	200
Hf	1·1	—	—	9·3‡	45	1
Nb	1·6	14·7†	20†	22·6‡	29	20
Ta	0·6	0·83†	3·5†	4·0‡	3·7	3·5
Tr+Y	—	220	210	420‡	500	185

*K, Na, and F, %; rest, ppm.
†Lyakhovich and Ovchinnikov.[5]
‡Kovalenko *et al.*[4]

rubidium, strontium and barium). The plagiogranites are acid differentiates of gabbroic magma and usually form massifs, which are small in size, and occur in association with rocks of basic composition. They differ in mineralogy and form part of a series of rocks ranging from diorite to plagiogranite and even alaskite. Geochemically, they are characterized by very low contents of potassium, lithium, rubidium, beryllium, tantalum and lead.

The ultra-metamorphic granites are usually developed over wide areas and, characteristically, have gradational or vague contacts with the host gneisses and migmatites. In spite of the differing composition of the metamorphic substrate from which these acid granitic melts form, they have a low content of volatiles, lithium, beryllium and tantalum, a high barium content and a predominance of potassium over sodium. Palingenic granites are a common geochemical type. They can form large magmatic bodies at a considerable depth, and are then frequently associated with a genetically related series of melanocratic to leucocratic rocks. Granodiorites and biotite granites are the most common types, and usually form the major part of palingenic granite complexes.

Table 1 shows that the rare-metal content is close to that of the 'average granite' of Vinogradov.[1]

In addition to normal palingenic granite, it is necessary to distinguish a sub-group of subalkaline palingenic granites, which form as a result of the melting of highly (or deeply buried) metamorphic rocks. Differences between the normal and subalkaline palingenic granites found in eastern Transbaikalia have been described elsewhere.[8]

obtained during a study of 20 magmatic complexes. For almost all the ore-forming elements studied there is not more than a twofold difference in their concentration in geochemically similar granites of different ages.

The ore deposits in the region are mainly related to the Mesozoic granites, whose ore-forming element contents are very similar to those of the Proterozoic granites and only slightly higher than those of the Palaeozoic granites.

Plumasitic leucogranites include leucogranites which have formed as hypabyssal intrusions at the terminal stage of orogenic events that involve eugeosynclinal zones or form during the latest phases of magmatism in zones undergoing remobilization. They are characterized by a predominance of potassium over sodium and are saturated with water and fluorine. Their fluorine content is almost five times higher than that of normal palingenic granites. Plumasitic leucogranites are also characterized by high lithium, rubidium, beryllium, tin, tungsten, niobium, tantalum and rare-earth contents. These intrusions are frequently associated with tin deposits and are often called 'tin-bearing'.

The small size of plumasitic leucogranite intrusions, their formation at a shallow depth (3–4 km), and the fact that they are saturated with volatiles (water and fluorine), which can considerably modify the silicate melts, predispose them to the intensive emanational differentiation of materials, which results in a very irregular (internal) trace-element distribution. This is especially striking in the upper parts of the intrusions, where additional, intense, high-temperature metasomatic processes (greisenization) may occur. These processes are not confined to

Table 2 Average ore content (ppm) in palingenic granites in eastern Siberia

Complex age	Complex number	Sn	Be	Mo	Pb	Zn
Proterozoic	5	5·6	4·3	1·8	22	66
Palaeozoic	6	3·9	2·3	1·6	16	57
Mesozoic	6	5·4	4·7	2·0	26	44
'Average granite' (after Vinogradov[1])		3·0	5·5	1·0	20	60

In spite of the ability of palingenic granites to form magmatic bodies of different sizes at different depths (abyssal, mesoabyssal and hypabyssal intrusions), they usually show only slight differences in trace-element content. Table 2 presents examples of data on the distribution of certain ore-forming elements in recently studied palingenic granite complexes in eastern Siberia. The data are the result of processing some 10 000 analyses

enrichment of the apical portions of the intrusions with a range of trace elements but also result in a considerable increase in the variance of these elements. This is illustrated in Table 3, in which the distribution parameters of a range of trace elements in hypabyssal intrusions of normal palingenic granite are compared with those of plumasitic leucogranites occurring in similar geological situations.

Table 3 Distribution of trace elements in hypabyssal intrusions of Mesozoic granites in eastern Transbaikalia

Geochemical type	Complex	F, %	Li		Rb		Be		Sn		Zn		Pb	
			$\bar{X}$	σ^2	$\bar{X}$	σ^2	$\bar{X}$	σ^2	$\bar{X}$	σ^2	$\bar{X}$	σ^2	$\bar{X}$	σ^2
Palingenic granites	Shakhtaminsky	0·08	32	71	150	211	3·0	0·28	2·9	0·41	41	150	25	68
	Amudjikano–Sretensky	0·14	66	357	190	2123	4·6	0·59	4·3	—	39	162	31	21
Plumasitic leucogranites	Kukulbeisko–Kharalginsky	0·30	97	571	400	3546	6·8	7·78	6·3	6·83	57	629	30	116

$\bar{X}$, arithmetic mean, ppm; σ^2, variance.

Agpaitic leucogranites are distinguished by being undersaturated in water and contain alkali-rich melanocratic minerals such as aegirine, members of the amphibole series, riebeckite–arfvedsonite, firelite, and alkali titano- and zircono-silicates. Agpaitic leucogranites are also peculiar in having a high zirconium and hafnium content, a high content of rare earths, yttrium, tin, niobium and tantalum, and a low content of strontium and barium.

The nature of the distribution of ore-forming trace elements in granite intrusions serves not only to distinguish different geochemical types but can also identify those having ore-bearing potential.

In central Transbaikalia there is tin mineralization which is related to Upper Jurassic hypabyssal intrusions of plumasitic leucogranite and mesoabyssal batholithic domes of Triassic palingenic granite. In Table 4 a comparison is made of the distribution of lithium and tin in the barren and potential ore-bearing facies of the Triassic palingenic granites and the Jurassic plumasitic leucogranites. It is evident that in areas of potential ore-bearing palingenic granite the concentrations of lithium and tin increase slightly, but the variances are 9 and 18 times as high, respectively. In the plumasitic leucogranites these dispersion parameters are even more important. The reason for such irregular distributions of elements in different portions of one and the same phase of palingenic granites is related to the fact that in mesoabyssal batholiths and hypabyssal intrusions, in particular, extensive migration of materials, volatiles and associated ore-forming elements occurs in the apical portions just prior to crystallization. As a result, the trace-element content of the apical and abyssal portions of the intrusion may be considerably different.

For example, observations by N. S. Kravchenko[2] on the Verkhneurmijsky granite massif showed that the tin content of the apical part was 25 ppm and that of the abyssal part (800 m beneath the roof of the intrusion) only 5·8 ppm. The tin content of the biotite also changed considerably: close to the roof of the intrusion the biotite contained 260 ppm of tin, but at the deeper level only 40 ppm. In an interval of 800 m the tin content of the biotite (which is the main concentrator of tin during the final stages of crystallization) increased sixfold.

Table 4 Distribution parameters of lithium and tin in barren and potential ore-bearing facies of granites in central Transbaikalia

Granite type	Ore content	Lithium		Tin	
		$\bar{X}$	σ^2	$\bar{X}$	σ^2
Palingenic	Barren Mesozoic biotite and leucocratic granites	47	100	4·3	0·7
	Potential ore-bearing Mesozoic biotite and leucocratic granites	64	710	7·4	12·6
Plumasitic leucogranites	Potential ore-bearing granites	74	880	9·4	54

$\bar{X}$, arithmetic mean, ppm; σ^2, variance.

A similar enrichment of the apical portions of mesoabyssal and hypabyssal intrusions with trace elements and a sharp decrease in concentration at depth have also frequently been observed in other regions. In addition to tin, lithium, beryllium, tungsten, niobium and tantalum, etc., are also concentrated in the apical portions of intrusions. This emanative concentration of a variety of trace elements in the apical portions of intrusions results not only in a simple increase in values, giving the impression that 'metallogenic specialization' of the magma has occurred, but also produces a proportionally much greater increase in the variance of the elements.

The passage of trace elements into the apical portions of intrusions occurs at the magmatic stage and during the early phases of the post-magmatic history. Associated with the concentration of volatiles and related rare elements in the apical portions of intrusive bodies may be widespread penetration of the exo-contact and intrusive zones by the accumulated emanations. This results in the formation of superintrusive emanation haloes for some elements in which the distribution parameters differ greatly from those to be found at some distance from the intrusions and those found in areas where high-temperature greisenization or hydrothermal mineralization have taken place. This can be illustrated by an example showing the distribution of lithium, tin and lead in the vicinity of the Tarbaldjeisky tin deposits, eastern Transbaikalia (Table 5). Since these elements are lognormally distributed, geometric means have been taken to represent the average concentrations, and the variances of the log-transformed data have been used to give a measure of the dispersion.

The concentrations of lithium and lead increase only slightly above background in areas in which mineralization haloes occur, but the variances of the log-transformed data differ by a factor of ten. In haloes surrounding zones of greisenization the concentrations of lithium and tin increase considerably, but the increase in the variance of the log-transformed data is proportionately smaller. The same holds for haloes surrounding zones of hydrothermal mineralization. If the emanation haloes for lead and lithium that surround greisen and hydrothermal mineralization are compared, there is a considerable decrease in the concentrations of lead and an increase in the concentration of lithium in the greisen haloes. It is, therefore, possible that the lithium/lead ratio may be useful in identifying the type of halo present in districts in which multi-element anomalies occur. The Li/Pb ratio in greisen haloes may be many times higher than in hydrothermal ones. In addition to symmetrical intrusion haloes, linear elongated emanation haloes (predating ore-formation) may occur along large tectonic fault zones.

An example of such a halo is the Shakhtaminsky deep fault in eastern Transbaikalia. The fault can be traced for a considerable distance, but the emanation halo occurs in a section 20–25 km in length which passes over a layered hypabyssal

Table 5 Distribution parameters of lithium, tin and lead in emanation, greisen and hydrothermal haloes, Tarbaldjeisky deposit, eastern Transbaikalia

District types of ore field	Li $\bar{X}$	σ^2	Sn $\bar{X}$	σ^2	Pb $\bar{X}$	σ^2	Li/Pb
Host sandstones (background)	28	0·01	3	0·04	10	0·02	2·8
District with emanation halo occurrence	41	0·09	10	0·15	16	0·25	2·5
District with greisenization development	505	0·46	55	0·23	6	0·10	84
District with hydrothermal vein mineralization development	115	0·25	34	0·23	30	0·52	3·8

$\bar{X}$, antilog of the geometric mean, ppm; σ^2, variance of the log-transformed data.

Table 6

	Mo $\bar{X}$	σ^2	Cu $\bar{X}$	σ^2	Pb $\bar{X}$	σ^2	Zn $\bar{X}$	σ^2
District of emanation halo of Shakhtaminsky structure zone	3·1	23·5	118	4200	60	440	65	180
Beyond halo districts of Shakhtaminsky structure zone	1·8	0·9	13	167	18	25	43	36

$\bar{X}$, arithmetic mean, ppm; σ^2, variance.

intrusion of palingenic granite of Jurassic age. The distribution parameters for molybdenum, copper, lead and zinc within the emanation halo vary considerably from the background (Table 6). Only copper and lead show a significant increase in concentration (nine and three times, respectively). The variances of molybdenum and copper, however, increase 25-fold and the variance of lead increases 18 times. This suggests that, in order to evaluate the ore-bearing potential of granite intrusions and the nature of the anomalies in ore-bearing regions, it is important not only to consider the absolute concentrations of the elements studied but also their variances. In many cases the latter have been found to provide more contrast and to be more useful.

In identifying geochemical types of granite and ore-bearing intrusions, in particular, it is also useful to use element concentration ratios (see Table 7).

as 'emanation concentration', is apparently the factor that is instrumental in deciding the ore-bearing potential of a magma.

To understand the part played by concentration deposition it is necessary to consider the different behavioural trends shown by the elements. The Ba/Rb ratio reflects an interesting trend in the development of granite magma, and Table 7 shows that the more widely used K/Rb ratio is less informative. As is known, barium tends to accumulate in high-temperature primary potassium minerals separating from granite magma, and it has a weak relationship with fluorine. Because of this it shows a strong tendency to be dispersed during crystallization. Rubidium also actively replaces potassium in the lattice of potassium minerals, but it shows a stronger relationship with fluorine and migrates with it. As a result, in magma differentiating with a high volatile content the barium concentration decreases and the rubidium content increases. The Ba/Rb ratio in such granites is, therefore, always minimal. Table 7 shows that the Ba/Rb ratio in palingenic granites differs from that found in plumasitic leucogranites by a factor of 10.

Table 7 Concentration ratios in geochemical granitic types

Concentration ratios	Plagiogranites	Ultra-metamorphic granites	Palingenic granites	Plumasitic leucogranites	Agpaitic leucogranites	'Average granite'[1]
K/Na	0·16	2·2	1·1	1·4	1·2	1·2
K/Rb	1250	385	240	100	290	165
Ba/Rb	45	11·5	5·3	0·5	0·34	4·1
(Li × 1000)/K	0·4	0·15	1·1	2·4	0·9	1·2
F/Li	75	16	16	31	29	20

When the geochemical history of elements in magmatic processes is being considered, it should be realized that their behaviour is governed by two basic tendencies. The majority of trace-element atoms are dispersed during crystallization in the lattices of rock-forming and accessory minerals and are effectively trapped in the crystallizing igneous rocks. A proportion of the trace-element atoms—in particular, those related to the volatile components—migrate with the volatiles during differentiation and crystallization processes. This may lead to concentration of volatiles and related trace elements in the residual melts of magma chambers. This tendency, which can be referred to

In some mineralized areas the difference between the Ba/Rb ratio of potential ore-bearing and barren granites is even greater. The examples noted in Table 8 are taken from the Hercynian granites of Czechoslovakia and the Mesozoic granites of Mongolia.[4] In the first case the Ba/Rb ratio in the ore-bearing granites is 50 times smaller than that in the barren ones, and in the second case it is 80 times smaller.

Table 8

		Average contents, ppm Ba	Rb	Ba:Rb
Czechoslovakia	Gorsky granites (barren)	2200	150	15
	Rudohorsky granites (ore-bearing)	150	580	0·26
Mongolia	Palingenic granites (barren)	640	250	2·5
	Plumasitic leucogranites (ore-bearing)	20	640	0·03

Table 9

Element	Complex	Phase one	Phase two (main phase)	Phase three	Vein aplites
F, %	Susamyrsky (central Tyan-Shan)	0·070	0·100	0·125	0·025
	Verkhneundinsky (east Transbaikalia)	0·045	0·068	0·058	—
	Jidinsky (west Transbaikalia)	0·082	0·098	0·079	0·015
	Kalbinsky (east Kasakhstan)*	0·070	0·150	0·036	0·018
Li, ppm	Susamyrsky	28	32	55	7
	Verkhneundinsky	25	42	39	8
	Jidinsky	34	35	32	10·5
	Kalbinsky*	85	123	77	26
F/Li	Susamyrsky	25	31	23	31
	Verkhneundinsky	18	16	15	—
	Jidinsky	24	28	25	14
	Kalbinsky	8	12	9	7

*Stavrov.[7]

The best indicator of the development of differentiation and emanation processes that have involved the migration of fluorine is lithium. For positive identification of this process it is necessary to consider the unusual ratio between lithium and potassium (Li·1000/K). The data given in Table 7 show that in plumasitic leucogranites where migration of material is related to that of fluorine the ratio (Li·1000/K) is highest where fluorine migration has been greatest. The plumasitic leucogranites differ from the agpaitic ones and the palingenic granites in this respect. There is a strong correlation between lithium and fluorine in the formation of haloes related to ores of different genesis. Table 5 shows that the Li/Pb ratio in greisen haloes is 20 times greater than that in those containing hydrothermal vein mineralization. The Li/Zn ratio is similarly useful.

A study of the Li/Zn ratio in granites from eastern Transbaikalia has shown that in ore-bearing complexes it is usually greater than 1 (Gudjirsky complex, 1·1; Amudjikano–Sretensky, 1·8; Kukulbeisko–Kharalginsky, 1·7), but in barren granite complexes it is less than 1 (Amanansky, 0·3; Kunaleisky, 0·6; Shakhtaminsky, 0·8). The differences are not very great, but they do show that lithium is more mobile than zinc, which is one of the least mobile of the ore-forming elements in magmatic processes.

The F/Li ratio is also very important to an understanding of the processes involved in the migration of emanations. The very similar behaviour of lithium and fluorine in magmatic processes is illustrated by the stability of their relationship in the intrusive portions of palingenic granite complexes (Table 9).

The relationship of lithium and fluorine to each other remains constant during successive stages of accumulation in late differentiates, as, for example, in the leucogranites of the Susamyrsky abyssal batholith, and also during degassing of late phases of mesoabyssal batholiths (Verkhneundinsky, Jidinsky and Kalbinsky). The constancy of the F/Li ratio in intrusive phases of complex magmatic complexes may be used to determine to which series the granites under investigation belong.

Summary

The data presented here have shown that element distribution functions, concentration levels, variances and concentration ratios can usefully be employed to distinguish differences in processes of crystallization and the dispersion and concentration of elements in granites of different geochemical types.

The use of distribution functions and ratios of trace-element concentrations in estimating the ore-bearing potential of granites is, at present, best understood for plumasitic leucogranites, which have tin, tungsten and other elements associated with them. Their use in estimating the ore-bearing potential of palingenic granites and agpaitic leucogranites is less well established, and for plagiogranites and ultra-metamorphic granites there are, as yet, insufficient data available.

References

1. VINOGRADOV, A. P. The average content of chemical elements in the main types of igneous rocks of the earth. *Geokhimiya*, 1962, 555–71; *Geochemistry*, 1962, 164–64.

2. GOVOROV, I. N. *et al.* Features and factors of geochemical specialization of granitoids of the Far

East. In *Geochemical criteria of potential orebearing of granitoids* (Irkutsk: Siberian Institute of Geochemistry, 1970), pt 1, 83–145. (Russian text); *Referat Zh., Geol.,* 1970, 11V 74.

3. KOVALENKO, V. I. and POPOLITOV, E. I. *Petrology and geochemistry of trace elements of alkaline and granitic rocks of north-east Tuva* (Moscow: Nauka, 1970), 258 p. (Russian text)

4. KOVALENKO, V. I. *et al. Raremetallic granitoids of Mongolia* (Moscow: Nauka, 1971), 240 p. (Russian text)

5. LYAKHOVICH, V. V. and OVCHINNIKOV, L. N. Some geochemical features of granitoids and their orebearing capacity. In *Geochemical criteria of potential orebearing of granitoids* (Irkutsk: Siberian Institute of Geochemistry, 1970), pt 1, 47–82 (Russian text); *Chem. Abstr.,* **75**, 1971, 131695.

6. PETROVA, Z. I. *et al.* Ultrametamorphism and granite formation in basic crystalline schists of Aldan. In *Problems of petrology and geochemistry of granitoids* (Sverdlovsk: Institute of Geology and Geochemistry, 1971), 148–58. (Russian text)

7. STAVROV, O. D. *Geochemical features of lithium, rubidium and caesium in the formation process of granite intrusives and their related pegmatites* (Moscow: Gosgeoltekhizdat, 1963), 142 p.; *Chem. Abstr.,* **59**, 1963, 15076.

8. TAUSON, L. V. *et al.* Geochemical particularities of extra-geosyncline granitoids of folded regions of different age. *Yb. Sib. Inst. Geochem. 1970,* 1971, 116–27. (Russian text)

550.42:546:553.067/.068(713)

Bedrock geochemistry as a guide to areas of base-metal potential in volcano-sedimentary belts of the Canadian Shield

P. H. Davenport, PH.D.

Ian Nichol, PH.D., M.I.M.M.

Both of the Department of Geological Sciences, Queen's University, Kingston, Ontario, Canada

Synopsis

Base-metal massive sulphide deposits associated with volcanic rocks are considered by many geologists to be of volcanic-exhalative origin. As such, they constitute an intrinsic part of these volcanic rocks, and the geochemistry of any particular group of these rocks might be expected to reflect the composition of the deposits developed in them. The definition of such a relationship would be valuable in base-metal exploration in this type of environment, since it would then be possible to predict the characteristics of any deposits which might occur in any particular group of volcanic rocks. The geochemistry of two mafic to felsic cycles of volcanic rocks and the base-metal deposits of each cycle were studied in the Birch–Uchi Lakes Belt, northwest Ontario. The distributions of metals in the volcanic rocks and sulphide deposits of each cycle are discussed and compared. The felsic volcanic rocks and sulphide deposits of the upper cycle, which contains the South Bay Mines, Ltd., Zn–Cu–Ag ore deposit, are higher in Zn and lower in Cu, Co and Ni than the equivalent rocks and deposits of the lower cycle, which contains no economic deposits. The distributions of Cu and Ni vary significantly along strike within the upper cycle, but the distribution of Zn is constant over a distance of 12 miles. Thus, in the study area the geochemistry of the felsic volcanic rocks of the two cycles does, in part, reflect the composition of the sulphide deposits.

It is suggested that the high Zn contents of felsic volcanic rocks containing significant mineralization may be a useful guide in distinguishing mineralized from unmineralized regions. The high average Zn contents of felsic volcanic rocks containing economically significant mineralization may be a useful guide in distinguishing these cycles from those which are unmineralized.

The Canadian Shield has long been a source of a diversity of mineral products, collectively making a major contribution to Canada's mineral wealth. Mineral exploration in the Shield is presented with many unique problems due to paucity of outcrop, glacial history and extensive muskeg cover. Exploration hitherto has largely been based on classical prospecting, geological mapping and geophysics. More recently, attention has been given to the development of geochemical exploration methods having regard to obtaining an understanding of factors affecting metal dispersion in this environment.

The results are presented here of a study of the geochemical characteristics of massive sulphide deposits and their host rocks in the Birch–Uchi Lakes area of northwest Ontario, which was aimed at establishing a method of identifying mineralized areas on the basis of bedrock geochemistry.[3]

Massive Zn–Cu sulphide deposits of the Canadian Shield are generally associated with volcano-sedimentary belts comprising well-differentiated mafic to felsic extrusive and pyroclastic rocks. The deposits are typically intimately associated with felsic volcanic rocks, which normally occur in the vicinity of ancient centres of volcanism. This association has led to the suggestion that these deposits are of volcanic-exhalative origin and possibly represent the late products of magmatic differentiation, although not all of the known volcanic centres have associated sulphide deposits. If there is a genetic relationship between the sulphide deposits and their volcanic host rocks, it might be expected that the mineralized sequences are geochemically distinct from their apparently similar, although unmineralized, equivalents.

In 1968 Selco Exploration Co., Ltd., located, on the basis of geophysics, geological mapping and drilling, the Grit orebody—a rich massive Zn–Cu–Ag deposit[1] now being mined by South Bay Mines, Ltd. The initial phase of airborne geophysical exploration located a number of anomalies, only one of which hitherto has proved to be economic. The follow-up procedure adopted involved ground geophysics, geological mapping and drilling. The purpose of the current investigation was to establish whether exploration geochemistry could assist in the screening of the geophysical anomalies.

Description of area

Geology

The Birch–Uchi Lakes volcano-sedimentary belt, of Archaean age, lies within the Superior Province

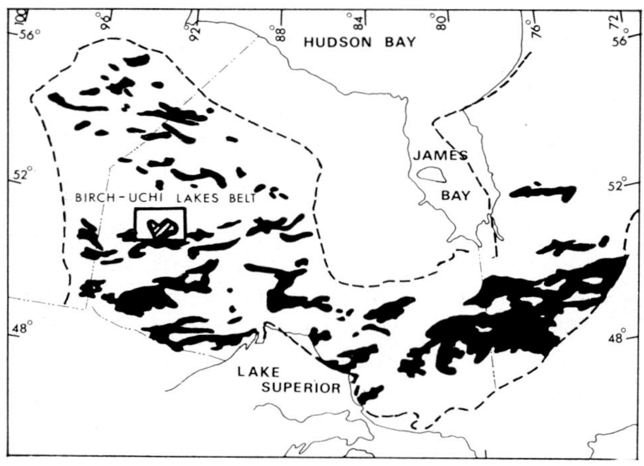

Fig. 1 Location of Birch–Uchi Lakes Belt in relation to other Archaean volcano-sedimentary belts of Superior Province of Canadian Shield

of the Canadian Shield and covers an area of some 300 square miles (Fig. 1). Geologically, the belt is typical of volcano-sedimentary belts in the Shield and is composed of predominantly volcanic rocks resting unconformably on an older predominantly sedimentary group[2, 7] and is surrounded by intrusive granites and granitic gneisses (Fig. 2).

The 72 square mile area, in Dent and Mitchell townships, on which this study was based, lies in the southwest part of the belt in an area underlain

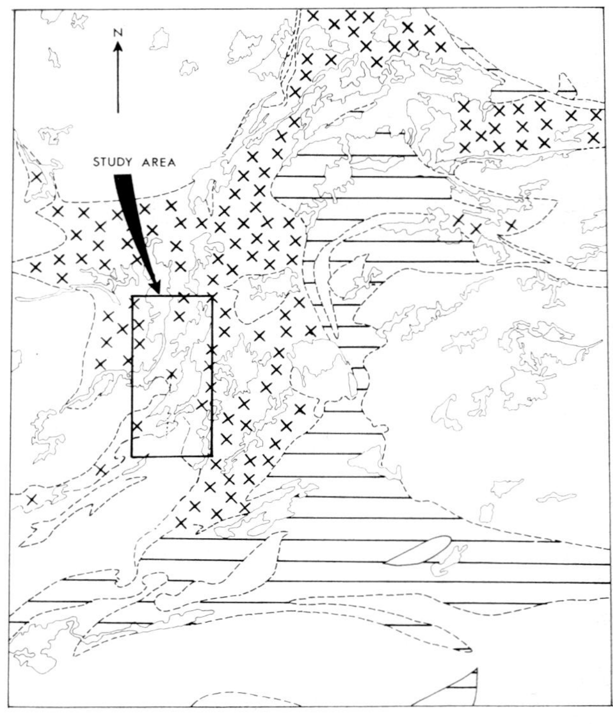

Fig. 2 Simplified geological map of Birch–Uchi Lakes volcano-sedimentary belt

by the predominantly volcanic group of rocks. The structure of this southwestern part of the belt is an isoclinal syncline or synclinorium, the study area lying mostly over the western limb and the synclinal axis lying near the eastern margin.[15]

The study area includes portions of two cycles of volcanism, which vary from basic to acid composition (Fig. 3). The older of the two cycles, the Woman Lake cycle, lies on the west side of the area and comprises predominantly basic volcanic flows, succeeded upwards by a series of predominantly intermediate pyroclastic rocks with minor flows, which, in turn, pass up into acidic tuffs and a thin sedimentary horizon. This cycle is cut by a granodioritic body towards the south.

The uppermost horizons of the Woman Lake cycle are overlain concordantly by the basic volcanic flows and minor interbedded greywackes of the Confederation Lake cycle. These basic volcanic rocks include minor intercalations of intermediate and acid units. A narrow formation of intermediate volcanic rocks overlies the basic volcanic formation in the southern part of the area, but apparently dies out to the north; thus, the acid volcanic rocks overlie the intermediate formation in the southern part of the area, and the basic volcanic rocks directly in the north. The acid volcanic rocks are cut by two major fault zones of unknown displacement. The acid volcanic rocks lying to the north of the northeast–southwest-

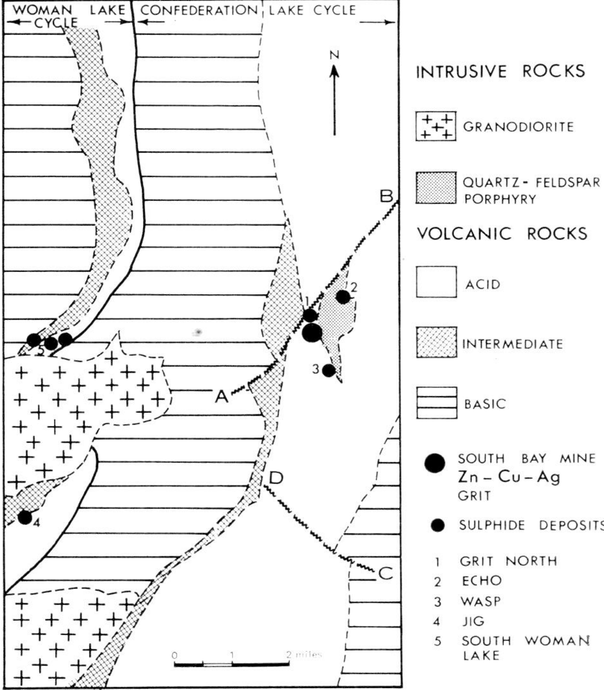

Fig. 3 Simplified geological map of study area, Dent and Mitchell townships, showing distribution of volcanic rocks and locations of base-metal deposits

trending fault are for the most part rhyodacite flows, with minor tuff units. Those in the central area are both pyroclastic rocks and flows of rhyodacitic composition, together with tuffs and agglomerates of rhyolitic composition. This central area is also characterized by bodies of quartz–feldspar porphyry, of which only the most extensive is shown in Fig. 3. These bodies are probably largely intrusive, but locally have the appearance of flows. These very high-level intrusive bodies, together with the widespread presence of agglomerates and volcanic breccias, suggest that this area is the site of a volcanic centre. The southern portion of acid volcanics, which occupies the area to the south of the southeast–northwest-trending fault, contains pyroclastic rocks predominantly of rhyolitic composition.

Gabbroic and dioritic intrusive bodies occur throughout the area, mainly as sills (not shown in Fig. 3). Similarly, several fingers of later intrusive granodiorite and minor sedimentary units have also been omitted.

Mineralization

Detailed investigations into the nature of the conductors revealed by the initial electromagnetic survey have located a number of discrete mineralizations. The deposits in the Confederation Lake cycle are associated with acid tuffs, in some places accompanied by minor argillitic units, and in many places with bodies of quartz–feldspar porphyry, whereas the deposits of the Woman Lake cycle are associated with acid to intermediate tuffs and clastic sediments. The form of the mineralization varies from pod-like lenses of massive sulphides in the Grit orebody to stringers and disseminations of sulphides in other occurrences. The mineralogy of the sulphides similarly varies from pyrite, sphalerite, chalcopyrite and minor galena at Grit to pyrite and pyrrhotite in the South Woman Lake occurrences. The mineralogy, form of mineralization and host rocks of the six individual deposits studied from the Confederation Lake cycle and the Woman Lake cycle, which are representative of the types of mineralization present in each cycle, are summarized in Table 1.

Sampling and analysis

Sampling

In order to determine the nature of the metal associations of the deposits of each cycle, samples of mineralization were collected from drill core. Sixty samples were collected from three drill cores from the ore zone of the Grit deposit, 29 samples from one core from the Wasp deposit, 9 from one core from the Grit North deposit, 6 from one core from the Echo deposit, 18 from one core from the Jig deposit and 5 from two cores from the South Woman Lake area. The drill core sampled was selected on the basis of available information to be as representative as possible of the individual deposits. This sampling is certainly not adequate as an indicator of the grade of the deposits, but it should suffice to identify characteristic metal associations of the deposits.

In order to identify any geochemical relationship between the base-metal sulphide deposits and the volcanic rocks, bedrock samples were collected from the entire succession of volcanic rocks in each cycle; 300 samples were taken from an area of 72 square miles, giving an average density of some 4 samples per square mile. Sample sites were preselected from available geological maps. The majority of the samples were taken along the shores of lakes, which provided easy accessibility and aided accurate location of the sample sites. Elsewhere, where additional samples were neces-

Table 1 Mineralogy and host rocks of the deposits

Deposits*	Nature of sulphides	Host rocks
1, Grit	Pod-like lenses of massive sulphides: pyrite, sphalerite, chalcopyrite, minor galena	Rhyolitic tuffs, tuff breccias, agglomerates and porphyritic flows; marginal to quartz–feldspar porphyry body
2, Wasp	Zone, up to 30 ft wide, consisting of stringers of sulphides: pyrrhotite, pyrite, sphalerite, minor chalcopyrite	Interbanded rhyolitic tuffs, with graphitic and argillaceous members present in the mineralized zone; marginal to quartz–feldspar porphyry body
3, Grit North	Zone, 10–20 ft in width, of stringers of sulphides: pyrrhotite, pyrite, minor sphalerite	Rhyolitic tuffs, sill or flow of quartz–feldspar porphyry
4, Echo	Zone about 10 ft in width of stringers and disseminations of sulphides: pyrrhotite, pyrite, minor sphalerite	Rhyolitic tuffs, marginal to quartz–feldspar porphyry body
5, Jig	Zone, up to 150 ft in width, of stringers and pods of massive sulphides: pyrrhotite, pyrite, chalcopyrite, minor sphalerite	Felsic tuffs and sediments (in places calcareous)
6, South Woman Lake	Number of thin zones, ½–10 ft in thickness, containing disseminated to massive sulphides: pyrite, pyrrhotite	Sediments, calcareous and graphitic argillites

*1–4, Confederation Lake cycle: 5 and 6, Woman Lake cycle.

sary, these were collected along a road traverse from South Bay Mines or traverses through the bush. This sampling procedure did not give an even sample distribution over the area, but it was hoped that it would suffice to delineate the presence of any metal associations or compositional trends. At each outcrop composites of chip samples (approximately 1 kg) were collected to reduce sample variability at the outcrop scale.

Sample preparation

The samples of mineralization and bedrock were prepared for analysis in separate batches to avoid contamination. The samples were crushed to ¼-in chips in a Braun jaw crusher with steel plates. The fines produced in crushing were passed through a 6-mesh nylon sieve and discarded in order to eliminate any steel chips which might have been introduced into the samples from the steel plates. These chips were then reduced to less than 6 mesh by use of a Lemaire jaw crusher equipped with 99% alumina plates. The samples were then reduced to −80 mesh in a Lemaire pulverizer with alumina plates. Tests have shown that, in general, less than 1% alumina is introduced into the samples by this procedure.

Analytical methods

The samples of mineralization were analysed variously by assay and atomic absorption and emission spectrographic methods. The Cu and Zn data for the Grit, Wasp, Grit North, Echo and Jig deposits were obtained by assay, made available by Selco Exploration Co., Ltd. The Fe content of all samples and the Cu and Zn content of the mineralization in the South Woman Lake area were determined by atomic absorption, an aqua regia attack being employed with a 6 M hydrochloric acid leach. The accuracy of the determinations was checked by including sulphide standards in the batch, and the precision of the analyses at the 95% confidence level was ±20% for Fe, ±15% for Cu and ±20% for Zn. The compatibility of the Cu and Zn data for the mineralization in the South Woman Lake area with the assay data was checked by re-analysis of samples from previously assayed core. Ag, Pb, Sn, Co and Ni were determined spectrographically by use of a modified method from that described by Nichol and Henderson-Hamilton.[14] A 2 : 1 : 1 mixture of sample, graphite and lithium carbonate is arced for 30 sec by use of anode excitation and the element concentration is determined with Ge as an internal standard.[3] The precisions of these determinations were in the range ±40–50% at the 95% confidence level.

Silica analyses of the bedrock samples were carried out by the determination of the refractive indices of the fused samples, a slightly modified version of the method described by Mathews[13] being used. A calibration curve was established by use of secondary standards selected from the samples, the SiO_2 content of these standards being determined by X-ray fluorescence. The overall precision of the method is estimated at ±7% at the 95% confidence level. Cu, Pb, Zn, Co, Ni, Fe and Mn contents of the bedrock samples were analysed by atomic absorption. The samples were heated to dryness in 5 ml of a 4:1 perchloric–nitric acid mixture, and leached in 2·5 ml of 6 M nitric acid to dissolve the Fe, followed by the addition of 10 ml of a 5% H_2O_2 solution in deionized water to reduce the MnO_2 to Mn^{2+}. The precision of these determinations at the 95% confidence level ranges from ±25 to 40%. The elements MgO, K_2O, Cr, V and Sn were analysed spectrographically, with precisions in the range ±40–50% at the 95% confidence level.

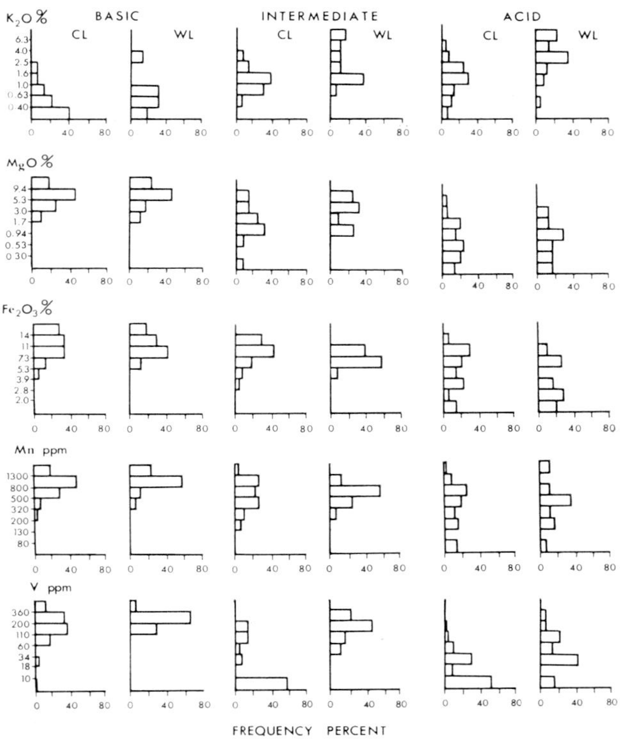

Fig. 4 Frequency distributions of K_2O, MgO, Fe_2O_3, Mn and V in basic, intermediate and acid volcanic rocks of Confederation Lake (CL) and Woman Lake (WL) cycles

Data processing

The frequency distributions of the elements in the volcanic rocks do not, in general, correspond to the Gaussian distribution (Figs. 4 and 5). The number of samples in each group is generally less than 50, and the distributions of a particular element in corresponding rock types in each cycle have markedly different variances in some cases. For these reasons, parametric statistical methods of comparison, such as Student's t test for small samples, are inappropriate, and the non-parametric Mann–Whitney U test was used for this purpose. This test requires no assumption to be made as to the shape of the distributions.

An excellent description of the U test has been given by Siegel.[19] Briefly stated, the test determines the probability that the bulk of two observed distributions were drawn from the same population. The procedure for carrying out the test may be summarized as follows: (1) consider two distributions, containing n_1 and n_2 samples, respectively, where $n_1 < n_2$, and $n_2 > 20$; (2) rank all the observations, from both distributions, from the lowest to the highest; (3) sum the ranks of the smaller data set to obtain R_1; (4) calculate the U statistic from the equation

$$U = n_1 n_2 + \frac{n_1 (n_1 + 1)}{2} - R_1$$

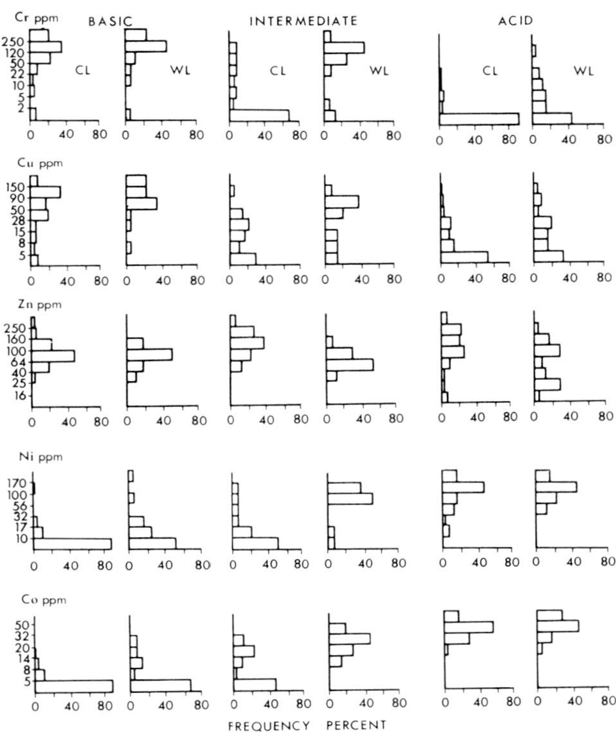

Fig. 5 Frequency distributions of Cr, Cu, Zn, Ni and Co in basic, intermediate and acid volcanic rocks of Confederation Lake (CL) and Woman Lake (WL) cycles

(5) calculate the z statistic from the equation

$$z = \frac{U - \frac{n_1 n_2}{2}}{\sqrt{\frac{(n_1)\ (n_2)\ (n_1+n_2+1)}{12}}}$$

(6) Determine the probability, p, that the two distributions are drawn from the same population from standard tables of z values.

If the two distributions are identical, the probability value would be unity. The greater the difference between the two observed distributions, the lower would be the probability that they were drawn from the same population. Three arbitrary significance levels were chosen, $\alpha_1 = 0{\cdot}05$, $\alpha_2 = 0{\cdot}01$, $\alpha_3 = 0{\cdot}002$, which correspond to the 95, 99 and 99·8% confidence levels, respectively.

When the probability, p, that two observed distributions were drawn from the same population is less than $\alpha_1 = 0{\cdot}05$, the distributions are significantly different at the 95% confidence level. If $p < \alpha_2 = 0{\cdot}01$, the distributions are significantly different at the 99% confidence level; and if $p < \alpha_3 = 0{\cdot}002$, they are significantly different at the 99·9% confidence level. A qualitative idea of the power of this test may be obtained by comparing the statistically significant differences in Fig. 9 with the frequency distributions in Figs. 4 and 5.

This test does not indicate which population is, on average, higher, since it is being used to determine the significance of differences between populations; but when the distributions differ significantly at the 95% confidence level, the population which is, on average, higher may be determined from the frequency distribution diagrams by inspection.

Metal associations of mineralizations

The mean metal contents of mineralized samples from the six deposits taken as being representative of deposits associated with the Woman Lake and Confederation Lake cycles show marked variations in level (Table 2). These differences, undoubtedly in large part, are due to variations in the proportions of sulphide minerals to silicate gangue present in the different deposits. From a consideration of the mean metal contents it is not possible to

Table 2 Mean metal content of mineralized material (ppm) from sulphide deposits

	Number of samples	Fe, %	Zn	Cu	Ag	Pb	Sn	Co	Ni
Grit	60	30·4	165 000	28 600	28	530	650	150	<5
Wasp	29	11·1	27 000	2 100	3·0	140	130	27	35
Grit North	9	8·9	14 000	1 500	2·6	71	61	22	26
Echo	6	7·5	2 600	340	<1·0	31	14	11	17
Jig	18	19·7	2 300	8 600	6·0	99	<10	75	390
South Woman Lake	5	16·0	124	167	3·0	55	<10	34	92

Table 3 Cation proportions (%) of Fe, Zn, Cu, Pb, Sn, Co, Ag and Ni in sulphide deposits

Deposit	Fe	Zn	Cu	Pb	Sn	Co	Ag	Ni
Grit	64	30	5·4	0·031	0·066	0·030	0·0031	‡
Wasp	82	17	1·4	0·028	0·045	0·019	0·0012	0·025
Grit North	87	12	1·3	0·019	0·028	0·020	0·0013	0·024
Echo	97	2·9	0·39	0·011	0·0087	0·014	†	0·021
Jig	95	0·95	3·7	0·013	*	0·034	0·0015	0·19
South Woman Lake	99·8	0·066	0·092	0·0094	*	0·020	0·0010	0·055

*Sn not detected in these deposits.

†Ag too low in this deposit to be reliably estimated.

‡Trace Ni (<5 ppm) only in this deposit.

identify differences in metal associations of the deposits.

Assuming that these elements Fe, Zn, Cu, Pb, Sn, Ag, Co and Ni will be strongly concentrated in the sulphide relative to the silicate phases, the variations caused by differing proportions of sulphides to silicate gangue will be largely eliminated by ratioing the mean content of each of these metals to the total content of these metals of each deposit. Table 3 gives the ratios of the atomic proportions of each metal (i.e. mean content divided by atomic weight) to the sum of atomic proportions of these metals for each deposit. In the deposits of the Confederation Lake cycle there is a negative correlation between the atomic proportions of Zn, Cu, Ag, Pb, Sn and Co with Fe. This feature is further illustrated in Figs. 6 and 7, where, for each deposit, the atomic proportions (from Table 3) of these elements are plotted as a percentage against the sum of the atomic proportions of all the elements less the atomic proportion of Fe. On this basis the deposits of the Confederation Lake cycle are higher in Zn and Sn and lower in Cu, Co, Ni and Ag than those of the Woman Lake cycle.

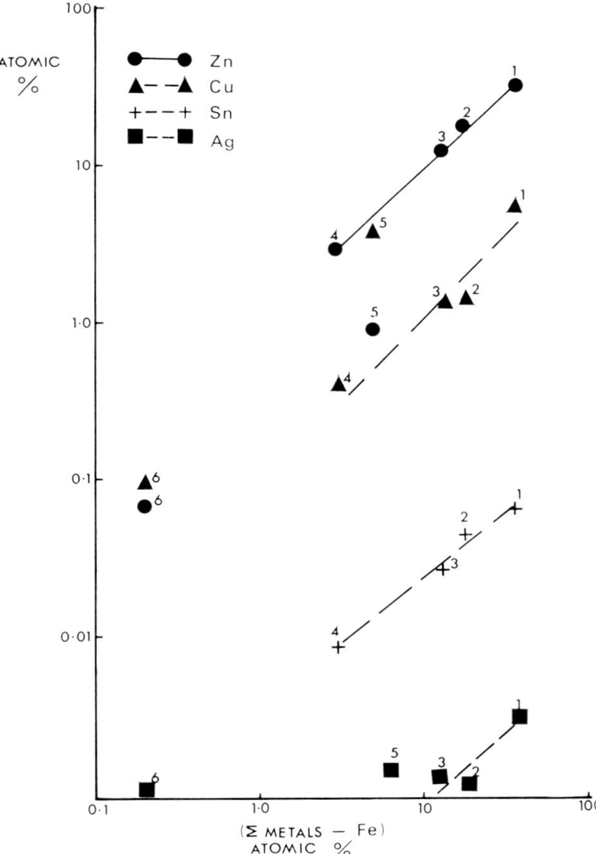

Fig. 6 Variations of atomic proportions of Zn, Cu, Sn and Ag in sulphide deposits

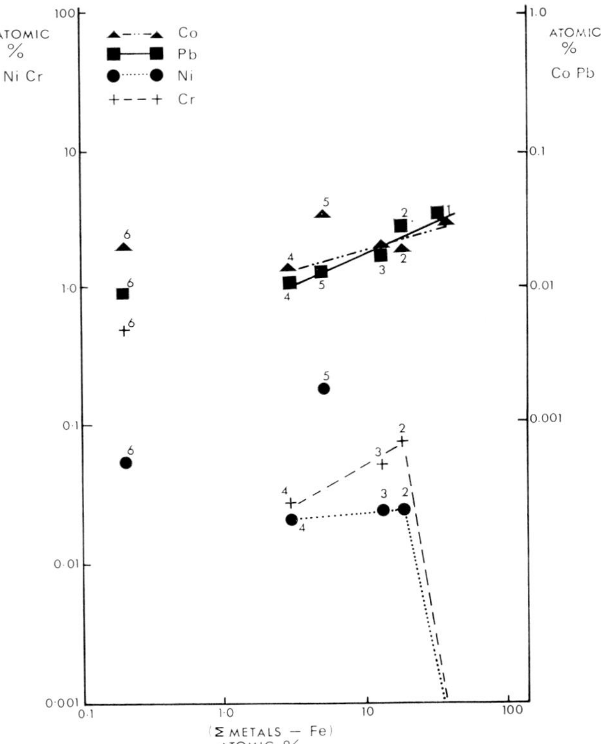

Fig. 7 Variations of atomic proportions of Co, Pb and Ni in sulphide deposits, and variations of Cr content of gangue material

Metal associations of volcanic rocks

Classification of rocks

The concentrations of a number of elements show variations through the basic to acid differentiation sequence of the volcanic rocks. From this it is clearly important, when considering variations in minor-element composition of rocks, to consider only comparable rock types. The bedrock samples have, therefore, been classified on the basis of their silica content. In this way it is possible to consider variations in metal contents of rocks which display only limited ranges of SiO_2 content.

The data have been divided into three arbitrary groups: <56% SiO_2 basic volcanics; 56–64% SiO_2 intermediate volcanics; and >64% SiO_2 acid volcanics. From a consideration of the frequency distribution of SiO_2 in rocks from the two cycles it is clear that basic, intermediate and acid varieties are not equally represented (Fig. 8). In the Confederation Lake cycle the small number of intermediate types reflects their paucity of occurrence, but the predominance of acid over basic varieties reflects a sampling bias conditioned by a greater accessibility of acid volcanics and the fact that they crop out over a proportionately greater area as they occupy the core of the syncline

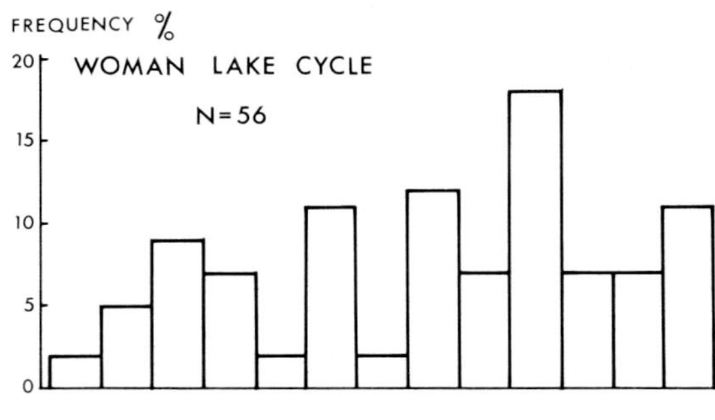

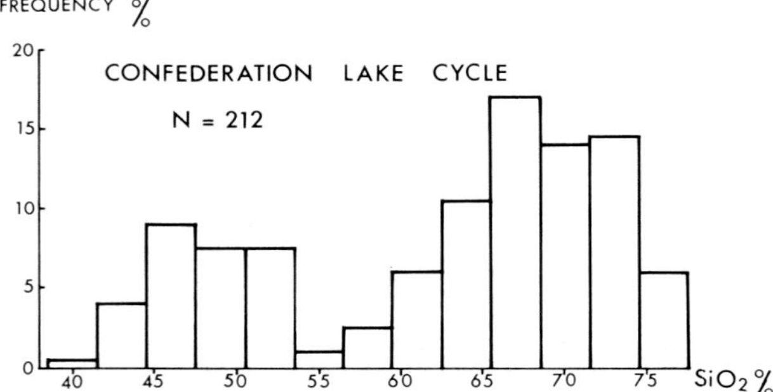

Fig. 8 Frequency distributions of silica in volcanic rocks of Confederation Lake and Woman Lake cycles

(Fig. 3). The samples from the Woman Lake cycle are more representative of the actual proportions of the three volcanic units.

Variation in geochemistry between volcanic cycles

The frequency distributions of K_2O, MgO, total Fe as Fe_2O_3, Mn, V, Cr, Co, Ni, Cu and Zn in the basic intermediate and acid volcanic rocks of each cycle are shown in Figs. 4 and 5. For some elements their distributions in certain rock types are obviously different, but in other cases the variance of the data precludes the identification of differences by casual inspection. Zinc is generally higher in the intermediate and acid rocks of the Confederation Lake cycle relative to similar rocks of the Woman Lake cycle. In addition, the Zn distribution within the Confederation Lake cycle displays a much higher variance than that in the Woman Lake cycle. K_2O appears to be generally higher in the intermediate and acid rocks of the Woman Lake cycle. Cu, Ni, Cr and V are generally higher in the intermediate volcanics of the Woman Lake cycle and Co is higher in the acid volcanics relative to the corresponding rocks of the Confederation Lake cycle. No variations in the distribution of MgO, Fe_2O_3 or Mn are apparent between comparable rocks of the two cycles. Pb and Sn are below the detection limits (10 ppm) of the analytical method for all samples.

These differences between the distributions of various elements between the two cycles of volcanism are qualitative observations and unavoidably subjective. In order to compare these distributions more objectively, the data were analysed statistically by the Mann–Whitney *U* test to determine the probability that the distributions of each particular element in corresponding bedrock types from each cycle were drawn from the same population.

As the distributions of many elements vary with the silica content, it is essential in any comparison of element distributions of rock units to ensure that their silica distributions are as similar as possible.

The probabilities that the SiO_2 distributions of the basic, intermediate and acid volcanics of the Woman Lake and Confederation Lake cycles are drawn from the same populations are 0·91, 0·13 and 0·21, respectively. The low values obtained for the intermediate and acid volcanics suggests that the SiO_2 contents for the intermediate and acid varieties are not entirely comparable. An

ELEMENT	BASIC	INTERMEDIATE	ACID
SiO_2			
K_2O		>>>	>>>
MgO		>	
Fe_2O_3		<<	
Mn			>
V	>	>>>	>>>
Cr		>>>	>>>
Co		>>>	>>
Ni		>>>	>>>
Cu		>	>
Zn		<<<	<<<

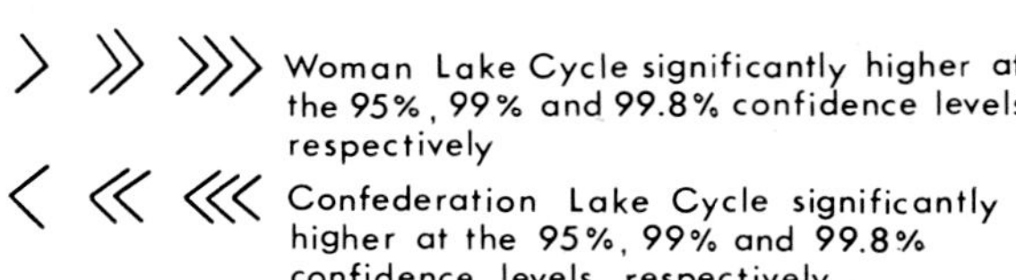

Fig. 9 Comparison of distributions of elements in corresponding volcanic rock types between Confederation Lake and Woman Lake cycles by Mann–Whitney U test

improved correspondence in populations was achieved by restricting the SiO_2 range of the intermediate volcanics to 57–64% and the acid volcanics to greater than 66%: the probabilities that the populations are the same are then, respectively, 0·65 and 0·84. Some of the data have, therefore, been neglected in the interests of achieving a comparison between rock units that contain SiO_2 contents that are as comparable as possible.

Fig. 9 summarizes the results of the *U* test; the direction of the significant differences is also indicated. The basic volcanic rocks of each cycle are similar in composition, with the exception of V, which is significantly higher in the Woman Lake cycle at the 95% confidence level. The intermediate rocks of the Woman Lake cycle are significantly higher in K_2O, Co, Ni, Cr and V at the 99·8% confidence level, and MgO and Cu at the 95% confidence level. The intermediate rocks of the Confederation Lake cycle are significantly higher in Zn at the 99·8% confidence level and in Fe_2O_3 at the 99% confidence level. The acid volcanic rocks of the Woman Lake cycle are significantly higher in K_2O, Ni, Cr and V at the 99·8% confidence level, and in Co at the 99% and Mn and Cu at the 95% confidence levels. The acid rocks of the Confederation Lake cycle are significantly higher in Zn at the 99·8% confidence level.

Thus, the Mann–Whitney *U* test essentially confirms the differences noted visually and, in addition, identifies as significant other differences not readily apparent before.

Variation of geochemistry within acid volcanic formation of Confederation Lake cycle

From a consideration of the areal element distribution of certain elements within the acid volcanics it is apparent that certain elements show marked changes in distribution along strike in the acid volcanic rocks of the Confederation Lake cycle. In particular, K_2O, MgO, V and possibly Ni show a general increase in level from north to south, whereas Fe_2O_3 and possibly Mn show a decrease southward. Cu, on the other hand, appears to attain maximum concentrations in the central part of the area.

The acid volcanic formation in the Confederation Lake area is cut by two major fault zones of unknown displacement. The northern area is composed predominantly of rhyodacitic flows with a silica content in the range 65–70%. The central area contains both rhyodacitic and rhyolitic rocks, with a silica content of 65–75%, and the southern area rhyolitic rocks with 70–77% silica—indicating a general increase of SiO_2 in passing from north to south. This variation in silica content precluded the direct comparison of all three areas over a single silica range by use of an analysis of variance technique. The northern segment was, therefore, compared with the central segment, samples in the range 65–69% SiO_2 being used, by the Mann–Whitney *U* test. Similarly, the central and southern segments were compared over the range 70–74% SiO_2. Thus, the two groups taken from the central portion are entirely distinct.

	NORTH		CENTRE		SOUTH
K_2O		<<			
MgO		<<<		<<	
Fe_2O_3					
Mn				>	
V		<<<		<	
Cr					
Co					
Ni		<<			
Cu		<			
Zn					

> >> >>> Distributions significantly different at the 95%, 99% and 99.8% confidence levels, respectively. Direction of arrows indicates direction of difference

Fig. 10 Comparison of distribution of elements between north, central and southern parts of acid volcanic rocks of Confederation Lake cycle by Mann–Whitney U test

The direction and significance level of changes in the distributions of elements in the acid volcanic rocks from north to south, after the effects of silica variation have been eliminated, are presented in Fig. 10. MgO and V increase consistently from north to south; K_2O, Ni and Cu are lower in the northern part, and have similar distributions in the central and southern parts; Mn is higher in the northern and central parts than in the southern portion. The apparent decrease in Fe_2O_3 from north to south is not statistically significant when samples of similar silica content only are compared. The apparent statistical similarity in the distributions of Cr and Co may be misleading, as the bulk of the data for these elements are below the detection limits of the analytical method used. The distribution of Zn in the three segments is very similar.

Discussion

The essential requirement for geochemistry to aid in regional mineral exploration in volcano-sedimentary belts is that the host rocks of economic deposits have some geochemically diagnostic feature that allows them to be distinguished from similar, but barren, rocks.

The Confederation Lake cycle, which contains the ore deposit, is characterized by a number of differences in the distributions of certain elements in the acid and intermediate volcanic rocks relative to the Woman Lake cycle. The mineral deposits in each cycle are also distinct on the basis of their metal content. In order to determine which of the observed differences between the volcanic rocks of each cycle are related to the differences in the metal content of the deposits of the cycles, it is necessary to consider the possible processes that affect the composition both of the volcanic rocks and the deposits.

Metal content of sulphide deposits

Variations in the compositions of sulphide deposits with increasing distance from volcanic centres have been noted in the Abitibi belt of the Canadian Shield.[10] In that belt pyritic Zn–Cu–Ag massive sulphide deposits occur adjacent to volcanic centres, whereas away from the centres the deposits are typically iron sulphides, essentially barren of base metals, commonly associated with volcanogenic clastic sediments. This spatial association of massive sulphide deposits may not be generally true, however, as is evidenced by the deposits of the Manitouwadge area of the Canadian Shield which are in metasediments.[16] Furthermore, Gilmour[5] considered massive, concordant base-metal deposits in general to be of volcanic-exhalative origin, regardless of whether volcanic rocks form a conspicuous proportion of their host rocks.

Within the deposits of the Confederation Lake cycle the elements Zn, Cu, Sn and possibly Ag show comparable relationships to one another in each of the deposits (Fig. 6). This is considered to indicate that these elements were derived from the same or similar exhalative sources. The depletion of these elements relative to iron may correlate with distance of the deposits from the original exhalative vent or vents, as is suggested for the Abitibi belt. From the present positions of the deposits, however, this is not apparent. In view of the absence of a breccia pipe in the vicinity of any of these deposits, such as that described by Roscoe,[17] the deposits may have been tectonically transported soon after their formation, as was suggested by Jenks.[11] Thus, the present positions of the deposits may be substantially different from their original positions.

The virtual absence (<5 ppm) of Ni in the Grit deposit, which, it is assumed, was formed adjacent to the exhalative vent, suggests that this element is not of exhalative origin in the other deposits of the Confederation Lake cycle, where it is present in higher concentrations, although it may be of exhalative origin in the Jig deposit. The similar patterns shown by Ni and Cr in the deposits (Fig. 7) suggest that these elements are associated with argillaceous material in the Wasp, Grit North and Echo deposits, since Cr is not associated with the sulphides. Co and Pb are associated with the sulphides of the Grit deposit, but do not decrease in relative content as rapidly as Zn, Cu and Sn in the other deposits. It is suggested that Co and Pb are associated both with the exhalations and the argillaceous material, and possibly the processes of deposition are such as to cause enrichment of Co and Pb relative to Zn, Cu and Sn.

If the exhalations were juvenile magmatic waters, the base-metal content of the late-stage magmatic liquids would, presumably, influence the base-metal concentrations present in these fluids. If, on the other hand, these fluids are recycled sea water which have leached metal from consolidated volcanic rocks, as Ferguson and Lambert[4] suggested from their study of hydrothermal spring waters associated with recent volcanism, the base-metal content of the exhalations will be controlled by the concentration of base metals in the volcanic rocks which they leach in their passage to surface. In the latter case, therefore, the presence of high concentrations of, for example, Zn in the late-stage volcanic rocks would favour Zn-rich exhalations. Thus, in either case, the composition of the exhalations might be expected to reflect, in part, the composition of the late-stage volcanic rocks.

Distributions of metals in volcanic rocks

The distribution of metals in the pyroclastic rocks will be controlled largely by magmatic processes, but may also be modified by sedimentary processes in their environment of deposition. The similarity in the geochemistry of the basic volcanic rocks, which are predominantly flows, suggests that the magma sources of the two cycles were similar and that the differences between the later differentiates, which are, in large part, pyroclastic, are due to factors such as the physico-chemical conditions of differentiation, and also to processes operating in the depositional environment of the pyroclastic rocks.

The volcanic centres occurred in sub-aerial or shallow-water environments, whereas in more remote areas deeper-water conditions prevailed. Away from volcanic centres intermediate and acid volcanics become subordinate to volcanogenic clastic sediments, and the thickness of pyroclastic rocks may decrease away from the centre. This suggests that equilibration between the volcanic material and the water may become increasingly important away from the centre in modifying trace-element distributions.

The portion of the Confederation Lake cycle sampled in the present study contains a volcanic centre with a preponderance of acid volcanics and high-level intrusives. In contrast, the area sampled within the Woman Lake cycle is apparently more remote from a centre with a smaller proportion of acid volcanics and a more extensive development of volcanogenic sediments.

The acid volcanic rocks of the northern segment of the Confederation Lake cycle are predominantly flows and, therefore, their composition will be largely unaffected by sedimentary processes during their deposition. Thus, the significant differences in the distributions of K_2O, MgO, V, Ni and Cu between these rocks and those of the central and southern segments may be caused, in part at least, by sedimentary processes which operated on the predominantly pyroclastic rocks of these two segments. The elements MgO and V are also significantly higher in the southern than in the central segment. The elements K_2O, V, Ni and Cu, together with Cr and Co, are significantly higher in the acid volcanic rocks of the Woman Lake cycle than in the Confederation Lake cycle, and it is suggested that these differences may be related, in part at least, to the greater effect of sedimentary processes in the Woman Lake cycle. This is supported by comparing the acid rocks of the southern part of the Confederation Lake cycle with those of the Woman Lake cycle. The distributions of V in the two data sets are not significantly different, and the confidence levels at which the differences in the distributions of Cr, Co, Ni and Cu are significant are lower in this case than when all of the data from the Confederation Lake cycle are used. The remaining differences may be related to initial differences in the composition of the volcanic rocks caused by magmatic processes, which have merely been exaggerated by sedimentary processes. Alternatively, by sampling areas at greater distances from the volcanic centre in the Confederation Lake cycle, the distributions of some or all of the elements may approach those in the portion of the Woman Lake cycle sampled in this study. The differences in the distributions of these elements between the intermediate volcanic rocks of the two cycles may also be due, in part, to the sampling of different segments of the two belts relative to the centres of volcanism in the two cycles.

The distribution of zinc in the acid volcanic rocks of the Confederation Lake cycle is remarkably constant: this suggests that the differences in the zinc distributions in the acid rocks of the two cycles are not a feature of the segments of the cycles, relative to the volcanic centre, which have been sampled. Rather, it is considered that the difference in the distributions of zinc between the acid and intermediate rocks of the Confederation Lake cycle from the equivalent rocks of the Woman Lake cycle is a fundamental geochemical difference between the two volcanic suites.

In the Confederation Lake cycle the mean zinc content increases from the basic rocks to the later intermediate and acid rocks, the means being 105, 134 and 128 ppm, respectively, whereas in the Woman Lake cycle the mean zinc content decreases from the basic and intermediate rocks to the later acid rocks, the means being 79, 74 and 63 ppm, respectively. The rather scanty data in the literature on the behaviour of zinc in suites of differentiated igneous rocks also indicate that in any particular suite zinc may show either an overall decrease or increase in level from the early to the late differentiates. [6, 12, 18, 20] It is considered that the difference between the zinc distributions in the intermediate and acid rocks of the two cycles may be reflecting different physico-chemical conditions under which differentiation occurred. It is suggested, therefore, that the increase of zinc in the later fractions of the Confederation Lake cycle favoured the formation of zinc-rich base-metal deposits.

Summary and conclusions

The purpose of this study was to investigate the geochemistry of the volcanic rocks of a portion of the Birch–Uchi Lakes greenstone belt in order to determine whether it is possible to distinguish the Confederation Lake cycle, which contains the Grit ore deposit and several similar sub-economic deposits, from the Woman Lake cycle, which contains no economic deposits. Insofar as the data presented refer only to a portion of a single volcano-sedimentary belt, any conclusions reached here may apply only to this particular area, although there is evidence to believe that these conclusions may be generally applicable to the belt as a whole.[3] In view of the similarities between Archaean volcano-sedimentary belts of the Canadian Shield,[8, 9] some of the conclusions may be more generally applicable, and similar studies in other belts where the volcanic stratigraphy and nature of the sulphide deposits are known would seem to be worth while on the basis of the present data.

Conclusions regarding the distributions of metals in the volcanic rocks and sulphide deposits and their interrelationships are outlined below.

(1) The relative contents of Zn, Cu, Sn and possibly Ag in the deposits are related directly to the composition of the volcanic exhalations from

which they were formed. The elements Pb and Co are related, in part, to the composition of the exhalations, but they may be also concentrated in argillaceous material. Ni is unrelated to the exhalations in the Confederation Lake cycle, and the nickel content of these deposits is due to the presence of argillaceous material. Ni in the deposits of the Woman Lake cycle may be derived, in part, from volcanic exhalations.
(2) The sulphide deposits of the Confederation Lake cycle are relatively richer in Zn and Sn, and poorer in Cu, Ag, Co and Ni, than the deposits of the Woman Lake cycle.
(3) The composition of the volcanic exhalations is considered to be related, directly or indirectly, to the composition of the late-stage volcanic rocks of each cycle. The acid and intermediate rocks of the Confederation Lake cycle contain significantly higher Zn, and lower Cu, Co and Ni, than the corresponding rocks of the Woman Lake cycle.
(4) The intermediate volcanic rocks of the Confederation Lake cycle are also significantly higher in Fe_2O_3, and lower in K_2O, MgO, V and Cr, than those of the Woman Lake cycle. The acid volcanic rocks of the Confederation Lake cycle are significantly lower in K_2O, Mn, V and Cr than those of the Woman Lake cycle.
(5) The significant variations in the distributions of V, Ni and Cu within the acid volcanic formation of the Confederation Lake cycle are considered to be related to sedimentary processes operating in the environment of deposition of the pyroclastic rocks. This being so, it might be expected that these elements, and also possibly Cr and Co, would show a general increase away from the volcanic centre in rocks deposited under deeper-water conditions. It is suggested, therefore, that the differences in the distributions of V, Cr, Co, Ni and Cu between the acid and possibly the intermediate volcanic rocks of the two cycles may be related, in part at least, to the position, relative to the volcanic centres, of the portions of the cycles sampled. These sedimentary processes may, however, have accentuated differences in the compositions of the volcanic rocks which are due to magmatic processes.
(6) The significant differences between the intermediate and acid volcanic rocks in the distributions of K_2O, MgO, Fe_2O_3 and Mn may possibly be related to magmatic processes. The relationship between the distributions of these elements, and V and Cr, and the nature of the sulphide deposits present in the two cycles cannot be assessed from the present study.
(7) The differences in the distributions of Zn in the acid and intermediate rocks of the two cycles and the lack of any variation along strike in the acid volcanic rocks of the Confederation Lake cycle suggest that this element is the most reliable indicator with respect to the type of mineralization developed in each cycle. Within the Birch–Uchi Lakes the cycle containing the South Bay Mines, Ltd., Zn–Cu–Ag ore deposit may be distinguished by the higher average Zn content throughout its intermediate and acid volcanic rocks.
(8) It is suggested that the distributions of certain elements, in particular zinc, in the intermediate and acid rocks of volcano-sedimentary belts may be useful in identifying favourable areas for economic base-metal mineralization. The sampling parameters required would vary in any particular area, being dependent on the degree of geological control available. In order to define adequately the distribution of zinc in any particular volcanic formation, it is suggested that at least 20, and preferably 50, samples are required. The determination of silica in all samples is a valuable aid in comparing only similar rocks from different areas, and is essential if the metamorphic grade of the rocks makes accurate field identification difficult.

Acknowledgment

Selco Exploration Co., Ltd., provided generous support for this work, which represents part of a Ph.D. research programme by P. H. Davenport at Queen's University supervised by Professor Ian Nichol. Grateful appreciation is expressed to Mr. J. S. Auston of Selco Exploration Co., Ltd., and Dr. O. W. Nicolls of Selection Trust, Ltd., for suggesting the investigations and for the benefit of their opinions throughout the study. The interest shown during the work by the geologists of Selco Exploration, their ready assistance and helpful comments are gratefully acknowledged. In particular, the authors wish to thank Dr. G. D. Pollock, Dr. I. G. Sinclair and Mr. V. Wierzbicki.

In addition, the staff of the Department of Geological Sciences, Queen's University, are thanked for assistance with analytical work and preparation of the paper.

References

1. Auston, J. S. Uchi Lake—a geophysical case history. Paper presented at Society of Exploration Geophysicists Annual Meeting, Calgary, Alberta, 1969.
2. Bateman, J. D. Geology and gold deposits of the Uchi–Slate lakes area. *Rep. Ont. Dep. Mines*, **48**, pt 8, 1939, 1–43.
3. Davenport, P. H. The application of geochemistry in base-metal exploration in the Birch–Uchi Lakes Volcano-Sedimentary Belt, northwestern On-

tario. Ph.D. thesis, Queen's University, Kingston, Ontario, 1972.
4. FERGUSON, J. and LAMBERT, I. B. Volcanic exhalations and metal enrichments at Matupi Harbor, New Britain, T.P.N.G. *Econ. Geol.*, **67**, 1972, 25–37.
5. GILMOUR, P. Strata-bound massive pyritic sulfide deposits—a review. *Econ. Geol.*, **66**, 1971, 1239–44.
6. GOLDSCHMIDT, V. M. *Geochemistry* (Oxford: Clarendon Press, 1954), 730 p.
7. GOODWIN, A. M. Volcanic studies in the Birch–Uchi lakes area of Ontario. *Misc. Pap. Ont. Dep. Mines* 6, 1967, 96 p.
8. GOODWIN, A. M. Evolution of the Canadian Shield. *Proc. Geol. Ass. Can.*, **19**, 1968, 1–14.
9. GOODWIN, A. M. and RIDLER, R. H. The Abitibi orogenic belt. In Basins and geosynclines of the Canadian Shield, BAER, A. J. ed. *Pap. geol. Surv. Can.* 70–40, 1970, 1–24.
10. HUTCHINSON, R. W. RIDLER, R. H. and SUFFEL, G. G. Metallogenic relationships in the Abitibi belt, Canada: a model for Archean metallogeny. *CIM Bull.*, **64**, April 1971, 48–57.
11. JENKS, W. F. Tectonic transport of massive sulfide deposits in submarine volcanic and sedimentary host rocks. *Econ. Geol.*, **66**, 1971, 1215–24.
12. LUNDEGÅRDH, P. H. Rock composition and development in central Roslagen, Sweden. *Ark. Kemi, Miner. Geol.*, **23**, nos 4–5 1947, 160 p.
13. MATHEWS, W. H. A useful method for determining the approximate composition of fine grained igneous rocks. *Am. Miner.*, **36**, 1951, 92–101.
14. NICHOL, I. and HENDERSON-HAMILTON, J. C. A rapid quantitative spectrographic method for the analysis of rocks, soils and stream sediments. *Trans. Instn Min. Metall.*, **74**, 1965, 955–61.
15. PRYSLAK, A. P. Geology of Dent and Mitchell townships, District of Kenora (Patricia portion). *Prelim. Maps Ont. Dep. Mines* pp. 592 and 593, 1970.
16. PYE, E. G. Geology of the Manitouwadge area. *Rep. Ont. Dep. Mines*, **66**, pt 8, 1957, 114 p.
17. ROSCOE, S. M. Geochemical and isotopic studies, Noranda & Matagami areas. *Trans. Can. Inst. Min. Metall.*, **68**, 1965, 279–85.
18. SANDELL, E. B. and GOLDICH, S. S. The rarer metallic constituents of some American igneous rocks. *J. Geol.*, **51**, 1943, 167–89.
19. SIEGEL, S. *Nonparametric statistics for the behavioral sciences* (New York: McGraw-Hill, 1956), 330 p.
20. WILSON, H. D. B. Geology and geochemistry of base metal deposits. *Econ. Geol.*, **48**, 1953, 370–407.

552.1:546.49:553.661.2(418.1)

Trace identification of mercury compounds as a guide to sulphide mineralization at Keel, Eire

R. J. Watling, B.SC.

Division of Mining Geology, Imperial College, London, England

G. R. Davis, B.SC., PH.D., F.I.M.M.

Division of Mining Geology, Imperial College, London, England

W. T. Meyer, B.SC. M.SC.

Applied Geochemistry Research Group, Imperial College, London, England

Synopsis

Primary dispersion patterns of mercury in the bedrock of the Keel zinc–cadmium prospect, County Longford, Eire, have been clarified by the identification of trace amounts of specific mercury compounds. These results have been obtained by quantitative measurement of the atomic absorption of mercury evolved during controlled heating of samples in a continuous air flow system. Mercury peaks released from natural samples at 150, 220, 300 and 520°C are correlated, respectively, with artificially prepared standards of mercurous chloride, mercuric chloride, mercuric sulphide and mercuric oxide, and two further peaks at 450 and 600°C are correlated with mercury released from pyrite and sphalerite lattices, respectively. Elemental mercury gave a release peak at 80°C and mercuric sulphate gave triple release peaks at 350, 420 and 510°C, but these two compounds are not common in samples analysed from the Keel area.

Mercuric sulphide was found to be concentrated in the vicinity of base-metal sulphides, mercuric chloride being the predominant phase in rocks more remote from sulphide ore. These distinct zones overlap to form a narrow mixed zone. By contrast, the total mercury content is not always directly related to the proximity of sulphide ore at Keel. Exploration drilling may therefore be aided by trace analysis of rock for mercury compounds rather than total mercury. Further case histories are required to test the general validity of this approach.

The bedrock geochemistry of the Keel zinc–cadmium prospect, County Longford, Eire, has been investigated as part of an interdisciplinary study of the deposit, financed by Barringer Research, Ltd. In view of the geochemical significance[1–4,6,14] of mercury as a pathfinder element for potential areas of sulphide mineralization, this metal received particular attention.

Previous work by Bradshaw and Köksoy[10,16] by use of a differential heating method[9] showed that it was possible to identify mercury compounds in rocks surrounding cinnabar and stibnite deposits in western Turkey. The research outlined in this paper has extended this approach to the Keel base-metal deposit by the use of a more precise method of trace compound identification[11] for examinations of dispersion patterns in wallrock. These compounds were subsequently found to have definite spatial relationships with the base-metal mineralization and their identification is, therefore, a useful indicator of base-metal build-up in this area.

Geology and mineralization at Keel

The geology of the Keel zinc–cadmium prospect has been described by Patterson.[15] The prospect occurs in Carboniferous marine sediments on the southern side of a Lower Palaeozoic inlier, mineralization occurring predominantly along northeast–southwest normal faults of Hercynian age. Finely disseminated base-metal sulphides do occur, however, throughout the area. The mineralogy of the deposit is quite simple: sphalerite is the predominant ore mineral, but galena does attain local significance, together with pyrite. Micromineralogy is somewhat more complex, the sphalerite containing up to 2% cadmium in solid solution and up to 180 ppm silver, held predominantly in the lattices of small inclusions of tetrahedrite. Up to 18% silver occurs in solid solution in the tetrahedrite at Keel.[15]

Mineralization is epigenetic and almost entirely confined to the clastic sediments which underlie the locally predominant reef limestones, and is probably the result of deposition from hot circulating waters in a sulphur-active area associated with these faults.

Primary dispersion of mercury

Sample collection and preparation

Wallrock samples from the underground workings, weighing in excess of 1 kg, were collected and stored immediately in sealed polythene bags and packed together in batches of 10. An effort was made to keep obviously mineralized and unmineralized samples apart. In the laboratory each sample was scrubbed with ordinary tap water to remove surface contamination, washed in deionized water and allowed to dry inside a large absorbent tissue at room temperature.

When dry, the sample was reduced to −1550 μm in a manganese–steel jaw crusher, and three size fractions were then sieved from this (−210 μm, −494 μm +210 μm, and −1550 μm +494 μm). After each sample had been taken, the jaw crusher was cleaned by scrubbing with warm water and sprayed with acetone to facilitate quick drying. Under normal procedures the jaw crusher is cleaned by running coarse sand through it, but as this sand was found to contain 1480 ppb (1480×10^{-9} g/g) mercury, its use as an abrasive cleaner was discontinued.

The three size fractions were stored in sealed Kraft paper bags, again in batches of 10, similar precautions with respect to mineralized and unmineralized samples being taken.

Drill core samples, which consisted of short lengths of whole core weighing about 1 kg each, were handled in the same way.

Initial results and orientation study

The mercury analysis was carried out initially by use of the sensitive mercury vapour analyser designed by James and Webb.[5] This instrument works on the atomic absorption principle by measuring the relative absorption of light of wavelength 253·7 nm by mercury vapour. A sample of material was heated to 700°C for 1 min and the evolved mercury was flushed through the apparatus as a single pulse, air being used as the carrier gas. Additional analyses were carried out by Barringer Research, Ltd., in Toronto with their own system of mercury analysis.[8]

*Table 1 Mercury content in different size fractions of rock samples from underground workings at Keel**

Sample	−210 μm	−494 μm +210 μm	−1550 μm +494 μm
K1	665	152	90
K2	2575	280	174
K3	2420	144	84
K4	300	140	106
K5	525	206	16
K6	240	170	130
K7	2855	1180	720
K8	1330	160	80
K9	705	184	116
K10	2460	390	49
K11	1075	216	139

*Results expressed as ppb.

The initial results obtained indicated that a mercury halo existed around Keel. When samples of the three mesh sizes were analysed it was found that the mercury content varied with the grain size (see Table 1). This variation could be due to a

*Table 2 Mercury contents of different size fractions of Keel drill core material**

Sample	−210 μm	−494 μm +210 μm	−1550 μm +494 μm
K12	998	1560	1020
K13	28	48	28
K14	22	32	36
K15	35	96	14
K16	166	210	110
K17	70	85	50
K18	14	42	15
K19	103	310	80
K20	60	130	52
K21	219	560	27
K22	169	700	150
K23	381	496	130
K24	44	210	27
K25	49	326	20
K26	160	172	110
K27	1675	2593	1020
K28	66	74	30
K29	44	236	25
K30	33	57	14
K31	141	280	100

*Results expressed as ppb.

combination of several factors: (1) the mercury occurs as minute grains, possibly discrete mercury minerals, which tend to concentrate in the finer fraction; (2) concentration of friable minerals, particularly sphalerite, occurs in the finer fraction: any ZnS lattice-held mercury would, therefore, also tend to concentrate in this fraction; and/or (3) delay in release of mercury from the coarser material due to a thermal gradient being set up, preventing release of mercury from the innermost parts of the larger grains during the 1-min heating cycle.

Analysis of the material from the underground workings of the prospect was followed by a survey of the drill core material from the west of the area (Table 2).

*Table 3 Comparison of mercury content of ground and unground −1550 μm +494 μm material from Keel**

Sample	Ground	Unground	Ratio Ground/unground
Material from underground workings			
K1	178	90	1·98
K2	1970	179	11·00
K3	280	84	3·23
K4	130	106	1·22
K5	320	16	20·00
K6	170	103	1·65
K7	1420	720	1·97
K8	380	80	4·75
K9	296	116	2·55
K10	483	49	9·85
K11	980	139	7·05
Material from drill core			
K12	2060	1020	2·0
K13	210	28	7·5
K14	240	36	6·7
K15	1040	14	74·3
K16	1840	110	16·7
K17	546	50	10·9
K18	480	15	32·0
K19	1880	80	23·5
K20	4200	52	80·7
K21	6800	27	251·8
K22	1560	150	10·4
K23	6900	130	53·1
K24	2100	27	77·8
K25	1520	20	76·0
K26	3400	110	30·9
K27	7400	1020	7·3
K28	3800	30	126·6
K29	4200	25	168·0
K30	1080	14	77·0
K31	1600	100	16·0

*Results expressed as ppb.

Comparison of Tables 1 and 2 shows that both for materials from the underground workings and for drill core the lowest mercury contents occur in the −1550 μm +494 μm material, and that mercury concentration increases with decreasing particle size for samples from the underground workings and increases with increasing particle size for drill core. This, in the light of information available at the time, was thought to be due to a build-up of pore- and fracture-held mercury in drill core material away from the known area of mineralization.

To test the possibility −1550 μm +494 μm material from the underground workings and from drill core was ground for 15 sec in a Tema mill and then re-analysed for mercury. The results obtained are shown in Table 3.

Grinding increased the apparent mercury concentration in both sets of samples, and the increase in mercury content of some samples of drill core material was substantially higher than the increase for the underground samples. The air inside the Tema mill was analysed for mercury after the grinding procedure in order to ascertain if the mine rock had lost more mercury than the drill core to the atmosphere inside the Tema mill. This experiment was carried out by substituting a

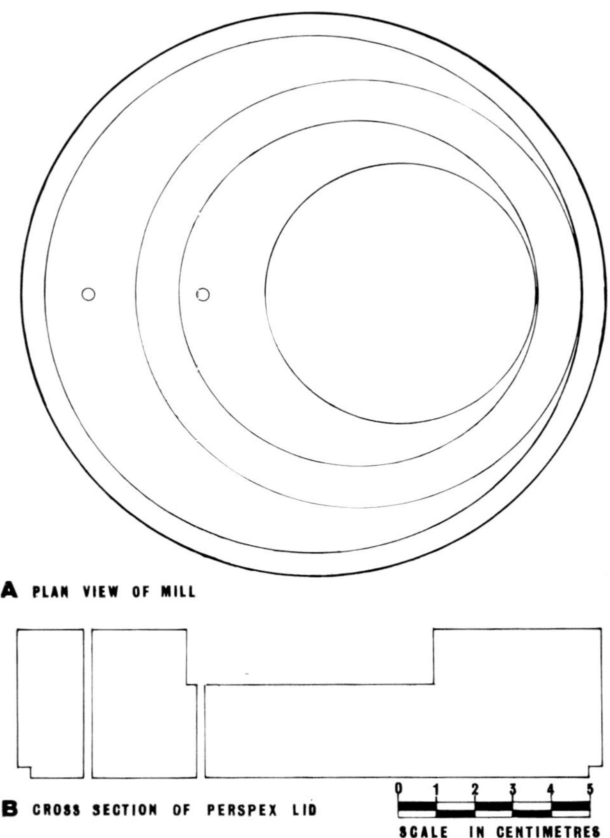

Fig. 1 Perspex lid design for gas extraction from a sealed grinding mill

4 cm thick perspex lid in place of the original Tema lid. Two holes were drilled through the lid in order to correspond to gaps between the grinding rings (Fig. 1). The holes were sealed with sellotape, both inside and outside, and the sellotape inside was pushed slightly into the hole to allow the inside to be continuously sealed throughout grinding. After grinding, a hypodermic needle was pushed through each hole and a 20-ml sample of air from inside the mill was drawn into a glass syringe. This air was then immediately transferred to the mercury analyser and the mercury was measured. Knowledge of the original sample weight and density of the specimen and the volume of the Tema mill made it possible to calculate that from 4 to 13 ng of mercury per gramme of rock was lost; hence, loss of mercury by volatilization during grinding was negligible. This fact indicated that the mercury in the drill core samples was bound up just as tightly as that in samples from the underground workings and that the distribution of mercury in rocks was governed by far more complex factors than a simple increase of elemental mercury in fractures and fissures.

Identification of mercury compounds in Keel area

Analytical method[11]

Consideration of the accumulated data led to the conclusion that a specific distribution of mercury compounds might be the cause of the apparent analytical discrepancies between samples from the underground workings and drill core material. To test this hypothesis it was necessary to adapt the available mercury analysis equipment into a continuous mercury release monitoring device. This was achieved by use of a thermostatically controlled tube furnace to heat the sample, which was contained within a silica tube, to 800°C over a period of 45 min. A thermocouple placed directly in the sample inside the heating chamber gave the temperature to within ±5°C, and a variable-scale multi-speed chart recorder enabled a continuous record of mercury release to be taken. Air, at a rate of 4 l/min, was continuously flushed over the 1-g sample during the course of this heating run (Fig. 2).

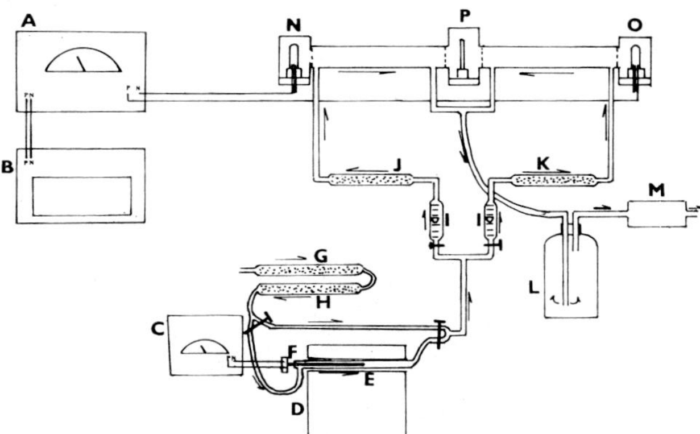

Fig. 2 Apparatus for continuous mercury thermal release analysis: A, volt meter; B, chart recorder; C, thermocouple; D, tube furnace; E, sample chamber; F, silica tube; G, silica gel; H, activated charcoal; I, flow meters; J, glass wool; K, $PdCl_2$ on glass wool; L, aspirator; M, pump; N, right photocell; O, left photocell; P, ultraviolet source; flow taps and flow directions (arrowed) are shown

Standardization

By use of this equipment a series of curves for mercury release from each rock was obtained. Synthetic standards containing 2·00, 1·00, 0·50, 0·25 and 0·125 ppm mercury in the form of solid mercurous and mercuric chloride, mercuric sulphide, mercuric oxide and mercuric sulphate were prepared for comparison with the mercury release curves obtained from the Keel rocks. It was not feasible to use elemental mercury in the preparation of standards of this type due to homogenization difficulties and oxidation, but a qualitative standard was prepared. As the thermal release characteristics of the various salts might be modified by the rock matrices, each set of standards was prepared in each of the five rock bases listed in Table 4.

Table 4 Composition of rock standard bases (%) used in mercury compound analysis

Base	SiO_2	$CaO/CaCO_3$	Fe_2O_3	Carbon
a	100			
b	90	10		
c	80	15	5	
d	80	15		5
e	75	15	5	5

Each base was heated to 900°C for 24 h and

Table 5 Average temperatures (°C) of mercury compounds release peaks

Compound	Peak 1	Peak 2	Peak 3
Mercury	80		
Mercurous chloride	170		
Mercuric chloride	210–220		
Mercuric sulphide	300–320		
Mercuric oxide	160	255–270	495
Mercuric sulphate	350	420	510

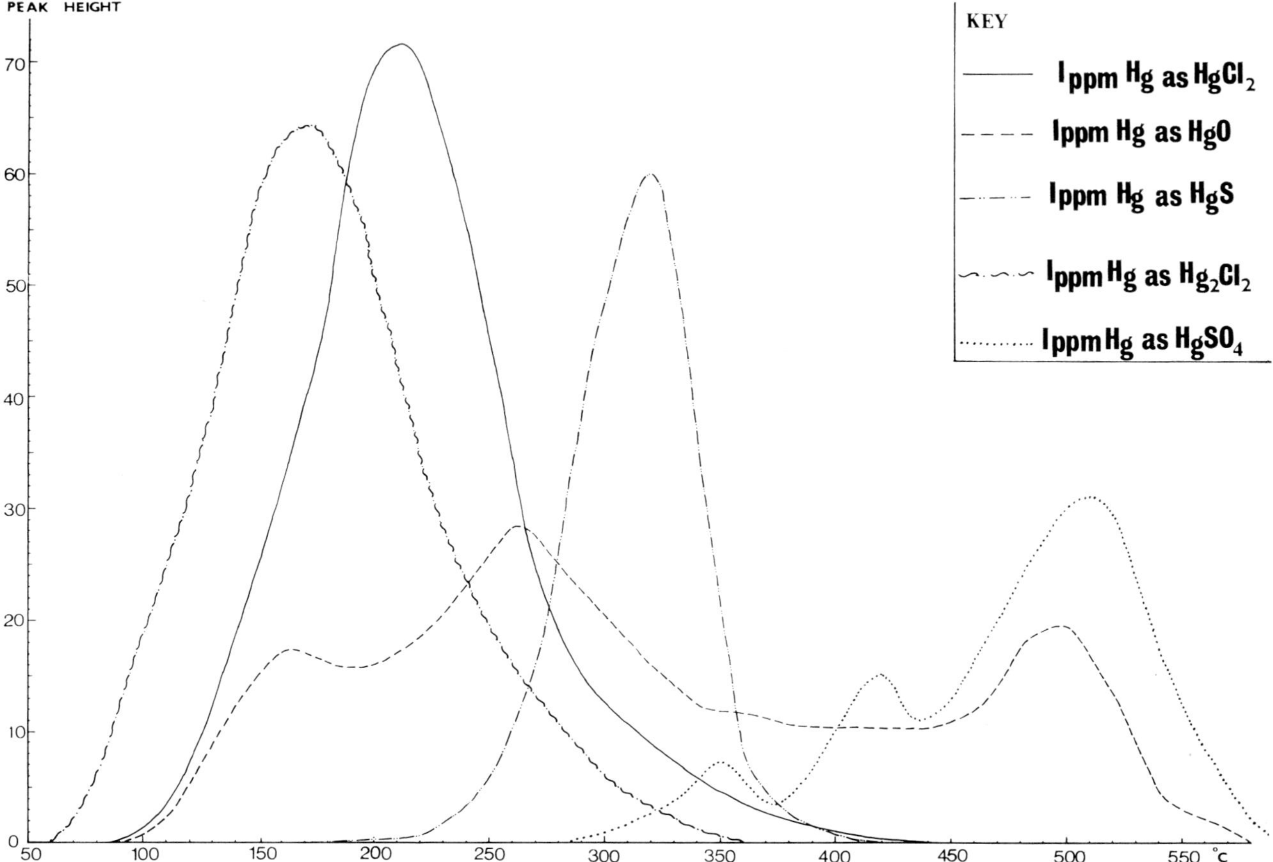

Fig. 3 Thermal release curves of mercury vapour from various mercury compounds

reground to $-53{\cdot}5$ μm before addition of each mercury compound. The mercury thermal release characteristics of these compounds are shown in Fig. 3, and peak temperatures are given in Table 5.

From these results the possibility of qualitative identification of these compounds in natural material is opened up. Calibration curves on the basis of the prepared standards were also calculated (Fig. 4). The temperature of maximum peak height and sub-curve area is specific for each mercury compound and comparison of the results given in Fig. 4 for mercuric sulphide and mercuric chloride shows that a sample containing 0·54 ppm mercury in the form of mercuric chloride will give an equal sub-curve area to one containing 1 ppm mercury in the form of mercuric sulphide. Thus, two samples containing equal amounts of mercury, one in the form of mercuric chloride and the other in the form of mercuric sulphide, will give different readings when analysed by the normal total volatilization method usually employed for mercury analysis, and misleading analytical data could result. The use of this kind of thermal dissociation technique thus offers the possibility of quantitative as well as qualitative mercury and mercury compound analysis.

Results

If the release graphs from natural materials are now compared with those from the synthetic standards (Figs. 3 and 5), two major patterns emerge for rocks from which mercury is volatilized below 400°C (in Fig. 5 vertical lines represent changes in chart recorder scale).

α Single release peaks at $\alpha 1$, 75–85°C
$\alpha 2$, 160–175°C
$\alpha 3$, 205–230°C
$\alpha 4$, 295–310°C

β Double release peaks at 205–230°C and 295–310°C

With reference to the synthetic standards, $\alpha 3$ can be seen to correspond to the release pattern of mercuric chloride and $\alpha 4$ to mercuric sulphide, $\alpha 1$ and $\alpha 2$ correlating with mercury and mercurous chloride, respectively. The β dissociation curve is a mixture of mercuric chloride and mercuric sulphide.

If the temperature scale to 900°C is considered, peaks at 450°C and 600–650°C emerge. These are formed by release of mercury from the lattices of iron pyrites and sphalerite, respectively. A triple-peaked oxide dissociation curve is primarily associated with a decomposed zone in which ferric

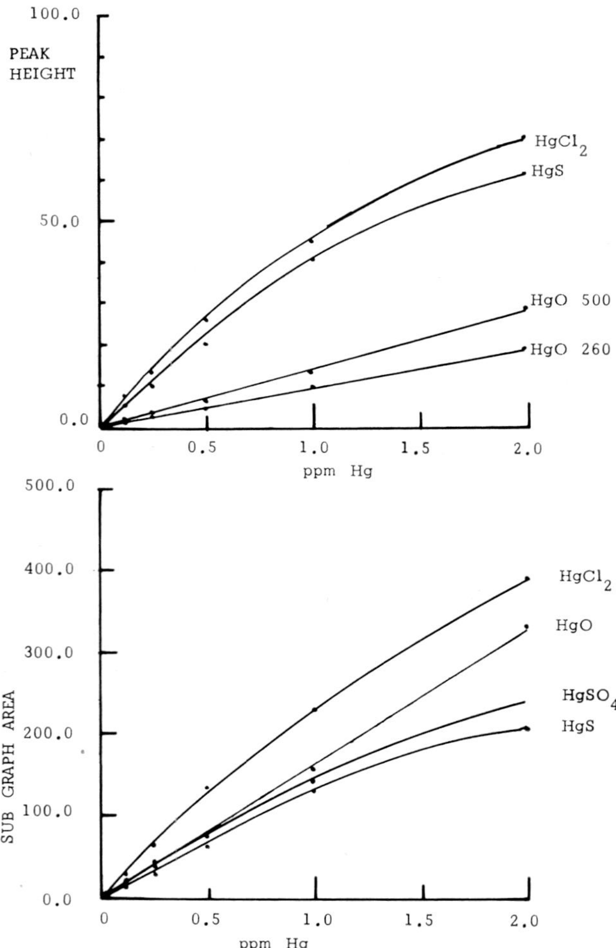

Fig. 4 Mercury compound calibration graphs

oxides are also present and give a mercury release peak at 420°C. Mercuric oxide is also present, associated with an area of faulting on the 200-ft level (Fig. 8). The relative magnitudes of the three peaks differ between natural and synthetic oxides, and this may be due to different crystallographic structures and, hence, different thermal release characteristics.

Fig. 5*(d)* shows the release curves of a rock from the mineralized zone at Keel in which the 600°C release peak appears. This peak varies between 600 and 650°C and is purely dependent upon the dissociation temperature of the particular sphalerite being analysed. No physical or chemical reason can be found for this difference in breakdown temperature. Multi-element analysis shows no significant difference in the trace-element composition, and major-element analysis of the rocks gives no indication as to why this temperature discrepancy occurs. When separated and cleaned, the sphalerites show exactly the same thermal release characteristics as they did in their original host rock.

The 450°C release peak is caused by mercury being released from the lattice of pyrites and is much more consistently temperature-dependent both for natural rock material and the separated mineral. At 420°C ±5°C mercury dissociates from ferric oxide to give a release peak. By taking these separate high-temperature peaks it is possible to know exactly how much mercury is actually associated with each mineral and, hence, to obtain far more accurate correlations between mercury and base-metal elemental concentrations (Fig. 6).

Subsurface distribution of mercury compounds at Keel[11]

The low-temperature sulphide, chloride and oxide peaks give information of a fundamental and meaningful nature. If their spatial distribution is plotted on plans of the underground levels of the mine area, a direct relationship to the distribution of the known areas of sulphide mineralization becomes apparent (Figs. 7–11).

Areas of significantly low concentrations of mercury exist to the north, southwest and south of the region. In the north the mercury, where present, is in the form of mercuric sulphide, whereas in the south and southwest it is in the form of mercuric chloride and is in direct association with the higher mercuric chloride areas. Mercuric chloride build-up is predominantly confined to the south and southwest of the area, and its obvious relationship to the extent of the reef limestone is very apparent, especially on the 200-ft level, where it conforms closely to the reef boundary.

Separating the mercuric chloride zone from the mercuric sulphide zone is an area of mixed mercuric sulphide and chloride. This zone appears to be continuous in the south, where the mercuric chloride halo is in close association with the reef limestone, and very abundant on the 100-ft level, where surface groundwater and possible bacterial effects have considerably modified the mercury compound distribution pattern (Fig. 7). In the southwest, however, the zone is far more sporadic in its distribution. This may be due in part to faulting which post-dates the mineralization and, hence, post-dates the emplacement of the mercury compound haloes. Lack of drill core material also plays an important part in this patchy distribution picture in the southwest. The mixed zone extends for some considerable distance into the underground workings on the 500-ft level, although the chloride component is extremely small. On the 200-ft level the zone is only apparent at the eastern end of the drive east.

The ratio of mercuric chloride to mercuric sulphide decreases as mineralization is approached and is very low in the underground workings,

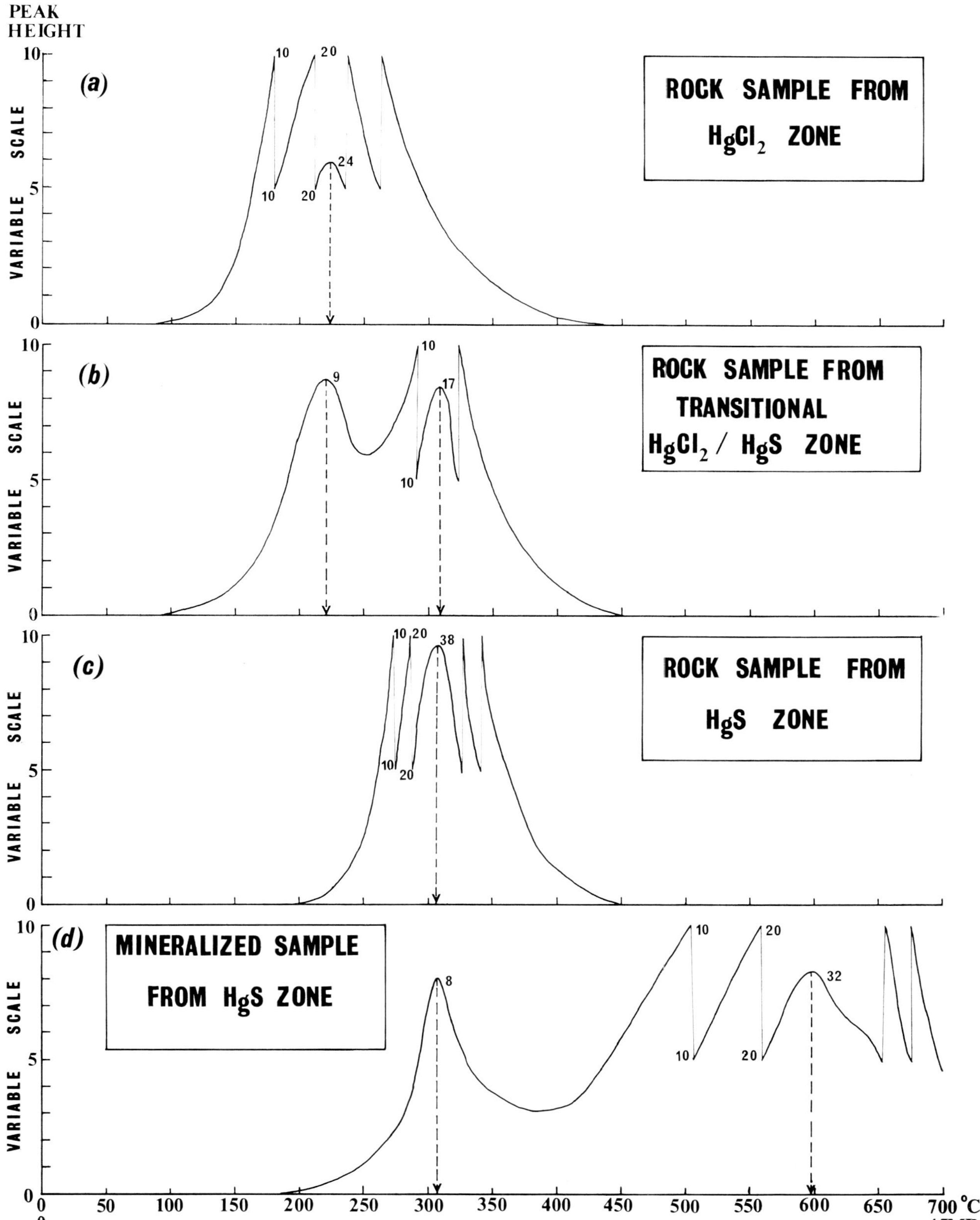

Fig. 5 Variation in mercury thermal release pattern in rock samples from different mercury compound zones, Keel (curves are traced from variable-scale chart recorder output: at each switchdown point scale is halved; true values shown against curves)

except in the southern end of the 500-ft level, where in one sample mercuric chloride makes up some 60% of the total rock mercury.

Mercuric oxide assumes prominence to the east of the area on the 200-ft level. This oxide zone, however, is very closely associated with faulting

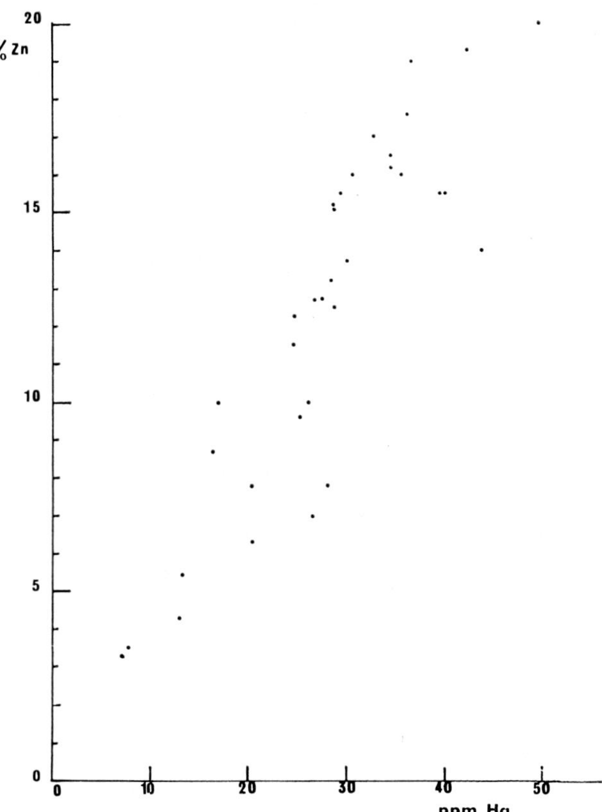

Fig. 6 Zinc–mercury relationship in Keel rock material

and is probably a direct result of oxidation of the existing mercuric chloride in that area. A greater area of mercuric oxide is associated with an area of decomposed rock to the northeast, where pyrite has also been oxidized.

Discussion

The total mercury content of the bedrock increases towards the underground workings and also with increasing depth. High concentrations occur where mineralization is encountered, and also in the mercuric chloride zone, where no mineralization is found. Quite commonly, very low levels of total mercury occur only ½ in away from massive mineralization, whereas high values can exist throughout a complete drill-hole in the unmineralized chloride zone. It is, therefore, of primary importance to identify the actual type of mercury compound giving rise to the anomaly.

At Keel the mineralization is characteristically fault-controlled, and as the distribution of mercury compounds appears to be aligned with the faults, it appears that the mercury compound distribution has been directly effected by the process or processes which gave rise to the mineralization in the area. Evidence as to whether the mercury came directly from the mineralizing solutions or

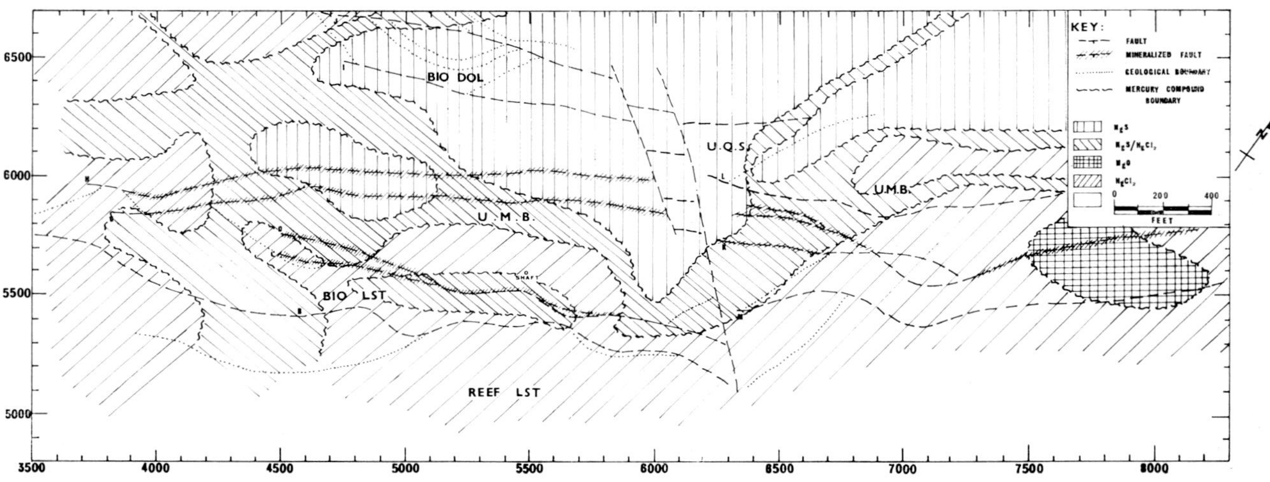

Fig. 7 Distribution of mercury compounds, 100-ft level, Keel

Table 6

Desiccator	Solution added	Compound formed at room temperature	Compound formed at 100°C	Compound formed at 140°C
A	1 ml 2% NaCl	$HgCl_2$ plus very minor HgO	$HgCl_2$	$HgCl_2$
B	1 ml 2% Na_2S	HgS	HgS	HgS
C	1 ml deionized water	Hg vapour only	Hg vapour	Hg vapour
D	0·5 ml 2% NaCl plus 0·5 ml 0·1% Na_2S	$HgCl_2$	$HgCl_2$	HgS

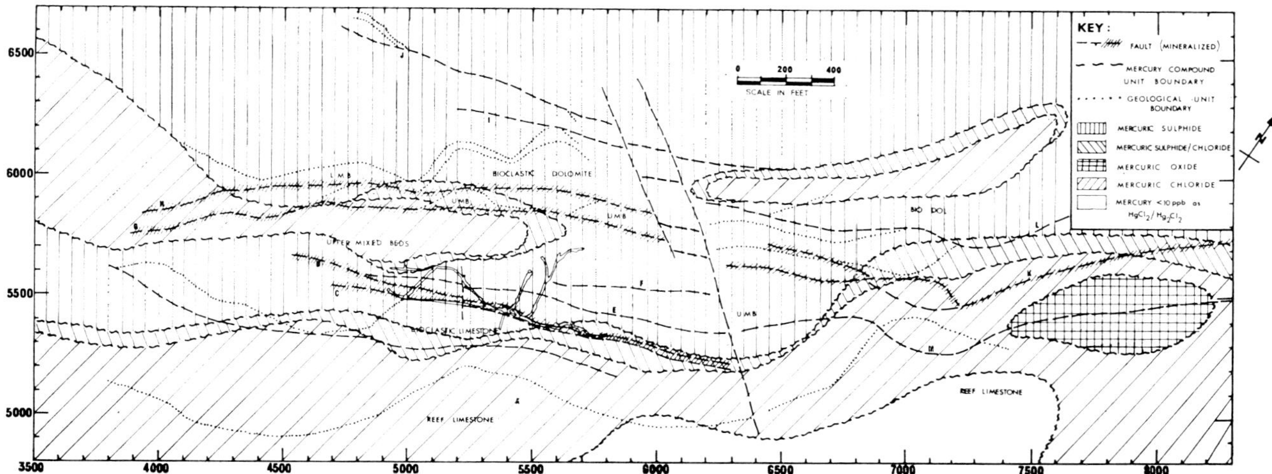

Fig. 8 Distribution of mercury compounds, 200-ft level, Keel

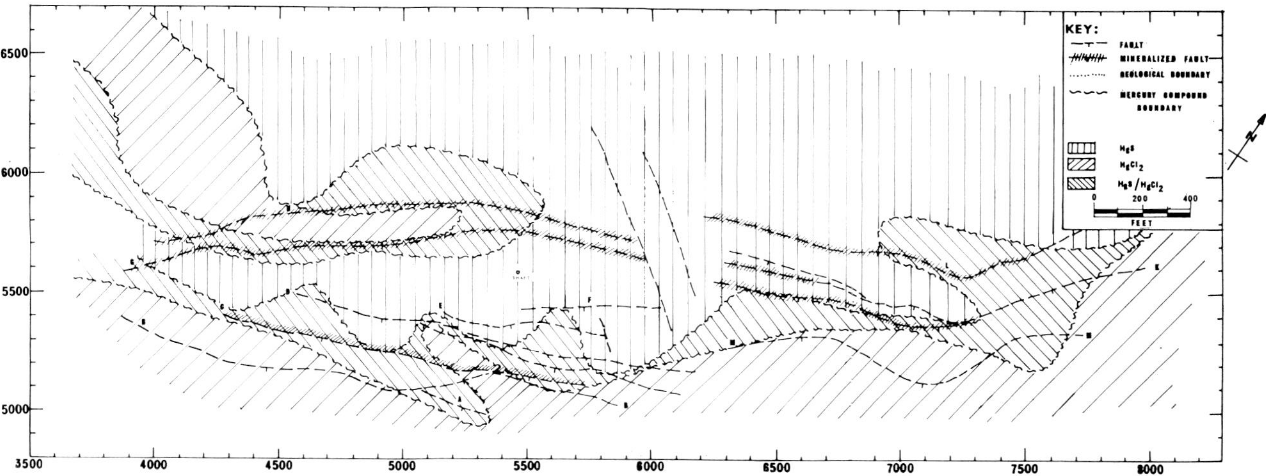

Fig. 9 Distribution of mercury compounds, 300-ft level, Keel

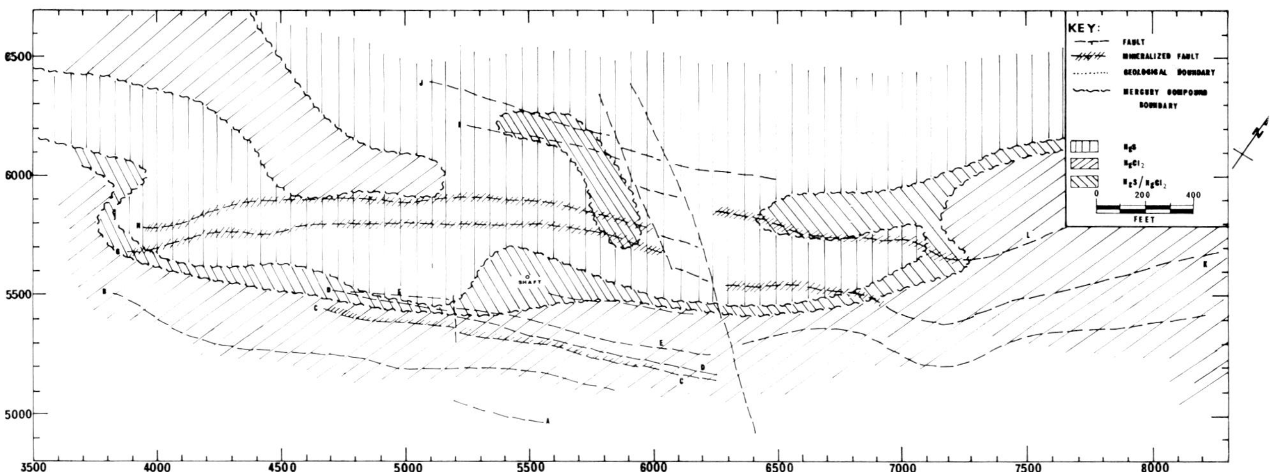

Fig. 10 Distribution of mercury compounds, 400-ft level, Keel

whether it built up in the area prior to mineralization and was later modified is purely tentative. The data suggest that waters charged with base metals in the form of chlorides and chloride complexes migrated towards the faults on the southern side of the inlier. The faulted area was an area of increased sulphur activity and temperature and, consequently, when the chloride-rich waters encountered the faulted region, base-metal sulphides were precipitated. This

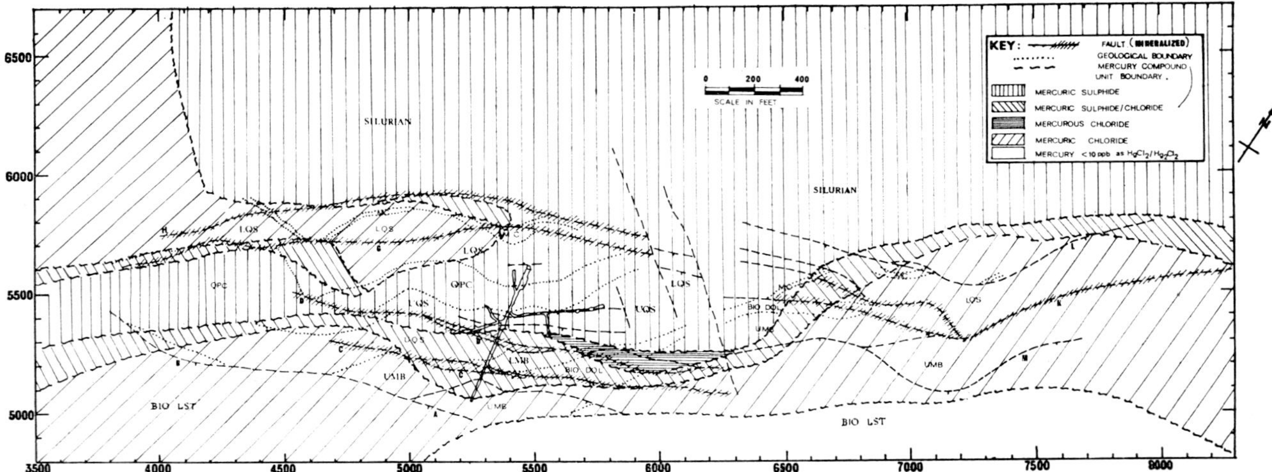

Fig. 11 Distribution of mercury compounds, 500-ft level, Keel

precipitation occurred predominantly in the faults that acted as natural channelways for the migrating solutions, but disseminated sulphides were also precipitated throughout this sulphur-active area. As the sulphides were insoluble, and, hence, stable, a gradual build-up of base-metal sulphides could occur by continuous precipitation from the migrating juvenile waters. The halo of mercury compounds around Keel is therefore interpreted as being due to migration of mercuric chloride into an area of sulphur activity and consequent conversion to insoluble mercuric sulphide. The greatest precipitation was in the area of base metal build-up, giving a high proportion of mercury in the lattice of sphalerite. When traced away from the faults, the mercuric sulphide zone merges into the mixed zone and finally back into the mercuric chloride zone. Traced farther away, true background areas of mercuric chloride are reached.

The marine Silurian to the north of the area is anomalous in that it contains mercuric sulphide but no mineralization. This is explicable by assuming that when the Silurian was deposited it contained trace mercury in the form of mercuric chloride, just as did the marine Carboniferous. The following experiments indicate that the mercuric chloride would be stable in this original environment, even though it would be in association with sulphur species.

Four samples (2 g each) of rock base *B* were placed on watch glasses, each inside a separate pyrex desiccator, referred to as desiccators *A, B, C* and *D*. 10 ml of deionized water was then poured into the base of each desiccator and three drops of redistilled mercury were added to the water. Each rock base was then spiked with a solution as noted in Table 6.

Each desiccator was then covered and allowed to equilibrate at room temperature for one week. The watch glasses were then removed and dried in separate desiccators for two days over silica gel. Each sample was then analysed by the differential thermal release technique to ascertain which mercury compounds had formed. Analyses were performed in triplicate on 500-mg samples.

The whole experiment was repeated twice more—once at 100°C and again at 140°C. At the latter temperature constant re-additions of mercury and water to the desiccator and water to the sample were required. Thereafter, the experiment had to be discontinued on account of the considerable volume of mercury being evolved and inadequate safety precautions.

At approximately 140°C in desiccator *D* HgS was formed in preference to the $HgCl_2$ formed in the experiments at room temperature and 100°C. This suggests that the degree of reactivity of the sulphide ion at 140°C is much greater than the chloride ion, even when the latter is in twenty times excess. As chloride is relatively more abundant than sulphide in the marine environment,[12, 13] it could be concluded that mercuric chloride initially would have been the stable mercury compound in the marine Silurian. Taking pressure effects into account, the conversion of this $HgCl_2$ to HgS in a sulphur-rich environment would take place at temperatures in excess of 150°C, and these conditions could have existed during the Caledonian orogeny.

This interpretation would also explain the fact that the Silurian is almost invariably uniformly low in mercury content, because under the conditions envisaged little migration of mercury would have occurred. It would also suggest that the mercuric sulphide in the Silurian rocks is older than that in the Carboniferous, and, hence, of a different genesis.

Summary and conclusions

The information recorded here appears to be of primary geochemical significance. Standard geochemical methods of mercury analysis are clearly of great use in defining relatively broad areas of potential base-metal mineralization.[17] At Keel bedrock zones containing mercuric chloride, which is not closely associated with base-metal concentrations, can give high total mercury values which may easily lead to misinterpretation of conventional geochemical data for total mercury. The method of differential mercury compound analysis can, however, be used to identify those mercury compounds which are specifically associated with base-metal mineralization, and the potential area of sulphide mineralization may thus be outlined with increased confidence.

This approach, however, is still in its infancy, and a great deal more work needs to be done before it can become a proven and worthwhile geochemical technique. Not only mercury but other elements and their compounds should be examined in both primary and secondary environments around many other ore deposits in both different geological and climatological environments.

Acknowledgment

Research into the trace-element geochemistry of the bedrock environment has been undertaken by the Mining Geology Division as part of a multidisciplinary research investigation at Keel. The authors express their gratitude to Dr. A. R. Barringer, Visiting Professor of Economic Geology at Imperial College, London, and President of Barringer Research, Ltd., who initiated the entire project and arranged the finance to support the costs and a research studentship to one of the present authors (R.J.W.). Sincere thanks are offered to Mr. A. J. Thompson of the departmental Geochemistry Laboratory for his invaluable help and advice throughout the analytical programme, and to other colleagues for helpful discussion and assistance—in particular, Dr. C. J. Morrissey.

References

1. Fursov, V. Z. Halos of dispersed mercury as prospecting guides at the Achisai lead-zinc deposits. *Geokhimiya*, 1958, 267–72; *Geochemistry, Ann Arbor*, 1958, 338–45.

2. Ozerova, N. A. The use of primary dispersion halos of mercury in the search for lead-zinc deposits. *Geokhimiya*, 1959, 638–45; *Geochemistry, Ann Arbor*, 1959, 793–802.

3. Hawkes, H. E. and Williston, S. H. Mercury vapor as a guide to lead-zinc-silver deposits. *Min. Congr. J.*, **48**, Dec. 1962, 30–2.

4. Friedrich, G. H. and Hawkes, H. E. Mercury as an ore guide in the Pachuca-Real del Monte district, Hidalgo, Mexico. *Econ. Geol.*, **61**, 1966, 744–53.

5. James, C. H. and Webb, J. S. Sensitive mercury vapour meter for use in geochemical prospecting. *Trans. Instn Min. Metall.*, **73**, June 1964, 633–41.

6. Barringer, A. R. Developments towards the remote sensing of vapours as an airborne and space exploration tool. In *Proc. 3rd Symp. remote Sens. Environm.* (Ann Arbor: University of Michigan, 1965), 279–92.

7. Vaughn, W. W. and McCarthy, J. H. Jr. An instrumental technique for the determination of submicrogram concentrations of mercury in soils, rocks, and gas. *Prof. Pap. U.S. geol. Surv.* 501–D, 1964, 123–7.

8. Barringer, A. R. Interference-free spectrometer for high-sensitivity mercury analyses of soil, rocks and air. *Trans. Instn Min. Metall. (Sect. B: Appl. earth sci.)*, **75**, 1966, B120–4.

9. Köksoy, M. Bradshaw, P. M. D. and Tooms, J. S. Notes on the determination of mercury in geological samples. *Trans. Instn Min. Metall. (Sect. B: Appl. earth sci.)*, **76**, 1967, B121–4.

10. Bradshaw, P. M. D. and Köksoy, M. Primary dispersion of mercury from cinnabar and stibnite deposits. W. Turkey. In *Rep. 23rd Int. geol. Congr.* (Prague: Academia, 1968), Sect. 7, 341–55.

11. Watling, R. J. Identification of trace mercury compounds in rocks as a guide to sulphide mineralization at Keel, Eire. *Trans. Instn Min. Metall. (Sect. B: Appl. earth sci.)*, **81**, 1972, B47–8.

12. Mellor, J. W. *A comprehensive treatise on inorganic and theoretical chemistry* (London: Longmans, 1929), vol. 4, 695–1049.

13. Rankama, K. and Sahama, Th. G. *Geochemistry* (Chicago, Ill.: University of Chicago Press, 1950), 94–263; 715–8.

14. Hawkes, H. E. and Webb, J. S. *Geochemistry in mineral exploration* (New York: Harper and Row, 1962), 415 p.

15. Patterson, J. M. Geology and mineralization of the Keel area, County Longford, Republic of Ireland. Ph.D. thesis, University of London, 1970.

16. Köksoy, M. Dispersion of mercury and other ore elements from mineral deposits in Turkey. Ph.D. thesis, University of London, 1967.

17. Evans, D. S. Secondary dispersion of mercury and associated elements at Keel, Eire. Ph.D. thesis, University of London, 1971.

18. James, C. H. The potential role of mercury in modern geochemical prospecting. *Min. Mag., Lond.*, **111**, 1964, 23–32.

552.1:546.49:553.43/.44(714)

Nature of mercury anomalies at the New Calumet Mines area, Quebec, Canada

E. C. Smith, M.SC.

Marine Sciences Centre, McGill University, Montreal, Quebec, Canada

G. R. Webber, PH.D.

Department of Geological Sciences, McGill University, Montreal, Quebec, Canada

Synopsis

To study the association of mercury with sulphur, zinc, copper, iron, magnesium and manganese in rock and ore, the concentrations of these elements were determined in 233 rock samples from New Calumet mine, Quebec (a zinc–lead deposit), and from a mineralized copper prospect in the Labrador Trough.

At New Calumet, mercury, zinc, sulphur and copper had anomalous concentrations over the ore. The anomalous concentrations of different elements complement one another because they are not always coincident. Significant correlations of mercury, zinc, sulphur and copper suggest that they were, in part, emplaced together. Mercury in surface rock samples was mostly readily extractable by heating at 500°C, whereas much lower proportions of mercury were released at this temperature from samples from the ore zone.

A reinterpretation of data on the iron content of sphalerite at New Calumet suggests that the ore deposit underwent regional metamorphism and annealing of sulphides at a pressure of 3·5–4·5 kb. There is an approximate correspondence in the location of zones of iron-rich sphalerite and readily extractable mercury which might be related to sulphur fugacities.

Correlations of mercury with iron, sulphur, copper and zinc were found in samples from the copper prospect in the Labrador Trough. Much of the mercury probably arrived in association with these elements.

A study has been made of the association of mercury with copper, zinc, iron, sulphur, magnesium and manganese in rocks and ore from New Calumet mine, Quebec, and from a base-metal showing in the Labrador Trough. Correlations and groupings of elements were derived from data by use of factor analysis and multiple-regression techniques. The principal objectives were to see whether the nature of mercury anomalies could be clarified by consideration of associated elements and to judge the relative advantages of mercury and the other elements in the search for ore in these particular environments.

Geology of New Calumet mine

The New Calumet mine property is situated on Calumet Island in the Ottawa River 60 miles northwest of Ottawa, and can be reached by bridge from the town of Bryson. Fig. 1 is a generalized map of the geology of the area. The deposit is found within a mass of biotite gneiss, lesser offshoots occurring in the bordering carbonates. Mining dates back to 1897, when a small, high-grade, lead–zinc tonnage was produced. Workings reached a depth of 2000 ft, and production eventually exceeded 1 000 000 tons of ore averaging about 0·51% lead, 8·17% zinc, 0·036 oz of gold and 5·70 oz of silver. The ore contains varying amounts of medium-grained sphalerite, pyrrhotite, marcasite, pyrite and galena, and minor amounts of chalcopyrite, arsenopyrite, tetrahedrite and molybdenite.[20, 23]

The principal rock types exposed on the property are 600 ft of coarse, massive, crystalline limestone, overlain by a body of granitized migmatite gneiss of varying thickness, overlain by a body of rusty-weathering biotite gneiss of variable thickness, overlain by from 100 to 300 ft of thinly layered or massive amphibolite. Osborne[23] favoured origin of the ore from hydrothermal solution associated with igneous activity, and suggested as controls of ore emplacement the presence of more permeable, argillaceous zones within the biotite gneiss, and the presence of a relatively impermeable ceiling of carbonate amphibolite. Moorhouse[22] also favoured hydrothermal origin for the ore. He referred to the resemblance of these ores to those at Edwards, New York. Brown[3] has suggested that the Balmat–Edwards deposits were emplaced as normal bedded deposits and later metamorphosed. This may be true at New Calumet also. Oreshoots were generally parallel to the attitude of major and minor fold axes, mineral lineation and diabase dykes, which all show a pitch of 20–30° to the east.[23] Cornwall[4] reported noticeable alteration up to 100 ft into the hanging-wall amphibolite, and detectable base-metal dispersion for up to 300 ft from the ore. The ore zone comes near the surface on the southwest side, but mineralization, except for pyrite, is not generally visible on the surface. Workings were 500–600 ft deep on the north side and 1300–1500 ft deep on the east side.

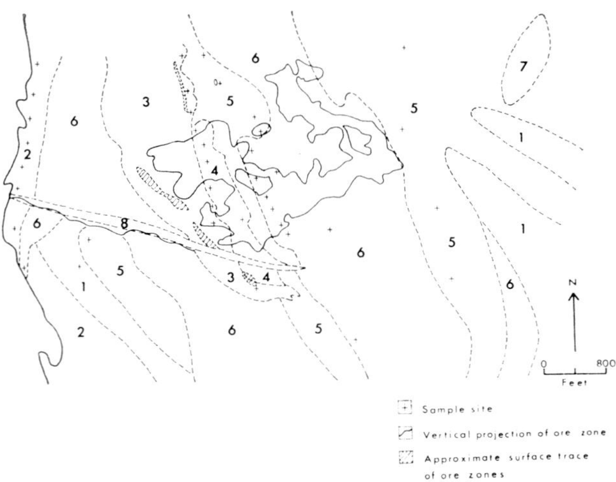

Fig. 1 General geology, surface sample site location and approximate position of ore zone at New Calumet property: 1, amphibolite; 2, limestone; 3, biotite gneiss; 4, carbonate amphibolite; 5, hornblende gneiss; 6, migmatite; 7, granite; 8, diabase. Geology after Moorhouse [22] and Osborne [23]

Geology of the Theta showing

Theta showing is a fictitious name for an actual area located within the middle-eastern part of the Labrador Trough. The private company which supplied the specimens from this property requested that the specific location not be disclosed at this time.

The region is characterized by abundant volcanics and metasediments, which are intruded by metagabbro sills. The effects of regional metamorphism are widespread, and increase toward the east. A front of new biotite formation runs about north–south through the showing. Contact metamorphism may often be recognized at the boundaries of sills. The property is located near a major synclinal axis, and schistosity and former bedding planes dip from 30 to 75° to the east. Several minor faults have been postulated to occur nearby.[25]

Exposures of the Hellancourt Formation cover an area which includes the Theta showing. The predominant lithologies are green aphanitic to fine-grained metabasalt flows, and pillow lavas containing scoria, quartz, carbonate and epidote fillings. All mineral occurrences are found close to, and appear to be related to, an extensive sill of gabbro with blotchy zones of clinozoite and albite. Most of the mineral showings in the region are pyrrhotite–chalcopyrite deposits with some nickel and gold, and are thought to be magmatic rather than hydrothermal, because of a lack of structural control, and because of their relation to the blotchy gabbro sill. A few other showings of pyrite and of copper–zinc mineralization of probable magmatic or syngenetic origin are also present.[25]

Previous work on mercury

Reviews of the chemistry of mercury in natural environments have already been given in various papers.[9, 14, 16, 21, 24] Several sources quoted concentrations between 10 and 80 ppb for acidic rocks and from 80 to 100 ppb for basic rocks.[11, 12, 24]

More recently, Ehmann and Lovering[6] arrived at a range of 4–7 ppb for mafic and ultramafic rocks. Marowsky and Wedepohl[19] estimated average concentrations of mercury at 77 ppb for granitic rocks and 35–70 ppb for basaltic–gabbroic rocks. High concentrations of mercury are found in sulphide minerals.[2, 13, 27]

Sample collection

Three sets of specimens were used in this study. At the New Calumet property 78 specimens of

rock were collected from outcrops by E. C. Smith in 1969. At each of the collection sites one sample was of comparatively fresh rock and another sample was of altered rock with limonite-stained fractures. Fig. 1 shows all the surface sample sites, except for one, which occurs well outside the area shown.

Seventy-eight ore specimens collected several years ago by L. A. Clark from pillars in underground workings of the New Calumet property were used for analysis. These had been stored in coarsely crushed form in glass bottles with friction tops.

Seventy-seven rock specimens from the Theta showing were obtained from a private company. These specimens consisted of fresh and altered samples. Some of the altered samples had limonite-stained fractures; others were bleached or had quartz-carbonate fillings.

Analytical procedure

Rock samples from New Calumet surface samples were pulverized in a small ceramic disc grinder. Underground and Theta showing rock samples were ground in a shatterbox. Mercury analysis was done with a Lemaire Bench Model 500B Mercury Detector, which functions on the basis of absorption of the 2537-Å radiation from a single beam supplied by a mercury lamp. Sample powders were heated in a small furnace to about 500 °C to free the mercury, which was collected on a gold wool trap, and other interfering gases were allowed to flow out of the system. The mercury concentration was then determined by heating the gold trap and introducing the mercury vapour into the absorption cell of the detector. The instrument was calibrated by use of known volumes of air saturated with mercury vapour.

Run times for the analyses were extended until mercury ceased to be given off by the gold trap. Specimens from New Calumet Mines were run in triplicate. Because only small amounts of sample were supplied from the Theta showing, only one run was possible.

On completion of the mercury analysis by use of the procedure outlined above, a series of tests was made with a Coleman Mercury Analyser MAS-50 and a sample preparation based on the procedure of Hatch and Ott,[10] as modified at the Geological Survey of Canada.[15] Tests showed that when samples were heated in the furnace, as previously described, about 80% of the mercury was driven off to be detected in the case of surface samples from New Calumet (69, 75, 78, 83 and 84% of 5 samples tested), about 85% for the Theta showing samples (10 samples tested), but for underground samples from New Calumet variable amounts of mercury could be driven off (1, 4, 7, 9, 24, 55 and 70% for 7 samples tested). Thus, the mercury values obtained here can be considered to be the readily extractable mercury. The bulk of the mercury in samples from the ore zone at New Calumet is probably bound tightly in sphalerite.

Analyses for magnesium, iron, sulphur, copper, zinc and manganese were performed on pelletized powders by use of a Philips PW 1220 semi-automatic X-ray fluorescence unit.

The average value of the relative deviations calculated for triplicate mercury analyses of individual samples of New Calumet ore was 18·5%. Corresponding values for New Calumet surface rock samples were 38·5 and 27·6% for the samples taken from along limonite-stained fractures and fresh rock samples, respectively.

Webber and Horska[30] reported a relative deviation for similar sulphur analyses of 8·3% for a concentration of 0·008% S, and of 1·25% for a concentration of 0·6% S. The average relative deviations for repeated analyses of the same pellets for copper, magnesium, iron, zinc and manganese are 1·9, 1·0, 0·5, 0·5 and 0·5%, respectively (calculated by running each pellet on three different occasions).

Results

Table 1 gives means, medians and relative deviations of element concentrations in fresh and altered rocks in both areas and for underground samples at New Calumet. Table 2 shows median concentrations of elements in different rock types at New Calumet. Rock classification was based on hand-specimen identification and an outcrop map by Osborne,[23] which differs in detail from Fig. 1. Note that median concentrations of mercury are about the same for all rock types except limestone.

Threshold concentrations of copper, zinc, mercury and sulphur were set at one standard deviation above the median and were determined from cumulative frequency plots drawn on logarithmic probability paper according to the method described by Lepeltier.[18] The relatively high proportion of samples over the ore zone would tend to give a strong representation of the population of anomalous samples and to introduce a bias to estimation of background, thus justifying the arbitrary choice of the relatively low threshold value of one standard deviation above the median. The straight-line character of these plots supports lognormal distributions. The presence of two populations of samples is suggested by slope changes for the New Calumet surface and Theta showing samples. Threshold

Table 1 Means, medians and relative deviations

	Mean	Median	Relative deviation, %	Mean	Median	Relative deviation, %
New Calumet						
		ppb Hg			% S	
Fresh	27	17	148	0·13	·016	330
Altered	50	25	204	0·28	·057	214
Underground	266	210	88	16·7	18	32
		ppm Zn			ppm Cu	
Fresh	162	100	121	142	69	181
Altered	735	100	477	137	84	121
Underground	14·9%	15·6%	39	1405	625	191
		Total Fe as % Fe_2O_3			% MgO	
Fresh	8·1	9·3	48	7·7	6·3	77
Altered	8·4	8·9	48	7·3	4·4	105
Underground	14·9	13·2	43	12·4	9·9	71
		ppm Mn				
Fresh	1078	1100	44			
Altered	1094	1100	43			
Underground	1642	1410	43			
Theta showing						
		ppb Hg			% S	
Fresh	55	34	144	0·90	0·030	622
Altered	45	30	120	1·07	0·027	532
		ppm Zn			ppm Cu	
Fresh	90	100	24	256	100	360
Altered	49	50	67	144	80	146
		Total Fe as % Fe_2O_3			% MgO	
Fresh	15·5	15·0	23	9·2	9·1	26
Altered	10·4	10·0	88	5·4	5·3	92
		ppm Mn				
Fresh	1743	1680	58			
Altered	934	970	67			

concentrations for mercury, copper and zinc are, respectively, 72 ppb, 230 ppm and 100 ppm for the Theta showing; 38 ppb, 200 ppm and 210 ppm for New Calumet rocks; and 510 ppb, 2100 ppm and 21% for New Calumet ore. Sulphur threshold for New Calumet surface rocks was set at 2100 ppm. Plan views showing mercury, zinc, copper and sulphur concentrations in New Calumet Mine surface samples are given in Figs. 2–5, and mercury and mercury/zinc ratios in the ore zone are included as Figs. 6 and 7. Anomalies of mercury, zinc, copper and sulphur in surface rocks are concentrated near and over the ore zone.

The data were analysed by use of available computer programs[29] for stepwise multiple regression and factor analysis in order to discover indications of correlations between mercury and the other measured elements. Data were transformed to $\log_e$ values for these calculations. Results from these numerical analyses are given

Table 2 Median concentrations of elements in different rock types of fresh New Calumet surface samples

	n	ppb Hg	% S	ppm Zn	ppm Cu	% Fe_2O_3	% MgO	ppm Mn
Limestone	7	12	0·014	74	22	0·21	18·2	590
Migmatite	8	21	0·014	96	54	8·35	3·3	865
Biotite gneiss	9	20	0·081	130	95	7·10	6·3	1030
Hornblende gneiss	7	22	0·014	110	170	11·10	4·8	1370
Amphibolite	8	17	0·053	120	135	10·80	9·6	1460

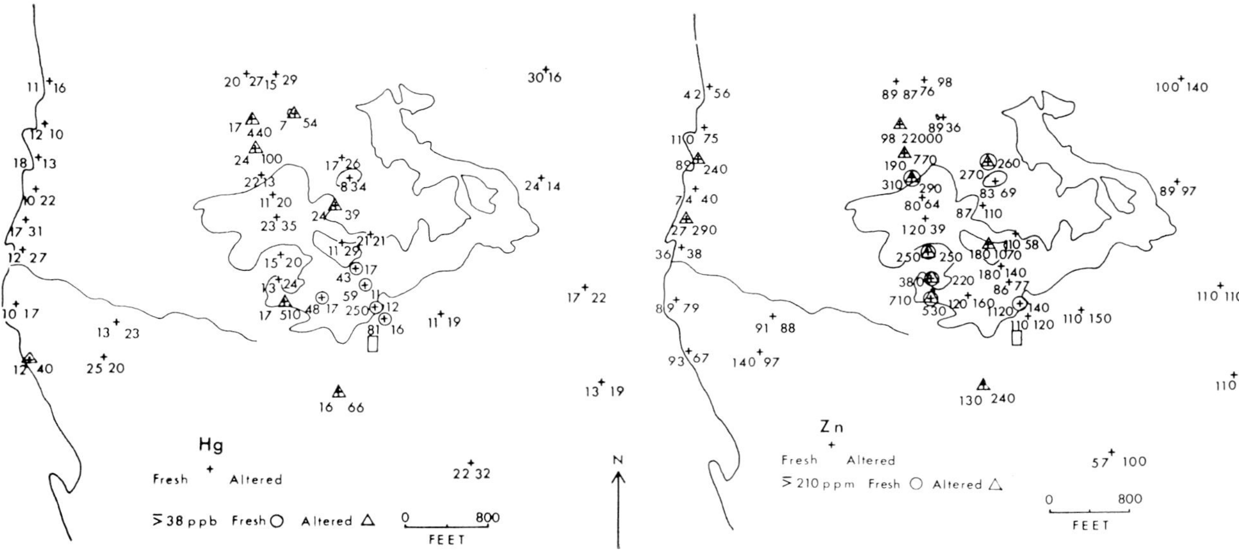

Fig. 2 Mercury concentrations (ppb) in surface samples at New Calumet. Small rectangle is site of mill

Fig. 4 Zinc concentrations (ppm) in surface samples at New Calumet

Table 3 Multiple-regression equations

		% of error reduced
New Calumet		
Fresh	Hg=4·21+0·26 S†−0·22 Mg†	36 (29 S, 7 Mg)
Altered	Hg=6·48+0·40 Zn†−0·79 Mn†+0·16 Fe*	47 (34 Zn, 10 Mn, 3 Fe)
Underground	Hg=2·01+0·23 Cu†+0·59 S†	11 (7 Cu, 4 S)
Theta showing		
Fresh	Hg=−1·46+1·54 Fe†+0·18 Cu†	35 (29 Fe, 7 Cu)
Altered	Hg=4·26+0·31 S†+0·05 Zn*	51 (48 S, 3 Zn)

In these equations Hg is $\log_e$ of ppb Hg; S is $\log_e$ of ppm S; Cu is $\log_e$ of ppm Cu; Mg is $\log_e$ of % MgO; Mn is $\log_e$ of ppm Mn; Fe is $\log_e$ of % Fe_2O_3; Zn is $\log_e$ of ppm Zn.

*Regression coefficient significant at 10% level.

†Regression coefficient significant at 5% level or better.

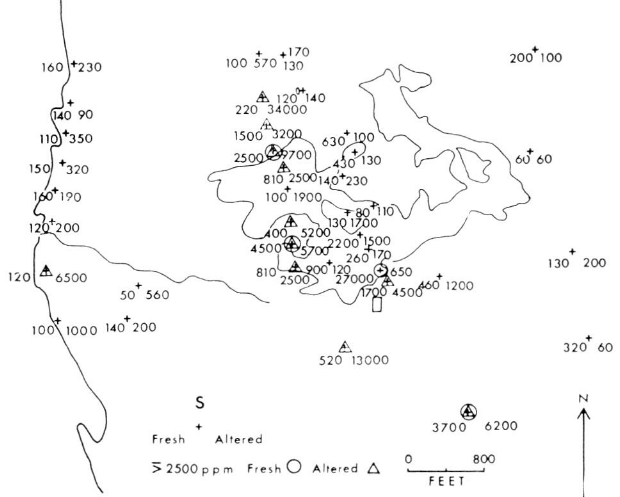

Fig. 3 Sulphur concentrations (ppm) in surface samples at New Calumet

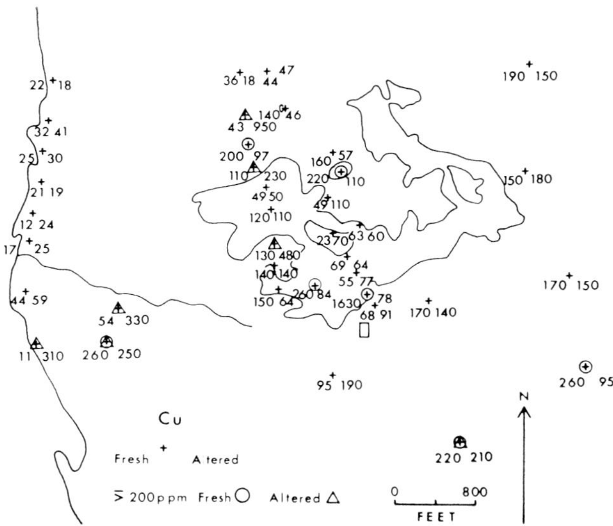

Fig. 5 Copper concentrations (ppm) in surface samples at New Calumet

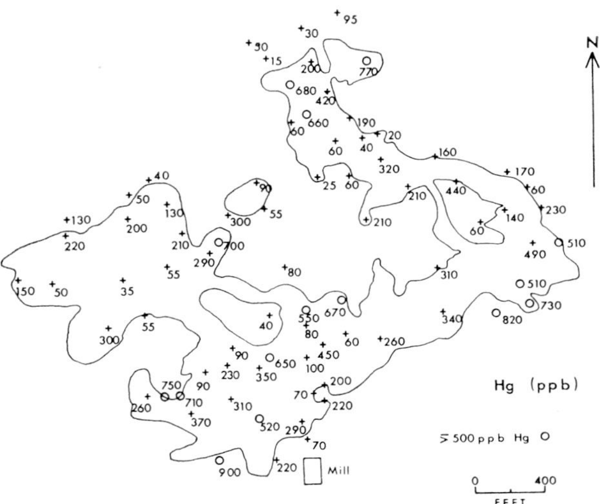

Fig. 6 Mercury (only portion readily removed by heating to 500°C) in underground samples at New Calumet. Solid line is generalized outline of orebody

in Tables 3, 4 and 5. Only factors with eigenvalues greater than unity were retained in factor analysis.

For the New Calumet surface samples Table 4 shows that mercury was found to be significantly correlated with zinc, sulphur and copper. The strongest correlation in the fresh rocks is with sulphur and in the altered rocks it is with zinc. This is also reflected in the multiple-regression equations (Table 3). Analysis of New Calumet ore specimens yielded a weak, but significant, positive correlation of mercury with copper and sulphur, but it should be noted that the regression accounted for only 11% of the variance of mercury, and, consequently, the six elements studied are not strong controls of the readily extractable mercury. Sears,[27] who used a higher temperature of extraction, found a strong correlation of mercury and zinc in samples from New Calumet.

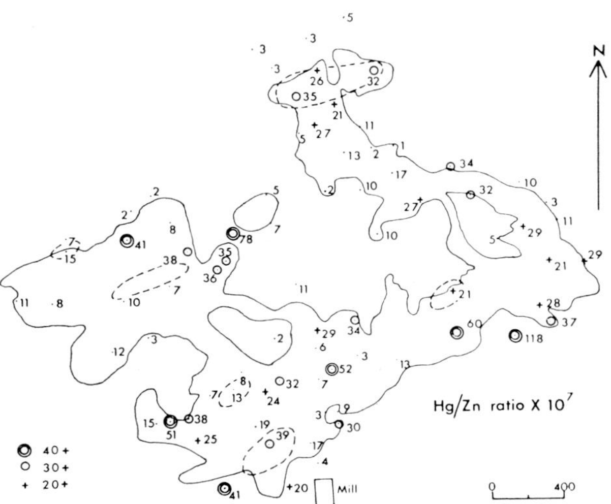

Fig. 7 Hg/Zn ratio in underground samples at New Calumet. Solid line is generalized outline of orebody; dashed lines represent zones of iron-rich sphalerite found by Erdosh[7]

Examination of mercury residuals from six-element multiple regression for the New Calumet samples showed that, for the fresh samples, about the same proportion of samples near ore and away from ore had an excess of mercury over that predicted by the regression (43 and 44%, respectively). Of the five anomalous fresh rock samples, however, only one would have been anomalous from prediction from six-element multiple regression, and four of these anomalous samples (and one background sample) contained more than 50% of their mercury content above the predicted value. Thus, mercury not predicted by the regression made an appreciable contribution to anomalous values in fresh rocks. For the altered samples there was an excess of mercury for 38% of the samples near ore and 61% of the samples away from ore. Of the eight anomalous altered samples, four would have been predicted from the six-element multiple regression. Two of these, two other anomalous samples and one background sample contained 50% or more of their mercury content above the predicted value. In the altered samples, therefore, both regression-predicted mercury and residual mercury contributed appreciably to anomalous samples.

Theta showing samples showed significant positive correlations for mercury with sulphur, copper and zinc by multiple regression, and a significant negative relation with manganese. For fresh samples, multiple regression indicates an association of mercury with iron and copper, and, for altered samples, with sulphur and zinc.

Of the Theta showing samples, two anomalous samples and four background samples contained 50% or more of their mercury content above the predicted value from five-element multiple regression. One-half of the anomalous Theta showing samples would have been predicted to be anomalous from the regression equations. Therefore, regression-predicted mercury and residual mercury both contributed appreciably to anomalous samples.

The three retained factors for factor analysis of New Calumet fresh and altered rock data (Table 5) grouped sulphur, mercury, zinc and copper; manganese, iron and copper; and magnesium and negative iron. The last two factors reflect differences in rock type. The three retained factors for New Calumet ore grouped zinc and sulphur; copper and mercury with negative magnesium; and iron with negative manganese.

Factor analysis of fresh rock data from the

Table 4 Correlation coefficients of mercury with other elements

	Mg	Fe	S	Cu	Zn	Mn
New Calumet						
Fresh	−0·27	0·16	0·54	0·44	0·41	−0·08
Altered	−0·10	0·03	0·42	0·26	0·58	−0·32
Underground	−0·13	0·04	0·21	0·25	0·15	−0·13
Theta showing						
Fresh	−0·05	0·54	0·46	0·48	0·19	−0·34
Altered	0·36	0·49	0·69	0·43	0·42	0·35

*Table 5 Varimax rotated factor loadings**

	Factor 1	Factor 2	Factor 3
New Calumet			
Fresh	Hg 0·80, S 0·89, Cu 0·67, Zn 0·72	Fe 0·64, Cu 0·58, Mn 0·90	Mg 0·95, Fe −0·65
Altered	Hg 0·82, S 0·76, Cu 0·54, Zn 0·85	Fe 0·83, Cu 0·65, Mn 0·90	Mg 0·97 (Fe −0·43)
Underground	S 0·90, Zn 0·94	Hg 0·53, Mg −0·70, Cu 0·85	Fe 0·87, Mn −0·60
Theta showing			
Fresh	Fe 0·84, S 0·82, Hg 0·77, Cu 0·63	Mg 0·89, Zn 0·79, Mn −0·70	
Altered	Zn 0·88, Mn 0·87, Mg 0·84, Fe 0·83, Cu 0·62	S 0·90, Hg 0·87	

*Only loadings greater than 0·5 are given, except for those shown in parentheses.

Theta showing set grouped iron, sulphur, mercury and copper in the first factor. For the altered rocks, factor analysis isolated mercury and sulphur in the second factor.

Discussion

Supposed temperature contours at the New Calumet deposit were developed by Erdosh[7] by use of the now discredited sphalerite geothermometer. The high spots which represent high concentrations of iron in sphalerite are indicated in Fig. 7 along with mercury/zinc ratios from the present study.

Scott and Barnes[26] have suggested that the FeS content of sphalerite can be used as a geobarometer when in equilibrium with pyrite and hexagonal pyrrhotite. From Erdosh's data on FeS content of sphalerite one can select samples having coexisting sphalerite, pyrite and pyrrhotite and calculate a mean concentration of 14·5 mol % FeS (standard deviation, 1·9 mol %). From the data of Scott and Barnes this would place the pressure in the range 2·8–4·5 kb, assuming temperatures of 300–600 °C, respectively. Scott and Barnes estimated pressure of 4·5 ± 1 kb at Balmat, New York, assuming a temperature of 475–500 °C. This suggests a rather similar history to that suggested for the ore at Balmat, New York;[17, 26] that is, an annealing of the sulphides during regional Grenville metamorphism. The temperature range involved at New Calumet would then narrow the pressure estimate to the range of about 3·5–4·5 kb.

Variations of FeS content of the sphalerite could be due to variations in sulphur fugacity (higher Fe content with lower sulphur fugacity) at the time of formation of the sphalerite. Perhaps there is a tendency for less mercury to enter the sphalerite structure when the fugacity of sulphur is lower. The remaining available mercury could then stay outside the sphalerite structure and might then be in a form more readily available for extraction by heating up to 500 °C. This relationship needs further study. The extractability of mercury may have a bearing on the nature of anomalies to be obtained near a mineral deposit.

The observation that readily extractable mercury constitutes a higher proportion of the total mercury in the surface rocks as compared with the ore zone rocks suggests a difference in mode of occurrence. This could be related to differences in the form in which mercury was deposited or might be related to grain size. For instance, medium-grained sphalerite in the ore zone might release mercury more slowly by diffusion than would fine-grained sphalerite in the surface rocks over appreciable time periods.

Previous workers have established that mercury can be transported as a vapour[1, 32] and in aqueous solution.[5, 32] It is of interest in geochemical prospecting to know which mode of transport is responsible for the emplacement of mercury in a given instance and to know whether mercury gives a different pattern from the other elements.

Table 6 Relationships between anomalous concentration of elements and ore zone

	Hg			Zn			Cu			S		
	F	*A*	*C*	*F*	*A*	*C*	*F*	*A*	*C*	*F*	*A*	*C*
% anomalies near ore	100	86	92	100	75	83	67	50	58	100	75	80
% samples near ore with anomalies	24	29	26	29	43	36	19	14	17	14	48	29

F, fresh rock; *A*, altered rock; *C*, combined.

For the New Calumet surface samples the correlation coefficients and factor analysis tend to support a common introduction for much of the mercury, zinc, sulphur and copper. Multiple-regression calculations showed that 29% of the variability of the mercury could be accounted for by sulphur concentration in fresh samples, and in the fracture samples 34% could be accounted for by zinc concentration. Multiple-regression results should be viewed with some caution when one is considering association of elements because precision of analyses can influence the appearance or disappearance of an element from the regression equation. If, for instance, mercury were associated with sphalerite in a rock where sphalerite was the essential sulphur mineral, the factor determining the appearance of sulphur or zinc in the multiple-regression equation could easily be the relative precision of the sulphur and zinc analyses. One could, however, still say that the element appearing in the equation was the best empirical predictor of mercury concentration with the analytical methods used.

For samples from the Theta showing the correlation coefficients, in particular, support an association of mercury, iron, copper and sulphur.

Where mercury is highly correlated with elements such as copper and zinc it is unlikely that the mercury was vapour-transported. An appreciable number of samples anomalous in mercury, however, were not predictable from regression relationships in both the New Calumet and Theta showing areas, and independent introduction of mercury may have taken place. Rather similar conclusions were reached in a study of two intrusive complexes.[31]

In the New Calumet rock types both the means and medians indicate that more mercury is contained in the samples with limonite-coated fractures, as was also noted at Getchell mine by Erickson *et al.*[8] This may be due to the increased exposure to circulating fluids along fractures or to adsorption of mercury on limonite. Five fresh rock samples were anomalous—as compared with eight altered rocks. At the Theta showing, however, means and medians were higher for fresh rocks, although five altered samples were anomalous and only three fresh rocks were anomalous in mercury.

The first row of data in Table 6 shows that the anomalies of mercury, sulphur and zinc at New Calumet occur almost exclusively over the ore zone, whereas the distribution of copper anomalies is more scattered. The second row of figures suggests that copper is the most likely of the four elements to miss ore.

Conclusions

(1) Mercury, zinc, sulphur and copper anomalies occur in surface rocks overlying the New Calumet orebody. Copper is not as good an indicator of the ore zone as are the other three elements.

(2) A reinterpretation of data of Erdosh[7] suggests that the New Calumet ore deposit underwent regional metamorphism and annealing of the sulphides at a pressure of 3·5–4·5 kb.

(3) Background concentrations of mercury, as given by the median values, are 17 and 25 ppb for fresh and altered New Calumet surface samples, 34 and 30 ppb for fresh and altered Theta showing samples and 210 ppb for ore specimens from the New Calumet area. These concentrations represent the mercury extractable by heating to about 500°C, and are about 80% of the total mercury concentration for the New

Calumet surface samples, about 85% for the Theta showing and from 1 to 70% for the New Calumet ore specimens. Causes for differences in extractability of mercury by heating from ore at New Calumet need further clarification. They may be related to fugacity of sulphur at the time of metamorphism of the ore deposit. Mercury is more readily extractable from surface samples than from ore samples at New Calumet. This might be related to diffusion processes in the minerals or to different forms of combination of mercury.

(4) Multiple regression and factor analysis established significant correlations of mercury with sulphur, zinc and copper in samples from New Calumet. The strongest element associations with mercury in the Theta showing samples were with iron, sulphur, copper and zinc. Much of the mercury probably arrived in association with these elements; however, an appreciable proportion of the mercury anomalies would not have been predicted from concentrations of other elements present and may represent independent introduction of mercury.

(5) Background mercury concentrations are higher in fracture samples from New Calumet than from fresh rocks and the number of anomalous samples is slightly larger. No fresh samples were anomalous away from the ore zone. Background mercury concentrations are higher in fresh samples at the Theta showing, but, again, slightly more of the fracture samples were anomalous.

Acknowledgment

Part of this paper represents material from an M.Sc. thesis by one of the authors.[28] Mrs M. L. Newbury and Mrs S. J. Horska provided assistance with analytical work. Mrs Newbury also made supplementary mercury analyses. The writers thank New Calumet Mines, Ltd., for their cooperation during the surface sampling on their property, L. A. Clark for providing underground samples at New Calumet, and W. H. MacLean for discussion of parts of the study.

This research was sponsored by funds from the National Research Council of Canada.

References

1. Aydin'yan, N. Kh. and Ozerova, N. A. Some genetic features of the formation of mercury mineralization based on data of a study of recent volcanic activity. In *Essays on the geochemistry of endogenous and supergene processes* (Moscow: Nauka Publishing House, 1966), 87–92. (Russian text)

2. Azzaria, L. M. A method of determining traces of mercury in geologic materials. *Pap. geol. Surv. Can.* 66–54, 1967, 13–26.

3. Brown, J. S. Ore deposits of the northeastern United States. In *Ore deposits of the United States, 1933–1967* Ridge, J. D. ed. (New York: A.I.M.E., 1968), vol. 1, 1–19.

4. Cornwall, F. W. D. Rock alteration and primary base metal dispersion at Barvue, Golden Manitou, and New Calumet Mines, Quebec. Ph.D. thesis, McGill University, 1956, 229 p.

5. Dickson, F. W. Solubility of cinnabar in Na_2S solutions at 50–250°C and 1–1,800 bars, with geologic applications. *Econ. Geol.*, **59**, 1964, 625–35.

6. Ehmann, W. D. and Lovering, J. F. The abundance of mercury in meteorites and rocks by neutron activation analysis. *Geochim. cosmochim. Acta*, **31**, 1967, 357–76.

7. Erdosh, G. Sphalerite and pyrrhotite geothermometry of the New Calumet sulphide deposit, Calumet, Quebec. M.Sc. thesis, McGill University, 1962, 67 p.

8. Erickson, R. L. *et al.* Geochemical exploration near the Getchell mine, Humboldt County, Nevada. *Bull. U.S. geol. Surv.* 1198–A, 1964, 26 p.

9. Grdenić, D. and Tunell, G. Mercury. In *Handbook of geochemistry* Wedepohl, K. H. ed. (Berlin, New York: Springer-Verlag, 1970), vol. II/2, 80–A–80–M.

10. Hatch, W. R. and Ott, W. L. Determination of sub-microgram quantities of mercury by atomic absorption spectrophotometry. *Analyt. Chem.*, **40**, 1968, 2085–7.

11. Hawkes, H. E. and Webb, J. S. *Geochemistry in mineral exploration* (New York: Harper and Row, 1962), 415 p.

12. James, C. H. A review of the geochemistry of mercury (excluding analytical aspects) and its application to geochemical prospecting. *Tech. Commun. Imp. Coll. geochem. Prosp. Res. Centre* 41, 1962, 42 p.

13. Jolly, J. L. and Heyl, A. V. Mercury and other trace elements in sphalerite and wallrocks from central Kentucky, Tennessee and Appalachian zinc districts. *Bull. U.S. geol. Surv.* 1252-F, 1968, 29 p.

14. Jonasson, I. R. Mercury in the natural environment: a review of recent work. *Pap. geol. Surv. Can.* 70–57, 1970, 39 p.

15. Jonasson, I. R. Personal communication.

16. Jonasson, I. R. and Boyle, R. W. Geochemistry of mercury and origins of natural contamination of the environment. *Trans. Can. Inst. Min. Metall.*, **75**, 1972, 8–15.

17. Lea, E. R. and Dill, D. B. Jr. Zinc deposits of the Balmat–Edwards district, New York. In *Ore deposits of the United States, 1933–1967* Ridge, J. D. ed. (New York: A.I.M.E., 1968), vol. 1, 20–48.

18. Lepeltier, C. A simplified statistical treatment of geochemical data by graphical representation. *Econ. Geol.*, **64**, 1969, 538–50.

19. Marowsky, G. and Wedepohl, K. H. General trends in the behavior of Cd, Hg, Tl and Bi in some major rock forming processes. *Geochim. cosmochim. Acta*, **35**, 1971, 1255–67.

20. Maufette, P. The geology of Calumet Mines

Limited. M.Sc. thesis, McGill University, 1941.
21. Mercury in the environment. *Prof. Pap. U.S. geol. Surv.* 713, 1970, 67 p.
22. Moorhouse, W. W. Geology of the zinc–lead deposit on Calumet Island, Quebec. *Bull. geol. Soc. Am.,* **52**, 1942, 601–32.
23. Osborne, F. F. Calumet Island area, Pontiac County. *Geol. Rep. Quebec Dep. Mines* 18, 1944, 30 p.
24. Saukov, A. A. Geochemistry of mercury. *Trudy Akad. Nauk. SSSR Inst. Geol. Nauk.*, 78 (Mineralogo-Geokhim. Seriya 17), 1946, 129 p. (Russian text)
25. Sauvé, P. and Bergeron, R. Gerido Lake–Thévenet Lake area, New Quebec, *Geol. Rep. Quebec Dep. nat. Resour.* 104, 1965, 110 p.
26. Scott, S. D. and Barnes, H. L. Sphalerite geothermometry and geobarometry. *Econ. Geol.*, **66**, 1971, 653–69.
27. Sears, W. P. Mercury in base metal and gold ores of the Province of Quebec. In *Geochemical exploration* (Montreal: CIM, 1971), 384–90. *(CIM Spec. vol. 11)*
28. Smith, E. C. The nature of mercury anomalies at the New Calumet Mines area, Quebec. M.Sc. thesis, McGill University, 1971, 129 p.
29. Tauchi, H. J. Remote access statistical system (RAX statistical system), 360D 13.0.003, IBM Corp., 1930 Century Park West, Los Angeles, California 90067, 1968, 146 p.
30. Webber, G. R. and Horska, S. J. Mercury and sulphur in two intrusive complexes, Rougemont and Lake Dufault, Quebec. *Proc. geol. Ass. Can.*, **22**, 1970, 11–8.
31. Webber, G. R. and Newbury, M. L. The nature of mercury anomalies of two intrusive complexes, Rougemont and Lake Dufault, Quebec, by multielement analysis. *Can. J. Earth Sci.*, **8**, 1971, 1197–202.
32. White, D. E. Mercury and base-metal deposits with associated thermal and mineral waters. In *Geochemistry of hydrothermal ore deposits* Barnes, H. L. ed. (New York: Holt, Rinehart and Winston, 1967), 575–631.

550.42:551.25:553.661.21

Differential secondary dispersion in transported soils and post-mineralization rocks: an electrochemical interpretation

G. J. S. Govett, B.SC., PH.D., D.I.C., F.I.M.M.

Department of Geology, University of New Brunswick, Fredericton, New Brunswick, Canada

Synopsis

Multi-element analysis of various residual soils, transported soils and post-mineralization rocks has shown antipathetic dispersion of some ore elements near mineralization. This is of particular interest with respect to measurable dispersion in soils derived from transported overburden and in post-mineralization rocks; it is suggested that the mechanism of such dispersion is strongly influenced by electrochemical processes.

Soils derived from glacial till overlie a sulphide orebody in northern New Brunswick, Canada; the sulphides consist of more than 80% pyrite, the other sulphides being galena, sphalerite, chalcopyrite and arsenopyrite; the average grade of the sulphide body is 2·0–3·0% Zn, 0·6% Pb and 0·2% Cu. The distribution of Pb shows a pronounced peak immediately over mineralization, with an abrupt decrease to background values each side; Zn and Mn are very low over mineralization, with very high and broad peaks each side; and Cu shows a similar, although much subdued, pattern. The ratio of Ni : Co is about 0·25 over mineralization and 2·0 to 4·0 either side of the orebody.

Some Ni–Co–Cu sulphide orebodies in mafic intrusions in southern New Brunswick are covered by a variety of transported overburden, including marine clay, beach sands, alluvium and glacial till; the sulphides are dominantly pyrrhotite, with lesser amounts of chalcopyrite, pentlandite and niccolite. Anomalous concentrations of Ni and Co (and, rarely, Cu) have been detected in these transported materials up to the surface. Even minor mineralization in bedrock is reflected by anomalous concentrations of Ni and Co in a 5-ft section consisting of beach sand, marine clay and glacial till in upward succession.

In Cyprus both pre- and post-mineralization pillow lavas adjacent to sulphide bodies are characterized by abnormally high concentrations of Zn and abnormally low concentrations of Cu. These abnormalities persist, in places, for more than a mile from the orebody.

The complex metal distribution related to these three sulphide deposits is clearly due to secondary processes; simple solution and movement in groundwater is not an entirely adequate explanation.

Preliminary laboratory tank experiments with an artificial mixed-sulphide 'orebody' set in a crushed basalt matrix covered with artificial sea water gave electrical potential differences consistent with normal self-potential anomalies, both laterally away from the 'orebody' and vertically through it. After a period of two months, analysis of the crushed basalt showed that the Zn concentration in the basalt had increased 1·2–1·4 times throughout, except at the top of the basalt, at the farthest distance from the sulphides, where no change had occurred. The maximum increase in Zn was close to the base of the 'orebody'. Lead showed a twofold increase in the basalt close to the base of the sulphide and a slight increase in the basalt close to the top of the sulphide; elsewhere in the tank the Pb concentration was unchanged or had decreased by as much as one-half. Copper concentration in the basalt was unchanged, except at the top, adjacent to the 'orebody', where it had increased 1·7–2·0 times.

These preliminary studies suggest that there can be significant differential secondary rearrangement of metals in soils and rocks related to sulphide orebodies under the influence of electrical potentials. Thus, the possibility arises that geochemical techniques may be used to prospect for deep deposits covered by transported overburden or post-mineralization cap rocks. Inasmuch as absolute changes in concentration of metals will decrease in magnitude with increasing distance from a sulphide orebody, differences in the behaviour of metals which have widely differing electrochemical characteristics should be sought.

The deliberate application of geochemical techniques to the search for blind and deeply buried ore deposits has been inhibited by the intellectual conditioning that geochemical dispersion conforms to the classical concepts of primary and secondary processes in rocks and surficial material, respectively. The traditional assumption is that rocks above a blind deposit cannot exhibit a geochemical response to the deposit unless it is of the leakage-halo type described by Hawkes and Webb.[10] Similarly, there is no great expectation of a direct response in transported overburden over an ore deposit. These beliefs arise largely from the lack of a reasonable mechanism to account for dispersion from deeply buried deposits.

The writer initiated a number of research programmes to investigate displaced anomalies in glaciated terrain in New Brunswick, Canada, but found that, in fact, geochemical response is precisely located above the ore deposit in each case;[13, 14, 17] at the time this was attributed to lack of movement of glacial material. A rock geochemical exploration programme was undertaken in Cyprus[8] on the initially false premise that the sulphide deposits post-dated the overlying rocks; a geochemical response was documented and it gradually became clear that the mineralization pre-dates the overlying rocks. The demonstrated fact that dispersion can occur over considerable distances in post-mineralization rocks prompted the initiation of another rock geochemistry programme, this time around a sulphide deposit in New Brunswick, where the rocks overlying the mineralization are, on the most likely interpretation, post-mineralization in age. Again, a distinctive and measurable dispersion of elements related to mineralization was found in the overlying rocks.*

At least in the case of the dispersion in rocks, solution and transfer in groundwater did not seem to provide a reasonable mechanism for the distances involved; it was speculated that the driving force of dispersion might be electrochemical in origin.[6, 8] Marked similarities in the behaviour of the same elements in soils derived from glacial till as in the rocks suggested that the same processes might be operative in both cases. A series of laboratory tank experiments was therefore designed to measure the effects of electrochemical reactions. The results of the first pilot experiment reported here lend some credence to the hypothesis proposed.

This paper describes metal dispersion patterns around a number of sulphide deposits; these patterns imply that deeply buried deposits may be found with geochemical techniques. An electrochemical interpretation for the mechanism of dispersion is offered; this is admittedly speculative at this stage, and this paper should be regarded as a preliminary report of ideas which may stimulate more intensive research into the problems of deep prospecting. To this end, laboratory experiments and field investigations of electrophysical and chemical properties of rocks and soils are being actively pursued at the University of New Brunswick.

*Details will be described in a separate paper by R. E. Whitehead, a Ph.D. candidate at the University of New Brunswick.

Electrochemical model of dispersion

There is a considerable body of evidence to show that dissolution and oxidation of sulphides is an electrochemical process.[3, 5, 9, 15, 18, 22, 23, 24] In the context of this paper an important characteristic is that when both electropositive and electronegative minerals are present, solutions are enriched in ions of the electronegative mineral, and the difference is greater for minerals with large differences in potential.[22] Generally, differences in electrode potential between different sulphides hasten solution of the more negative and retard solution of minerals with more positive electrode potentials. The actual potential of different sulphides is difficult to determine for, apart from the normal variation due to concentration of its metal ions and the pH and Eh of the surrounding solution, it varies as a function of other metal ions in solution[1] and is also dependent on the degree of sulphur saturation in the sulphide.[20] The greater the sulphur saturation, the more electropositive is the sulphide.

Therefore, it is not possible to draw up an electrochemical series of sulphides of universal applicability. Some comparative lists from a number of workers are given in Table 1; it will be noted that the data of Agócs[1] differ from the rest by showing galena to be more electronegative than sphalerite. Since the theoretical half-cell potential, at unit activity for the cation in solution, can vary from +280 mV to −763 mV for ZnS and from +372 mV to −126 mV for PbS simply as a function of sulphur activity of the sulphide,[20] such reversals in experimentally determined data are to be expected.

The most generally applicable series of electrode potentials of common sulphides is probably pyrite > chalcopyrite > galena > sphalerite. Interpreting the data of Sato[20] and Venkatachalam and Mallikarjunan,[26] it seems probable that NiS, CoS, MnS and stoichiometric FeS have potentials lower than that of sphalerite.

Experiments reported by Sveshnikov and his co-workers[22, 23, 24] indicate that in the dissolution

Table 1 Comparative list of relative electrode characteristics of some common sulphides according to various authors

Increasing electropositive character ↑	Rasskazov[18]	Mikhailov[15]	Sveshnikov and Dobychin[22]	Agócs[1]
	Marcasite			
	Pyrite	Pyrite	Pyrite	Pyrite
	Chalcopyrite	Chalcopyrite	Chalcopyrite	Marcasite
	Pyrrhotite			Pyrrhotite
	Arsenopyrite			
	Pentlandite			
	Galena	Galena	Galena	Sphalerite
	Molybdenite	Molybdenite		
	Sphalerite		Sphalerite	Galena

of sulphide pairs the sulphide with the lower potential shows the greater dissolution; the amount which dissolves increases as the difference in potential between the sulphides increases. Thus, in the pairs galena–sphalerite, galena–chalcopyrite and galena–pyrite in distilled water, the maximum amount of Pb in solution was obtained from the galena–pyrite pair and the least amount (15 times less) from the galena–sphalerite pair. The concentration of Cu in solution of mixed sulphides is characteristically low, except in monomineralic chalcopyrite or chalcopyrite–pyrite mixtures—in the latter case chalcopyrite is electrically negative with respect to pyrite. The extreme importance of pyrite in increasing the dissolution of sulphides more electronegative than itself—which includes most of the common primary sulphides—is obvious from these results. It should also be noted that iron and sulphate do not appear in the solutions. These conclusions are valid in both acid and basic solutions, and in the presence or absence of air;[23] the electrochemical control was proved by obtaining negative results in a system where electrochemical activity was prevented by a buffered oxidation–reduction aqueous medium.

There is another important electrochemical feature of a sulphide body whereby the sulphide acts as a chemically inert conductor of electrons in the characteristic self-potential mechanism. According to Sato and Mooney[21] (and supported by Sveshnikov and Ryss[24] and Becker and Telford[2]), the potential differences which cause the natural electric currents around sulphide bodies ('self-potential') are due to differences in Eh in solutions in contact with the orebody. The substances in solution are relatively oxidized above the water-table and relatively reduced below it. Electrons travel through the orebody from the reducing agents at depth, which are therefore oxidized, to the oxidizing agents near the surface, which are therefore reduced. Thus, the upper part of the orebody acts as a cathode and the lower part of the orebody acts as an anode. Sato and Mooney[21] suggested that the potentials are maintained by Fe^{2+}–Fe^{3+} and H_2O_2–O_2 couples (and possibly also Cu^{2+}–Cu_2O) above the water-table and the Fe^{2+}–$Fe(OH)_2$ couple below the water-table. It should be noted that they also stated that although a water-table across the orebody provides the most favourable conditions, it is not essential; this is supported by the existence of self-potential anomalies at depth in working mines. These reactions necessarily imply that at the surface positive ions must be added or negative ions removed and *vice versa* at depth.

A number of Soviet workers have claimed that heavy metal aureoles occur over a blind or deeply buried deposit due to electrochemical reactions;[22, 24] this has been specifically noted over ore deposits where oxygen is unlikely to penetrate.[18] These aureoles, however, are explicitly stated to occur in waters, and there appears to have been no consideration given to the logical extension of the idea to account for the occurrence of aureoles in rocks or surface overburden (in the translated literature available).

It is to be expected that ions within an electric field will move. In fact, experiments by Noritomi[16] showed that an applied high voltage to dry, solid basaltic andesite at elevated temperatures caused a migration of positive ions to the negative electrode. Similarly, the technique of clay stabilization by direct current shows a considerable movement of material.[25] A series of experiments is being conducted at the University of New

Brunswick to investigate the possible electrochemical mechanism for secondary dispersion from a sulphide deposit. The results from the first small-scale pilot experiment are described below.

Tank experiment results

A glass tank was filled with crushed basalt, and a tube of ground sulphide wrapped in 200-mesh nylon screen was placed in the centre. Platinum electrodes were placed in the basalt and the sulphide as shown in Fig. 1, and the tank was filled with artificial, trace-element-free sea water to about 1 in above the top of the basalt;* the tank was covered with a lid, but was not sealed from the air. Potentials were measured between the platinum electrodes and two calomel reference electrodes with a Leeds and Northrup potentiometer. Readings were taken daily for 24 days and then weekly. The potentials of all electrodes on the first day were about −200 mV; they gradually separated in an erratic fashion during the first two weeks until the pattern illustrated in Fig. 1 was established.

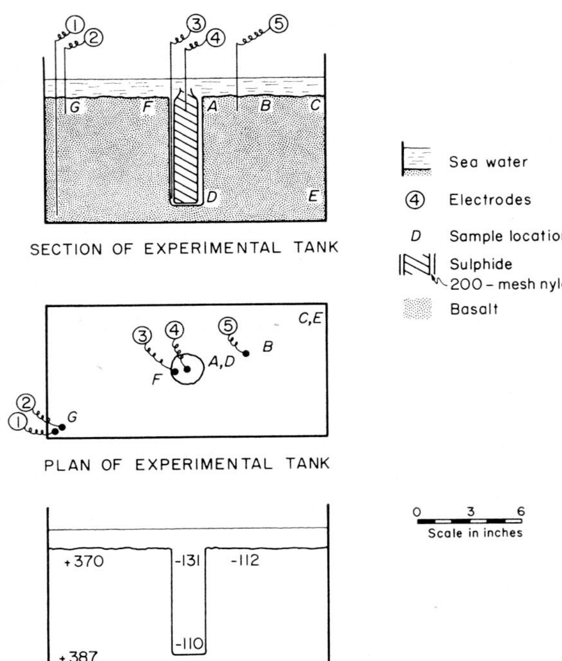

Fig. 1 Diagrammatic representation of sulphide–basalt tank experiment and potentials developed

The most important characteristic of the experiment was that the top of the sulphide was more negative than the base. This is consistent with the self-potential model of Sato and Mooney.[21] Interestingly, this effect persisted in the basalt at its greatest distance from the sulphides, although both top and bottom had positive potentials.

The experiment was terminated after two months, when, over a period of several days, the differential in the potential between the top and the bottom of the sulphide disappeared and the potentials in the basalt decreased rapidly towards that of the sulphide. It is presumed that the disappearance of the potentials is a reflection of chemical equilibrium due to the static nature of the experiment. Duplicate samples of the basalt were taken at the locations shown in Fig. 1, washed thoroughly in deionized water, ground by hand to pass a 100-mesh screen, and analysed for Cu, Pb and Zn.* The results obtained are presented in Table 2.

Obviously, there was considerable solution and movement of material. The maximum combined changes occurred at the bottom closest to the sulphides, with the minimum changes at the bottom farthest from the sulphides. The absolute differences in concentration compared with 'background' (i.e. basalt before the experiment) were small, but a distinct pattern exists of differences which are beyond the limits of analytical and sampling error (Table 2). Within the limitations of the data, the following general trends may be noted: *(a)* an apparent movement of Zn from the sulphide, having its greatest magnitude at the bottom next to the sulphide; *(b)* an apparent movement of Pb from the basalt towards the sulphide at the top, and a movement from the sulphide to the basalt at the bottom closest to the sulphide; and *(c)* movement of Cu from the sulphide to the basalt at the top, which is restricted to the zone adjacent to the sulphide.

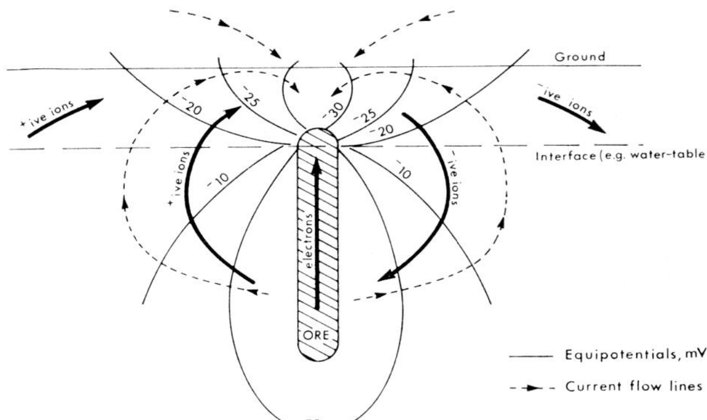

Fig. 2 Theoretical electrochemical model for development of electrical potential differences around a sulphide body. After Sato and Mooney[21] and Becker and Telford[2]

*Sea water was used because of a specific interest in determining in the same experiment the possible effects of submarine leaching of sulphides.

*All analytical results reported here are for hot, concentrated nitric acid digestion, with estimation by atomic absorption spectrophotometer.

Table 2 Conductivity and content of Cu, Pb and Zn in basalt samples from experimental tank

Sample location	Number of samples	Conductivity	ppm Cu	Pb	Zn
A	2	450*	43*	35*	159*
B	2	295*	20	20†	147*
C	2	183	21	22†	140*
D	2	800*	30	58*	172*
E	2	530*	21	26	120
F	1	270*	37*	25	137*
G	1	350*	27	15†	120
Basalt before	20 mean:	156	21	29	121
experiment	mean $+2s$:	206	33	34	128
	mean $-2s$:	106	9	24	114

*Positive anomaly.

†Negative anomaly.

The resultant dispersion pattern is consistent with the electrochemical model illustrated in

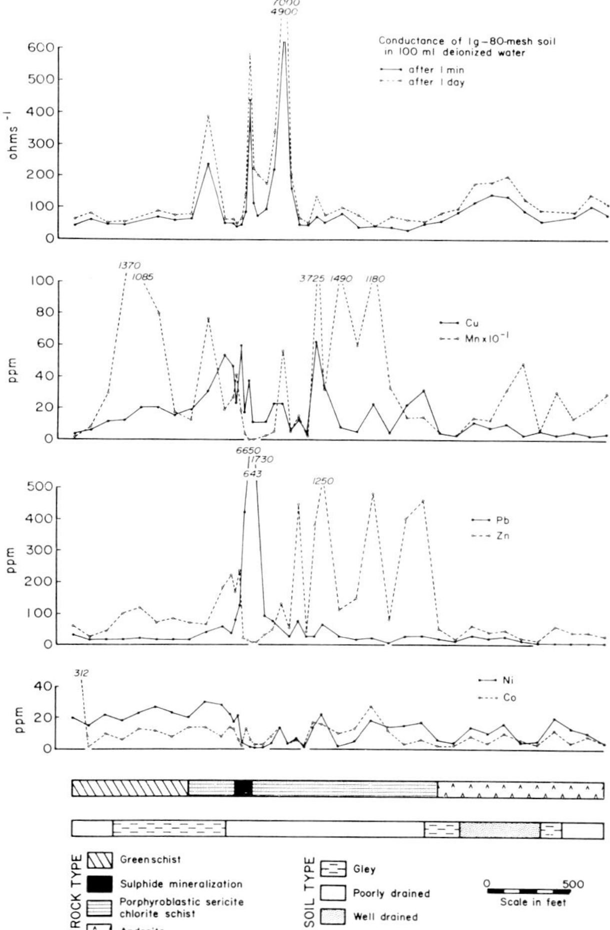

Fig. 3 Distribution of Ni, Co, Pb, Zn, Cu and Mn in soils, and conductance of soil slurries at Armstrong A deposit, New Brunswick. Metal distribution data from Pilch[17]

Fig. 2, although the results of this single experiment cannot be regarded as conclusive proof that the mechanism is electrochemical. By virtue of the high concentration and wide dispersion of Zn throughout the tank, ZnS has suffered the most dissolution. It is presumed that this is an anodic process towards the base of the sulphide mass, and that Zn ions move upwards and outwards; there may be a movement towards the sulphide mass at the top of the basalt. Lead is similarly released at the base of the sulphide, but is not, apparently, dispersed (possibly it is immobilized as the sulphate). An interesting feature is the apparent removal of Pb from the basalt at the top. The very small amount of Cu released is consistent with the electrochemical model of dissolution of polymetallic sulphides described above.

An important secondary aspect of the passage of electric current through rocks (and soils) is that it may accelerate solution,[11] thus promoting migration of ions in solution by all processes as well as under the influence of electric potentials. For these reasons it may reasonably be expected that the intrinsic physical properties of such soils and rocks will differ from those not affected by electrical processes. For example, Sakovtsev[19] has reported that rocks overlying ore zones are better conductors than those some distances away. To test this, 1·0 g of ground rock sample from the experimental tank was placed in 100 ml of deionized water, stirred for 1 min, and the conductivity of the suspension was measured. The results, recorded in Table 2, show significant deviations from background conditions; the greatest increases in conductivity occurred nearest to the sulphide, and the conductivity of the bottom is generally greater than at the top.

Description of possible electrochemical dispersion in soils and rocks

Armstrong *A* sulphide deposit

The following brief description of the Armstrong *A* sulphide deposit of Anaconda American Brass Company of Canada is taken from Pilch.[17] The deposit is located 17 miles west of Bathurst in northeastern New Brunswick. The rocks in the vicinity are pre-Silurian in age, metamorphosed, and have a northerly strike and a vertical to steep easterly dip. The host rock is porphyroblastic quartz–feldspar chlorite schist, with greenschists to the west and andesitic volcanic rocks to the east of the deposit.

The northern lens is about 450 ft long, and has a maximum width of 50 ft; the southern lens is about the same length, but has a maximum width of about 150 ft. Both lenses persist to a depth of more than 800 ft from the surface, but decrease in width downwards. The two lenses are separated by a narrow zone of disseminated pyrite.

Fine-grained pyrite accounts for about 85% of the sulphides, the remainder being sphalerite, galena, chalcopyrite and arsenopyrite. The average grade per cent of the deposit is:

	North lens	South lens
Cu	0·2	0·3
Pb	0·6	0·6
Zn	2·2	3·2

The soils of the region are podzols, which, in the vicinity of the deposit, are poorly drained; they are developed on glacial till, which appears to be locally derived. The absence of clay minerals (other than chlorite and sericite) in the clay size fraction suggests that the soils are in a youthful stage of development and that there has been little formation of secondary minerals due to surficial processes. The direction of ice movement was west to east; this is coincident with the direction of the local drainage.

The general character of the dispersion of Cu, Pb, Zn, Ni, Co and Mn in the *B* horizon of soils in the vicinity of the deposit is shown by a west–east traverse across the southern lens illustrated in Fig. 3. Only Pb shows a simple positive anomaly over the deposit; Cu, Zn and Mn have pronounced negative anomalies immediately over the deposit, with broad lateral positive anomalies. The absolute variation in Co and Ni is small, but a small trough in the values occurs over the mineralization. The concentration of Ni is normally greater than the concentration of Co, but there is a reversal in their relative concentrations over, and laterally to the east of, the deposit where the content of Co exceeds that of Ni.

This dispersion pattern in soils should be compared with data in Table 3: the concentration of all these elements is greater in the sulphides than in the host rocks by a factor of four to 400.

The orebody is probably near enough to the surface to admit the possibility of simple solution and dispersion by groundwater. Such an explanation would imply a control of the elements by sulphate solubility, precipitating Pb immediately over the orebody under acid conditions while allowing the other elements to disperse laterally until precipitated by change in pH to more basic conditions. This does not, however, explain the lateral anomalies for all elements (except Pb) on the up-drainage as well as the down-drainage side. Moreover, although the concentration of Cu in the sulphides is about 100 times greater than the concentration of Ni, their concentration in the soils is similar.

The probable ease of dissolution for the elements in the sulphide body based on sulphide potentials is (Co, Ni, Mn) $>$ Zn $>$ Pb $\gg$ Cu—thus accounting for the exceptionally low Cu concentrations in the surface environment. Assuming the model of Sato and Mooney[21] for potentials developed in a sulphide body, there is a strong probability that ions released by dissolution from

Table 3 Summary analysis of drill core through Armstrong A deposit

	Footwall schists	Sulphides	Hanging-wall schists
Number of samples	6	15	12
Footage represented	125	175	221
Cu, ppm	17	3330	8
Pb, ppm	44	5250	39
Zn, ppm	62	27 200	980*
Ni, ppm	8	36	10
Co, ppm	12	62	11
Mn, ppm	332	3220	520

*Range 43–3250 ppm; general trend of increasing concentration towards sulphide body.

the sulphide body would migrate according to the model shown in Fig. 2; this would naturally lead to lateral concentration peaks on each side of the sulphide. The failure of Pb to appear as lateral concentrations must be attributed to its being immobilized close to the sulphides as a sulphate (see tank experiment above). Normal surface oxidative reactions and groundwater dispersion must also occur at this locality, and it is reasonable to suppose that metals other than Pb (which is initially immobilized by its sulphate insolubility) will be leached away from above the ore zone to be precipitated downslope. This interpretation would explain the Pb peak over the ore and the tendency for the other metals, especially Zn, to have greater lateral dispersion downslope. It is possible that the pronounced asymmetry of the Zn lateral peaks is partly influenced by the anomalous concentrations of this metal in the hanging-wall rocks which underlie the downslope soils (see Table 3).

Conductivity measurements were made on the soils. Results for 1-g –80-mesh soil slurries in 100 ml of deionized water after 1 min, and after one day, are shown in Fig. 3. There is a pronounced concentration of high conductivities over the ore zone. The only significant effect of soil type is a broad zone of moderately high conductivities over the well-drained soils.

St. Stephen Ni–Cu–Co deposits

Ni–Cu–Co sulphide mineralization is associated with ultrabasic intrusions around the town of St. Stephen in southern New Brunswick. The following account is taken from Mersereau[14] and Lahti.[13]

The general geology conforms to part of the Appalachian geosyncline with Palaeozoic sedimentary rocks and acid and basic intrusions. Vein-type and massive mineralization occurs in various types of gabbro and norite (possibly associated with metasedimentary lenses), and disseminated mineralization occurs in peridotites. In both types pyrrhotite accounts for about 90% of the sulphides; pentlandite, chalcopyrite and niccolite are the most important of the other sulphides. The general grade of the deposits is indicated by the average of four sulphide samples: 0·9% Ni, 0·2% Cu and 0·1% Co.

The soils are podzols, which are developed on glacial till. The rock fragments are generally angular, unsorted and unstratified, and correlate with local geology. The recent history is obviously complex; in places marine clay and beach deposits occur, sometimes interstratified with the till. The thickness of the overburden is normally 3–12 ft, although it is more than 50 ft thick in places.

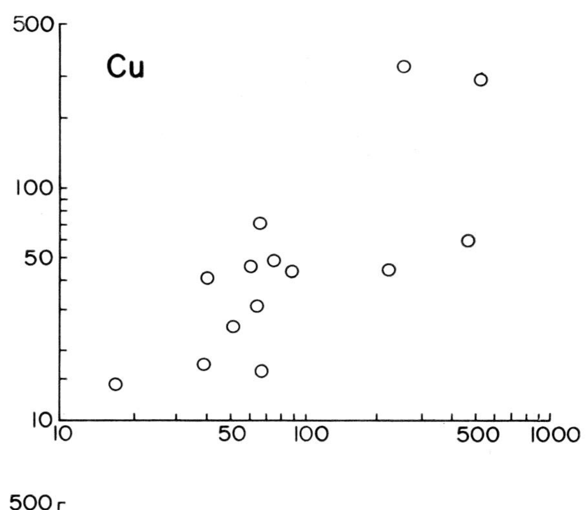

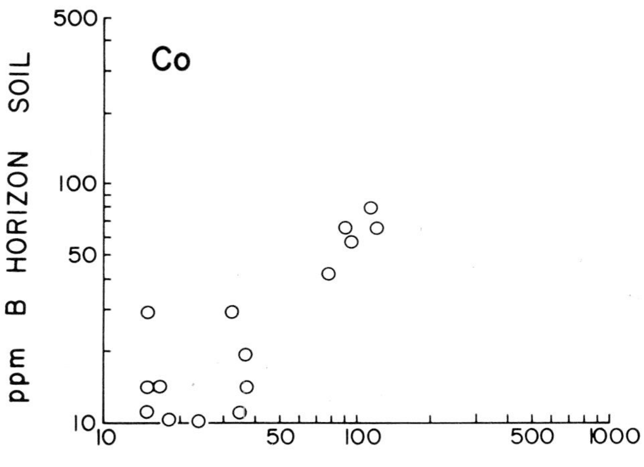

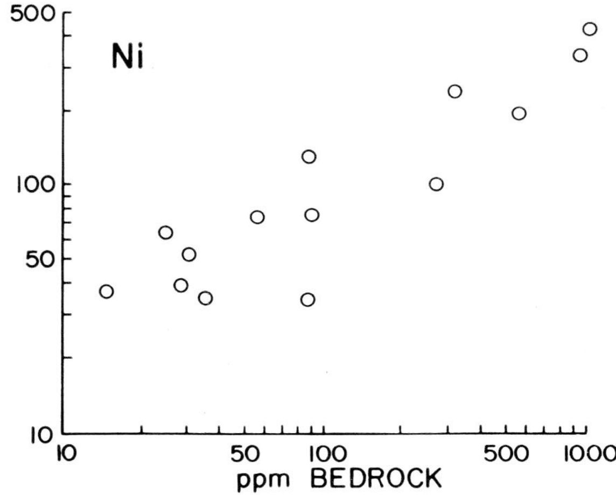

Fig. 4 Correlation between metals in B-horizon soils and underlying bedrock, St. Stephen, New Brunswick. Data from Lahti[13]

Notwithstanding the transported and diverse nature of the overburden, there is a definite positive (though erratic) correlation between the concentration of Ni, Co and Cu in soils and the underlying bedrock (Fig. 4). Furthermore, Cu, Ni, Co and Mn concentrations over mineralized zones are 3–10 times higher than background, and the vertical distribution in the profile does not appear to be significantly affected by soil processes.[13]

The metal distribution in a particularly complex profile is illustrated in Fig. 5. Here, anorthositic gabbro has weak disseminated mineralization, and

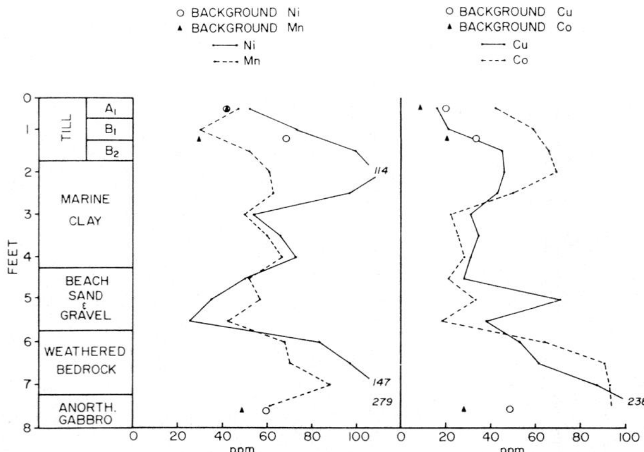

Fig. 5 Distribution of Ni, Mn, Cu and Co in a soil profile at St. Stephen, New Brunswick. Data from Lahti[13]

therefore contains anomalously high concentrations of Ni, Cu and Co. The gabbro is overlain by beach deposits, marine clay and till in upward succession. The concentration of metals in the *B* horizon of the podzol soils is generally greater than in the underlying materials; compared with background for soils overlying anorthositic gabbros, Co is distinctly anomalously high and Ni is slightly anomalously high, whereas Cu and Mn are present in background concentrations. The important point illustrated here is that, although Cu is present in bedrock in similar concentrations to Ni, and some two and a half times the concentration of Co, in the *B* horizon of the soil its concentration is only one-third that of Co and Ni. This clearly confirms the conclusion that chalco-

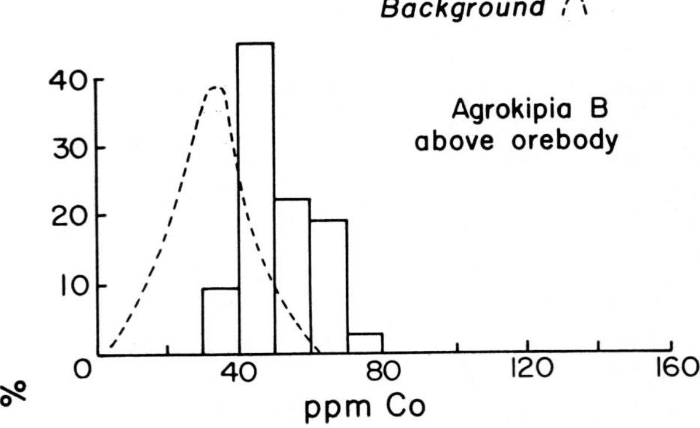

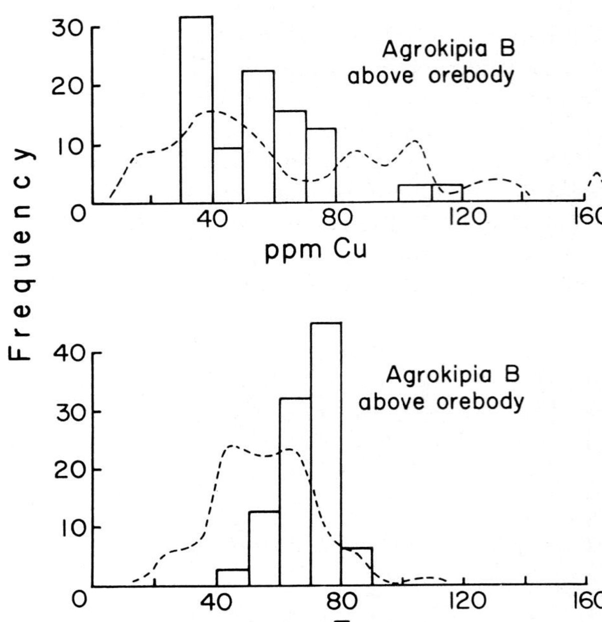

Fig. 6 Frequency distribution of Zn, Cu and Co in surface rocks 450 ft above Agrokipia B orebody, Cyprus

pyrite will remain relatively insoluble in the presence of the much more electronegative—and soluble—sulphides of Ni and Co. Movement of metals through the transported overburden is

Table 4 Distribution of Cu, Zn and Co in drill-hole A–103 through Agrokipia B orebody

Rock type	Depth, ft	Number of samples	Mean, ppm Cu	Zn	Co
Lavas	39– 99	2	105	78	36
	131–230	2	88	70	34
	262–328	2	44	54	22
	335–360	1	81	89	33
	365–386	1	81	1400	33
	386–424	1	69	2200	31
	426–455	1	400	1900	22
Massive ore	460–488	4	38 875	34 025	29
Mean background		111	89	64	30

probably aided by development of differential potentials related to the sulphides, although at this particular location the relatively shallow depth of the overburden allows the possibility of simple groundwater dispersion. There is also the possibility of generation of small local electrical fields by the marked physical differences in the various horizons of the material. The higher concentration of metals at the marine clay–till interface suggests that the latter may be important.

Cyprus pyritic sulphide deposits

Pyritic sulphide deposits in Cyprus were emplaced in depressions in basic pillow lavas during a pause in volcanicity, and are overlain by rather more basic pillow lavas, which are post-mineralization in age.[4] In rocks adjacent to and overlying the sulphides the concentration of Cu tends to decrease abruptly (over several inches) from ore grade to below background levels, whereas the concentration of Zn is anomalously high for distances of more than one mile from some deposits; the concentration of Co is also anomalously high in rocks near sulphide deposits which are relatively rich in Co. The details of metals distribution have been described elsewhere,[8] and only one illustrative example is given here.

In Table 4 the distribution of Cu, Zn and Co in 460 ft of post-mineralization lavas lying above the Agrokipia *B* orebody is shown; the frequency distribution of these metals in surface lavas overlying the deposit is illustrated in Fig. 6. The characteristically abnormally high concentrations of Zn and Co and the relatively low concentrations of Cu in rocks overlying mineralization are shown by the frequency distributions; in the particular drill-hole illustrated, only the abnormal Zn distribution is evident in the overlying lavas. Very clearly, there has been a significant positive dispersion of Zn which, because of the demonstrable geological relationships, has to be due to post-mineralization secondary processes.

The sulphides are dominantly pyritic, with varying amounts of chalcopyrite and sphalerite. According to the electrochemical dissolution sequence, solutions around the sulphide bodies should be preferentially enriched in Zn and the concentration of Cu should be extremely low. The major problem here, as in similar situations in New Brunswick, is to account for the movement of metal over large distances. Diffusion in response to simple concentration gradients seems to be totally inadequate as an explanation. It is suggested that a model based on electrochemical gradients provides a reasonable solution (similar to the model shown in Fig. 2). The apparent lack of movement of Cu is readily explained by its relative insolubility under the conditions of the sulphide mineralogy; the apparent deficiency in Cu in rocks adjacent to the sulphides may be due to its being leached and moving towards the orebody in response to its strong electropositive nature and its participation in the potential-producing reactions (the Cu deficiency in rocks adjacent to the sulphide deposit in Cyprus should be compared with the Pb deficiency noted in an analogous situation in the tank experiment described above).

Discussion

The approximate correlation in metal content between soils derived from diverse transported sources and underlying bedrock at St. Stephen, the presence of pronounced anomalies in glacial till associated with underlying mineralization at Armstrong *A* and the characteristic metal association in post-mineralization rocks overlying sulphide deposits in Cyprus all imply a significant movement of material. In general, dispersion will be controlled by dissolution of the source, solubility of the dissolved species (which is related to mechanisms of fixation) and a mechanism of movement. Interpretation of controlling processes is obviously extremely difficult: a low concentration of dispersed Cu can equally well be interpreted as a very low dissolution of the source or as extreme solubility and mobility of dissolved Cu.

In the three case histories cited above, Cu occurs as chalcopyrite in orebodies with sulphides of Pb and Zn, and its dispersed concentration in soils and rocks is very low; this differs from a disseminated copper porphyry type deposit in the Philippines, where chalcopyrite is the only important sulphide apart from pyrite, and the dispersed concentration of Cu in soils is very high.[7] It is suggested that this is due to the fact that in the Philippine example chalcopyrite is the most electronegative sulphide relative to pyrite and is readily leached, whereas in the former cases chalcopyrite is electropositive relative to other sulphides and therefore remains relatively inert. Obviously, the fact that the soils overlying the Philippine deposit are residual, the climate is tropical and the nature of the mineralization is different must cause some differences in the nature of dispersion. The writer has, however, noted a depression of Cu dispersion in residual soils over polymetallic deposits in Greece similar to that found in glacial soils in New Brunswick.

It is important to distinguish the macro-galvanic effect by which self-potential currents are generated from the micro-galvanic effect in sulphide bodies, which significantly influences dissolution of the minerals, and also rearrangement and concentration of metals and formation of secondary minerals within an orebody.[4] Thus, self-potential current flows *because* of reactions at the sulphide–country rock interface and not *vice versa*. The existence of imposed potential differences in the vicinity of sulphide bodies must, however, have some influence on the migration of ions; the sense of the movement will presumably depend on the disposition of lines of equi-potential, and these depend on the dimensions and configuration of the sulphide body. Moreover, it should be realized that electric currents may also be generated by structural disturbances,[12] temperature differentials[27] and other causes; these potentials may be small in magnitude, but they may have important long-term effects on metal movement.

The degree of influence of electrochemical reactions on ion migration in individual instances is difficult to determine: for example, in any particular case it is uncertain whether a potential difference exists because there is a concentration gradient of a particular metal or *vice versa*. This is, however, largely irrelevant for the complete system, since ions of elements having different redox characteristics from the elements causing the potential will behave in a different fashion. It is to be expected, therefore, that the character of dispersion of different elements from a sulphide body will vary depending on both the mineralogy of the sulphides and the electrochemical nature of the ion relative to the electrical field around the sulphide body. It is this differential dispersion which is important in geochemical exploration; it has been effectively used through the application of metal ratios to design an exploration procedure in Cyprus.[6] This procedure has the advantage of partly overcoming one of the difficulties in detecting the very small differences likely to be encountered in metal concentrations due to deeply buried deposits, i.e. lack of analytical precision and accuracy. Another fruitful approach to the problem of small differences in metal concentration is likely to be through the application of differential analytical extraction techniques.

The pronounced anomalous conductivities of rock and soil slurries for samples near to sulphide bodies are clearly a potentially powerful aid to exploration. Whether these high conductivities are related to actual movement and relocation of material, or whether they are due to physical changes in fixation of metals, is not clear from the data now available. Moreover, the changes in conductivity are not necessarily related to the trace elements—they are more likely due to variations in the major elements. Similarly, if secondary dispersion of the type described is electrochemical in nature, other electrical—or combined electrical and chemical measurements—of soils and rocks may be expected to reveal evidence of the processes. Furthermore, they should provide the basis of additional exploration techniques.

This discussion of electrochemical processes is not intended to imply that surface processes—such as soil-forming mechanisms and oxidative secondary enrichment of orebodies—are unimportant; they may enhance, modify or completely obscure the effects of electrochemical reaction and dispersion. It is suggested, however, that electrochemical processes may control the amount and kind of ions released from a sulphide and play a major role in the dispersion of elements into surrounding rocks and soils.

Conclusions

There is empirical evidence that metals may be dispersed from polymetallic sulphides through transported overburden and through post-mineralization cap rocks. A characteristic of such dispersion patterns is the marked differential movement of the elements involved—this can be exploited to advantage to detect the subtle anomalies due to deeply buried deposits.

It is suggested that the dispersion of elements is influenced by natural electrical currents, generated primarily by Eh differences in solutions surrounding sulphide bodies. The differential dispersion is related to varying dissolution of sulphides due to their differing electrode potentials, and possibly also to the varying electrochemical behaviour of elements in solution.

The case for electrochemical dispersion is not regarded as proved; however, it is believed that there is enough evidence to warrant extensive research. If it is a significant factor, there are a number of new approaches to exploration through various electro-physical and chemical measurements on soils and rocks. Regardless of the interpretation of the mechanism of dispersion, it is believed that there is enough evidence that geochemical methods can detect deeply buried deposits in some situations to justify a major research effort to develop techniques and determine mechanisms.

Acknowledgment

The writer gratefully acknowledges the financial support from the Geological Survey of Canada Grant Number 17–68 for the work at St. Stephen, and from the National Research Council of Canada Grant Number A5585 for the work at

Armstrong *A* and for analytical and experimental programmes.

The work of T. G. Mersereau and H. R. Lahti at St. Stephen and of P. G. H. Pilch at Armstrong *A* on the research projects which formed part of their M.Sc. studies at the University of New Brunswick is acknowledged.

The tank experiment was conducted by Miss Kirsty Clark under the supervision of Mr. R. E. Whitehead (who was supported by an International Nickel Company of Canada Fellowship). Mr. F. Klidaras has been responsible for the analyses performed specifically for this paper.

Grateful acknowledgment is given to Dr. G. Constantinou of the Cyprus Geological Survey for the data presented in Table 4 and especially for originally drawing the writer's attention to the importance of electrochemical reactions in secondary enrichment processes, and, hence, having stimulated the development of the hypotheses presented here.

Professor K. B. S. Burke of the University of New Brunswick is thanked for reading the manuscript and making pertinent suggestions. Mr. R. E. Whitehead is also thanked for his assistance and for critical discussion of the manuscript.

References

1. Agócs, M. Investigations on the electrode potential of sulfide ores. *Acta Mineral.-Petrog.*, **18**, no. 2 1968, 61–72.

2. Becker, A. and Telford, W. M. Spontaneous polarization studies. *Geophys. Prospect.*, **13**, 1965, 173–88.

3. Butler, B. S. and Burbank, W. S. Relation of electrode potentials of some elements to formation of hypogene mineral deposits. *Trans. Am. Inst. Min. Engrs*, **85**, 1929, 341–59.

4. Constantinou, G. and Govett, G. J. S. Genesis of sulphide deposits, ochre and umber of Cyprus. *Trans. Instn Min. Metall. (Sect. B: Appl. earth sci.)*, **81**, 1972, B34–46.

5. Douglas, G. V. Goodman, N. R. and Milligan, G. C. On the nature of replacement. *Econ. Geol.*, **41**, 1946, 546–53.

6. Govett, G. J. S. Interpretation of a rock geochemical exploration survey in Cyprus: statistical and graphical techniques. *J. Explor. Geochem.*, **1**, 1972, 77–102.

7. Govett, G. J. S. and Hale, W. E. Geochemical orientation and exploration near a disseminated copper deposit, Luzon, Philippines. *Trans. Instn Min. Metall. (Sect. B: Appl. earth sci.)*, **76**, 1967, B190–201.

8. Govett, G. J. S. and Pantazis, Th.M. Distribution of Cu, Zn, Ni and Co in the Troodos Pillow Lava Series, Cyprus. *Trans. Instn Min. Metall. (Sect. B: Appl. earth sci.)*, **80**, 1971, B27–46.

9. Habashi, F. The mechanism of oxidation of sulfide ores in nature. *Econ. Geol.*, **61**, 1966, 587–91.

10. Hawkes, H. E. and Webb, J. S. *Geochemistry in mineral exploration* (New York: Harper and Row, 1962), 415 p.

11. Karasev, A. P. *et al.* Influence of electrophysical properties of sulfides and electrochemical phenomena on hydrothermal ore formation. *Int. Geol. Rev.*, **12**, 1970, 952–8.

12. Kurdyukov, A. A. Galvanic effect of minerals and rocks. *Dokl. Akad. Nauk SSSR*, **176**, 1967, 668–71; *Dokl. Acad. Sci. USSR, Earth Sci. Sect.*, **176**, 1967, 124–6.

13. Lahti, H. R. Factors contributing to secondary dispersion of trace elements in glacial soils, St. Stephen area, New Brunswick. M.Sc. thesis, University of New Brunswick, 1971.

14. Mersereau, T. G. Secondary dispersion of the metals nickel, cobalt and copper near the St. Stephen gabbro, Charlotte County, New Brunswick. M.Sc. thesis, University of New Brunswick, 1969.

15. Mikhailov, A. S. Experiments in electrochemical oxidation and solution of molybdenite. *Geokhimiya*, 1962, 818–25; *Geochemistry*, 1962, 940–8.

16. Noritomi, K. Migration of charged carrier in the case of electric conduction of rocks. *Sci. Rep. Tohoku Univ. Geophysics*, **9**, 1957–8, 120–7.

17. Pilch, P. G. H. Dispersion of some trace elements in soils of glacial origin near the Armstrong 'A' sulphide deposit. M.Sc. thesis, University of New Brunswick, 1970.

18. Rasskazov, N. M. Experimental study of electrochemical solution of arsenide–sulfide mixtures. *Int. Geol. Rev.*, **7**, 1965, 202–4.

19. Sakovtsev, G. P. The search for deep copper pyrites deposits in Ural. *Int. Geol. Rev.*, **12**, 1970, 787–91.

20. Sato, M. Half-cell potentials of semiconductive simple binary sulphides in aqueous solution. *Electrochim. Acta*, **11**, 1966, 361–73.

21. Sato, M. and Mooney, H. M. The electrochemical mechanism of sulfide self-potentials. *Geophysics*, **25**, 1960, 226–49.

22. Sveshnikov, G. B. and Dobychin, S. L. Electrochemical solution of sulphides and dispersion aureoles of heavy metals. *Geochemistry*, 1956, 413–9.

23. Sveshnikov, G. B. and Kedrinskiy, I. A. Electrochemical solution of sulphide ores. *Int. Geol. Rev.*, **7**, 1965, 225–32.

24. Sveshnikov, G. B. and Ryss, Yu. S. Electrochemical processes in sulphide deposits and their geochemical significance. *Geokhimiya*, 1964, 208–18; *Geochemistry*, 1964, 198–204.

25. Transformation of the properties and composition of clays through the action of direct electric current. In *Electrochemical induration of weak rocks* Titov, N. I. *et al.* (New York: Consultants Bureau, 1961), 1–6.

26. Venkatachalam, S. and Mallikarjunan, R. Anodic potentials and anodic polarization of sulphides. *Trans. Instn Min. Metall. (Sect. C: Mineral Process. Extr. Metall.)*, **79**, 1970, C181–8.

27. Yamashita, S. The electromotive force generated within the orebody by the temperature difference. *J. Min. Coll., Akita Univ., Ser. A*, **1**, no. 1 1961, 69–78.

550.422:546.59:553.43(729.5)

Gold—a useful pathfinder element in the search for porphyry copper deposits in Puerto Rico

Robert E. Learned, PH.D.

U.S. Geological Survey, Denver, Colorado, U.S.A.

Rafael Boissen, B.S.

Center of Environmental Investigations, Department of Public Works, Santurce, Puerto Rico

Synopsis

Geochemical studies of residual soils in selected areas of the copper belt of west-central Puerto Rico indicate that porphyry copper deposits that have been enriched by supergene processes may easily be missed in soil surveys that utilize copper or other acid-leachable elements as indicators. Gold in soils, however, serves as a reliable indicator of both primary and supergene-enriched deposits. Gold and copper data evaluated together may serve to distinguish soils overlying primary deposits from soils overlying supergene-enriched deposits.

Within the copper belt of Puerto Rico the residual soils overlying five major deposits contain strongly anomalous amounts of gold. The gold content of these soils is uniform from deposit to deposit and is about equivalent to the gold content of the primary deposits themselves.

The silver and base-metal contents of the soils show much less consistent relations to the deposits than does gold. In the Rio Vivi district, the area studied most thoroughly, the copper concentrations in the soils of two adjacent deposits differ greatly. The median copper concentration in soils over the Piedra Hueca deposit is 2300 ppm; at the Cala Abajo deposit it is only 300 ppm.

The gold concentration in soils over the two deposits is similar. Median concentrations in soils are 0·19 ppm over the Piedra Hueca deposit and 0·15 ppm over the Cala Abajo deposit.

The Cala Abajo deposit is a supergene-enriched deposit overlain by a thick, intensely leached capping, whereas the Piedra Hueca deposit is simply a hypogene deposit. The much lower copper content, as well as the lower content of zinc, silver and manganese, in soils over the Cala Abajo deposit as compared to the amounts of these elements in soils over the Piedra Hueca deposit is attributed to acid leaching.

Recognition of the foregoing relations is based on analyses of the −0·25-mm fraction of samples from the *B* soil horizon; but, in fact, the parameters of soil horizon and grain size are not critical.

Although it is well known that gold is a common trace constituent of most porphyry copper deposits, its use as a pathfinder element has not been emphasized. Previous investigators have detected gold in soils over porphyry copper deposits (see, for example, Archer and Main[1]), but their emphasis has been directed toward the use of other elements, particularly copper and molybdenum, as indicators.

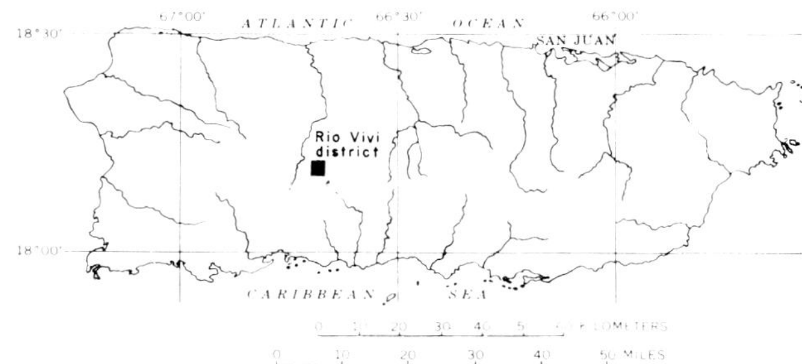

Fig. 1 Index map of Rio Vivi copper district, Puerto Rico

The usefulness of gold as a pathfinder for porphyry copper deposits that are covered by intensely leached cappings is emphasized in this paper by contrasting the geochemical character of the soils over two adjacent deposits in the Rio Vivi district of west-central Puerto Rico (Fig. 1). The two deposits of interest are the Cala Abajo deposit, a deposit exhibiting extensive effects of leaching and supergene enrichment, and the Piedra Hueca deposit, a deposit exhibiting negligible effects of leaching and supergene enrichment.

The work presented here represents part of a continuing geochemical research project undertaken jointly by the U.S. Geological Survey and the Puerto Rico Department of Public Works. All chemical and spectrographic analyses reported in this paper were done at the Geological Laboratory of Natural Resources of the Puerto Rico Department of Public Works, San Juan, Puerto Rico.

Historical background

The Rio Vivi copper district was discovered in 1957 by the A. D. Fraser Company during reconnaissance stream-sediment sampling throughout Puerto Rico.[2] Ponce Mining Company, a subsidiary of American Metal Climax, Inc., acquired the concession in 1960 and soon initiated a drilling programme based primarily on the results of geochemical soil surveys for copper. The Piedra Hueca deposit and the northwestern part of the Cala Abajo deposit were drilled in the early stages of the programme, but the major supergene-enriched part of the Cala Abajo deposit remained undrilled for several years because of poor geochemical indications there.

The decision to drill at a central Cala Abajo site in 1964 was based on petrologic identification and geologic mapping of the highly weathered porphyry (R. A. Bradley, personal communication, 1972).

Both the Piedra Hueca and the Cala Abajo deposits have now been extensively drilled, but exploitation of the deposits awaits the favourable outcome of current negotiations between American Metal Climax, Inc., and the government of Puerto Rico.[6]

Geographical setting

Puerto Rico is the easternmost island of the Greater Antilles of the West Indies. It lies in the trade-wind belt approximately 18° north of the equator and 66° west of the Greenwich meridian (Fig. 1). The general climate of the island is tropical marine, but the extremes in annual rainfall range from 600 cm in the northeast to 90 cm in the southwest. The Rio Vivi area, in the western third of the island on the northern slope of the Cordillera Central, receives about 200 cm of rainfall annually, mostly between July and November.

The Rio Vivi area is situated in strongly dissected uplands between two mountain streams, the Rio Vivi and the Rio Pellejas. The slopes in this area of rugged topography are predominantly 20°–30°, but slopes in excess of 30° are common. Elevations range from 400 to 750 m above sea level.

Geologic setting

The geology of the Rio Vivi area has been described by Mattson.[7] The area studied is composed principally of Cretaceous and lower Tertiary lavas and tuffs, but includes significant amounts of Upper Cretaceous and Eocene intrusive rocks. The Utuado batholith, a granodiorite body of Late Cretaceous age, is the principal intrusive unit of the area. Its maximum exposed dimensions are about 35 km by 10 km, its long axis trending westerly. All the known porphyry copper deposits occur along the southern margin of this batholith in bodies of quartz-diorite porphyry of Eocene age that have intruded the batholith or the volcanic rocks adjacent to the batholith.

A major fault zone parallels the southern margin of the batholith in the volcanic rocks adjacent to the batholith. It is along this fault zone that the Cala Abajo and Piedra Hueca stocks occur. In the Rio Vivi area, displacements along the fault system are mainly vertical; but, elsewhere, the movement has been left-lateral.[7] The rocks in the fault zone are extensively hydrothermally altered.

Geology of the copper deposits

Bradley[3] has recently described the geology of the porphyry copper deposits of the Rio Vivi area. The following description is taken mainly from his paper. Both the Piedra Hueca and the Cala Abajo deposits occur in quartz-diorite porphyry stocks. The stocks are separated by only 600 m and are located in similar topographic environments. Their physical and mineralogic characteristics are, however, considerably different.

The Piedra Hueca stock, 250 m × 450 m in outcrop, is weakly fractured. Leaching and secondary enrichment are negligible. The hydrothermal alteration suite, characterized by potassium feldspar and biotite, is a potassic assemblage.

The Cala Abajo stock, 200 m × 1200 m in outcrop, is cut by a major fault and is strongly fractured. Leaching and secondary enrichment are extensive. The hydrothermal alteration suite, characterized by albite and chlorite, is a propylitic assemblage. Hydrothermal biotite and potassium feldspar are absent.

The two deposits together contain approximately 100 000 000 tons of ore averaging 0·8% copper.

A third body of quartz-diorite porphyry, the Sapo Alegre stock, occurs in the Rio Vivi area, but its copper mineralization is of such limited extent that it has little economic potential.

Sampling procedures

Because of rugged topography, dense vegetation and limited time and manpower, soil sampling was accomplished by use of ridge and spur traverses rather than a grid. Although spacing between ridges is not always optimum, ridge and spur surveys do have some advantages over grid surveys: (1) the characteristics of the soils sampled are apt to be more uniform inasmuch as the soils have developed under similar topographic conditions and (2) the soils on ridge crests are more likely to be residual soils than the soils on ridge flanks or in topographic lows because very little transported material moves directly down crests.

The specific spacing of sample sites along the traverses was selected to ensure several samples falling within the boundaries of the deposits on each traverse. With few exceptions, the samples were collected from the upper *B* horizon.

A limited amount of drill core was analysed so that geochemical comparisons could be made between rocks of the ore deposits and soils over them. The drill core was provided by Ponce Mining Company.

Preparation and analysis of samples

The soil samples were oven-dried at about 110°C, passed through a jaw crusher to break the aggregates, and sieved into −0·25- and +0·25-mm fractions. The two fractions were then ground by use of a vertical pulverizer with ceramic plates. Both were retained for analysis. The core samples were crushed and then pulverized.

The samples were analysed for 30 elements by the semiquantitative spectrographic method developed by Grimes and Marranzino.[5] In the present report only the spectrographic data for molybdenum, manganese and silver are discussed.

Copper, lead and zinc contents were determined from a hot nitric acid extraction by atomic absorption spectrophotometry by use of the method of Ward *et al.*[12] Gold content was also determined by atomic absorption spectrophotometry. The samples were ignited at 700°C for 2 h, and then analysed by the method developed by Thompson, Nakagawa and VanSickle.[11]

Sampling and analytical errors

Thirty-five randomly selected sites were sampled in duplicate. These samples were prepared and analysed as described above. The method outlined by Garrett[4] was used to determine the variability caused by sampling and by analytical errors. Overall data variance, combined sampling and analytical variance, and variance ratio were estimated for copper and gold. An arbitrary value of 0·01 ppm was assigned to gold analyses below the 0·02 ppm lower limit of analytical detection. The results obtained (Table 1) indicate that the sampling and analytical errors are very small as compared to total elemental variation at both the 95% confidence level and the 99% level. Results are shown only for copper and gold, the elements of principal importance in the present discussion.

Table 1 Estimation of sampling and analytical errors in soil samples, Rio Vivi area, Puerto Rico

	σD^2	σSA^2	$F_{calc.}$	$F_{tabulated}$	
				$\alpha=5\%$	$\alpha=1\%$
Cu	0·3019	0·0094	32·12	≈4·0	≈7·0
Au	0·2620	0·0296	8·85	≈4·0	≈7·0

σD^2, overall data variance; σSA^2, combined sampling and analytical variance; F, variance ratio; α, significance level.

Studies of soil profiles

General characteristics

The soils of the Rio Vivi area have developed on the steep slopes of strongly dissected, humid uplands. Characteristically, these soils are immature and well drained and of yellow-brown hue. The pH ranges from 4·0 to 6·2 and averages 5·5. The average thickness of the solum is 0·9 m, except in the vicinity of the Piedra Hueca deposit, where the average is 0·7 m.

The *A* horizon in profiles of the Rio Vivi area is characterized as follows: (1) thickness, 0·3 m; (2) colour, hues 5–10*YR*, values 2–3, chromas 3–4; (3) structure, medium granular; (4) texture, clay (see reference 8 for descriptive notations).

The *B* horizon is characterized as follows: (1) thickness, 0·6 m, except at Piedra Hueca, where the average thickness is 0·4 m; (2) colour, hues 5–10*YR*, values 4–8, chromas 4–6; (3) structure, medium sub-angular blocky; (4) texture, clay. The *B* horizon in most of the area is a cambic horizon and the soils are classified as inceptisols.[9, 10] A small fraction of soils in the area, developed on the gentler slopes, contains argillic *B* horizons and are classified as ultisols.[9, 10]

The *C* horizon in most of the area is saprolite that exhibits the relict textures of the parent rock. The rock at Piedra Hueca is less intensely altered than that at Cala Abajo, but fresh rock in the area studied is confined to the immediate vicinity of major stream channels. The effects of weathering are deep; they are observed in exposed cuts to depths of 50 m.

Metal content as a function of grain size and depth

The relationships between metal content, grain size, depth and soil horizon are summarized graphically in Figs. 2 and 3. These data are a compilation of information from 110 vertical soil profiles in the Rio Vivi area.

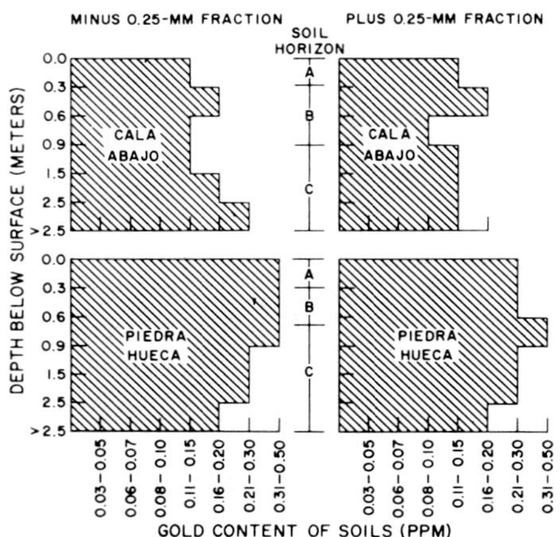

Fig. 2 Gold content of soils in vertical profiles, Rio Vivi district. Average background content is less than 0·02 ppm, the lower limit of analytical detection

Fig. 2 indicates that over both deposits the gold content of soils in all horizons and in both sieve fractions is highly anomalous. The background gold content of soils outside the deposits is generally less than 0·02 ppm—the lower limit of analytical detection. The –0·25-mm fraction of the upper *B* horizon yields high gold concentrations more consistently than do other combinations of horizon and grain size.

Fig. 3 indicates that over the Piedra Hueca deposit the copper content of soils in all horizons and in both sieve fractions is highly anomalous as compared to the copper content of background soils. The greatest contrast is found in parts of the *C* horizon.

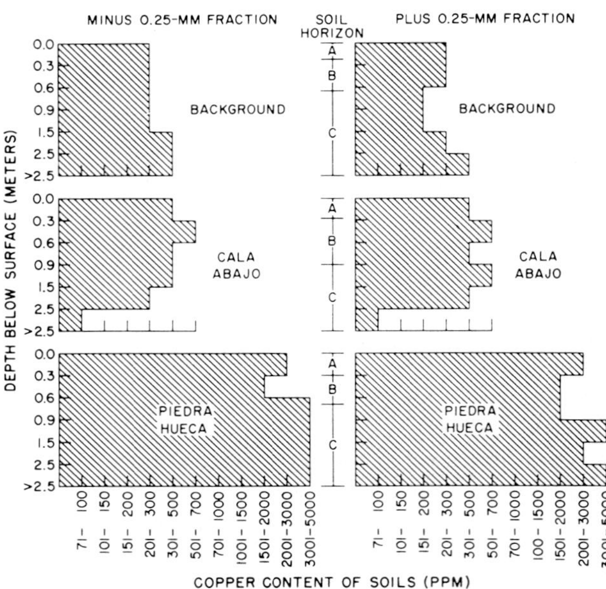

Fig. 3 Copper content of soils in vertical profiles, Rio Vivi district. Background refers to sites outside limits of known deposits

Over the Cala Abajo deposit the copper content of the soils is only weakly anomalous as compared to the copper content of background soils. The strongest contrast is found in the upper *B* horizon and in the uppermost *C* horizon.

The –0·25-mm fraction of the upper *B* horizon most consistently yields the optimum concentrations of gold and copper, the elements of principal interest. In addition, the *B* horizon is easily accessible, contains relatively little organic matter and is generally free from surface contamination.

Graphic presentation of results of the survey

To depict the areal distribution of concentrations of the elements in soils the sample population of each element was divided into four concentration intervals, represented by four sizes of circles (Figs. 4–9). The approximate limits of the four intervals are as follows: (1) the highest 5% of the population, represented by the largest circles; (2) the next lower 15% of the population, represented by the second largest circles; (3) the following lower 25% of the population, represented by the third largest circles; and (4) the remaining fraction of the population, comprising the median and lower concentrations, represented by the smallest circles.

The scheme employed is an approximation of one in which the four intervals described above represent (1) concentrations greater than two standard deviations, (2) concentrations between one and two standard deviations, (3) concentrations between the mean and one standard deviation and (4) concentrations comprising the mean and lower concentrations. Strict adherence to this system proved undesirable because of higher sample density in areas of the deposits.

Copper distribution

The most striking feature of the distribution pattern of copper in soils of the Rio Vivi district is the contrast between concentrations over the Piedra Hueca deposit and those over the major (southeastern) part of the Cala Abajo deposit (Fig. 4). This sharp contrast of concentrations can also be seen in Table 2.

Table 2 Median concentrations, ppm, of selected elements in soils of Rio Vivi district, Puerto Rico

	Cala Abajo deposit	Piedra Hueca deposit	District background
Copper	300	2300	160
Gold	0·15	0·19	<0·02
Molybdenum	7	<5	<5
Zinc	7	36	15
Manganese	3	30	15
Lead	7	11	6
Silver	<0·5	0·7	<0·5

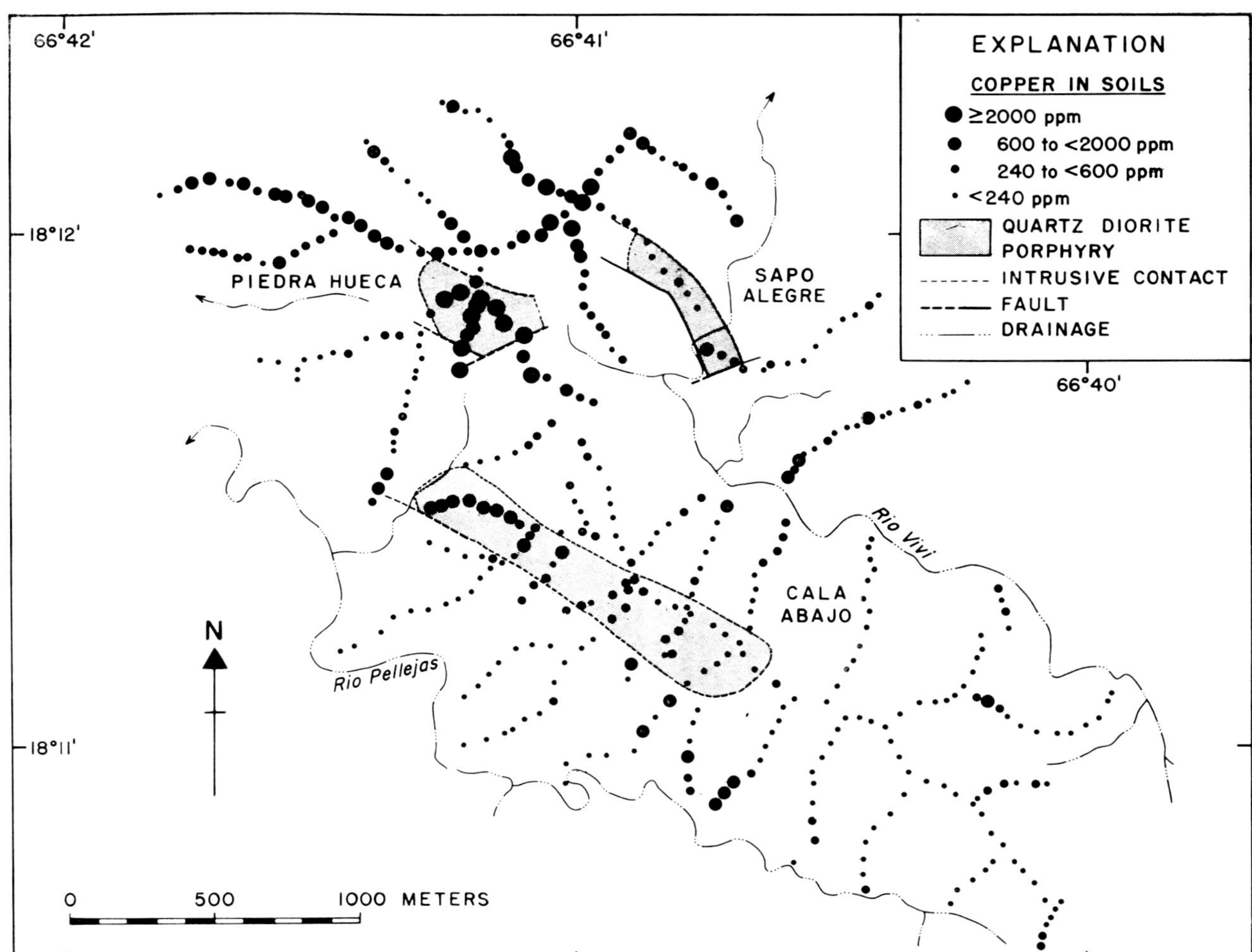

Fig. 4 Distribution of copper in soils, Rio Vivi district (—0·25-mm fraction, upper B horizon). Geology from Mattson[7]

Within the confines of the two ore deposits, high copper concentrations occur in soils over the entire Piedra Hueca deposit and the northwestern part of the Cala Abajo deposit, whereas low copper concentrations occur in soils over the major (southeastern) part of the Cala Abajo deposit. The effects of supergene leaching and enrichment are negligible in the Piedra Hueca deposit and in the northwestern part of the Cala Abajo deposit, and are extensive in the southeastern part of the Cala Abajo deposit. Therefore, it is concluded that the low copper concentrations in soils over the southeastern part of the Cala Abajo deposit are the result of supergene leaching processes that accompanied secondary enrichment.

The copper content of soils over the Cala Abajo deposit is somewhat higher than that of background soils; the ratio of median copper concentrations over the Cala Abajo deposit to median background concentrations outside the deposits is less than 2 to 1. Inasmuch as the ratio for the Piedra Hueca deposit is 15 to 1, the distribution of copper concentrations in soils does not indicate a deposit of ore grade at Cala Abajo.

The cluster of circles representing concentrations greater than 2000 ppm, shown north of the Piedra Hueca deposit (Fig. 4), presents an attractive exploration target; but its significance, in terms of known mineralization at depth, is unknown. Further investigation seems warranted.

Gold distribution

The feature of greatest importance in the pattern of gold distribution (Fig. 5) is the high gold content of soils directly over both ore deposits. The median concentrations of gold in soils are similar over both deposits—0·19 ppm for Piedra Hueca and 0·15 ppm for Cala Abajo. By use of concentration data given by Lutjen[6] it is calculated that the average gold concentration for the two deposits is about 0·20 ppm.

High gold concentrations in soils have also been found over three other known porphyry copper deposits in the same structural belt about 15 km northwest of the Rio Vivi district. The range of gold concentrations in the soils over these three deposits is similar to that over the Rio Vivi deposits.

The consistent relationship between the gold content of soils and the known porphyry copper deposits of Puerto Rico suggests that gold will be a useful pathfinder for porphyry copper deposits elsewhere.

The cluster of circles, representing concentrations between 0·06 and 0·19 ppm, shown north of the Piedra Hueca deposit (Fig. 5) presents an attractive target, especially in the light of high copper and molybdenum concentrations in soils of the same area. The significance of these clusters is

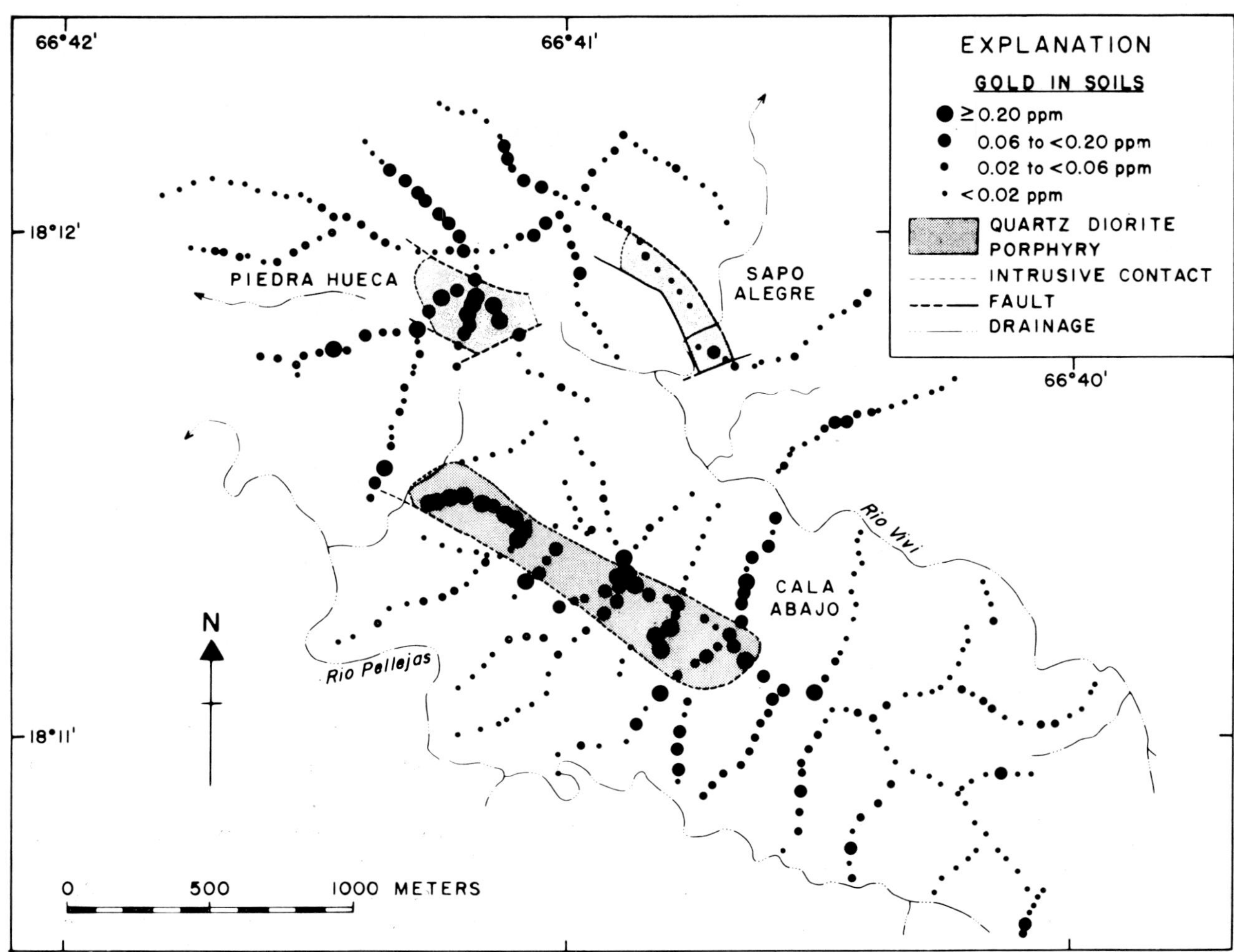

Fig. 5 Distribution of gold in soils, Rio Vivi district (—0·25-mm fraction, upper B horizon). Geology from Mattson[7]

not known, but further investigation seems warranted.

Molybdenum distribution

The feature of principal interest in the distribution pattern of molybdenum (Fig. 6) is the contrast between concentrations in soils over the Piedra Hueca deposit and those over the major (southeastern) part of the Cala Abajo deposit.

The actual molybdenum content of the deposits is not known, but it is apparently very low. Assays of the core were not made because extremely little molybdenite was observed in the drill core (R. A. Bradley, personal communication, 1972). The small amount of drill core that has been analysed does not contain molybdenum in concentrations above the lower limit of detection (5 ppm). Thus, inferences regarding mobility of molybdenum are at present inconclusive. Analyses

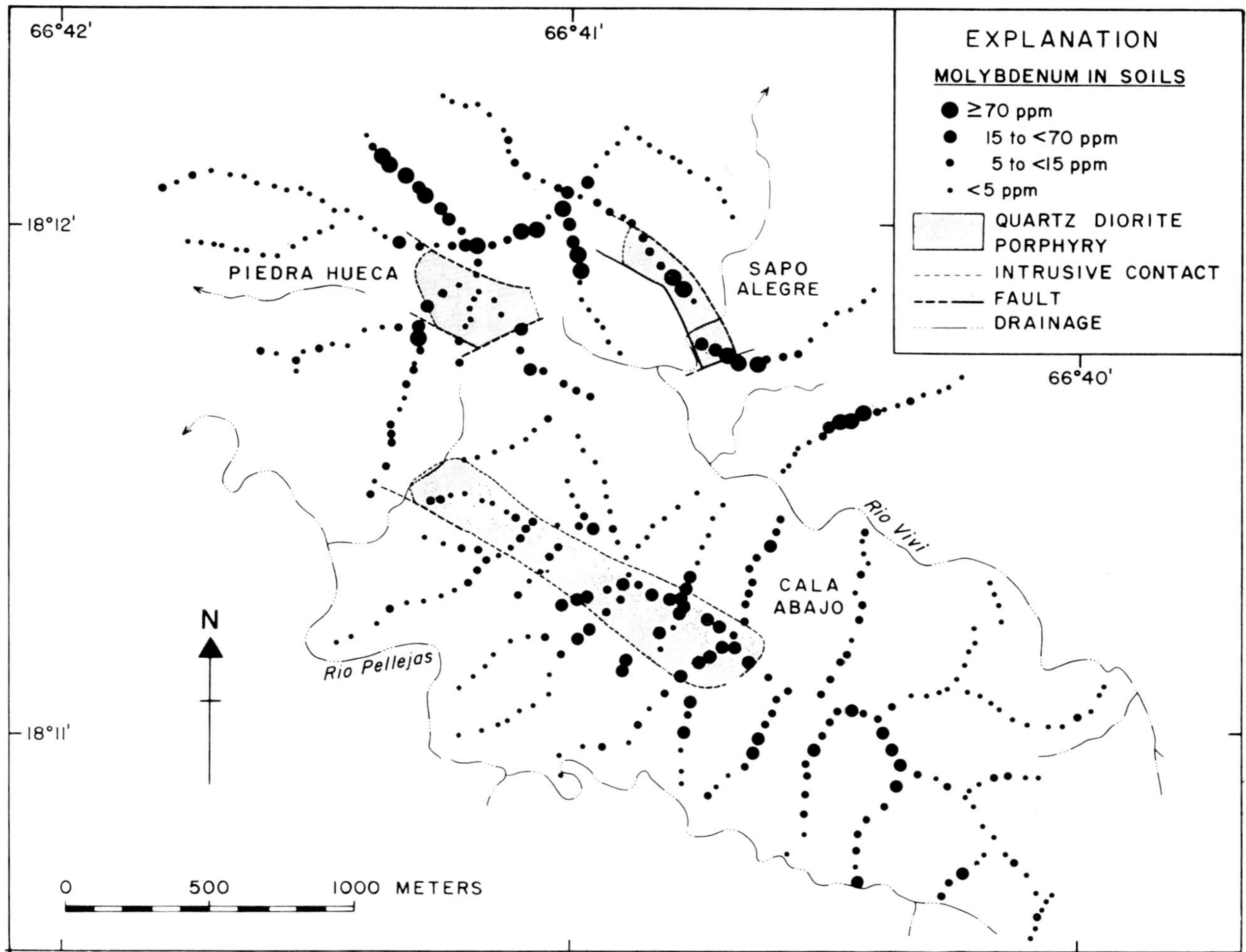

Fig. 6 Distribution of molybdenum in soils, Rio Vivi district (−0·25-mm fraction, upper B horizon). Geology from Mattson[7]

Within the confines of the deposits the molybdenum content of the soils exhibits an inverse relationship to the copper content of the soils; the relatively low molybdenum concentrations occur in the soils over the Piedra Hueca deposit and the northwestern part of the Cala Abajo deposit, whereas the higher molybdenum concentrations occur in the soils over the major (southeastern) part of the Cala Abajo deposit. This relationship suggests that molybdenum was immobile under the strongly acid conditions that accompanied supergene leaching and was mobile under less acid conditions.

of additional drill core need to be made.

The significance of the clusters of high molybdenum concentrations in soils north of the Piedra Hueca deposit is not known, but further investigation seems warranted.

Zinc distribution

The distribution pattern of zinc in soils of the Rio Vivi district (Fig. 7) shows the highest zinc concentrations to be in areas peripheral to the copper ore deposits. Similar patterns are common to other porphyry copper districts and are attributed to

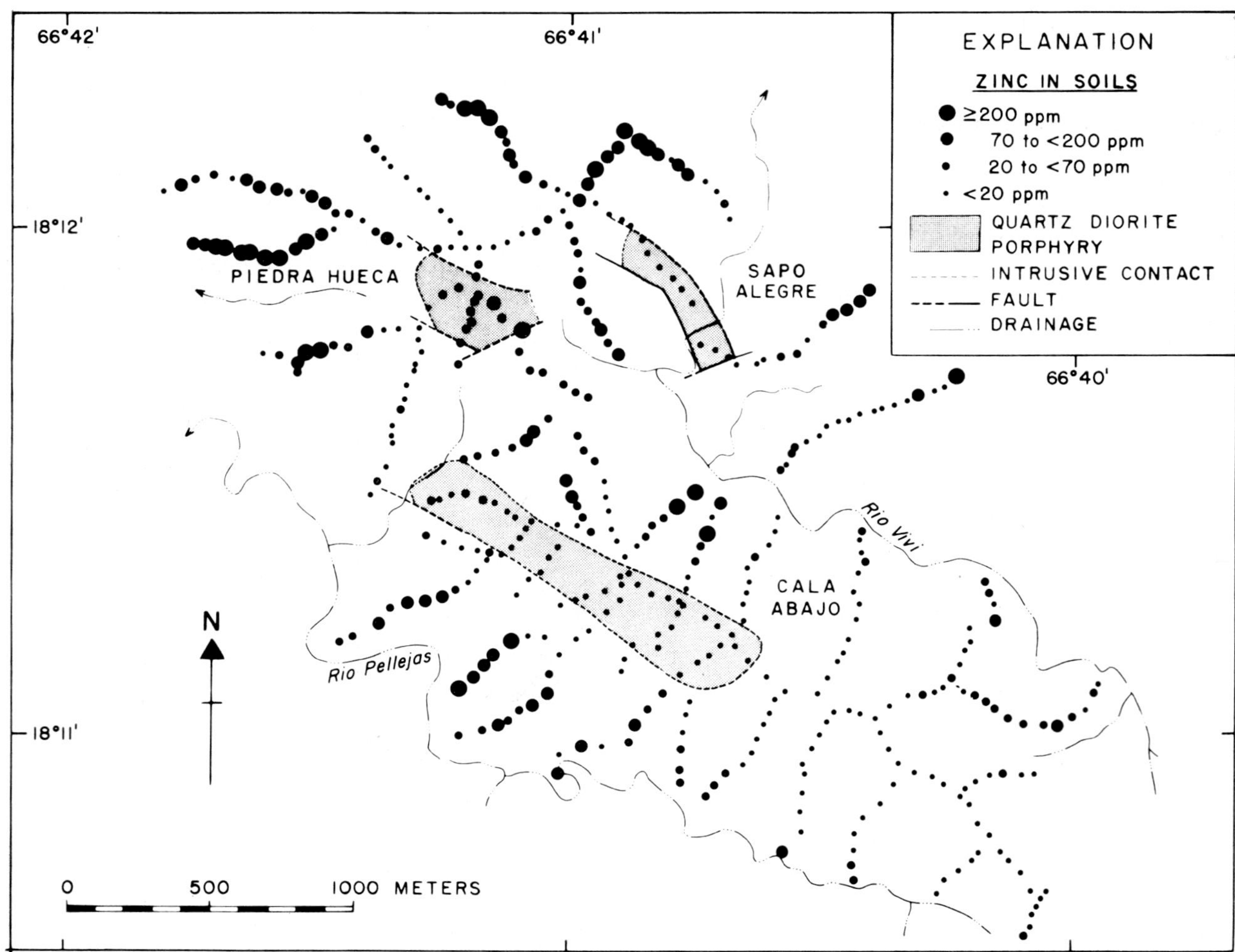

Fig. 7 Distribution of zinc in soils, Rio Vivi district (−0·25-mm fraction, upper B horizon). Geology from Mattson[7]

primary zoning.

Within the limits of the ore deposits, the distribution pattern of zinc in soils is similar to that of copper; the relatively high zinc concentrations occur in the soils over the Piedra Hueca deposit and the northwestern extremity of the Cala Abajo deposit, whereas the lower zinc concentrations occur in soils over the major (southeastern) part of the Cala Abajo deposit. The median concentrations (Table 2) reflect this distribution pattern.

It is concluded that zinc, like copper, has been leached from the capping over the major (southeastern) part of the Cala Abajo deposit. Limited analytical data of drill core corroborate the conclusion.

Manganese distribution

The distribution pattern of manganese in soils of the Rio Vivi district (Fig. 8) is similar to that of zinc; the areas of highest concentration are peripheral to the copper ore deposits. The pattern probably reflects primary zoning.

Within the limits of the copper deposits, the distribution of manganese in soils is like that of copper and zinc; the relatively high manganese concentrations occur in the soils over the Piedra Hueca deposit and the northwestern extremity of the Cala Abajo deposit, whereas much lower manganese concentrations occur in the soils over the major (southeastern) part of the Cala Abajo deposit. This distribution pattern is shown in Table 2. The conclusion that this relationship is a result of supergene leaching of the southeastern part of the Cala Abajo deposit is strengthened by limited analyses of drill core; manganese concentrations increase markedly below the zone of oxidation.

Lead distribution

The distribution pattern of lead in the soils of the Rio Vivi district (Fig. 9) is like that of zinc and manganese; the areas of highest lead concentrations are peripheral to the copper ore deposits. This pattern in the soils of the district probably also reflects primary zoning.

The lead concentrations in the soils over the Piedra Hueca deposit are somewhat higher than those over the Cala Abajo deposit (Table 2), but this difference may reflect a difference in primary concentrations. Limited analyses of drill core show similar lead concentrations in both the zone of leaching and the zone of secondary enrichment at Cala Abajo.

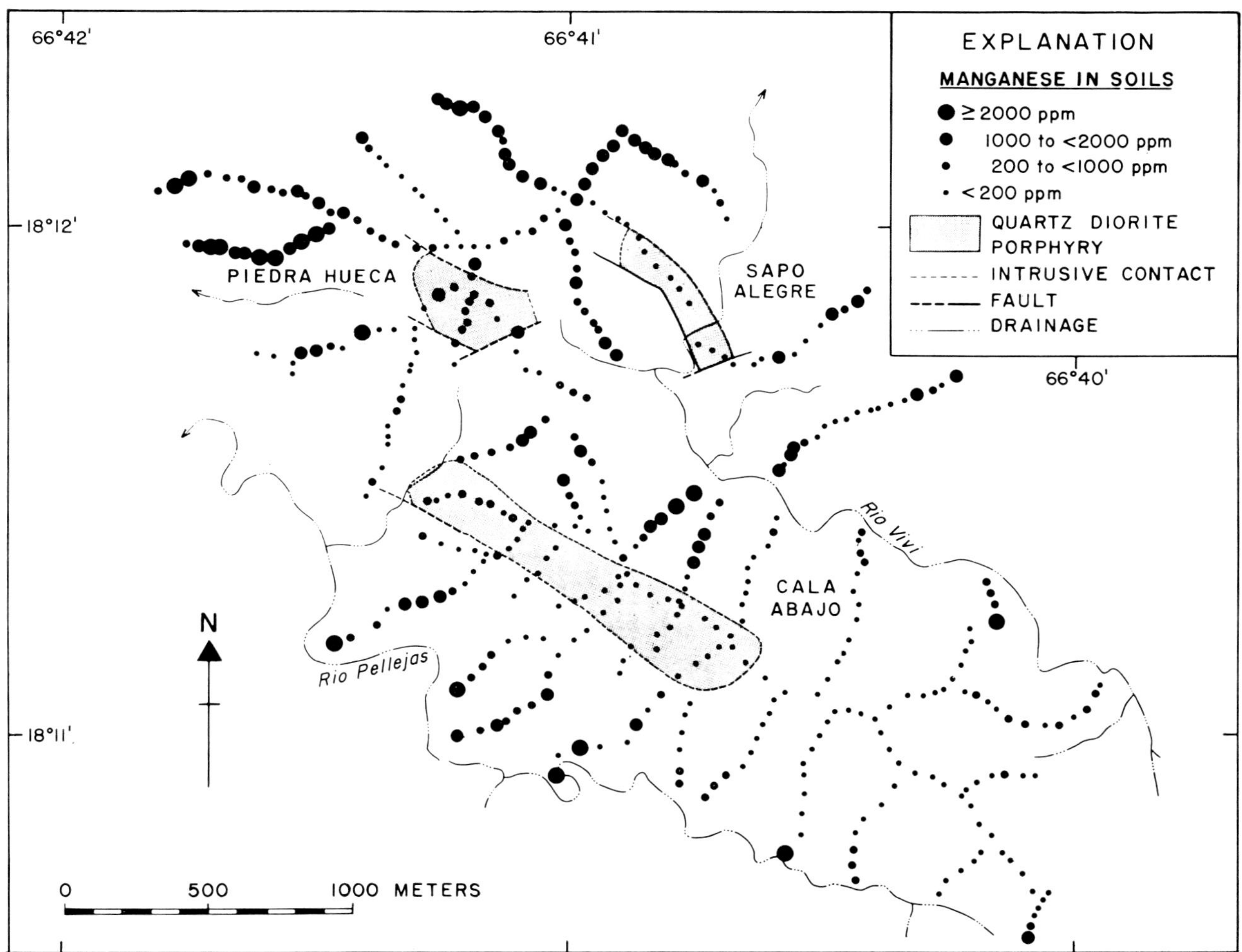

Fig. 8 Distribution of manganese in soils, Rio Vivi district (—0·25-mm fraction, upper B horizon). Geology from Mattson[7]

Silver distribution

The distribution of silver in soils is not shown in map form because most of the samples did not contain silver in detectable concentrations (≥0·5 ppm). Lower silver concentrations in soils over the Cala Abajo deposit relative to those over the Piedra Hueca deposit, however, also (Table 2) probably resulted from supergene leaching of the Cala Abajo deposit. Limited analyses of drill core show appreciably higher silver concentrations below the zone of oxidation at Cala Abajo.

Summary and conclusions

In the Rio Vivi district the copper content of soils over the known porphyry copper deposits is not a reliable guide to the location and grade of underlying mineralization because the leaching process responsible for secondary enrichment can result in low copper content of soils overlying copper deposits of economic grade.

In contrast, the gold content of soils over porphyry copper deposits in the copper belt of Puerto Rico does not differ greatly from deposit to deposit, and it apparently is similar to gold concentrations in the deposits themselves. These relationships hold for deposits covered by thick, intensely leached cappings as well as for primary deposits. Therefore, the gold content of soils should serve as an especially useful pathfinder for the location of porphyry copper deposits, at least under conditions similar to those in the areas studied.

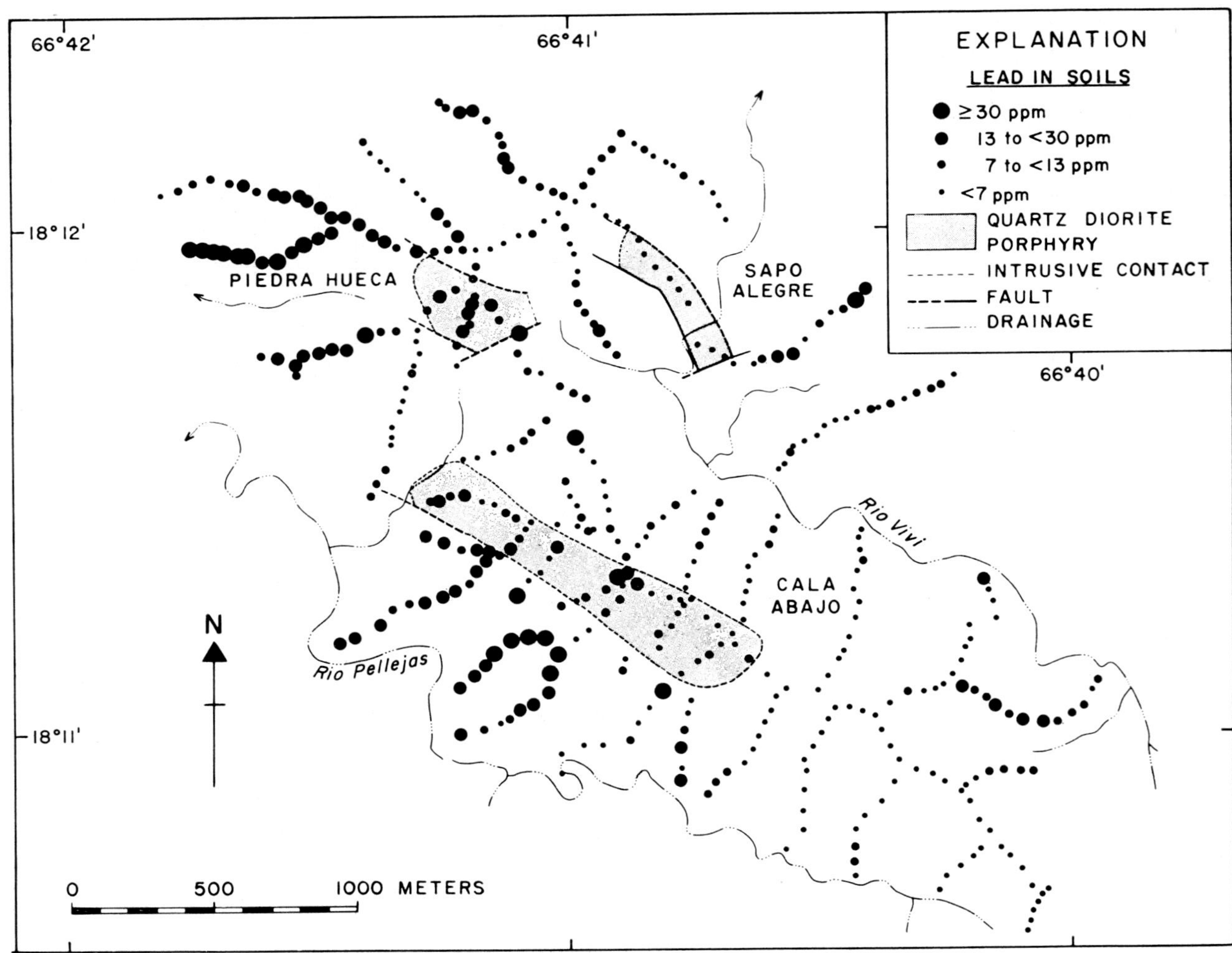

Fig. 9 Distribution of lead in soils, Rio Vivi district (—0·25-mm fraction, upper B horizon). Geology from Mattson[7]

Gold and copper content of soils, considered together, may help to distinguish supergene-enriched deposits from primary deposits in known districts.

Knowledge of the molybdenum, zinc, manganese, lead and silver content of soils may play a useful role in the search for porphyry copper deposits in Puerto Rico and in similar environments, but their role seems to be subsidiary to that of the gold and copper contents of soils.

Acknowledgment

The authors gratefully acknowledge the valuable contributions made by many persons to their investigation. The analytical support was provided by the Geological Laboratory of Natural Resources, San Juan, Puerto Rico. Its staff, directed by the second author, includes Carmen Anna Abrahamson, Sara Cortés, Ruth D. González. Lisbeth Hyman, Juan B. Martínez and Isidro Toro.

Francisco García and Richard B. Tripp assisted the first author in the field. Maurice A. Chaffee, Dennis P. Cox and Ralph L. Erickson critically read the manuscript. George H. Loring and Lamont O. Wilch gave assistance with the computer programs applied to the investigation.

Sergio Rojas drafted the illustrations.

Special thanks are due to Roger A. Bradley of American Metal Climax, Inc., who provided valuable geological information and logistical assistance.

This paper is published by permission of the Director, U.S. Geological Survey.

References

1. Archer, A. R. and Main, C. A. Casino, Yukon—a geochemical discovery of an unglaciated Arizona-type porphyry. In *Geochemical exploration* (Montreal: CIM, 1971), 67–77. (*CIM Spec. vol. 11*)
2. Bergey, W. R. Geochemical prospecting for copper in Puerto Rico. *Publ. Jamaican geol. Surv.* 95, 1966, 113–9.
3. Bradley, R. A. The geology of the Rio Vivi porphyry copper deposits, Puerto Rico. *Geol. Soc. Am. Abstr.*, **3**, 1971, 511.
4. Garrett, R. G. The determination of sampling

and analytical errors in exploration geochemistry. *Econ. Geol.*, **64**, 1969, 568–9.
5. Grimes, D. J. and Marranzino, A. P. Direct-current arc and alternating-current spark emission spectrographic field methods for the semiquantitative analysis of geologic materials. *Circ. U.S. geol. Surv.* 591, 1968, 6 p.
6. Lutjen, G. P. The curious case of the Puerto Rican copper mines. *Engng Min. J.*, **172**, Feb. 1971, 74–84.
7. Mattson, P. H. Geologic map of the Adjuntas quadrangle, Puerto Rico. *U.S. geol. Surv. Misc. geol. Invest. Map* I–519, 1 : 20,000, 1968.
8. Soil Survey Staff. Soil survey manual. *Handb. U.S. Dep. Agric.* 18, 1951, 503 p.
9. Soil Survey Staff. *Soil classification—a comprehensive system, 7th approximation* (Washington, D.C.: Soil Conservation Service, U.S. Department of Agriculture, 1960), 265 p.
10. Soil Survey Staff. Soil survey of the Adjuntas–Utuado area, Puerto Rico. *Unpubl. Rep. U.S. Dep. Agric.*, 1969.
11. Thompson, C. E. Nakagawa, H. M. and VanSickle, G. H. Rapid analysis for gold in geologic materials. *Prof. Pap. U.S. geol. Surv.* 600-B, 1968, 130–2.
12. Ward, F. N. *et al.* Atomic-absorption methods of analysis useful in geochemical exploration. *Bull. U.S. geol. Surv.* 1289, 1969, 45 p.

550.42:551.311.14:553.43/.44(481+429+711)

Some aspects of geochemical dispersion from base-metal mineralization within glaciated terrain in Norway, North Wales and British Columbia, Canada

M. B. Mehrtens, B.SC., PH.D.

Rio Tinto Canadian Exploration, Ltd., Toronto, Canada

J. S. Tooms, B.SC., PH.D., M.I.M.M.

Applied Geochemistry Research Group, Imperial College, London, England

A. G. Troup, B.SC.

Rio Tinto Canadian Exploration, Ltd., Vancouver, British Columbia, Canada

Synopsis

Some results are presented of geochemical research and exploration within glaciated terrain over and in the vicinity of base-metal mineralization in Norway, North Wales and the Central Interior of British Columbia, Canada. These study areas are characterized by siliceous overburden and a topography of rolling hills and broad U-shaped valleys. The climates of the regions, however, are somewhat dissimilar, varying from cold and dry to temperate and wet.

The results of the investigations show that ore elements are dispersed, from their bedrock source beneath glacial till, dominantly in shallow groundwaters and to a lesser extent by mechanical (ice) transport and biochemical processes.

It is concluded that secondary metal dispersion patterns related to sulphide mineralization in these and similar environments may be readily detected on a broad regional scale by sampling groundwater seepage sites. Where seepages occur in lakes, as, for example, in the Central Interior of British Columbia, the existence of bedrock mineralization in the general vicinity of the lakes can be detected by sampling the organic-rich bottom sediments in the central deeper parts of these lakes. Interpretation of these seepage anomalies, as well as of anomalies which may occur as a result of seepage in streams or rivers, is dependent on an assessment of groundwater dispersion trains.

In south-central Norway detailed research into the secondary dispersion patterns of ore elements from massive sulphide deposits sub-outcropping beneath ground moraine provided an opportunity for the obtaining of an understanding of dispersion mechanisms in a strongly glaciated region. The results of subsequent geochemical exploration for base metals within similar glaciated environments in North Wales and within the Central Interior of British Columbia have indicated that the conclusions of the Norwegian study may have a more widespread applicability.

Observations and conclusions which resulted from work undertaken in Norway, North Wales and the Central Interior of British Columbia are presented, together with some of the implications of these results to the planning of geochemical reconnaissance programmes.

Sampling and analytical procedures

In the Norwegian study the distribution of copper, zinc, lead and iron was investigated in bedrock, glacial overburden and soils from barren and mineralized areas.

Profile samples from the overburden to depths of 2 m were normally obtained by hand by use of a spiral auger. Over one of the orebodies (Tverrefjellet), however, a bulldozer and backhoe were employed to excavate trenches to bedrock across the strike of the ore zone. Rock samples were crushed to 2 mm in a jaw crusher and then ground to less than 200 μm in a pestle and mortar.

All other sample materials were dried at 80°C and then sieved through an 82-mesh screen (204 μm), the oversize material being discarded. Previous size analysis had determined that the −82-mesh fraction gave the greatest contrast between anomalies and background.[5]

A variety of sample attacks was employed, but, unless otherwise stated, the data refer to results after fusion with $KHSO_4$. Except for cold extraction methods (by use of a buffer, dithizone–benzene), where the precision[2] was ±30% at the 95% confidence level, a precision of better than ±25% at the 95% confidence level was obtained during colorimetric estimation.[9,10,11]

The Welsh and British Columbian samples were sieved to −82 mesh and their metal contents determined by atomic absorption spectrophotometry following a hot oxidizing acid attack. The precision of these data is better than ±25% at the 95% confidence level.

The base-exchange capacity was determined by use of the method of Bray and Willhite.[12]

The pH of stream waters was determined *in situ* by use of a portable meter and that of the sediments and overburden by determination of the pH of a 1:5 solid/water suspension equilibrated by shaking for 1 h on an automatic shaking machine. Deionized boiled water was used in these experiments.

Carbonate in soil was determined by a modification of the titration method described by Piper.[13] Sulphate in soil and water was determined turbidimetrically as barium sulphate.[14]

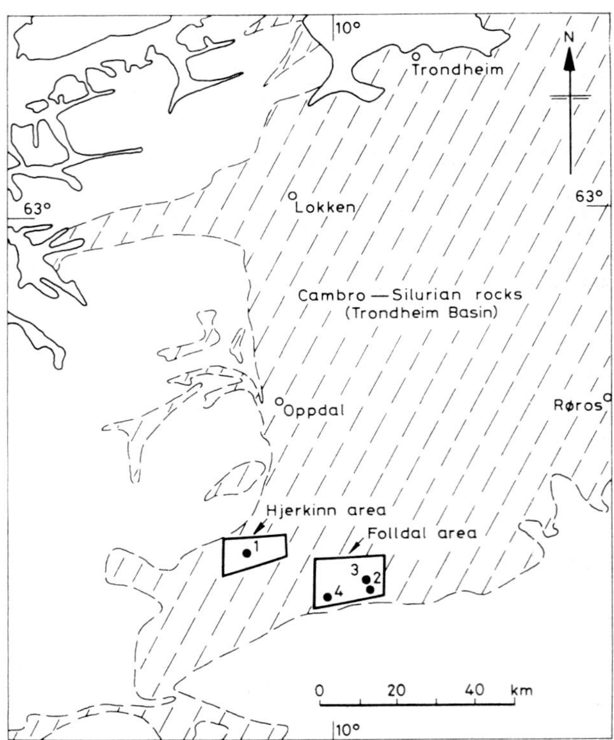

Fig. 1 Simplified geological map of Trondheim Basin showing field areas: 1, Tverrefjellet; 2, Søndre Gjeitteryggen; 3, Nordre Gjeitteryggen; 4, Grimsdal. Geology after Holtedahl[15]

Norway

The area investigated is located in south-central Norway, 100 km south of Trondheim in the Hjerkinn–Folldal district.

The detailed study was carried out over the Tverrefjellet massive sulphide orebody near Hjerkinn and less detailed comparative studies were undertaken in the Folldal area in the vicinity of the Søndre and Nordre Gjeitteryggen mines and Grimsdal mine (Fig. 1).

General description of the field areas

The massive sulphide deposits of the Hjerkinn–Folldal area occur as conformable bodies within a steeply dipping sequence of metamorphosed spilite–keratophyre submarine volcanics and related sedimentary rocks of the Lower Palaeozoic Støren Series.[7] The orebodies range in size from 1 000 000 to 15 000 000 tons and have an average grade of 1% Cu, 2·5% Zn, 0·5% Pb and 35% S.[8]

Quaternary glacial deposits overlie the region, and in the study areas above the 1000-m topographic contour (i.e. the tree line) periglacial features such as solifluxion benches, polygons and frost tussocks are well developed.[3]

The thickness of the glacial till is highly variable, but commonly exceeds 4 m, and was deposited by the last glacial ice, which moved toward the north and northeast. Soil profile development is minimal, the soils being characterized by an indistinct profile similar to podzols.

The present-day climate is cold and dry; mean annual temperature is −0·8°C and precipitation 230 mm. The vegetation above the tree line consists of 'scrub-like' tundra flora dominated by dwarf species of willow, birch and juniper.

Contamination

At the time of the study the Tverrefjellet deposit was being explored from an inclined shaft situated downstrike and to the east of the ore zone. Development work was at a very early stage and, therefore, contamination of the surface was confined to the immediate vicinity of the shaft and did not seriously restrict the investigations reported here.

The mines at Søndre Gjeitteryggen, Nordre Gjeitteryggen and Grimsdal had been in existence for a number of years. Geochemical sampling was therefore confined to strike extensions of the ore in locations remote from disturbed and contaminated ground.

Results

Geochemical characteristics of background moraine

In the Hjerkinn area two locations underlain by barren rocks of the greenstone group were selected for a background study. The areas are freely drained and situated on gentle hill slopes above the tree line. The results obtained are summarized in Table 1. There is no enrichment of ore elements in any of the soil horizons compared with metal values in the parent material. Moreover, base-metal values increase to a depth of 1 m, below which the metal concentrations are essentially constant. The increase in metal content with depth is accompanied by a rise in pH and an increase in the carbonate content.

Table 1 Geochemical values in moraine and overlying soil horizons, Hjerkinn

Depth, cm	Description	$KHSO_4$ metal, ppm Cu	Zn	Pb	cxCu	Base-exchange capacity, meq/100 g	Carbonate, %	pH
0–5	Black to brown A_0	5	70	10	0·2	10	0·1	4·3
5–15	Grey leached loam	10	60	10	0·5	8·0	0·1	4·6
15–25	Brown/medium grey loam	20	100	10	0·7	6·0		4·2
25–30	Semi-weathered moraine	20	100	10	0·6	6·0		4·8
30–50	Grey unweathered moraine	30	100	10	0·6	6·0		5·1
50–90	Grey unweathered moraine	50	150	10	1·0	4·3	0·2	6·0
90–120	Grey unweathered moraine	60	160	10	1·1			
120–150	Grey unweathered moraine	60	140	10	1·2			
150–170	Grey unweathered moraine	65	160	10	1·2			
170–180	Grey unweathered moraine	65	160	10	1·2			
180–200	Grey unweathered moraine	70	160	10	1·2	4·0	0·2	6·3

Mean values of 66 samples from 10 profiles. Data from 1 profile.

Table 2 Comparison of average base-metal content (ppm) of ground moraine from Hjerkinn area with mean values of unmineralized bedrock from which it was derived

	Cu	Pb	Zn	No. of samples
Mean values for unmineralized bedrock	50	10	140	15
Average values in background ground moraine	60	10	150	66

The depletion in the Cu content of the soil horizons compared with Zn is possibly due to the addition of Zn-rich plant debris to the soil from *Betula nana*—a so-called Zn accumulator plant.[4, 5]

Table 3 Calculated threshold values (ppm) for solum and moraine in Hjerkinn–Folldal area

	Cu	Zn	cxCu
Solum (50 samples)	30	140	1·0
Moraine (70 samples to a depth of 1 m)	80	180	1·4

In general, therefore, the overburden is partially leached of its base-metal and carbonate content to a depth of 1 m, below which the moraine is essentially unweathered. Comparison of the base-metal content of moraine from a depth of 1 m with mean values for the unmineralized greenstone from which it was derived (Table 2) confirms that there is little weathering of the deeper moraine.

Threshold values for base-metals in the soil horizons taken as a group (the solum) and the moraine below the soil horizons to a depth of 1 m are given in Table 3. These thresholds are the upper limit of normal background variations at the 95% confidence level. Pb in background samples of soil and moraine was found to vary between 5 and 20 ppm. At this concentration, which is near the analytical detection limit, the variations cannot be considered to be significant. Accordingly, an arbitrary upper background level of 20 ppm Pb for all levels of the overburden is used here to comment on the data.

Distribution of ore elements in overburden at Tverrefjellet

The detailed surface features at Tverrefjellet are shown in Fig. 2. The ground underlain by the orebody is drained by the Porg stream, which rises in a large mass of iron oxides in the moraine downslope of a collapse structure and swallow-hole located immediately above the sub-outcrop of the ore. These features occur at the western extremity of the ore zone. At the eastern end of the ore sub-outcrop, near to the Porg stream, there is

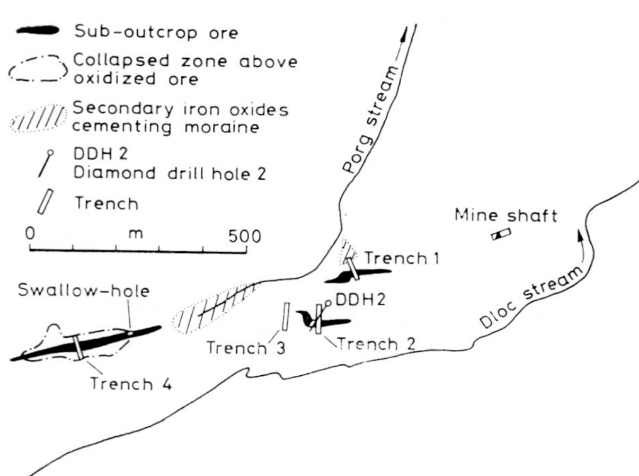

Fig. 2 Map showing surface features at Tverrefjellet

a second patch of ferruginous moraine. This Fe-rich overburden consists of moraine cemented with Fe oxides and marks the emergence of strongly acid groundwaters, which contain very anomalous concentrations of dissolved base metals and iron, as has been reported elsewhere.[5, 6] In general, however, the overburden is freely drained and consists of grey moraine, overlain, in places, by a thin layer of sand and gravel. The directions of drainage and latest ice transport coincide.

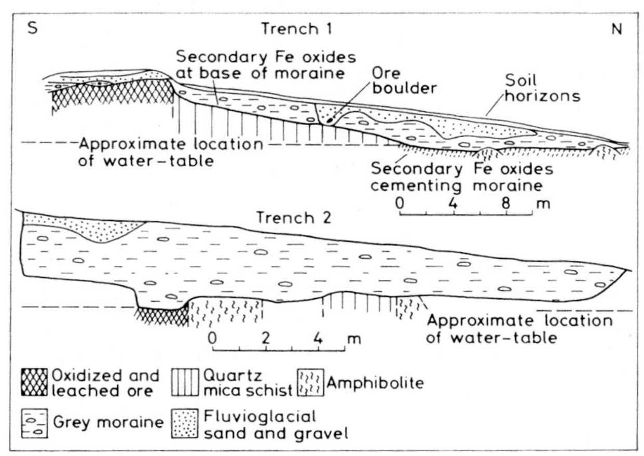

Fig. 3 Geology of trench 1 and 2, Tverrefjellet

The overburden above the mineralization was investigated in a number of trenches excavated across the strike of the ore zone (Fig. 2). These trenches revealed the basal 10 cm of the overburden, overlying and downslope of mineralization, to be cemented with secondary iron oxides, and the upper part of the ore to be thoroughly leached. In these locations the moraine has a maximum thickness of 4 m (Fig. 3).

Continuous channel samples were collected

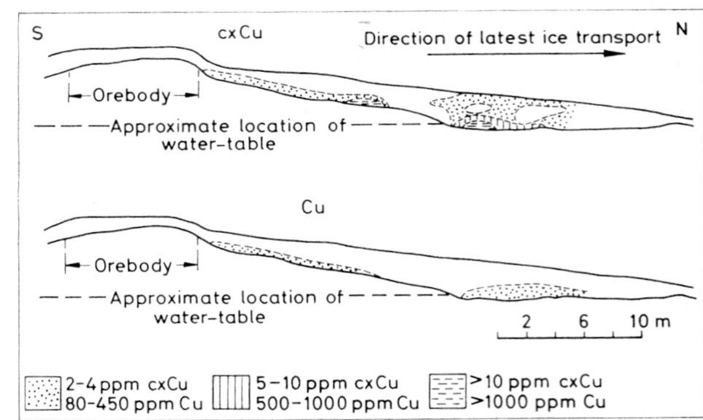

Fig. 4 Distribution of anomalous cxCu and Cu in overburden, trench 1, Tverrefjellet

from vertical profiles, approximately every 2 m along the length of the trench face. Additional profiles were sampled at critical locations following analysis of the systematic samples. The length of each channel sample was related to visual variations in the soil or till, but did not normally exceed 40 cm.

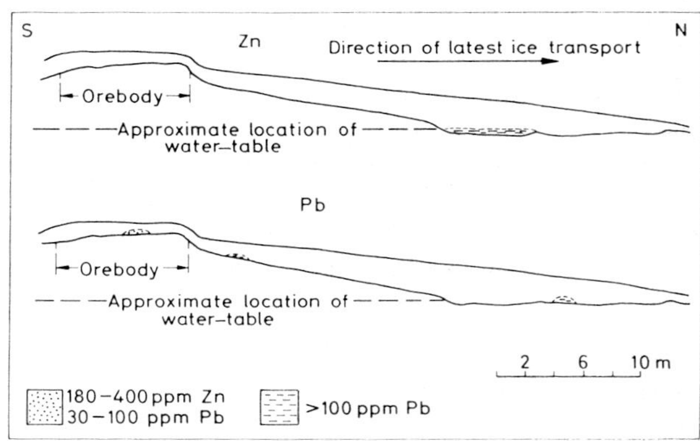

Fig. 5 Distribution of anomalous Zn and Pb in overburden, trench 1, Tverrefjellet

The distributions of anomalous metal in two of the trenches are shown in Figs. 4–7. Only the basal 50–100 cm of the moraine downslope of ore contain anomalous concentrations of base metals, and the metal values tend to decrease in intensity upwards. In trench 2, however, the highest cxCu and Cu values (up to 26 and 15 times threshold,

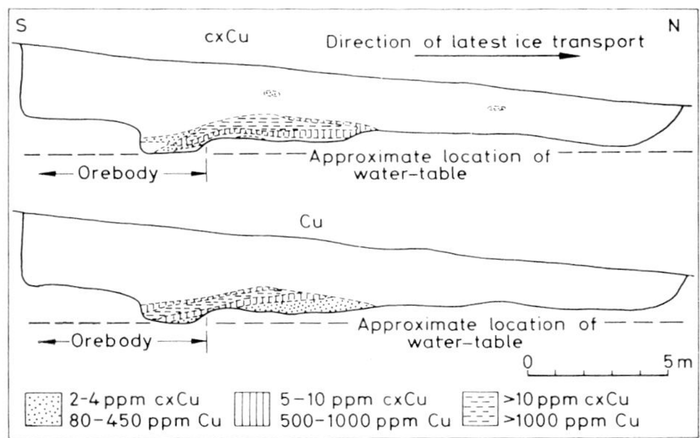

Fig. 6 Distribution of anomalous cxCu and Cu in overburden, trench 2, Tverrefjellet

respectively) occur in the upper layer of the strongly anomalous zone and are immediately overlain by barren moraine (Fig. 6). A detailed investigation of a vertical profile taken from this location (Table 4) shows that the vertical cutoff occurs at different elevations for the various ore elements studied. These observations tend to suggest, therefore, that the ore elements have been introduced into the moraine via groundwaters, resulting in differential migration and fixation of the various ionic species.

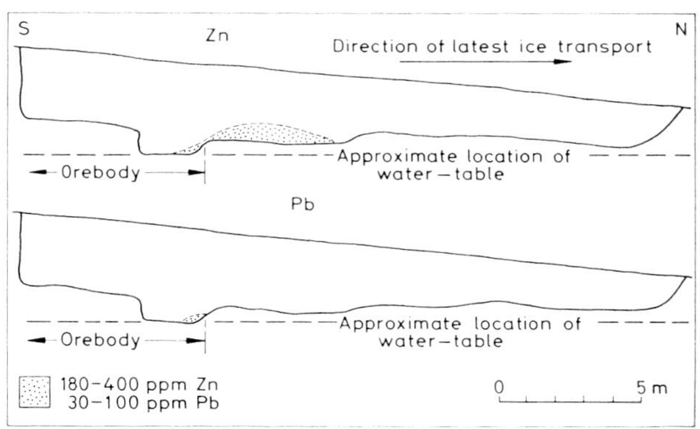

Fig. 7 Distribution of anomalous Zn and Pb in overburden, trench 2, Tverrefjellet

A comparison of the Cu, Zn and Pb ratios in fresh ore, gossan and anomalous basal moraine in trench 2 and DDH2 (Table 5, Fig. 2) indicates that the metal content of the mineralized moraine bears no direct relationship to the fresh ore or to the gossan. Of particular significance are the low Pb values of the anomalous moraine, since the mixing of sulphide ore particles or gossan in the moraine would be expected to result in the secondary dispersion of Pb as well as the other ore elements.

Table 5 Cu:Zn:Pb ratios in unoxidized ore, gossan and anomalous basal moraine

Material	Ratios of base-metal content Cu : Zn : Pb
Unoxidized ore	4 : 6 : 1
Gossan	2·3 : 1 : 1
Anomalous basal moraine	40 : 25 : 1

The data are, therefore, consistent with the ore elements having been introduced into the moraine via groundwaters, the two most striking features of the base-metal dispersion patterns being the very much greater extent of the anomalous cxCu values, compared with Cu and Zn, and the rare occurrence of Pb.

Table 4 Vertical distribution of ore elements in moraine overlying mineralized zone in trench 2, Tverrefjellet

Depth, cm	Description	cxCu, ppm	Cu, ppm	Zn, ppm	Fe, %	Pb, ppm	SO_4, ppm
120–140	Grey moraine	0·2	40	65	3·4	15	n.d.
140–160	Grey moraine	0·4	25	50	2·0	12	n.d.
160–180	Grey moraine	1·0	65	90	3·7	12	n.d.
180–200	Grey moraine	18·4	550	160	3·6	18	n.d.
200–220	Grey moraine	37·0	1150	340	4·0	20	n.d.
220–230	Grey moraine	14·0	525	320	5·0	15	25
230–240	Grey-yellow moraine	6·6	325	130	7·5	12	4250
240–250	Red-brown moraine	0·8	500	320	10·0	12	250

n.d., not detectable.

Evidence for ore boulder train at Tverrefjellet

Three ore boulders have been discovered at Tverrefjellet—at 20, 100 and 5000 m northeast of the ore sub-outcrop. Continued search failed to discover additional macro-size ore fragments. A study of the coarse-grained (−10- +80-mesh sizes) and heavy mineral fraction of the moraine over and down-ice of the ore has not produced any evidence to indicate the presence of a significantly developed indicator train.

Comparative studies, Folldal area, Norway

Less detailed comparative studies were made over the ore zones at the Søndre and Nordre Gjeitteryggen and Grimsdal mines (Fig. 1) in locations where oxidation of the mineralization is very limited—in contrast to the Tverrefjellet deposit.

Søndre Gjeitteryggen

The ore occurrence is located near the brow of a hill in an area occupied by a mature coniferous forest. The overburden is freely drained and consists of grey moraine overlain by a thin podzol. Samples were collected from along the traverse at 10- to 20-m intervals. Channel samples were collected through the depth of the profile, the length of each sample relating to the thickness of soil horizons and any variations in the nature of the till or soil. Interpretation of the geochemical data is helped by the fact that at this locality the transport direction of the latest glacial ice is opposite to that of the drainage.

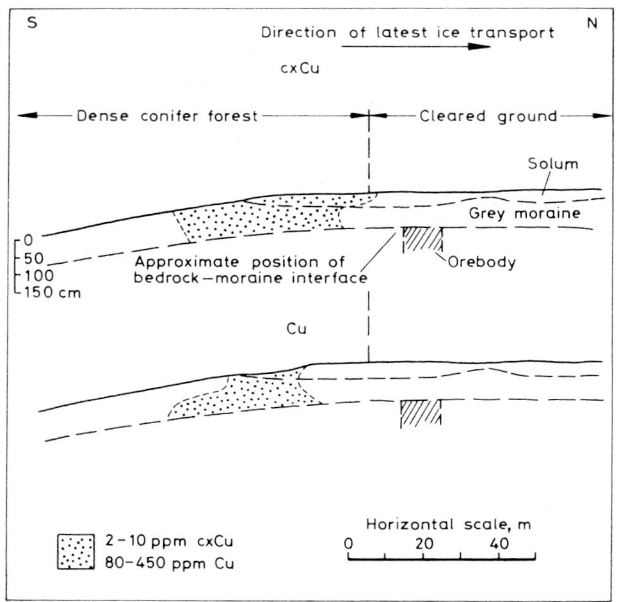

Fig. 8 Distribution of anomalous cxCu and Cu in overburden, Søndre Gjeitteryggen

The salient features of the metal distribution patterns in the overburden (Figs. 8 and 9) are (*a*) that the metal anomalies are displaced downslope from the bedrock source and (*b*) that there is a slight, but significant, increase in the cxCu

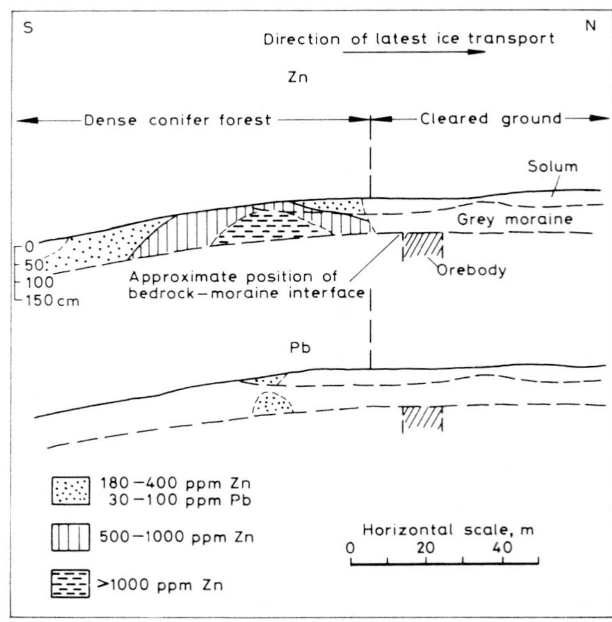

Fig. 9 Distribution of anomalous Zn and Pb in overburden, Søndre Gjeitteryggen

content of the soil horizons as compared with the underlying moraine.

These observations would be in agreement with the view that the metals have been dispersed,

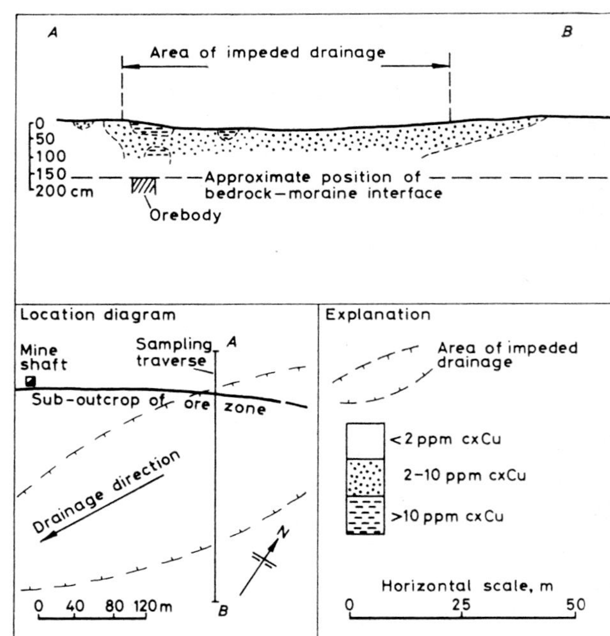

Fig. 10 Distribution of anomalous cxCu in overburden, Nordre Gjeitteryggen

initially, via the groundwaters, supplemented by biochemical processes.

Nordre Gjeitteryggen

In this location above the timber line, mining activities have resulted in the contamination of a wide area around the mine shafts. Consequently, the sample traverse was located outside the contaminated area and to the east, where the ore grade is reported to be low. The traverse is located in an area of impeded drainage through which the groundwater moves to the southwest (Fig. 10). Both ends of the traverse, however, are in freely drained ground. The overburden consists of grey waterlogged ground moraine overlain by a thin layer of semi-decomposed humus, except on the freely drained margins of the transect, where a brown *B* soil horizon is developed above

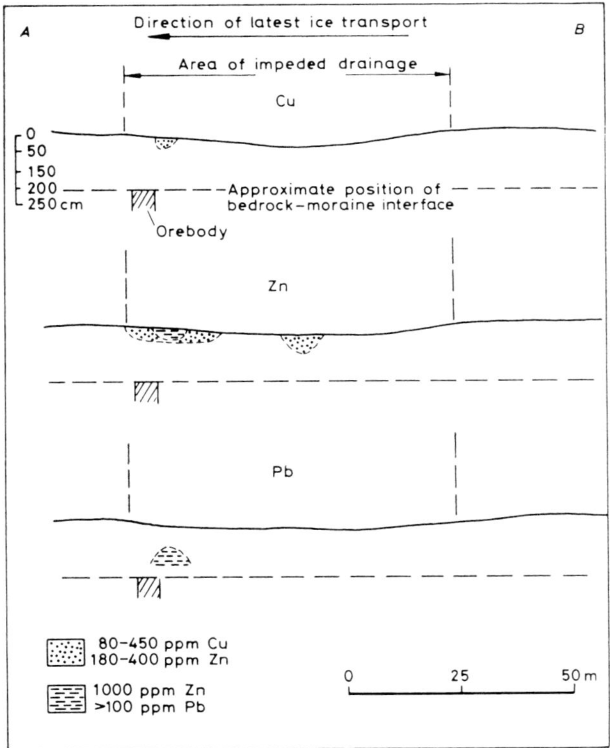

Fig. 11 Distribution of anomalous Cu, Zn and Pb in overburden, Nordre Gjeitteryggen

the moraine. The base-metal dispersion patterns (Figs. 10 and 11) are similar to Tverrefjellet in the major respect of the very much greater extent of the anomalous cxCu values compared with those for Cu, Zn or Pb. In addition, it is a matter of observation that the cxCu anomaly is related to the groundwater being dispersed downslope in a direction opposite to that of the latest glacial ice. The cxCu values over the ore increase upwards, however, to a maximum in the organic-rich surface layer, which is also the location of the very restricted anomalous Cu and Zn values. These features were not observed at Tverrefjellet. The accumulation of metal in the organic-rich horizon may be due to biochemical processes. In this connexion it is important to note that the water-table is very near surface, which makes it possible for the shallow-rooted vegetation to tap metal-rich groundwater. The absence of anomalous Pb values from the surface layer of the overburden above the ore tends to confirm that groundwater is the medium by which the ore elements are dispersed.

Grimsdal

At Grimsdal, which is above the tree line, an extension of the massive sulphide zone sub-outcrops beneath a small bog, which drains to the north (Fig. 1). pH measurements of the peat along a north–south transect indicate a trend towards a low pH (pH <4·0) in the subsurface material in the vicinity and downslope of mineralization, whereas in the surface layers over the entire transect and in the subsurface up-drainage of the ore pH values are generally greater than 4·0 (Figs. 12 and 13). It is thought

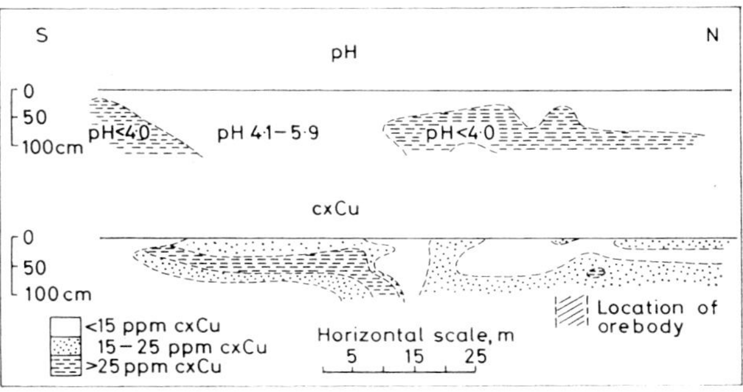

Fig. 12 pH and cxCu peat values, Grimsdal bog

to be significant that the zone of relatively higher pH values contains the strongly anomalous cxCu and Cu values, whereas in the low pH material the metal content is substantially lower. In

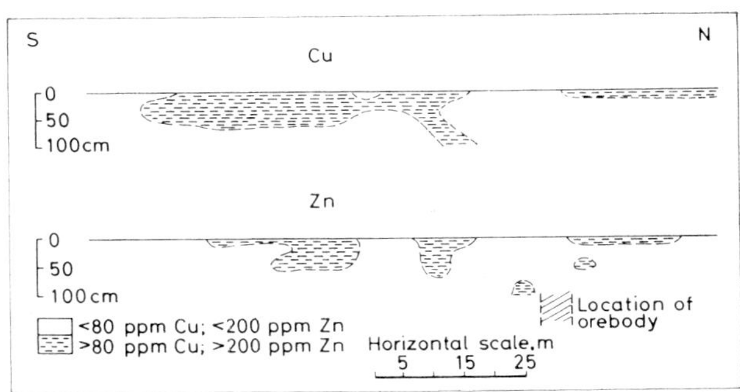

Fig. 13 Distribution of anomalous Cu and Zn in peat, Grimsdal bog

addition, the anomalous Zn values (i.e. greater than 200 ppm, as indicated by a bog study reported elsewhere[5]) are confined to small erratic zones in the surface peat where the pH is at a maximum value for the bog. The peak cxCu and Cu values occur in a flat-lying zone at a depth of 50 cm below the surface up-drainage of the ore. Below the strongly anomalous zone the bog material has a pH of 4·1–5·9, but it contains much lower metal values. The reason for this is not understood.

These data imply that, at least in this area, the metal content of the peat is closely related to the grade of the ore, the distance from the sub-outcrop and the pH of the peat. This would be consistent with the upward migration of mineralized groundwaters followed by fixation of the anomalous metal in zones of high pH.

Summary of results of Norwegian study

Geochemical dispersion from base-metal mineralization sub-outcropping beneath siliceous ground moraine takes place predominantly in solution in shallow groundwater following oxidation of the ore. The assemblage of anomalous ore elements developed within the overburden generally does not include Pb, but contains the loosely bonded form of Cu as one of the more prominent components. This is considered to be a reflection of the relative solubilities of the ore elements (and their secondary mineral species) under the conditions of pH and Eh prevailing in the groundwater.

In well-drained areas above the tree line (i.e. Tverrefjellet) metal anomalies occur only in the basal 50–100 cm of the overburden in the down-drainage vicinity of the ore, whereas metal anomalies in the surface layers of the overburden develop only at seepage sites receiving mineralized groundwater, and these may be well removed from the bedrock metal source. In localities where ore oxidation is sufficiently active, the anomalous seepage areas may be indicated by an accumulation of secondary iron oxides.

Displaced anomalies were also observed in areas under coniferous forest cover, but here the displacement downslope from the ore sub-outcrop is less than at Tverrefjellet, which may be attributable to the capability of the vegetation to tap metal-rich groundwaters. Surface metal anomalies that directly overlie sub-exposed mineralization were found to be restricted to areas of impeded drainage and to bogs. Under these conditions the strongest metal anomalies can occur in the top organic horizon—probably as a result of biochemical enrichment as well as by sorption of metal on to organic material as a consequence of a very near-surface ground water-table.

North Wales and Central Interior of British Columbia

The observations and conclusions on the secondary dispersion mechanisms in the glaciated terrain of south-central Norway have been found to apply to other similar environments. Brief details of such regions are given here, together with some of the implications of the results obtained to the planning of geochemical reconnaissance and follow-up programmes.

In *North Wales*, within an area of moderate relief and relatively high annual rainfall (approximately 2000 mm) low-grade chalcopyrite mineralization was discovered by Rio Tinto Finance and Exploration, Ltd., within altered and fractured hornblende quartz diorite. The mineralized intrusive is probably Lower Ordovician in age and is intruded into Upper Cambrian greywackes and argillites. The overburden is dominantly grey ground moraine deposited during the last Quaternary glaciation by ice which moved towards the south. The moraine has an average thickness of 6 m, and is overlain by a poorly developed soil 15 cm thick.

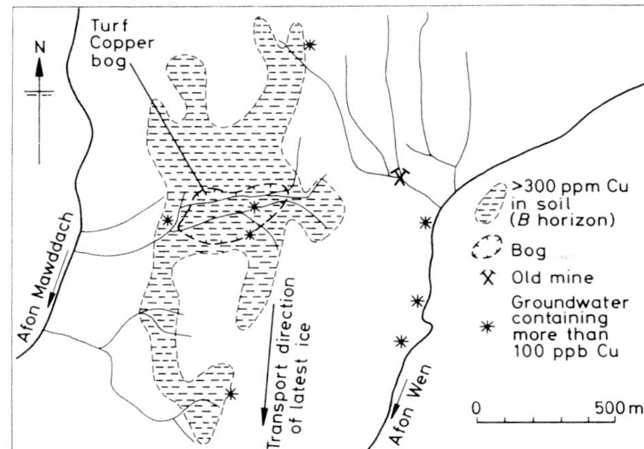

Fig. 14 Distribution of anomalous Cu in B horizon soils and near-surface groundwaters in vicinity of 'Turf Copper bog', North Wales

The most striking evidence of mineralization in the area is a small bog, containing up to 2% Cu in the dry peat, known locally as 'Turf Copper'. A broad soil anomaly (Fig. 14) elongated north–south (in the latest ice transport direction) surrounds the Cu-rich bog. Outcrops of weakly mineralized intrusive and sedimentary rocks occur within the soil anomaly, together with weakly mineralized rock fragments. The mineralized rock fragments appear to have been transported no great distance (i.e. tens rather than hundreds of metres) and to have been derived from bedrock within the northern part of the anomaly.

These observations suggest that the metal soil anomaly is developed from the mineralized rock fragments it contains and is related to the underlying weakly mineralized bedrock. The Cu contents of spring and seepage waters, however, indicate that substantial quantities of Cu are being dispersed in solution in groundwater. Moreover, the distribution of strongly anomalous spring and seepage water (Fig. 14) indicates the existence of a concealed mineralized bedrock zone somewhere to the east of 'Turf Copper'.

The concealed mineralization was eventually discovered by a drilling programme, which also revealed the existence of a gravel-filled stream channel, 30 m deep, buried beneath 6 m of ground moraine. This old stream course underlies the 'Turf Copper' bog (as indicated by isopachytes for the overburden, Fig. 15) and must be a channelway for the metal-rich groundwaters, which emerge within the bog after passing through the mineralized ground. The drilling also showed that in the critical area the bedrock 'watershed' is located 200 m east of the surface watershed (Fig. 15).

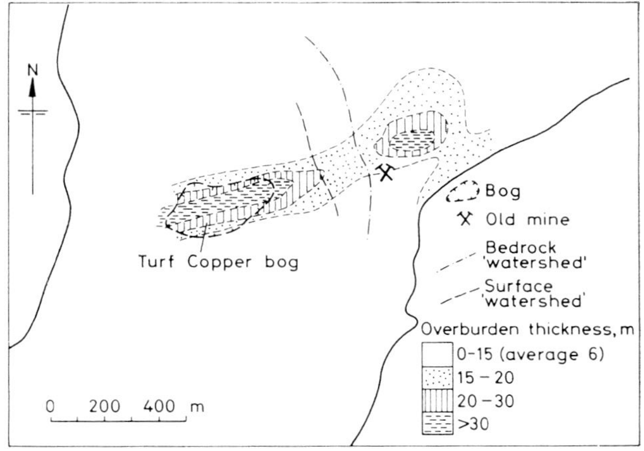

Fig. 15 Thickness of overburden in vicinity of 'Turf Copper bog', North Wales

The evidence from the drilling explains the displacement of the metal soil anomaly from the significantly mineralized bedrock zone as being due, in part at least, to subsurface flow of metal-rich water along a course that would not normally have been suspected.

Geochemical characteristics similar to those of the previously mentioned areas are also typical of the *Central Interior of British Columbia*, where the relief is generally subdued and the precipitation moderate (i.e. approximately 500 mm per year). Metal anomalies are frequently well displaced from their bedrock source and are developed in

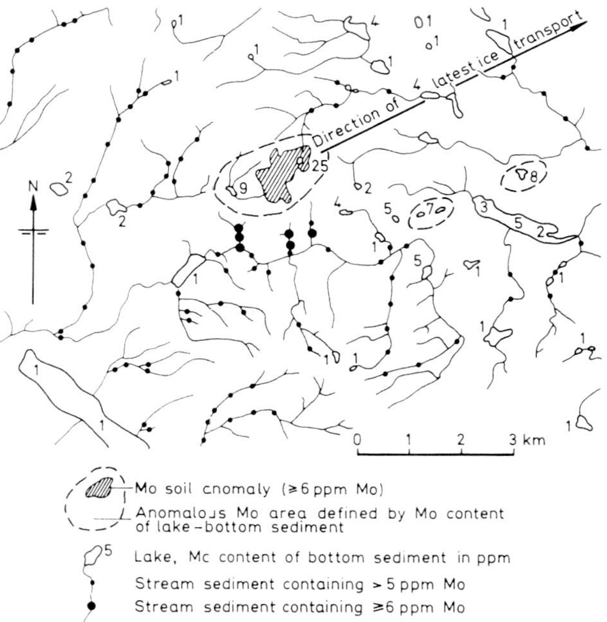

Fig. 16 Drainage and soil geochemistry of an area within Central Interior of British Columbia

bogs, swamps and in the components of the drainage, except in areas where the bedrock mineralization occurs in low ground.

Some of the characteristics mentioned above are illustrated in Fig. 16, which shows the results of a geochemical study carried out by Rio Tinto Canadian Exploration, Ltd. The area shown was investigated by stream-sediment sampling, lake-bottom sediment sampling and *B* horizon soil sampling. All three media were of use in defining an area of molybdeniferous bedrock later shown to carry up to 0·06% Mo.

An extensive Mo anomaly developed in the *B* soil horizon is broadly coincident with the bedrock source. The soil anomaly has a tendency to be elongated in the transport direction of the latest glacial ice, but the degree of transportation is probably quite small. The drainage anomalies, however, first appear over barren granodiorite at a distance of 800 m south and downslope of the mineral occurrence, which is the direction opposite to that of the latest ice transport.

These data indicate that dispersion is dominantly saline and illustrate the degree of separation that can occur between a bedrock metal source and its associated drainage anomalies. They further imply that in a region such as the Central Interior, in which the frequency of lake distribution approximates to 1 per 10 km^2, the sampling of lake-bottom sediment may provide an ideal geochemical reconnaissance medium. In particular, the organic-rich ooze commonly found in the central parts of these lakes may be expected to be an excellent geochemical trap for metals entering the lake environment via groundwater seepage. The results of lake-bottom sampling in the area mentioned above (Fig. 16) demonstrate that variations in the trace-metal content of organic-rich lake sediments can disclose the presence of bedrock mineralization. In the example quoted detailed stream sampling of the tributary drainage would have indicated the presence of mineralization, but at a higher cost.

Practical implications

The following observations refer specifically to glaciated districts similar to those described in this paper.
(1) The dispersion patterns are controlled to a very important degree by groundwater movements.
(2) The mineralized groundwaters are thought to be generated at or near the interface between moraine and the mineralized bedrock and to travel at or just below the interface.
(3) The assemblages of anomalous ore elements developed within the overburden from base-metal sulphide deposits generally exclude Pb, but contain the weakly bonded form of Cu as one of the more prominent components.
(4) In well-drained overburden metal anomalies develop in the surface layers of the overburden only at seepage sites that receive metal-rich groundwaters, and these may be well displaced from the bedrock metal source.
(5) Where mineralized bedrock is overlain by poorly drained overburden, a metal anomaly is commonly found in the surface soils directly above the mineralization. It is, however, rarely possible to determine whether a metal-rich bog overlies or is remote from mineralization without relatively extensive investigations.

Recommendations on prospecting procedures and interpretation can be summarized as follows.
(1) Reconnaissance sampling should be directed to obtaining sediment from seepage areas on a broad regional scale. In the Central Interior of British Columbia lake-bottom sediments are ideal, whereas in some other areas stream sediments and bank material should be collected.
(2) In following up anomalous areas detected by a reconnaissance survey, all seepages, bogs and other localities of impeded drainage should be investigated in some detail, particular attention being given to obtaining systematic profile samples of the overburden at each site.
(3) At each seepage the pH and metal content of the water (mainly Cu, Zn and Mo) should be determined by use of *in situ* measurements, where these are possible.
(4) Following (2) and (3), systematic soil sampling grids can be planned.
(5) In all interpretations consideration should be given to the influence of topography, etc., on groundwater movements, it being borne in mind that groundwater may not always drain in the same direction as the surface water in any given locality.

Acknowledgment

During the course of a geochemical study in Central Norway (1958–59) the Norges Geologiske Undersøkelse (NGU) demonstrated the 'anomalous' character of the drainage in the vicinity of the Tverrefjellet deposit.[1] Following this early work, a research project aimed at obtaining a better understanding of geochemical dispersion processes in the glaciated terrain near Hjerkinn was initiated by NGU in collaboration with Professor J. S. Webb of the Applied Geochemistry Research Group at the Royal School of Mines, London. The research was financed jointly by NGU and the Department of Scientific and Industrial Research (United Kingdom).

The authors gratefully acknowledge the financial assistance which made the research possible and express their appreciation of the technical help and guidance received from NGU geochemists—in particular, Dr. A. Kvalheim, B. Bølviken and R. Krog, and the assistance afforded by their colleagues at the Royal School of Mines, London.

They also wish to thank Folldal Verk A/s, Rio Tinto Finance and Exploration, Ltd., and Rio Tinto Canadian Exploration, Ltd., for permission to publish data on the various areas, and acknowledge contributions from employees of these companies—in particular, G. J. Sharp of Rio Tinto Finance and Exploration, Ltd.

References

1. Bølviken, B. Recent geochemical prospecting in Norway. In *Geochemical prospecting in Fennoscandia* Kvalheim, A. ed. (New York: Wiley, 1967), 225–53.
2. Craven, C. A. U. Statistical estimation of the accuracy of assaying. *Trans. Instn Min. Metall.*, **63**, Sept. 1954, 551–63.
3. Holmsen, P. Quaternary geology of the Hjerkinn area, central Norway. *Unpubl. Rep. Norges geol. Unders.*, 1963.
4. Lounamaa, J. Trace elements in trees and shrubs growing on different rocks in Finland. In *Geochemical prospecting in Fennoscandia* Kvalheim, A. ed. (New York: Wiley, 1967), 287–317.
5. Mehrtens, M. B. Geochemical dispersion from base-metal mineralisation, central Norway. Ph.D. thesis, University of London, 1966.
6. Mehrtens, M. B. and Tooms, J. S. Geochemical drainage dispersion from sulphide mineralization in glaciated terrain (central Norway). In *Prospecting in*

areas of glacial terrain (London: IMM, 1973), in press.
7. STRAND, T. and HENNINGSMOEN, G. Cambrio–Silurian stratigraphy. In *Geology of Norway* HOLTEDAHL, O. ed. *Norges geol. Unders.* no. 208, 1960, 128–284.
8. WALTHAM, A. C. Classification and genesis of some massive sulphide deposits in Norway. *Trans. Instn Min. Metall. (Sect. B: Appl. earth sci.)*, **77**, 1968, B153–61.
9. Determination of lead in soil, sediment and rock samples. Abbreviated operating instructions. *Tech. Commun. appl. Geochem. Res. Group, Imperial College, Lond.* no. 16, 1962, 2 p.
10. Determination of copper with 2,2′-diquinolyl in soil and sediment samples. *Tech. Commun. appl. Geochem. Res. Group, Imperial College, Lond.* no. 23, 1962, 2 p.
11. Determination of zinc in soil, sediment and rock samples. Abbreviated operating instructions. *Tech. Commun. appl. Geochem. Res. Group, Imperial College, Lond.* no. 30, 1962, 2 p.
12. BRAY, R. H. and WILLHITE, F. M. Determination of total replaceable bases in soils. *Ind. Engng Chem. (analyt. Edn)*, **1**, 1929, 144.
13. PIPER, C. S. *Soil and plant analysis: a laboratory manual of methods by the examination of soils and the determination of the inorganic constituents of plants* (Adelaide: The University, 1942), 368 p.
14. Determination of sulphate in soil and sediment samples. Abbreviated operating instructions. *Tech. Commun. appl. Geochem. Res. Group, Imperial College, Lond.* no. 33, 1962, 2 p.
15. HOLTEDAHL, O. ed. Geology of Norway. *Norg. geol. Unders.* no. 208, 1960, 540 p.

550.84:553(411)

Detection of concealed mineralization in northern Scotland

U. McL. Michie, B.SC.

M. J. Gallagher, PH.D., M.I.M.M.

A. Simpson, B.SC.

All of the Institute of Geological Sciences, London, England

Synopsis

In the search for uranium in northern Scotland geochemical sampling of stream sediment, water and soil, supplemented by scintillation counter surveys, the use of gamma and radon probes and mapping of mineralized boulders, has led to the identification of uranium and other mineralization in Devonian sediments and Caledonian granites. Over much of the region there is 2 m or more of overburden comprising peat, soil, glacial till and locally derived 'head' material. Although geochemical patterns in peat and soil overlying mineralization are subject to modification by secondary factors, metal enrichments in a peat bog have provided a useful guide to the presence of mineralization in nearby granite. Concentrations of uranium in stream water and sediment and of lead in sediment and peaty soil have also served to outline mineralized sites in granite.

From a study of the metal content of peaty soil, the gamma radiation at surface and in depth and the radon content of soil air a coherent picture has been obtained of metal dispersion from mineralized Devonian arkose concealed by a few metres of overburden. Surface gamma anomalies are displaced 100 m or more downslope from uraniferous till, and a similar dispersion is shown by uranium, copper, lead and zinc in the soil. The distribution pattern of radon in overburden appears to be displaced downslope from the mineralization by some tens of metres, extending to more than 100 m. Most effective in this environment have been close-interval gamma-probe surveys to determine radioactivity at or near the base of the peat. In more flat-lying areas of thin soil cover overlying uraniferous disseminations and structures in Middle Devonian sediments the radon and gamma-probe patterns largely coincide, serving to delineate and extend sub-outcropping uranium mineralization more effectively than the pattern of gamma radiation at surface.

Between 1968 and 1970 small field parties of the Geochemical Division, Institute of Geological Sciences, were engaged in a reconnaissance for uranium in the Northern Highlands of Scotland, especially in eastern Sutherland and Caithness, where numerous new occurrences of uranium and other mineralization were soon discovered.[1] Because of the pronounced association of uranium with Old Red Sandstone (Devonian) sediments in Caithness the investigations were extended in 1971 to the Orkney Islands, where significant uranium enrichments were found.[2] The rocks of eastern Sutherland, Caithness and to a lesser extent Orkney are generally buried beneath 'head' (locally derived material), glacial moraine or fluvioglacial material and covered by peat or soil; peat is especially widespread in southwest Caithness as well as in upland districts of granitic and metamorphic rocks. This paper describes some of the problems encountered in exploring for uranium and other metals and in assessing and extending radiometric and geochemical anomalies in this region of scarce rock outcrop.

The general geology of the region explored is shown in Fig. 1, together with the locations of the mineral prospects referred to in the paper. Of the six principal localities investigated, four occur in ORS sediments and two in Caledonian granitic masses. It can also be seen that most other

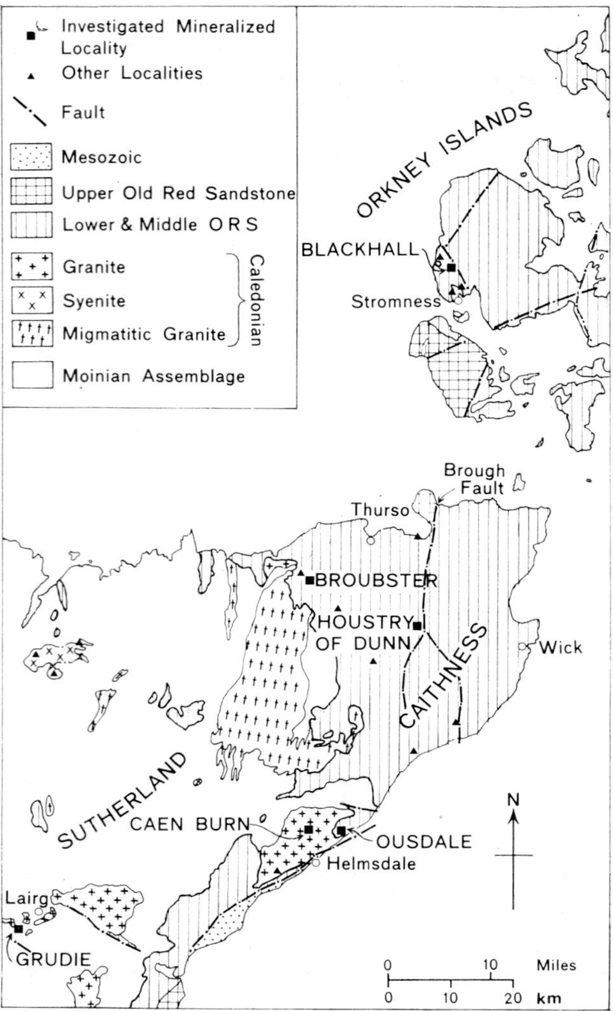

Fig. 1 General geology of part of north of Scotland showing investigated localities and other mineralized sites of note

mineralized localities of note are associated with the ORS area of Caithness and Orkney. Primary reconnaissance of the region was based on vehicle-mounted scintillation counter surveys and on-foot radiometric traversing of coastal and stream sections, the multi-element analysis of −100-mesh (150-μm) stream-sediment samples[3] and hydrogeochemical sampling for uranium. Stream water sampling was particularly effective in Caithness and Orkney, where the ORS-derived waters are enriched in bicarbonate. The distribution of uranium in the stream waters of Caithness is illustrated elsewhere (reference 4, Fig. 5) as are the uranium anomaly patterns in stream sediment and water related to Caledonian granite at Helmsdale (reference 1, Fig. 8).

The region has a cool temperate climate, the annual sea-level temperature variation being from 0 to 2°C in winter to 13 to 15°C in summer. Annual rainfall is generally 800–1000 mm, rising to more than 1000 mm on the higher ground and also to the west of the region. Heather is the dominant vegetation, but an Arctic–Alpine flora is found on the higher hills. During the period of maximum glaciation shelly till derived from the Moray Firth moved northwest across the ORS platform to fill preglacial hollows and partly overlie and mix with earlier local till. Although generally only 1–3 m in depth, it may attain 30 m. Shell fragments have been destroyed by leaching to depths of 2 m. This dark grey-blue clayey till is markedly different in character from the local red-brown stony till, and in Caithness both may be covered by later moraines and gravels derived from the local Sutherland ice cap. The natural dominant surface deposit is blanket peat formed during the post-glacial period, but pastoral and agricultural activity has produced loamy brown forest soils. Peat is generally more than 1 m deep in southern Caithness and Sutherland, reaching a maximum of about 8 m. In western Caithness, at Altnabreac, a peat bog covers 8500 hectares to an average depth of 2·2 m.[5] Although the peat may occur unbroken over both till and sandy drift, the lower layers being compressed and quite dry with surface peat drainage, over sandy drift the peat usually cracks, permitting the development of sub-peat drainage. The underlying drift is intensely leached by the acid waters and the lower layers of the peat are sodden. Deep peat in hollows may be moist throughout, the lower layers being completely saturated. In this condition the peat can reflect the bedrock geochemistry, except in cases where clay-rich lacustrine sediments form an impermeable barrier.

Some of the most promising radiometric and geochemical anomalies were assessed by soil sampling and by the use of gamma and radon probes prior to trenching and sampling drilling. The principles underlying the determination of the alpha activity of ^{222}Rn in soil air by the use of recently developed radon probes have been described by Miller and Loosemore.[6] In practice, the alpha-activity measurement is made immediately after withdrawal of the hole-making tool (a sliding-hammer steel bar, see Fig. 6) designed for this purpose, and is followed by measurement of the gross gamma activity near the bottom of the hole with a probe having a 25 mm × 25 mm NaI(Tl) detector mounted 18 mm from its end.

Mineralization in Caledonian granite

Traces of mineralization are associated with many of the Caledonian granitic masses of Scotland, including those of Helmsdale and Grudie. K–Ar ages of 401 ± 14 m.y. and 397 ± 14 m.y. have been obtained on the Helmsdale granite,[7] and R. Harding (personal communication) has recently obtained a provisional Rb–Sr age of 380±170 m.y. on the Grudie granite, the large error being attributed to the weathered character of the surface samples. These ages support the classification of these granites as members of the Newer Caledonian group.[8]

The nature of the disseminated pyrite–fluorite–molybdenite mineralization in the Grudie granite and of fracture-bound galena–fluorite–baryte veins located chiefly in metamorphic rocks (Moinian) at the southeast margin of the granite boss has already been described,[1,9] and attention drawn to the high metal values present in peat and glacial materials downslope from the exposed mineralization. Anomalous values of uranium, molybdenum, lead and other metals have also been recorded in sediment samples from streams draining the granite and associated vein mineralization.[3] Although the soils overlying the granite contact and adjacent vein mineralization are heavily enriched in metal, particular interest lies in the presence of up to 200 ppm U, 100 ppm Mo and 2000 ppm Pb in the dried residues of peat and peaty soil from a bog southeast of the granite. The distribution patterns of molybdenum, lead, zinc and copper are in broad agreement with that of uranium and are attributed to the fixation of metals on peat from waters draining the granite and vein mineralization upslope. Anomalous metal values persist for a distance of approximately 400 m from the granite contact and exposed veins, so the conclusion can be drawn that reconnaissance sampling of peat bogs in similar environments elsewhere should be a useful method of exploration.

Previous work on the Helmsdale pluton[1,3]

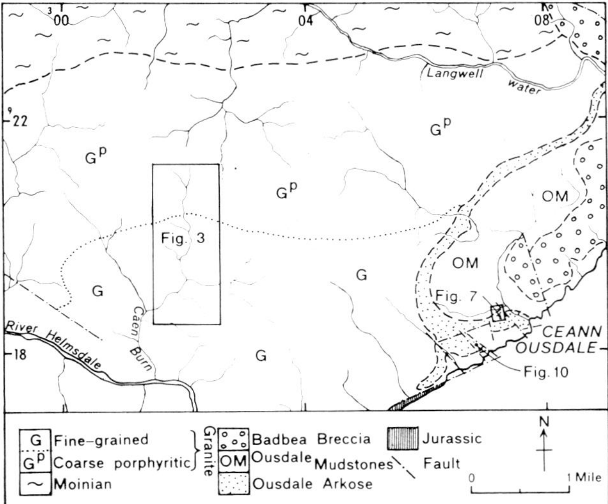

Fig. 2 Sketch map of geology of eastern Sutherland and southern Caithness. Investigated locality in drainage basin of Caen burn (Fig. 3 inset) lies within Helmsdale granite, and two other localities (Figs. 7 and 10) are associated with Lower ORS sediments

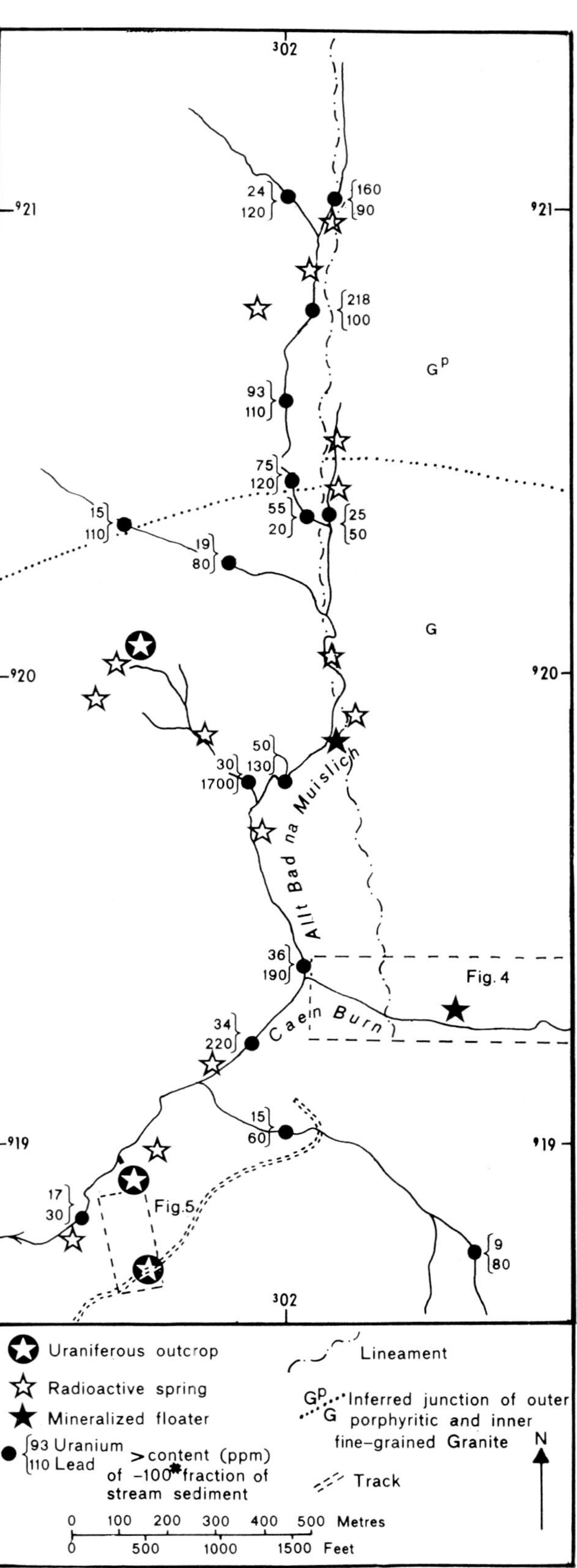

Fig. 3 Distribution of uranium and lead in −150-μm stream sediment in upper drainage basin of Caen burn, near Helmsdale, Sutherland (located in Fig. 2)

located several small uranium occurrences in the drainage basins on Caen burn (Fig. 2) and, to the north, a large area of weakly uraniferous (10–40 ppm U) and deeply disaggregated granite which may be related, in part at least, to the chemical weathering during the last interglacial noted by Synge[10] and Phemister and Simpson.[11] In this district outcrop is confined to incised stream channels and isolated knolls. Humus iron podzols in valley bottoms pass up through peaty podzols with iron pan on valley sides to blanket peat on the high level moorland (*ca* 300 m elevation) carrying *Calluna, Trichophorum* and *Eriophorum.* The valleys are presently being exhumed from thick glacial drift transported from the northwest. Highly acid radioactive springs are common on the valley floors and since their waters contain up to 150 μg U/l they constitute a major influx of uranium into the streams.

The northernmost tributary of Caen burn cuts the apparently transitional boundary between the outer coarse porphyritic member and the inner fine-grained member of the adamellite mass, and in places follows a physiographic lineament, probably a fault (Fig. 3). From south to north up this tributary there is a substantial increase in the uranium content of stream sediment (Fig. 3), which is reflected in the stream water values—although the highest of these (45 μg U/l) occurs 400 m downstream from the maximum sediment value. The distribution of lead in the same drainage is erratic, one very high value occurring downstream from an isolated outcrop of silicified granite containing 0·3% Pb and 100 ppm U. Both lead and uranium occur at anomalous levels within the area of Fig. 4, with cutoff peaks 30 and 100 m, respectively, downstream from the extrapolated line of the structure shown. These peak values in stream sediment correspond to a broad zone of lead and uranium anomalies in *B/C* horizon soils collected beneath peat cover on the north side of Caen burn (Fig. 4). The very high uranium content of the drainage illustrated in Fig. 3 can, therefore, be related in part to the inferred presence of a mineralized structure some 2 km in length and in part to the presence of a large-volume source of uranium represented by weakly uraniferous and altered granite occurring in the north of the area (cf. Szalay and Sámsoni[12]).

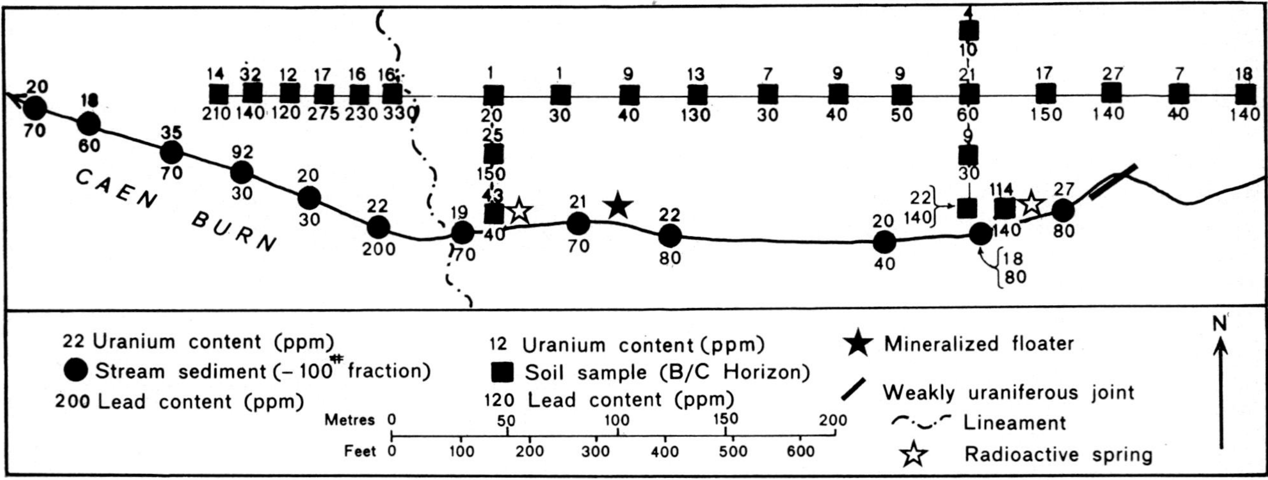

Fig. 4 Distribution of uranium and lead in —150-μm stream sediment and in soils from section of Caen burn, near Helmsdale, Sutherland (located in Fig. 3)

A pronounced association of uranium with the light organic fraction was found on partial separation of the stream-sediment sample of maximum uranium content:

Size fraction, μm	sp. gr.	ppm U
−150 (standard sediment)	—	218
63–125	<2·5	*1410*
	2·5–2·9	35
	>2·9	350
−63	—	740

The light fraction consists almost entirely of weathered peat containing uranium markedly in excess of its daughter products (U/*e*U *ca* 3). Enhancement of uranium in the organic fraction of stream sediment has also been noted by Dall'Aglio,[13] who found that entirely spurious anomalies were produced, and by Dyck *et al.*,[14] who observed that 'organic matter in sediments is no hindrance in uranium exploration'. Our own evidence suggests that uranium adsorbed on the organic fraction of stream sediment obscures the detailed pattern of uranium dispersion; but, because of the enhanced contrast and longer dispersion train it provides, this feature is advantageous in surveys on a regional scale.

In a further investigation of concealed uranium mineralization in the Caen burn district radon

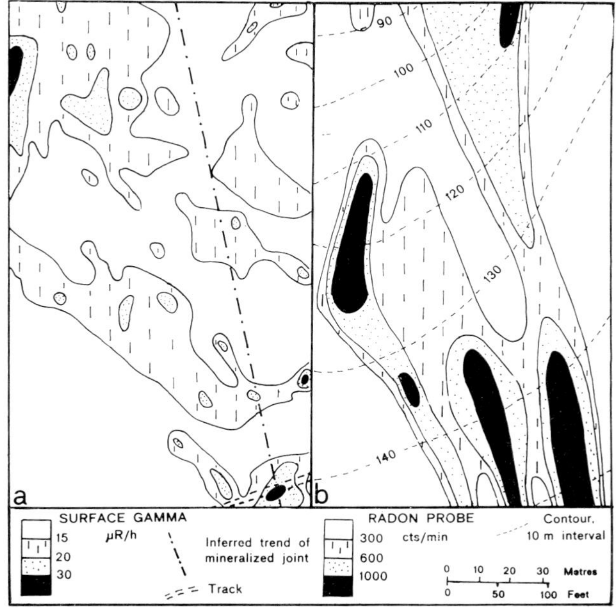

Fig. 5 Surface gamma radiation (a) and radon-probe measurements (b) over trend of mineralized structure extrapolated from its outcrop in Caen burn, near Helmsdale, Sutherland (located in Fig. 3)

probe profiles were made across the projected trend of a minor mineralized structure in ground obscured by peat and till. The uphill cutoff to the pronounced radon anomalies shown in Fig. 5*(b)* defines a well-marked linear feature, which is interpreted as the strike extension of the uraniferous structure. The overall pattern suggests that downslope dispersion and reconcentration of radon has taken place and the ill-defined surface radiometric pattern (Fig. 5*(a)*, based on work by Dr. L. Haynes) can be interpreted on the same basis, representing as it does the accumulation of gamma-emitting decay products.

Mineralization in Devonian arkose

The eastern margin of the Helmsdale granite is overlain by arkose of Lower ORS (Devonian) age, which passes upwards into mudstones (Figs. 1 and 2). Following the discovery of uranium mineralization in stream and coastal exposures near the eastern limit of this arkose,[1] a combina-

Fig. 6 Prototype radon probe in use to detect concealed uranium mineralization in Lower ORS arkose at Ousdale, southern Caithness. Photograph taken looking north from centre of area depicted in Fig. 7 with broch (Pictish ruin) on left

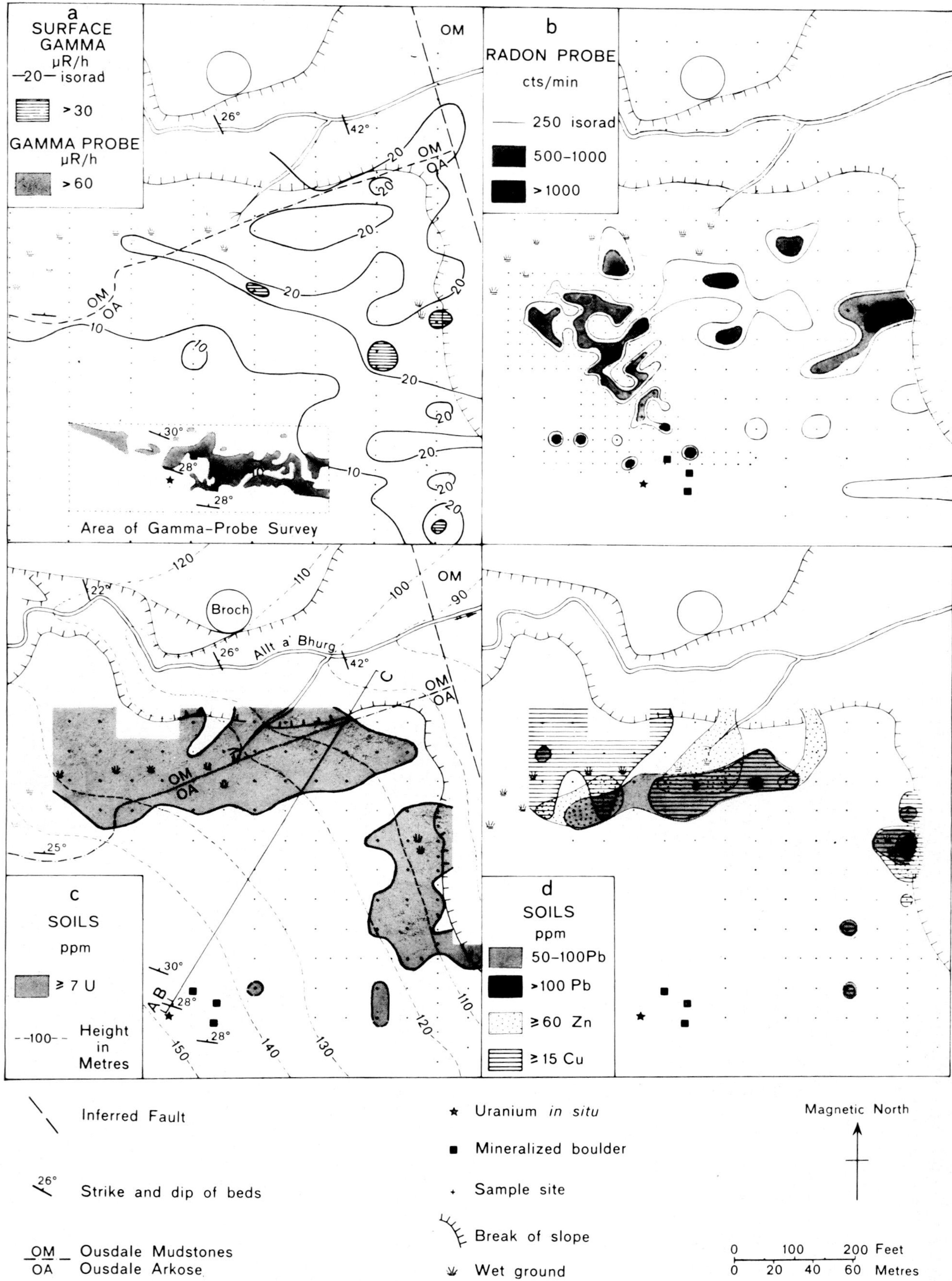

Fig. 7 Surface gamma radiation and gamma-probe measurements (a), radon-probe measurements (b) and distribution of uranium (c), lead, zinc and copper (d) in soils associated with uranium mineralization in Lower ORS arkose underlying mudstones at Ousdale, Caithness (located in Fig. 2)

tion of soil-sampling and radiometric techniques was used to extend the mineralization westwards for more than a kilometre in ground generally obscured by 2 m or more of peat and till.

The nature of the terrain is illustrated in Fig. 6, which also shows a prototype of the radon probe employed. The terrace feature in the foreground of the photograph represents the northern half of the site depicted in Fig. 7, located some 300 m west of the discovery outcrops. Weak surface gamma anomalies at this site were initially investigated by soil sampling. The uranium distribution in poorly developed *B* horizon peaty soils corresponds broadly with the surface radiometry, although the main anomaly lies a little downslope from the principal arcuate gamma anomaly (cf. Fig. 7*(a)* and *(c)*). Weak copper, lead and zinc anomalies in soil (Fig. 7*(d)*) are located within the zone of anomalous uranium concentration, which probably reflects no more than a favourable environment for metal accumulation at sites of groundwater discharge and organic enrichment. This view is reinforced by the strong concentration of radioactivity in a decomposed organic horizon within the soil profile above weathered till. The soil sampling was followed up by radon-probe measurements on a 15-m grid spacing which defined broad anomalies coinciding, for the most part, with surface gamma anomalies, but a few scattered high readings were also obtained upslope to the southwest.

On the results of the soil sampling and the preliminary radon survey, investigations were extended southwestwards through increasingly thick peat. Gamma probing on a 2- to 3-m spacing to near the base of the peat located blocks of uraniferous arkose within radioactive till having no gamma expression at surface and describing a linear anomaly occurring nearly 100 m upslope from the main surface gamma anomaly (Fig. 7*(a)*). Close-interval radon-probe measurements revealed only isolated high readings within the area of the gamma probe anomaly, but a complex pattern of pronounced radon anomalies occurs in ground immediately to the northwest (Fig. 7*(b)*). Trenching on the gamma-probe maxima exposed a zone of anomalous radioactivity, perched within about 2 m of sandy till, which exhibits downslope solifluxion and rests on arkose containing only a small enrichment of uranium (Fig. 8). This comprises a zone of uraniferous boulders in the till and is regarded as having been displaced mechanically downhill from bedrock mineralization some 10–20 m southwest of point *A* in Fig. 7*(c)* in ground where trenching was inhibited by deeper cover.

The spatial distribution of the several types of anomaly is summarized in the section in Fig. 9.

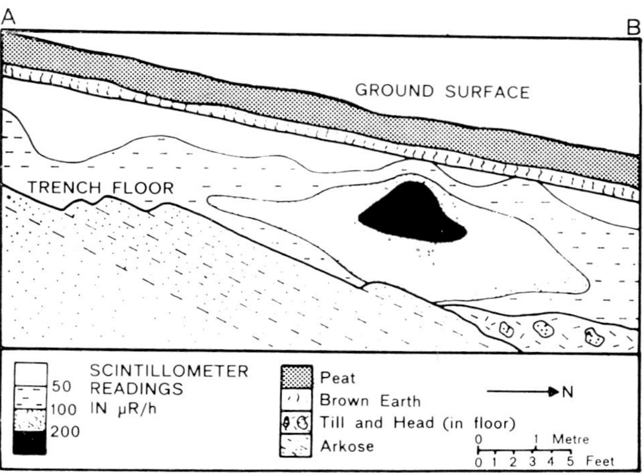

Fig. 8 Radiometric profile through till exposed in trench wall above weakly uraniferous Lower ORS arkose at Ousdale, Caithness (see Fig. 7(c) for location of trench A–B)

Although gamma-probing proved to be the most effective method of tracing concealed mineralization, attention was first drawn to the site by the detection of surface gamma anomalies. Recognition of downslope dispersion as the mechanism producing the geochemical soil and surface radiometric anomalies led to orientation of the close-interval gamma-probe work upslope beneath peat cover.

Attention was drawn to a second site in the Ousdale district (Fig. 10, locality shown in Fig. 2) by the discovery of a large boulder containing secondary uranium minerals in ground otherwise uniformly covered by thin peat and free of radiometric anomalies. Probe investigations demonstrated the presence of a pronounced sub-peat gamma anomaly some 30 m uphill from the boulder and a much more restricted radon anomaly (Fig. 10) was also defined. Trenching on the gamma-probe maxima through 1 m of dry cracked peat with some thin soil exposed shattered and leached arkose carrying uranium mineralization.[1] The presence of radioactive seepages more than 130 m downslope from the uranium locality and at 30 m lower elevation is related to groundwater leaching of this highly permeable arkose.

Mineralization in other Devonian sediments

In Caithness, reconnaissance exploration of the Devonian rocks by scintillation counter surveys and hydrogeochemical sampling[15] led to the discovery of numerous phosphatic and black shale horizons containing a maximum of 1500 ppm U over 30 cm and enrichments of Pb, Zn, Cu, Mo and Ag. These Middle ORS sediments consist of a calcareous and bituminous cyclic sequence of grey sandstones, siltstones, mudstones

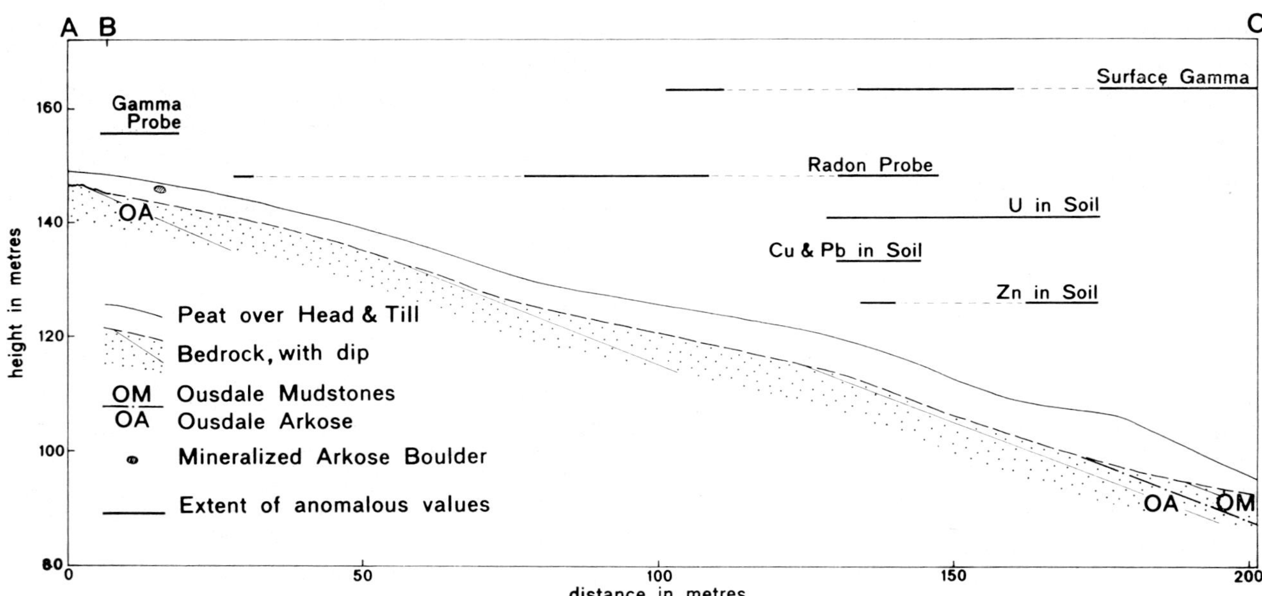

Fig. 9 Relationships between gamma-probe, radon-probe, surface gamma and soil anomalies in peaty soils covering head and till above Lower ORS sediments at Ousdale, Caithness (see Fig. 7(c) for location of section ABC)

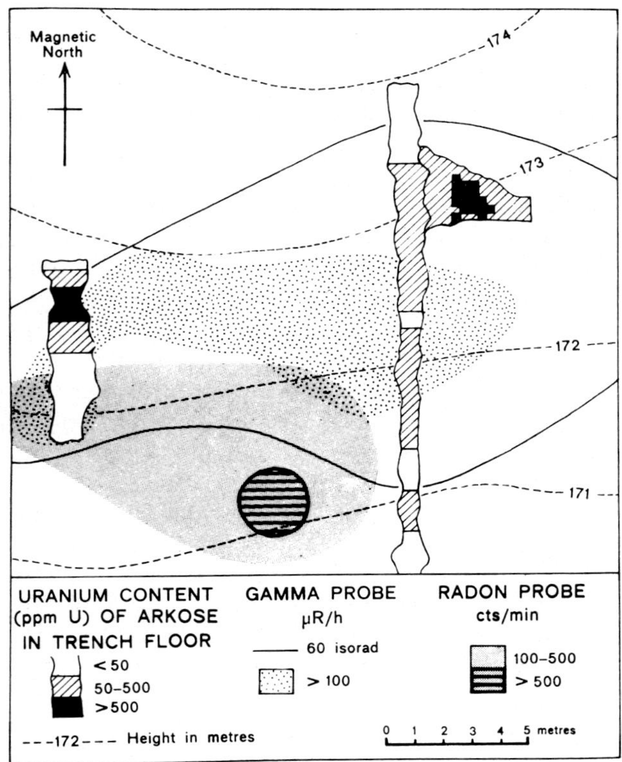

Fig. 10 Radon- and gamma-probe anomalies associated with uranium mineralization exposed by trenching beneath peat in Lower ORS arkose at Ousdale, Caithness (see Fig. 2 for location)

and shales. Fracture-controlled uranium mineralization also occurring in the Middle ORS probably represents material remobilized from the penecontemporaneous enrichments, as does associated galena, sphalerite, chalcopyrite and molybdenite.

One group of radiometric and hydrogeochemical anomalies occurs at Houstry of Dunn in north-central Caithness in undulating farmland underlain by the Middle ORS Caithness Flags. The anomalies are located on either side of a major north–south fracture, known as the Brough fault, which post-dates Upper ORS rocks (see Fig. 1). Early work showed the uranium to be restricted to phosphatic siltstone and shale horizons.[1] Subsequent gamma- and radon-probe surveys, followed by trenching and sample drilling, point to the movement of uranium along faults and its concentration in phosphate and hydrocarbon along with lead and zinc.

An example of uraniferous bedrock giving rise to a distinct radon anomaly and a modest gamma-probe anomaly, yet having only a weak gamma expression at surface, is shown at the eastern end of profile *E* in Fig. 11. Other radon anomalies in profiles *E* and *F* remain to be investigated, but the well-defined peak in profile *D* is located in the clay fill of an old agricultural drain indicating the downslope dispersion of radioelements in groundwater. A similar origin is attributed to the radon anomaly in the central section of profile *C*. Results obtained in profile *A* suggest that uranium mineralization dies out to the north, and near the western limit of profiles *B* and *C* a clear association of radon- and gamma-probe anomalies with known uranium mineralization can be seen.

The patterns produced by close-interval probe measurements over one small area (Fig. 12) differ only in detail from those of the surface gamma radiation, but they exhibit greater contrast. There is a small uranium-in-soil anomaly displaced a

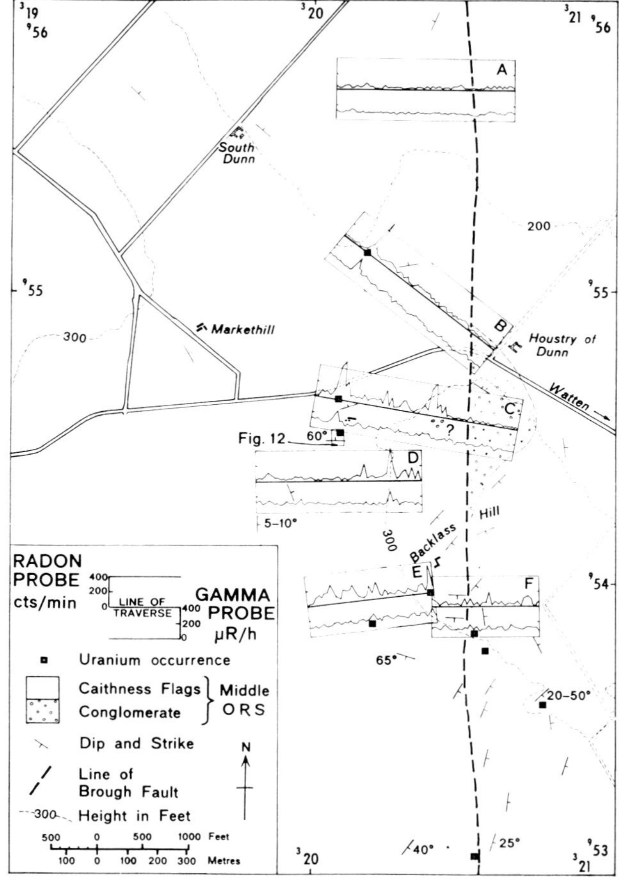

Fig. 11 Radon- and gamma-probe profiles across uranium occurrences in Middle ORS Caithness Flags adjacent to Brough fault at Houstry of Dunn, north-central Caithness (see Fig. 1 for location)

drained bedrock occurs beneath about 20 cm of soil. A sample from a prominent lenticle of radioactive 'clay' material in the steeply tilted Caithness Flags gave the following results:*

	%		ppm
P_2O_5	27·9	U	1190
Fe_2O_3	3·2	Zn	770
		Pb	550
		Cu	40
		Ag	5

Examination of this unusual material by I. R. Basham shows it to be composed largely of irregular aggregates of apatite. Some of the uranium, however, with appreciable amounts of thorium, is located in nodules of hydrocarbon up to 7 mm across. The characteristics of this occurrence suggest that phosphate and hydrocarbon, together with uranium, lead and zinc, were mobilized from horizons within the Caithness Flags which are known to be enriched in all of these constituents,[1] and were concentrated in structures associated with the Brough fault.

At the Broubster uranium occurrence in north-west Caithness (Fig. 1) a radiometric anomaly occurs over a small zone of brown earth soils underlain by a bituminous limestone of the Caithness Flags. The limestone is cut by calcite veins containing uraniferous hydrocarbon, galena and sphalerite,[1] and is the site of an old quarry. There is groundwater movement along the shallow excavation from west to east, where a waterlogged

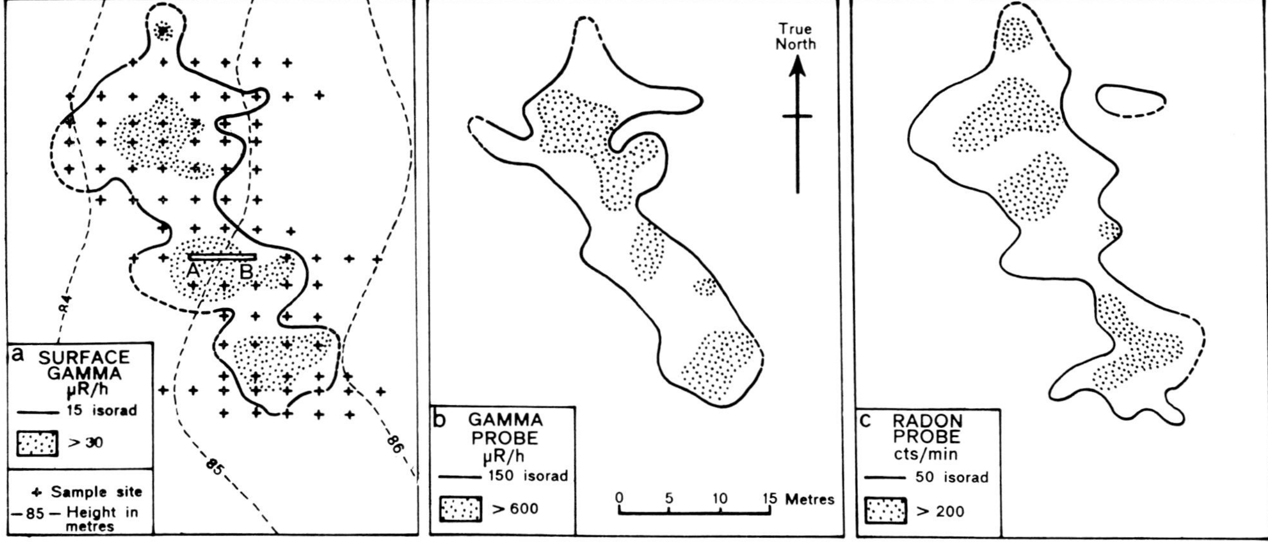

Fig. 12 Surface gamma radiation (a), gamma-probe measurements (b) and radon-probe measurements (c) over uraniferous mineralization in Middle ORS Caithness Flags beneath thin soil cover at Houstry of Dunn, Caithness (located in Fig. 11)

little downslope from the main concentration of uranium (480 ppm U over 1·8 m) in the lower wall of a trench at this site (Fig. 13), where well-

area of peaty gleys is developed. The level ground

*Analyst, D. Peachey.

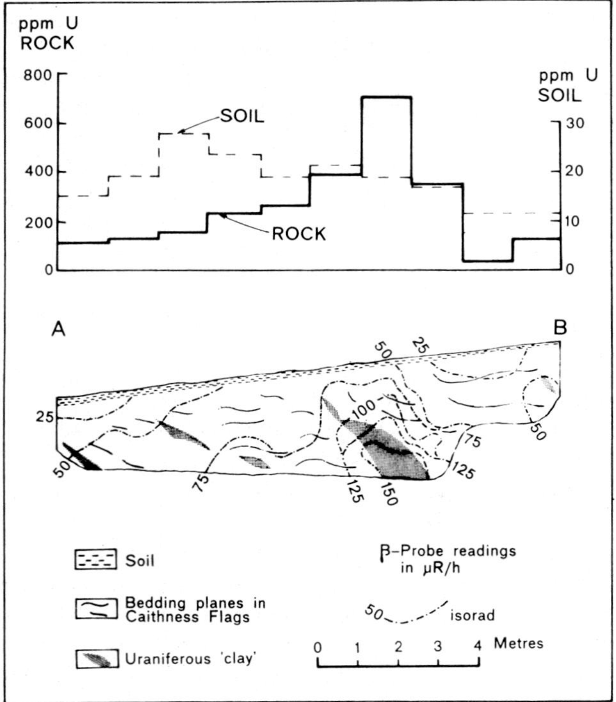

Fig. 13 Distribution of uranium in soil and rock overlying uraniferous phosphatic 'clay' lenticles in disturbed Caithness Flags at Houstry of Dunn, Caithness. Radiometric pattern produced by uraniferous zones shown in trench section (A–B in Fig. 12)

surrounding the excavation is obscured by 1 m or more of peat and till and its waterlogged condition at the time of investigation inhibited the use of radon probes. Preliminary work indicated that sampling of *A* horizon surface peat was effective only where the profile permitted upward metal migration. The metal content of surface peat is generally lower than in *B* or *C* horizons, particularly where surface drainage is developed over compacted peats, and local anomalies present in surficial peaty gley are usually displaced with respect to those in underlying till. To the west of the area the presence of secondary iron oxides as bog iron ore near the surface of waterlogged deep peat produces weak, but widespread, anomalies for copper, zinc and, to a lesser extent, uranium in the peats.

On the basis of sampling of the *B* and/or *C* horizon (Fig. 14), both uranium and zinc serve to outline the presumed vein direction, except in the east, where their dispersion has been modified by groundwater movement. Highest uranium values are found not in the brown earth soils west of the road, where surface radioactivity is at a maximum, but in peaty gleys east of the road, where there is probably an interface between bicarbonate-rich and bicarbonate-poor groundwaters at which uranium would be sorbed on to peat.[16] Similarly, erratic radiometric anomalies in surficial peat are usually developed at the boundaries between waterlogged peat and slightly elevated drier areas. Poor correlation between surface gamma activity and uranium in peat is not unexpected and can be explained by the differing geochemical mobilities of uranium and its gamma-emitting daughter products.[17,18] Lead is concentrated only in the reworked brown earth soils, where it is probably present in a clastic component.

Near Blackhall in Orkney (Fig. 1) surface gamma anomalies occurring at the foot of a steep hill are due to tabular enrichments of uranium within sandstone units near the local base of the ORS and are masked by up to 2 m of head. Pronounced radon anomalies coincide with the surface anomalies, but there is also a line of anomalies along the fault shown at the centre of Fig. 15. These anomalies have no associated gamma expression and are thought to be related to leakage from down-dip extensions of the mineralization. The radon pattern also shows that there is no continuity between the northern surface anomaly and the main anomaly.

The basal ORS sediments also exhibit minor surface gamma activity some 600 m to the west of the Blackhall occurrence at Sowa Dee (Fig. 16). Here cover is thicker, consisting of 30–60 cm of peat above 1–1·3 m of local till. Gamma-probe readings beneath the peat provide a more coherent pattern of radioactivity than the isolated surface expression and define a northwest–southeast trend corresponding to the strike of the underlying beds. This trend is also shown by the radon pattern, although it exhibits a greater degree of downslope dispersion. In this instance the close coincidence of the radon and gamma measurements implied the presence of uraniferous bedrock, and shallow trenching revealed uraniferous shaly horizons within the sequence.

Conclusions

The results of the work in northern Scotland on the detection and evaluation of concealed mineralization show that different techniques have to be employed for reconnaissance and assessment work. Generally, a combination of techniques is more effective than any one method in isolation. At the reconnaissance level, for uranium exploration, surface gamma measurements and high-density sampling of waters are the best approach. Apart from zinc, the limited solubility of many other ore metals precludes the wider application of hydrogeochemical sampling, and for these stream sediments were employed. Over the calcareous Caithness flagstones uranium is carried almost exclusively in solution, so stream sediments

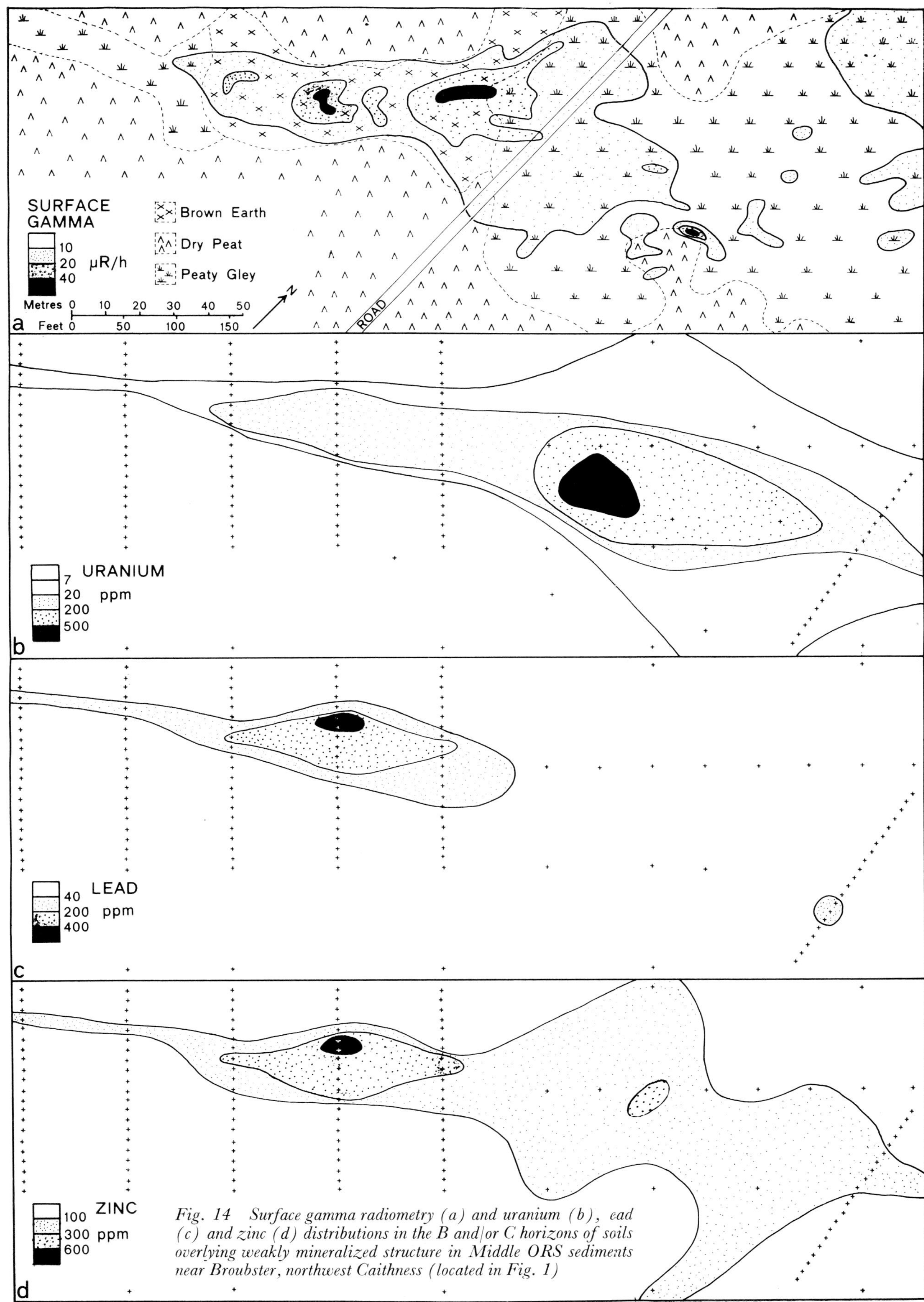

Fig. 14 Surface gamma radiometry (a) and uranium (b), ead (c) and zinc (d) distributions in the B and/or C horizons of soils overlying weakly mineralized structure in Middle ORS sediments near Broubster, northwest Caithness (located in Fig. 1)

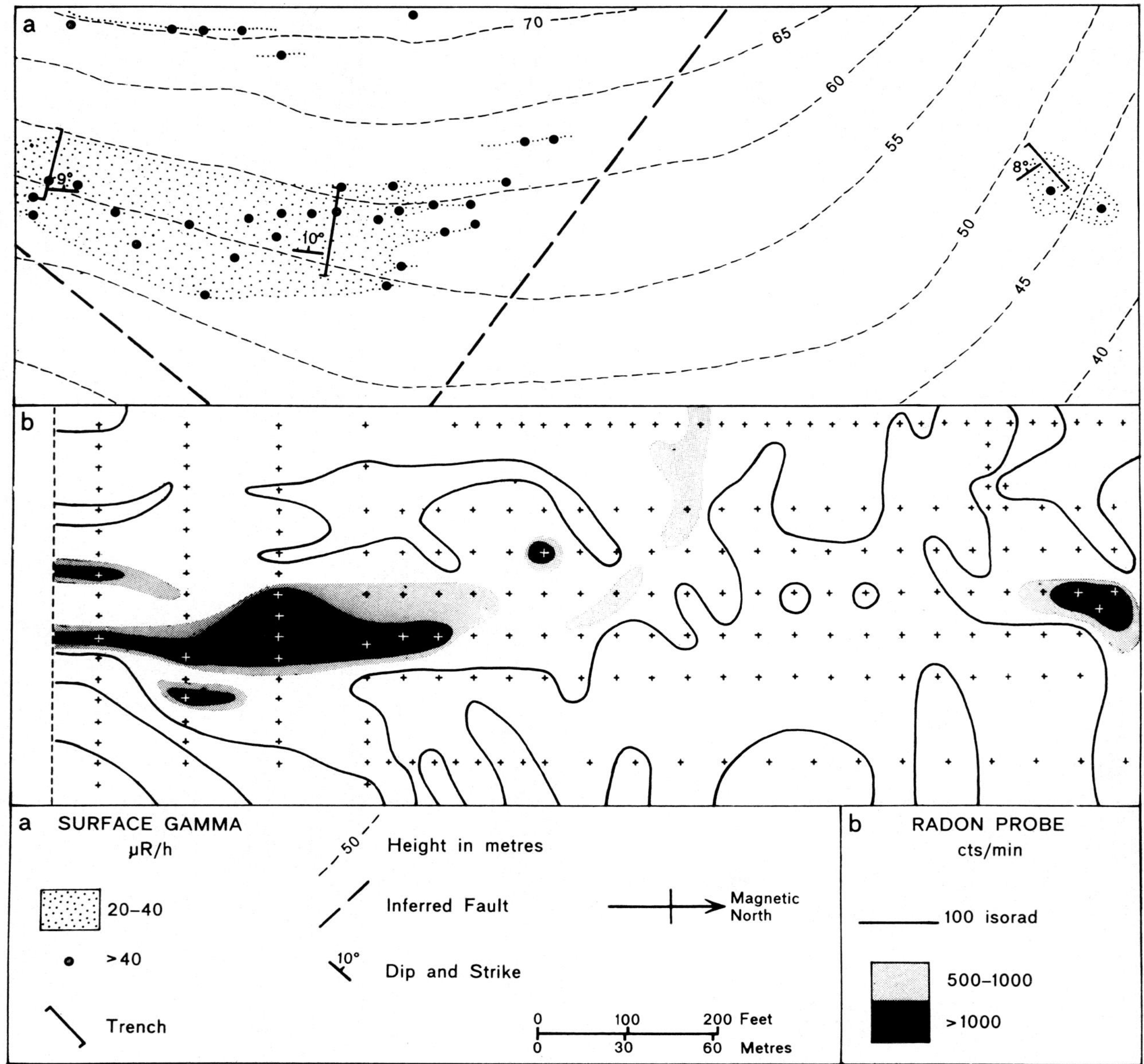

Fig. 15 Comparison of surface gamma radiation (a) with radon-probe measurements (b) over uranium occurrences in ORS sediments near Blackhall, Orkney (see Fig. 1 for location)

are of little value; but over the metamorphic and granitic rocks the uranium in stream-sediment patterns are very similar to that in the stream water. In this latter case, because of the useful multi-element data they provide, stream sediments are preferred. Both water and stream-sediment metal values are affected in areas where peat is a significant component of the catchment drainage. The organic material not only removes uranium and other metals from solution but also gives rise to spurious anomalies in sediment, especially when hydrous oxides of manganese and iron are produced.[3]

Reconnaissance based on soil or peat sampling was not attempted, but the appreciable extent of dispersion exhibited by several metals has shown that this approach could be used. Sampling at points of metal accumulation, such as seepages at breaks of slope or boggy areas, during regional stream-sediment surveys can be expected to provide much useful complementary information.

Follow-up investigations require the recognition of the surficial geochemical environment to eliminate spurious anomalies and to determine the migration pattern of hydromorphically transported metals and radioelements. The results presented here indicate that the in-field determination of alpha activity in soil air (radon monitoring) can be used in the assessment of radioactive anomalies, but that it is best employed in conjunction with gamma measurements. These permit discrimination between radon anomalies

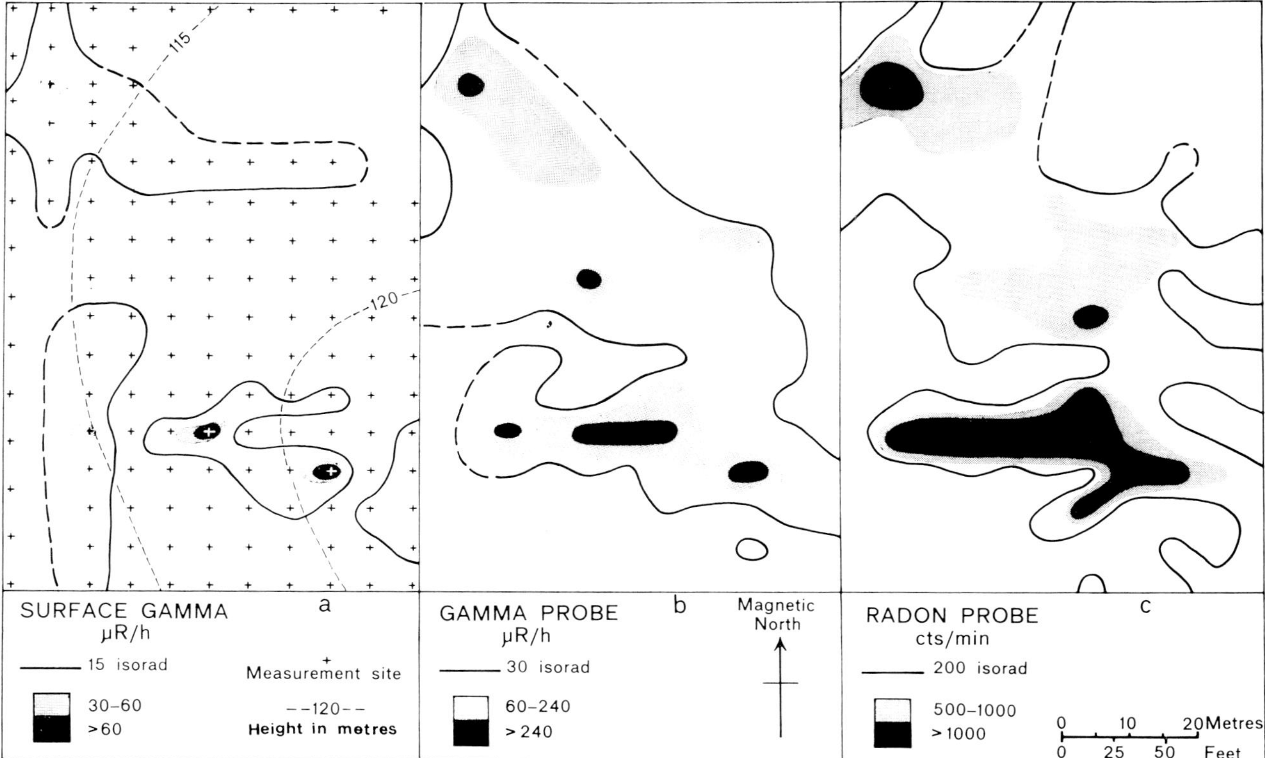

Fig. 16 Surface gamma radiometry (a) gamma-probe measurements (b) and radon-probe measurements (c) over uranium occurrence in ORS sediments at Sowa Dee near Blackhall, Orkney (see Fig. 1 for location)

due to near-surface uraniferous occurrences, local transported radioelement accumulations and deep-seated radon leakage anomalies.

Work in areas of thick peat at Ousdale and elsewhere has shown that the most effective results are obtained where the probes penetrate the peat cover. Often, the generally waterlogged nature of the peat produces a seal beneath which radon can accumulate, and which may give rise to minor associated gamma anomalies. Filling of soil air holes by groundwater in boggy ground is a serious problem and can prevent the use of the radon-probe technique.

These in-field determinations were supplemented by soil sample analyses—particularly for uranium, but also for associated elements with differing geochemical mobilities such as lead. If the soil sample is taken from a suitable soil horizon within the air hole, this helps in the recognition of the nature of the anomalies.

Due to the greater contrast and wider dispersion exhibited by the radon anomalies, the technique could be extended to the regional scale, particularly in areas of moderately deep uniform cover having free drainage. Specifically, the application of probe techniques in assessment work has served rapidly to outline, extend and evaluate surface indications of uranium and associated mineralization prior to more expensive trenching or drilling operations.

Acknowledgment

This work formed part of a uranium reconnaissance programme on behalf of the United Kingdom Atomic Energy Authority directed by Dr. S. H. U. Bowie and D. Ostle. It was made possible by the instrumentation development carried out by J. M. Miller and R. Williamson, which included early trials in northern Scotland by J. M. Miller and R. D. Beckinsale. D. C. Cooper, R. T. Smith and R. N. Aitken also contributed to the field work.

The paper is published by permission of the Director, Institute of Geological Sciences, London, and Dr. M. Davis, U.K.A.E.A.

References

1. Gallagher, M. J. *et al.* New evidence of uranium and other mineralization in Scotland. *Trans. Instn Min. Metall. (Sect. B: Appl. earth sci.)*, **80**, 1971, B150–73.

2. Michie, U. McL. Further evidence of uranium mineralization in Orkney. *Trans. Instn Min. Metall. (Sect B: Appl. earth sci.)*, **81**, 1972, B53–4.

3. Plant, Jane. Orientation studies on stream-sediment sampling for a regional geochemical survey in northern Scotland. *Trans. Instn Min. Metall. (Sect. B: Appl. earth sci.)*, **80**, 1971, B324–45.

4. Davis, M. *et al.* United Kingdom research and development work in aid of future uranium resources.

Paper presented at 4th International Conference on Peaceful Uses of Atomic Energy, Geneva, 1971, P/489.
5. DEPARTMENT OF AGRICULTURE AND FISHERIES FOR SCOTLAND. *Scottish peat. Second report of the Scottish Peat Committee* (Edinburgh: HMSO, 1962), 222 p.
6. MILLER, J. M. and LOOSEMORE, W. R. Instrumental techniques for uranium prospecting. In *Uranium prospecting handbook* (London: IMM, 1972), 135–46.
7. MILLER, J. A. and BROWN, P. E. Potassium–argon age studies in Scotland. *Geol. Mag.*, **102**, 1965, 106–34.
8. READ, H. H. Aspects of Caledonian magmatism in Britain. *J. Liverpool Manchester geol. Soc.*, **2**, 1961, 653–83.
9. GALLAGHER, M. J. Galena–fluorite mineralization near Lairg, Sutherland. *Trans. Instn Min. Metall. (Sect. B: Appl. earth sci.)*, **79**, 1970, B182–4.
10. SYNGE, F. M. The glaciation of north-east Scotland. *Scot. geogr. Mag.*, **72**, 1956, 129–43.
11. PHEMISTER, T. C. and SIMPSON, S. Pleistocene deep weathering in north-east Scotland. *Nature, Lond.*, **164**, 1949, 318–9.
12. SZALAY, S. and SÁMSONI, Z. Investigation of the leaching of uranium from crushed magmatic rock. *Földtani Közlöny*, **97**, no. 1 (undated), 60–72; *Geochem. intn.*, **6**, 1969, 613–23.
13. DALL'AGLIO, M. A study of the circulation of uranium in the supergene environment in the Italian Alpine Range. *Geochim. cosmochim. Acta*, **35**, 1971, 47–59.
14. DYCK, W. *et al.* Comparison of regional geochemical uranium exploration methods in the Beaverlodge area, Saskatchewan. In *Geochemical exploration* (Montreal: CIM, 1971, 132–50. *(CIM Spec. vol. 11)*
15. GALLAGHER, M. J. Exploring for uranium in the Northern Highlands of Scotland: a case-history. In *Uranium prospecting handbook* (London: IMM, 1972), 313–8.
16. DEMENT'YEV, V. S. and SYROMYATNIKOV, N. G. Conditions of formation of a sorption barrier to the migration of uranium in an oxidising environment. *Geokhimiya*, 1968, 459–65; *Geochem. intn.*, **5**, 1968, 394–400.
17. WHITEHEAD, N. E. and BROOKES, R. R. A comparative evaluation of scintillometric, geochemical and biogeochemical methods of prospecting for uranium. *Econ. Geol.*, **64**, 1969, 50–6.
18. TITAYEVA, N. A. Association of radium and uranium with peat. *Geokhimiya*, 1967, 1493–9; *Geochem. intn.*, **4**, 1967, 1168–74.

550.84:546:551.312.4(71)

Lake geochemistry—a low sample density technique for reconnaissance geochemical exploration and mapping of the Canadian Shield

R. J. Allan, PH.D.

E. M. Cameron, PH.D.

C. C. Durham

All of the Geological Survey of Canada, Ottawa, Ontario, Canada

Synopsis

In order to test methods of geochemical exploration and mapping by use of lake sediments for large-scale reconnaissance surveys of the Canadian Shield 176 samples of inorganic lake sediment were collected from seven areas within the Bear and Slave geological provinces, at a sampling density of one per 10 square miles. Two of these areas—High Lake and Hackett River—contain Archaean massive sulphide ore deposits, and a third—Terra mine—contains Cu and Ag ores. In order to relate the chemistry of the lake sediments to that of the surrounding bedrock, 447 samples of various rock types were collected from five of the above areas. The rock powders and the −250-mesh fraction of the lake sediments were analysed for Si, Al, Fe, Mg, Ca, Ti, Na, K, Mn, Ba, Zn, Cu, Pb and Hg. The lake sediments were further analysed for Ni, Co and Ag, and the rocks for As.

The data obtained show that the −250-mesh fraction of the lake sediments is a particularly homogeneous sampling medium, which closely reflects the average composition of the surrounding bedrock. It provides a uniform material for the sorption of metal ions dispersed by the weathering of ore deposits and smaller, more abundant, sulphide masses contained in the rocks enclosing the ore deposits. Ore metal anomalies are readily distinguished at this sample density, and the presence of Mg alteration zones associated with the deposits and areas of favourable rock type can be outlined. It is suggested that this economical method of geochemical exploration and mapping may be applied over the 70% of the Canadian Shield which is likely to contain inorganic lake sediments.

Canada is the second largest country in the world. More than half of its area is underlain by the complex Precambrian block of the Canadian Shield. Permafrost conditions affect more than two-thirds of the Shield. For more than one-third of Canada's land area there are, therefore, three dominant conditions: size, complex Precambrian geology, and permafrost. Any economical method of performing a reconnaissance geochemical assessment of trace-element variations within this area must accommodate these three conditions. An outstanding feature of the drainage map of the Canadian Shield, relative to most other parts of the world, is the prominence of lakes. These lakes are the key to reconnaissance geochemistry in this vast, flat, Precambrian Shield.

The frequency of lakes ranges up to more than 7000 in a 5000 square mile area of northern Saskatchewan. In nearly all parts of the Shield there are sufficient lakes that lake muds or waters may be sampled at a density of one sample per square mile or greater. The materials sampled from lakes for geochemical investigations have generally been surface waters and inorganic sediments. Trace-element variations in drainage basins of lakes are reflected in these media by processes depicted in a very simplified form in Fig. 1. Rock types are normally characterized by certain trace-element concentration ranges.[47] Oxidation and leaching of the bedrock and the overlying Quaternary deposits releases these trace elements into soils, streams and, finally, lakes.

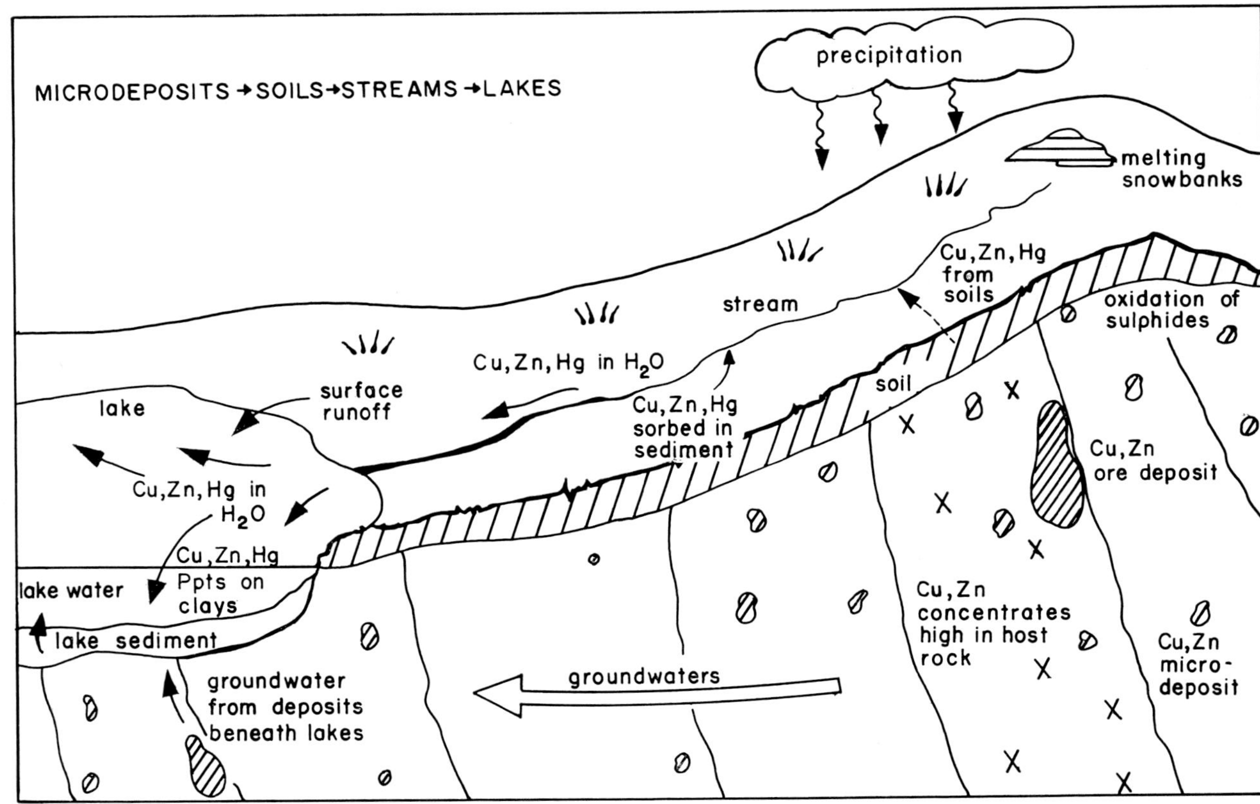

Fig. 1 Dispersion of metals from sulphides into lakes and lake sediments

Such geochemical dispersion in permafrost areas of Canada has only recently been described for Cu,[2] Zn and Ni[4,5] and Hg.[24] In the U.S.S.R., however, documentation of geochemical dispersion due to chemical weathering processes in permafrost areas has been available for many years and appeared in major works on geochemical exploration methods,[37] hydrogeochemical methods,[22] geochemistry of landscapes[33] and geochemical regionalization surveys.[23] Detailed studies on oxidation and dispersion processes for sulphides in permafrost zones are common in the Soviet literature.[29,31,36,44] Elements in permafrost zones are dispersed into drainage systems following migration through the overlying Quaternary deposits and soils.[30,34,35] Groundwater, surface water and flash runoff in the spring are important factors in this dispersion process.[26,27,28,41,42] Trace elements may also reach streams and then lakes in a sorbed form on the surface of clays or as included ions and complexes in the iron, manganese and organic coatings of fine sediment particles. Also, in rare cases where massive sulphide deposits are in the immediate vicinity, movement of sulphides in a particulate form may occur. These means of dispersion of trace elements into lakes have been more fully discussed elsewhere.[3]

A lake in the impeded and disorganized drainage system of the Shield normally has several small streams entering it from different points, but it has either no obvious exit stream or only one larger exit stream. Samples of sediment taken from such a lake either at the outflow stream bay, or at least away from any inflow bay, will reflect the trace-element content of the entire lake drainage basin. Through natural processes of mixing and averaging the lake sediment becomes a composite sample of the rocks within a drainage basin. The size of such drainage basins may commonly be one mile square or larger. In an area of the Shield where the rock type is uniformly basaltic it has been found that the variation in the bedrock trace-element content of Cu and Zn could be adequately shown at a lake sediment site density of one per 10 square miles.[3] It appears that the averaging and smoothing effect on metal distribution produced by the drainage system will allow regional changes in bedrock trace-element geochemistry to be assessed at very low sample densities. Such an approach enables many of the problems of regional geochemistry in the Canadian Shield to be overcome.

The fact that lake sediments represent an average of the composition of the surrounding bedrock means that samples that are collected at wide—and therefore economic—intervals can adequately reflect the chemical composition of a wide area of rocks. Before such wide-interval sampling can be applied for mineral exploration, however, the effect of the distribution of elements within the rock mass must be assessed.

If anomalous concentrations of ore elements within a given area were confined to one or a few ore deposits, it is improbable that the resulting geochemical signal could be detected at a sample density of one per 10 square miles, for if the dispersion of elements from these deposits along the drainage system is restricted, it is unlikely that the small anomaly would be intersected. Alternatively, if these elements are widely dispersed, their signal will be too weak to be detected. Cameron and Baragar,[10] however, have suggested that the condition is rare where ore-related mineralization within an area is confined to a few economic-size deposits. Instead, they have proposed that ore masses are distributed by size in a probabilistic fashion, and that orebodies are but the upper portion of more extensive populations of ore masses truncated by economic—not geological—considerations. These populations are continuous through non-economic 'showings' to very small masses and grains ('micro-deposits'). In terms of mass per cubic mile of rock these micro-deposits and showings probably exceed by many orders the mass of economic orebodies. It is these widely distributed showings and micro-deposits, as well as the ore deposit, that influence the composition of sediment in a lake with a drainage basin several miles in extent and allow anomalous areas to be detected at a wide sampling interval. By way of warning, it should be added that there must be many populations of micro-deposits that are truncated before the critically economic size is reached, thus leading to spurious anomalous results.

History of development and present status of low density lake geochemistry

The possible use of lake materials for geochemical exploration within Canada was first considered in the late 1950s.[21, 38] In the 1960s there were numerous investigations of the radon and uranium contents of surficial waters, including lake waters from various parts of Canada.[14, 18, 45] Recently, lake waters were used with various degrees of success in regional exploration for base metals.[2, 5, 24] Up to this time water had remained the principal lake sample medium. Only in rare cases were lake sediments collected—and then usually only over small areas and at high sample densities for uranium exploration.[19] In 1970 the first regional lake sediment survey[3] was conducted

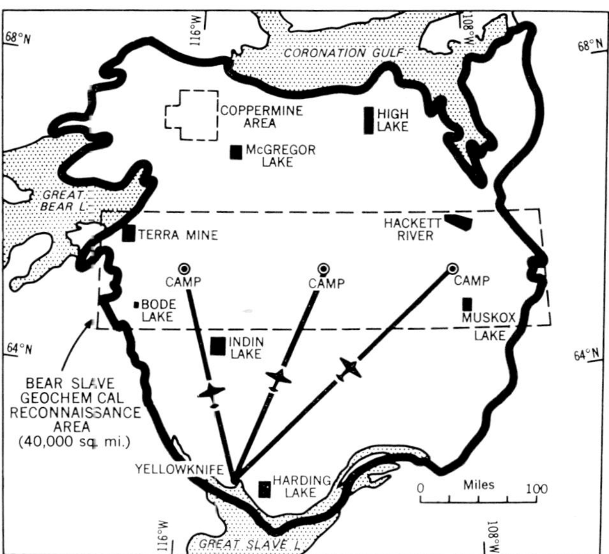

Fig. 2 Study areas within Bear and Slave geological provinces

for Cu in the Coppermine area (Fig. 2), lake sediments being shown to be a superior sample medium to lake waters. In part, this may have been attributable to analytical difficulties relating to the low element contents in waters. Lake sediments are, however, inherently closer to the original bedrock or surficial sediment compositions of the drainage area than waters. Weathering of rocks to sediments and sorption mechanisms on sediments are both complicated processes. Waters, which may have concentrations of trace elements dependent on environmental and other non-geological variations, are even more steps in the geochemical cycle away from the original rock material

While the above methods were being tested at universities and by government agencies, many companies in Canada were using lake waters for uranium exploration, and a limited number had begun to use lake sediment in the late 1960s in detailed exploration for base metals. None of this work is currently available.

When the Geological Survey of Canada was carrying out the regional lake sediment survey in the Coppermine area, the Geology Division of the Saskatchewan Research Council was conducting detailed lake sediment and water sampling in the Cu–Ni-rich Flin Flon area,[7] their conclusions confirming that sediment was the better sample medium. Also in the same year, one company which had conducted a regional gyttja sampling programme outside the Shield recommended gyttja as an excellent medium for mobile elements such as Mo. This raises the possibility of a third lake sample medium that could be used in certain organic-rich lake areas of the southern Shield. The use of organic gyttjas has been proposed in the

U.S.S.R. as a means of prospecting in swamp regions.[42]

What is planned in Canada for 1972? Government agencies will be involved in at least three regional lake sediment and water surveys. The Geological Survey of Canada will conduct a geochemical reconnaissance of a 40 000 square mile area of the Bear and Slave Provinces of the Shield at a site density of one per 10 square miles. The Department of Indian Affairs and Northern Development will cover several greenstone belts in the vicinity of Yellowknife at site densities of greater than one per square mile. The total area covered will be in the order of 1000 square miles. Lake sampling programmes will form part of more general geochemical surveys planned for parts of Newfoundland and Nova Scotia by the Department of Regional Economic Expansion. The newly formed Department of the Environment may carry out lake sampling in parts of Quebec to evaluate natural contamination by toxic elements. The Geological Survey renders technical advice on these programmes of other federal government departments. The Saskatchewan Research Council, Geology Division, will conduct a detailed geochemical sampling programme, including lake media, over a 1500 square mile area of the Canadian Shield in northern Saskatchewan. Several companies now use detailed lake sampling in base-metal exploration. Some of the larger companies will use this method in 1972 to assess numerous electromagnetic conductors outlined by airborne surveys. At least one Ph.D. thesis is planned on the subject of lake geochemical processes as they apply to mineral exploration. If successful, the 40 000 square mile survey (Fig. 2) to be carried out by the Geological Survey may eventually result in a geochemical reconnaissance of a large part of the Canadian Shield being conducted in this way.

Location and physiography of study areas

Eight study areas were selected on the basis that they are, or have been, sites for considerable prospecting activity. Five of the areas are in the Slave Province and three in the Bear Province. All eight areas contain rocks of possible economic potential, with the exception of the Harding Lake area, which is underlain by granitic rocks. Two of the areas contain proven ore deposits, as yet undeveloped. One area contains several small operating silver mines and another area contains abandoned gold mines.

One-third of the Bear Province and two-thirds of the Slave Province lie above the treeline in the so-called Barrens. The treeline runs approximately from the northeast corner of the Great Bear Lake to the northeast corner of the Great Slave Lake (Fig. 2). The treeline also very closely approximates the boundary of the discontinuous and continuous permafrost zones. Bode Lake, Indin Lake and Harding Lake are regions of low hills with a relief of a few hundred feet, forested mainly by black spruce and underlain by discontinuous permafrost. The remaining five areas have very low relief, a tundra vegetation predominantly of mosses and lichens, and they are affected by continuous permafrost conditions.

Geology and metallogeny of study areas

Slave Province, 75 000 square miles in area, shares with the much larger Superior Province to the south the distinction of being the oldest portion of the Canadian Shield. Except for Proterozoic sediments and volcanic rocks along the margins of the province, it is composed of rocks of Archaean (>2400 m.y.) age. The Archaean rocks comprise a 'matrix' of granitic and high-grade metamorphic rocks enclosing north-trending belts of volcanic and sedimentary rocks metamorphosed to lower greenschist to amphibolite grade. The most common sequence found in the latter belts are mafic lavas, succeeded by much less abundant intermediate and acid volcanic and pyroclastic rocks. The volcanic sequences are up to 40 000 ft thick and generally occur along the margins of the belts. They are overlain, often conformably, by typical flysch-facies greywacke–shale sediments that occupy the axial zone of the belts. Sedimentary rocks are volumetrically much more abundant (4:1) than the volcanic rocks—a feature which is the converse of that in the volcanic-sedimentary belts of the Superior Province and which has an important bearing on the relative mineral potential of the two areas.

Very few ore-grade deposits have been located in the Slave Province. The mineral showings have been classified by Thorpe[46] into 11 types. Of these the massive sulphide Zn–Cu and Zn–Pb–Ag–Cu deposits of volcanic origin are by far the most economically attractive. Efforts in the Slave Province have, therefore, been very largely devoted to developing a low sample density technique to locate this type of deposit. The massive sulphide Archaean deposits do not in themselves provide a large target for surficial geochemical methods. It is believed, therefore, that if the methods described here are successful in locating these deposits, they will certainly be effective for a number of other, larger, lower-grade deposits.

The massive sulphide ores of the Canadian Shield are generally considered to have formed subaqueously by the precipitation of metals brought in to the sea by fumarolic solutions. These solutions were released around centres of acidic volcanism. The deposits are thus associated with acid, often fragmental, volcanic rocks and with sediments of exhalative origin, such as chert, sulphide-facies iron formation and carbonate sediments. The deposits are often found at the interface between the intermediate and acid volcanic rocks or near the contact of the acidic rocks at the top of a volcanic pile and the overlying sediments. Unpublished studies by Cameron and others of the chemistry of the rocks associated with the massive sulphide deposits show that mineralization with ore elements is largely confined to the feeder pipe, to the deposit itself, and to a thin stratigraphic horizon in which the deposit lies.

The areas sampled within the Slave Province for this study are noted below.

High Lake

The High Lake Cu–Zn deposit is situated within a north-trending belt of intermediate and acid volcanic rocks near to their contact with apparently conformable tuffaceous sediments. It contains 5 200 000 tons of 3·5% Cu and 2·5% Zn, and is composed of pyrite, sphalerite, chalcopyrite and minor galena. Many details of the deposit are unknown because of insufficient drilling, but the character of the deposit and its host rocks clearly mark it as a typical Archaean exhalative sulphide occurrence. Gossans resulting from the oxidation of masses of pyrite, with minor pyrrhotite and arsenopyrite, are common within this belt of volcanic rocks.

Hackett River

The Hackett River Zn–Pb–Ag–Cu deposit is also a clear example of an Archaean massive sulphide deposit of exhalative origin, situated near to the contact of acid volcanic rocks with tuffaceous sediments.

The volcanic rocks around the deposit, although highly silicified and metamorphosed, indicate the presence of a vent structure or alteration pipe. One of the most interesting features of the deposit is the presence of limestone horizons of possible exhalative origin. Early drilling results indicated more than 10 000 000 tons of 8% Pb–Zn and 9 oz/ton Ag for the main zone, but it appears that there is more than one orebody in the general area.

Muskox Lake

This area is largely underlain by sedimentary and volcanic rocks of the Yellowknife Supergroup, although there are also a number of small intrusions of quartz diorite. No ore deposits have been discovered here, but there are a number of gossans resulting from the oxidation of sulphide bodies and there are lenses composed of pyrite with minor pyrrhotite and chalcopyrite. These occurrences are located in the centre of the sampled area in acid and intermediate volcanic rocks.

Harding Lake

The Harding Lake area is largely underlain by rocks of the granitic 'matrix' of Slave Province—chiefly biotite granites and muscovite granites. The granitic rocks are fringed to the northwest, north and east by Yellowknife Supergroup sediments that have been metamorphosed to knotted quartz–mica schists and hornfels near the intrusive granite contact. No significant sulphide mineralization has been noted in this area.

Indin Lake

The Indin Lake region was selected to represent a 'background' situation for both the rock and lake sampling because it contains all the rock types typical of the Archaean of Slave Province, but no known base-metal deposits. Basic, intermediate and acid volcanic rocks and sediments of the Yellowknife Supergroup underlie most of the area, and granitic rocks occur in the northwest portion. The sediments and volcanics have been metamorphosed to varying degrees up to granulite facies. The only important mineral occurrences are gold deposits, with some chalcopyrite and pyrite in quartz veins. These occur commonly in shear zones in the volcanics and sediments.

The three remaining areas sampled are in the Bear Geological Province. This province, to the west of Slave Province, is composed of rocks of Proterozoic age. In the southern part of the province miogeosynclinal sediments of Aphebian age occur in a zone flanking the margin of the Slave craton. To the west the eugeosynclinal facies of these sediments has been metamorphosed and migmatized. Farther to the west the lithology changes to a complex of andesitic to trachytic volcanic and pyroclastic rocks and molasse sediments which form the Echo Bay and Cameron Bay Groups. They are intruded by granites and are, in places, highly metamorphosed. North of Great Bear Lake these deformed rocks are overlain by sandstones and dolomites of the Hornby Bay and Dismal Lakes Groups, and these, in turn, by the Coppermine River Group. The Coppermine River Group is a thick series of basalts overlain by red beds with intercalated flows, and these rocks are overlain by quartzites, dolomites and shales of the Rae Group.

In Bear Province there is a considerable variety of mineral deposits, although few are of economic importance. Their metallogeny has recently been summarized by Thorpe.[46] The most pertinent to this study are (1) native copper and chalcocite in the Coppermine basalts; (2) native silver associated with nickel–cobalt–iron arsenides and pitchblende in faults and fractures in the Echo Bay Group; (3) uranium and/or copper in fault-controlled massive quartz stockwork zones (these zones may be in excess of 1000 ft wide and 10 miles long); (4) fracture-controlled chalcopyrite mineralization in porphyritic lavas of the Echo Bay Group; and (5) stratiform copper mineralization in tuffaceous Echo Bay sediments.

Geochemical studies of the rocks and surficial materials associated with the Coppermine basalt ores have already been reported.[3, 10] Studies in other areas of the Shield[14, 18] indicated that lake geochemical methods are likely to be effective for delimiting uraniferous areas of Bear Province. The present studies have, therefore, been limited within Bear Province to two areas that contain deposits of types (2), (4) and (5) above and to one area that contains a distinctive (ultramafic) rock type.

Terra mine

Deposits of types (2) and (5) above are found at this mine, which has been operating since 1970. The rocks in the vicinity of the mine are mainly tuffaceous sediments and intermediate to acid volcanic rocks, dominantly porphyries, of the Echo Bay Group. Native silver, argentite, matildite, nickel–cobalt arsenides and sulphides, native bismuth, bismuthinite, chalcopyrite, bornite, pyrite, sphalerite and galena are found along faults and fractures. In the mine and outcropping on the adjacent hillside is a zone of chalcopyrite and pyrite with some galena and sphalerite. A further interesting feature of the mine area is a mass of magnetite–apatite.

Bode Lake

The rocks of the area sampled around Bode Lake are almost entirely porphyries: in the main, these appear to be lavas, but some may be intrusive. Chalcopyrite with pyrite and quartz is found in fracture zones in these lavas a short distance north of Bode Lake.

McGregor Lake

The geological feature of greatest interest in this region is the Muskox Intrusion—a layered ultrabasic body of Neohelikian age. It is intruded into a basement complex of metamorphosed Aphebian sediments, migmatites, granite–gneiss and granodiorite. Non-economic concentrations of copper and nickel sulphides occur in the body and around its margin.

Sample collection, preparation and analysis

Bedrock sampling

Bedrock samples were collected at High Lake, Hackett River, Indin Lake, Terra mine and Bode Lake.

At High Lake 94 samples were taken along three traverses across the strike of the volcanic rocks and tuffaceous sediments (Fig. 3). Two of these traverses *(C1, C2)* are close to the sulphide deposit; the other *(C3)* is about two miles to the south. At Hackett River one traverse was made across the sequence of sediments and of volcanic rocks that lies 3–4 miles along strike from the main

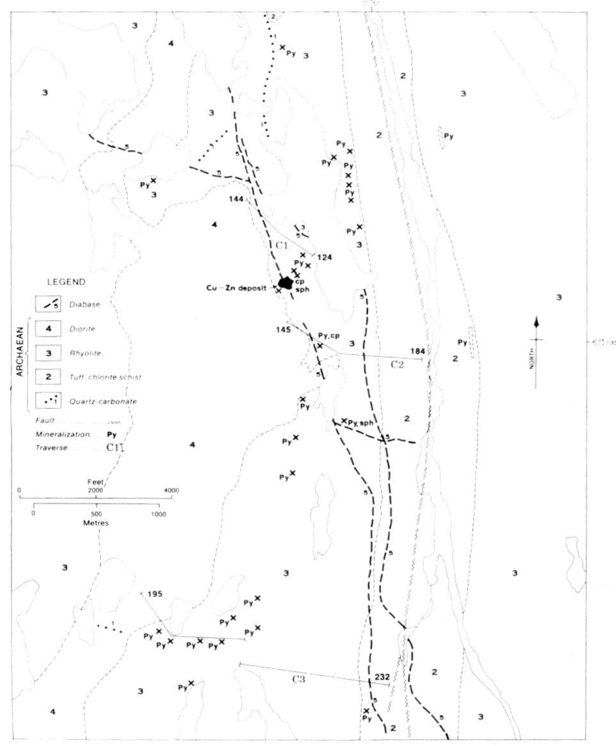

Fig. 3 Rock sampling traverses, High Lake, N.W.T. Geology after E. O. Dearden and B. A. Bradshaw, Kennarctic Explorations, Ltd.

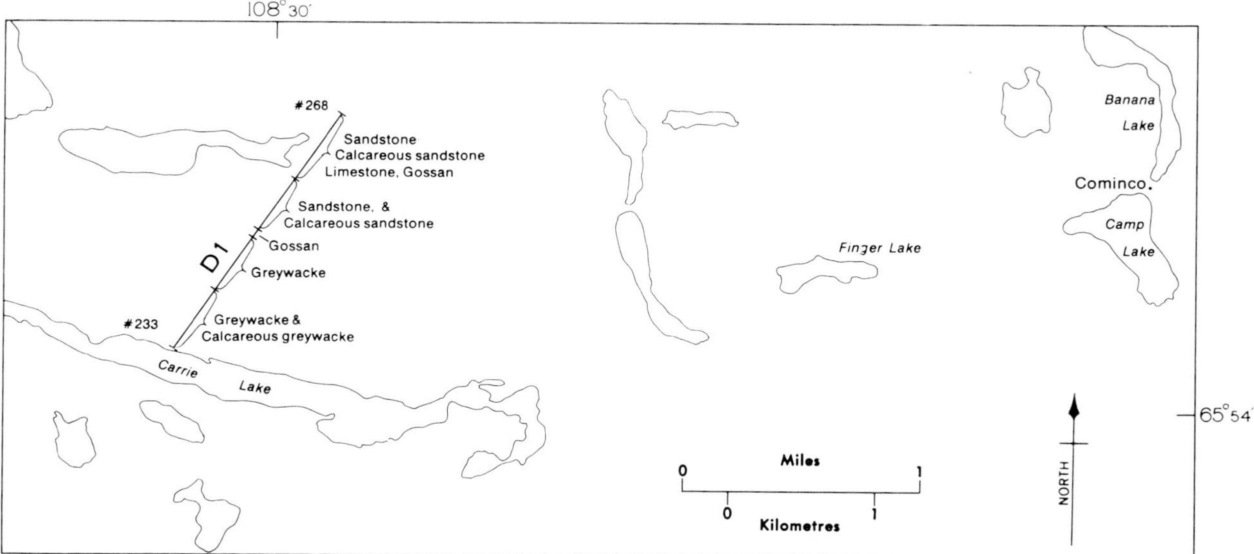

Fig. 4 Rock sampling traverse, Hackett River, N.W.T. Zn–Pb–Ag deposit is situated on north side of Camp Lake

ore deposit (Fig. 4). Here, 37 samples were collected. At Indin Lake 195 samples were taken along 12 traverses. These samples represent the basic, intermediate and acid volcanic rocks and the sediments of the Yellowknife Supergroup, and also the granitic rocks (Fig. 5). At Bode Lake three traverses were made across the feldspar porphyries, and 57 samples were collected (Fig. 6). One of the traverses passed over the mineralized zone. At Terra mine (Fig. 7) three traverses were made across the strike of the sedimentary-volcanic sequence that contains the two types of ore deposit referred to above. Some 40 samples were collected on these traverses. A further 25 samples were collected from porphyries that contain no obvious mineralization on the northeast bank of the Camsell River, opposite the mine. On most of these traverses fist-size samples were collected at 100-ft intervals. Outcrop is generally excellent. On the Hackett River traverse and on traverse *C3* at High Lake samples were taken at 200-ft intervals.

Lake water and sediment sampling

Water and sediment samples were collected regionally at a density of approximately one per 10 square miles at the five locations where bedrock samples were collected and also at the McGregor Lake, Muskox Lake and Harding Lake areas. The lakes were reached by small fixed-wing aircraft and, because landing was only possible on large lakes, site distributions are uneven. By helicopter, a regular grid can be easily maintained.

Water samples were always taken from just below the lake surface since this avoids layering effects in lake water related to redox changes.[7] The organic and iron oxide coated sediments found for 1 or 2 cm below the water–sediment interface were discarded as these are expected to show variations in trace-element content due to other than geological factors.[39, 40] Sediment samples came from immediately below this discarded layer. Sediments came from near the lake shore in 3–8 ft of water.

Preparation of lake water and sediment samples

The sediments were air-dried and sieved to −250 mesh (63 μm): this is the approximate boundary, in most surficial sediments, between the sand fraction, containing many rock fragments, and the silt fraction, that is largely composed of grains of individual minerals. In most surficial sediments from temperate and Arctic environments the dominant minerals in the silt fraction are quartz and feldspar.[1, 9, 17] Although the quartz and feldspar may be derived from a variety of rocks, these components are often supplemented by other minerals, such as amphiboles and micas, that are more specific to certain rock types. This must be the case for the samples described here, for their major-element chemistry responds well to changes in the surrounding bedrock (see later).

Sediment less than 250 mesh was chosen because it has a relatively high surface area, which provides a fairly uniform absorbing medium.[25] Such a medium might be expected to reflect the chemical dispersion of elements carried in solution from surrounding mineral deposits. This has previously been discussed by Allan,[3] who pointed out that the sorption capacity of lake sediments exceeds the amount of trace-element available in the drainage system. Major-element analysis (see

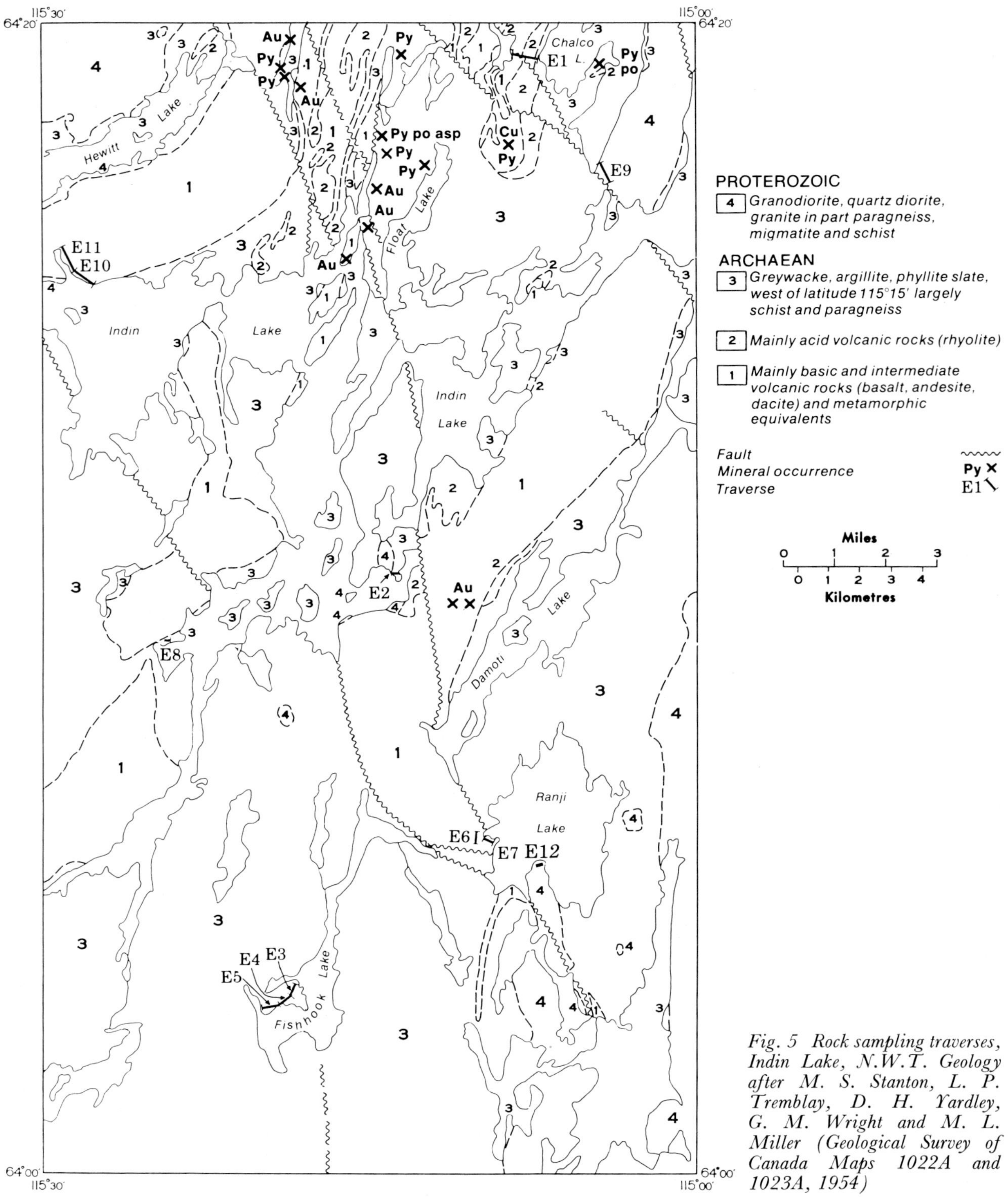

Fig. 5 Rock sampling traverses, Indin Lake, N.W.T. Geology after M. S. Stanton, L. P. Tremblay, D. H. Yardley, G. M. Wright and M. L. Miller (Geological Survey of Canada Maps 1022A and 1023A, 1954)

later) shows the sieved samples to be homogeneous, with variation from area to area and within areas that is largely the product of changes in the composition of the surrounding bedrock rather than variation in the silt/clay ratio. These sieved lake sediments may be compared to an enormous, flat, and relatively uniform resin bed that receives an input of various elements at different places across its surface.

The quantity of trace elements removed from the drainage waters will mainly be dependent on both the concentration of the elements in the water and the available surface area of the sediment. Depending on these two major factors, an equilibrium amount of trace element will be removed by the sediment of the lakes. This process

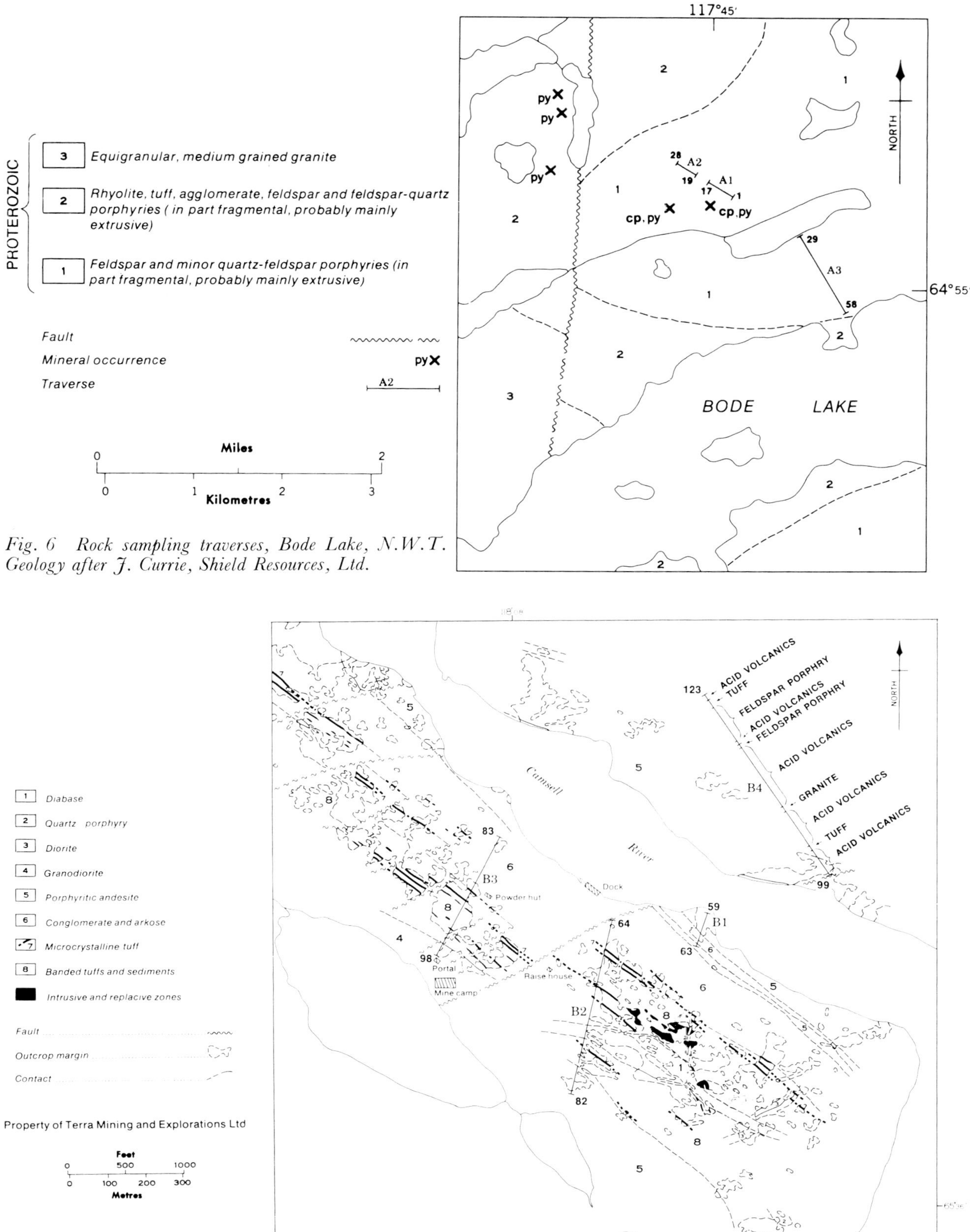

Fig. 6 Rock sampling traverses, Bode Lake, N.W.T. Geology after J. Currie, Shield Resources, Ltd.

Fig. 7 Rock sampling traverses, Terra mine, N.W.T. Geology after N. Badham

is repeated at each lake in the drainage system until background levels are reached in both waters and sediments.

There are a number of variations which should be added to this rather simplistic explanation. A major one is that ions are probably sorbed on to particles during transport of the latter in streams feeding the lakes. Another is that the analytical methods may leach not only the sorbed ions but also dissolve some of the ions that are bound within the structure of the mineral, and which have been carried in this form from the source rock.

Water samples were not filtered as there was no visible evidence of suspended loads.

The rock samples were reduced to $\frac{1}{4}$-in chips between the steel jaws of a Chipmunk crusher, after which they were further reduced to about 20 mesh in a Braun pulverizer equipped with ceramic plates. 10 g of this material was sampled and ground to approximately −150 mesh in an alumina ceramic mill.

Analytical methods

The −250-mesh lake sediment samples and the rock samples were analysed for Si, Al, Fe, Mg, Ca, Ti, Mn and Ba by direct-reading optical emission spectrometer. After fusion of the sample with lithium tetraborate, containing cobalt oxide and strontium tetraborate as stabilizers and internal standards, the powder was sparked on a 'Tape Machine'. Na[16] and K were analysed by adding 10 ml of 1 N HNO_3 to 210 mg of the above fused powder and heating until dissolved. 10 ml of 10 000 μg/ml Sr was added and the solution was made up to 100 ml, then being analysed by atomic absorption spectrometry. For Zn, Cu, Pb, Ni, Co and Ag 3 ml of 4 N HNO_3 and 3 drops of concentrated HCl were added to 0·2 g of the sample and heated to 95°C for 90 min. This was then diluted to 10 ml and analysed by atomic absorption spectrometry. For the Pb, Ni and Co a deuterium background correction was used. For the As determination the sample was fused with potassium pyrosulphate. The material was dissolved in 6 N HCl and the As was determined by the cold spot Gutzeit method. The cold vapour method for Hg has been described by Cameron and Jonasson.[12]

All of the water samples were acidified with concentrated HNO_3 on arrival in Ottawa. They were extracted with ammonium pyrrolidine–dithiocarbamate as chelating agent and methyl isobutyl ketone as solvent, and then analysed by atomic absorption spectrometry for Zn and Cu. The limit of detection for these two elements is 1 ppb.

At the time of writing most of the rock samples had not been analysed for Ni and Co, and none for Ag; none of the lake sediment samples had been analysed for As.

Presentation of results

The lake sediment and water results are presented in the form of computer-drawn contour maps for each area. The contours are drawn from a regular grid computed from the original values. To compute this regular grid a uniform circular search radius is used: this is large enough to include several original values, which reduces the effect of an extremely high concentration. Because the search area is circular, and the original points are close to being on an even grid, any linear trends observed in the contours are due to trends in the bedrock or to other geological factors, and are independent of the contouring process.

For the various elements contour levels are:
Zn (Fig. 11) 20, 40, 60, 80, 100, 120, 160, 200 ppm
Cu (Fig. 12) 10, 20, 30, 40, 50, 60, 80, 100 ppm
Ni (Fig. 13) 10, 20, 30, 40, 50, 60, 80, 100 ppm
Co (Fig. 14) 5, 10, 15, 20, 25, 30, 40, 50 ppm
Ag (Fig. 15) 0·2, 0·4, 0·6, 0·8, 1·0, 1·2, 1·6, 2·0 ppm
Pb (Fig. 16) 5, 10, 15, 20, 25, 30, 40, 50 ppm
Hg (Fig. 17) 15, 20, 25, 30, 40, 50 ppb

In Figs. 11–17 the upper two contour levels are hachured; all numbers shown that are not labels for contour levels are actual sample values.

Bedrock geochemistry of sampled areas

As was demonstrated for the Coppermine area, the broad Cu anomalies in lake materials, miles in extent, are not the product of metal dispersion from ore deposits. Rather, they are the product of dispersion from 'micro-deposits' of copper minerals which are contained in the basalts, and which are genetically related to the orebodies. The objective of the rock sampling programme described here was to determine whether comparable occurrences of ore elements are associated with other important types of deposits in the Bear and Slave Provinces.

Areas in Slave Province

Where possible, the samples are classified into a number of groups. At Indin Lake they have been grouped into volcanic, sedimentary and granitic types. The volcanics have been further subdivided into basic (40·0–54·9% SiO_2), intermediate (55·0–67·9% SiO_2) and acid ($\geqslant$68% SiO_2). The grouping into sedimentary or volcanic has not been rigid; where a tuffaceous sediment is intercalated in an overwhelmingly volcanic sequence, it too is classified as a volcanic rock—conversely for a volcanic sample in a dominantly sedimentary sequence. In a number of cases assignment to one or other of these classes is difficult. At High Lake extrusive and pyroclastic rocks and tuffaceous sediments are intimately related. The samples have, therefore, not been subdivided into sediments and volcanics; neither have those from Hackett River, where the succession sampled is dominantly sedimentary. At High Lake there are sufficient samples to make it practical to subdivide them into basic, intermediate and acid classes. It should be noted, however, that metasomatism of the rocks around the alteration pipe and ore deposit has modified their silica content, changing the 'classification' of some samples compared with that at the time of formation of the rocks.

Major elements

It would be inappropriate in a paper of this nature to make any extensive discussion of the major-element chemistry of the rocks; this will be done elsewhere. Much of this information is conveniently summarized by plotting the data for the different oxides in the different classes against SiO_2. This is done in Fig. 8, where the three classes of volcanic rocks from Indin Lake and from High Lake are compared with data for average Canadian Shield volcanic rocks of Archaean age. These data are for Baragar and Goodwin's[8] basalt (49·7% SiO_2) and salic fraction (66·8% SiO_2). The intermediate plot at 55·2% SiO_2 is a weighted average of their andesite, type *A* volcanic and type *B* volcanic in the abundance ratio 23:12·2:5·1 determined by Baragar and Goodwin. Major-element analyses for the different areas and rock types are shown in Table 1.

It may be seen from Fig. 8 that there is little difference for most elements in the plots for Baragar and Goodwin's data and those for Indin Lake. There is a definite enrichment, however, in alumina for the intermediate volcanic rocks at Indin Lake. Differences in the content of MnO and Ba may, at least in part, be due to analytical bias at this low concentration. The same degree of conformity does not apply to the data from High Lake. The plots for most of the oxides for this locality show what is believed to be the effects of Mg metasomatism of the rocks that surround the alteration pipe and ore deposit. MgO is more abundant in the rocks from High Lake than in rocks of equivalent silica content at Indin Lake or in those represented by the averages of Baragar and Goodwin. The plots in Fig. 8 suggest loss of Na_2O and Ba, particularly from the more acid varieties. Many of the plots of the other elements appear to have been displaced to the left as a result of a reduction in the SiO_2 content caused by the addition of Mg, and also possible loss of Si. The Al_2O_3 content of the acid volcanics at High Lake is very much lower than the general trend shown in Fig. 8.

The introduction of Mg into the rocks can also be readily detected by examining the sequence of analyses for individual samples. This introduction of Mg is accompanied by irregular loss of Na and Ba, and sometimes by K and Ca. Fe also seems to have been introduced or deposited with a number of the rocks sampled, mainly as sulphide. All of these metasomatic changes are largely confined to sampling traverses *C1* and *C2* in the area of the alteration pipe and ore deposit; traverse *C3* (Fig. 3), some distance to the south, does not show these changes.

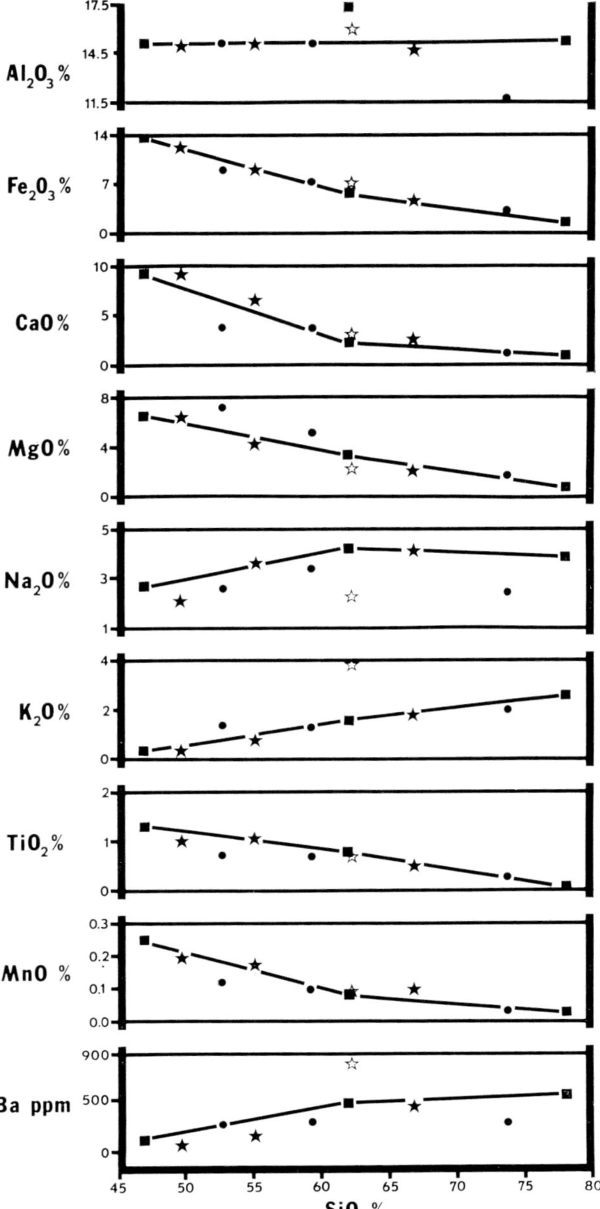

Fig. 8 Variation diagram: SiO_2 versus Al_2O_3, total Fe as Fe_2O_3, CaO, MgO, Na_2O, K_2O, TiO_2, MnO and Ba: ■, basic, intermediate and acid volcanic rocks, Indin Lake; ●, basic, intermediate and acid volcanic rocks, High Lake; ☆, volcanic rocks, Bode Lake; ★, basalts, andesites (with type A and type B) and salic volcanic rocks from Archaean of the Canadian Shield. From Baragar and Goodwin[8]

Table 1 Major-element composition, % (as arithmetic means), for rock types of study areas

Area and rock type	Number of samples	SiO_2	Al_2O_3	Fe_2O_3*	MgO	CaO	Na_2O	K_2O	TiO_2	MnO	Ba
High Lake, all samples	94	59·8	14·6	7·11	4·98	3·29	2·78	1·42	0·60	0·091	0·028
High Lake, basic	22	52·7	15·2	9·41	7·41	3·83	2·50	1·27	0·69	0·118	0·025
High Lake, intermediate	54	59·4	15·2	7·08	4·70	3·75	3·39	1·31	0·67	0·096	0·029
High Lake, acid	16	73·6	11·5	3·29	1·76	1·06	1·42	2·01	0·25	0·031	0·028
High Lake, traverses *C1–C2*	57	60·0	13·8	6·86	5·58	2·75	1·96	1·66	0·50	0·079	0·027
High Lake, traverse *C3*	37	59·4	15·7	7·49	4·06	4·11	4·03	1·07	0·75	0·110	0·029
Hackett River, all samples	37	59·8	13·1	4·66	2·25	8·94	1·99	1·90	0·36	0·211	0·031
Indin Lake, all volcanics	109	58·5	15·5	9·01	4·18	5·67	3·24	1·21	0·89	0·156	0·030
Indin Lake, basic	55	47·0	15·1	13·92	6·35	9·51	2·59	0·36	1·33	0·254	0·014
Indin Lake, intermediate	21	62·1	17·2	6·62	3·40	2·46	4·18	1·61	0·79	0·084	0·044
Indin Lake, acid	31	77·9	15·2	1·48	0·89	0·94	3·83	2·50	0·16	0·025	0·050
Indin Lake, sediments	81	61·8	16·7	6·81	3·08	1·81	2·65	2·05	0·71	0·076	0·047
Indin Lake, granites	5	58·6	17·6	7·79	5·24	4·96	3·71	1·59	0·87	0·112	0·060
Bode Lake, all samples	57	62·2	15·9	6·13	2·35	3·00	2·22	3·80	0·62	0·097	0·082
Terra mine, all samples	65	56·1	15·4	9·56	3·25	3·95	3·09	3·31	0·60	0·215	0·076

*In this and all subsequent tables total Fe as Fe_2O_3.

The section sampled at Hackett River contains an abundance of limy beds and sediments with calcitic cement. The rock association in this area, and particularly the relation of these horizons to the alteration pipe near Camp Lake (Fig. 4), suggests that the carbonate material is of exhalative origin. This contention is supported by the manganiferous nature of the carbonate. The three samples with the highest amount of CaO, 48·4, 35·0 and 29·8%, contain, respectively, 1·23, 0·65 and 0·85% MnO.

The Indin Lake sediments are the chemically highly immature types so characteristic of the Archaean greenstone belts.[6]

Trace elements

Before the distribution of the trace elements in the Slave rocks is discussed, some mention should be made of the degree of extraction of these metals from the samples during analysis. Zn, Cu and Pb were analysed by a partial leach in order that the component of the metal most subject to solution and dispersion within the drainage system would be analysed. In Table 2 there is a comparison of results for this attack and for a total leach of 10 representative samples from High Lake. These data indicate that the HNO_3–HCl attack has removed most of the Cu and Pb present in the rock, and less of the Zn. For As and Hg the extraction used is believed to remove virtually all of the metal in the majority of rock samples (J. J. Lynch, personal communication).

Zinc For Zn the area with the highest arithmetic mean content is High Lake (Table 3). There are two causes for this high level: first, the background level of abundance of Zn, shown by the geometric mean of 74 ppm and the median of 70 ppm, is

Table 2 Comparison of analytical data for Zn, Cu and Pb on 10 representative rock samples from High Lake (first column for each element gives results for total extraction ($HF–HNO_3–HClO_4$); second column is for the partial leach ($HNO_3–HCl$) method used to obtain data on Zn, Cu and Pb given in this paper)

Sample number	Zn, ppm (HF)	Zn, ppm (HNO_3)	Ratio HF/HNO_3	Cu, ppm (HF)	Cu, ppm (HNO_3)	Ratio HF/HNO_3	Pb, ppm (HF)	Pb, ppm (HNO_3)	Ratio HF/HNO_3
710124	144	142	0·99	29	32	1·10	<8	<8	—
710133	87	82	0·94	13	14	1·08	8	10	1·25
710143	157	102	0·65	37	37	1·00	20	22	1·10
710153	89	72	0·81	28	25	0·89	16	18	1·12
710163	74	61	0·82	23	18	0·78	<8	<8	—
710173	103	80	0·78	76	78	1·03	14	14	1·00
710183	29	24	0·83	57	61	1·07	<8	<8	—
710203	131	107	0·82	14	12	0·86	<8	<8	—
710213	86	43	0·50	71	66	0·93	<8	<8	—
710223	72	61	0·85	157	169	1·08	8	8	1·00
Mean ratio			0·80			0·98			~1·0

greater than for the other areas; secondly, the frequency distribution is skewed to high values, reflecting the presence of scattered mineralization along the two traverses, *C1* and *C2*, close to the ore deposit. Because of this mineralization along *C1* and *C2*, the arithmetic mean for these traverses is much higher than that for *C3*—which does, however, show a background level similar to that of the other two, with a geometric mean and median close to that for the whole area.

For the three classes of rock at High Lake the median value for Zn decreases from basic to acid, but there is a corresponding increase in the arithmetic mean. These features may be related to the generally higher background level of Zn within the major minerals composing basic rocks, and by the tendency for Archaean Zn sulphide mineralization to occur in acid volcanic rocks.

For the samples from the Hackett River traverse the background level, given by the geometric mean and median values of 42 and 43 ppm, respectively, is only slightly more than half the background at High Lake. The higher percentiles have values that are, however, similar for the two areas. This indicates the presence of scattered mineralization at both High Lake and Hackett River.

For the Indin Lake volcanics the background, shown by the median of 38 ppm and the geometric mean of 33 ppm, is even less than that found at Hackett River. There are no high Zn values at the higher percentile levels, confirming the general absence of mineralization along the sampled traverses. There is a general decrease in the arithmetic and geometric means and medians for Zn on passing over the sequence basic, intermediate and acid volcanic rocks.

The sedimentary rocks sampled within the Indin Lake area have a quite different distribution of Zn. Background, shown by a geometric mean of 76 ppm and a median of 81 ppm, is higher than even the High Lake area. The statistical dispersion of the data is, however, very low, the highest value recorded being 112 ppm. These rocks, therefore, form a relatively high and uniform source of Zn.

Copper The arithmetic mean of Cu (71 ppm) at High Lake is approximately double the geometric and median values of 33 and 37 ppm. This is caused by scattered mineralization along all three traverses skewing the frequency distribution. At 46 ppm the arithmetic mean of traverse *C3*, most distant from the ore deposit, is less than that of *C1* and *C2*. As for Zn, however, the background level of Cu is relatively high along all three traverses.

Again, as for Zn, the geometric mean and median for Cu at Hackett River (17 and 23 ppm) are approximately half of those at High Lake. Cognizance should be paid to the facts that neither are the rock types in this area the same as for High Lake nor is the distance from the traverse to the most important deposit the same as at High Lake. Also, the main deposit at Hackett River contains little Cu. The Hackett River samples show Cu values at high percentile levels that are much smaller than those for High Lake.

The data for the Indin Lake volcanics reflect the very marked tendency for Cu to associate with basic rocks. The basic group have a median of 93 ppm Cu and very similar values for the arithmetic and geometric means. In contrast, the medians for the intermediate and acid classes are 16 and 4 ppm, respectively. Turning to High Lake, there is very little difference between the basic, intermediate and acid rocks there. There may be two reasons for this: first, originally acid and

Table 3 Statistical data for Zn, Cu, Pb and As (ppm) and Hg (ppb) in rock samples from Bear and Slave geological provinces

Area and rock type	Number of samples	Arithmetic mean	Standard deviation	Geometric mean	Percentile 2·5	5	10	20	30	40	50	60	70	80	90	95	97·5	99
Zinc																		
High Lake, all samples	94	247·4	1591·8	74·2	25	31	41	48	56	64	70	80	83	91	142	192	656	15 500
High Lake, basic	22	85·6	36·2	80·1	45	45	55	67	70	75	80	81	83	102	182	190	190	190
High Lake, intermediate	54	71·4	30·0	65·8	28	31	42	52	57	62	68	78	81	87	100	142	192	192
High Lake, acid	16	1037·4	3857·3	80·5	24	25	31	37	46	47	52	70	82	142	307	15 500	15 500	15 500
High Lake, traverse *C1–C2*	57	364·1	2042·7	81·4	25	31	37	46	55	69	78	81	89	102	190	307	15 500	15 500
High Lake, traverse *C3*	37	67·6	19·6	64·2	28	43	47	52	57	62	68	75	80	85	92	101	107	107
Hackett River, all samples	37	67·7	83·3	42·1	7	9	11	23	28	34	43	53	60	101	146	370	370	370
Indin Lake, all volcanics	109	44·4	31·5	33·4	5	9	11	17	23	30	38	46	50	76	93	103	111	161
Indin Lake, basic	55	48·2	31·4	39·8	14	15	17	23	29	32	42	49	56	77	98	105	161	161
Indin Lake, intermediate	21	50·4	33·4	38·4	12	12	13	18	24	30	50	70	83	91	96	100	100	100
Indin Lake, acid	31	32·3	27·6	21·4	3	4	5	9	11	17	32	38	46	49	89	94	111	111
Indin Lake, sediments	81	78·2	14·8	76·4	56	58	61	67	70	75	81	84	87	91	96	99	102	112
Indin Lake, granites	5	66·8	14·7	65·5	47	47	47	62	63	63	77	77	77	85	85	85	85	85
Bode Lake, all samples	57	82·9	25·4	79·8	54	56	64	67	70	75	79	83	88	93	105	156	172	172
Terra Mine, all samples	65	117·7	163·3	74·0	20	21	30	36	43	51	72	86	103	146	273	383	787	1004
Copper																		
High Lake, all samples	94	71·0	125·1	33·1	3	4	5	14	25	33	37	44	58	76	174	404	544	840
High Lake, basic	22	73·4	99·0	40·6	5	5	8	23	34	37	44	61	65	98	181	462	462	462
High Lake, intermediate	54	62·9	123·2	31·6	3	4	5	18	28	33	39	43	50	60	83	183	840	840
High Lake, acid	16	72·1	134·5	25·0	3	3	5	6	12	14	30	43	61	101	182	544	544	544
High Lake, traverse *C1–C2*	57	87·2	156·4	32·0	3	3	4	6	18	29	39	46	62	93	183	481	840	840
High Lake, traverse *C3*	37	46·0	36·3	35·0	8	8	12	28	31	33	37	43	51	61	79	169	174	174
Hackett River, all samples	37	28·4	29·7	17·1	3	3	4	5	7	16	23	35	40	47	59	63	163	163
Indin Lake, all volcanics	109	57·6	58·4	24·6	2	2	3	3	8	18	42	65	88	109	138	160	170	280
Indin Lake, basic	55	98·2	52·6	80·8	6	29	42	57	78	85	93	107	118	138	160	170	280	280
Indin Lake, intermediate	21	23·0	31·1	10·3	2	2	3	3	3	5	16	23	31	36	84	127	127	127
Indin Lake, acid	31	6·7	6·2	5·0	2	2	2	3	3	3	4	6	8	12	15	18	30	30
Indin Lake, sediments	81	42·2	25·9	34·8	9	12	17	21	27	31	35	41	50	67	83	94	113	115
Indin Lake, granites	5	24·2	13·3	21·1	10	10	10	13	24	24	32	32	32	42	42	42	42	42
Bode Lake, all samples	57	16·3	60·8	5·3	1	1	2	2	3	3	4	6	9	16	31	35	462	462
Terra Mine, all samples	65	89·6	247·4	10·7	1	1	1	1	3	6	7	13	30	92	282	824	919	1522
Lead																		
High Lake, all samples	94	21·1	70·8	11·2	4	4	4	7	8	9	11	12	14	18	24	42	87	687
High Lake, basic	22	14·7	8·7	12·7	7	7	7	7	11	12	14	15	18	18	32	42	42	42
High Lake, intermediate	54	13·6	14·7	10·3	4	4	4	7	8	9	10	11	13	16	21	36	87	87
High Lake, acid	16	55·1	169·4	11·7	4	4	4	4	8	8	9	10	11	18	75	687	687	687
High Lake, traverse *C1–C2*	57	30·1	90·0	16·0	7	7	8	10	11	12	14	16	18	21	36	75	687	687
High Lake, traverse *C3*	37	7·2	3·1	6·5	4	4	4	4	4	7	7	8	9	10	12	14	14	14
Hackett River, all samples	37	15·0	20·5	10·5	4	4	4	7	8	9	11	11	13	14	31	55	122	122
Indin Lake, all volcanics	109	8·2	4·6	7·2	4	4	4	4	4	7	7	8	10	11	16	18	20	25
Indin Lake, basic	55	8·8	4·9	7·6	4	4	4	4	6	7	8	9	10	12	18	18	25	25
Indin Lake, intermediate	21	8·0	4·0	7·1	4	4	4	4	4	7	8	9	11	12	16	16	16	16
Indin Lake, acid	31	7·5	4·6	6·5	4	4	4	4	4	4	7	8	8	10	14	18	23	23
Indin Lake, sediments	81	13·6	4·1	12·9	8	8	9	10	12	12	13	15	15	17	18	20	26	27
Indin Lake, granites	5	11·2	4·1	10·3	4	4	4	12	13	13	13	13	13	14	14	14	14	14
Bode Lake, all samples	57	16·4	10·0	14·6	8	9	9	11	12	13	14	15	16	20	25	32	69	69
Terra Mine, all samples	65	27·2	67·0	15·7	8	8	8	10	12	13	13	14	15	20	49	75	184	522
Mercury																		
High Lake, all samples	94	20·8	57·6	9·7	2	2	5	5	6	7	10	12	14	15	25	56	310	458
High Lake, basic	22	15·9	29·6	8·5	2	2	2	5	6	7	9	10	12	15	43	143	143	143
High Lake, intermediate	54	11·0	7·1	9·3	2	5	5	5	6	7	10	12	14	15	17	19	47	47
High Lake, acid	16	41·3	112·7	10·2	2	5	5	5	5	6	6	8	9	56	67	458	458	458
High Lake, traverse *C1–C2*	57	26·4	73·6	8·7	2	2	4	5	5	6	7	7	10	17	47	143	458	458
High Lake, traverse *C3*	37	12·2	3·5	11·6	5	6	6	10	12	12	13	14	14	15	16	17	19	19
Hackett River, all samples	37	17·5	18·6	14·6	9	10	10	12	12	13	14	14	15	16	29	45	120	120
Indin Lake, all volcanics	109	10·5	5·9	9·2	2	5	5	6	7	8	10	11	12	14	16	17	32	38
Indin Lake, basic	55	8·5	5·9	7·4	2	5	5	5	6	7	8	8	9	10	12	17	38	38
Indin Lake, intermediate	21	12·3	6·9	11·1	6	6	6	7	10	10	12	12	15	16	16	38	38	38
Indin Lake, acid	31	13·3	2·7	13·0	8	10	11	11	12	13	14	14	14	15	17	17	22	22
Indin Lake, sediments	81	10·3	3·7	9·8	5	6	7	8	9	9	10	10	11	12	14	15	21	30
Indin Lake, granites	5	4·0	2·8	3·3	2	2	2	2	2	2	6	6	6	8	8	8	8	8
Bode Lake, all samples	57	6·0	2·0	5·6	2	2	4	5	5	5	6	7	7	7	9	10	10	10
Terra Mine, all samples	65	7·4	2·7	6·9	4	4	4	5	6	7	7	8	8	9	10	13	16	17
Arsenic																		
High Lake, all samples	94	5·5	7·9	3·8	3	3	3	3	3	3	3	3	3	3	10	30	40	40
High Lake, basic	22	5·0	6·2	3·8	3	3	3	3	3	3	3	3	3	5	15	30	30	30
High Lake, intermediate	54	4·7	7·1	3·5	3	3	3	3	3	3	3	3	3	3	5	10	40	40
High Lake, acid	16	6·9	8·5	4·7	3	3	3	3	3	3	3	3	5	15	15	35	35	35
High Lake, traverse *C1–C2*	57	6·3	8·8	4·2	3	3	3	3	3	3	3	3	3	5	15	35	40	40
High Lake, traverse *C3*	37	4·3	6·2	3·4	3	3	3	3	3	3	3	3	3	3	3	10	40	40
Hackett River, all samples	37	4·9	11·0	3·3	3	3	3	3	3	3	3	3	3	3	3	5	70	70
Indin Lake, all volcanics	109	3·8	3·6	3·3	3	3	3	3	3	3	3	3	3	3	3	15	15	30
Indin Lake, basic	55	4·0	4·5	3·3	3	3	3	3	3	3	3	3	3	3	3	15	30	30
Indin Lake, intermediate	21	3·9	3·0	3·4	3	3	3	3	3	3	3	3	3	3	10	15	15	15
Indin Lake, acid	31	3·5	2·2	3·2	3	3	3	3	3	3	3	3	3	3	3	5	15	15
Indin Lake, sediments	81	16·9	11·7	12·4	3	3	3	5	10	10	15	20	20	30	30	40	45	50
Indin Lake, granites	5	3·4	0·9	3·3	3	3	3	3	3	3	3	3	3	5	5	5	5	5
Bode Lake, all samples	57	3·1	0·4	3·0	3	3	3	3	3	3	3	3	3	3	3	3	6	6
Terra mine, all samples	65	6·2	9·1	4·1	3	3	3	3	3	3	3	3	3	5	25	30	40	45

intermediate rocks in the High Lake volcanic sequence may have become more 'basic' by metasomatic introduction of Mg, and reduction in the SiO_2 content, but with no associated introduction of Cu; secondly, the much higher Cu levels of the acidic rocks at High Lake in comparison with those from Indin Lake may reflect the association of a Cu–Zn deposit with the acid volcanics at this location.

The sediments from Indin Lake are moderately rich in Cu (arithmetic mean, 42 ppm) and, again, show a low dispersion of values around the mean.

Lead The background level of Pb in the rocks differs little from area to area, as is indicated by the median values of 11, 11, 7 and 13 ppm for, respectively, the High Lake area, the Hackett River area, the Indin Lake volcanic rocks and the Indin Lake sediments. There is also very little difference between the geometric mean and median values for acid, intermediate and basic lavas at either High Lake or Indin Lake. The most notable feature of the Pb distribution is a marked skewing to high values of Pb for the higher percentiles of the High Lake and Hackett River distributions, which reflects the presence of Pb mineralization, associated with ore deposits, in these samples. This is particularly marked for the acidic volcanics at High Lake.

Mercury Hg shows a rather uniform distribution, around the 10 ppb Hg level, in the rocks sampled from Slave Province (Table 3). The marked skewing to high values of Hg in the High Lake and Hackett River samples indicates the association of this element with sulphide mineralization.

Arsenic Arsenic also shows a uniform distribution. The determinations for most samples fall close to or below the detection limit of 3 ppm. In performing calculations this value is allocated to all samples showing some colour in the final solution. There is no marked association between the sulphide mineralization and the arsenic content of the rocks at either High Lake or Hackett River. The most interesting feature of the results is the much higher level of the values for the Indin Lake sediments. The median of 15 ppm is five times greater than that of any other group. The arithmetic mean of 16·9 ppm is three to four times that of the other groups. These differences would possibly be greater if this element could have been measured precisely at lower detection levels. If all samples assigned values of 3 ppm are, instead, assigned values of 0 ppm, the averages become: High Lake (94), 3·0 ppm; Hackett River (37), 2·0 ppm; Indin volcanic rocks (109), 1·0 ppm; Indin Lake sediments (81), 16·3 ppm. These values are in good agreement with Onishi's[32] average for igneous rocks of 1·5 ppm As and for shales of 13 ppm.

Areas in Bear Province

Major elements

In Fig. 8 the major-element data for the 57 samples from Bode Lake are plotted. All of these samples fall within the limits 55–67·9% SiO_2 used above to define intermediate igneous rocks. The rocks at Terra mine are a rather heterogeneous mixture of sediments and volcanics, and are, therefore, not shown. The Bode Lake data for Al_2O_3, total Fe as Fe_2O_3, MgO, CaO, TiO_2 and MnO show no marked deviation from the Archaean trends. K_2O and Ba are, however, much higher in the Bode Lake rocks, and Na_2O is lower. Eade and Fahrig[20] have previously documented the higher level of K_2O in Proterozoic rocks of the Shield in comparison with those of Archaean age. They did not note any significant difference for Ba.

Trace elements

Zinc For Zn the median and geometric mean for Bode Lake and Terra are at approximately the same levels (Table 3) as for High Lake. At higher percentile ranges the Bode Lake and Terra frequency distribution curves distinctly diverge from each other. The former is rather flat—indicative of a general absence of zinc mineralization along the traverses. The Terra data curves sharply up to higher Zn values in a similar fashion to the data from High Lake.

Copper The median values for Cu at Bode Lake and Terra (4 and 7 ppm, respectively) and the geometric means (5 and 11 ppm) are only a fraction of the same parameters at High Lake (37 and 33 ppm). The Bode Lake frequency distribution curve continues to be low up to the highest percentile levels. Only one sample, at 462 ppm, contains moderate Cu; the next highest is 36 ppm. This confirms the data for Zn on the very weakly mineralized nature of the rocks at Bode Lake. The mineral prospect at Bode Lake has not, to the present, been shown to be other than minor. The Terra distribution curve shows a rapid escalation in Cu values at higher percentile values. Thus, at Terra the zone containing the ore deposits contrasts much more sharply with the background level of Cu in rocks of the area than was the case with High Lake. Although this feature will be of assistance in making detailed geochemical surveys by use of soils, stream sediments or lake muds, it is not helpful for reconnaissance surveys. At Terra, more so than in the other areas, the anomalous results for Zn, Cu, Pb and As are mainly confined to a zone extending several hundred feet on either side of the stratiform copper ore horizon. Traverse *B4* on the opposite side of the Camsell River from traverses *B1–B3* shows few high results for any of these elements.

Lead The median Pb values for Bode Lake and Terra (14 and 13 ppm) and the geometric mean values (15 and 16 ppm) are slightly higher than the equivalent values for the Archaean areas. The distribution curve for the element at Terra resembles that at High Lake because of the effect of mineralization in the two areas. The Bode Lake curve shows moderate escalation.

Mercury For Hg the featureless nature of its distribution, noted for the Archaean areas, persists for the areas in Bear Province. The median values of 7 ppb at Terra and 6 ppb for Bode Lake are slightly smaller than those for the Archaean areas. At High Lake and Hackett River some high Hg values were associated with the mineralization;

at Bode Lake and Terra there are none. At Terra the highest Hg value is 17 ppb; at Bode Lake 10 ppb. For these two areas this element would appear to be of no value for either reconnaissance or detailed geochemical surveys.

Arsenic One sample from Bode Lake contains an As concentration greater than the detection limit of 3 ppm. At Terra there are a number of values greater than 3 ppm in the zone around the stratiform Cu horizon and silver arsenide veins. The arithmetic mean and geometric mean values and the distribution curves for As are very similar for the High Lake and Terra data.

Major-element composition of lake sediments

One of the more interesting aspects of this study is the major-element composition of the lake sediment samples. These data allow an assessment to be made of the effectiveness of lake sediments as a composite sample of the surrounding bedrock. The results obtained suggest that lake sampling may be useful for geological mapping as well as for mineral exploration. In Table 4 are listed the average major- and minor-element compositions of the lake sediments for each area. A comparison is made of the means for different areas, of the means for Slave Province samples with those from the Bear, and of the sediment data with the estimated composition of the underlying bedrock. In order to provide a meaningful comparison of the sediment data with that for bedrock all data given in Table 4 are corrected to 100% in terms of the oxides listed. This eliminates variation caused by changes in the moisture and organic content of the sediment samples.

The average of the five areas sampled for sediments within Slave Province is compared with Eade and Fahrig's[20] average of the Archaean for the Canadian Shield; and the average of the Terra and Bode Lake areas is compared with their average Proterozoic rock. The McGregor Lake area results have been excluded from the Bear Province average because these sediments were, in part, derived from an uncommon (ultramafic) rock type.

In comparison with the Archaean bedrock, the Slave Province lake sediments show a large percentage decrease in Ca, Na and Ba. The former two components are the major elements most readily lost during rock weathering.[11,15] The apparent loss in Ba may be due to analytical bias between laboratories for this minor element. The next largest decrease is for Fe and Mn. It is possible that some of these components move to the surface of the sediment during diagenesis; the iron-rich layer seen at the sediment–water interface at many sample sites was excluded from all the samples taken. For all other elements the difference between the sediment and rock averages is quite small in proportion to the amount present. The largest absolute difference is for silica. The amount of silica and alumina in the samples

Table 4 Comparison of major-element data for lake sediments

Area	Row	SiO_2	Al_2O_3	Fe_2O_3	MgO	CaO	Na_2O	K_2O	TiO_2	MnO	Ba
Indin Lake	1	69·0	15·5	4·23	2·39	2·29	3·19	2·62	0·66	0·072	0·062
High Lake	2	71·3	14·0	4·33	2·33	1·96	2·53	2·69	0·69	0·067	0·049
Muskox Lake	3	71·4	13·8	4·38	2·59	1·95	2·68	2·52	0·56	0·074	0·055
Harding Lake	4	70·6	15·8	3·89	1·93	1·42	2·72	3·07	0·46	0·055	0·060
Hackett River	5	78·2	11·7	2·84	1·33	1·05	2·37	1·99	0·40	0·044	0·037
Average R1–R5	6	72·1	14·2	3·93	2·11	1·73	2·70	2·58	0·55	0·062	0·053
Average Archaean	7	65·7	16·1	4·8	2·3	3·4	4·1	2·72	0·50	0·080	0·080
$\frac{(R6-R7) \times 100}{R7}$	8	+10%	−12%	−18%	−8%	−49%	−34%	−5%	+10%	−23%	−34%
Bode Lake	9	72·0	14·5	2·93	1·72	2·04	3·11	3·08	0·49	0·062	0·063
Terra mine	10	68·7	15·1	4·78	2·67	1·51	2·39	3·95	0·75	0·091	0·073
Average R9–R10	11	70·4	14·8	3·86	2·20	1·78	2·75	3·52	0·62	0·077	0·068
Average Proterozoic	12	65·5	16·1	5·0	2·1	3·3	3·5	3·54	0·62	0·091	0·082
$\frac{(R11-R12) \times 100}{R12}$	13	+7%	−8%	−23%	+5%	−46%	−21%	−1%	0%	−15%	−17%
$\frac{(R11-R6) \times 100}{R6}$	14	−2%	+4%	−2%	+4%	+3%	+2%	+36%	+13%	+24%	+28%
McGregor Lake	15	66·7	13·4	6·81	6·20	1·48	1·55	3·08	0·69	0·116	0·048

Data in rows 7 and 12 from Eade and Fahrig.[20] Row 8 is difference between Slave Province lake sediments and average Archaean rocks; row 13 is difference between Bear Province sediments and average Proterozoic rocks; row 14 is difference between lake sediments of Bear and Slave Provinces. All data converted to 100% in terms of oxides given above.

characterizes them as silty rather than argillaceous. The differences between the Bode Lake and Terra mine means and the average Proterozoic rock are almost precisely the same as for the Slave comparison.

Comparison of the Bear with the Slave lake sediment averages show that there are negligible differences for Si, Al, Fe, Mg, Ca and Na. There is a relative enrichment of 36% for K in the Proterozoic lake sediments and 13% for Ti, 24% for Mn and 28% for Ba. For K and Ti Eade and Fahrig[20] reported differences of this magnitude for the Archaean and Proterozoic crust of the Shield. They did not detect any marked difference for Mn or Ba.

Comparing individual areas, the most striking differences are observed for McGregor Lake. The sediments of this area reflect the contribution of ultramafic rocks in much higher amounts of Fe, Mg and Mn and a higher amount of Ti, with lower contents of Si, Al, Ca and Ba. Variation among the averages for other areas is much less. Indin Lake, High Lake and Muskox Lake—areas principally underlain by Archaean volcanic and sedimentary rocks, with some granitic material—are remarkably similar in their major- and minor-element composition. The Hackett River lake sediments are more siliceous: this is associated with the siliceous nature of the acid volcanics and sedimentary rocks of the area. If the calcitic component of the analyses is removed from the rock data for Hackett River, the sampled rocks are more siliceous than those from either High Lake or Indin Lake. The sediment data from

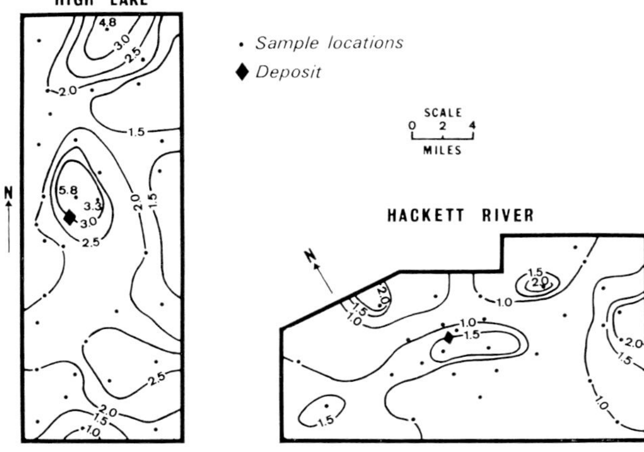

Fig. 9 Distribution of MgO, %, in lake sediments, High Lake and Hackett River areas

Harding Lake suggest that the underlying 'granitic' material is of granodioritic composition.

On the basis of the above information it can be suggested that major- and minor-element analyses of lake sediments can produce maps of the bedrock geology and geochemistry that will complement present types of mapping.

It is not proposed here to make any extensive discussion of the mapping of the major-element variations within the areas studied—this will be covered in a subsequent paper that will describe the multivariate statistical treatment of the major- and trace-element data. As an example of the variation that is obtained, however, contour maps for MgO in the lake sediments of the High Lake and Hackett River areas are shown in Fig. 9. In the discussion on the geochemistry of the rocks it was noted that there was Mg metasomatism of the rocks surrounding the vent and ore deposit at High Lake. The contoured plots for this area show an anomaly in the MgO content of the sediments near to the ore deposit. At Hackett River the MgO content of the sediments is low along the band of acid volcanics, because of their silicic character. There is, however, a MgO anomaly close to the known vent and ore deposit. In the typical geological environment of the Archaean massive sulphide ore deposit this pattern of low MgO values might be expected over the acid volcanic horizons, with anomalies outlining the metasomatized vent structures. There are at both High Lake and Hackett River other MgO anomalies, most of which can be eliminated as possible mineralized vents by comparing the trace-element levels of Zn and Ag at these locations. Nevertheless, at least one unexplained MgO anomaly exists at both locations.

Trace-element composition of lake sediments

Each trace element is discussed below separately. Variation between and within the areas can be considered in relation to their geology (Fig. 10) and to known or suspected mineralization. Because of space considerations, only seven of the eight sampled areas are shown as contour diagrams. *Not included in the contour diagrams are the data for the highly mineralized lakes adjacent to the High Lake and Hackett River ore deposits. These data are presented in Table 5.* Statistical data on the major- and trace-element content of sediments from all eight areas are given in Table 6. In comparing the sediment with the rock data median values have generally been used in order to avoid the untoward effects of a few mineralized samples.

Zinc

The area with the highest and most extensive Zn concentrations is High Lake (Fig. 11), with a median content of 124 ppm. The background

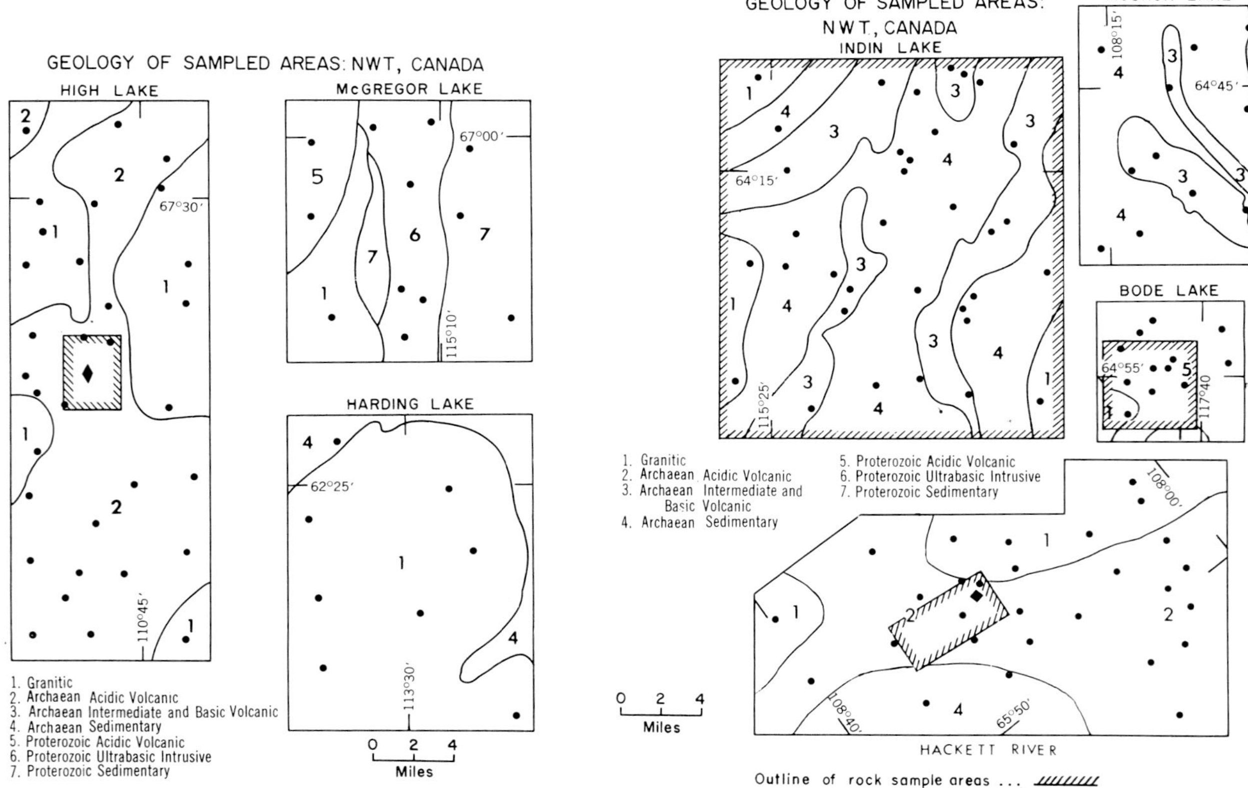

Fig. 10 Geology of areas sampled for lake sediments. Cu–Zn deposit located by diamond symbol at left centre of High Lake map; Pb–Zn–Ag deposit by diamond symbol at centre of Hackett River map; dots represent sample locations

level of abundance at Hackett River is much less, with a median of 44 ppm Zn. The Pb–Zn–Ag deposit at Hackett River, however, is clearly defined by a sharp anomaly, which, at its peak, has Zn values as high as those at High Lake. From these two examples it can be seen that areas of ore potential may be outlined by broad areas of high values or by relatively less extensive and sharper anomalies in the lake sediment results. After a reconnaissance survey perhaps 50% of the area sampled at High Lake might have been selected for detailed follow-up survey (say, within the 160 ppm Zn contour); but much less for Hackett River.

It was shown above that there is a similar trend in the background for Zn in the rocks of the two areas; the lake sediments in the High Lake area do, however, have a higher median value of Zn than the rocks that were sampled. At High Lake the 100 ppm contour very approximately marks the granite–greenstone contact.

All of the other areas sampled show median values rather higher than those at Hackett River; but only two, McGregor Lake and Indin Lake,

*Table 5 Major- and minor-element concentrations in −250-mesh sediments from lakes adjacent to Archaean massive sulphide deposits**

	Major elements: arithmetic means, %									Minor elements: arithmetic means, ppm							
Sample type	SiO_2	Al_2O_3	Fe_2O_3	MgO	CaO	Na_2O	K_2O	TiO_2	MnO	Cu	Pb	Zn	Ag	Ni	Co	Hg	Ba
High Lake†	62·9	14·0	5·00	2·49	1·54	2·43	2·29	0·67	0·07	3700	88	1791	1·1	47	50	0·042	40
Camp Lake‡	69·1	12·1	3·00	1·66	1·12	2·02	1·69	0·50	0·03	624	140	1419	0·9	32	11	0·048	40
Iron-rich sample§	41·5	32·5	26·5	0·99	0·00	0·54	2·77	0·83	0·01	278	319	213	25·6	6	5	0·075	40

*The concentrations for High Lake and for Camp Lake given above were not used in any of the regional diagrams (Figs. 11–17). They are presented to show the great concentration difference between the regional trace-element levels and those in lakes adjacent to orebodies.

†Average of nine samples evenly distributed over High Lake which is the closest lake to the 5 000 000-ton Cu–Zn deposit in High Lake area (Fig. 3).

‡Average of three samples evenly distributed over Camp Lake which is the closest lake to the 10 000 000-ton Pb–Zn deposit in Hackett River area (Fig. 4). Compare with analysis of Hackett River ore given in text.

§This sample from the High Lake area is anomalous in Ag (25·6 ppm; Fig. 15) and is referred to in the text as the 'iron-rich sample'. It is not incorporated in the statistical treatments, but is included in Figs. 11–17. The values are for one sample.

Table 6 Statistical data on major- and trace-element content of lake sediments

Area		SiO_2	Al_2O_3	Fe_2O_3	MgO	CaO	Na_2O	K_2O	TiO_2	MnO	Ba	Zn	Cu	Pb	Ni	Co	Ag	Hg
High	$\bar{X}$	66·9	13·1	4·06	2·19	1·84	2·37	2·53	0·65	0·063	0·046	138·5	55·5	22·5	31·7	15·9	0·67	25·2
Lake	S	5·2	1·1	1·47	0·96	0·67	0·36	0·48	0·16	0·021	0·008	75·7	28·2	43·2	16·3	10·3	0·18	14·2
$N=31$	G	66·7	13·1	3·83	2·05	1·73	2·34	2·48	0·63	0·060	0·045	120·0	46·9	14·1	27·5	13·1	0·63	22·4
	M	68·5	13·0	3·53	1·99	1·82	2·45	2·55	0·65	0·061	0·050	124·0	51·0	11·0	28·0	14·0	0·70	20·0
Hackett	$\bar{X}$	73·5	11·0	2·67	1·25	0·99	2·22	1·87	0·37	0·042	0·035	71·3	34·1	29·3	24·0	9·2	0·60	19·7
River	S	7·5	2·0	1·36	0·55	0·36	0·67	0·55	0·11	0·016	0·013	71·5	32·5	95·1	27·4	11·6	1·16	11·7
$N=28$	G	73·2	10·8	2·37	1·15	0·91	2·14	1·81	0·35	0·039	0·033	50·5	24·7	11·9	16·8	6·3	0·30	17·4
	M	74·5	11·0	2·14	1·16	1·12	2·04	1·69	0·38	0·039	0·030	44·0	22·0	11·0	16·0	5·0	0·20	17·0
Indin	$\bar{X}$	65·7	14·7	4·03	2·28	2·17	3·04	2·49	0·63	0·068	0·059	75·8	37·7	12·1	35·1	15·2	0·37	14·3
Lake	S	5·6	0·8	1·46	0·80	0·38	0·22	0·54	0·13	0·026	0·013	47·4	26·5	2·8	16·3	8·7	0·19	5·0
$N=35$	G	65·4	14·7	3·79	2·17	2·14	3·03	2·43	0·62	0·065	0·057	66·5	31·5	11·7	32·3	13·3	0·33	13·8
	M	66·4	14·7	3·93	2·32	2·14	3·06	2·59	0·67	0·066	0·060	70·0	31·0	12·0	32·0	14·0	0·30	14·0
Muskox	$\bar{X}$	66·5	12·9	4·08	2·41	1·82	2·50	2·35	0·53	0·069	0·051	97·6	41·8	9·4	45·7	19·7	0·42	54·9
Lake	S	7·9	1·2	1·01	0·69	0·30	0·20	0·37	0·14	0·017	0·009	33·4	16·5	2·4	13·0	6·3	0·13	16·4
$N=11$	G	66·0	12·8	3·94	2·30	1·80	2·49	2·32	0·51	0·067	0·050	93·4	38·6	9·1	44·0	18·9	0·40	52·8
	M	69·1	13·8	4·63	2·65	1·96	2·44	2·43	0·52	0·075	0·050	96·0	41·0	9·0	50·0	21·0	0·50	59·0
Harding	$\bar{X}$	63·7	14·3	3·51	1·74	1·28	2·46	2·77	0·41	0·049	0·054	55·2	25·4	9·8	27·7	10·9	0·50	24·0
Lake	S	8·3	1·8	1·84	0·93	0·24	0·75	0·65	0·14	0·018	0·022	26·9	10·0	4·2	9·7	5·4	0·19	11·9
$N=8$	G	63·2	14·2	3·01	1·50	1·26	2·36	2·70	0·39	0·046	0·050	49·0	23·6	9·0	26·2	9·5	0·47	21·7
	M	63·8	14·7	4·46	2·16	1·26	2·43	3·16	0·45	0·050	0·060	56·0	27·0	11·0	29·0	12·0	0·50	17·0
Bode	$\bar{X}$	68·3	13·8	2·78	1·63	1·94	2·95	2·92	0·46	0·059	0·060	59·3	23·2	12·3	23·5	11·4	0·37	44·7
Lake	S	5·7	1·3	1·59	0·79	0·35	0·34	0·58	0·16	0·021	0·009	32·5	12·6	6·5	15·6	8·5	0·24	17·3
$N=12$	G	68·1	13·7	2·46	1·48	1·90	2·93	2·88	0·43	0·056	0·059	51·4	19·8	10·7	19·2	8·6	0·30	42·2
	M	69·6	13·4	2·43	1·33	1·96	3·10	2·81	0·45	0·052	0·060	61·0	30·0	13·0	18·0	9·0	0·30	37·0
Terra	$\bar{X}$	66·0	14·5	4·60	2·57	1·45	2·30	3·79	0·72	0·088	0·070	95·9	31·6	20·4	34·4	18·3	0·50	31·6
Mine	S	7·4	1·9	1·81	1·09	0·36	0·26	0·48	0·14	0·036	0·012	31·8	11·1	6·2	17·2	10·5	0·24	19·7
$N=39$	G	65·6	14·4	4·25	2·34	1·41	2·28	3·76	0·71	0·082	0·069	90·3	29·3	19·4	29·7	15·6	0·44	28·3
	M	67·2	14·4	4·82	2·65	1·40	2·29	3·92	0·75	0·076	0·070	95·0	34·0	21·0	34·0	18·0	0·50	31·0
McGregor	$\bar{X}$	57·1	11·4	5·83	5·31	1·27	1·33	2·63	0·59	0·100	0·041	114·0	48·6	20·1	185·0	26·5	0·58	21·0
Lake	S	9·0	2·6	2·02	3·38	0·87	0·37	0·71	0·14	0·036	0·010	48·6	23·3	7·6	236·0	23·6	0·17	6·6
$N=12$	G	56·4	11·2	5·55	4·54	1·04	1·28	2·52	0·58	0·094	0·039	105·0	44·0	19·0	92·0	20·7	0·54	20·1
	M	56·9	11·7	5·29	4·31	0·84	1·35	2·72	0·57	0·093	0·040	129·0	43·0	18·0	132·0	18·0	0·60	20·0

$\bar{X}$, arithmetic mean; S, standard deviation; G, geometric mean; M, median.
SiO_2 to MnO,%; Ba to Ag, ppm; Hg, ppb.

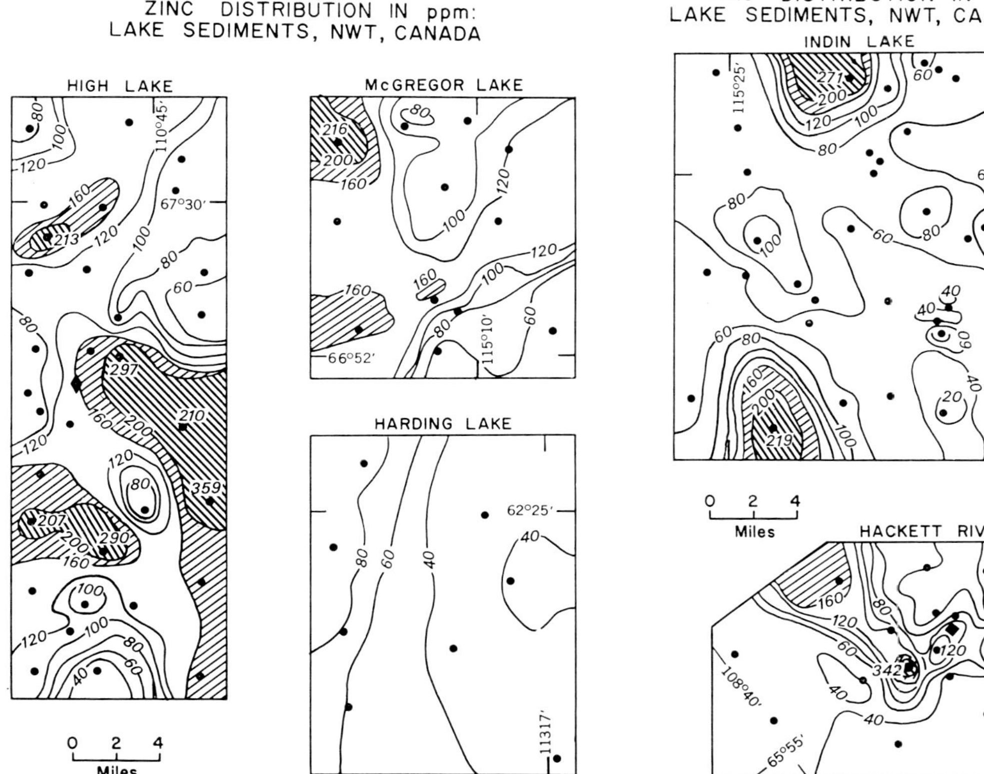

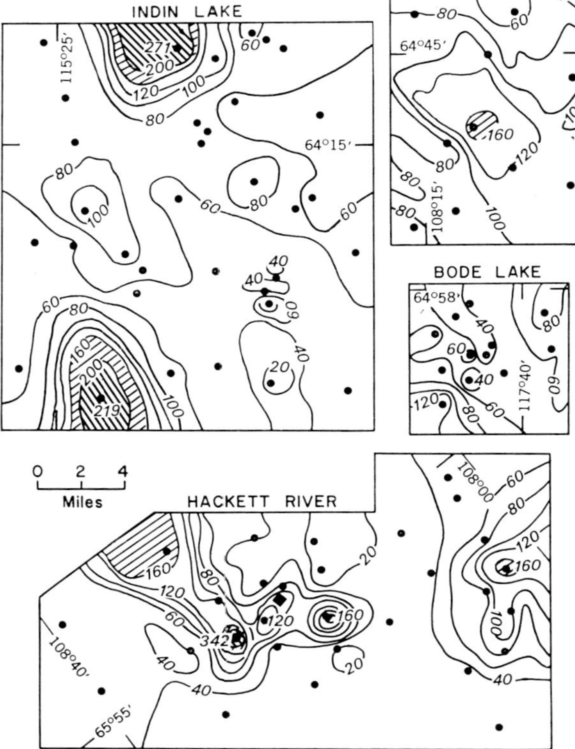

Fig. 11 Distribution of Zn in lake sediments of study areas

show sharp single sample anomalies. For all of these areas, apart from High Lake, the level of abundance in the rocks and lake sediments is in broad agreement.

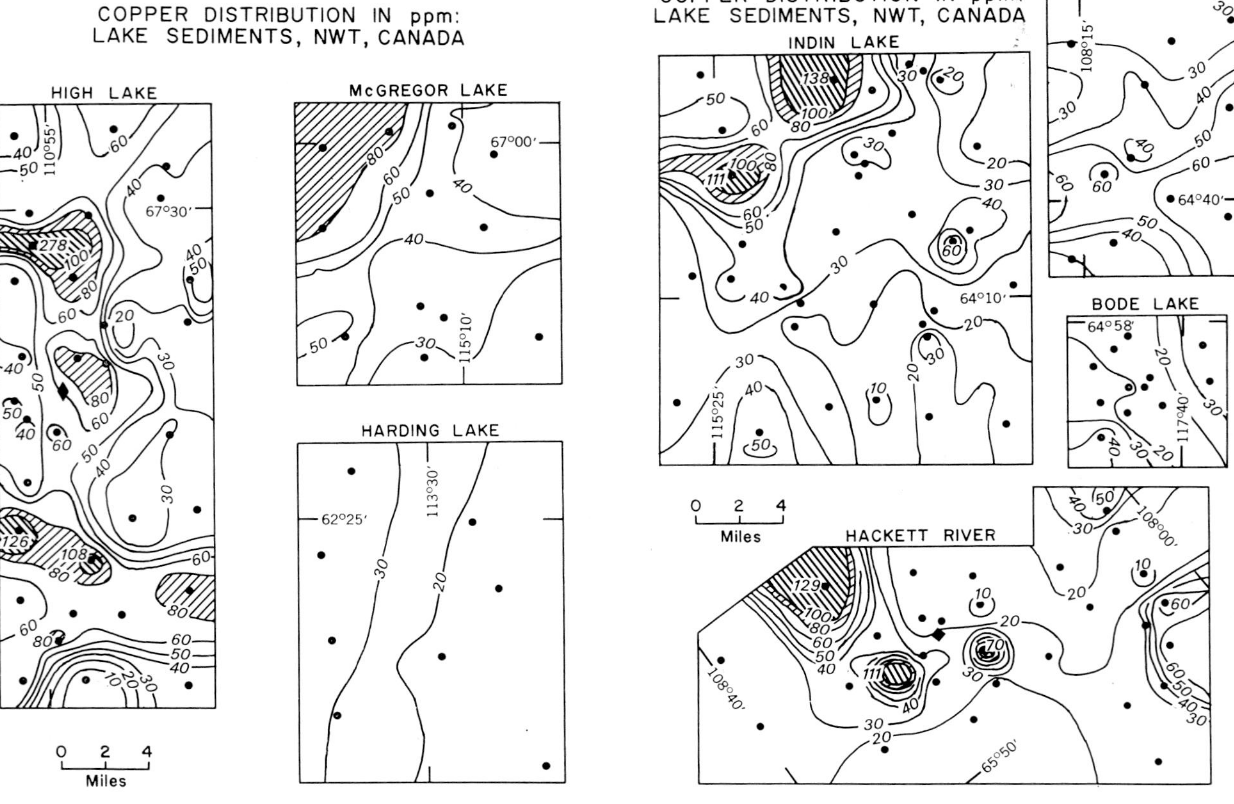

Fig. 12 Distribution of Cu in lake sediments of study areas

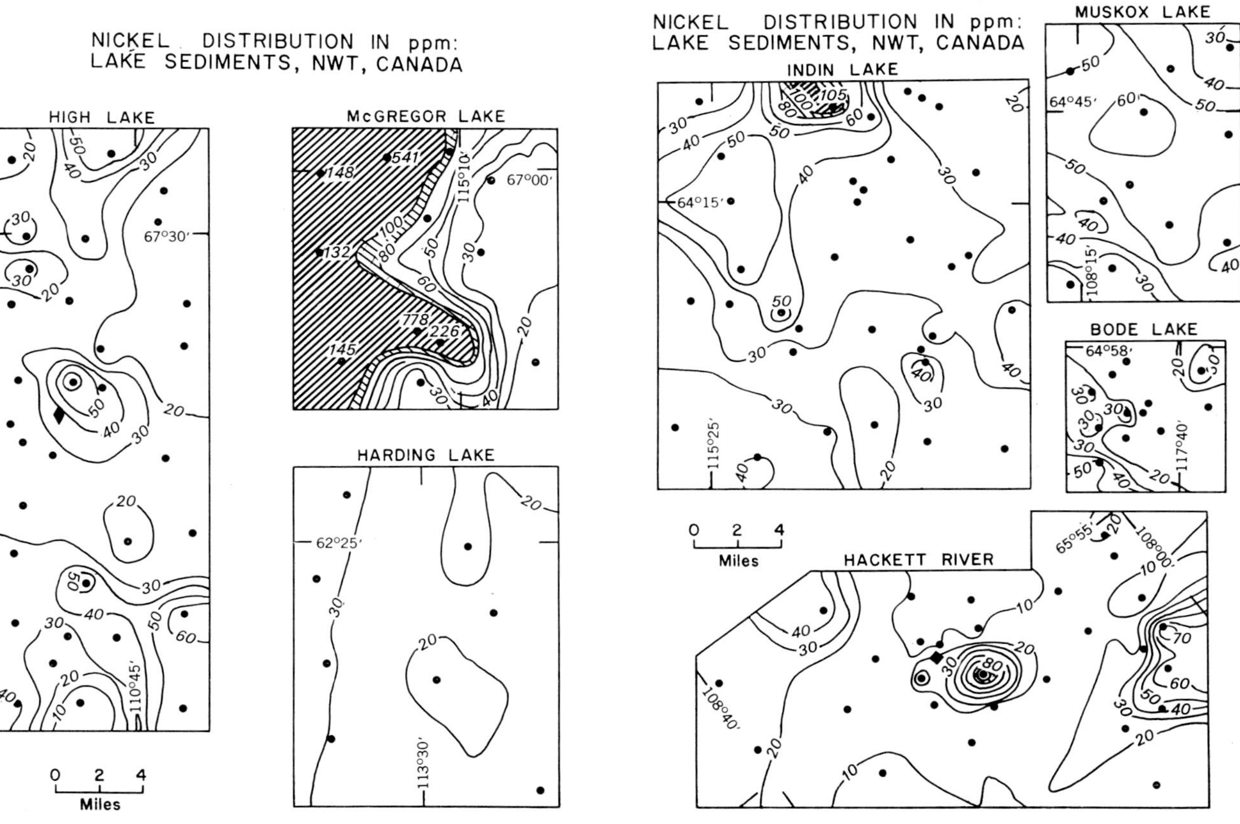

Fig. 13 Distribution of Ni in lake sediments of study areas

Copper

Concentrations of Cu in lake sediments (Fig. 12) produce somewhat similar results to those for Zn. The areas which appear to have the most favourable anomalies are High Lake, Indin Lake, Hackett River and McGregor Lake. High Lake again has the highest background level. Since this is the one area known to contain a large Cu–Zn deposit, this is an entirely satisfactory result. At Hackett River the higher Cu concentrations are found in the same area as the high Zn results. This may indicate the same type of ore association in this area as is present at High Lake. The anomalous Cu and Zn concentrations at Hackett River appear to lie along a zone in the acid volcanics.

At Indin Lake there is an interesting anomaly for Cu in the northwest corner of the area, where it coincides with a Zn anomaly. Airborne electromagnetic surveys in the Indin Lake area have been confined to the eastern volcanic belts. On the basis of the analyses, the northwestern volcanic belt might be more thoroughly investigated in order to explain the high Cu and Zn results.

Aside from these anomalous zones, the concentration of Cu in the lake sediments correlates well with the geology and the concentration of Cu in the different rock types. The 40 ppm contour at High Lake roughly parallels the granite–greenstone contact. The Hackett River area, the rocks of which are silicic, is low in Cu in the non-anomalous parts of the area. The area of granodiorites at Harding Lake is low and shows little contrast. Within the Indin Lake region the basic and intermediate volcanic areas are richer in Cu than those parts of the area underlain by acid volcanics. The low median Cu values of the rock samples from Bode Lake and Terra mine are also matched by similarly low results for the sediments, although in the former case the geometric mean Cu concentration in the sediment (Table 6) is unusually high relative to the rock mean Cu concentrations (Table 3). One area that has not been discussed so far is McGregor Lake. The Cu and Zn concentrations here are best reviewed with nickel.

Nickel

Lake sediments in the McGregor Lake area, especially where the sample sites lie within the Muskox Intrusion, have by far the highest Ni concentrations of any area sampled (Fig. 13), with up to 778 ppm. To the west of the intrusion, in country underlain by acidic volcanics and gneiss, there are moderately high Ni values in the range 132–148 ppm. These and the higher values over the intrusion show a good correlation with the MgO content of the sediments; it therefore appears likely that the enhanced Ni content of these sediments has principally come from the Ni contained in the silicate minerals of the ultrabasic rock. The continental ice sheet, which moved from the southeast to the northwest in this region, may have carried ultrabasic rock detritus to the west of the Muskox Intrusion.

Nickel sulphide mineralization of the Shield, which is associated with ultrabasic rocks, is almost invariably accompanied by copper sulphides.[13] In such areas as the Ungava nickel belt, where there are large Ni sulphide zones, the Cu concentrations in lake sediments substantiate this concept.[4] Here, Ni concentrations of 100–200 ppm in lake sediment are accompanied by Cu concentrations of 80–100 ppm Cu, with a high value of 693 ppm Cu in the sediment closest to the ore zones of the Donaldson mine at the Raglan property. In many of the high Ni lake sediments from the Muskox Intrusion the Cu is not particularly high (40–50 ppm)—confirming the above interpretation of the origin of the high Ni values. In the northwest quadrant of the sampled area, however, relatively high Cu and Zn values accompany the enhanced Ni levels (Figs. 11 and 12). This region may merit further investigation.

Although no other areas have Ni concentrations approaching those at McGregor Lake, there are lower, but sharp, Ni anomalies at Hackett River, Indin Lake and High Lake. The latter is in the general area of the ore deposit and vent and coincides with the MgO anomaly (Fig. 9). Analyses of the High Lake rock samples for Ni that became available as this paper was being prepared show scattered (to 713 ppm) Ni mineralization in the metasomatized rocks around the vent. A similar explanation may also account for the Ni anomaly near the ore deposit at Hackett River. Analyses of the sediments from the lakes immediately adjacent to the ore deposits at High Lake and Hackett River, which have not been included in the contour diagrams for any element, do not show high levels of Ni (Table 5). The actual ore at Hackett River, however, also has a low Ni concentration (47 ppm). These Ni anomalies are thus associated with vent metasomatism rather than with the actual ore. (A sample of ore-grade material from Hackett River analysed 28% Zn, 210 ppm Cu, 97 ppm Pb, 21 ppm Ag, 47 ppm Ni and 103 ppm Co.)

The Ni levels in non-anomalous parts of the different areas correlate reasonably well with the geology; sediments overlying acidic rocks are generally low, those overlying basic rocks generally high. Thus, the median value for Hackett River (Table 6) is lowest of all, confirming that this area

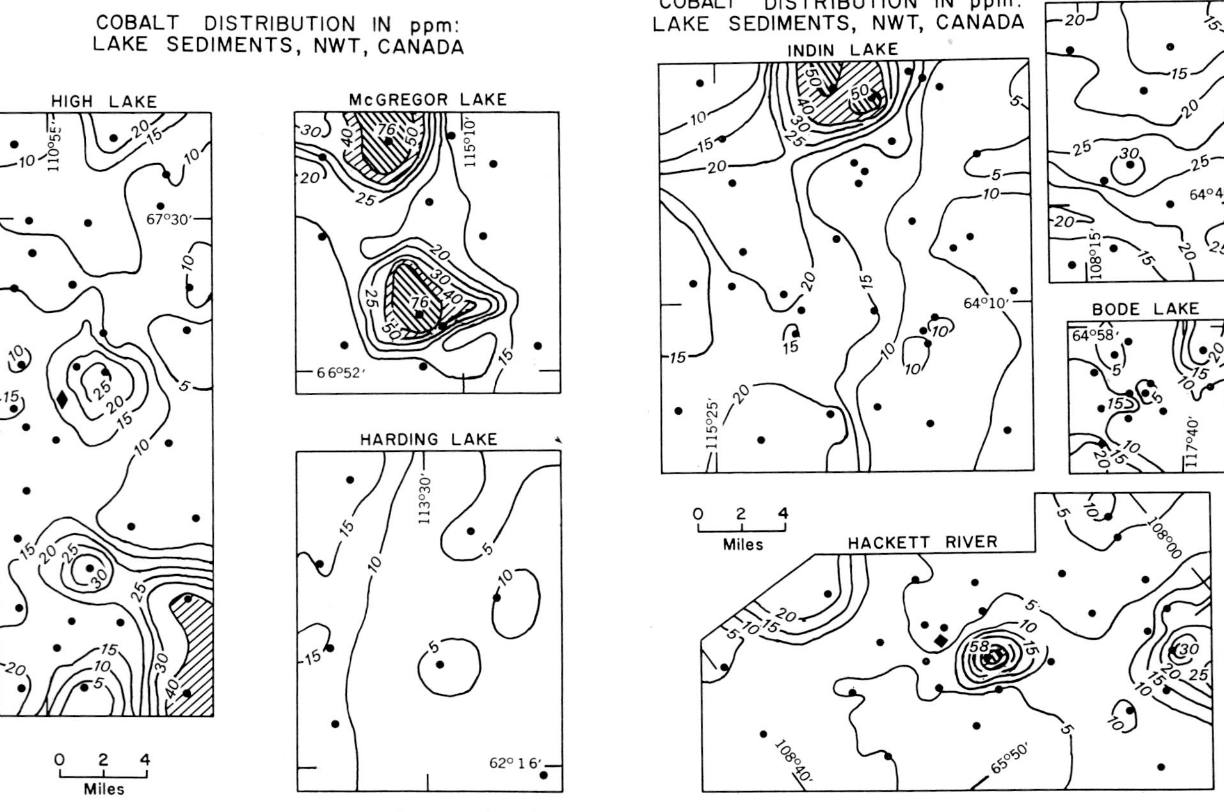

Fig. 14 Distribution of Co in lake sediments of study areas

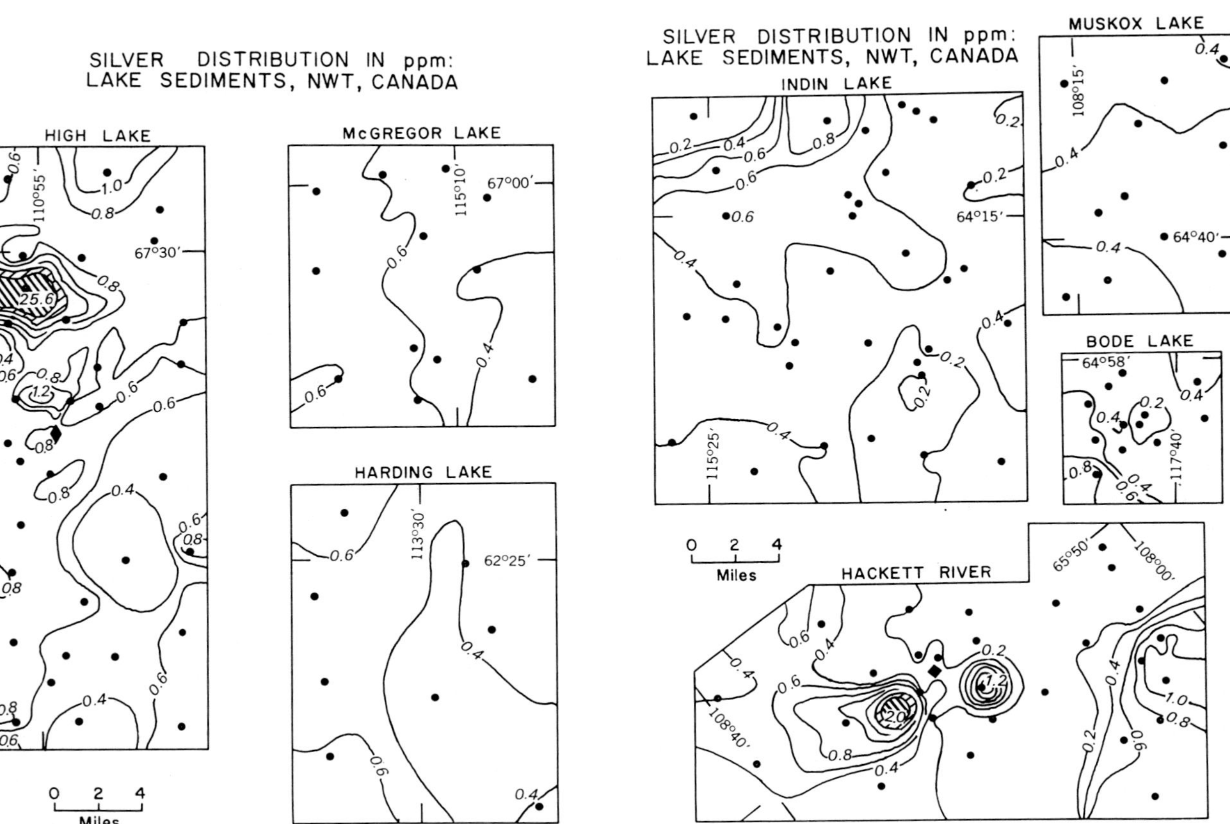

Fig. 15 Distribution of Ag in lake sediments of study areas

is largely underlain by silicic volcanic rocks and sediments that are poor in the base metals. The levels are also rather low at Harding Lake and Bode Lake, but higher at Muskox Lake, where there are a good deal of basic volcanic rocks. To what extent the moderately high Ni values at different points within the High Lake region are associated with sulphide mineralization is undecided, since the levels are rather higher than one might expect from the geology of the area.

Cobalt

The distribution patterns for Co (Fig. 14) are very similar to that for Ni. At McGregor Lake, however, the high Co values are more closely restricted to the area underlain by the ultrabasic Muskox Intrusion than are the high Ni values (Fig. 13). Again, at Hackett River, High Lake and Indin Lake the general pattern of Co anomalies parallels the Ni anomalies, and is presumably also caused by mineralization around a vent—at least for the first two examples. Away from the anomalous areas the distribution of Co closely follows that of Ni, being low in areas of acidic rocks and higher in areas of basic volcanics. Again, the Hackett River area has the lowest Co median, followed by Bode Lake and Harding Lake. As with Ni, the Muskox Lake area is intermediate in Co concentration.

Silver

The McGregor Lake, Harding Lake, Muskox Lake and Bode Lake areas have uniformly low Ag contents in their lake sediments (Fig. 15). In detail, such as at Indin Lake, the higher Ag values can be related to areas of basic rocks. There is no marked enhancement of Ag in lake sediments in the area of the Ag deposits at Terra mine (Table 6). The most striking areas in terms of Ag anomalies are Hackett River and High Lake. Two Ag values greater than 1 ppm were found in the former area. They are about five miles apart, straddling this Pb–Zn–Ag deposit. They stand out strongly against the low background of the silicic rocks of the surrounding area. These anomalies are clearly related to the massive sulphide mineralization of the Hackett River camp, and thus confirm the pattern given by the other elements. At High Lake there is an anomaly of the same magnitude immediately north of the location of the ore deposit. Farther north there is the unusually high value of 25·6 ppm Ag in iron-rich sediments taken from a lake that has a large gossan at its northern edge. This anomalous sample is listed in Table 5 rather than in the compilations given in Table 6. Zn and Cu are also higher at this site, but not Ni and Co. Apart from these two anomalies, a number of other samples from High Lake have fairly high values in comparison with other areas. This is further evidence that this entire area is rather rich in base metals and, as such, is suitable for further prospecting.

Lead

The pattern of Pb distribution is very similar to that for Ag (Fig. 16). High Lake and Hackett River again provide most of the interest with a number of prominent anomalies. At High Lake the Fe- and Ag-rich sample is also enriched in Pb. There is a prominent Pb anomaly to the east of the deposit at High Lake and another in the southeast quadrant. At Hackett River there is a strong anomaly to the west of the deposit, in the same location as anomalies for Ag, Cu and Zn. Another, weaker, anomaly is close to the southeastern border of the area. This location contains anomalous levels of a number of other metals; it would appear to be worth further study. The background level of Pb in the lake sediments of the different areas corresponds well with the median values previously established for the rocks. Because there is very little difference in the Pb content of acid, intermediate or basic rocks, this element shows little contrast in its distribution across many areas, such as Indin Lake, Muskox Lake, McGregor Lake and Harding Lake.

Mercury

There are clear anomalies for Hg in the lake sediments from the Hackett River sites (Fig. 17). They flank either side of the ore deposit, and coincide with other base-metal anomalies. The higher value at the northwest site at Hackett River may be related to a rather higher content of free iron in this sample, which contains anomalous amounts of Cu, Zn, Co and Ni. At High Lake there are no Hg anomalies close to the ore deposit, but there are a number of others in different parts of the area, the origin of which is unknown. They are apparently close to the granite–greenstone contact (Figs. 10 and 17).

For the three areas in Slave Province where rock and lake sediment samples were collected there is good agreement in the background level of Hg measured in each material. For the Bode Lake and Terra mine areas the levels reached in the lake sediments are a good deal higher than those in the rocks. A possible explanation for this is that the rock sampling was restricted to igneous rocks or to tuffaceous sediments. Cameron and Jonasson[12] have shown that Aphebian shales from

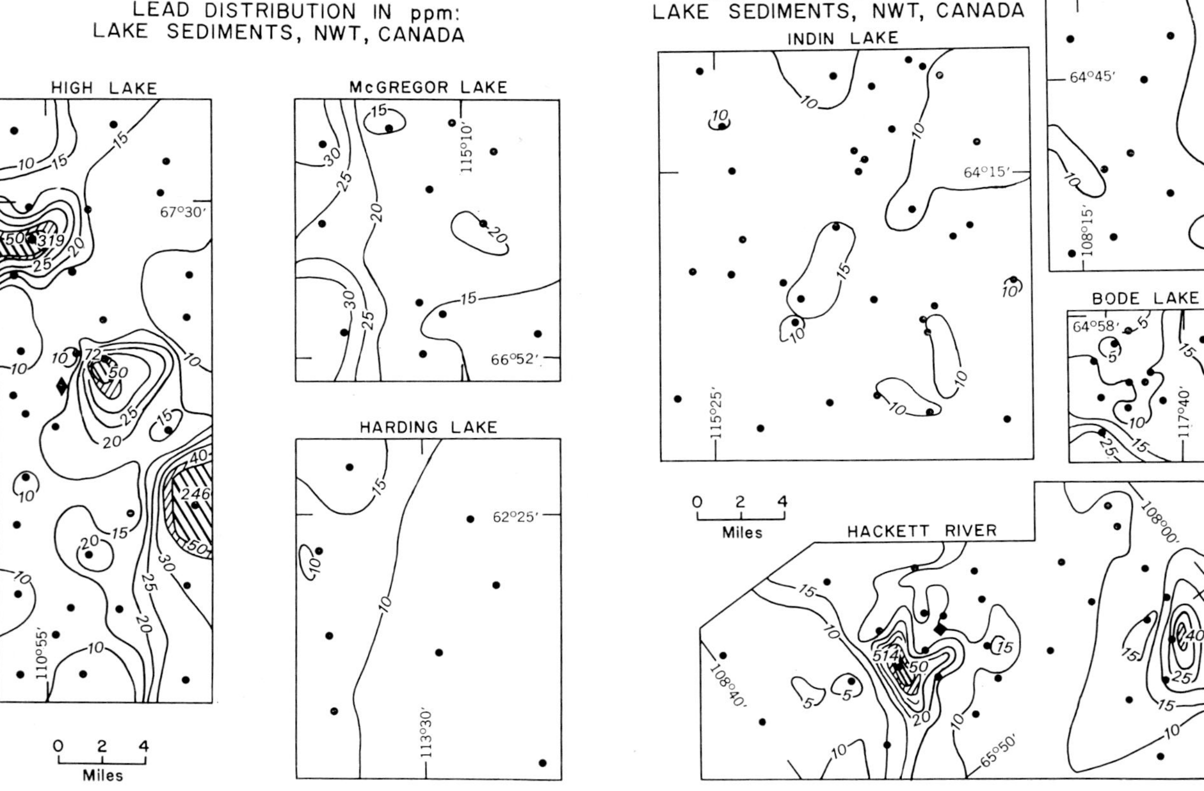

Fig. 16 Distribution of Pb in lake sediments of study areas

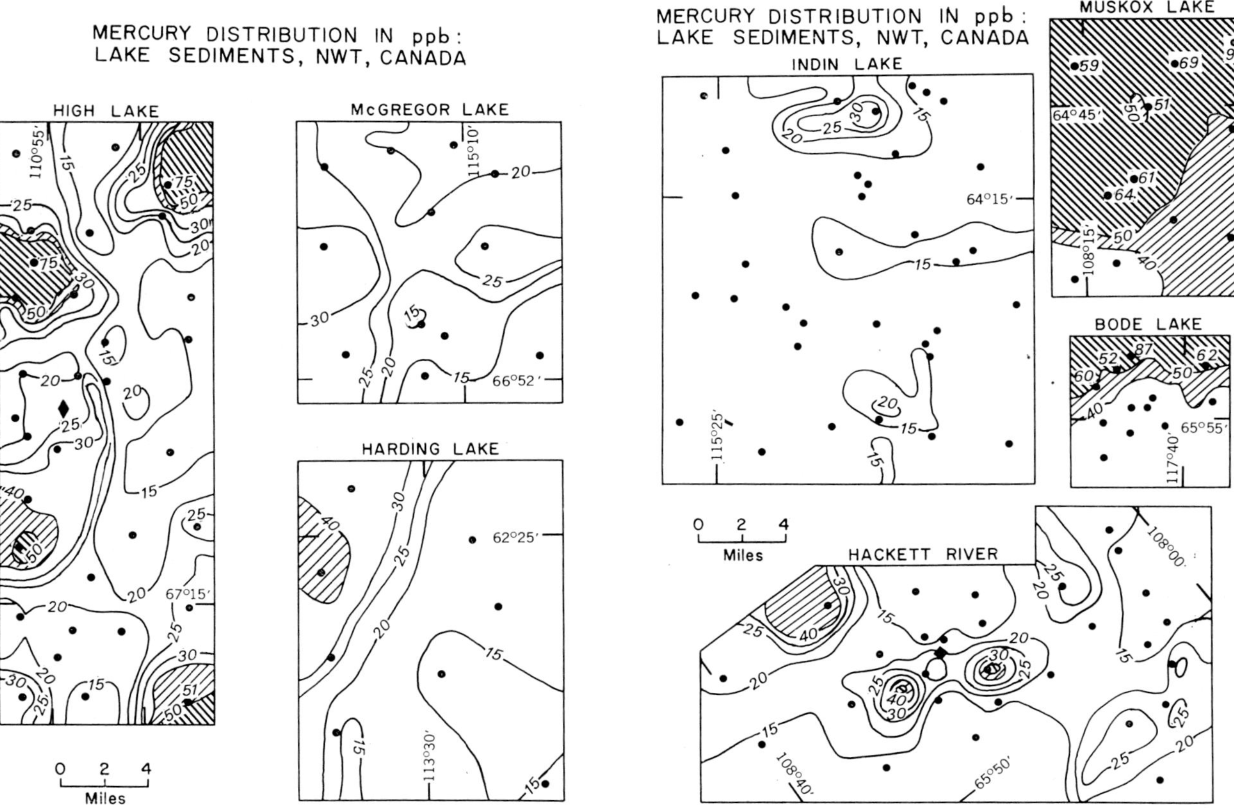

Fig. 17 Distribution of Hg in lake sediments of study areas

the southern part of the Shield have a notably higher level of Hg than do either Archaean or Phanerozoic shales and a very much higher content than igneous rocks; they attributed this to a greater degree of mantle degassing during the Proterozoic. It is possible that the Terra and Bode Lake areas contain Hg-rich sedimentary rocks of Aphebian age that contribute to the high content of this element in the lake sediments. The lake sediments at Muskox Lake have a uniformly higher level of Hg than those from any other part of Zn. None of the waters from Bode Lake or from Harding Lake reaches these levels, and at Terra mine only four very scattered sites reached this minimum level, so these areas are not illustrated. The distribution patterns for Cu and Zn in the waters is quite similar to that of the lake sediments, although there are some obvious differences—the Hackett River deposit is very clearly outlined; the High Lake deposit rather less clearly at a lower level of metals. There are a number of other anomalies on the Hackett River (the one along

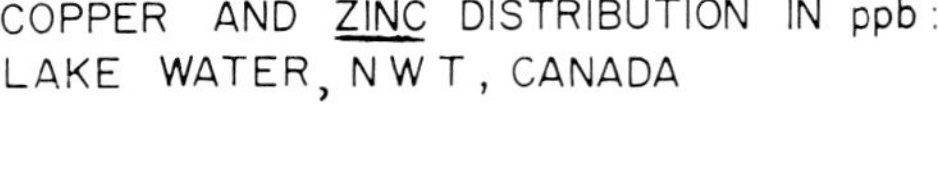

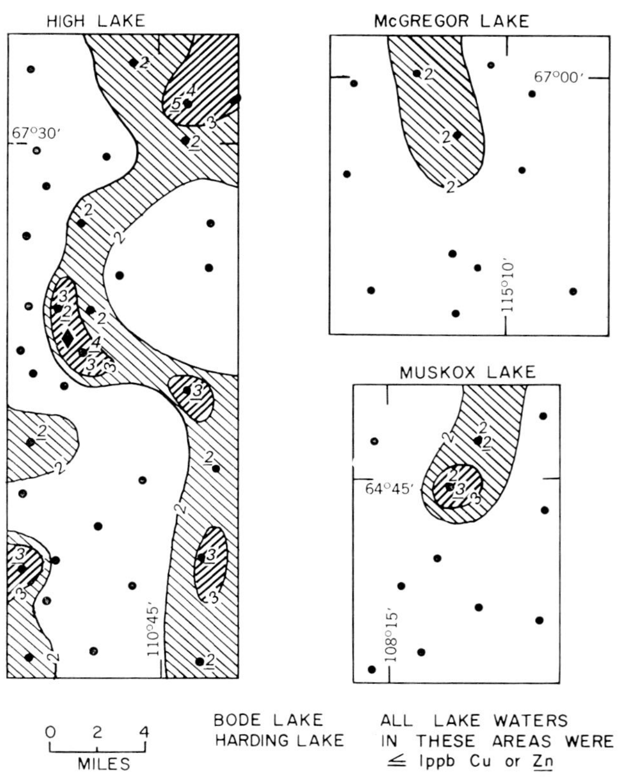

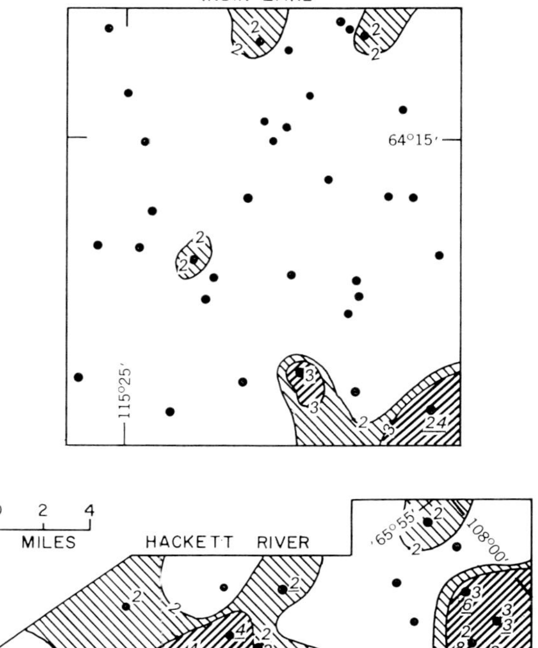

Fig. 18 Distribution of Cu and Zn in lake waters of study areas

of Slave Province. Presumably, this reflects a greater concentration of Hg in the Yellowknife Supergroup sedimentary rocks of this region than for the equivalent rocks in the other sampled areas, as the highest values are in the area underlain by such rocks (Fig. 10).

Trace-element composition of lake waters

Cu and Zn concentrations in lake waters from the sampled areas are shown in Fig. 18. Contours are drawn to enclose those areas where the waters contain 2 ppb and more of Cu *or* 3 ppb and more the southeast margin of the map looks particularly interesting), High Lake, Muskox Lake and Indin Lake maps, where no deposits are known.

Water analyses give the promise of faster and cheaper geochemical surveys because it is much easier to obtain a water sample than it is to obtain a lake sediment sample. The ability to utilize waters as a sampling medium is, however, handicapped on two counts. First, present analytical methods have a limit of detection for the ore indicator elements that is close to or above the natural abundance of these metals in lake waters; more sensitive methods must be devised. Secondly, knowledge of the behaviour of trace metals in lake waters of the Shield is virtually non-existent. Since it is believed that trace metals in waters are much more subject to seasonal and environmental influences than are the same metals within lake sediments, it is imperative that this information be available before lake water data can be properly interpreted. The cause of some of the anomalies shown in Fig. 18 must for the present remain unexplained; the more definite ones are probably related to mineralization, the others to environmental effects.

Table 7 Comparison of geometric means for Zn, Cu, Pb and Hg in rocks and lake sediments

Area and sample material	Zn, ppm	Cu, ppm	Pb, ppm	Hg, ppb
High Lake				
Basic volcanics	80·1	40·6	12·7	8·5
Intermediate volcanics	65·8	31·6	10·3	9·3
Acid volcanics	80·5	25·0	11·7	10·2
Lake sediments	120·0	46·9	14·1	22·4
Hackett River				
Rocks	42·1	17·1	10·5	14·6
Lake sediments	50·5	24·7	11·9	17·4
Indin Lake				
Basic volcanics	39·8	80·8	7·6	7·4
Intermediate volcanics	38·4	10·3	7·1	11·1
Acid volcanics	21·4	5·0	6·5	13·0
Sedimentary rocks	76·4	34·8	12·9	9·8
Granites	65·5	21·1	10·3	3·3
Lake sediments	66·5	31·5	11·7	13·8
Muskox Lake				
Lake sediments	93·4	38·6	9·1	52·8
Harding Lake				
Lake sediments	49·0	23·6	9·0	21·7
Bode Lake				
Rocks	79·8	5·3	14·6	5·6
Lake sediments	51·4	19·8	10·7	42·2
Terra mine				
Rocks	74·0	10·7	15·7	6·9
Lake sediments	90·3	29·3	19·4	28·3
McGregor Lake				
Lake sediments	105·0	44·0	19·0	20·1

Summary of relationships between rocks and lake sediments for trace elements

Because the lake sediments have been shown to be relatively homogeneous within each area (Table 6), the samples obtained are probably representative of all lake sediments present within these areas. The same cannot be said for the rock data because of their relative inhomogeneity. In Table 7 the geometric mean values for Zn, Cu, Pb and Hg for the lake sediments and for the rock types of the different areas are compared. The geometric mean (and median) values are more stable to sampling imprecision than the arithmetic mean, particularly when the data are skewed, as they are for many of the rock populations.

Hg is consistently higher in the lake sediments than it is in the rocks. A possible explanation that might account, in part, for the enrichment of Hg in the sediments at Bode Lake and Terra mine has already been discussed, but it appears from the data in Table 7 that there may be a general tendency for Hg to be more abundant in lake sediments than in the surrounding rocks. Discounting possible analytical bias, which it is planned to investigate further, possible explanations for this difference are (1) that some of the Hg in the lake sediments did not come from the rocks but from the atmosphere, or through faults and fractures[26] in the crust; (2) that diagenetic or biological processes have concentrated the Hg near to the surface of the sediments of lakes; and (3) that there is loss of Hg from near-surface rock samples.

Turning now to the other elements, Zn, Cu and Pb, for each of the areas containing significant mineral deposits—High Lake, Hackett River and Terra mine—the geometric mean content of each of these elements in the lake sediments is greater than that of any of the rock types measured in the same area. For the areas that are not known to contain large mineral deposits the reverse is more frequent: thus, at Indin Lake, one or more of the rock types contain a greater amount of these elements than do the lake sediments.

At Bode Lake, only Cu, perhaps significantly, has a higher geometric mean in lake sediments than in rocks. There are similar trends if the lake sediments of the Muskox Lake area are compared with the Indin Lake rocks, or if the Harding Lake sediment data are compared with the Indin Lake granitic rocks. Although the sampling methods have not been rigorous, and thus do not allow one to place statistical bounds of reliability on these differences, they do seem to be consistent. They are explained in the following terms.

In rocks intimately associated with mineral deposits there are at least two populations for each of the ore elements—the 'normal' rock population and the population due to micro-deposits of the ore minerals. This leads to a combined frequency distribution that is right skewed to a greater or lesser extent. The arithmetic mean of this population is often a good deal higher than the geometric mean or median. In the course of rock weathering and the processes of dispersion and, ultimately, sedimentation of these elements in lake bottoms, there is a general tendency to smooth the very irregular distribution that is characteristic of these elements in their host rock. In statistical terms there will be a tendency for the arithmetic mean and standard deviation of the ore element to decrease in a population of lake sediment samples and the median and geometric mean to increase relative to that of the original, unweathered rock population. The background level of abundance of an ore element, as defined by its geometric mean or median, will increase relative to the source rocks. All of these tendencies may be observed in the data presented here.

Even considering the above, the differences between the geometric means for Zn and Cu in the lake sediments at High Lake and the geometric means of these elements in the rocks are surprisingly high. Another general tendency should, therefore, be considered. This tendency is for the elements contained in sulphide masses to weather preferentially and be dispersed in the drainage system. At High Lake and Hackett River there are several visible gossans. It has been shown in permafrost areas of the U.S.S.R. that sulphide zones are oxidized to considerable depth.[29, 31] In special cases this weathering may have been preglacial, as was proposed for the Norilsk deposit.[48] It is reported that in these areas of the U.S.S.R., due to the heat of the oxidation processes, these sulphides may form thawed zones in the permafrost, thus assisting in the processes of leaching and dispersing the elements of the sulphide masses. The presence of gossans, however, whether they be visible or concealed by vegetation or drift, is probably not critical to the development of anomalous metal values. Indeed, at Coppermine[3] the area with the most visible gossans is not the area with the highest level of mineralization, as revealed by lake sediment analysis.

Conclusions

The −250-mesh fraction of inorganic sediments from lakes in the northern part of the Canadian Shield has been shown to be a homogeneous sampling medium that closely reflects the average composition of the surrounding bedrock. It provides a uniform material for the sorption of metal ions that are derived from ore deposits and from smaller, but more numerous, sulphide masses that are contained in the rocks around the ore deposits. By combining these two important characteristics of lake sediments it should be possible, in reconnaissance surveys, to outline simultaneously anomalies for the ore metals that are directly related to the ore deposits, together with other geochemical features associated with alteration zones and areas of favourable geology. Thus, for a typical Archaean massive sulphide ore deposit it may be possible to develop a hierarchical sequence of favourable geological–geochemical indicators on the basis of lake sediment analysis: (1) the occurrence of acid volcanic rocks; (2) within (1), zones underlain by exhalative facies rocks, such as carbonates and iron-rich sediments, that are related to the processes of ore deposition; (3) Mg or other alteration zones around the volcanic vent; and (4) anomalies of the ore metals.

The lake sediments are sufficiently homogeneous that the data derived from their analysis are eminently suitable for multivariate statistical treatment. In fact, the objective of geochemical mapping, as described in the preceding paragraph, is probably only completely attainable when such methods are employed. In a sequel to this paper an integrated statistical interpretation will be presented of the major- and trace-element data described here.

The methods of geochemical exploration and mapping described can probably be applied over the 70% of the Shield that is underlain by continuous or discontinuous permafrost, and which, therefore, likely contains inorganic lake sediments. There remains 30% of the Shield that is outside the permafrost zone and is covered by thick forests. In this region trace metals in solution in the lakes may be scavenged by organic gyttjas and thus not reach the inorganic sediment. A programme of geochemical mapping and exploration over this region may require analysis of both organic and inorganic lake materials.

Analysis of lake waters provides a faster and cheaper method of geochemical exploration within the Shield than is possible with lake sediments. Before this can be realized, more sensitive analytical methods must be devised and a better understanding gained of the effects of seasonal and environmental influences on the trace-element contents of lake waters. Lake waters cannot, however, provide the broad range of opportunities provided by lake sediments for geochemical mapping of rock types.

One of the questions which remains largely unresolved is the optimum interval for lake sediment sampling within the Shield. The density of one sample per 10 square miles, which was used here and which will be again used in the survey of 40 000 square miles of the Bear and Slave geological provinces in the summer of 1972, is largely an arbitrary choice influenced by economic considerations. Previous studies of the lake sediment and bedrock chemistry of the Coppermine region showed that small ore masses and micro-deposits were so widely distributed and abundant in the rocks enclosing the Coppermine ores that a distinctive geochemical anomaly could be easily outlined at a sample density of one sample per 10 square miles. For the massive sulphide ore deposits at High Lake and Hackett River the distribution of micro-deposits and small sulphide masses is less widespread than at Coppermine. The evidence presented here has shown that this sample interval is still adequate for detecting areas of interest, and even the presence of massive sulphide ore deposits, in a greenstone belt. The sample density of one per 10 square miles is, however, no more than adequate, and where actual deposits are sought, rather than areas of interest, this density should be increased. Archaean massive sulphide deposits are probably one of the smallest base-metal targets in geochemical terms. For other types, particularly the large, low-grade, disseminated ores, a reconnaissance site density of one per 10 square miles should give very satisfactory results.

Acknowledgment

The authors wish to express their thanks to the analysts of the Geochemistry Section, Geological Survey of Canada, under the direction of Mr. J. J. Lynch. The individual analysts involved are Mrs. A. MacLaurin, trace elements in rocks and sediments; Mr. G. Gauthier and Mr. P. Belanger, trace elements in waters; Mr. J. C. Pelchat and Mr. P. Belanger, As and Sb in rocks; Mr. L. Trip, mercury in rocks, sediments and waters; and Mr. R. Horton and Mr. W. Nelson, major elements in rocks and sediments. Mr. L. Trip, Mrs. A. MacLaurin and Mr. R. T. Crook assisted with the major-element analyses of the rocks and sediments.

They would also like to thank several individuals who assisted with various parts of the operation: Mr. N. G. Lund, who gave help with all of the lake sampling; Dr. I. R. Jonasson, who brought the authors' attention to most of the Soviet literature referred to here; Mr. M. Showalter, the pilot of the Beaver aircraft on all rock and lake sampling trips; Mr. D. Hobbs, for assistance on the computer processing; Mr. G. M. Thomas and Mr. R. T. Crook, for drafting and compilation; Mr. R. Hornal, Yellowknife Office, Indian Affairs and Northern Development, for cooperation during the field season; Mr. N. Badham, for a geological map of Terra mine; and Mrs. R. Chaffey, who typed the manuscript.

They also wish to acknowledge the use of a helicopter made available by Cominco in the Hackett River area. Thanks are due to the Cartographic Section, including Mr. C. McNeil, Mr. L. Babcock and Mr. R. Daugherty, of the Geological Survey of Canada.

Finally, the authors wish to note the contribution of Dr. C. S. Lord and Dr. S. C. Robinson. Their interest in new approaches to the study of the Canadian Shield was instrumental in this type of programme being commenced at the Geological Survey of Canada. Dr. R. Thorpe and Dr. W. W. Shilts kindly read the manuscript and offered critical comment.

This paper is published by permission of the Director, Geological Survey of Canada.

References

1. Allan, R. J. Brown, J. and Rieger, S. Poorly drained soils with permafrost in interior Alaska. *Proc. Soil Sci. Soc. Am.*, **33**, 1969, 599–605.

2. Allan, R. J. and Hornbrook, E. H. W. Development of geochemical techniques in permafrost, Coppermine River region. *Can. Min. J.*, **91**, April 1970, 45–9.

3. Allan, R. J. Lake sediment: a medium for regional geochemical exploration of the Canadian Shield. *CIM Bull.*, **64**, Nov. 1971, 43–59.

4. Allan, R. J. *et al.* Geochemical methods of exploration in permafrost areas. *Pap. geol. Surv. Can.* 72–1, pt A, 1972, 62–8.

5. ALLAN, R. J. LYNCH, J. J. and LUND, N. G. Regional geochemical exploration in the Coppermine River area, district of Mackenzie; a feasibility study in permafrost terrain. *Pap. geol. Surv. Can.* 71–33, 1972, 52 p.
6. ANHAEUSSER, C. R. *et al.* A reappraisal of some aspects of Precambrian shield geology. *Bull. geol. Soc. Am.*, **80**, 1969, 2175–200.
7. ARNOLD, R. G. The concentrations of metals in lake waters and sediments of some Precambrian lakes in the Flin Flon and La Ronge areas. *Circ. geol. Div. Sask. Res. Council* 4, 1970, 30 p.
8. BARAGAR, W. R. A. and GOODWIN, A. M. Andesites and Archean volcanism of the Canadian Shield. *Proc. Andesite Conf., Int. Upper Mantle Project Sc. Rep.* 16, 1969, 121–42.
9. BORCHARDT, G. A. JACKSON, M. L. and HOLE, F. D. Expansible layer silicate genesis in soils depicted in mica pseudomorphs. In *Proc. Int. Clay Conf.* (Jerusalem: Israel Program of Scientific Translations, 1966), vol. 1, 175–85.
10. CAMERON, E. M. and BARAGAR, W. R. A. Distribution of ore elements in rocks for evaluating ore potential: frequency distribution of copper in the Coppermine River Group and Yellowknife Group volcanic rocks, N.W.T., Canada. In *Geochemical exploration* (Montreal: CIM, 1971), 570–6. *(CIM Spec. vol. 11)*
11. CAMERON, E. M. and BAUMANN, A. Carbonate sedimentation during the Archean. *Chem. Geol.*, **10**, 1972, 17–30.
12. CAMERON, E. M. and JONASSON, I. R. Mercury in Precambrian Shales of the Canadian Shield. *Geochim. cosmochim. Acta*, **36**, 1972, 985–1005.
13. CAMERON. E. M. SIDDELEY, G. and DURHAM, C. C. Distribution of ore elements in rocks for evaluating ore potential: nickel, copper, cobalt and sulphur in ultramafic rocks of the Canadian Shield. In *Geochemical exploration* (Montreal: CIM, 1971), 298–313. *(CIM Spec. vol. 11)*
14. CHAMBERLAIN, J. A. Hydrogeochemistry of uranium in the Bancroft–Haliburton Region, Ontario. *Bull. geol. Surv. Can.* 118, 1964, 19 p.
15. CONWAY, E. J. Mean geochemical data in relation to oceanic evolution. *Proc. R. Irish Acad.*, **B48**, 1942, 119–59.
16. DANIELSSON, A. Spectrochemical analysis for geochemical purposes. In *XIII Colloquium spectroscopicum internationale, Ottawa* (London: Hilger, 1968), 311–23.
17. DEMENT, J. A. The morphology and genesis of the sub-Arctic brown forest soils of Central Alaska. Ph.D. thesis, Cornell University, Ithaca, 1962.
18. DYCK, W. *et al.* Comparison of regional geochemical uranium exploration methods in the Beaverlodge area, Saskatchewan. In *Geochemical exploration* (Montreal: CIM, 1971), 132–50. *(CIM Spec. vol. 11)*
19. DYCK, W. Lake sampling vs. stream sampling for regional geochemical surveys. *Pap. geol. Surv. Can.* 71–1, pt B, 1971, 70–1.
20. EADE, K. E. and FAHRIG, W. F. Geochemical evolutionary trends of continental plates—a preliminary study of the Canadian Shield. *Bull. geol. Surv. Can.* 179, 1971, 51 p.
21. GLEESON, C. F. Studies on the distribution of metals in bogs and glaciolacustrine deposits. Ph.D. thesis, McGill University, Montreal, 1960.
22. GOLEVA, G. A. *Hydrogeochemical prospecting for hidden ore deposits* (Moscow: Nedra, 1968), 291 p; *Chem. Abstr.*, **72**, 1970, 135264.
23. GRIGORYAN, S. V. Efficiency of geochemical methods of prospecting for ore deposits and the broad application of these methods by the Geological Survey of the U.S.S.R. In *Theses presented at an All-Union seminar, Moscow*, 1967 GRIGORYAN, S. V. ed. (Moscow: Ministry of Geology, 1967), 71 p. (Russian text)
24. HORNBROOK, E. H. W. and JONASSON, I. R. Mercury in permafrost regions: occurrence and distribution in the Kaminak Lake area, N.W.T. *Pap. geol. Surv. Can.* 71–43, 1971, 13 p.
25. JACKSON, M. L. *Soil chemical analysis—advanced course, 3rd edn* (Maidison, Wisc.: Department of Soil Science, University of Wisconsin, 1967), 894 p
26. JONASSON, I. R. and ALLAN R. J. Snow—a sampling medium in hydrogeochemical prospecting. In *Geochemical exploration 1972* (London: IMM, 1972), 161–76. *(Proc. 4th intn. geochem. Explor. Symp., London, 1972)*
27. KARLOVA, V. P. Experience in applying hydrochemical methods of exploration in the south Krasnoyarsk region. In *Materials on the geology and minerals of the Krasnoyarsk region* (Krasnoyarsk: 1962), 209–14; *Referat. Zh., Geol.*, 1964, 3D85.
28. KONTOROVICH, A. E. SADIKOV, M. A. and SHVARTSEV, S. L. Abundance of certain elements in the surface and ground waters of the northwestern part of the Siberian platform. *Dokl. Akad. Nauk SSSR*, **149**, 1963, 179–80; *Dokl. Acad. Sci. USSR, Earth Sci. Sect.*, **149**, 1963, 168–9.
29. KRAVTSOV, E. D. Minerals from the permafrost oxidation zone of the D'yakhtardakh ore deposit. *Zap. Vses. Mineral. Obshchest.*, **100**, no. 3 1971, 282–90; *Chem. Abstr.*, **75**, 1971, 99824.
30. KRITSUK, I. N. and KVYATKOVSKII, E. M. Distribution of metals in the various classes and fractions of talus haloes of deposits in eastern Transbaikal. *Zap. leningrad. gorn. Inst.*, **51**, no. 2 1966, 51–61; *Chem. Abstr.*, **67**, 1967, 56088.
31. OLEINIKOV, B. V. and SHVARTSEV, S. L. Recent sulphate formation in the oxidation zones of pyrrhotite–chalcopyrite hydrothermal ore manifestations of the N.W. Siberian platform. *Geol. Geofiz.*, no. 6 1968, 15–24. (Russian text)
32. ONISHI, H. Arsenic. In *Handbook of geochemistry* WEDEPOHL, K. H. ed (Berlin, etc.: Springer, 1969), 33.
33. PEREL'MAN, A. I. *Geochemistry of landscape* (Moskow: University Press, 1966), 392 p. (Russian text)
34. PITUL'KO, V. M. Geochemical survey methods under permafrost conditions on the Siberian platform. *Zap. leningrad. gorn. Inst.*, **51**, no. 2 1966, 42–50; *Chem. Abstr.*, **67**, 1967, 45879.
35. PITUL'KO, V. M. Soil formation in permafrost-taiga regions in the northern part of the Siberian platform and methods of lithochemical soil sampling.

Zap. leningrad. gorn. Inst., **56**, no. 2 1969, 58–62. (Russian text)
36. Pitul'ko, V. M. and Shilo, N. A. Geochemistry of frozen terrain and prospecting for ore deposits. *Geol. Geofiz.*, no. 11 1969, 21–8; *Chem. Abstr.*, **72**, 1970, 81379.
37. Saukov, A. A. *Geochemical methods of prospecting for useful mineral deposits* Valyasko, M. G. ed. (Moskow: University Press, 1963), 247 p. (Russian text)
38. Schmidt, R. C. Adsorption of Cu, Pb and Zn on some common rock forming minerals and its effect on lake sediments. Ph.D. thesis, McGill University, Montreal, 1956.
39. Shimp, N. F. Leland, H. V. and White, W. A. Studies of Lake Michigan bottom sediments—no. 2, distribution of major, minor and trace constituents in unconsolidated sediments from southern Lake Michigan. *Illinois St. geol. Surv. envir. Geol. Notes* 32, 1970, 19 p.
40. Shimp, N. F. *et al.* Studies of Lake Michigan bottom sediments—no. 6, trace element and organic carbon accumulation in the most recent sediments of southern Lake Michigan. *Illinois St. geol. Surv. envir. Geol. Notes* 41, 1971, 25 p.
41. Shvartsev, S. L. Some results of hydrogeochemical investigations under permafrost conditions. *Geol. Rudn. Mestorozhd.*, **5**, no. 2 1963, 100–10; *Chem. Abstr.*, **59**, 1963, 6136.
42. Shvartsev, S. L. Hydrogeochemical method of prospecting in northern swamp areas. *Geol. Geofiz.*, no. 7 1965, 3–10. (Russian text)
43. Shvartsev, S. L. Physico-chemical processes in a series of permafrost rocks. In *Cryogenic processes in soils and rocks* (Moskow: Nauka, 1965), 132–40; *Chem. Abstr.*, **64**, 1966, 17282.
44. Shvartsev, S. L. and Lukin, A. A. Hydrogeochemical zonality of groundwaters of some sulfide ore deposits in permafrost rocks. In *Cryogenic processes in soils and rocks* (Moskow: Nauka, 1965), 141–8; *Chem. Abstr.*, **64**, 1966, 15567.
45. Smith, A. Y. and Dyck, W. The application of radon methods to geochemical exploration for uranium. *CIM Bull.*, **62**, March 1969, 215. (Abstract)
46. Thorpe, R. I. Preliminary report on the metallogeny of the Bear and Slave structural provinces, N.W.T., Canada. Paper presented at Exploration Symposium, Yellowknife, 1972.
47. Turekian, K. K. and Wedepohl, K. H. Distribution of the elements in some major units of the Earth's crust. *Bull. geol. Soc. Am.*, **72**, 1961, 175–91.
48. Zontov, N. S. The Wurmian oxidation zone in the Norilsk copper–nickel sulphide deposits. *Dokl. Akad. Nauk SSSR*, **129**, 1959, 405–7; *Dokl. Acad. Sci. USSR, Earth Sci. Sect.*, **129**, 1959, 1057–9.

550.846:551.578.46

Snow: a sampling medium in hydrogeochemical prospecting in temperate and permafrost regions

I. R. Jonasson, PH.D.

R. J. Allan, PH.D.

Both of the Geological Survey of Canada, Ottawa, Ontario, Canada

Synopsis

It has been demonstrated that snow may be used as a sampling medium for hydrogeochemical prospecting. Such metals as Hg, Zn, Cu, Pb, Ni, Cd and Mn have been determined in snow samples overlying buried mineralization. In particular, the metal ions which constitute the bulk of the sulphide ore in any given deposit are the most useful indicators in snow. Dispersion aureoles are usually quite broad, especially near massive sulphide deposits. Samples were taken of clean snow from 2 to 6 in above the soil-ground ice layers. The mechanism by which metal ions are dispersed from subsoils to snow is considered to be ionic migration via capillary and pellicular water, which encases the packed snow crystallites. Studies of metal-content variations with depth of snow reveal a distinct increasing gradient of concentration, often up to fifty-fold, from snow surface down to the soils. Comparison of such snow horizons in mineralized locations with those from unmineralized locations confirms that airborne particulates contribute an insignificant amount of metals to the snow strata.

Typical trace-metal contents (ppb) found in the snow are: Hg, ⩽0·01–1·69; Cu, 2·8–161; Zn, ⩽1–53; Pb, ⩽2–15; Ni, 3·6–40·8; and Cd, ⩽0·4–7·2.

The relationships between such metal migration in snow and similar processes in permafrosted soils are discussed, along with the applications of these ideas to spring thaw and fall runoff water sampling for concealed mineralization. Advantages and hazards of the use of snow-melt waters for geochemical exploration and mapping are described with particular reference to the Canadian Arctic.

Recent work on the geochemistry of soil gases above hidden sulphide ore deposits has provided much information on the migration patterns of mercury from such orebodies. In particular, detailed studies by Karasik and Bol'shakov,[19] Fursov and co-workers,[11] and publications by McCarthy and co-workers,[25, 26] have shown that mercury in soil gases can be used as an effective prospecting tool in arid zones with sparse vegetation cover. Current interest in mercury geochemistry is not confined to its usefulness or otherwise in mineral exploration. Considerable attention is now being focussed on mercury, the environmental contaminator. A number of comprehensive review articles documented these aspects of mercury geochemistry.[17, 18, 42, 48] Volatile organic compounds, such as dimethylmercury, have been shown to be present in lake muds and sediments (see, for example, Jensen and Jernelöv[15]), and are thought to be generated in humic soils as well. But whatever the chemical forms of volatile mercury in soils may be, there becomes a definite possibility that mercury may also move into fallen winter snows from underlying base-metal mineralization.

There is limited information only on metal contents of fresh snow, and practically none at all on the trace-metal chemistry of fallen snow overlying mineral deposits.

Articles published in Japan have discussed the hydrogeochemistry of molybdenum,[39] copper and zinc[29] and vanadium.[38] In each paper a few figures were given for the metal contents (Mo, Cu, Zn, V) of fresh snow. These data, however, reflect industrial, volcanic and some mining activity because falling snow scrubs the atmosphere clean of floating dust particles. Mellor[27] summarized the chemical properties of snow and described the nature of organic and mineral impurities in snow from polar regions. He considered most inorganic impurities to derive from aerosols, dust and gases from the earth's surface, further contamination being due to wind-blown dust settling into fallen snow.

Similar information has been detailed by Murozumi and co-workers,[30] who made an extensive study of the lead content of arctic and antarctic snow strata, relating their findings for deep core samples to advances in man's industrial development over about 2500 years. Brocas and Picciotto[8] collected similar samples and analysed them for nickel, which was subsequently considered to be of extra-terrestrial origin. Both groups of workers also published extensive data on alkali and alkaline-earth metals and halogens. Numerous other publications have presented similar results for these same elements. It is not within the scope of this article to adequately summarize all such work. In a recent paper[45] the increase in mercury content in a Greenland ice sheet over several decades was measured, it being concluded that the permanent snow fields of the ice caps preserve a record of the introduction of mercury into the atmosphere both by man's industrial activities and by natural degassing processes from within the earth's crust. The samples collected had been analysed previously for lead, sulphate and selenium.[44]

Brief mention was made by Johnels *et al.*[16] of the mercury content of snow lying on a frozen lake in Sweden. The writers noted that the mercury content of the fresh surface snow was about one-third of that of a lower layer, one month old. It is quite likely that the mercury is moving up into the snow from within the lake itself, possibly as an organometallic vapour. Zitko *et al.*,[47] in a paper which discussed similar mercury pollution problems in New Brunswick, found measurable mercury in two snow samples collected near the town of St. Andrews. But perhaps the most interesting paper is that of Kolotov and co-workers,[21] who measured secondary dispersion aureoles of Zn, Cu and Pb (as total heavy metals) in snow overlying a tin ore deposit. Their explanation was that falling snow carried down dust clouds which exist over the ore deposit. Weiss[46] has demonstrated that direct sampling of such airborne dust particles is itself a potentially useful prospecting technique. Table 1 summarizes data described in the work cited above.

Table 1 Trace metals in snow: some previous work

Sample location	Element	Range, ppb	Average, ppb
Nagoya, Japan	Mo	0·12–0·30	0·20
Nagoya, Japan	Al	⩽70	
Nagoya, Japan	Fe	⩽120	
Sweden	Hg	0·07–0·21	
Nova Scotia	Hg	0·15–0·20	
Greenland	Hg	0·030–0·075	0·060
Greenland, after onset of pollution	Hg	0·053–0·230	
Japan	V	0·33–2·8	
Japan	Cu	0·4–2·3	1·3
Nagoya, Japan	Zn	3·5–12	6·5
Antarctica	As	4	
Greenland	Se	0·005–0·025	
U.S.S.R., near tin deposit	(Zn, Cu, Pb)	1·5–3·0	2
Antarctica	Ni	0·2–13·5	1·5
Greenland	Pb	<1–200	
Antarctica	Pb	<1–20	

Samples taken near cities are undoubtedly contaminated by local industrial sources. Table is not intended as a complete summary of all work done.

There is, however, no information in the literature concerning migration of individual trace elements from buried ore deposits into overlying fallen snow. Moreover, little attention has been paid to the further possibility that such elements as antimony, arsenic, copper, tellurium, selenium, cadmium, zinc, lead and mercury may also occur naturally as volatile organometallics* in humic soils. If this is the case, it might be expected that they would be recoverable if the solid atmosphere were to be sampled. It was also considered that the collection of metals by snow and ground ice and a study of the mechanisms by which this would occur would be most relevant to elucidation of the old problem of metal ion migration in permafrost regions. Numerous Russian workers and several Canadian scientists (e.g. Hornbrook and Allan[14] and Boyle[6,7] have shown that hydrogeochemical techniques are applicable to prospecting in permafrost, but that till or soil sampling is somewhat less effective. If the source of the metals in lake and stream systems arises because of a flushing-out of the active layer of permafrost soils during the spring snow melt, which leaves the soils relatively clean of trace metals, then the success or otherwise of geochemical techniques in permafrost regimes is understandable.

With these considerations in mind a preliminary snow sampling programme was devised for the winter of 1970–71, which was continued throughout the winter of 1971–72. It is relevant to note that the first winter was characterized by virtually continuous deposition of snow in the test areas, and at no time was the snow cover subjected to mid-winter freeze–thaw cycles. After four months of snowfalls the cover of lightly packed snow was about 5 ft deep. The winter conditions of 1971–72 were quite different, although total precipitation was close to that of 1970–71. The difference was that snow deposition was interrupted by freezing rainstorms and several thaws, which resulted in the soil surface being directly covered with at least $\frac{1}{2}$ in of ice, which itself was covered by interlayed ice and snow up to 3 ft deep in all test areas. The two winter periods are to be regarded as close to extreme snow deposition conditions for the Ottawa Valley region. Therefore, comparisons of trace-element contents of snow in each winter season might be expected to be quite interesting.

Test sites

The principal test site was chosen to be at a known mercury prospect near Clyde Forks, Eastern Ontario. There the mineralization consists of narrow veinlets and blebs of disseminated tetrahedrite, chalcopyrite and pyrite in barite–calcite gangue. Traces of cinnabar are visible in association with each of the sulphides. The tetrahedrite contains up to 4% mercury. The occurrence itself lies in Precambrian sediments of Grenville age and appears to be related to a number of northeast-striking parallel faults.

A second test area was located at the site of an old gold mine close to Joe Lake, which is also in the Clyde Forks area, the mineralization again being predominantly tetrahedrite and pyrite. The trace-element content of the ore differs in that there is much less mercury present and considerably more Au, Ag, Sb and As.

An area near to a recently closed lead–zinc mine on Calumet Island, Quebec, was the third main test site. Whereas the first two sites were vein deposits, the Calumet ore is massive and is predominantly sphalerite, galena and pyrrhotite. It also lies in Precambrian sediments of Grenville age. One of the interesting characteristics of this deposit is that the sphalerite contains around 30–40 ppm Hg.

Table 2 Ore mineralogy and trace-element chemistry of test sites

Test site	Principal sulphides present	Metals in ore
Clyde Forks: mercury prospect	Tetrahedrite, chalcopyrite, pyrite, cinnabar	Cu, As, Hg, Zn, Ni, Cd, Mn, Sb, Fe
Joe Lake mine: gold	Tetrahedrite, pyrite, sphalerite (?)	Cu, As, Hg, Zn, Ni, Cd, Mn, Sb, Fe, Au, Ag
New Calumet mine: lead–zinc	Sphalerite, galena, pyrrhotite, pyrite	Zn, Pb, Cd, Sb, Fe
Donaldson mine, Ungava: copper–nickel	Pentlandite, chalcopyrite, pyrrhotite	Cu, Ni, Co, Zn, Fe

*Organometallic: a compound in which a covalent metal to carbon bond exists. Metal-organic: a coordination complex compound in which the metal to ligand binding is essentially ionic in character, and in which the metal is exchangeable.

Snow samples were also collected over extensive nickel mineralization near to Donaldson mine on the Ungava Peninsula of Quebec, where the principal sulphides, chalcopyrite and pentlandite, are won from Proterozoic serpentinite sills. Grab samples of bottom snow were collected just prior to the spring thaw in that region. Table 2 summarizes the mineralogy and trace-element chemistry of all four test sites.

Analytical techniques and sample treatment

Snow samples were kept frozen for about 24 h in well-tied polyethylene bags. They were filtered (Whatman No. 31) to remove particulates, usually small pieces of twig, leaves, etc., after melting at room temperature. 1 ml of 18 M sulphuric acid was then added for each 500 ml of water sample. Filtration through such coarse papers was considered adequate for the purposes of these experiments, although, obviously, colloidal aggregates will not be removed from the samples in this way. The study was designed, however, to develop new techniques in hydrogeochemical exploration, and, therefore, it would be unwise to attempt to remove, for example, humate or silica colloids, which probably are carriers of metals, from the system, and so render the sampling programme useless. The question as to what is in true solution and what is colloidal is, therefore, not relevant to this work, although it does remain of considerable interest.

Mercury was determined by a cold vapour atomic absorption method, which utilized a cell of 1-m path length; limit of detection was 0·01 ppb. Copper, lead, zinc, cadmium and nickel were extracted as complexes of pyrrolidine dithiocarbamate (as APDC) and then determined by atomic absorption spectrometry (aas), where the lowest concentration measurable was about 1 ppb (ng/ml) depending on the volume of sample extracted. Manganese was determined directly by aas without prior extraction, with a limit of detection of 15 ppb. Reproducibility of all determinations was within 10%.

Table 3 Trace metals in snow horizons, ppb, Clyde Forks, 1970–71

Snow horizon, depth, ft, from snow surface	Hg	Cu	Zn	Cd	Pb	Mn
Samples from an unmineralized site						
Surface, 0–1	⩽0·01	3	8	⩽1	5	⩽15
Middle, 1–2	⩽0·01	4	15	1	15	⩽15
Lower, 3–4·5	⩽0·01	8	16	⩽1	12	⩽15
Total depth, 5 ft						
Samples from a mineralized site close to location C–4 (Fig. 1)						
Surface, 0–1	⩽0·01	3	12	⩽1	9	⩽15
1–2	0·40	27	28	1	11	⩽15
Middle, 2–3	0·19	73	14	⩽1	10	⩽15
3–4	0·30	74	182	2	2	15
Lower, 4–5	0·52	98	18	⩽1	4	20
Total depth, 5 ft						
Ice samples from within adit						
Icicles hanging from oxidized sulphide veinlets	0·87	73	13	1	3	⩽15
Ice stalagmites on floor	0·56	58	6	⩽1	⩽2	⩽15
Frost crystals on roof	0·23	23	6	⩽1	⩽2	⩽15
Trace elements in soils underlying sample locations, ppm						
Soil description						
Humic soil overlying mineralized site	278	370	65	⩽0·5	12	910
Mineralized gravels	1200	11 800	400	3·5	20	1030
Humic soil from unmineralized site	0·38	31	220	2·0	55	1600

Mineralization is covered by 1–2 ft of soils.

Results of snow sampling surveys

Clyde Forks mercury prospect

A preliminary survey in 1970–71 was designed as a feasibility study simply to check on whether or not trace elements did penetrate into snow overlying mineralization. Samples were collected from two surface locations carefully selected to represent a typical unmineralized site and a site close to some buried tetrahedrite veinlets. The whole area is covered in hardwood forest and underbrush, except on sharp ridges, which are bare of trees.

Up to five samples were taken at different depths of snow at each location (these are described as 'snow horizons' or 'snow strata' for the purposes of this discussion). An adit has been driven beneath some of the surface-exposed tetrahedrite mineralization. Groundwaters leaching through the sulphides and enclosing carbonate rocks formed icicles and ice stalagmites on the roof and floor of the adit, respectively. Frost crystals which covered the roof and walls were

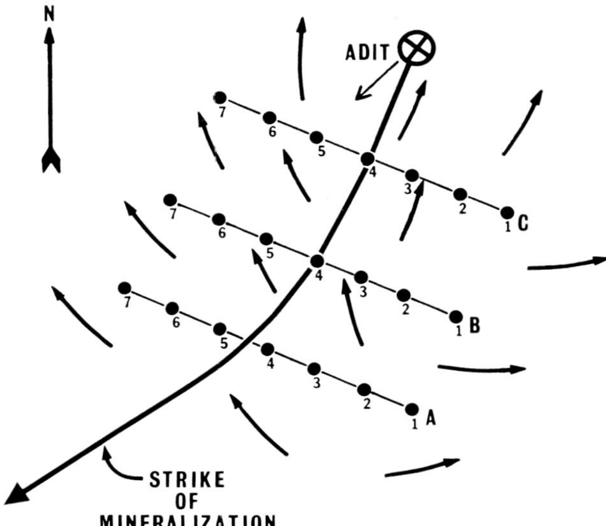

Fig. 1 Sample locations at Clyde Forks Hg prospect (sample interval, 50 ft; line interval, 100 ft; arrows indicate drainage flow)

probably formed from evaporation of the same groundwaters and icicles: for comparative purposes

Table 4 Trace elements in snow, Clyde Forks, 1971–72

Sample location	Trace metals, ppb					
	Hg	Cu	Zn	Cd	Pb	Ni
A–1	0·04	9·7	29	1·1	10·0	7·3
A–2	0·08	4·2	20	1·1	10·0	10·0
A–3	0·08	8·1	53	1·5	12·0	5·7
A–4	0·12	10·3	46	1·1	12·0	14·2
A–5	0·06	2·8	18	⩽0·4	10·0	3·6
A–6	0·07	5·0	20	0·8	11·6	29·4
A–7	0·15	5·6	46	1·1	12·0	15·8
B–1	0·05	3·1	12	3·0	12·0	7·3
B–2	0·04	1·9	23	0·4	14·0	4·7
B–3	0·04	2·5	4	1·1	5·8	4·2
B–4	0·10	27·6	⩽1	1·8	3·6	40·8
B–5	0·03	1·9	3	0·7	5·8	4·2
B–6	0·03	5·6	25	7·2	4·0	7·3
B–7	0·06	2·5	2	0·4	14·0	5·2
	Snow horizon samples					
B–4 at 2 in	0·10	27·6	⩽1	1·8	3·6	40·8
B–4 at 8 in	0·03	5·0	20	2·6	5·8	6·3
B–4 at 12 in	0·05	3·3	⩽1	2·6	5·8	5·7
B–4 at 24 in	0·03	3·3	⩽1	2·2	5·8	4·7
C–4 at 2 in	1·69	161·0	19	0·4	5·8	4·7
C–4 at 12 in	0·45	72·5	25	2·2	5·8	6·3
C–4 at 24 in	0·09	10·6	⩽1	0·7	5·8	4·2
	Single sample of frost crystals from adit					
D–1	0·52	84·6	⩽1	⩽0·4	4·6	13·6

All samples were collected from 2 to 6 in above the soil.

samples of each were collected in order to ascertain the influences carbonates would have on the hydrochemical migration of the trace metals of interest—Cu, Hg, Zn, Pb, Ag, Mn and Cd. Table 3 shows the trace-element data from these trials and information on the trace-element content of underlying soils.

In 1971–72 a repeat study of the snow horizons was undertaken and a small geochemical sampling grid (Fig. 1) was set up across the strike of the mineralization. In general, the trace-metal contents for snows taken at site *C–4* in both seasons are much the same, despite the different winter conditions which prevailed. Table 4 describes the results of the second survey for six trace metals, including nickel. Prior to the study it had been shown that the Clyde Forks area tetrahedrites contain about 0·7% Ni. Of the six metals utilized in the study, only Hg, Cu and Ni clearly outlined the mineralized veins. These are also the only metals which show any clear trends in behaviour through different snow horizons. In general, the highs occur, as expected, around sample locations 4, but it is also noteworthy that highs are found near to locations 7, where much of the drainage from the deposit seems to pass through. Thus, there is a source of metals which is transported from the zone of origin several hundred feet away.

From the data presented it is apparent that carbonate ions have a severe restrictive effect on trace-element migration in solution, most of the metals being precipitated from the icicles with calcium carbonate as the ice forms. It is curious to note that the maximum metal contents attained by the snows were generally greater than those measured for the adit ice samples. Perhaps one interpretation is that the mechanisms of transport in each case may be quite different, the transfer of metals from soil to snow probably being achieved by ionic diffusion, whereas that from rock to water (ice) is controlled by purely solution processes. It is demonstrated that in both back-

Table 5 Trace elements in snow and soils from Joe Lake gold mine

Snow sample location	Trace metal found, ppb						
	Hg	Zn	Cu	Pb	Ni	Cd	As
JL 01	0·05	16	3·8	3·2	6·4	⩽0·4	—
02	0·05	4	2·1	⩽2·0	2·8	⩽0·4	—
03	0·03	10	2·1	⩽2·0	1·7	⩽0·4	—
04	0·04	16	2·9	5·5	2·8	⩽0·4	—
05	0·02	8	2·9	3·2	2·8	⩽0·4	—
06	0·05	39	4·7	4·5	6·4	⩽0·4	—
07	0·04	⩽1	13·5	3·2	3·9	⩽0·4	—
08	0·05	25	13·2	3·2	4·5	⩽0·4	—
09	0·05	30	2·6	10·3	5·7	⩽0·4	—
10	0·03	12	2·6	8·0	3·9	⩽0·4	—
11	0·03	8	4·4	5·5	3·4	⩽0·4	—
12	0·03	18	2·1	8·0	3·9	1·3	—
All samples were collected from 2 to 6 in above the soil.							
	Snow horizon samples collected between 07 and 08						
+2 in above soil	0·05	16	12·9	10·0	6·3	1·5	—
+12 in above soil	0·05	12	2·8	7·9	6·8	0·4	—
+24 in above soil	0·05	8	3·9	12·0	5·2	0·4	—
Soil location	Trace metals in soils, ppm						
JL 01	110	86	67	93	15	—	3
02	78	86	50	85	12	—	3
03	129	131	32	93	10	—	15
05	146	169	37	127	12	—	3
08	105	165	23	78	8	—	20
09	108	110	21	107	10	—	20
10	90	107	18	71	13	—	15
12	42	67	16	33	15	—	40

ground and anomalous snow horizon samples a definite concentration gradient exists—which decreases from ground level snow up to surface snow.

Two deductions appear probable: that the source of the metals, especially Hg, Cu and Ni, is from below the snow and not from atmospheric particulates; and that these metals are derived mainly from the weathered mineralization beneath the soils rather than from the soils proper or from dead vegetation. One further point of interest is that the trace metals Ag, Cd, Mn and Pb, and perhaps Zn, are present in the background snow and anomalous snow at about the same concentrations: this is because these elements are not of great significance in the ores or the mineralized soils of the deposit and, therefore, do not produce measurable dispersion aureoles in the soils around the sulphides. The contents of Ag and Cd in the snows are very close to detection limits and so are of little use as indicators of nearby mineralization.

Joe Lake gold mine

The general layout of the sample locations at this site is depicted in Fig. 2. Although the ultimate

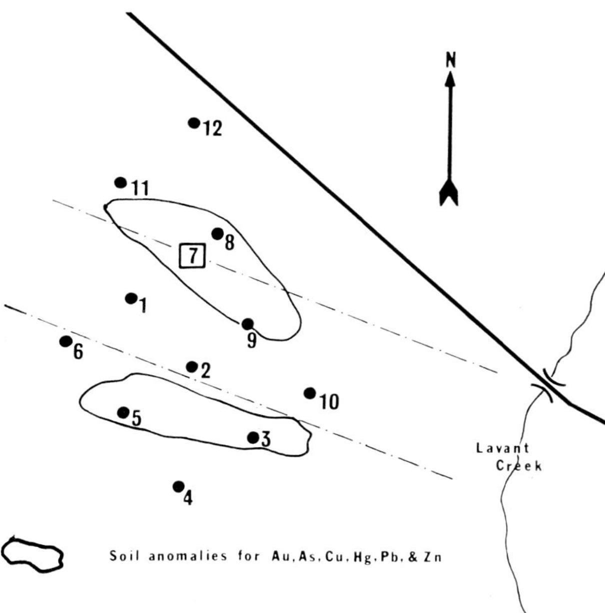

Fig. 2 Sample locations at Joe Lake gold mine (dotted line indicates strike of mineralization; line and sample intervals, 100 ft; sample point 7 located an old filled shaft; drainage is generally towards road and creek)

course of drainage in the area is towards the creek and road, there is some channelling to be observed between ridges which mark the strike of the mineralized rocks. Table 5 summarizes the analyses for a variety of elements found in snow samples taken at the grid points, and also in some soils underneath the snow cover. Although Hg, Cu, Zn and Pb outline the mineralization quite satisfactorily in soils, only Cu and Zn show any anomalous trends in snow samples collected over the same locations. Although the mineralization here is much the same as that at Clyde Forks, the showing is very much smaller and the forest cover is quite different. Joe Lake mine lies in a pine plantation and so, beneath the snow, there is a thick pine needle mat which was cemented solid

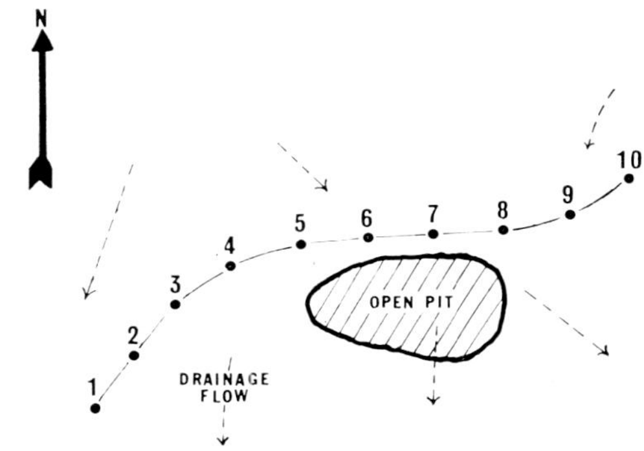

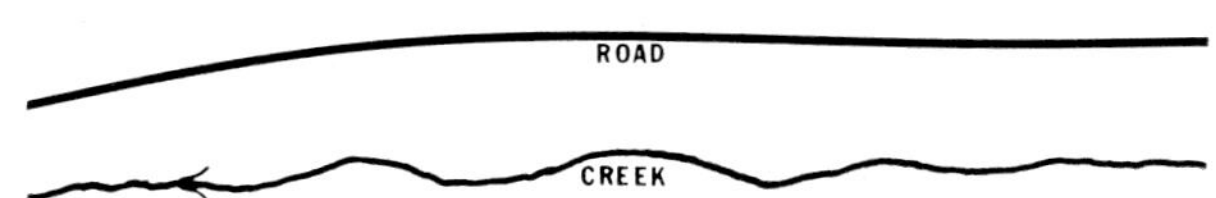

Fig. 3 Sample traverse line, Calumet Island, P.Q. (sample interval, approximately 100 ft)

with ice. The ice, which represents sub-snow drainage water, was not sampled since the objective of the programme was to take clean snow, where possible, and to ignore samples obviously contaminated by soil or vegetation debris.

Table 6 Trace elements in snow from New Calumet mine area

Snow sample location	Trace metals, ppb				
	Hg	Zn	Cu	Pb	Ni
CAL 01	0·03	77	10·5	3·2	5·0
02	0·02	75	13·5	8·0	10·0
03	0·02	75	7·3	8·0	5·0
04	0·02	35	3·2	5·5	2·8
05	0·03	46	5·0	15·5	5·4
06	—	35	3·2	19·0	5·0
07	0·02	120	3·4	8·0	7·9
08	0·02	275	10·8	42·9	6·4
09	0·02	92	4·4	15·5	6·1
10	0·02	28	4·4	15·5	6·4

All samples were collected from 2 to 6 in above the soil.

The snow horizon samples show similar characteristics to those at Clyde Forks, only the important trace metals, Cu and Zn, showing any distinctive patterns. The magnitude of the anomalies at the showing itself is not greater than four times local background, which undoubtedly reflects the very small amount of sulphides present.

New Calumet mine area

New Calumet mine, which has been closed down for about five years, was a lead–zinc producer. Soil contamination in the immediate vicinity of the old shafts and pits is quite extensive, so a traverse line was selected to cross a mineralized area on the edge of fairly thick, scrubby bush. An open-pit in which some blocks of sulphides are still visible lies down-drainage from the sample line. The mineralization is thought to underlie locations 6–9. Fig. 3 is a representation of the sample traverse line; Table 6 presents the analytical data for snows taken along this line.

It is apparent from these data that only Zn and Pb show anomalous patterns in samples 6–9. These metals, being the most abundant in the sulphides, are obviously the best indicators of such ore, although Cu may possibly be useful. No data are currently available for trace metals in soils along the traverse line. Comparisons, as in the other test areas, would be useful. The effect of the drainage pattern in the area on the snow contents is not known since soil tests have not been made up-drainage from the line.

Donaldson mine area, Ungava

Fig. 4 depicts the geochemical sample grid over an area where copper–nickel sulphide mineralization in serpentinite sills sub-outcrops into the till cover. Drainage is as shown and is generally east towards the Povungnituk River. Vegetation is sparse and consists of tundra mosses and lichen.

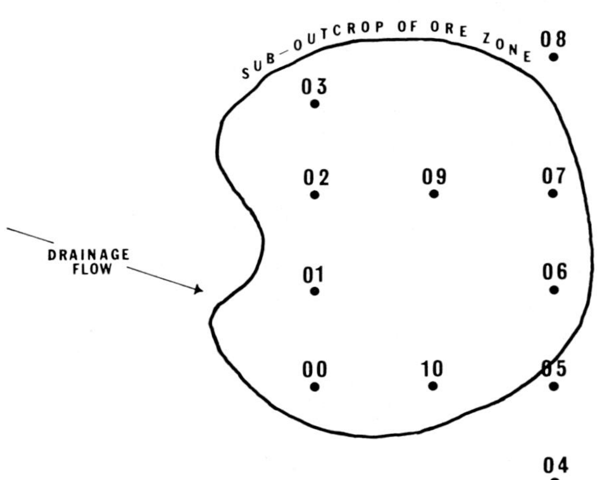

Fig. 4 Sample locations, Donaldson mine area, Ungava (sample interval, 100 ft)

Table 7 presents analytical data for Ni, Cu, Zn and Hg in wet snows. Once again the snows are shown to carry elevated quantities of the metal ions in which the ore itself is enriched (Ni and Cu), but in these samples some high values are found for both Zn and Hg, although there are no obvious relationships between these elements

Table 7 Trace-metal accumulation in snow at Ungava, Quebec

Sample location	Trace metals, ppb			
	Ni	Cu	Zn	Hg
00	35	34	220	0·51
01	24	20	117	0·84
02	44	18	40	1·01
03	53	10	4	0·69
04	14	10	130	—
05	54	10	4	0·64
06	24	11	77	0·48
07	5	5	36	0·55
08	15	14	44	0·86
09	29	3	5	1·10
10	37	17	89	—
11	8	5	28	0·48
12	8	18	155	1·04
13	6	6	45	0·77
14	4	3	16	0·54

Sampling interval, 100 ft.
Samples 11–14 lie to the north of and outside the known ore zone.

Table 8 Trace metals in silty frost boils and rocks outcropping in ore zone of Ungava deposit

Sample and type		Trace-metal content, ppm			
		Ni	Cu	Zn	Hg
Rock	*RSW3*	4146	7294	48	0·010
	RSW5	3000	510	18	0·010
	RSW6	1883	424	25	0·008
	RSW7	1393	518	29	0·009
Frost boil	*224*	454	208	39	0·005
	220	678	266	41	0·012
	219	590	227	48	—
	218	338	204	43	0·012

Metals were determined by aas methods after a hot, dilute nitric acid leach.

and the ore materials. It is possible that extensive formations of shales in the same area are contributing significant quantities of Zn and Hg to the snows. Unfortunately, these rocks were not sampled in the course of the geochemical survey conducted in the area by one of the authors (R.J.A.). Table 8 presents analytical results for some samples of serpentinites, and soils from frost boils close to the snow sample locations. It is quite obvious that the sources of Zn and Hg are not associated with these materials. The implications of these and other findings in the four test areas are discussed later.

Significance of snow sampling in geochemistry

The results presented for trace-element analyses of snows in all four test areas are encouraging from the viewpoint of the use of snows in hydrogeochemical exploration. In each case the major metals in the sulphide ores show distinctive patterns in the snows collected above ore zones—sufficiently distinctive to outline the immediate sources of mineralization. The anomalous factors range from three to fifty times threshold values, depending on the element under consideration. Tests have been made for the presence of antimony in snows above tetrahedrite-rich rocks. At the time of writing, the analytical technique is still under development and existing results are not yet sufficiently reliable to be reported in detail. Sb is, however, present in some of the snow—perhaps as much as 10 ppb in some samples. Evidence available at this stage of the investigation points to a rapid migration mechanism common to all of the elements tested. It seems likely that the metals move into fallen snow in dissolved form. At all times throughout the winter months the frozen soils showed the presence of free moisture. This was especially apparent along slight depressions, to which such moisture would gravitate and where thicker ground ice was always observed. For instance, at Clyde Forks two anomalous zones were detected—one immediately above the mineralization and one on a slope down-drainage from the ore zone, the latter reflecting drainage away from the ore. Metals apparently move upwards into snow cover via capillary and pellicular water across and between ice-crystal faces. It may well be that the metals are in complex form as sulphates, fulvates or analogous materials. In some situations such complexes may become uncharged or anionic in form and are not so easily adsorbed by soil-iron and soil-manganese oxides, for example. Metals which form the strongest associations with humic materials might be expected to concentrate to the greatest levels in water and snow. It is not possible to draw any firm conclusions as to the chemical forms of the metals in snow, since all cations of the metals studied—Hg, Pb, Cu, Zn, Ni, Cd and Ag—are known to form strong metal-organic coordination complexes with organic ligands.[34]

The fact that they can be extracted, under conditions of mild acid hydrolysis, into solutions where they can be complexed by pyrrolidine dithiocarbamate ligand, indicates that they occur bound (or unbound) in ionic linkages. The same observation does not apply for Sb, which could not be detected in a Brilliant Green extract—the conventional colorimetric method of analysis for antimony. After treatment with strong sulphuric acid, however, the Sb was converted to a form in which it was so extracted. It is suggested, therefore, that Sb migrates in anionic form in aqueous solution, perhaps as antimonite ion or some hydrolytic derivative. Further work is necessary to clarify this point. With the analytical techniques at our disposal it has not been possible to find evidence whether or not organometallics are present in snows or soil gases. As yet, no evidence exists to show positively that they are, in fact, generated in humic soils.

The question of mercury migration is worthy of further discussion. There is no doubt at all that mercury vapour is mobile under all conditions which might be expected to obtain in the test areas. Since elemental mercury itself is water-soluble to about 40 ppb at $0^{\circ}C$, however, movement of mercury into snow as vapour can probably be discounted. Experiments in our laboratory with soil-sniffing equipment have shown that wet soils greatly deplete the amount of mercury available in soil gases. Once the soils dry out, the mercury again volatilizes. Thus, it is considered that if elemental mercury is present in snows, it is in dissolved form and has, therefore, probably been transported by the same route as taken by zinc, for example. But, whatever the mechanism, it is contended that snow sampling may have its uses in hydrogeochemical prospecting methodology. It could be used to map extensions of vein systems or buried ore pods, particularly if care were taken to collect samples in gentle drainage depressions.

Interpretation of the drainage pattern for the area would then yield useful information. In these circumstances subsurface (close to ground) snow sampling could pick up anomalies within several hundred feet of the source of the metals. Once the snow melts, rapid runoff probably removes the anomalous pattern completely from the soils, especially in the drainage depressions which are well flushed in the springtime. The presence of metals in snow depends on a slow build-up over the long winter months resulting in their being trapped either in the soil ice layers or in the snows themselves. It is considered that the metals contents of the soils just before the spring thaw are significantly higher than may be found later in the summer season.

In areas where the freeze is very deep and lasts several months—for example, in Ungava and similar permafrost areas of the north or in Siberia—the trace-element build-up can be quite impressive.[32,35] In the U.S.S.R. some unusual minerals, composed of metal sulphates, have been reported which have crystallized at the ground surface during winter.[23,28,31] Undoubtedly, the snows in these regions will also contain abnormally high contents of mineral salts and complexes (especially near buried ore zones) at the winter's end. The relationship between such build-up and its application to spring-thaw hydrogeochemistry can now be discussed.

Spring-thaw hydrogeochemistry and related phenomena

The build-up of trace metals in fallen winter snows has been described above. It follows that once these snows melt, a flush runoff of these metals and associated anions, such as HCO_3^-, $SO_4^=$ and Cl^- will occur, resulting in a sharp increase in measured metal content of stream and lake waters. The areal size over which this increase is observed will depend on the magnitude of the source of the metals, e.g. sulphides, shales, carbonates or other rock types, in a given drainage basin. The runoff effects may persist over several weeks. These influences have been observed and described on a number of occasions, particularly in Soviet geochemical and hydrological literature (e.g. Polikarpochkin *et al.*[33]). During the spring thaw not only will metals in snow and ground ice be washed free but considerable leaching of the upper soil layers will also take place. Contributions due to radioactive atmospheric fallout have also been recorded.[41] Leaching of soils constituting the active layer of permafrosted ground is a particularly important process in the establishment and replenishment of hydrogeochemical anomalies in Arctic regions.[33,36,41]

A schematic representation, which summarizes the mechanism by which hydrogeochemical anomalies are established in permafrost, is presented in Fig. 5.

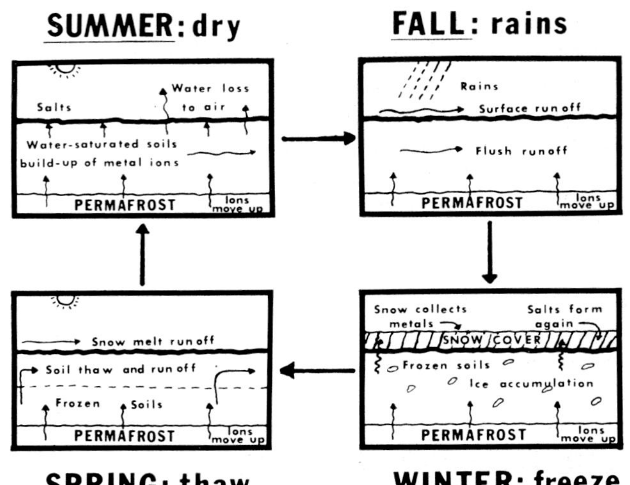

Fig. 5 Metal ion migration in permafrost: effects of season changes. The basic process by which ions move from permafrost upwards is probably ionic diffusion in solution. There is sufficient moisture in permafrost to allow capillary type migration to take place. In winter the advance of the freeze front down from the surface and up from the permafrost base causes metal cations (and anions) to concentrate in the frozen soils and subsequently to be moved to the surface, where carbonate and sulphate crusts may form; and to be moved into overlying snow. There is a continuous accumulation over the winter months in the upper soil layers. In spring the thaw flushes out a pulse of ions in the early runoff. Thaw occurs only from the surface down, so these processes may take several weeks to leach the frozen soils. In summer surface runoff has ceased, but subsurface migration may continue and carry dissolved material to the surface, where an ion build-up occurs again if the drought is long. Fall rains complete the cycle, resulting in a flush runoff which may be as intense as the spring thaw under certain conditions

Some previous work

A number of excellent papers have described various aspects of these and analogous processes, particularly for the winter–spring sequences. For example, Stremyakov[37] recorded increases in the sulphate content of surface waters on the Chukotsk Peninsula after the spring thaw of upper layers of permafrost, which he attributed to winter oxidation of galena, chalcopyrite and pyrite in the area. Shvartsev and Lukin[36] discussed the hydrochemical properties of the different water layers found in permafrost ground and described the sequence of events which begins with cryogenic weathering of ores and rock deep in the permafrost and which ends with the seasonal spring runoff flushing the metals into the surface water systems. Shvartsev[35] has made very thorough chemical analyses for Pb, Cu, Zn, Ag, Cr, Ni, V, Sn and Mn in permafrost ice of mineralized areas of the Siberian Platform. He was able to construct a relative mobility series for the metals in question in permafrost ice and water. His findings have been applied to hydrochemical investigations of other areas, including the vast Noril'sk copper–nickel deposits. Oleinikov and Shvartsev[31] observed heavy incrustations of a wide variety of sulphates, including gypsum, melanterite, epsomite and several alums above the oxidation zones of pyrrhotite–chalcopyrite showings in Siberia. These remain in abundance until rains disperse them. Miroshnikov[28] described the presence of free sulphuric acid on the surface of permafrost, where free metal sulphates also abound. Apparently, the acid forms as a result of hydrolysis of the hygroscopic sulphates formed earlier.

These papers, and many others, demonstrate the nature of the vigorous chemical activity which occurs in permafrost throughout the winter and which leads to a build-up of salts at or in the surface soil layers.

Two further sample papers illustrated the effects of runoff on the mineral content of surface waters. Chernyshev and Vlasov[10] described the chemical characteristics of four lake waters and ices, which showed marked increases in sulphate and bicarbonate content during the spring as compared with the rest of the year. Uryvayev[40] measured the increases in total mineralization in river waters of the northeastern U.S.S.R. during early summer, ascribing these changes to the increased accumulation of salts of spring runoff waters in the catchment basins. A recent paper by Gosling and co-workers,[13] who were studying hydrochemical dispersion aureoles of gold in Colorado, made comment on the increase in gold content of waters during 'spring flush-out' of groundwaters. In this case, however, the authors could see little use in this type of sampling for gold (cf. Goleva[12]).

The phenomenon of flush runout of metal ions by sudden precipitation or snow melt has been well observed under varied climatic conditions. For example, Atkinson[3] measured seasonal variations in total heavy metals in waters in tropical Angola. He noted large increases in content when heavy rains fell after a dry period. Webb and Millman[43] observed similar metal variations in Nigeria, which could be related to increased groundwater movement. Boyle and co-workers[4, 5] pursued similar studies in the Yukon Territory in terrain underlain by permafrost. Zn, Pb, Fe and sulphate were measured in snow-melt and spring waters derived from the permafrost active layer, and it was concluded that surface waters in permafrost areas were best collected for hydrogeochemical sampling in late summer because of the variability in metals content caused by snow melt earlier in the summer. This work was the first of its kind to be applied to permafrosted regions. Kozhara[22] made analogous studies with the same conclusions in eastern Siberia, and Karlova[20] described the effects of autumn downpours of rain on trace-metal contents of runoff waters in the permafrost of Krasnoyarsk.

All of these papers serve to illustrate that a considerable amount of study on runoff waters has been made. The *present* work, however, is a first attempt to show that fallen snow can play a far more important role in thaw and runoff hydrogeochemistry than to act simply as a source of metal-transporting waters. It is contended that the build-up of metal ions (or complexes) in snow is merely an extension of the migration processes which persist (Fig. 5) in frozen soils and rocks. The latter processes are, in turn, analogous to the movement of moisture and ions which have been shown to take place in soils beneath snow cover in more temperate regions of Canada.

Present work on runoff sampling

During 1970 and 1971 a number of areas in the Northwest Territories and northern Quebec were selected for hydrogeochemical sampling. One at Ungava has already been described because snow samples were also taken there. The others were taken at Little Cornwallis Island, where a lead–zinc prospect in Palaeozoic carbonates is known, and at Coppermine River, where a number of chalcocite prospects in Proterozoic basalts have been mapped. All sites are characterized by deep, continuous *permafrost*, which is subject to surface summer thaw and runoff. For comparative purposes waters from lakes and streams were collected later in the summer, and in one instance, at Coppermine River, just after an autumn rainstorm.

It became obvious that the waters from streams and lakes, collected during the spring and late autumn runoffs, held considerably higher quantities of trace elements than those collected in summer—indicating that oxidation processes and elemental migration continue unabated throughout the year. One consequence of the effects of runoff was the dramatic increase in the mercury content of waters. This phenomenon has been observed and recorded in three widely separated areas of Canada, and it probably reflects the high mobility of mercury in runoff waters.

Spring runoff at Ungava, Quebec

Table 7 presents data on trace metals in snow overlying a nickel orebody in a serpentinite sill. It was noted at the time that the mercury values were unusually high in view of the low amounts present in the ore and host rocks (Table 8). Table 9 compares some of the snow data with analyses for runoff water samples taken up- and down-drainage from the sill. It should be noted that although the ore-indicating elements Cu, Ni and probably Zn decrease considerably, the quantities of Hg remain much the same over a wide area. It is considered that metals are lost from waters derived from melting snow as these

Table 9 Runoff water analyses up- and down-drainage from a nickel-bearing sill

Sample number	Sample type	Trace metals, ppb			
		Ni	Cu	Zn	Hg
2000	Lake (down)	29	7	4	0·94
2001	Lake (down)	6	1	1	0·88
2002	Lake (down)	6	1	2	0·88
2003	Lake (down)	6	1	2	0·88
2004	Stream (down)	20	3	2	0·60
2005	Lake (down)	9	4	4	0·81
2006	Lake (down)	7	3	2	0·70
2007	Lake (down)	3	2	3	0·81
2008	Lake (down)	5	4	4	0·78
2009	Lake (down)	2	0	1	1·10
2025	River (up)	4	1	2	0·55
2026	River (up)	2	1	2	0·94
2027	Stream (up)	4	1	2	0·70
2028	Stream (up)	3	0	1	1·18
2029	Stream (up)	3	1	1	1·23
2030	Lake (up)	2	1	0	0·73
2031	Lake (up)	3	0	1	0·61
2032	Lake (up)	4	1	0	0·83
2033	Stream (up)	4	2	1	0·82
Means for 13 samples	Snow near ore (from Table 7)	24	12	68	0·73

Samples were taken at irregular intervals for several thousand feet up- and down-drainage from the ore zone.

flow away from their source. Adsorption on to till and soils is the probable mechanism by which these processes occur. In this way till and sediment dispersion haloes are regenerated by runoff waters. Table 9 also shows that the quantities of trace metals, Ni, Cu and Zn, are higher, on average, by at least a factor of two down-drainage than up-drainage. It is apparent that the snow-melt waters exert an influence several hundred metres down-drainage from their source. By way of comparison, waters sampled over nickel-bearing rocks and till later in the summer contained, on average, 1 ppb Ni and 0·03 ppb Hg. Nickel contents of some waters collected some distance from such rocks were not measurable, i.e. $<0{\cdot}5$ ppb.

Autumn runoff on Little Cornwallis Island, N.W.T.

These water samples were collected in August, 1970. At that time early snowfalls in the area had followed a dry spell, resulting in rapid melting and causing the streams and rivers, previously dry, to flow again. The runoff was not of the magnitude observed in the spring at Ungava, or later in the fall at Coppermine. It should be emphasized that prior to the runoff there were *no* surface waters to be sampled in the vicinity of the ore showings on Little Cornwallis Island. Table 10 summarizes the results of the survey. The waters related to the ore zone definitely carry higher amounts of Pb and Zn, the major metals of the deposit, but very little mercury by comparison. Zinc is the best indicator of mineralization, perhaps reflecting its higher mobility over lead. The results obtained indicate quite clearly that the runoff after dry weather rapidly mobilizes soluble salts which collect around the ore zone.

Table 10 Autumn runoff in high Arctic: trace-element concentrations in surface waters from Little Cornwallis Island

	Sample number	Trace metals, ppb		
		Pb	Zn	Hg
A	*111*	13	46	0·08
Related to	*112*	3	41	0·14
ore zone	*113*	⩽3	16	0·10
	114	⩽3	18	0·12
B	*104*	5	3	0·24
Not directly	*107*	3	4	0·10
related to	*108*	⩽3	2	0·12
ore zone	*109*	⩽3	2	0·11
	115	3	2	0·12
	116	3	3	0·16
	117	3	9	0·13

Sample type, stream.

Summer and autumn runoff at Coppermine River

Several lakes in the vicinity of copper mineralization were sampled regularly over the whole summer field season, most remaining fairly constant in Cu content throughout the sampling period. The exceptions, those which were quite close to mineralization, showed a sharp increase in Cu when feeder streams resumed flow after an early winter rainfall (Table 11).

The most striking demonstration of the effects of a flush runoff on metal contents of surface waters is with mercury. The samples collected in September, 1970, were very high in mercury (normal lake water content is 0·01–0·05 ppb). Snow had fallen for several days previously, causing the streams and lakes to flood. In 1971 samples were collected from the same areas, but in July, the driest period of the year. Most streams had dried up or were quite stagnant. Some results of this study are presented in Table 12.

Table 11 Seasonal variations in copper concentrations (ppb) related to drying up of streams after spring runoff and return of high concentrations after autumn runoff, Coppermine basalt area, N.W.T.: changes during July and August, 1970, in A, a lake near no known mineralization and B, lake closest to 47-Zone deposit

Sample	Lake system near no known mineralization									
A	July 5	July 11	July 17	July 23	July 29	Aug. 4	Aug. 10	Aug. 16	Aug. 22	Aug. 28
Stream 1	3	5	3	4	4	3	3	3	4	6
Stream 2	4	6	3	3	4	4	4	5	3	6
Stream 3	3	3	1	3	3	3	3	6	4	4
Lake site 1	2	4	2	3	3	3	6	4	4	6
Lake site 2	1	3	2	3	3	3	4	5	4	3
Lake site 3	1	3	3	3	4	3	3	4	3	5
B	Lake system near mineralization									
Stream 4	8	18	16	13 stag.*	12 stag.*	17 stag.*	18	18	18	25
Stream 5	8	s.d.u.†	s.d.u.†	s.d.u.†	s.d.u.†	s.d.u.†	12	s.d.u.†	s.d.u.†	16
Stream 6	6	8	6	6	4	5	10	7	8	13
Lake site 4	7	8	4	6	3	4	6	9	9	8
Lake site 5	9	8	8	7	4	4	8	9	9	10
Lake site 6	8	8	8	6	5	4	6	9	8	8
Lake site 7	8	8	8	5	4	5	8	8	8	8

Colorimetric analyses in the field. Beginning of autumn rains about 6 August. Snowfalls and melts between 22 and 28 August.
*stag., stagnant stream, no visible flow.
†s.d.u., stream dried up completely.

Table 12 Effect of autumn runoff on mercury contents of waters: Coppermine area: change in mercury content in those waters with highest content of mercury in early winter, 1970, and content in the middle of the summer, 1971

Sample number	Date of collection Early Sept, 1970 Mercury content, ppb	Late July, 1971	Sample type
100	2·14	0·08	Lake
101	0·72	0·02	Stream
104	1·53	0·02	Lake
105	4·80	0·03	Stream
109	4·60	0·01	Stream
112	0·62	0·02	Lake

Waters were collected from exactly the same sample sites both years.

Discussion

The effect of precipitation and flush runoff is clearly to increase the concentration of mercury in the waters of lakes and streams by factors up to 500 times the concentrations measured during dry periods. The mercury contents of nearly 5 ppb are exceptionally high. Such levels are usually found only with mercury mineralization where a stream actually passes through a deposit. In the Coppermine basalt area it is obviously quite misleading to regard these values as indicating proximity to a rich ore deposit. Although mercury in runoff waters would be a confusing tool to use, to say the least, there is a possible compensating benefit involved. The question 'Where does the mercury come from?' has arisen earlier in this paper when the Ungava nickel deposit was being discussed. It seems probable that mercury does not come from igneous rocks or ores at Ungava, and the same is true at Coppermine. Analyses of chalcocite ore, pyrites and basaltic rocks reveal no more than 500 ppb Hg, on average.

Because of the known mobility of mercury (during dry periods) as a vapour, it was considered of interest to plot the distribution of mercury highs found in the runoff waters in the Coppermine area (Fig. 6). The high mercury values tend to outline the major fault structures in the area (except in the vicinity of the Cu anomaly), particularly in the regions covered by the Coppermine basalts.[2] Several proposed extrapolations of faults have been added which also fit the data fairly well, and it may be that others could be drawn in too. There is one section, in the north of the area, where the mercury values all exceeded 1 ppb (see Table 12 for some examples). Here the basalts have given way to sediments which are

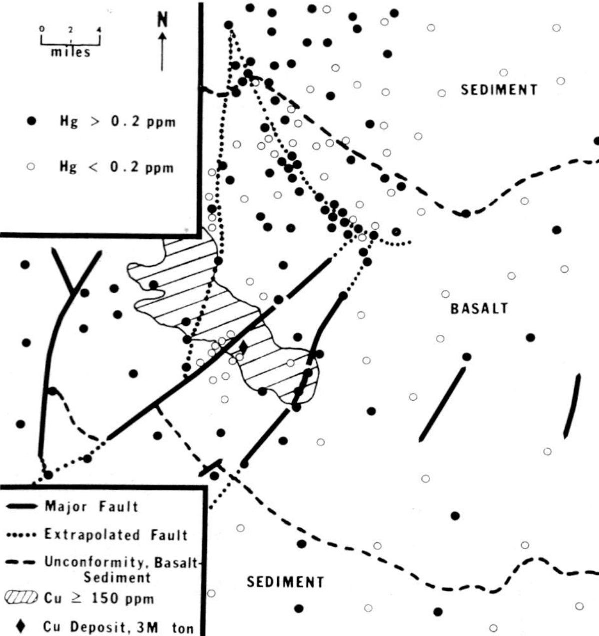

Fig. 6 Anomalous mercury in lakes and streams, Coppermine basalt area, N.W.T.

primarily Proterozoic dolomites and shales. The higher mercury contents of stream waters in the sediments probably derive from the rocks themselves, with considerable amplifying effects due to the runoff processes. Analysis of the sediments will be necessary in any follow-up study. Cameron and Jonasson,[9] in a recent study, have shown that Proterozoic shales have an elevated content of mercury when compared with both older and younger shales. It may well be that herein lies a possible explanation for the high mercury found in runoff waters at Ungava and Coppermine, but not on Little Cornwallis Island.

Mercury in water may prove to be useful in mapping faults, unconformities and other similar structures. Kuzmin and Posokhov[24] have used Ag and Cr in waters to similarly map tectonic fractures. It could well be that mercury, moving as vapour up an easy pathway along joints or fissures of some kind, can accumulate under snows, in frozen soils or soils in general, where its concentration will reflect the underlying structure. Such a reflection may show through a considerable depth of loose overburden. The data in Table 12 indicate that the phenomenon is a transient one, occurring only during a flush runoff of streams crossing, flooding or following the faults.

Conclusions

(1) Snow has been shown to gather trace metals from beneath soils in the vicinity of sulphide ore showings. The rate of accumulation is relatively rapid, and measurable concentrations may be built up over two or three months. Hg, Cu, Zn, Cd, Mn, Ni and Pb have been determined in snows and ground ice.

(2) The mechanism by which snows collect metals and their associated anions is considered to be by ionic migration along water films covering ice-crystal faces and filling micro-cracks and fissures in the snow strata. The same process probably accounts for ionic movements in frozen soils, which are related to cryogenic weathering processes taking place in permafrost regions.

(3) The use of snow as a sampling medium is probably limited to detailed prospecting in known mineralized areas. The best elements to analyse for seem to be the metals which would constitute the economic ore of a prospect.

(4) The accumulation of metals in snow has been shown to be part of a cycle by which metals move from below soil surfaces, from orebodies, whether in permafrost or not, and into the surface water systems of the locality in question.

(5) It could well be that by demonstrating the presence of metals in surface snows overlying permafrost, a measure of the rate of metal ion movement in frozen ground can be made. Moreover, the usefulness of snow sampling in the arctic further supports the proposal, long accepted in the U.S.S.R., that the hydrochemical method of prospecting in permafrost is reliable.

(6) Some aspects of spring-thaw hydrogeochemistry and fall runoff hydrogeochemistry have been considered in detail. Although there is obvious value in the fact that these flush-out processes are responsible for replenishing lakes and streams with trace metals for summer water and sediment sampling, caution should be exercised in interpreting the magnitude of anomalies found for some metals, e.g. mercury, in the runoff waters themselves.

(7) Mercury in snow and snow waters may find some use as a mapping aid, particularly in faulted areas covered by till or sediments. Consideration should be given, however, to the nature of the rock types in the area under study, particularly if shales occur.

Acknowledgment

The authors most gratefully acknowledge the contributions of Mr. C. C. Durham, who assisted in the field sampling programme and who made many helpful suggestions regarding collection techniques. They also commend Mr. G. Gauthier, Mr. P. G. Belanger, Mr. L. J. Trip and Mrs. A. MacLaurin for their excellent analytical work, and thank Mr. R. T. Crook and Mr. G. M. Thomas, who drafted the illustrations. Thanks are also due to Mrs. R. Chaffey, who prepared the manuscript.

This paper is published by permission of the Director, Geological Survey of Canada.

References

1. ALLAN, R. J. and HORNBROOK, E. H. W. Development of geochemical techniques in permafrost, Coppermine River region. *Can. Min. J.*, **91**, April 1970, 45–9.

2. ALLAN, R. J. LYNCH, J. J. and LUND, N. G. Regional geochemical exploration in the Coppermine River area, District of MacKenzie; a feasibility study in permafrost terrain. *Pap. geol. Surv. Can.* 71–33, 1972, 52 p.

3. ATKINSON, D. J. Heavy metal concentration in streams in North Angola. *Econ. Geol.*, **52**, 1957, 652–67.

4. BOYLE, R. W. ILLSLEY, C. T. and GREEN, R. N. Geochemical investigation of the heavy metal content of stream and spring waters in the Keno Hill–Galena Hill area, Yukon Territory. *Bull. geol. Surv. Can.* 32, 1955, 34 p.

5. BOYLE, R. W. PEKAR, E. L. and PATTERSON, P. R. Geochemical investigation of heavy metal content of streams and springs in the Galena Hill–Mount Haldane area, Yukon Territory. *Bull. geol. Surv. Can.* 36, 1956, 12 p.

6. BOYLE, R. W. Native zinc at Keno Hill. *Can. Mineralogist*, **6**, 1961, 692–4.

7. BOYLE, R. W. Geology, geochemistry, and origin of the lead–zinc–silver deposits of the Keno Hill–Galena Hill area, Yukon Territory. *Bull. geol. Surv. Can.* 111, 1965, 302 p.

8. BROCAS, J. and PICCIOTTO, E. Nickel content of Antarctic snow—implications of the influx ratio of extraterrestrial dust. *J. geophys. Res.*, **72**, 1967, 2229–36.

9. CAMERON, E. M. and JONASSON, I. R. Mercury in Precambrian shales of the Canadian Shield. *Geochim. cosmochim. Acta.*, **36**, 1972, 985–1005.

10. CHERNYSHEV, L. A. and VLASOV, N. A. Physicochemical characteristics of Tazheran lakes during the winter. *Izv. Nauch.—Issled. Inst. Nefte.—Uglekhim. Sin. Irkutsk. Univ.*, **12**, 1970, 175–7; *Chem. Abstr.*, **75**, 1971, 25068.

11. FURSOV, V. Z., VOL'FSON, N. B. and KHVALOVSKIY, A. G. Results of a study of mercury vapour in the Tashkent earthquake zone. *Dokl. Akad. Nauk SSSR*, **179**, 1968, 1213–5; *Dokl. Acad. Sci. USSR, Earth Sci. Sect.*, **179**, 1968, 208–10.

12. GOLEVA, G. A. Hydrogeochemistry of the gold deposits of the Baley district. *Int. Geol. Rev.*, **12**, 1970, 195–203.

13. GOSLING, A. W. JENNE, E. A. and CHAO, T. T. Gold content of natural waters in Colorado. *Econ. Geol.*, **66**, 1971, 309–13.

14. HORNBROOK, E. H. W. and ALLAN, R. J. Geochemical exploration feasibility study within the zone of continuous permafrost; Coppermine River region, Northwest Territories. *Pap. geol. Surv. Can.* 70–36, 1970, 35 p.

15. JENSEN, S. and JERNELÖV, A. Biological methylation of mercury in aquatic organisms. *Nature, Lond.*, **223**, 1969, 753–4.

16. JOHNELS, A. G. *et al.* Pike (*Esox lucius* L.) and some other aquatic organisms in Sweden as indicators of mercury contamination in the environment. *Oikos*, **18**, 1967, 323–33.

17. JONASSON, I. R. Mercury in the natural environment: a review of recent work. *Pap. geol. Surv. Can.* 70–57, 1970, 39 p.

18. JONASSON, I. R. and BOYLE, R. W. Geochemistry of mercury and origins of natural contamination of the environment. *CIM Bull.*, **65**, Jan. 1972, 32–9.

19. KARASIK, M. A. and BOL'SHAKOV, A. P. Mercury vapour at the Nikitovka ore field. *Dokl. Akad. Nauk SSSR*, **161**, 1965, 1201–4; *Dokl. Acad. Sci. USSR, Earth Sci. Sect.*, **161**, 1965, 204–6.

20. KARLOVA, V. P. Experience in applying the hydrochemical method of exploration in the south Krasnoyarsk area. *Material. Geol., Polez. Iskop. Krasnoyarsk Kraya. Sb.* no. 3, 1962, 209–14; *Referat. Zh., Geol.*, 1964, 3D83.

21. KOLOTOV, B. A. KISELEVA, YE. A. and RUBEIKIN, V. Z. On the secondary dispersion aureoles in the vicinity of ore deposits. *Geokhim.*, 1965, 878–80; *Geochem. intn.*, 1965, 675–7.

22. KOZHARA, V. L. Some features of the hydrologic migration of chemical elements in permafrost regions in connection with hydrochemical prospecting. *Tr. Inst. Geol. Rudn. Mestorzhd., Petrogr. Mineral. Geokhim.* no. 99, 1963, 122–35; *Chem. Abstr.*, **59**, 1963, 7245.

23. KRAVTSOV, E. D. Minerals from the permafrost oxidation zone of the D'yakhtardakh deposit. *Zap. Vses. Mineral. Obshchest.*, **100**, no. 3 1971, 282–90; *Chem. Abstr*, **75**, 1971, 99824.

24. KUZMIN, E. E. and POSOKHOV, E. V. Experiment on the use of a hydrogeochemical method in prospecting for ore deposits in the Noril'sk area. *Sb. Statei Gidrogeol. Geoterm.*, no. 1 1969, 54–61; *Chem. Abstr.*, **75**, 1971, 90159.

25. MCCARTHY, J. H. Jr. GOTT, G. B. and VAUGHN, W. W. Distribution and abundance of mercury and other trace elements in several base- and precious-metal mining districts. *Circ. New Mex. Bur. Mines Miner. Resour.* 101, 1969, 99–108.

26. MCCARTHY, J. H. Jr. *et al.* Mercury in soil, gas and air—a potential tool in exploration. *Circ. U.S. geol. Surv.* 609, 1969, 16 p.

27. MELLOR, M. Properties of snow. *U.S. Army Cold Regions Res. Engng Lab. Rep.* No. 111-A1, 1964, 102p.

28. MIROSHNIKOV, L. D. Mineral acids beyond the Polar Circle. *Priroda*, **52**, no. 3 1963, 76–7. (Russian text)

29. MORITA, Y. Distribution of copper and zinc in various phases of the earth materials. *J. Earth Sci., Nagoya Univ.*, **3**, no. 1 1955, 33–7 (English text); *Chem. Abstr.*, **49**, 1955, 12233.

30. MUROZUIMI, M. CHOW, T. J. and PATTERSON, C. Chemical concentrations of pollutant lead aerosols, terrestrial dusts and sea salts in Greenland and Antarctic snow strata. *Geochim. cosmochim. Acta.*, **33**, 1969, 1247–94.

31. OLEINIKOV, B. V. and SHVARTSEV, S. L. Recent

sulphate formation in the oxidation zones of pyrrhotite–chalcopyrite hydrothermal ore manifestations of the N.W. Siberian Platform. *Geol. Geofiz.*, no. 6 1968, 15–24. (Russian text)

32. Pitul'ko, V. M. Soil formation in permafrost-taiga regions in the northern part of the Siberian Platform and methods of lithochemical soil sampling. *Zap. Leningrad Gorn. Inst.*, **56**, no. 2 1969. 58–62. (Russian text)

33. Polikarpochkin, V. V. *et al.* Changes in the chemical composition of waters of small rivers of eastern Transbaikalia in relation to their regimes and questions of hydrochemical exploration methods. *Oreoly Rasseyan. Mestorzhd. Vost. Sib.*, 1971, 178–99. (Russian text)

34. Saxby, J. D. Metal-organic chemistry of the geochemical cycle. *Rev. pure appl. Chem.*, **19**, 1969, 131–50.

35. Shvartsev, S. L. Physical-chemical processes in a series of permafrost rocks. In *Cryogenic processes in soils and rocks* (Moskow: Nauka, 1965), 132–40 (Russian text); *Chem. Abstr.*, **64**, 1966, 17282.

36. Shvartsev, S. L. and Lukin, A. A. Hydrogeochemical zonality of groundwaters of some sulfide ore deposits in permafrost rocks. In *Cryogenic processes in soils and rocks* (Moskow: Nauka, 1965), 141–8 (Russian text); *Chem. Abstr.*, **64**, 1966, 15567.

37. Stremyakov, A. Ya. Application of the hydrochemical method to exploration of ore deposits under permafrost conditions. *Razved. Okhr. Nedr.*, **24**, no. 3 1958, 46–7; *Chem. Abstr.*, **52**, 1958, 16132.

38. Sugawara, K. Naitô, H. and Yamada, S. Geochemistry of vanadium in natural waters. *J. Earth Sci., Nagoya Univ.*, **4**, no. 1 1958, 44–61 (English text); *Chem. Abstr.*, **50**, 1956, 13340.

39. Sugawara, K. Okabe, S. and Tanaka, M. Geochemistry of molybdenum in natural waters. 11. *J. Earth Sci., Nagoya Univ.*, **9**, 1961, 114–28 (English text); *Chem. Abstr.*, **57**, 1961, 590.

40. Uryvayev, A. P. Quantitative relationships between mineralization and chemical composition of river waters in the northeastern USSR. *Sov. Hydrol.*, 1968, 626–31.

41. Vodovozova, I. G. Horizontal aqueous migration of strontium-90. *Tr. Inst. Eksper. Meteorol.* no. 21, 1971, 87–90. (Russian text)

42. Wallace, R. A. *et al.* Mercury in the environment. The human element. *U.S. A.E.C.* ORNL–NSF–EP–1, 1971, 61 p.

43. Webb, J. S. and Millman, A. P. Heavy metals in natural waters as a guide to ore; a preliminary investigation in West Africa. *Trans. Instn Min. Metall.*, **59**, 1949–50, 323–36.

44. Weiss, H. V. Koide, M. and Goldberg, E. D. Selenium and sulfur in a Greenland ice sheet: relation to fossil fuel combustion. *Science, N.Y.*, **172**, 1971, 261–3.

45. Weiss, H. V. Koide, M. and Goldberg, E. D. Mercury in a Greenland ice sheet: evidence of recent input by man. *Science, N.Y.*, **174**, 1971, 692–4.

46. Weiss, O. Airborne geochemical prospecting. In *Geochemical exploration* (Montreal: CIM, 1971), 502–14. (*CIM Spec. vol. 11*)

47. Zitko, V. *et al.* Methylmercury in freshwater and marine fishes in New Brunswick and in the Bay of Fundy and on the Nova Scotia Banks. *J. Fish. Res. Bd Can.*, **28**, 1971, 1281–91.

48. Mercury in the environment. *Prof. Pap. U.S. geol. Surv.* 713, 1970, 67 p.

550.847

Gaseous geochemical methods in structural mapping and prospecting for ore deposits

L. N. Ovchinnikov, ACADEMICIAN

The late V. A. Sokolov, PROF.

A. I. Fridman, DR.

I. N. Yanitskii, DR.

All of the Institute of Mineralogy, Geochemistry and Crystallochemistry of Rare Elements, Moscow, U.S.S.R.

Synopsis

As the search for new ore deposits continues, the proportion of such deposits which are either exposed at the surface or easily found by conventional geological methods decreases. For this reason it is becoming increasingly necessary to adopt exploration methods which are capable of detecting blind orebodies. At the present time, most of the geochemical prospecting methods used in the U.S.S.R. that are capable of finding such bodies are of the lithochemical type, and, as such, are only effective where dispersion aureoles are both present in the rock and accessible for sampling. It is suggested that gaseous emanations provide a reliable alternative method of searching for blind deposits.

Large volumes of different gases escape from the earth's crust into the atmosphere continually: however, neither the quantity nor the composition of such gaseous emanations is regular over the surface of the lithosphere. Areas of especially high activity appear to be related to zones of deep tectonic fracturing and the accompanying jointing in which mineralization is sometimes located.

The migration of elements in the gaseous phase is a widely distributed phenomenon. The phase itself constitutes an important component in the formation of endogenous deposits, and also remains active in the post-ore formation period. In the supergene conditions of the immediately subsurface parts of orebodies, gaseous components connected with the oxidation of ores may also be formed. Thus, gases of different genesis and composition accompany ore deposits during the various stages of their evolution. Such gases are either preserved in the gaseous and gaseous–liquid micro-inclusions of rocks and ores, or migrate along joints related to ore-controlling fault systems. Endogenous and exogenous gas aureoles around ore deposits can therefore be used as highly sensitive geochemical indicators, and in structural mapping they can be used to classify fracture systems of different ages, and also to investigate the origin and depth of bedding. The same methods may also be employed to outline those parts of an orebody which have the highest economic potential.

In the U.S.S.R. gaseous methods of geochemical prospecting are being developed along several lines. Many of these are concerned with the study of gases in micro-inclusions and occluded in rocks and ores. The object of such studies is to obtain a better understanding of the conditions of ore formation, such as temperature and fluid composition. The effect of the geological age of the material is also of relevance in this connexion. A great deal of information has now been obtained on the component composition, gas content, regularities of isotopic ratios and physical properties of such gases. Recent studies include detailed investigations of the helium, argon, hydrogen, carbon dioxide, nitrogen and other gaseous components in metasomatic contact zones, vein sulphides and both pyrite and mercury ores. Helium, neon, argon and nitrogen have also been studied by isotopic methods. Thus, Volkova *et al.*[1] have shown that considerable losses of ^{40}A occur where helium, lithium and rubidium are abundant, and high helium is found to coincide with concentrations of the ore metals such as tin, lead and zinc.

It is also possible to determine the nature and trend of superimposed processes of ore formation. For example, Zhirov *et al.*[4] studied variations in the concentrations of helium, argon, nitrogen and other gases in the sulphide ores of the Kola Peninsula. Here, the presence of abnormal concentrations of occluded gases in ores and host rocks is thought to be related to the presence of hidden orebodies. Research into gases contained in pyrite, polymetallic titano-magnetite and gold ore deposits of the Urals has also indicated that nitrogen, carbon dioxide, hydrogen, methane, argon and helium all play a major role in the ore-forming process.[5] A dependence of the quantity of occluded gases in pyrite on the age of the deposits in which they occur has, however, been statistically proved, high gas contents being

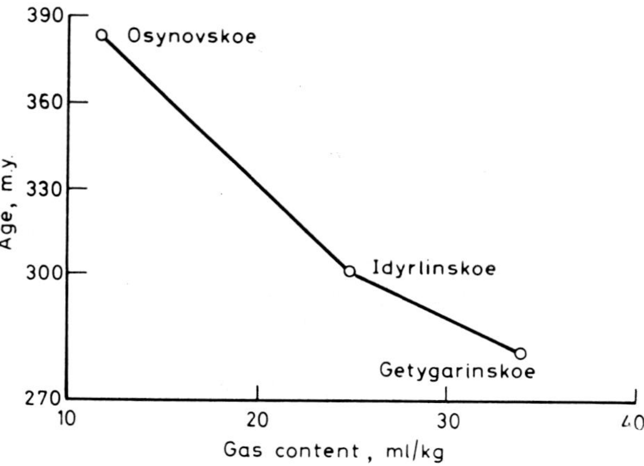

Fig. 1 Dependence of gas content on age in pyrites from gold deposits

typical of younger deposits and low contents being associated with older ones (Fig. 1). This feature may be explained by a gradual loss with time of the occluded components in primary aureoles, such a gradual loss of mobile gases giving rise to a seepage of gas which may be used to detect such deposits. Thus, mercury deposits of the Caucasus and Donbass areas are characterized by a high hydrogen content (up to 144 cm^3/kg) and carbon dioxide (up to 220 cm^3/kg).[6] These components are present not only in ores but also in hydrothermally altered host rocks; thus, widespread gaseous aureoles, which are intimately related to the ores, occur in the vicinity of ore-controlling fractures.

The study of mobile gaseous aureoles has recently been initiated. Such studies can be divided into two different types according to the genesis of the component gases: *(a)* those of polygenetic origin formed at large depths; and *(b)* those originating from oxidized ores in the zone of weathering. The latter are products of chemical reactions between sulphides and other unstable endogenous compounds with atmospheric oxygen in the zone of leaching. Below the water-table such reactions may still occur with oxygen dissolved in vadose water. Among the gases formed by these reactions are carbon dioxide, sulphur dioxide and hydrogen sulphide, and these, together with the halides and other volatile combinations of metals such as arsenic, antimony, caesium and rubidium that are released from ores, may be identified as secondary products.[7]

Ores containing mercury are characterized by mercury vapour, and radioactive ores by radon and occasionally hydrogen. In some cases gaseous aureoles derived by subsurface chemical or radioactive processes are related to corresponding mineralization and can be used in the search for hidden orebodies.[9] Such local anomalies associated with 'ore' are, however, usually connected with the orebody itself by means of faults in the surrounding rocks.[10] The problems of using gaseous aureoles of subsurface origin in prospecting are therefore many and complex. For example, the aureole of mercury vapour in the Chonkoi deposit is shown in Fig. 2.[8] The less intensive anomaly is associated with ore, whereas the larger anomaly is controlled by post-ore fracturing in the hanging-wall of the deposit. Similar distributions of gases of both subsurface and deep origin are also typical of carbon dioxide and radon.

Gases generated at depth are mobile and closely related to systems of deep tectonic faulting. In variety of form, intensity of aureole and abundance they greatly prevail over gases of subsurface origin. Among such gases carbon dioxide, hydrogen, helium, methane, nitrogen and mercury are the most typical, and they are quantitatively proportioned by their abundance in the earth's crust. Among them are gases distributed in great concentrations—carbon dioxide, nitrogen, hydrogen and methane—whereas the rarer components are present as admixtures—helium, mercury, radon, etc. Gases of mixed composition are typical of mobile gaseous aureoles, although in some cases a single component may comprise 70–95% of the total. Such dominant components may be nitrogen, carbon dioxide, hydrogen or methane. The actual

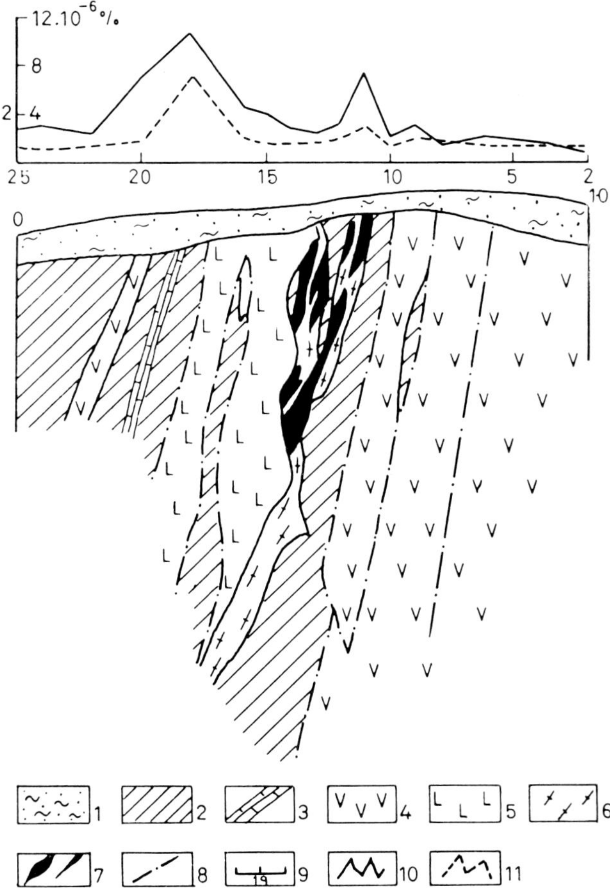

Fig. 2 Mercury concentrations in gaseous and solid phase in the Chonkoi deposit (profile III): (1) allochthonous clayey soil; (2) shales; (3) limestones; (4) effusive rocks; (5) serpentines; (6) listvenites; (7) orebodies; (8) tectonic faults; (9) measurement points; (10) concentration of mercury vapours; (11) concentration of mercury in solid phase

dominant gas varies from region to region. Thus, in east Transbaikalia there is a dominance of carbon dioxide, in the north Caucasus carbon dioxide and hydrogen, and in the large alkaline massifs of the Kola Peninsula methane is prevalent. Nitrogen is widespread in both abnormal geothermal regions and in most platform areas. Gases present in smaller quantities appear to be independent of the nature of the main component gas. Only in rare cases (as in the so-called 'dry carbon dioxide current', where the concentration of CO_2 is 99·9% by volume) is the role of other components, including minor constituents, sharply decreased.

The composition and intensity of the gas flow depends dominantly on geothermal conditions. Here, tectonic activity, stage of volcanic and magmatic activity and the rate of opening of the fluid-bearing systems are the main factors. Accordingly, the weaker gaseous patterns are observed in platform and shield areas, that is, consolidated blocks of the earth's crust, whereas stronger patterns are encountered in areas of late orogenic and tectono-magmatic activity.[3, 6]

The most effective form of gaseous prospecting is based upon sampling subsoil air. It aims at detecting gaseous aureoles of dispersion over the permeable fault structures which frequently control endogenous ore mineralization. For some components, however, the investigation of gases which are soluble in groundwater (but which may itself be controlled by faulting) is found to give better results. Within the earth's crust different processes of gas migration take place simultaneously, i.e. diffusion, filtration (effusion) and transportation by groundwater. Under normal conditions these processes are closely interrelated. In massive rocks and others with low permeability (i.e. most sedimentary, igneous and metamorphic rocks) diffusion is the main mechanism of gas migration, although the patterns formed in this way vary considerably in character. In the presence of faults and associated open joints very much larger filtrational aureoles are formed, and the actual shapes and sizes of such patterns are determined by the localization of such structures. In such cases the concentrations of the mobile gaseous components are much higher and the anomalies formed show good contrast with background.

A variety of sampling methods are employed when mobile gases are being investigated. Boreholes and probes can be operated within the limits of the aeration zone (i.e. boring to 0·5–1·5 m and probes down to 5–10 m), and springs, wells and boreholes may be used to provide access to the zone of saturation. In the first case carbon dioxide, hydrogen, methane, mercury, radon and, rarely, helium are sampled for analysis, whereas in the second helium and, less usually, radon or methane are employed.

There is some degree of overall regularity in the distribution of the mobile gaseous components inasmuch as they are fixed in the most permeable parts of tectonic fracture systems. Such places are often related to conjugate fractures of different orientation (Fig. 3), and only to those elements which are mobile. For these reasons consolidated parts of faults as well as fracture systems contained wholly within large massive blocks of rocks are unlikely to be associated with abnormal gas concentrations.

Another specific feature of mobile gases of deep origin is their indifference to ore deposition, flow patterns being present in all of the higher penetration zones regardless of whether endogenous ore formation has taken place or not. This characteristic is true not only of the main gaseous components but also gases such as helium, radon and mercury vapour, which are always present

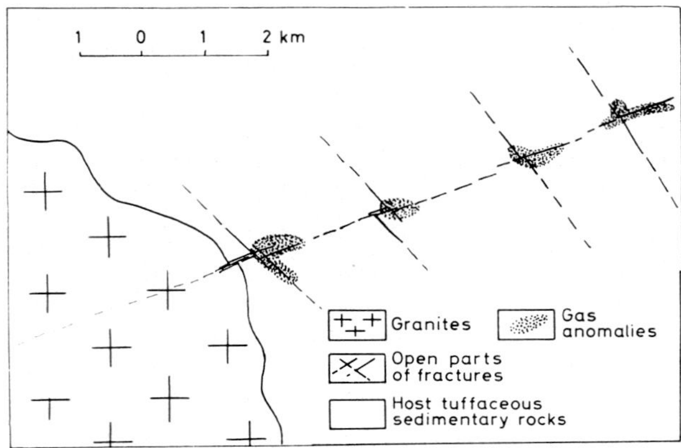

Fig. 3 Gas anomalies, Baksensk granite massif, north Kazakhstan

in rocks through which gas has flowed, even when no ore deposition has taken place. Thus, mobile gases which are solely related to tectonic factors overlap the hydrothermal ore process during its early stage and accompany ores during subsequent stages of evolution of the deposits so formed. Mobile gases of deep origin therefore have a direct connexion with fault tectonics and an indirect one with endogenous ore deposits. Accordingly, gas surveys which investigate mobile constituents of deep origin represent an effective method of locating ore deposits regardless of the nature of the indirect connexion between the anomalies and the ore itself. Such surveys are likely to be particularly effective during the early stages of an exploration programme. Furthermore, since the association of zones of ore deposition with deep fractures and the so-called 'feather joint' is established, the paragenetic relationship described may also allow the prognostication of hidden ore mineralization within the limits of known ore fields.

At the present time carbon dioxide, helium and mercury vapour are the gases most investigated as indicators of endogenous ore in depth.

Examples of regions studied

North Caucasus

Aureoles of carbon dioxide in subsoil air have been studied over a number of mercury deposits in the north Caucasus. The background content of the gas in this area is 0·3–0·7%, but over ore-controlling structures it reaches concentrations of 5–11% (Fig. 4). The anomalies reflect long narrow zones of such structures, which strike in both northwesterly and northeasterly directions. These anomalies reflect the conjugate fractures which constitute structural nodes in this area.

Central Kazakhstan

Local aureoles of helium in groundwaters associated with lead–zinc and tungsten–molybdenum deposits in central Kazakhstan have also been investigated. Here, ore mineralization is related

Fig. 4 Distribution of carbon gas concentrations in subsoil air, Shabsugskay mercury-bearing region, northwest Caucasus

to zones of crushed Devonian volcanic and tuffaceous rocks and, rarely, to hydrothermally altered granitoids. Over the Shalgiya molybdenum ore deposit a local helium anomaly with an intensity of 0·054% is controlled by a system of post-ore joints in the hanging-wall of the ore stockwork. On the flanks of the deposit the helium content decreases to a local background level of only 0·0006% (Fig. 5). The contrast of the geochemical anomaly associated with the structural node is therefore 90. For larger deposits contrasts in excess of 100 have been discovered.

Chukotka

Over the Plamennoe mercury deposits at Chukotka the mercury content of subsoil air varies from a background of 1×10^{-6} mg/l to anomalies of $1{\cdot}1 \times 10^{-3}$ mg/l.[8] The anomaly with the maximum intensity is related to a large zone of crushed host rocks which contain the ore, the concentration of mercury vapour being lower over the main orebody (Fig. 6).

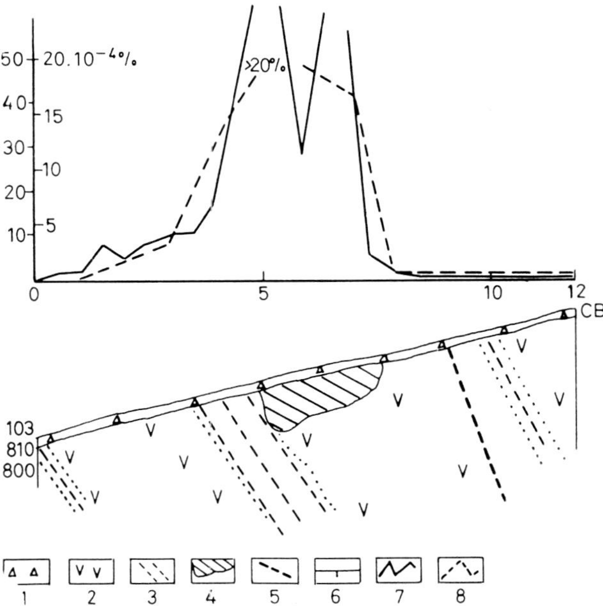

Fig. 6 Mercury concentrations in gaseous and solid phase, Plamennoe deposit, profile III: (1) eluvium–colluvium of orebody and host rocks; (2) Lower Cretaceous liparites; (3) crushed zone; (4) orebody; (5) tectonic fault; (6) measurement points; (7) concentration of mercury in gaseous phase; (8) concentration of mercury in solid phase

General discussion

An important feature of aureoles derived from mobile gases of deep origin is their stability in times of relative tectonic quiescence. Long-term observations on a number of components (carbon dioxide, methane, helium, mercury and radon) have shown that variations are within the limits of experimental error over periods of time. Both the volume of flow and the elemental and isotopic composition of the gases, however, display rapid and intensive change at times of tectonic activity.[7] Studies on a carbon dioxide anomaly over a mercury deposit in the north Caucasus showed a threefold increase in flow during the Anapa earthquake of 12 June, 1966 (Fig. 7), together with an isotopic change in the carbon, the proportion of ^{13}C changing from the normal 2·92 to 2·37% on the day of the earthquake.[6] The intensity of the earthquake at the observation point was 4–5 on the Richter scale.

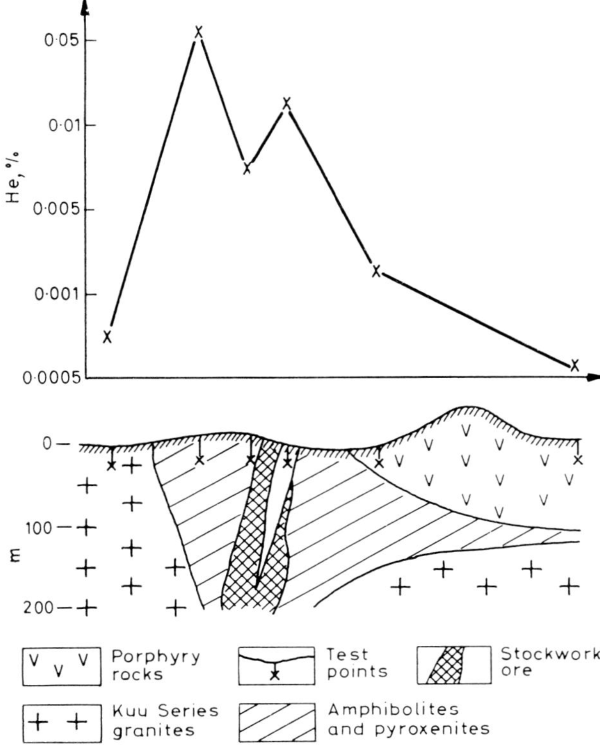

Fig. 5 Helium concentrations in groundwater, Shalgiya molybdenum ore deposit, central Kazakhstan

During the Tashkent earthquake of 1966 the helium content of the water in the observational bore rose by a factor of 12, and that of radon increased threefold.[2]

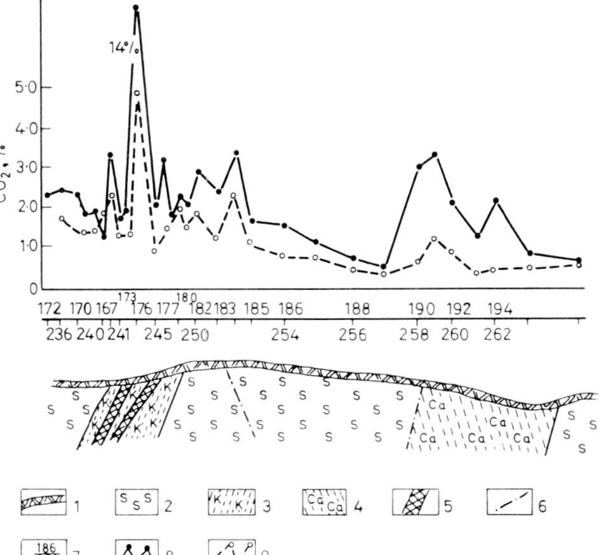

Fig. 7 Concentrations of carbon gas in subsoil air, Sakhalinsk deposit, northwest Caucasus: (1) eluvial and colluvial sediments; (2) aleurolitic clays; (3) zone of folding and intensive fracture of rocks with dickite mineralization; (4) as (3) with calcite mineralization; (5) orebodies; (6) faults; (7) test points; (8) and (9) concentrations of carbon gas in subsoil air 8–12 h before Anapa earthquake (12 July, 1966) (8) and two months after earthquake (10 September, 1966) (9)

These investigations show that the gas flow begins to increase with the growth of the dynamic load on the rocks and reaches a maximum at the time of seismic break. Constant, systematic observations on the gas flows and component compositions of gases emanating from seismic zones might therefore be used as a means of predicting both the times and locations of earthquakes.

Conclusions

Natural gases of differing composition and genesis are widely distributed within the earth. They are either syngenetic with the ore process, or present a product of subsurface chemical reaction or reflect the constant and general process of degassing of the earth. An extraordinarily wide range of gas distribution allows them to be used for a variety of different geochemical tasks, but the importance of further research into mobile gases of deep origin must be stressed.

Information on the distribution of mobile gases can be employed in structural mapping, especially in regions of sedimentary rocks and those areas overlain by loose sedimentary cover. Gas surveys allow the recognition of fracture systems with deep tectonic penetration. The results of such surveys can also be used to locate endogenous mineralization. Systematic observations of gas flows should be included in systems designed to forecast earthquakes.

These aspects show that geochemical studies of natural gases constitute an important branch of geological science.

References

1. Volkova, N. V. *et al.* Geochemistry of argon and helium in metasomatic aureoles at the contact of granitoid and diopside schists (Aldan) as indicator of trend of geological processes. In *Trudy tretjevo vsesoyuznovo simpoziuma po promeneniyu stabilnykh izotopov v geokhimii* (Moskow: Geokhi, 1970), 18–9. (Russian text)

2. Gorbushina, L. V. *et al.* On effect of geological–tectonic factors on the content of gases in ground water of Tashkent artesian basin. In *The Tashkent earthquake, 26 April, 1966* (Tashkent: FAN, 1971), 198–200; *Referat. Zh., Geol.*, no. 8 1971, 8E92.

3. Eremeev, A. N. Ershov, A. D. and Yanitskii, I. N. Some aspects of helium surveying in structural–geological mapping and prediction of endogenous mineralization. *Sb. 'Geokhim. metody pri poiskakh i razvedke rudnykh mestorzhdenii'* pt 5, 1971, 49–65; *Referat. Zh., Geol.*, no. 12 1971, 12D70.

4. Zhirov, K. K. *et al.* On the geochemistry of gaseous components (atmosphere)—helium, argon-40, nitrogen—in endogenous sulphide process. In *Trudy tretjevo vsesoyuznovo simpoziuma po promeneniyu stabilnykh izotopov v geokhimii* (Moskow: Geokhi, 1970), 52–4. (Russian text)

5. Ovchinnikov, L. N. *et al.* On the characteristics of gases containing in ores of some deposits of the Ural. *Sb. 'Geokhim. metody pri poiskakh i razvedke rudnykh mestorozhdenii'* pt 5, 1971, 100–18 (Russian text)

6. Fridman, A. I. *Natural gases of ore deposits* (Moskow: Nedra, 1970), pp. 173–81. (Russian text)

7. Fursov, V. Z. Vol'fson, N. B. and Khvalovskiy, A. G. The results of the study of mercury vapours in the zone of the Tashkent earthquake. *Dokl. Akad. Nauk SSSR*, **179**, 1968, 1213–5; *Dokl. Acad. Sci. USSR, Earth Sci. Sect.*, **179**, 1968, 208–10.

8. Fursov, V. Z. Possibility of using vapours of elements for search for ore deposits. *Sb. 'Geokhim. metody pri poiskakh i razvedke rudnykh mestorozhdenii'* pt 5, 1971, 30–49; *Referat. Zh., Geol.*, no. 12, 1971, 12D73.

9. Elinson, M. M. Application of gas surveying for search and prospecting of ore deposits. *Sb. 'Geokhim. metody pri poiskakh i razvedke rudnykh mestorozhdenii'* pt 5, 1971, 85–100; *Referat. Zh., Geol.*, no. 12, 1971, 12D70.

10. Khairetdinov, I. A. *et al.* Application of gas test for search for sulphide bodies in faults of East Saya. *Geologiya Geofiz.*, no. 10 1965, 135–7. (Russian text)

543.51:546.291:550.8

Application of helium surveying to structural mapping and ore deposit forecasting

A. N. Eremeev, PROF.

The late V. A. Sokolov, PROF.

A. P. Solovov, PROF.

I. N. Yanitskii, DR.

All of Moscow State University and the Scientific Council for Geochemical Prospecting, Moscow, U.S.S.R.

Synopsis

The survey described was based on the investigation of the natural field of free-filtering helium in the saturation zone, samples being taken at a depth of 20–30 m. At such depths the effect of atmospheric factors is eliminated and the high sensitivity of the method is assured. Direct analysis of the helium concentration is done by use of mass spectrometry (sensitivity above 5.10^{-5} equivalent %) or ion pump-type detectors (sensitivity above 5.10^{-4} equivalent %). The results of the analyses of samples from different depths are attributed to the most informative horizon by interpolation.

The number of sampling points required for the first stage of the study varies from 1 to 4 per 100 km^2 (survey scale, 1 : 1 000 000). Investigations to date have mainly been carried out in areas where the basement is exposed or lies not far from the surface (less than 500 m). A helium field determined by this method is characterized by high contrast relative to geological background. It varies in the order of from 1 to 6 at a horizon 30 m below surface. Helium flow observations carried out over a number of years show that the helium field is stable in time: significant variations arise only as a result of seismic activity.

It appears that only rock permeability affects the helium field, and no influence arises from factors related to geological composition near the surface. Permeability zones through which the helium flow is transported to the surface extend to significant depths, and they are of tectonic origin. Helium migration is mainly effected by means of filtration with other gases and fluids.

The intensity of helium anomalies depends directly on the depth at which the fracture zone is located. The main helium source would appear to be below a granitic sedimentary layer.

The earth's crustal structure in the helium field resembles a combination of impervious solid blocks, separated by permeable zones. Granitic masses usually occur inside these solid blocks, and the helium field within the blocks has the minimum intensity. Junctions and intersections of the zones between the blocks are regarded as maximum permeability areas. The helium-bearing zones are in agreement with the grid lines of high horizontal gradients and seismic zones. In general terms, the earth's helium field reflects the position of small, vertical-amplitude, diagonal and orthogonal fractures and consolidated blocks separating them.

Helium-bearing anomalies occur in geological formations directly or indirectly associated with mobile diagenetic fractures. Diamond-bearing kimberlite pipes, carbonatites, hydrothermal gold, lead–zinc, iron and mercury deposits are mostly found in association with these anomalies.

Further development and the application of helium surveying in conjunction with geophysical and geochemical studies offer considerable promise in structural mapping, prospecting for ore deposits and earthquake forecasting.

1969 marked the one-hundredth anniversary of the discovery of helium. Much has been published on the geology of helium, work by Vernadsky[3] and Rogers[12] being of considerable importance. Later, on account of the complexity of the problem, contradictory views have been put forward. The migration forms of helium, its source, connexion with radioactivity, etc., remained completely unknown.

Special importance was attributed to granites and granitic rocks in the production of large amounts of helium, the role of abyssal spheres in the earth's crust either being considered unimportant or being ignored because of the low radioactivity level of basic and ultrabasic rocks.

The discovery of unique helium-bearing regions in the U.S.A. led to calculations of the quantity and possible production of helium, and it was shown that the practically admissible amount of radioactive element dispersed in the crustal mass cannot account for such a large production of helium in some areas. Thus, suggestions were made with regard to the role of 'original helium'[17] and the presence of large radioactive sources at depth.[18] Attempts were also made to relate regional helium-bearing zones with belts of high radioactivity.[11,16] The accumulation of further data, however, did not support the existence of a regular relationship between helium anomalies and increased radioactivity, but a more general connexion was detected between high helium-bearing areas and geothermal and geochemical anomalies.[9,14] High radioactivity values were not recorded in this instance.

Forms of helium

There are three forms of helium in the earth's crust: (1) *solid solution*—He ions (or alpha-particles) with an electron captured at the breaking point; (2) *atomic inclusions*—neutral atoms in crystal lattices or in the closed pores of minerals and rocks; and (3) *free atomic*—neutral atoms in exposed breaks and connecting pores.

These three forms of helium determine its migration through the earth's crust. Helium in the solid solution form is practically immobile. For included helium diffusion migration is possible only when the diffusion factor is sufficiently low (about 10^{-7} cm^2/sec). The constants of this migration were determined by Newton and Round[10] and were acknowledged to be less than the average speed of sedimentation. The main migration form of free helium is by filtration, the speed of underground liquid phases being greater than the diffusion speed by several orders of magnitude.[5] Thus, diffusion migration can be ignored for practical purposes.

A literature study (including fundamental contributions by Kharley[15] and Gerling[4]), as well as our own experimental work, made it possible to draw diametrically opposite views of the distribution of 'immobile' and 'mobile' helium. If helium is held by solid, low-permeability rocks, the flow of the mobile component is absent; in areas of tectonic crushing, fissuring and high permeability the flow of free helium reaches its maximum (Fig. 1).

The present paper gives the results of a study of the distribution of natural mobile helium, the subject of the investigation being the earth's outer crust to a depth of 100 m. This choice was made on the basis of efficiency of work (probing at limited depth is less expensive) and larger contrast of the helium field in subsurface zones.[6,7]

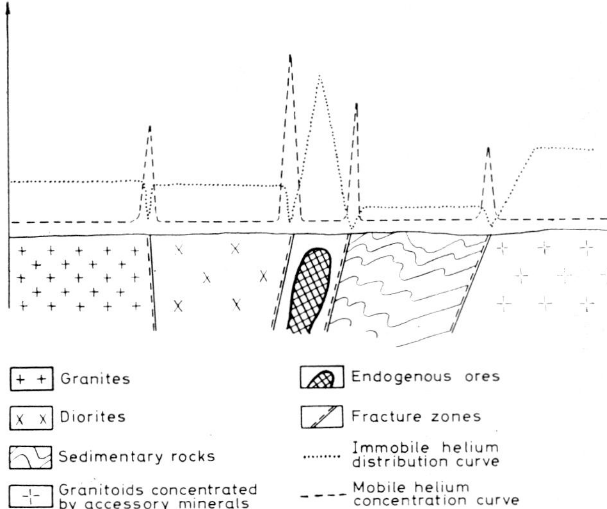

Fig. 1 Distribution of mobile and immobile helium in outer part of earth's crust

Natural mobile helium

Natural mobile helium can be detected with the aid of core sampling, water or gas surveying. Although they are almost equally informative, these methods are very different in terms of efficiency. The most simple and efficient method is that of helium surveying by underground waters (ground, fissure and strata waters) flowing from springs or tapped in wells and boreholes. Surface waters (rivers, lakes and swamps) have helium concentration in equilibrium with that in the atmosphere and are not suitable for testing.

The most convenient means of expressing the helium concentration in water is as a volume percentage of the equivalent gaseous phase. According to Henry's law, the quantitative ratio of helium when water and gas phases are in equilibrium is approximately equal to 1 : 100.

In the uppermost part of section many tectonic fissures are open joints, owing to the general fall in rock pressure and the influence of weathering zones. Hence, the highest possible water content is characteristic of the upper 30–50 m, particularly for the debris of crustal weathering. Here the helium-bearing fluid, lifted from depth by fissures, mingles with the subsurface waters and is dispersed. The investigation of water enables the area of helium coming to the surface to be located, as well as the area of its dissemination.[5]

Instrumentation

The helium content in water was measured by use of specially designed ion pumps (Fig. 2). Technical details of the instruments, which are suitable for field work, are given in Table 1.

Fig. 2 INGEM–1 helium analyser

Table 1 Technical characteristics of INGEM–1

Gas phase analytical range	From 5.10^{-3} to 15 volume %
Water (sensibility) analytical range	From 1.10^{-5} to 0·15 equivalent %
Weight of measurement assembly, kg	3·8
Overall size, mm	248 × 122 × 190
Electric power and required capacity	Accumulator or dry battery; 2·5 W
Weight of power assembly with energy supply for 50 working hours, kg	1·4
Warm-up time	<20 min
Productivity	Up to 30 tests per hour

The possibility of measuring the helium concentration directly in water eliminates the intermediate operation of degassing, which was previously unavoidable in mass-spectrometric analysis; an additional advantage of the method is the absence of any influence by hydrocarbons, which are widely distributed within the earth's crust, on the results of the analysis. The general applicability of the instruments in geological investigations is suggested.

Distribution of helium

The distribution of helium has been studied in vertical section. Its concentration due to dispersion changes greatly, generally being lowest in the upper part and increasing with increasing depth. The borderline of a saturation area can be marked at a certain depth, beneath which concentration growth ceases. Fig. 3 shows the helium distribution in abnormal sites specified by relatively stagnant conditions of underground water (curve *A*), these hydrogeological peculiarities prevailing in most areas with flat or slightly rugged surfaces. The borderline of the saturation zone usually runs at a depth of 100–200 m. In areas of relief and ascending underground water flow the distribution of helium assumes a different character (curve *B*). The increase of concentration with depth depends on the speed of ascending water flow: if it is sufficiently high, large concentrations can be found at surface—close to that in the saturation area. On the other hand, changes in helium concentration with depth in areas of background values are not dependent on conditions in the underground waters—background concentrations prevail in areas of relatively stagnant waters as well as in relief areas.

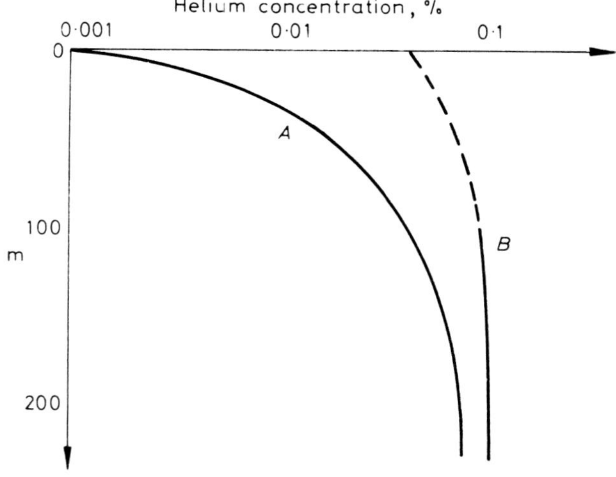

Fig. 3 Helium dispersion in subsurface zone: A, under stagnant conditions of underground water and B, under conditions of ascending water flow

Field investigations

The areal distribution of helium was investigated mainly in areas of Kazakhstan and the Russian platform. Several thousand water sources were tested—wells, springs and different types of boreholes. The helium contents measured were noted on test charts, background and abnormal values

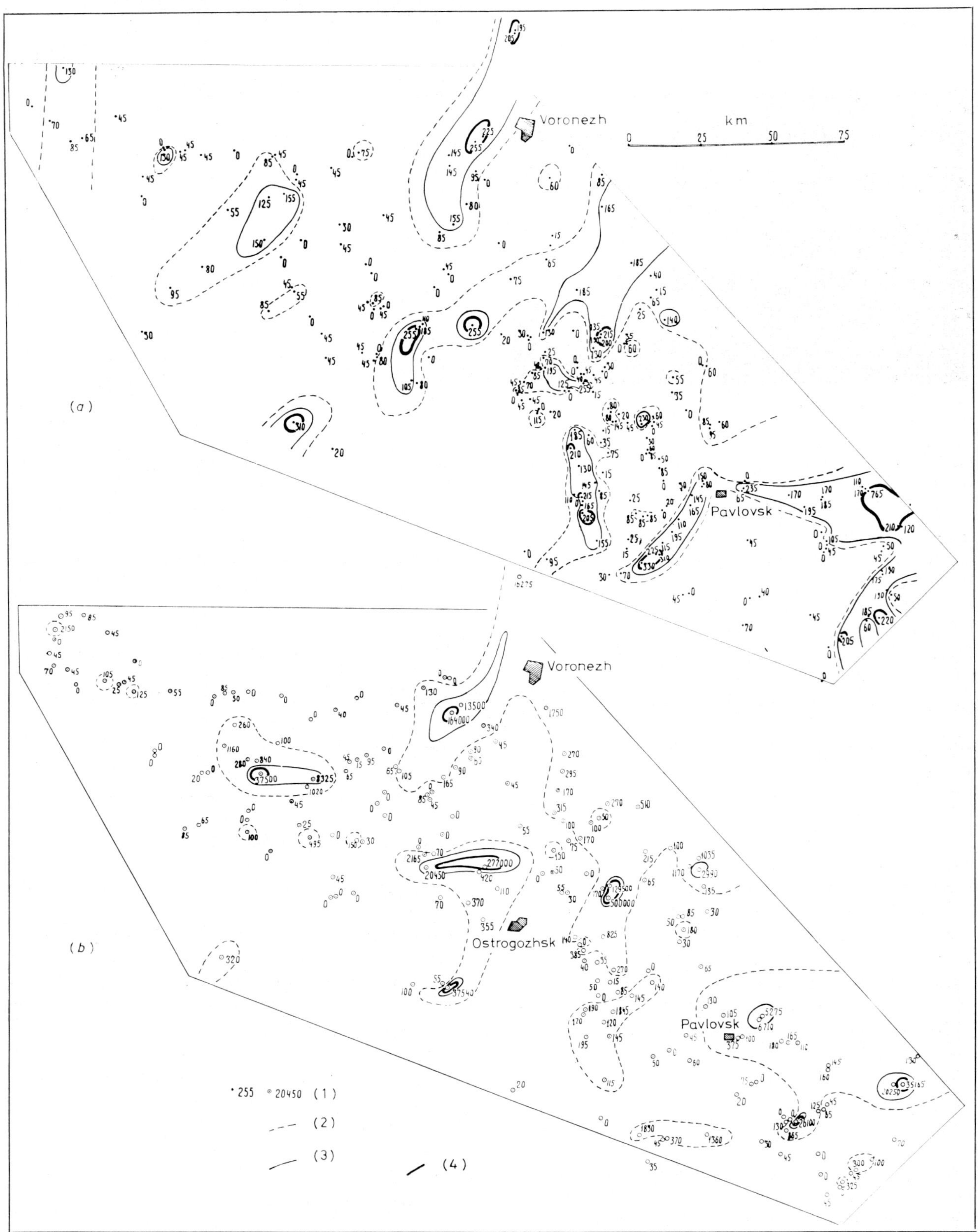

Fig. 4 Map of helium-bearing Voronezh crystalline core area: (a) upper water-bearing level; (b) basic water-intake level. (1), Point of probing and helium concentration; (2)–(4), lines of iso-concentrations of helium with values, respectively, 0·00025 and 0·0005, 0·0005 and 0·0125, and 0·001 and 0·125 volume % (recalculation of data according to formula $5n \times 10^{-6}$; helium concentration in atmosphere subtracted)

being determined by use of well-known statistical methods. From the values obtained, iso-concentration lines were drawn.

Difficulties arose in drawing iso-concentration maps for water sources from different testing depths. Correction factors were obtained statistically by comparison of the helium concentration in samples taken at different depths for each water source. For regions in which there were large differences (more than 10) between the helium content at the top and bottom levels separate maps had been drawn. Examples of the separate drawings are given in Fig. 4 (*(a)* and *(b)*) for subsurface (5–10 m) and deeper (100–150 m) levels. The space coincidence of background and anomalous fields for these levels is satisfactory, though the value of helium content differs from 30 to 100 times.

An important feature of the helium field is its high contrast, variations of several orders of magnitude being apparent. For north Kazakhstan, for instance, at the 30- to 50-m level the concentration changes from 5.10^{-4} to 5 volume %, i.e. 10^4 times.

Table 2 Results of multiple observations of control water areas in Kazakhstan

Location of water area and summary of test conditions	Date of test	Helium content, 5.10^{-6} volume %
Ruzaev district, borehole: water-table, 11 m; test depth, 100 m	July, 1963	188 000
	August, 1964	196 000
	August, 1965	177 000
	September, 1966	220 000
	August, 1967	200 000
	July, 1968	210 000
Adaibul district, spring: flow, 1·8 l/sec	July, 1966	100
	August, 1967	110
	July, 1968	120
	September, 1969	100
Shortandin district, Gelambet mine: 390-m level; borehole 639	August, 1964	2 000 000
	June, 1965	1 668 000
	July, 1965	1 750 000
Krasnoarmeisky district, Roshinskoje village: outflowing borehole N24; flow 12 l/sec	July, 1968	140 000
	August, 1969	144 000
	September, 1969	140 000

The high stability of helium field, as far as the time factor is concerned, should be emphasized. Prolonged observations of some water sources in Kazakhstan revealed slight fluctuations in concentrations which can be partially explained as being due to the methods employed (see Table 2 for results). In the preliminary investigations the average frequency of testing was 1 point per 100 km². As a result of such surveys it was possible to divide large fields with low and higher helium concentrations satisfactorily. Detailed testing (up to 1 point per 10–25 km²) was applied only for areas with high helium concentrations (not more than 15–20% of all regions investigated).

Fig. 5 shows the map of helium concentration in north Kazakhstan for an area in excess of 150 000 km², approximately half the area being taken up by fields with the minimum (5×10^{-4} volume %) concentration. A slightly higher helium content (up to 0·0125 volume %) was recorded for 35% of the area; for about 10% it was high (from 0·0125 to 0·125 volume %); and the remaining 5% was very high (more than 0·125 volume %).

It was found that helium content does not, in practical terms, depend on the radioactivity of the rocks, the content of mobile helium in basic and alkaline and acid intrusions very seldom exceeding its atmospheric concentration. The connexion between higher helium concentrations and fracture distortions is obvious, increasing significantly with increases in depth. Thus, if this connexion could be traced only as a trend for a generally spread network of subsurface fissures, it was distinct for zones of heterogeneous rocks obtained in interpretations of geophysical fields at depths of 3–5 km.

The coincidence of helium-bearing zones with the lines of higher horizontal gradients of gravity (Δg) has been recorded. This coincidence increased regularly with passage from observed anomalies, reflecting breaks up to 3–5 km deep, to those interpreted at depths of 10, 20 and 30 km. Fig. 6 shows the zones of maximum permeability in helium fields. The main zones correlate well with the lines of higher horizontal gradients (Δg)

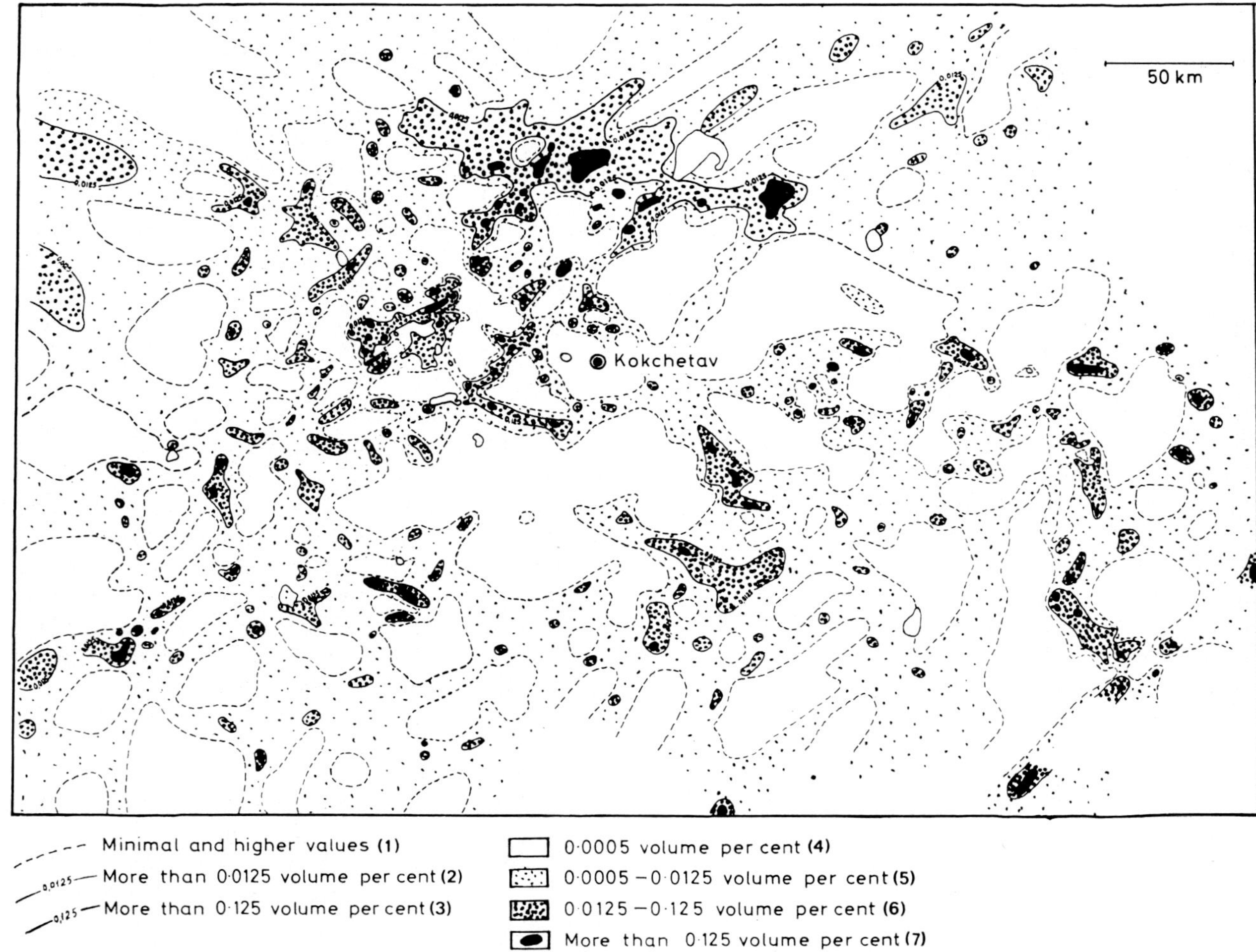

Fig. 5 Map of helium-bearing zone, north Kazakhstan: (1)–(3), borderlines of helium fields; (4)–(7), helium fields

recalculated at 20–30 km, the considerable depth of formation of permeable fracture systems being indicated. Their mobile condition was confirmed from the transfer of helium surveying data to profiles of deep seismic soundings,[2] a specific correlation being observable between intensity of helium anomalies and depth of fracture formation (Fig. 7).

Fig. 8 is a map of recent movements in north Kazakhstan.[1] Comparison of Figs. 5, 6 and 8 clarifies the common character of lines, the main directions spreading northwest and northeast to build up a network of diagonal fractures. Such a network is stipulated by the system of planetary fractures due to the rotation of the earth.[8] The system is 300–400 m.y. in age, as is reflected in the rock facies, beginning with Cambro–Ordovician rocks, and in the disposition of metallogenic zones. Fig. 7 illustrates this, the heavy lines showing general fractures and metallogenic zones.[13]

Metallogenic zones

Metallogenic zones in a helium field are depicted by use of contrasting intricate lines. They consist of a combination of small, low-permeability solid blocks (small intrusions, hydrothermally altered rocks) and permeable areas of renewed fractures, which delineate these blocks. Evidence of such zones is a high-contrast helium field severely differentiated, which makes it possible to forecast the most favourable places for exploration for endogenous deposits.

Structural junctions are characterized by multicomponent specialization with a set of different minerals. Cases are noted when in one structural block endogenous deposition of a number of metals occurs—for instance, gold, iron, tantaloniobates and tin (Kazakhstan), lead, zinc and mercury (Caucasus), etc. The time of formation of these deposits is obviously different; as far as their location in structural junctions is concerned, they are separated both vertically and horizontally and, therefore, they usually seem to be completely isolated. Nonetheless, the presence of helium anomalies in ore-bearing contours gives cause to expect the existence of common feeder channels (Fig. 9). The fact that these channels were being

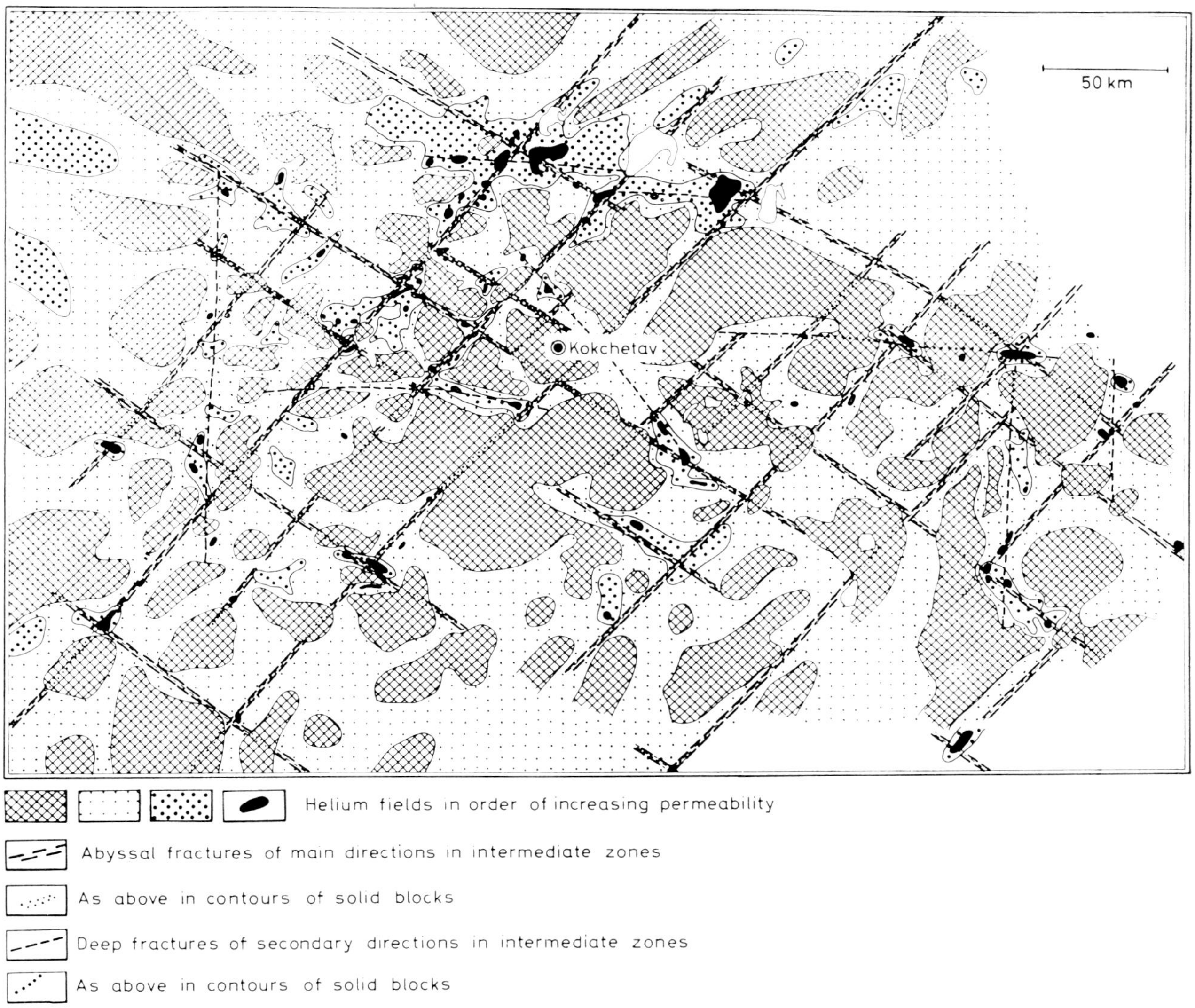

50 km
Kokchetav
Helium fields in order of increasing permeability
Abyssal fractures of main directions in intermediate zones
As above in contours of solid blocks
Deep fractures of secondary directions in intermediate zones
As above in contours of solid blocks

Fig. 6 Maximum permeable zones

preserved tens and hundreds of millions years after the process of ore deposition illustrates the prolonged and multiphase renewal of permeable structural junctions.

10^{-3}%
He
20
10
0
1 2 3 4 5 6
Number of peaks anomalies
a b
Number of peaks on section of profile of GSZ
100 300 500 700 1000
Depth, km
0 10 20 30 40 50
d_3^K
d_4^K
d_0^M
(1) (2) (3) (4) (5)

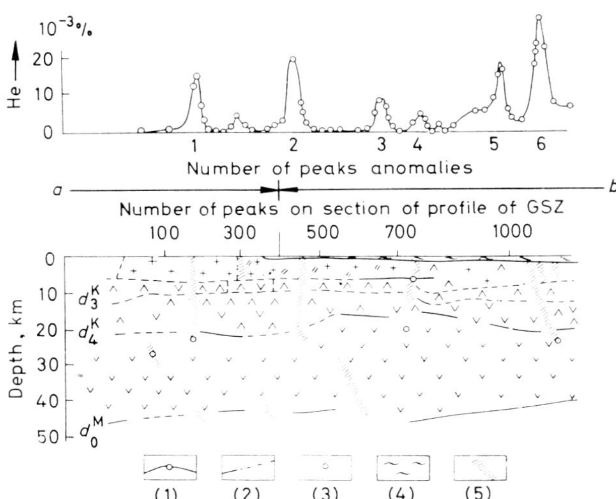

Fig. 10 presents another kind of structural ore-bearing junction with an intrusion of the central type. The helium anomaly has a circular form, whereas endogenous ore deposition is usually connected with exo-contact intrusions.

The direct relationship between endogenous deposition and weakened tectonic zones of deep fracturing would explain the great number of cases of the coincidence of helium anomalies and ore deposits, examples being available from many parts of the world. A very large list could be compiled of endogenous formations occurring in

Fig. 7 Helium surveying profile and GSZ. After Bulashevich and Bashorin.[2] (1) Observed values of helium concentrations; (2) layer borders: d^M, Mohorovicic; $d_4{}^K$, Conrad, $d_3{}^K$ base of granite–gneiss complex; (3) points of diffraction according to GSZ; (4) Mesozoic–Cenozoic sediments, Zauralje; (5) zones of tectonic distortion in earth's crust according to seismic data. (a) East Ural flexure; (b) Zaural lifted block

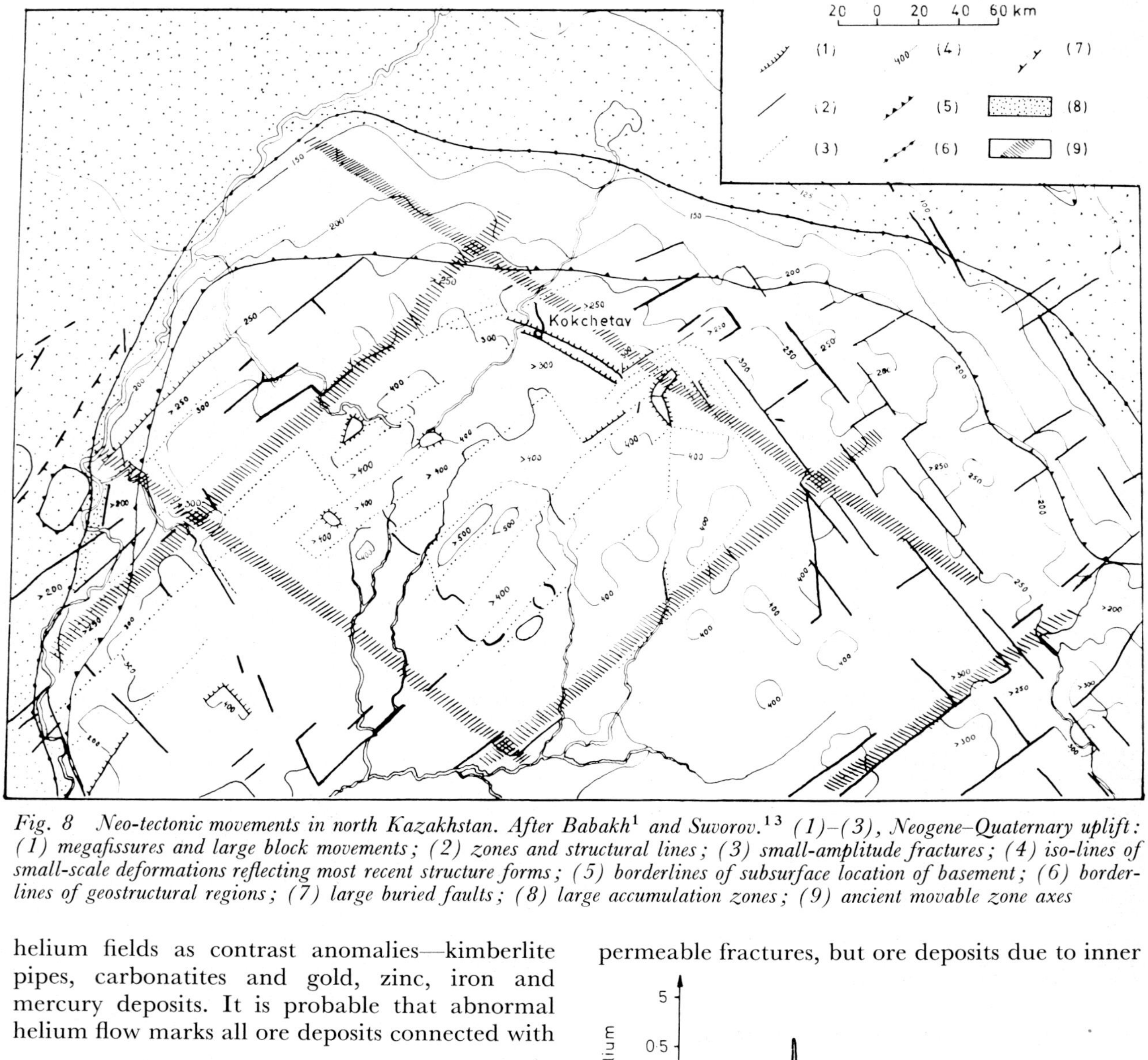

Fig. 8 Neo-tectonic movements in north Kazakhstan. After Babakh[1] *and Suvorov.*[13] *(1)–(3), Neogene–Quaternary uplift: (1) megafissures and large block movements; (2) zones and structural lines; (3) small-amplitude fractures; (4) iso-lines of small-scale deformations reflecting most recent structure forms; (5) borderlines of subsurface location of basement; (6) borderlines of geostructural regions; (7) large buried faults; (8) large accumulation zones; (9) ancient movable zone axes*

helium fields as contrast anomalies—kimberlite pipes, carbonatites and gold, zinc, iron and mercury deposits. It is probable that abnormal helium flow marks all ore deposits connected with

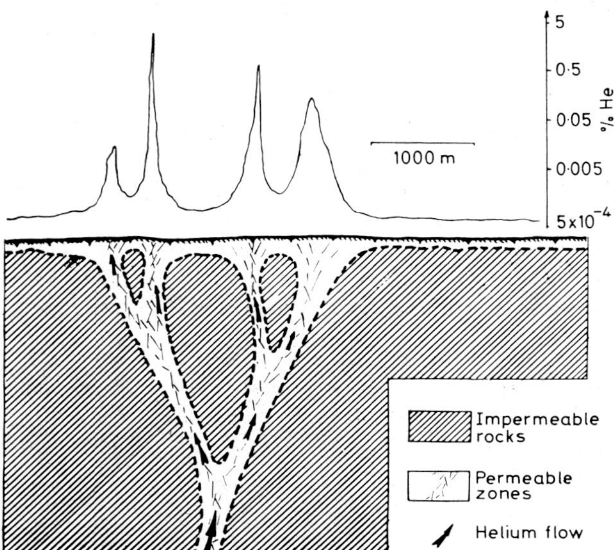

Fig. 9 Distribution of helium in structural junction

permeable fractures, but ore deposits due to inner

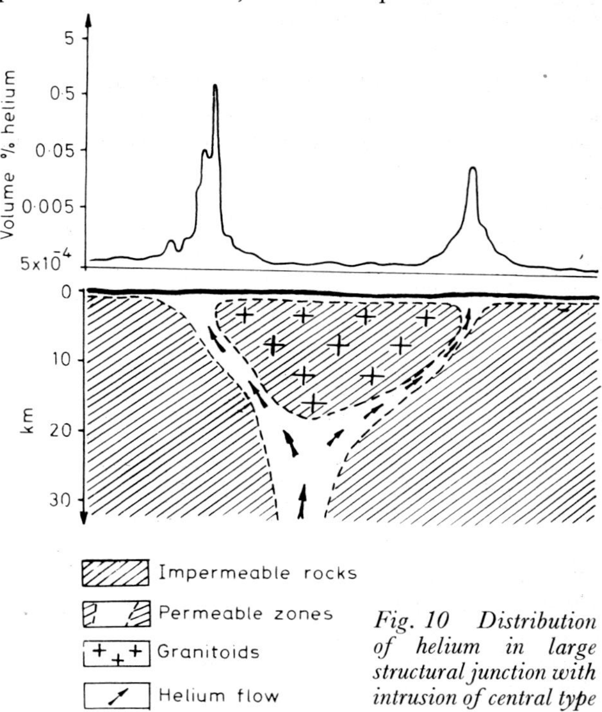

Fig. 10 Distribution of helium in large structural junction with intrusion of central type

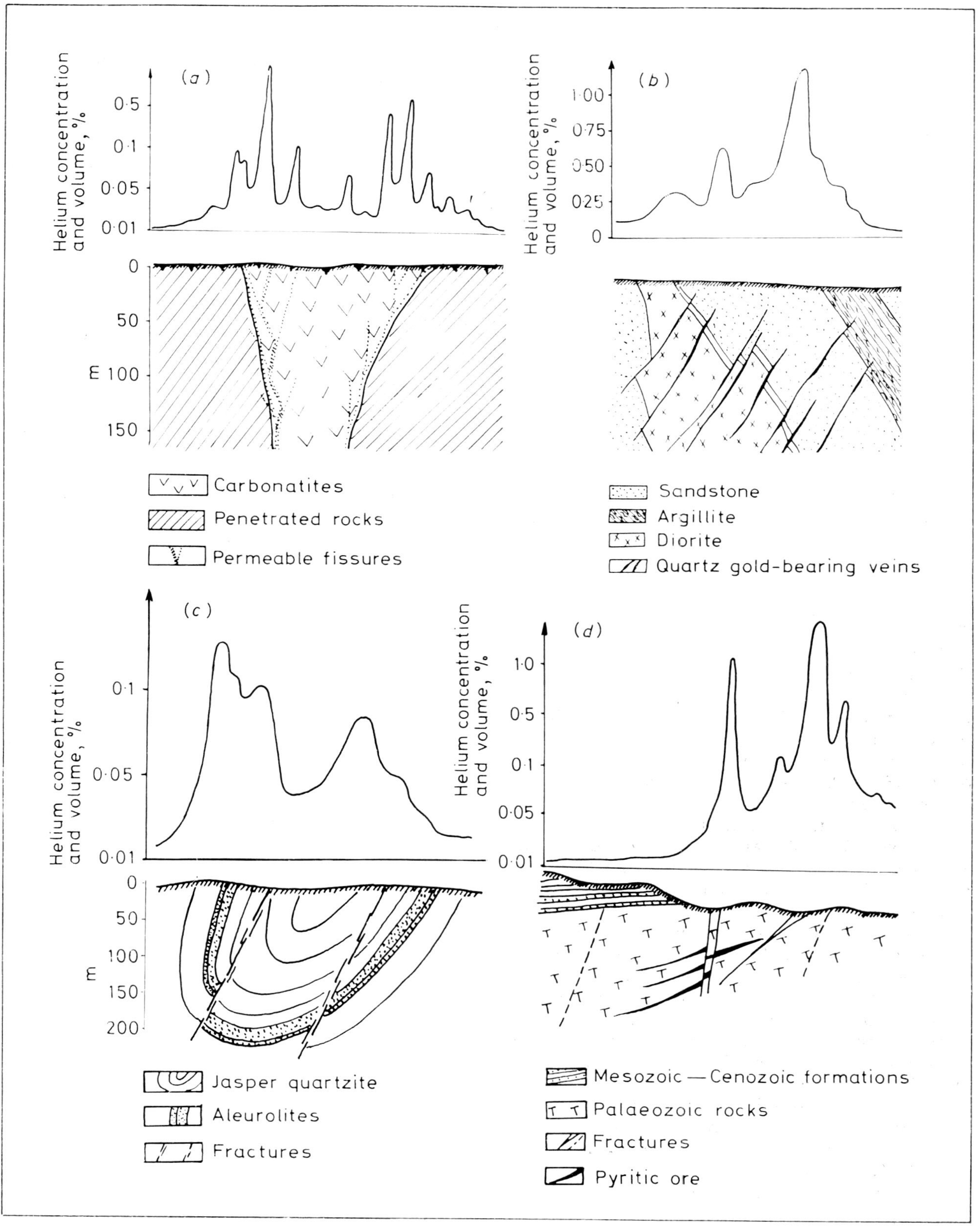

Fig. 11 Helium haloes in regions of ore deposits: (a) Barchensk carbonatite deposit, north Kazakhstan; (b) gold-bearing placer deposit, Bestube, north Kazakhstan; (c) lead–zinc–baryte deposit, Bastoba, central Kazakhstan; (d) Urup copper–pyrite deposit, Caucasus (after A. I. Fridman)

redistribution of matter are not reflected in the helium field—these are magmatic liquation and metasomatic mineral deposits. Fig. 11 (*(a)–(d)*) illustrates anomalous helium fields in the upper part of some ore deposits. The survey was performed in the upper part of underground waters (10–30 m below the surface).

Conclusions

A natural field of mobile helium is connected with durable deep fractures and reflects the permeability of the earth's crust. The correspondence of endogenous ore deposition and deep fractures enables the data obtained by the helium survey to be used to locate areas favourable to exploration for ore deposits. The practical significance of the helium method lies in deep fracture mapping and forecasting the most suitable sites for ore deposits. The use of helium surveying is advisable also in combination with the methods of abyssal geophysics and, in particular, of deep seismic sounding.

The distribution of mobile helium in permeable zones of the earth's crust reflects the general picture of deep gas migration, with which helium is connected genetically, and is a very suitable component for observation purposes.

References

1. Babakh, V. I. The most recent structure of the Kazakhsky Shield. *Geotektonika*, no. 6 1969, 86–99. (Russian text)
2. Bulashevich, B. P. and Bashorin, V. N. Helium in subterranean waters on the profile of the subsurface seismic soundings in Zaural. *Dokl. Akad. Nauk SSSR*, **193**, 1970, 573–5. (Russian text)
3. Vernadsky, V. I. *Outline of geochemistry* (Moscow: Gosgeolnefteizdat, 1934), pp. 669–72. (Russian text)
4. Gerling, E. K. Migration of helium from minerals and rocks. *Trudy Radievogo Inst., izd. Akad. Nauk SSSR*, **5**, no. 2 1957, 64–87; *Chem. Abstr.*, **52**, 1958, 10827.
5. Golubev, V. C. Osipov, Yu. G. and Yanitskii, I. N. Some peculiarities of the migration of helium in permeable systems of the upper part of the earth's crust. *Geokhimiya*, 1970, 1341–8. (Russian text)
6. Eremeev, A. N. *et al.* The pattern of dispersion of the concentration of helium in the upper part of the earth's crust with its subsurface formation and endogenic mineralization. *Bull. Izobr. otkrit.* no. 35, 1969, 1. (Russian text)
7. Eremeev, A. N. *et al.* Some aspects of helium survey during structural–geological mapping and forecast of endogenic mineralization. In *Trudy IMGRE*, **5**, 1971, 49–65. (Russian text)
8. Kosygin, Yu. A. *Tektonika* (Moscow: Nauka, 1968), pp. 170–5. (Russian text)
9. Mazur, N. N. The temperature of the upper layers of the earth—an important factor of helium accumulation in sedimentary stratum. In *Aspects of the increase of resources of the Ukraine* (Kharkov: Ukrtekhknniga, 1969), pp. 68–72. (Russian text)
10. Newton, R. and Round, G. F. The diffusion of helium through sedimentary rocks. *Geochim. cosmochim. Acta*, **22**, 1961, 106–32.
11. Osipov, Yu. G. and Yanitskii, I. N. On the mechanism of the formation of regional helium accumulation (Abstr.). *Geokhimiya* no. 1 1966, 36–47; *Goechem. int.*, **3**, 1966, 13.
12. Rogers, D. Sh. Natural helium gases. M–L., (ONTI: 1935). (Russian text)
13. Suvorov, A. I. *Mechanism of the formation of subsurface fractures* (Moscow: Nauka, 1968), 211–64. (Russian text)
14. Tikhomirova, V. G. and Tikhomirov, V. V. Helium content of waters occurring below the petroleum layer of the young platform of the USSR. *Izv. Akad. Nauk SSSR, Ser. geolog.*, no. 2 1971, 116–20. (Russian text)
15. Kharley, P. M. Helium method of the determination of age. *Yadernaya geologiya. M., In. Lit.*, 1956. (Russian text)
16. Yakutseny, V. P. *Mechanism of formation of deposits of helium-bearing gases* (Leningrad: Gostoptekhizdat, 1963), 132 p. (Russian text)
17. Cook, G. A. ed. *Argon, helium and the rare gases* (New York Interscience: 1961), p. 111 (Biblio.)
18. Deaton, W. M. Helium, another natural gas *Mines Mag.*, **49**, no. 11 1959, 46–9; 54.

550.42:551.462(267.5)

Dispersion from a submarine exhalative orebody

R. Holmes, B.SC.

J. S. Tooms, B.SC., PH.D., M.I.M.M.

Both of the Applied Geochemistry Research Group, Imperial College, London, England

Synopsis

Along the length of the median valley of the Red Sea are layered metalliferous deposits, which vary in thickness from a few centimetres to several metres or tens of metres. The best documented occurrence is that in the Atlantis II Deep off Jeddah, Saudi Arabia, where layered manganese and iron oxides and sulphides contain variable quantities of copper and zinc.

Occurrences of metalliferous precipitates compositionally similar to certain layers in the Atlantis II Deep have been recorded in the first 1 or 2 m of sediment in other deeps elsewhere in the median valley. Whether these deposits persist at depth and whether they are layered in a similar manner to those in the Atlantis II Deep has yet to be determined.

Thin (few cm) metalliferous layers observed at numerous localities in the median valley are commonly highly manganiferous, but with lower concentrations of copper and zinc than were observed in the major deposits referred to above.

Apart from the Atlantis II Deep occurrence, the existence of brines at other widely separated localities has recently been established—off Suakin, Sudan, and the Brothers Lighthouse, Egypt. The composition of these brines and the associated sediments is considerably different from that in the Atlantis II Deep. This is consistent with the brines being meteoric water whose compositions are closely related to the temperature of reaction with rocks through which they have passed.

Evidence of the existence of the exhalative deposits in the median valley is widespread in the sediments of the Red Sea. The ore metal content of the calcareous sediment varies with depth, being greatest within the median valley and lowest near the coast. There is also a variation in ratios of the ore metals with distance from the median valley.

Analysis of suspended particulate matter has revealed variations in composition and metal ratios, which can be related, at least in part, to the occurrence of the brines and to metal-rich precipitates not known to be still associated with brines. Preliminary (continuing) studies of the filtered Red Sea waters, collected both within and outside the median valley, have revealed variations which are considered to be also related to the deposits in the median valley.

The results obtained have important implications to prospecting for present and past submarine exhalative deposits.

The existence of an orebody, in process of formation, which is available for investigation and detailed study is a matter of very great significance. Incontrovertible evidence is provided of the origin of at least one type of orebody, and the information obtained can be applied to the interpretation of orebodies formed in the geological past and in the selection of areas where conditions in the past will have been conducive to the emplacement of ores of similar origin. In addition, the effect of the ore-forming processes on the local and general environment of the orebody can be determined prior to modifications which result from diagenetic and other changes. This latter information can be applied to the search for other similar deposits which are being formed at this time or which have been formed in the recent past. The paper describes the results obtained to date during a continuing investigation of the effects of the existence of metalliferous brines and associated metalliferous precipitates on the overall environment of the Red Sea and of the dispersion mechanics. The data are considered to be of sufficient interest and significance to justify their presentation as an interim report.

Indications of the existence of anomalously high temperatures in the bottom waters of the Red Sea were obtained in 1948 by the Swedish research ship *Albatross*.[1] Not until the mid-1960s was the fact established that there were high-temperature, highly saline brines in deeps on the sea-floor in the general area of the Swedish observations. Two brine-filled deeps were found in close juxtaposition in the median valley of the Red Sea opposite Jeddah as a result of sampling in the area by the R.R.V. *Discovery*,[2] R.V. *Meteor*[3] and R.V. *Atlantis II*[4] and these deeps were named Discovery and Atlantis II. In both deeps the brines contained high metal contents and were underlain by metalliferous (Fe, Mn, Zn and Cu, in particular) sediments. The first relatively detailed investigations of the two deeps were undertaken by the R.V. *Chain* in 1966, the results obtained being published in 1969.[5] More recently (1969 to date), much more intensive studies, particularly of Atlantis II Deep, have been made from the ships M.V. *Wando River* and M.V. *Valdivia* to determine the economic potential of the deposits. In addition, the scientists on these two ships have explored most of the length of the Red Sea median valley and have located a considerable number of other brine deposits. Unfortunately, few data on this most recent work, which was financed by Preussag, A.G., Metallgesellschaft, A.G., and the government of the German Federal Republic, have appeared in print. In addition, a further research cruise was undertaken by Woods Hole Oceanographic Institute scientists in R.V. *Chain* during 1971.

In 1969 the Natural Environment Research Council approved a grant to the second author, in part to investigate the effect of the metal deposits and their associated brines on the overall environment of the Red Sea with a view to developing prospecting methods. This research, which is still continuing, was undertaken in collaboration with the Geological Survey of the Sudan (1969) and with the Directorate General of Mineral Resources, Kingdom of Saudi Arabia (1970 and 1971). Furthermore, much useful information has been exchanged with the American and German scientists working in the Red Sea and one of the writers (J.S.T.) participated in one leg of the *Valdivia*'s 1972 Red Sea cruise.

Although this paper is concerned with dispersion of elements from the deposits rather than with the deposits themselves, some mention must be made of the nature of the brines and their precipitates. It should be emphasized, however, that the following paragraphs are only a very brief summary of the salient relevant features of these deposits which have a significant bearing on the dispersion data. For more detailed information on Atlantis II and Discovery Deeps the reader is referred to the book edited by Degens and Ross.[5] Data on the deposits and brines in Nereus Deep, discovered during this work, and on Oceanographer Deep[6] are currently being prepared for publication, and it is understood that Dr. H. Backer and his co-scientists intend to publish information in the near future on Suakin, Sudan, Albatross and the other brine deeps they have discovered.

Salient features of selected hot brine deposits

The only brine deposit discovered to date which can be considered at this time to be a potential ore deposit is that in Atlantis II Deep (21° 33′ N, 38° 04′ E), and by far the most information is available about this occurrence. Thus, attention is concentrated in this paper on the nature of the Atlantis II brines and precipitates.

Atlantis II Deep, at a depth of more than 2200 m, is in a very steep-sided graben forming part of the median valley of the Red Sea. Within the deep there is a discontinuous median ridge. The boundary walls of the deep are in places overhanging and appear to represent faults. It is along one of these postulated faults that the brines are entering the deep, as indicated by brine temperature variations.[7]

The temperature of the brines was recorded in 1965 as 56°C, but in the area of the brine springs it had increased to 59·2°C in March, 1971,[7] and to more than 60°C in 1972, reflecting the continued addition of brine to the pool. Overlying the 56°C brine are a number of layers of lower-temperature, more dilute brines, which Turner[8] has suggested are due to development of distinct convection cells as a result of mixing of the brines with normal Red Sea waters. Similar multiple-layering has been observed in at least a number of other Red Sea brine pools.

Compositionally, the hot waters of the deep are near saturated sodium chloride brines (Table 1) containing abnormally high concentrations of several metals. Deuterium–oxygen isotope data provide strong evidence that the waters were originally normal Red Sea water which has taken

Table 1 Composition of Atlantis II brines

	Atlantis II 56°C brine,[5] g/kg	Sea water, g/kg
Na	92·6	10·76
K	1·87	0·39
Ca^{++}	5·15	0·41
Mg^{++}	0·76	1·29
Sr^{++}	0·04	0·008
Cl^-	156·03	19·35
Br^-	0·13	0·066
SO_4^{--}	0·84	2·71
HCO_3^-	0·14	0·72
Si	0·03	0·004
$I \times 10^3$	0·03	0·06
$F \times 10^3$	0·05	1·3
$Fe \times 10^3$	80	0·02
$Mn \times 10^3$	80	0·01
$Zn \times 10^3$	5	0·005
$Cu \times 10^3$	0·3	0·01
$Pb \times 10^3$	0·6	0·004
Salinity	257·6‰	
Temperature	56·5°C	

up additional salt and trace elements during their passage through the rocks.[9] Craig suggested that on present-day evidence the source of the waters was in the southern Red Sea, but increasing information on the occurrences of brines in the Red Sea places this proposition more open to question. The salt content of the brines can be readily explained as due to solution from the extensive salt deposits known to occur along either side of the Red Sea and close to the median valley. The uptake of metals by brine solutions from unmineralized rocks through which they pass has been shown to be largely related to the brine concentration and temperature.[10] At least in several of the lower-temperature brines (e.g. in Oceanographer Deep) the metal content of the waters is very considerably lower than in Atlantis II Deep.

The sediments below the brines in Atlantis II Deep are highly metalliferous and evidently were formed by precipitation from the brines, presumably as a result of oxygen uptake by the anioxic brines from normal Red Sea water and by decreasing stability of the chloride complexes as the brines cool. The minerals are commonly very finely grained or amorphous.[11] Apart from detrital or normal Red Sea calcareous sediment layers, the major mineralogical facies are iron montmorillonite, goethite, sulphides, manganite and anhydrite. The metalliferous precipitates are extremely finely banded, both in Atlantis II Deep and in those other deeps where there are similar precipitates. This is in marked contrast to certain other deposits, which are considered to have been formed in similar fashion in the past, as, for example, the Cyprus ores.[12] In several of the other brine deeps where the temperature of the brines is low (e.g. Oceanographer Deep) the composition of the underlying sediments does not (on present evidence) differ greatly from that of normal Red Sea sediments. The full range of mineralogical facies observed in Atlantis II Deep has only been noted elsewhere, as far as the writers are aware, in Nereus Deep, where the temperature of the brines (27°C) is only a few degrees above that of the normal Red Sea waters. Whether this reflects the fact that these brines are old extrusions which have cooled from original higher temperatures is not known. The composition of the Nereus Deep brines has not yet been determined.

As might be expected, the metal contents of the

Table 2 Composition of mineralogical facies in Atlantis II Deep. After Bischoff[11]

	Detrital per cent (salt-free)	Fe montmorillonite	Goethite, amorphous	Sulphide	Manganite
SiO_2	27·3	24·4	8·7	24·7	7·5
Al_2O_3	8·4	1·7	1·1	1·5	0·7
Fe_2O_3 (total)	6·5	37·1	64·2	24·3	30·5
FeO	1·4	11·7	2·7	13·4	0·4
Mn_3O_4	0·6	2·1	1·1	1·1	35·5
CaO	23·6	4·8	3·4	2·5	2·9
ZnO	0·08	3·2	0·7	12·2	1·4
CuO	< ·01	0·8	0·3	4·5	0·1
CO_2	23·1	8·6	3·6	5·7	2·2
S	0·3	3·9	0·6	16·8	0·6

different mineralogical facies varies markedly, the highest contents of Cu and Zn being observed in the sulphide layers (Table 2).

Sample collection and preparation

Sediment samples were collected by use of either standard or modified UMEL gravity cores on 4-mm wire. The corers had a plastic liner. Water samples were collected in National Institute of Oceanography sampling bottles. These are nylon bottles, with rubber end caps, which were coated with polytetrafluorethylene before use. The only metallic parts are corrosion-resistant metal clamps, for attachment to steel hydrographic wire, and springs to hold the sample in the bottle by the end caps. The bottles are tripped in depth series down the wire by brass messengers started from the ship. The water-sampling bottles were used in random sequence and stored on deck prior to filtration. Filtration was completed within one hour of collection of the sample.

Filtration was carried out by use of a borosilicate glass–stainless steel assembly under vacuum; the stainless steel was not in contact with the water. 'Millipore' 0·45-μm pore diameter membrane filters were used to filter particulate from dissolved species in sea water. Storage bottles for the filtered water were high-density polythene bottles prewashed with 'Analar' hydrochloric acid and deionized water. The interiors were kept wet after cleaning. Immediately before use the bottles were rinsed with filtered sea water. Filtrates were stored at pH 2 or less by addition of redistilled nitric acid to the sample. The membrane from each filtration was retained folded in a sealed polythene bag for particulate analysis.

Sediment analysis

Unwashed sediment samples were dried at 80°C and ground prior to analysis. They were then analysed by use of a nitric–perchloric acid attack, and the estimations were carried out with a Perkin-Elmer 403 atomic absorption spectrophotometer.

Water analysis

Copper, zinc, manganese and iron dissolved species in sea water were concentrated by solvent extraction to levels suitable for subsequent atomic absorption analysis.

Iron, zinc, copper and manganese were extracted by buffering the sample to pH 3·5, chelating with ammonium pyrrolidine dithiocarbamate and extracting into chloroform, followed by chelation with excess sodium diethyl dithiocarbamate at pH 5·8–6·1 (to extract the remaining bulk of the manganese) and extraction into chloroform. The two chloroform extracts were then combined. The chelate complexes in the chloroform were destroyed by addition of one drop of redistilled nitric acid, and the chloroform was evaporated off on a water-bath. The residue was taken up in a suitable volume of molar hydrochloric acid and the element was determined by atomic absorption spectrophotometry. For iron and zinc determinations a deuterium source background correction was used to reduce the interference effects of sodium and calcium extracted with the other metals. To avoid contamination, buffer and chelates were prepared and purified daily.

Particulate species (>0·45 μm in diameter) retained from water filtration were analysed for zinc, manganese and iron. The particulate matter was separated from the Millipore filter by a microanalysis technique. The filter membrane was extracted in boiling acetone and, after centrifuging, the extracted Millipore was removed with the supernatant liquid. After three extractions the residue, after evaporation of the acetone, was ignited and taken up in acid prior to estimation of the elements by atomic absorption spectrophotometry. The results were recorded in terms of the ignited sample.

Dispersion from metalliferous brine deposits

There is a sharp interface between the brines and sea water, which can be detected as an acoustic reflector. There is, however, no doubt that there is some interchange of material across this interface, oxygen entering the brines and metals and other elements entering the sea water. Moreover, the lower-temperature and less saline upper brine layers have almost certainly formed due to a mechanical mixing of sea water and brine, during which some loss of brine to the normal water mass can be expected.

Thorough investigation of the dispersion of elements from the brines in the Red Sea is difficult owing to a variety of factors. In particular, only a limited number of samples can be collected in a given time (at depths of approximately 2000 m it would be exceptional to complete six hydrographic stations in a 24-h day). Furthermore, the sea water tends to be complexly layered, the layers often being very thin, and, therefore, erratic variations can be expected. In addition, the concentrations of most ore elements are extremely low in sea water, and complete avoidance of contamination from the wire on which the sampling bottles are suspended is difficult. Accordingly, it should be emphasized that although single-sample variations may be relevant to an understanding of the dispersion processes, only systematic variations in element concentrations can normally be taken to be significant.

Attention is concentrated here on dispersion from the Nereus and Atlantis II Deep brines. Far less information is available on the Nereus brines at the present time, but there are known to be a number of other brine deeps of differing composition immediately south and west of Atlantis II Deep, as well as possibly to the north. This complicates the dispersion patterns and makes interpretation difficult. Accordingly, much of the following illustrative information is based on the Nereus Deep area.

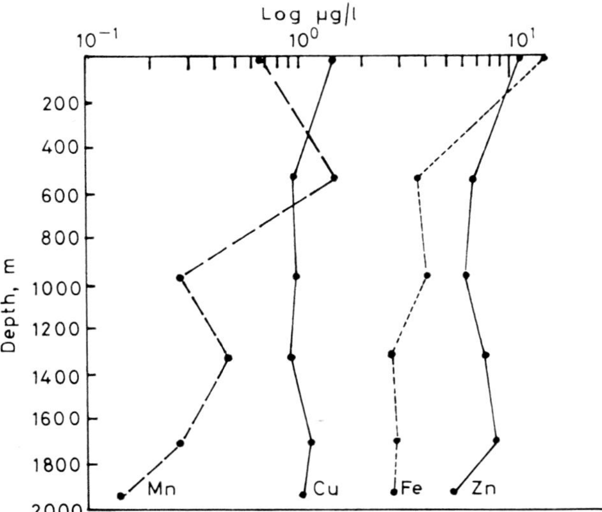

Fig. 1 Variation of Mn, Fe, Cu and Zn with depth at station 71–74

The distribution of metals in the water column is strongly related to depth. With increasing depth there is, in general, a decrease in the dissolved metal contents. This is most marked in the case of zinc and manganese (Fig. 1). Only near the bottom in the median valley is this trend commonly reversed, although there may be minor reversals at various depths in individual profiles. A similar decrease is observed in the case of metal in particulate matter. At least in part this is due to the effect of biological (plankton) and physico-chemical changes with depth. This depth effect complicates the study of the dispersion. Other things being equal, there will tend to be an increase in metal content away from the deeps in samples taken a given distance above the sediment–water interface due to the decreasing depth effect.

The effect of the Atlantis II brines on the composition of the dissolved species in the overlying near-bottom sea water and to the east and west is readily apparent (Fig. 2). Marked maxima

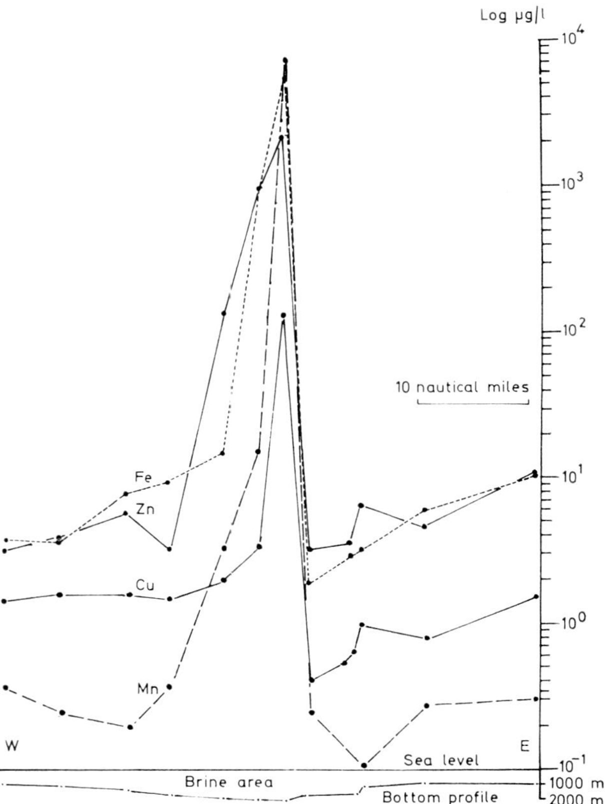

Fig. 2 Variation of dissolved metal species in near-bottom water, west–east across Atlantis II Deep (no samples from brine)

in all four of the metals considered occur immediately over the brine deposits and can be observed to extend several miles laterally, particularly to the west. East of Atlantis II Deep the effect of decreasing depth rapidly overrides the dispersion effect. A somewhat similar metal distribution is observed in the particulate matter in the bottom waters.

In the surface waters over Atlantis II Deep the dissolved species metal distributions are erratic and no definite trend can be observed.

The data for near-bottom waters over Nereus Deep reveal much less marked lateral (east–west) variations in the dissolved species than over Atlantis II Deep (Fig. 3). Strong iron and manganese anomalies can be observed, but the zinc and copper concentrations do not markedly increase. The slight decrease in copper and zinc, however, despite the decreased depth, is highly significant.

By contrast with Atlantis II Deep, in the near-bottom waters over Nereus Deep the particulates disclose a much more marked dispersion than

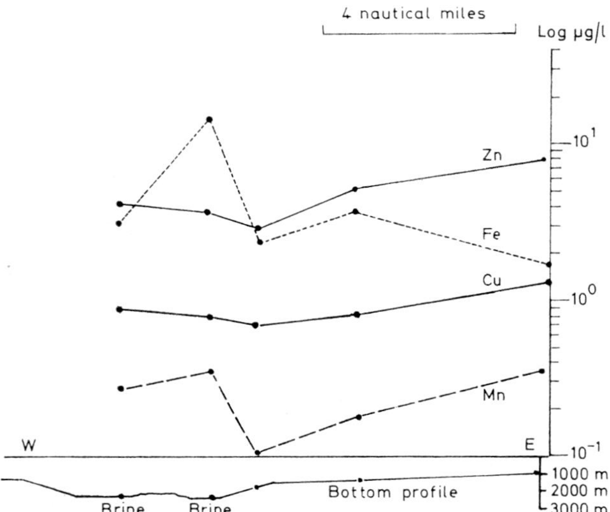

Fig. 3 Variation in dissolved metal species in near-bottom water, west–east across Nereus Deep (no samples from brine)

do the dissolved species. Strong Fe, Mn and Zn maxima occur over the brine area, and to the east the metal contents of the particulates decrease over a distance of a mile of more, despite the rapidly decreasing water depth.

If the variations along the axis of the median valley are considered, the effect of the Nereus brines can be much more readily observed in that no marked changes in depth complicate the dispersion pattern (Fig. 4). Along the axis of the

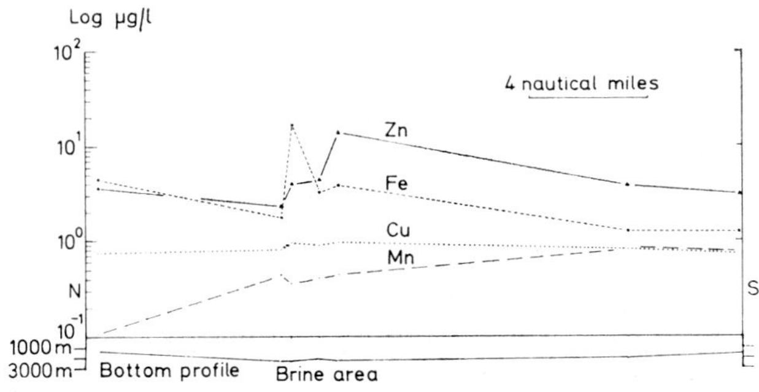

Fig. 4 Variation in dissolved metal species in near-bottom water, north–south across Nereus Deep (no samples from brine)

median valley the iron and zinc dissolved species contents decrease to the south from the brine area for the whole distance of the detailed sampling—a distance of more than 15 nautical miles. The manganese concentrations, however, increase southwards. The reasons for this are not fully understood, but are considered to be possibly related to the differing stability of the metal chloride complexes in normal sea water and the mixing of bottom water with the near-bottom water. There is also a well-marked metal anomaly in the particulates which extends for a distance of at least 10–15 miles to the south along the axis of the median valley (Fig. 5).

Both the particulate and dissolved species data strongly indicate dispersion of elements from Nereus Deep to the south.

Consideration of the metal contents of the waters and suspended particulate matter in the

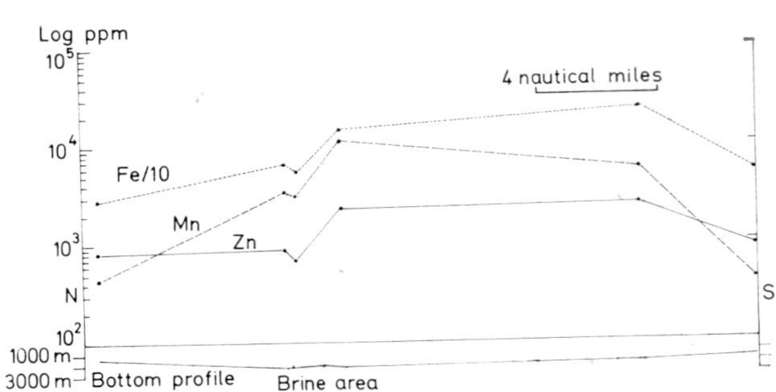

Fig. 5 Variation in particulate metal contents in near-bottom water, north–south across Nereus Deep

near-bottom waters along the median valley on a broader scale emphasizes the effect of the known brine deeps. A strong maximum exists over Nereus Deep both for particulates and dissolved species. Over Atlantis II Deep only a single sample anomaly peak was recorded, but to the south there is a broad anomaly, which covers an area where a number of brines are known (Fig. 6). This latter anomaly is even more significant if the water-depth variations are considered.

To the north of Nereus Deep the metal values tend to increase: this is probably largely due to the effect of decreasing depth. The maxima some 120 nautical miles north of Nereus Deep might, however, be indicative of the existence of another (previously unknown) brine deposit. This is an area in which very little exploration has been undertaken.

It should be emphasized that the number of samples involved in this study is extremely small when the distances involved are considered. Particularly on a broad scale the location of individual stations may have a disproportionate effect on the overall distribution pattern.

Very little is known about currents in the Red Sea. At surface it has been observed that currents of several knots velocity may exist in the central areas of the Sea, often flowing in a direction at right angles to the wind directions, but no systematic study of the currents has been made. Even less information is available on bottom currents. It is known that upwellings occur adjacent to the reefs, and there is an inflow of Indian Ocean water during much of the year. It may be assumed from the surface information

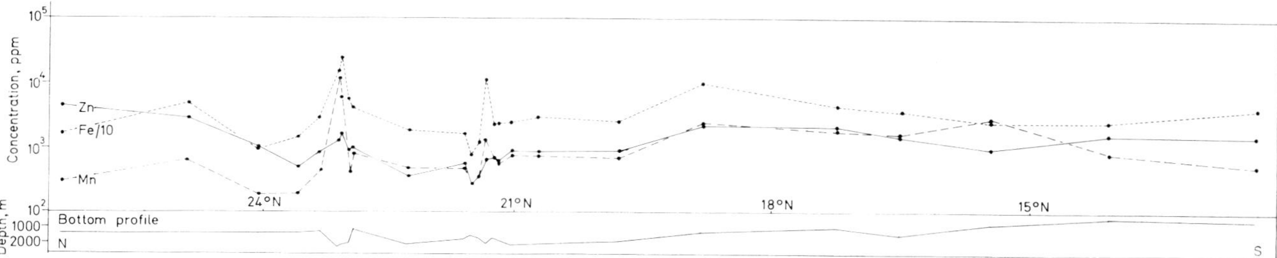

Fig. 6 Metal content of particulates in near-bottom water along median valley, Red Sea

that the bottom current directions are highly variable, although the dominant direction in the median valley could be expected to be southwards along the axis of the Red Sea.[13] When these facts are considered together with the markedly layered nature of oceanic waters and the depth effects previously mentioned, the detection of strong metal anomalies related to the location of the brines must be considered highly significant. The effect of dispersion on the composition of the bottom sediments will, of course, be related to the movements and mixing of water masses over a prolonged period. All the indications obtained from the water data, however, indicate that the major direction of dispersal is along the median valley, with a lesser amount of lateral dispersion.

Metals dispersed from the brines through the normal sea water will be precipitated either inorganically or due to biological processes. The overall effect should be that the composition of the sediment will reflect the net dispersion effects. The iron, manganese, copper and zinc contents of the first few centimetres of cores have been determined, therefore, to enable that effect to be assessed (Figs. 7 and 8). It should be emphasized, however, that the surface sediment often has a very high water content and may be displaced or lost during coring. Accordingly, the age of the sediment analysed almost certainly varies from one site to another.

Along an east–west traverse across Atlantis II Deep the concentration of all four elements is greatest over the median valley in the immediate

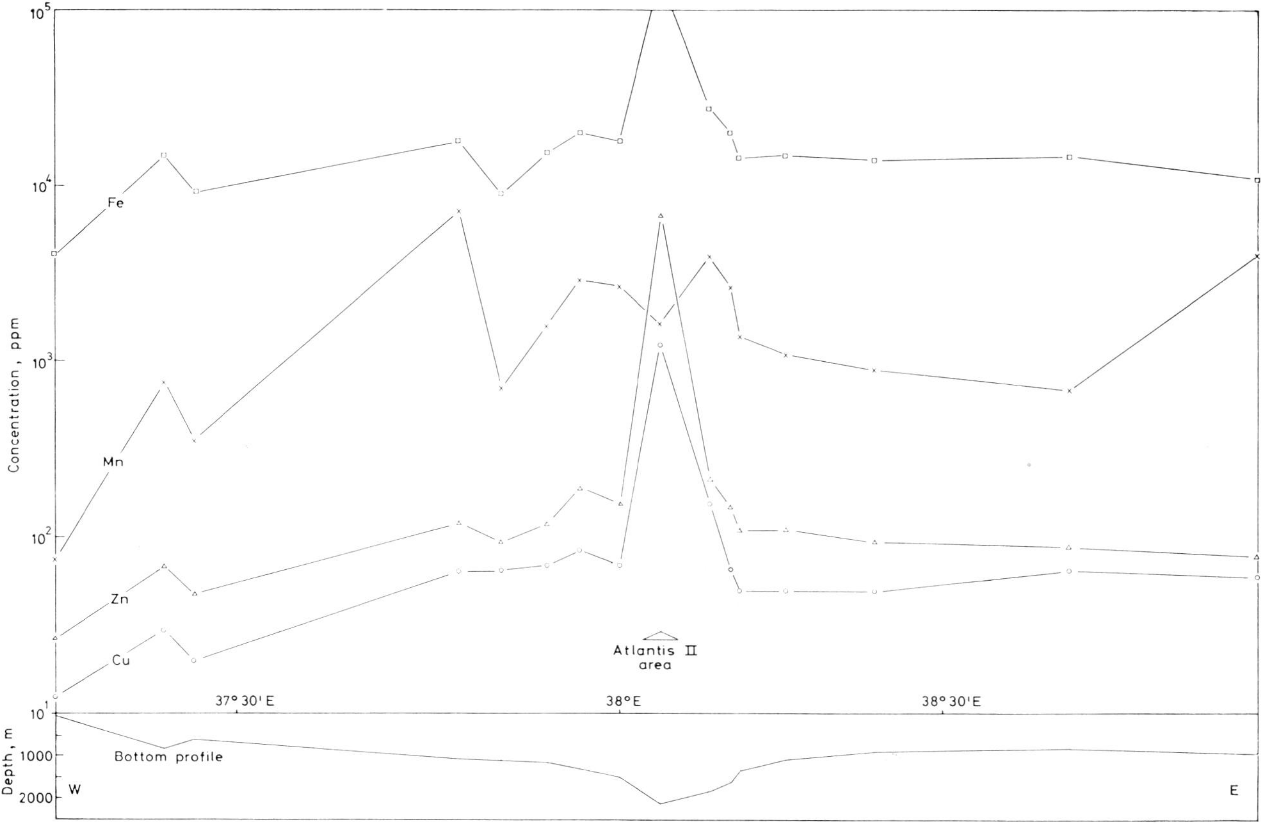

Fig. 7 Cu, Zn, Fe and Mn contents of surface sediments along traverse across Atlantis II Deep

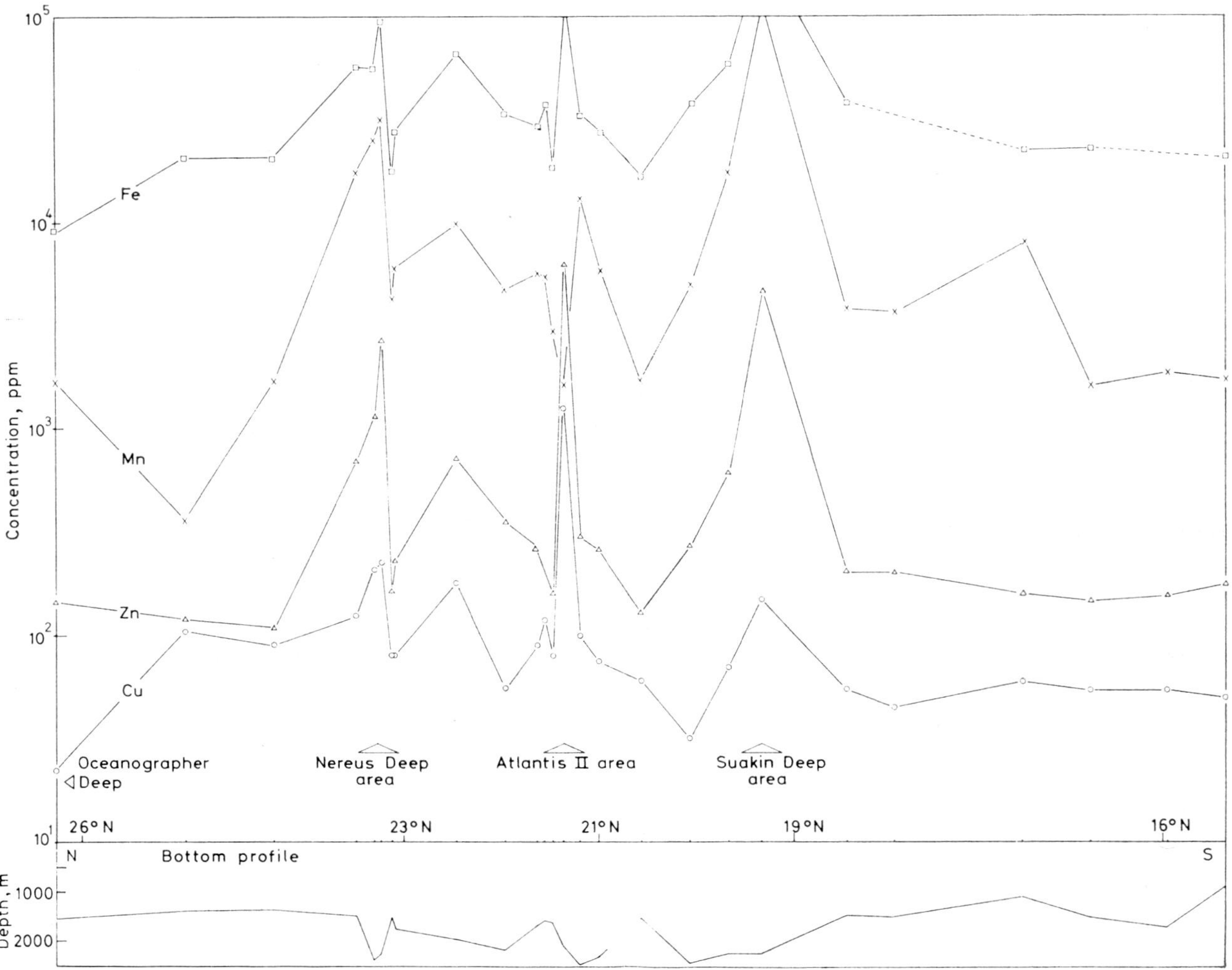

Fig. 8 Cu, Zn, Fe and Mn contents of surface sediments along axis of median valley, Red Sea

vicinity of Atlantis II Deep (no samples have been included from beneath the brines). Away from the median valley the concentrations of the metals decrease towards both the Saudi Arabian and Sudanese coasts. The secondary metal maxima immediately west of the median valley are probably related to the brine occurrences in this area. A minor increase in all metals is also observed in a small deep near the Sudanese coast which cannot be explained by pollution. The cause for this increase is not known.

The very limited apparent contribution of land-derived metal to the sediments of the Red Sea is emphasized by the absence of any reversal in the concentration trend on approaching the coasts. It is probable that land-derived material is largely trapped behind the reefs, which parallel much of the coasts. The few samples collected within the reefs (e.g. most westerly sample on the traverse), however, have extremely low concentrations of the metal, and it is probable that the land contribution is swamped by biogenic carbonates. It must be concluded, therefore, that the major-element concentration control is the dispersion from the brine areas and/or sedimentation variation related to depth. Study of the results obtained along the median valley indicate that, here, a sample depth mechanism control of sediment composition is most unlikely to be capable of explaining the data.

Along the whole length of the median valley the metal contents are high compared with shallower-water sediment sampled on sections across the Red Sea. Variations in the metal contents of surface sediment in the median valley are not related to depth, although the highest results are recorded in samples from basins (not necessarily the deepest parts of the median valley) in which brines have accumulated. Furthermore, comparison of samples collected from the main axis of the median valley with samples from similar depths in deeps to one side of the axis reveals that the metal content of the latter samples is considerably lower than that of the main axis samples. It is concluded, therefore, that the overall high metal contents along the median valley are due to dispersion of metals from the brine deeps.

In detail (on present evidence), higher metal contents in the surface sediments of the median valley are confined, in general, to the central portion, south of 18° 30′ N and south of 24° N. This includes the area in which all the known brines (except Oceanographer Deep) occur. Oceanographer Deep is a low-temperature, metal-poor brine, which would not be expected to have a significant effect on the sediment composition. The reason for the relatively high copper contents north of 24°N is not known, but it does correlate with the dissolved species water data from this general area.

The zinc, iron and manganese contents of the three samples north of 24°N are markedly lower than those to the north of 18° 30′ N. It is possible that this reflects the dominant movement of the bottom waters southwards.

Strong metal maxima are observed in the vicinity of Nereus and Atlantis II Deeps and to the south of Atlantis II Deep in the vicinity of Suakin Deep, where the water data are also anomalous. The much greater areal extent of high values in the vicinity of Nereus Deep as compared with Atlantis II Deep is probably due to the enclosed configuration of the basin within which the brines occur; the basin enclosing Atlantis II and several other brine deeps is open to the south.

Maxima in all four metals occur at 22° 28′ N, between Nereus and Atlantis II Deeps, which is again an area where anomalous water data have been obtained. In Fig. 8 this is shown as a single sample, but similar analytical results were obtained on another core a few miles to the east. No brine has been detected in this area, but there are strong indications in cores of the probable existence of metalliferous brines in the past.

Both on traverses across and along the median valley the metal concentrations decrease rapidly from the peak values near or at the brine sources. On the lateral traverses, however, the metal content continues to decrease with distance from the median valley, whereas in the median valley itself the metal contents remain strongly anomalous and relatively constant, even one or two hundred miles from a metalliferous brine. It is possible that minor seepages have occurred in the recent past, or are occurring now, throughout the length of the median valley. Alternatively, the consistent high values in the median valley, away from brine deep, may reflect the lack of mixing, and dilution, of the deep bottom waters as indicated by the water data. By contrast, any waters moving laterally will tend to be diluted from the shallower-water masses. In this connexion it should be noted that the Red Sea is unusual in that it is long, very narrow and only open at one end, but reaches relatively great depths. It is possible that under more open-sea conditions the dispersion patterns observed here would be considerably modified and that a more definite dispersion decay would be observed along the brine source fracture zone.

Conclusions

The results obtained to date demonstrate that dispersion of metals from the metalliferous brines is taking place through the normal sea water. Furthermore, this dispersion can be detected in both the dissolved species and in the particulate matter. The net effect of the dispersion processes is reflected by the surface sediments.

The main direction of dispersion appears to be along the axis of the median valley in a southward direction.

Application of even the present incomplete data would permit similar deposits to be located by relatively broad-scale sampling. The application of the data to exploration for similar deposits formed in the more or less distant geological past must, however, await the completion of work on the form of the elements in the sediments in order that a reasonable assessment may be made of the probable effects of diagenetic processes.

References

1. Bruneau, L. Jerlov, N. G. and Koczy, F. F. Physical and chemical methods. In *Reports of the Swedish deep-sea expedition, 1947–48* (Göteborg: Elanders, 1953), vol. 3, Appendix, Table 2, XIV–XXX.

2. Swallow, J. C. and Crease, J. Hot salty water at the bottom of the Red Sea. *Nature, Lond*, **205**, 1965, 165–6.

3. Dietrich, G. *et al.* Reisbericht der Indischen Ozean Expedition mit dem Forschungsschiff METEOR 1964–1965. *METEOR-Forschungsergebnisse, Reihe A*, **1**, 1966, 53.

4. Miller, A. *et al.* Hot brines and recent iron deposits in deeps of the Red Sea. *Geochim. cosmochim. Acta.*, **30**, 1966, 341–59.

5. Degens, E. T. and Ross, D. A. eds. *Hot brines and recent heavy metal deposits in the Red Sea* (Berlin, New York: Springer-Verlag, 1969), 600 p.

6. Ostapoff, F. A fourth brine hole in the Red Sea? Reference 5, 18–21.

7. Ross, D. A. Red Sea hot brine area: revisited. Directorate General of Mineral Resources, Ministry of Petroleum and Mineral Resources, Jiddah, Saudi Arabia (U.S. Department of the Interior—Geological Survey) SA(IR)-140. *Saudi Arabian Project Rep.* 140. (Unpublished)

8. Turner, J. S. A physical interpretation of the observations of hot brine layers in the Red Sea. Reference 5, 164–73.

9. Craig, H. Geochemistry and origin of the Red Sea brines. Reference 5, 208–42.

10. Ellis, A. J. Natural hydrothermal systems and experimental hot-water/rock interaction: reactions with NaCl solutions and trace metal extraction. *Geochim. cosmochim. Acta*, **32**, 1968, 1356–63.

11. Bischoff, J. L. Red Sea geothermal brine deposits: their mineralogy, chemistry, and genesis. Reference 5, 368–401.

12. Constantinou, G. and Govett, G. J. S. Genesis of sulphide deposits, ochre and umber of Cyprus. *Trans. Instn Min. Metall. (Sect. B: Appl. earth sci.)*, **81**, 1972, B34–46.

13. Siedler, G. General circulation of water masses in the Red Sea. Reference 5, 131–7.

14. Brewer, P. G. and Spencer, D. W. A note on the chemical composition of the Red Sea brines. Reference 5, 174–9.

550.42:551.763(712):550.8

Regional geochemical study of Cretaceous acidic rocks in the northern Canadian Cordillera as a tool for broad mineral exploration

R. G. Garrett, A.R.S.M., PH.D., D.I.C., M.I.M.M.

Geological Survey of Canada, Ottawa, Ontario, Canada

Synopsis

A constant problem in the interpretation of geochemical exploration data is the varying background levels encountered in different sample materials. This problem is most severe with certain trace elements present in the common rock-forming minerals, e.g. copper and zinc in hornblende and biotite, and lead in potash feldspar. Variations in rock type, and, hence, mineralogy, lead to wide variations in whole-rock trace-element content.

The Geological Survey of Canada has undertaken a rock geochemistry programme in east and central Yukon, and, as part of this study, an approach to the interpretation of base-metal trace-element data has been evolved. Principal-component analysis is used to orthogonalize the major- and minor-element data, which leads to a new set of variables that tend to be geological-process-orientated rather than chemical-response-orientated. The principal-component scores are computed and selectively used in a multiple-regression model with the trace-element data. Attention is thereby focussed on samples having abnormal contents of the trace elements in terms of the geological processes represented in the sample.

The resulting screened data are compared with the raw data, both sets being interpreted in the light of the distribution of known mineral occurrences. The interpretative procedure evolved delineates certain bodies of acid plutonic rocks as being of interest in terms of mineral potential, whereas an inspection of the raw data either failed to indicate the nearby mineral occurrences or, more importantly, was not considered selective enough. On the basis of these findings it is concluded that the interpretative technique is locally of value and may have broader application to other problems and other areas.

The use of rock geochemistry as an exploration tool has not been extensive outside the U.S.S.R. Until recently, studies in the west have been relatively few and have tended to be of an academic nature. The last five years have seen a resurgence of interest in the methods by western geochemists—for a variety of reasons. Many of the early attempts in the use of rock geochemistry had limited success due to an inadequate understanding of the problems of acquiring truly representative samples. Adequate analytical techniques are now available, in terms of precision and productivity, to handle the large number of samples derived from well-planned sampling programmes. Rock geochemistry also offers a detour around the problems of modification of trace-element patterns imposed by the secondary environment, and leads to data which may be used for more basic studies related to regional petrogenesis and the geochemical relationship of the rocks to associated mineral occurrences.

The Geological Survey of Canada commenced a study of the application of the sampling and analysis of granitoid rocks in the Yukon and adjoining areas of the Northwest Territories in 1969. The objectives of the programme are twofold: first, to investigate the use of bedrock geochemistry as a tool for broad mineral reconnaissance—that is, to design a geochemical filter that could be used at an early stage of a planned exploration effort to select those areas of increased mineral potential out of a much larger area of general interest; and, secondly, by means of the data, to establish correlations between discrete plutonic bodies and investigate both the regional petrogenesis of the area within the tectonic framework of the Northern Cordillera and the geochemical relationships of the plutonic rocks to the mineral occurrences associated with them.

Description of field area

The field area forms part of the Cordilleran system, which can be traced from Antarctica, through the Americas and into the northeastern U.S.S.R. The granitoids of the Selwyn Fold Belt extend arcuately on the northeastern side of the Tintina Trench for some 500 miles between Watson Lake and Dawson. The data presented in this paper relate to the northern half of this belt between 60° 20′ N and 64° 40′ N (Fig. 1). The acid intrusive rocks to Cretaceous age. The oldest rocks are Helikian and are found in the northern parts of the study area: these were deformed by the Racklan orogeny, and a great thickness of Hadrynian impure clastic sediments was then deposited. No definite unconformity has been observed between the Hadrynian sediments and the overlying Lower Cambrian miogeosynclinal rocks. Both Middle and Upper Cambrian rocks were deposited, but are only observed in the eastern part of the field area. The Selwyn Basin is composed mainly of Silurian graptolitic shales, siltstones and cherts; some Ordovician rocks may be present, and this whole assembly lies unconformably on Hadrynian rocks and grades upwards into eugeosynclinal Devono–Mississippian shales and chert-pebble conglomerates. In the northwest of the field area, near Dawson, Triassic shales occur, but apart from some sandstones and shales north of Ross River, Triassic strata are not found. Both Jurassic and Cretaceous sediments are found only in the northwest and north of the field area, and some

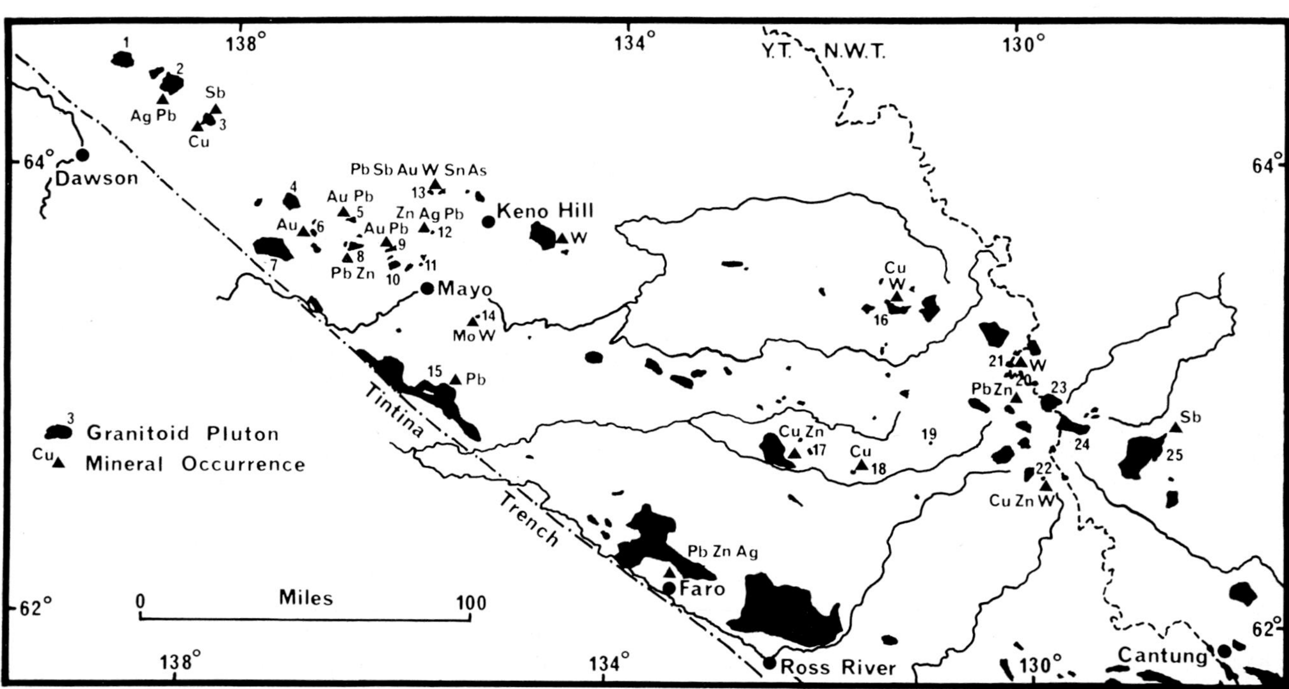

Fig. 1 Location of granitoids and major mineral occurrences

exhibit K–Ar ages in the range 74–110 m. y. (Middle to Upper Cretaceous), and it is considered that the deformation just prior to the intrusion of the granitoids took place during the Columbian orogeny.[5] A wide range of compositions is present, extremes being alaskite, granite, syenite and quartz-diorite; however, the vast majority of the rocks are granodiorites or quartz-monzonites. The granitoids are intruded into sedimentary and metamorphic rocks of Proterozoic Lower Cretaceous basaltic flows occur northeast of Ross River.

The intrusion of the Selwyn Belt granitoids was not accompanied by strong compressional activity, evidence for this being the lack of strong folding over much of the central part of the field area. Along the eastern margin and in the northwest of the field area folding was more intense, but the general tectonic situation in late Columbian times must have been one of platformal uplift.

Only to the south of the area discussed in this paper do the granitoids seem to have been involved in post-intrusion folding. One possible explanation could be that those intrusives were emplaced early and were competent before the close of the Columbian tectonic event.

The Selwyn Fold Belt hosts a number of mineral deposits, those of greatest economic importance being the lead–zinc deposits and silver–lead–zinc deposits. The massive strata-bound lead–zinc deposits at Faro and Macmillan Pass (20, see Fig. 1) are probably totally unrelated to the acid intrusive rocks. The silver–lead–zinc deposits of Keno Hill are vein-type and their origin is still open to conjecture; however, the only probable link with the granitoid intrusives is that they may have provided a heat source to stimulate a plumbing system that mobilized the ore elements from the surrounding country rock and transported them to depositional sites in the veins. There are two distinct areas of tungsten mineralization, both of which have a direct relationship to the granitoid intrusives. The first, known as the Selwyn Belt, lies in the Selwyn Mountains and consists of scheelite skarns with associated chalcopyrite developed in the country rock at or close to the contacts of the intrusives. The two most important occurrences are at Cantung and Mount Allan (21) near Macmillan Pass. The second, named the McQuesten Belt, consists of gold–tungsten mineralization with associated lead and antimony minerals in some instances. The tungsten occurs both in skarns as scheelite and also in quartz veins as scheelite and wolframite, and the gold occurs in arsenopyrite veins forming part of the same mineralization sequence. Important occurrences of this type occur at Potato Hills (13) and Scheelite Dome (9), north and west of Mayo. Throughout the Selwyn Fold Belt there are many scattered base-metal showings of varying significance, some of which also contain silver and antimony in important amounts.

Sampling and analytical methods

The granitoids were sampled by use of duplicates collected at an ideal minimum of 15 sites per pluton. This minimum number of sites yields sufficient data for the statistics computed on each pluton to be meaningful. The same general approach and number of samples collected is adopted in the U.S.S.R. in exploration programmes aimed at defining granitoids associated with tantalum and beryllium mineralization.[21] The duplicate sampling allows the use of analysis of variance to ascertain if any single pluton is internally zoned and to assess the overall significance of observed patterns in the data.[7] In fact, the number of sample sites per pluton varies considerably. In large intrusions many more than 15 sites are sampled so that zoning may be defined at any predetermined scale—usually one site per 1–2 square miles; and in small rugged intrusions it is often impracticable to sample more than six sites owing to the constraints of time set in a large regional sampling programme.

The samples weigh between 2 and 3 lb each and are comprised of fresh chips collected over a small area, the second sample collected at a site usually being removed from the first by some 20 ft. Care was taken at all times to ensure that unaltered rocks were sampled for the purposes of the statistical analysis of the data. In certain instances disseminated sulphides were observed in the rocks and, where these were widespread and not a feature of secondary alteration, those rocks were sampled. Separate samples were collected of altered rocks and vein mineralization, and these have not been included in the data under discussion. The problems of obtaining a random unbiased sample for geochemical programmes have been discussed by Miesch.[13] All but a hand specimen and material for thin-section work are reduced to -60 mesh, and a split of this material is ground to -100 mesh. The analyses for the major and minor elements were carried out after a lithium tetraborate fusion; Si, Al, Mg, Ca, Ti and Mn were determined by optical spectroscopy[4] and Na, K and Fe were determined by atomic absorption spectrophotometry after dissolution of the fusion product with dilute nitric acid. The trace-element determinations were carried out by atomic absorption spectrophotometry after a $HF–HClO_4$ attack; for Pb a deuterium arc lamp was used to correct for background variations in Ca and Mg.

Investigation of sampling and analytical variability: precision and accuracy

The analytical variances and precisions were determined by both replicate analysis of a bulk sample and a large number of repeat analyses of randomly chosen samples. These data, together with the overall sampling variability data derived from the duplicate sampling, were used in an analysis of variance (Table 1). All the elements exhibit a regional variability that is significantly higher, at the 99% level, than the local variability, the critical value for F being approximately 1·1 for 2135 and 2136 degrees of freedom at the 99% level. Of the major elements, it is noticeable that Al has the lowest F value: this reflects, first, the relatively high analytical variability for Al

Table 1 Analysis of sampling and analytical variance

	σ^2_D	σ^2_{SA}	F σ^2_D/σ^2_{SA}	σ^2_A	σ^2_S	σ^2_S/σ^2_A	$\bar{x}_1$	P	$\bar{x}_2$
Si	6·48	0·97	6·68	0·75	0·22	0·29	31·39	5·7	29·91
Al	0·56	0·17	3·29	0·17	0·00	0·02	8·72	8·9	9·27
Mg	0·53	0·03	17·67	0·01	0·02	3·82	1·03	11·9	1·33
Fe	1·26	0·12	10·50	0·01	0·12	23·59	2·59	6·9	2·16
Ca	1·12	0·12	9·33	0·02	0·10	4·06	2·62	12·3	2·49
Na	0·20	0·03	6·67	0·01	0·03	5·76	1·97	4·9	2·75
K	1·34	0·13	10·31	0·01	0·12	20·51	3·75	5·3	2·83
Ti	0·086	0·007	12·29	0·002	0·005	2·49	3·415	2·4	3·419
Mn	0·041	0·006	6·83	0·003	0·003	0·94	2·760	4·1	2·537
Zn	0·044	0·008	5·50	0·002	0·006	3·01	1·772	4·7	1·900
Cu	0·113	0·028	4·04	0·004	0·024	6·09	0·888	9·1	1·372
Pb	0·050	0·013	3·85	0·017	−0·004	−0·22	1·506	18·4	1·386

σ^2_D, Overall data variance. σ^2_{SA}, Combined sampling and analytical variance. σ^2_A, Analytical variance. σ^2_S, Estimate of on-site sampling variance. $\bar{x}_1$, Level at which analyses of variance were carried out. P, Analytical precision at 95% confidence level. $\bar{x}_2$, Level at which analytical precision was determined. Data for major elements recorded in per cent; minor and trace elements in ppm. Data for Ti, Mn, Zn, Cu and Pb have been $\log_{10}$-transformed.

and, second, the rather small total variability for Al in the granitoid environment. The percentage precision figures given in Table 1 were determined at the 95% confidence level at the analytical levels shown. These analytical levels are not identical to the overall data means at which the main test of variability is carried out. When the overall range of the data is considered, however, the differences are small and are not viewed as significant.

As variances are additive, it is possible to determine the local geological or sampling variability at the site for each element from the combined sampling and analytical variance and the analytical variance. Only in the cases of Fe and K is the sampling variability markedly higher than the analytical variance—indicating that only in the instance of these two elements could the sample sites not be considered homogeneous in terms of the measurement errors.

With regard to Pb, the analytical variance is greater than the combined sampling and analytical variance—thus leading to a negative site variance. This negative variance has no meaning and only emphasizes that the analytical variance is high with respect to the combined variances. Of the twelve elements determined, eight show sampling and local geological variability to be more than double the analytical variability. This feature serves to illustrate the importance of good sampling practice; in a general sense analysis is no longer the limiting factor that it was in geochemical surveys. Rather our comprehension of, and success in handling, sampling problems will govern our ability to successfully undertake effective surveys.

On the basis of these analyses of variance it was considered that the data were worthy of more rigorous analysis.

An investigation of the accuracy of the major- and minor-element data was undertaken. Ten samples selected from across the range of the data for each element were submitted for special analysis as sub-standards to the Analytical Chemistry Section of the Geological Survey of Canada. Determination of the major and minor elements was made after the method of Abbey[1] by use of the appropriate international standard rocks to bracket the unknowns. If it is assumed that the analyses so provided are a reasonable estimate of the actual amounts present, the percentage precision figures derived from the ten pairs of samples would be an estimate of the absolute percentage accuracy of the Yukon granitoid rock data (Table 2). It must be pointed out, however, that the accuracy figures are only an estimate, as at least 30 samples should be carefully treated in this fashion. Notwithstanding these limitations, the accuracy estimates do give a measure of the uncertainty involved in comparing the data with those derived by other workers.

Table 2 Inter-laboratory accuracy test

	$\bar{x}_1$	$\bar{x}_2$	Δx	$\bar{X}$	σ^2	σ	A
Si	29·02	29·24	−0·22	29·13	0·3632	0·6027	4·68
Al	8·96	8·51	0·45	8·74	0·2272	0·4767	12·34
Fe	3·99	4·09	−0·10	4·04	0·0466	0·2159	12·09
Mg	1·47	1·27	0·20	1·37	0·0326	0·1806	29·82
Ca	2·97	2·90	0·07	2·94	0·0471	0·2170	16·70
Na	1·76	1·70	0·06	1·73	0·0029	0·0539	7·05
K	3·76	3·61	0·15	3·69	0·0189	0·1375	8·43
Ti	3·4728	3·4472	0·0256	3·460	0·0023	0·0480	3·14
Mn	2·8015	2·7654	0·0361	2·784	0·0041	0·0640	5·20

$n = 10$. $t(9, 0{\cdot}95) = 2{\cdot}262$. $\bar{x}_1$, Determinations by Geochemistry Section. $\bar{x}_2$, Determinations by Analytical Chemistry Section. σ^2, Mean sum of squares of differences. A, Estimate of accuracy, made at 95% confidence level. Data for Ti and Mn have been $\log_{10}$-transformed.

Distribution of base metals in granitoids

The base-metal content of granitoids varies with position in the differentiation series, quartz-diorite to granite, and diorite to syenite. Clarkes computed for the base metals, together with the results of this study, are given in Table 3. Additionally, figures published by Tauson[22] are quoted which indicate the between-phase variations of base-metal content for three granitoid complexes in the U.S.S.R. (Table 4). Similar data have been published by Kuzmin[12] and Ivanov.[9] That the base-metal content may vary commonly by factors of two or more is quite apparent.

Table 3 Mean base-metal content (ppm) of acid plutonic rocks

	Zn	Cu	Pb
Granitoids (Vinogradov[25])	60	20	20
Granites (Krauskopf[11])	40	10	20
Granites (Taylor[23])	40	10	20
Low Ca granites (Turekian and Wedepohl[24])	39	10	19
High Ca granites	60	30	15
Syenites	130	5	12
This study	64	10	35

Table 4 Mean base-metal content (ppm) of some U.S.S.R. granitoids. After Tauson[22]

	Zn	Cu	Pb
Susamir batholith			
Dioritic phases	96	33	9
Porphyritic granodiorites	56	12	25
Leucocratic granites	30	9	34
Vein granito-aplites	12	10	34
Shakhtaminski massif			
Diorites and monzonites	54		20
Granodiorites and quartz-monzonites	40		24
Late-stage granites	33		28
Soktuy massif			
Biotite granites	61		30
Late-stage granites	42		19
Grano-syenites	85		22

Many workers in the field of bedrock geochemistry have pondered over the problems of whether they should be looking for areas of anomalously high or low trace-element content. Some maintain that areas containing mineralization of a particular element will be characterized by the rocks of the area having higher than normal amounts of that element, whereas others claim that mineralized areas will have lower levels of the trace elements in the rocks, as the elements in question will have been depleted at some stage to form the mineralizing solutions. Both these views may be true—the former on a regional scale and the latter on a more local scale intimately associated with the mineral deposits in question. A third alternative arises in the cases of mineral deposits which are essentially strata-bound and pre-intrusive in age. In such cases anomalously high trace-element values may be found due to the assimilation of sedimentary material which itself is anomalously high in such trace elements, or to the involvement of the hot intrusive in some plumbing system that draws metal-rich water from mineralized host rock. There are obviously problems of mass transfer and subsequent dilution in the intrusive body; however, the writer feels that the possibility should not be ignored. In this particular study, which is of a broad regional nature and aimed at delineating areas of increased mineral potential rather than individual mineral deposits, it was decided to look for plutons associated with anomalously high base-metal contents.

In the field area significant deposits of tungsten (scheelite) occur, together with gold, which is placer-mined by small operators. An interpretation of the tungsten data has been published elsewhere,[8] but it should be noted that these deposits contain varying amounts of the base-metal sulphides, and in the current case the base metals may be considered pathfinder elements for areas of tungsten or gold–tungsten mineralization.

Initial investigation of the data

In order to assess the gross geochemical features of the data and the degree to which the data could be considered to be drawn from a single magmatic series a *Q*-mode principal-component analysis was undertaken. The 2136 samples involved in the study area are drawn from 74 distinct plutonic bodies: thus, for each of these plutons the arithmetic and geometric means, together with Fisher's t estimator, based on the theory of maximum likelihood,[20] were computed. The value of the t estimator lies above the geometric mean, but below the arithmetic mean. It has a higher reliability than the arithmetic mean, i.e. fewer samples are required to attain the same precision of determination, and is less sensitive to occasional high values. In the case of the South African gold deposits this estimator yields, in most cases, a value lying nearer the true mean grade as determined by mining.

The input data to the *Q*-mode principal-component analysis was therefore 74 sets of maximum-likelihood estimators of the major, minor and trace elements. No transformations, other than for range, were applied to the estimators: thus, on computation of the similarity matrix all the elements were scaled to between zero and one, so giving them equal weight in the analysis. The first two varimax-rotated principal components account for 87·8% of the total variability, and a plot of the loadings on these two components is shown in Fig. 2. The arcuate distribution of the points coincides with a trend in the rocks from basic in the upper left to more acidic in the lower right. This general trend was taken as evidence that the rocks could be viewed as being drawn from one, or several overlapping, magmatic series. One drawback with the presentation of *Q*-mode principal components is that they tend to emphasize the common features, but do not lead to a presentation which allows the data to be clustered into geologically meaningful groups. To achieve this end a technique known as non-linear mapping, which was developed by Sammon,[19] was employed. Two matrices of Euclidian distances are computed—one from the initial transformed data and the other from the two principal-component loadings. Corrections are made iteratively to the second matrix until the Euclidian distances of that matrix most closely match those of the initial data. In doing this a two-dimensional projection of the n-dimensional initial data is found which most closely preserves the data structure as exhibited by the inter-point

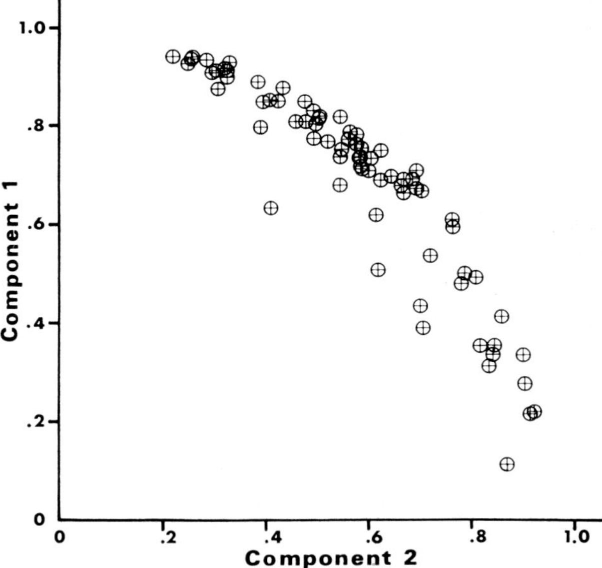

Fig. 2 Plot of rotated Q-mode principal component 1 versus principal component 2

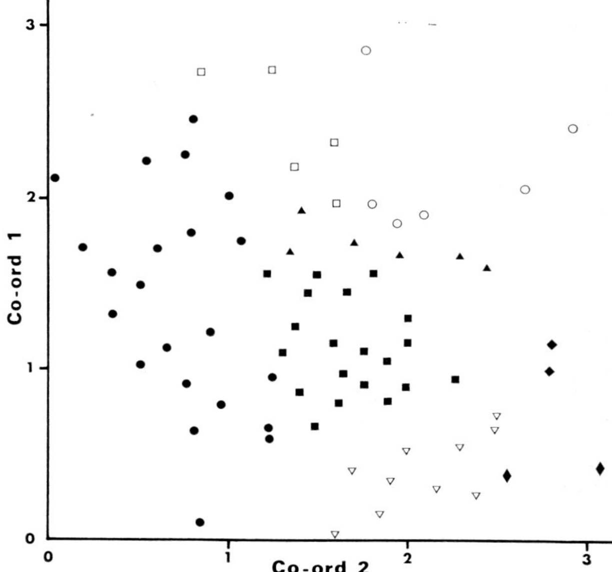

Fig. 3 Non-linear mapping presentation of Q-mode data (see Fig. 4 for description of symbols)

distances. The resulting computation leads to the presentation shown in Fig. 3. The plutons represented on the new data presentation were divided into groups on the basis of geological similarity and areal proximity, the areal distribution of these groups being shown in Fig. 4 and the average base-metal content for each group in Table 5. The data indicate twofold variations between different groups for zinc and up to fourfold and threefold variations for copper and lead, respectively. Thus, at this point the specific magnitude of the problem in the Selwyn Belt granitoids has been established and the data have been shown to be drawn from a single, or several overlapping, magmatic series.

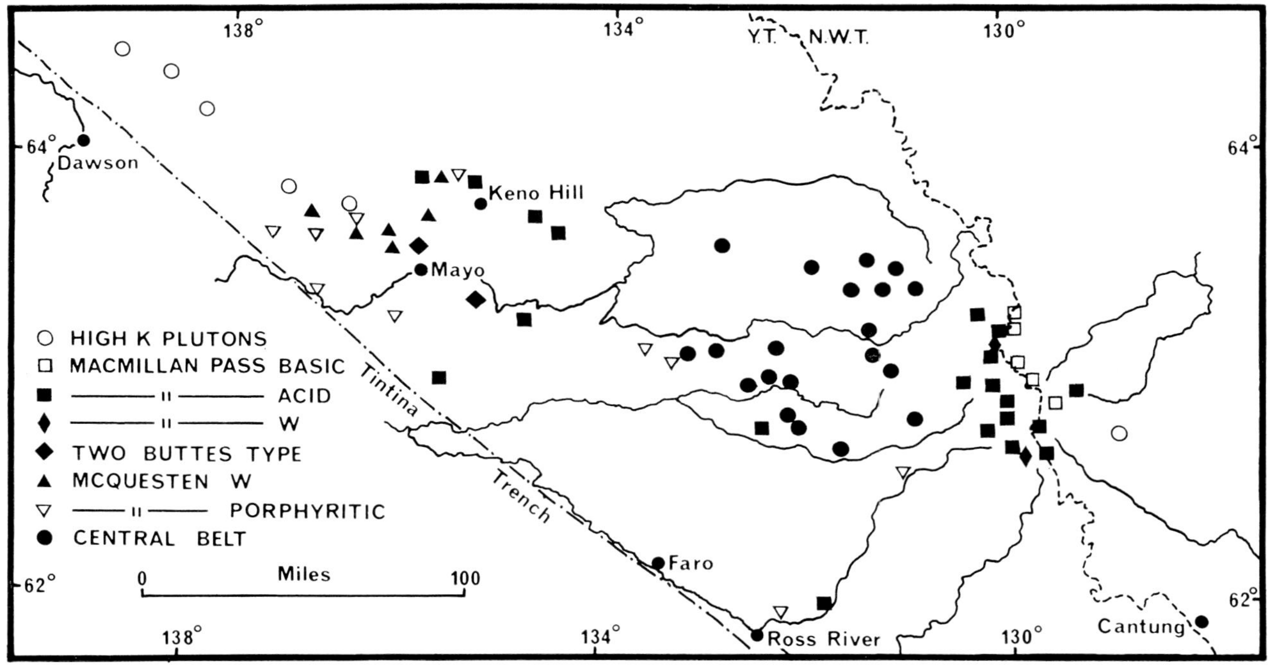

Fig. 4 Areal distribution of sub-groups derived from the non-linear mapping presentation

Table 5 Mean base-metal content (ppm) of acid plutonic rocks of various sub-groups

	Zn	Cu	Pb
High K granitoids	90	18	49
Macmillan Pass basic phase	61	17	38
Macmillan Pass acid phase	58	7	33
Macmillan Pass tungsten belt	47	7	59
Two Buttes type	73	4	36
McQuesten Porphyry tungsten belt	111	8	34
McQuesten porphyritic type	47	6	35
Central belt	63	11	23

Statistical analysis to identify anomalous plutons and samples

Multiple regression has been shown to be a useful tool in allowing for variations in background in stream sediments by Rose and co-workers[17,18] and by Nichol.[16] Previously, the independent variables have been elemental data, sometimes combined with a quantification of the relative importance of lithological units in the stream catchment areas. Because of the inherent redundancy in the major- and minor-element data for plutonic rocks, it was decided to minimize these effects by computing a new set of uncorrelated orthogonal variables by the use of *R*-mode principal-component analysis, i.e. where a correlation matrix with unities in the diagonal is used as a starting point. A beneficial product of this approach is the generation of new independent variables that are more closely related to the geological processes dominant in the environment. The computation of the *R*-mode principal-component analysis followed standard procedures. No attempt was made to allow for closure in the data such as was demonstrated by Miesch and co-workers,[14] as it was considered that Al would have been the most suitable element to use as a denominator for ratio generation, and in this data set the error variance associated with the Al data was large. The principal-component analysis was followed by a Varimax rotation and the computa-

Table 6 Varimax-rotated principal-components matrix

	1	2	3	4	5	6	7	8	9
Si	−0·56	−0·11	0·13	0·03	−0·28	0·76	0·05	−0·01	0·03
Al	0·13	0·01	−0·98	0·09	0·09	−0·07	−0·02	0·01	−0·01
Fe	0·73	−0·03	−0·15	−0·20	0·44	−0·29	−0·01	0·04	−0·34
Mg	0·91	−0·05	−0·01	−0·26	0·18	−0·18	0·06	−0·19	0·07
Ca	0·80	−0·14	−0·15	−0·13	0·26	−0·25	−0·41	−0·01	−0·01
Na	−0·27	0·09	−0·10	0·95	−0·19	0·02	0·02	−0·01	0·02
K	−0·07	0·99	−0·01	0·08	0·08	−0·05	0·02	0·00	0·00
Ti	0·86	−0·02	−0·17	−0·21	0·27	−0·16	0·04	0·32	−0·04
Mn	0·40	0·13	−0·11	−0·01	0·88	−0·18	−0·04	0·01	−0·01
Sum of squares	3·31	1·04	1·07	1·09	1·23	0·82	0·17	0·14	0·13
% of variability	36·7	11·6	11·9	12·1	13·7	9·1	1·9	1·6	1·4
Cumulative %	36·7	48·3	60·2	72·3	86·0	95·1	97·0	98·6	100·0

tion of the component scores for each of the 2136 samples on the nine orthogonal axes.[10] The Varimax-rotated principal-component loadings are given in Table 6. Component 1 accounts for some 36·7% of all the major- and minor-element variability, and appears to be inversely related to the degree of differentiation of the rocks. Larsen's differentiation index was computed for the 2136 rocks and the correlation coefficient between these and the scores on component 1 was found to be −0·79, which yields a value of Student's t of 58·7 with 2135 degrees of freedom and is considered highly significant. Component 2 is almost entirely related to the K content of the rock and is related to the presence of porphyritic microcline, and, as such, is considered to be a measure of porphyricity of the rocks. This latter component and the following four account for approximately 11% of the total data variability each, whereas the last three account for less than 2% each. All these components, with the exception of the fifth, are essentially single-element loaded, and no ready explanation is available for them in common petrologic terms or relationship to mineralization. Only component 5, which is a Mn–Fe association, is possibly explainable in terms of colour index or magnetite content of the rock. The correlation coefficients of the logarithmically transformed base metals with the rotated principal-component scores are given in Table 7. Component 1 is one of the largest correlations in all three elements, and component 2 is prominent in the cases of Cu and Pb, especially the latter. Zinc does not conform, having an important correlation with component 5, which may have some petrologic significance (see above).

Stepwise multiple regression was undertaken by use of the forward addition of independent variables as described by Efroymson.[6] It was noted that all rotated principal components with absolute correlation coefficient values in excess of 0·03 were included as contributing to a significant decrease in the error sums of squares of the regressions. This is due to the very small value the mean sums of squares of the residuals will assume when large numbers of samples are used in the regression. Although this situation is statistically valid, it is not satisfying to a geologist undertaking an interpretation as it is doubtful if petrological processes, or significant meanings, can be attached to any other than the first two principal components with confidence. It was therefore decided to use only the first two rotated principal-component scores in the multiple-regression equation. Thus, samples that yielded residuals within ±2 standard deviations of the regression could have their base-metal content explained in terms of the differentiation of a magmatic series

Table 7 Correlation coefficients between logarithmically transformed base metals and rotated principal-component scores

	1	2	3	4	5	6	7	8	9
$\log_{10}$ Zn	0·31	0·07	−0·20	−0·13	0·38	−0·24	0·03	−0·06	−0·11
$\log_{10}$ Cu	0·45	0·23	0·04	−0·16	0·12	−0·18	0·00	−0·14	−0·02
$\log_{10}$ Pb	−0·27	0·45	0·03	0·05	0·06	0·07	0·03	−0·15	0·08

Table 8 Percentage of variability explained by regressions

	Zn	Cu	Pb
Variability explained by all 9 elements	37·50	34·66	32·04
Variability explained by principal components 1 and 2	9·82	25·31	27·84
Percentage of explainable variability that is contained in principal components 1 and 2	26·19	73·02	86·89

and the amount of porphyritic feldspar present in the rock. Attention would therefore be focussed on samples which contain some component of non-standard rock-forming processes, one of which could be a mineralization process. Although only two variables are included in the regression equation, these contain 48·3% of the total data variability (Table 6) and account for varying amounts of the regression variability explainable by all the major and minor elements (see Table 8). For Cu and Pb the first two components account for approximately 75% of the variability explainable by the major and minor elements, whereas in the case of zinc only some 25% of the explainable variability is contained in the first two components.

The standardized residuals, i.e. the regression residuals divided by the square root of the mean sums of squares of the residuals, were computed for all samples for each of the base metals. These score values for each of the 74 plutons of the field area. Maps have been prepared for each element showing the maximum-likelihood value (hereafter referred to as the mean), the highest observed result, the mean score and maximum observed standard score for each pluton (Figs. 5–16).

Description of results

The data for zinc are presented in Figs. 5–8. The most noticeable feature of the regional zinc distribution is the concentration of plutons with high mean zinc content in the northwest of the field area (Fig. 5). A less conspicuous area of high mean zinc content lies in the Central belt (17, see Fig. 1). The distribution of means is positively skewed, and plutons with means in excess of 90 ppm must be considered anomalous. The map of the maximum observed zinc values (Fig. 6) reveals generally similar features as the means.

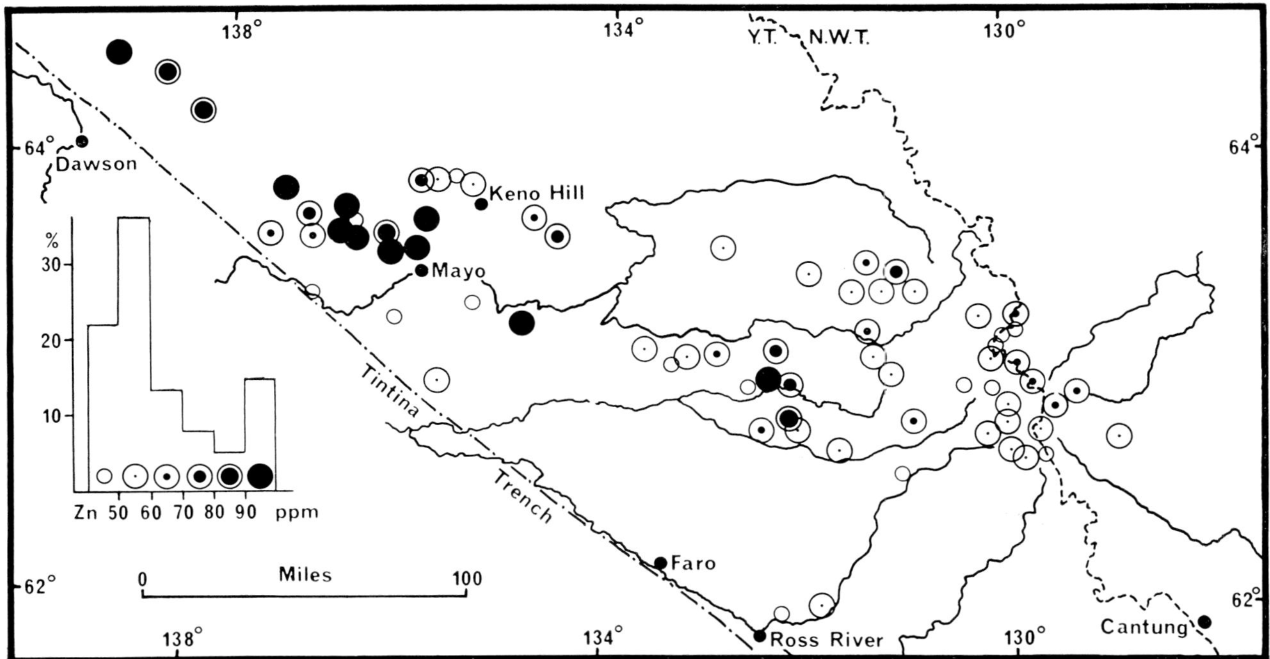

Fig. 5 Mean zinc content of granitoids

data were then used to determine the arithmetic mean score and maximum and minimum standard

Where relatively low maximum values are found in conjunction with high means, a low data

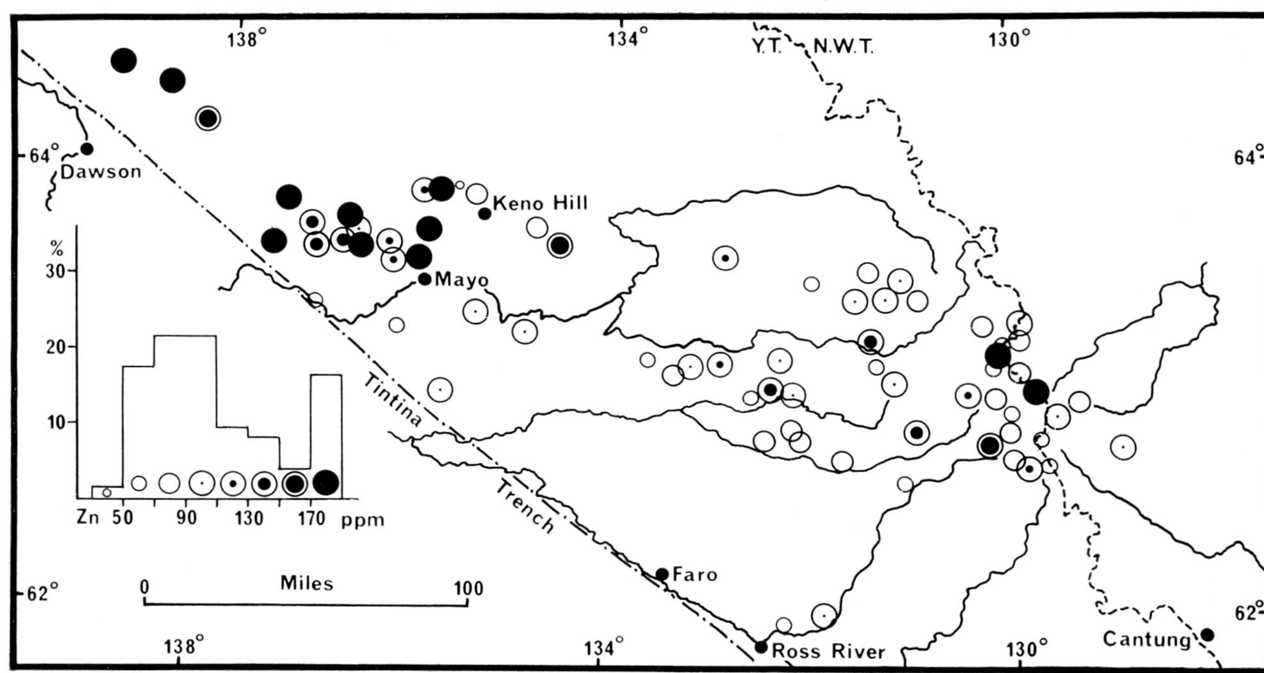

Fig. 6 Maximum zinc content of granitoids

variance is indicated. Conversely where high maximum values are found associated with either high or low means, a higher data variance or many cases only very small changes in means may be observed, but the data variance and positive skewness are distinct diagnostic features.[2,3,8,12,22]

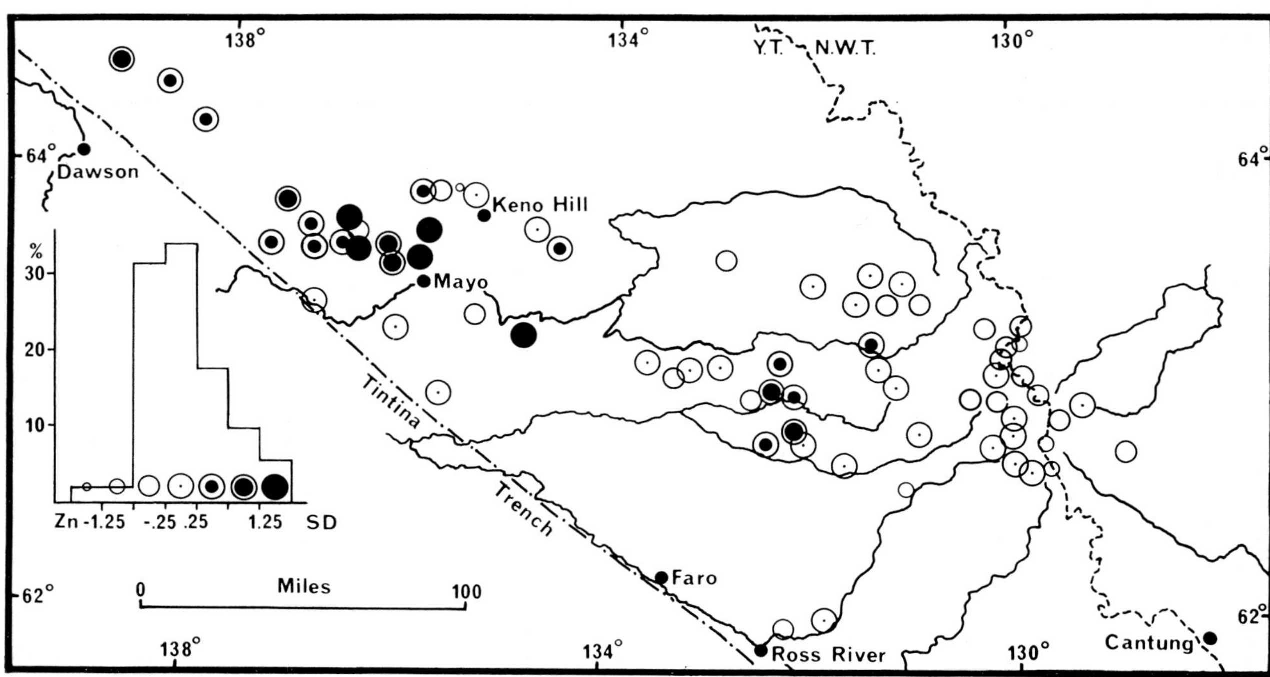

Fig. 7 Mean residual scores for zinc in granitoids

positive skew in the data is indicated. It is now generally recognized that the variance, or dispersion, and positive skewness of geochemical data sets are criteria of great importance in recognizing data derived by sampling mineralized areas. In Thus, the plutons of the Central belt (16, 17) become less interesting and others to the east (21, 23) more attractive. Similar general findings are apparent in the northwestern parts of the field area.

Fig. 7 depicts the mean residual scores for each pluton and indicates the mean residual zinc in the pluton after allowance for the effects of crystallo-chemical differentiation (principal component 1) and the presence of porphyritic microcline (principal component 2). The majority of the plutons in the eastern half of the field area appear to have near the expected zinc contents. The exceptions to this general trend are a reduced number of plutons showing higher than expected means in the Central belt (16, 17). The number of anomalous (>90 ppm Zn) plutons in the north-western part of the field area are reduced by almost half and those remaining with means in excess of 0·75 standard deviation units are, with one exception, associated with the McQuesten

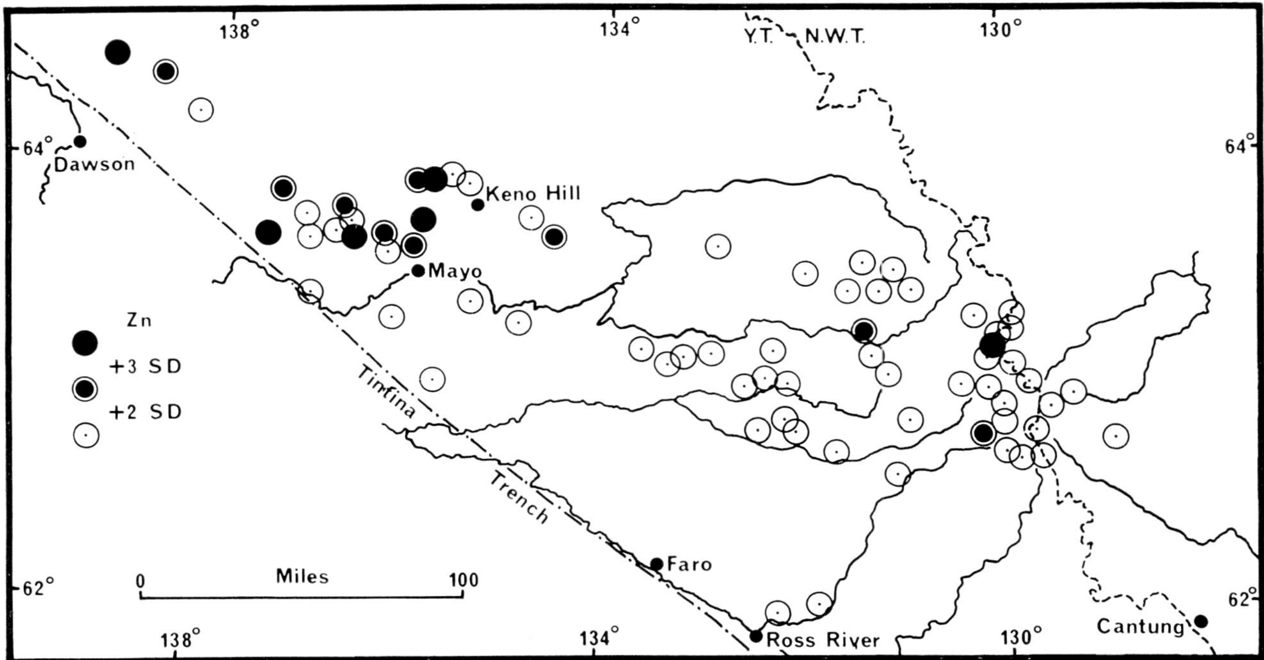

Fig. 8 Maximum residual scores for zinc in granitoids

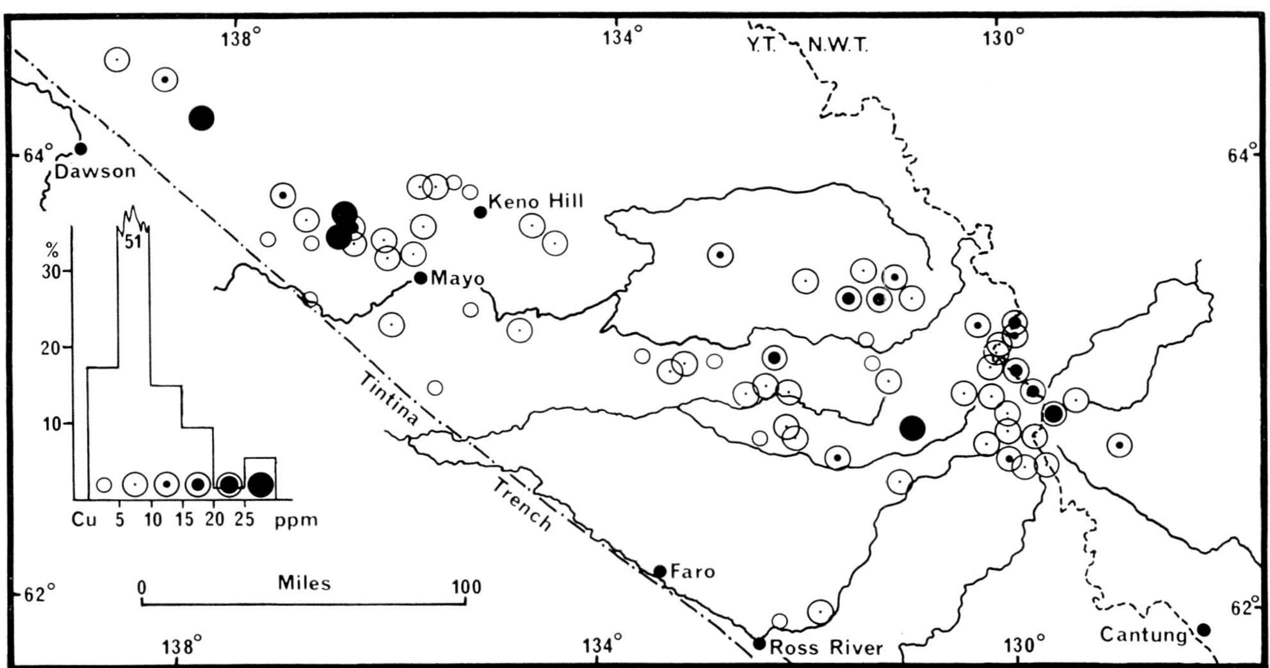

Fig. 9 Mean copper content of granitoids

mineralized belt. The whole of the northwestern part of the field area is characterized by small sulphide showings of lead, zinc and antimony minerals, and it is conjectured that the generally high zinc levels in the area are a reflection of these mineral occurrences. The maximum observable scores are presented in Fig. 8: of the six plutons with maximum scores in excess of 3 standard deviation units, four are known to be the mineral occurrences associated with the syenites and monzonites (2, 3) are reflected, and farther to the southeast mineral occurrences in the Red Mountain area (5) and on nearby Sunshine Creek are associated with plutons carrying anomalous copper contents (>100 ppm Cu). In the eastern half of the field area three anomalous plutons are present; the northernmost of these contains disseminated chalcopyrite

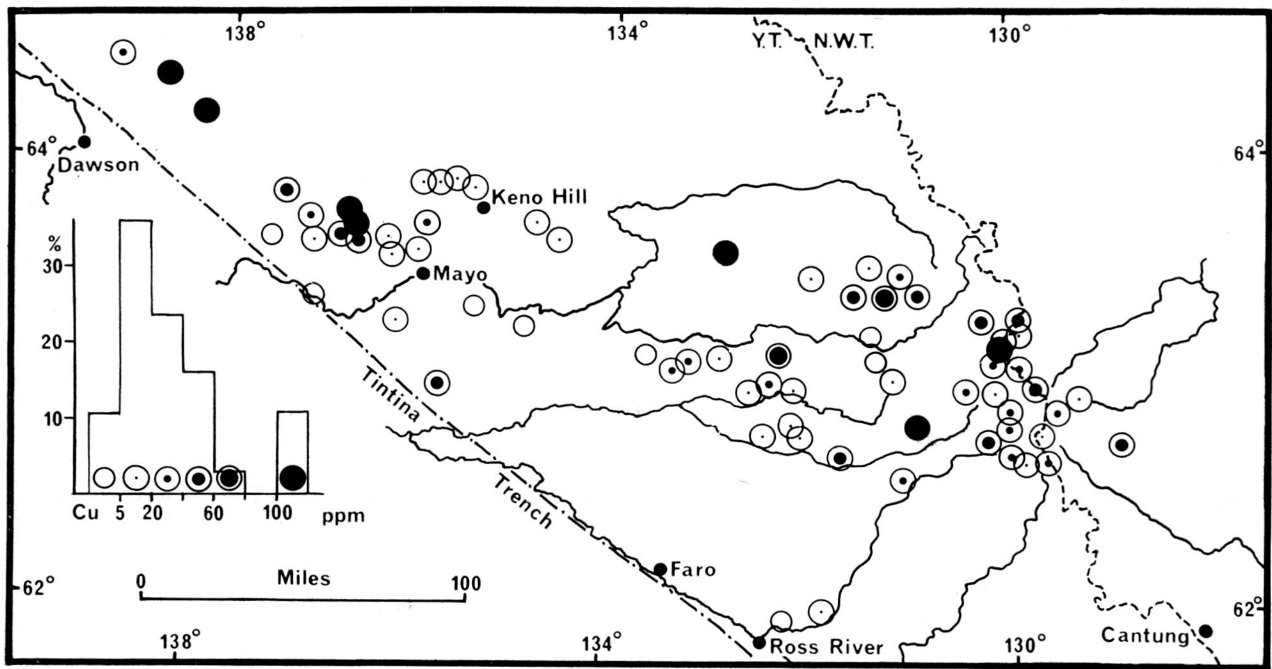

Fig. 10 Maximum copper content of granitoids

associated with mineralization (8, 12, 13, 21). Nine more plutons exhibit maxima between 2 and 3 standard deviation units: of these, five are known to have mineralized showings in the vicinity. Three plutons associated with mineralization (3, 6, 22) do not have anomalous zinc contents—contrary to expectations based on the mineralogy of the associated occurrences; however, these deposits are reflected in the results of at least one of the remaining base metals.

The data for copper are presented in Figs. 9–12. The frequency distribution of mean copper contents (Fig. 9) is less skewed and not so prominently bimodal as those for zinc. Copper minerals are not as widespread in the study area as are those of zinc, and the distribution reflects this fact. The area of high mean copper content in the northwest of the field area is less prominent than that for zinc. In the east, however, there are two clusters of higher than average copper content plutons and a single isolated anomalous pluton. The maximum observable values (Fig. 10) define a number of areas of interest. Northeast of Dawson

and pyrrhotite, and Mount Allan (21) is genetically related to a large skarn tungsten deposit. The third (19) is not known to be associated with any mineral occurrence, but a pluton to the southwest (18) contains chalcopyrite in a fractured zone in the apical part of the pluton. Two plutons contain high background mean levels—both in the Central belt (16, 17)—and one of these (17) is associated with a contact pyrometasomatic copper–tungsten mineral occurrence. The second lies above an area of limonitic staining on the north shore of Fairweather Lake; extensive soil sampling was undertaken in this area, but no definable target was found.

The mean residual scores for copper in plutons are presented in Fig. 11, the majority of the higher than to be expected copper levels lying in the eastern part of the field area. Three mineralized areas (3, 5, 8) in the west, however, are characterized by high mean scores. In the east there is a well-defined area of higher copper content (17–21) and two more isolated areas (16 and 23, 24). The majority of the known

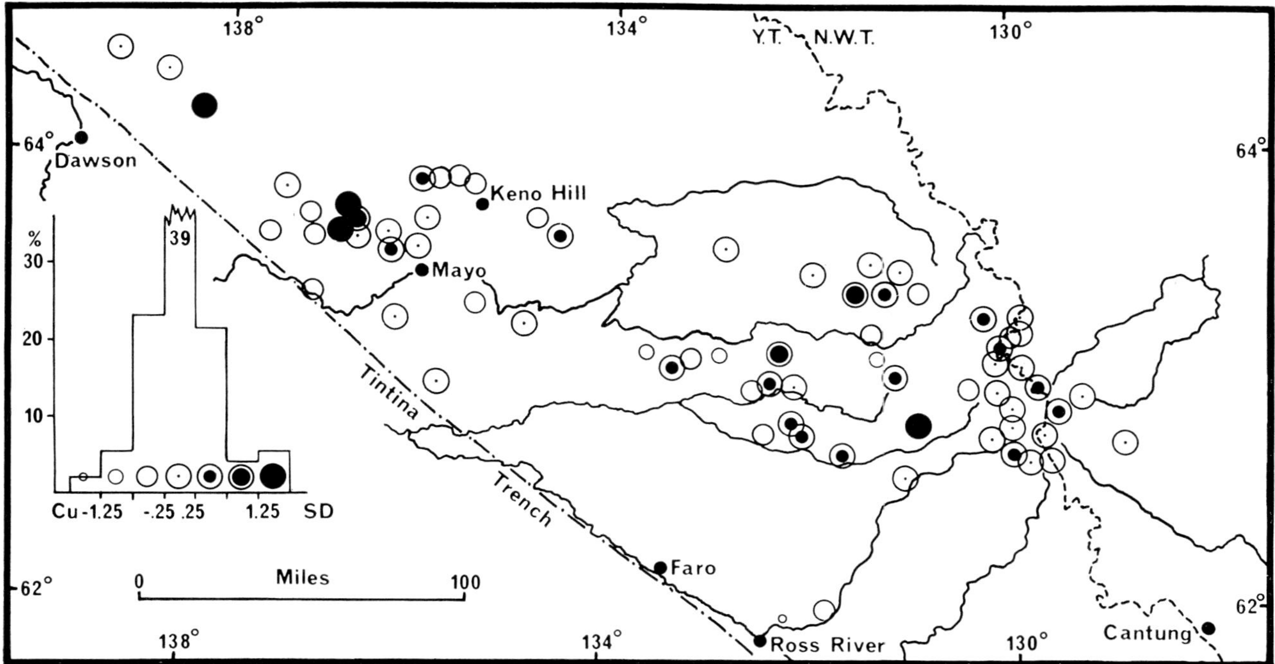

Fig. 11 Mean residual scores for copper in granitoids

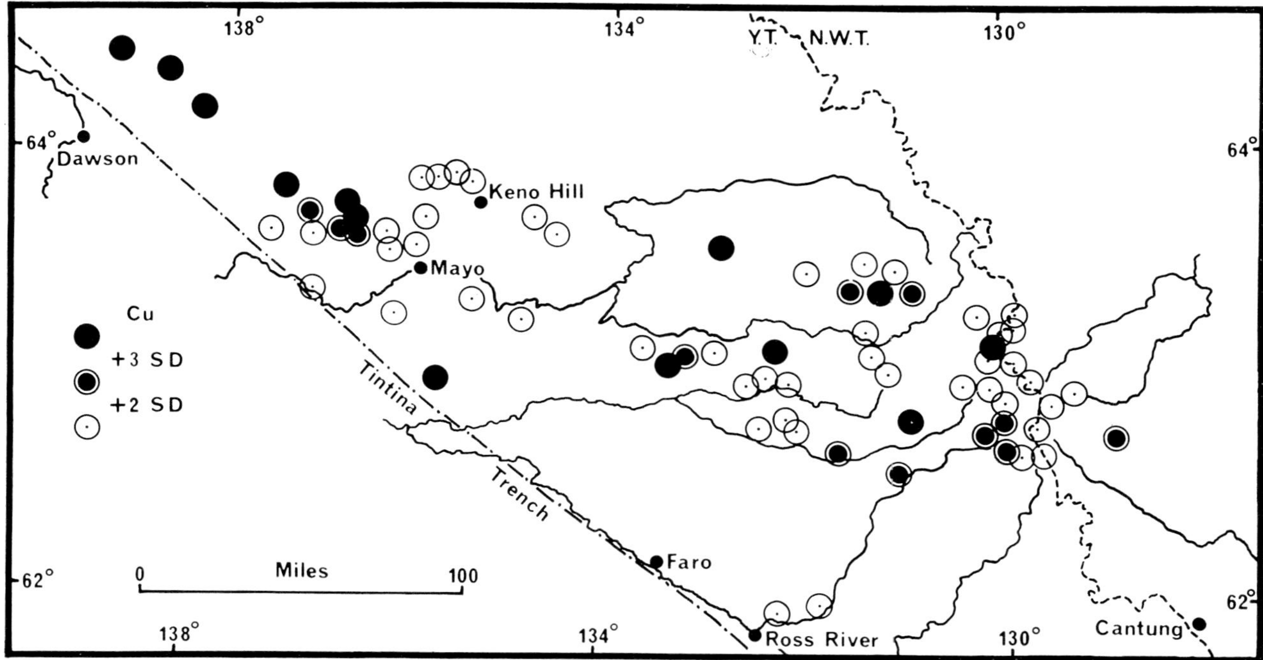

Fig. 12 Maximum residual scores for copper in granitoids

copper occurrences fall within these areas. The maximum observed residual scores are shown in Fig. 12. Thirteen plutons contain anomalously high scores, and eight of these have verified mineral occurrences in their vicinities (including 2, 3, 4, 5, 16, 21). Of the five remaining plutons, mineralization can be reasonably expected to be in the vicinity of three on the grounds of local geology (1, 19) or other geochemical data. The other two (15, northwest of 17) have completely unknown potential; however, there are old reports of mineralization having been found in both areas. Those plutons exhibiting maximum scores between 2 and 3 standard deviation units above the regression are all spatially associated with known mineral occurrences or are close to similar plutonic bodies intruding similar strata. Certain notable exceptions are again present—the most marked in the copper–zinc–tungsten skarn (22) in the southeast of the field area, which

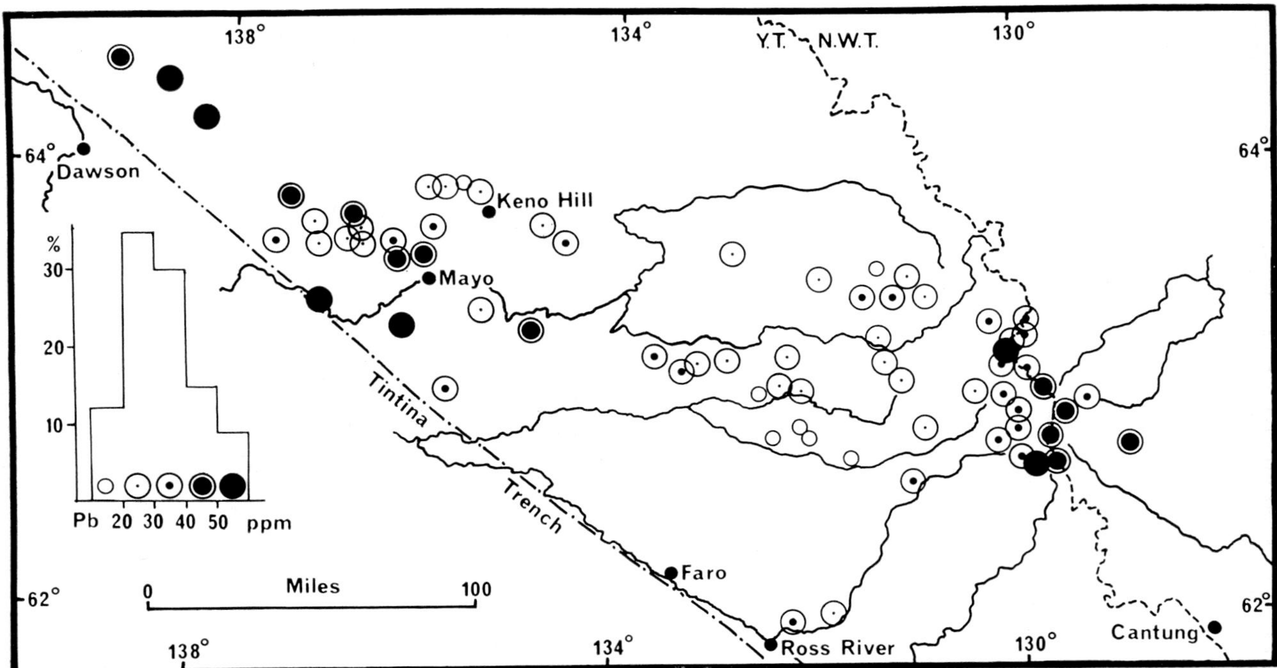

Fig. 13 Mean lead content of granitoids

is not reflected in any way by the copper data. It is perhaps not surprising that the plutons of the eastern part of the McQuesten Belt (9–13) are not anomalous, as copper is not a significant element in these deposits.

The data for lead are presented in Figs. 13–16. Unlike the previous two elements, which are geochemically related to the femic components of the rocks, lead is closely correlated with potassium content. The frequency distribution of the means is not significantly skewed and is unimodal, areas of high lead content being confined to the eastern and western extremities of the field area (Fig. 13). In the eastern area all plutons with major mineral occurrences are reflected by high mean contents (21, 22, 25). In the western area the high mean lead contents fall into two areas—one in the extreme northwest and the other near and south of Mayo. The former of these contains a number of lead-bearing

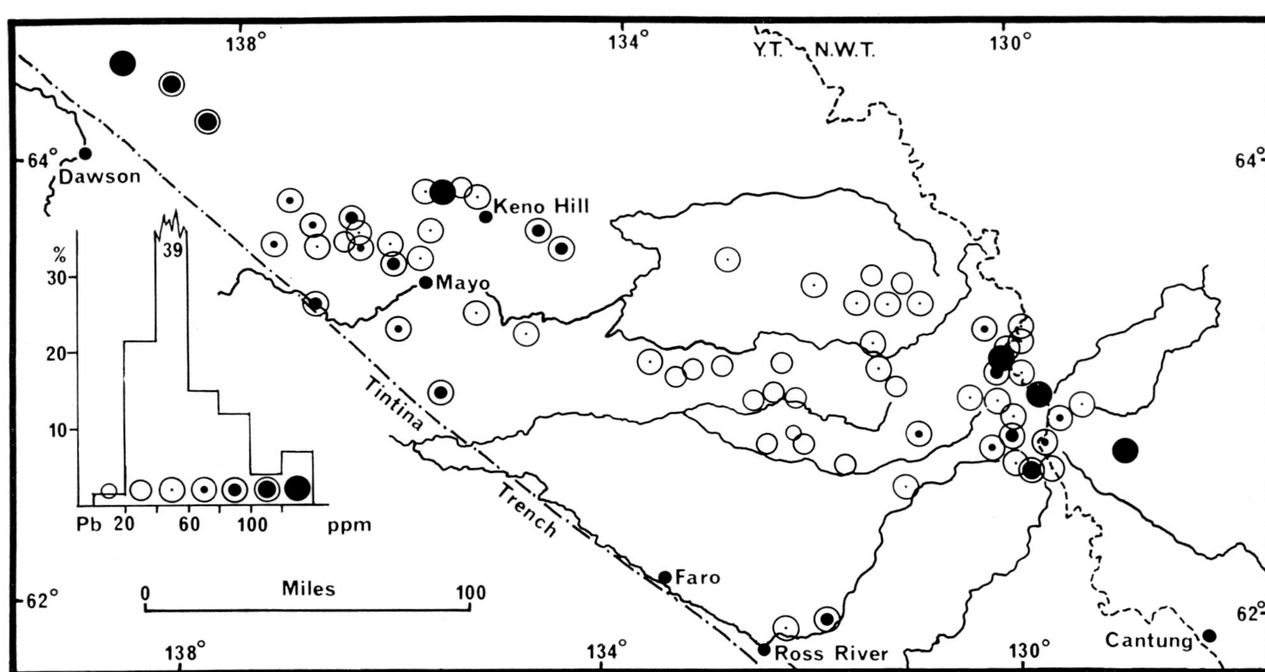

Fig. 14 Maximum lead content of granitoids

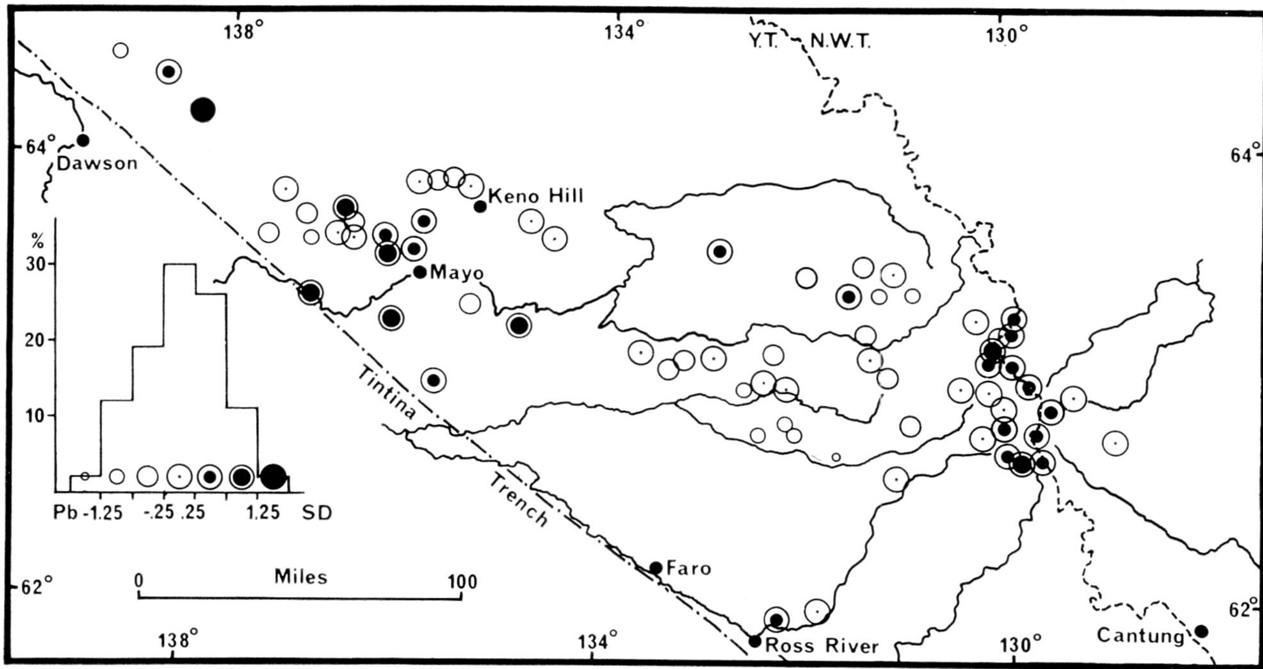

Fig. 15 Mean residual scores for lead in granitoids

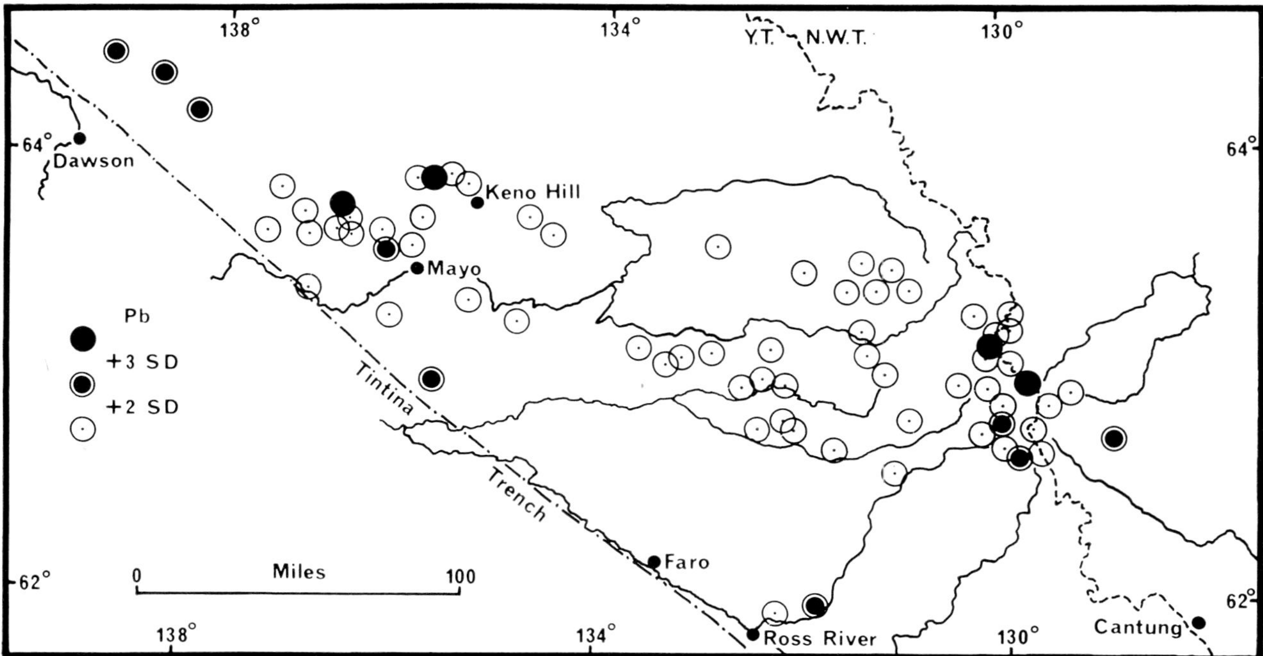

Fig. 16 Maximum residual scores for lead in granitoids

mineral occurrences, but the latter group with plutons south of Mayo has no known relationship to mineral occurrences. The maximum observable lead levels are presented in Fig. 14, and attention is drawn to all plutons containing in excess of 100 ppm Pb. Eight plutons are so characterized and, of these, six have associated mineral occurrences (2, 3, 13, 21, 22, 25), the remaining two (1, 23) being in generally favourable areas and perhaps associated with further mineral occurrences.

The regression approach to interpretation allows the effects of potassium to be taken account of, and the residual score means for each pluton are presented in Fig. 15. Three areas of higher than normal lead are apparent: one in the northwest, one around Mayo and one in the eastern part of the field area. The Central belt is generally characterized by low lead levels. The maximum residuals for each pluton are presented in Fig. 16.

The anomalous (>3 S.D. units) plutons are associated with mineral occurrences (5, 13, 21), with one exception (23), and that area must be considered favourable for exploration. Nine plutons exhibit samples with residual scores in the range 2–3 standard deviation units. Of these, six are associated with mineralization (2, 3, 11, 15, 22, 25); of the remaining three plutons, two are considered favourable target areas (1, north of 22), and the third, near Ross River, is of unknown potential. Of considerable interpretational importance is the elimination of the previously considered anomalous plutons south of Mayo. These plutons, although characterized by high lead levels, do not exhibit the skewness typical of plutons associated with mineralization.

Regional implications

Before concluding, some general points may be made about the regional geochemistry of the base metals in the northern half of the Selwyn Belt. First, the area northwest of Mayo is characterized by higher than average base-metal content, and the fact that this general area contains the largest concentration of mineral occurrences cannot be fortuitous. Secondly, the eastern edge of the field area coincides with the western margin of the miogeosynclinal rocks, and therefore lies close to a hinge-line, the eugeosynclinal rocks of the Selwyn Basin lying to the west. This area along the eastern edge of the field area is characterized by high lead levels, significantly higher copper levels and marginally higher zinc levels. These features could be due to a variety of reasons—the tapping of deeper sources of the elements due to the influence of hinge-line faults, regional differences in magma composition related to original differences in the composition of the crust subjected to palingenesis, and, lastly, the assimilation of miogeosynclinal rocks in contrast to eugeosynclinal rocks assimilated by plutons westward and farther out in the Selwyn Basin. Finally, the central part of the Selwyn Basin is characterized by higher than normal zinc levels, and the reasons for this could be one of those mentioned above with reference to the eastern part of the field area.

Conclusions

The data presented illustrate the application of a number of mathematical–statistical techniques in a planned manner to solve a series of interpretational problems. The final data presentation is shown to have advantages in delineating those plutons known to be associated with mineralization, at the same time eliminating others which appear to be anomalous by use of other presentation methods but which are not considered favourable on other grounds.

The technique of combining *R*-mode principal-components analysis with multiple regression also has a conceptual approach which is naturally appealing to the geologist. First, a transformation of the major and minor elements is made from a response-orientated, and conceptually difficult, form to a process-orientated form more easily understood in broad geological rather than geochemical terms. Secondly, the multiple-regression analysis allows the selection of samples which do not fit within certain bounds (e.g. 2 or 3 S.D. units) of the selected *R*-mode process model. Those samples lying outside the selected bounds must be assumed to have been influenced by some process other than those selected from the *R*-mode model for inclusion as independent variables in the multiple regression. If the processes, as in the present case, are selected on simple petrochemical and petrological grounds, samples outside the bounds must have been influenced by more complex processes, one of which could be related to the formation of mineral deposits.

This approach to data interpretation may be thought of as an extension to very large numbers of samples of *Q*-mode principal-components analysis and the use of communality to indicate the lack of fit of any particular sample to a process model chosen on the basis of *Q*-mode principal components.[15] It does, however, have the advantage of being specific to chosen elements, and there is no reason why elements included in the principal-components analysis could not be used as dependent variables in the regression analysis.

The interpretational scheme outlined contains no radically new data-manipulation technique; rather, it is a logical amalgamation of a number of well-tried methods in an attempt to improve the objectivity of the interpretational step the geochemist must make in moving from response-orientated chemical data to a geological process model. It is hoped that the methodology will have applications to other problems in other areas where the step from response to process is even more difficult than it is with bedrock geochemical data.

Acknowledgment

The author would like to express his sincere thanks to the many individuals who have participated in this study, both in the field and in the laboratory. The enthusiasm and diligence of students and aircraft crew made the field work in mountainous terrain possible, and the laboratory support provided was of the highest order. In this latter respect special thanks are due to the Geochemistry Section analytical team, who provided more than 60 000 determinations for the study, and the Analytical Chemistry Section, who carried out the whole-rock analyses for the accuracy investigation.

This paper is published by permission of the Director, Geological Survey of Canada.

References

1. Abbey, S. Analysis of rocks and minerals by atomic absorption spectroscopy. Part 3. A lithium-fluoborate scheme for seven major elements. *Pap. geol. Surv. Can.* 70–23, 1970, 20 p.
2. Bolotnikov, A. F. and Kravchenko, N. S. Criteria for recognition of tin-bearing granites. *Dokl. Akad. Nauk SSSR*, **191**, 1970, 678–80; *Proc. Acad. Sci. USSR, Earth Sci. Sect.*, **191**, 1970, 186–7.
3. Cameron, E. M. and Baragar, W. R. A. Distribution of ore elements in rocks for evaluating ore potential: frequency distribution of copper in the Coppermine River Group and Yellowknife Group volcanic rocks, N.W.T., Canada. In *Geochemical exploration* (Montreal: CIM, 1971), 570–6. *(CIM Spec. vol. 11)*
4. Danielsson, A. Spectrochemical analysis for geochemical purposes. In *XIII Colloquium spectroscopicum internationale, Ottawa* (London: Hilger, 1968), 311–23.
5. *Geology and economic minerals of Canada* Douglas, R. J. W. ed. (Ottawa: Geological Survey of Canada, 1970), 838 p. (*Econ. Geol. Rep.* no. 1)
6. Efroymson, M. A. Multiple regression analysis. In *Mathematical methods for digital computers* Ralston, A. and Wilf, H. S. eds. (New York: Wiley, 1960), vol. 1, 191–203.
7. Garrett, R. G. The determination of sampling and analytical errors in exploration geochemistry. *Econ. Geol.*, **64**, 1969, 568–9.
8. Garrett, R. G. Molybdenum, tungsten and uranium in acid plutonic rocks as a guide to regional exploration, S.E. Yukon. *Can. Min. J.*, **92**, April 1971, 37–40.
9. Ivanov, L. B. Intrusive complexes of the Batystau mineral field in Central Kazakhstan: their geochemical and metallogenic specialization. *Sov. Geol.*, no. 9 1969, 120–33. (Russian text)
10. Kaiser, H. F. Formulas for component scores. *Psychometrika*, **27**, 1962, 83–7.
11. Krauskopf, K. B. *Introduction to geochemistry* (New York: McGraw-Hill, 1967), 721 p.
12. Kuzmin, M. I. The function of the rare element distribution in granitoids and their parameters. In *Origin and distribution of the elements* Ahrens, L. H. ed. (Oxford, etc.: Pergamon, 1968), 641–8.
13. Miesch, A. T. The need for unbiased and independent replicate data in geochemical exploration. In *Geochemical exploration* (Montreal: CIM, 1971), 582–4. (*CIM Spec. vol. 11*)
14. Miesch, A. T. Chao, E. C. T. and Cuttitta, F. Multivariate analysis of geochemical data on tektites. *J. Geol.*, **74**, 1966, 673–91.
15. Nichol, I. Garrett, R. G. and Webb, J. S. The role of some statistical and mathematical methods in the interpretation of regional geochemical data. *Econ. Geol.*, **64**, 1969, 204–20.
16. Nichol, I. The role of computerized data systems in geochemical exploration. *CIM Bull.*, in press.
17. Rose, A. W. Dahlberg, E. C. and Keith, M. L. A multiple regression technique for adjusting background values in stream sediment geochemistry. *Econ. Geol.*, **65**, 1970, 156–65.
18. Rose, A. W. and Suhr, N. H. Major element content as a means of allowing for background variation in stream-sediment geochemical exploration. In *Geochemical exploration* (Montreal: CIM, 1971), 587–93. (*CIM Spec. vol. 11*)
19. Sammon, J. W. Jr. A nonlinear mapping for data structure analysis. *Trans. IEEE*, **C-18**, 1969, 401–9.
20. Sichel, H. S. New methods in the statistical evaluation of mine sampling data. *Trans. Instn Min. Metall.*, **61**, 1951–52, 261–88.
21. Sitnin, A. *et al.* Evaluation of ore bearing potential of granitoid complexes with respect to tantalum and beryllium on the basis of geochemical specialization of the granitoids. In *Efficiency of geochemical methods of prospecting for ore deposits and the broad application of these methods by the Geological Survey of the USSR* Grigoryan, S. V. ed. (Moscow: Ministry of Geology, 1967), 9–10. (Russian text)
22. Tauson, L. V. Distribution regularities of trace elements in granitoid intrusions on the batholith and hypabyssal types. In *Origin and distribution of the elements* Ahrens, L. H. ed. (Oxford, etc.: Pergamon, 1968), 629–39.
23. Taylor, S. R. Abundance of chemical elements in the continental crust: a new table. *Geochim. cosmochim. Acta*, **28**, 1964, 1273–85.
24. Turekian, K. K. and Wedepohl, K. H. Distribution of the elements in some major units of the Earth's crust. *Bull. geol. Soc. Am.*, **72**, 1961, 175–92.
25. Vinogradov, A. P. Average contents of chemical elements in the principal types of igneous rocks of the Earth's crust. *Geokhimiya*, 1962, 555–71; *Geochemistry*, 1962, 641–64.

519.272:552.3:550.42(944)

Cluster analysis of rocks in the New England igneous complex, New South Wales, Australia

W. R. Hesp, D.SC.TECH.

D. Rigby, A.R.I.C.

Both of CSIRO Division of Mineralogy, North Ryde, New South Wales, Australia

Synopsis

The new England igneous complex—with an area of about 20 000 km^2—consists of numerous outcrops of granitoid rocks (granites, adamellites, granodiorites) and also porphyrites. The cluster analysis of 72 samples (taken from 34 outcrops) based on the major-element geochemistry of the rocks resulted in good general agreement between the previous geological mapping and the present statistical grouping of the rocks. Most of the samples were in one of two distinctive groups, which comprised adamellites–granodiorites–porphyrites (46 samples) on the one hand, and granites–leucoadamellites (24 samples) on the other. The mean silica contents were 67·2 and 75·0%, respectively. Geochemical differences within certain outcrops and similarities between outcrops bearing different names suggest the necessity of some modification of rock-type nomenclature within outcrop boundaries.

In the different types of rocks the distribution of trace elements is, in general, more uniform than that of the major elements—suggesting that post-magmatic alteration processes were responsible for trace-element concentrations and that these processes had a regional character.

The preference of major and trace elements for the two main types of rocks was generally in accord with findings elsewhere. Of the trace elements, Li, Mo, Y, Pb and Ga were present in higher concentrations in the granites, and Co, Cr, Cu, Ni, Sr and V in the adamellites–granodiorites. The behaviour of Sn, Ba and Zr was unexpected, since these elements were more concentrated in rocks of low silica content.

Several linear relationships between elements were found which may be of importance concerning mineralization. These are highlighted by the correlation of Pb with Y and Li with Mo and Sr in the granites–leucoadamellites. In the adamellite–granodiorite group good correlations of Mo with Sn and Cu, Sn with Y, Co with Ni, and Cu with Ni and Sc were found.

The results suggest that the technique of cluster analysis is a useful tool not only in that it improves the mapping and geochemical characterization of rocks but also yields information which may be relevant to the processes of ore formation and exploration.

The New England batholith covers a large area, approximately 20 000 km^2, of northeastern New South Wales and southeastern Queensland (Fig.1) and outcrops discontinuously from Tamworth in the south to Warwick in the north.[1] To the east and west it is bordered by fault systems defining the Eastern and Western Serpentine Belts. The granitoid rocks of the batholith intrude into Palaeozoic rocks of the Central Complex.[2] Wilkinson[1] observed that, in general, the intrusions belonging to the batholith possess sharply defined contacts and well-delineated metamorphic aureoles with local evidences of metasomatism. He considered that the epithermal nature of mineralization,[3] the development in some areas of breccia veins and stockworks filled with pegmatite[4] and the presence of trydimite in bostonite dykes[5] indicate relatively high levels of emplacement.

Although clearly possessing a complex history of intrusion, the New England batholith is generally regarded as Permian in age.[6]

It was thought earlier that the different types of rocks were formed in a series of intrusions of

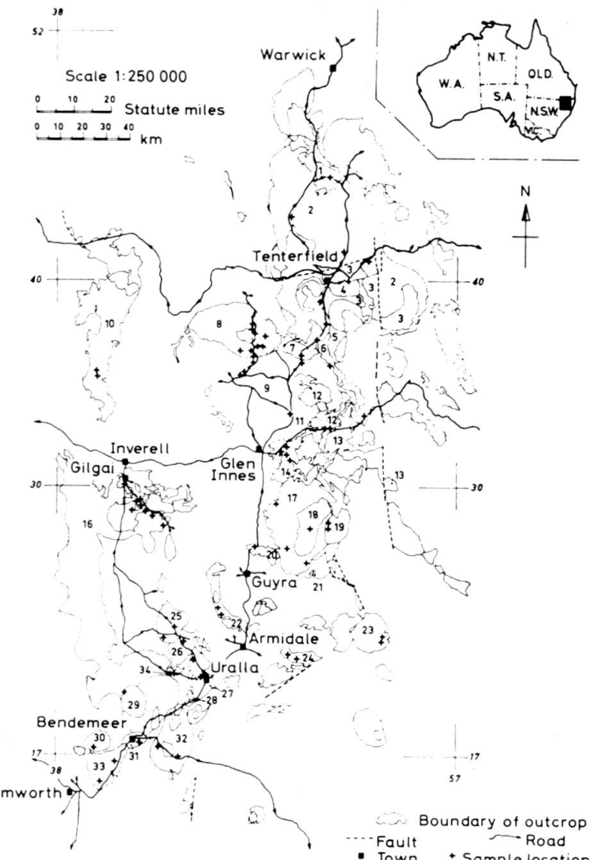

Fig. 1 Map of the New England igneous complex and sample locations. After Flinter.[24] Numbers refer to list of outcrops (see Table 1)

decreasing basicity with the following sequence: 'blue granite' (adamellite–porphyrite) 64% SiO_2; sphene–diorite–porphyry, 65–69% SiO_2; coarse-grained acid granites and adamellites, 70–73% SiO_2; and acid dykes and pegmatites. Later work has indicated that this scheme was oversimplified. Wilkinson[8] found that gabbros from which such rocks could have been derived by fractional crystallization are virtually absent in the New England batholith. He suggested that processes involving fractional crystallization of basaltic magma modified by sialic contamination also possess limitations in view of the relatively minor development of intermediate and basic rock types, and that it is more likely that partial melting of sialic crustal material was responsible for the formation of these rocks. Also, he found evidence for hybridism in several locations, which is inconsistent with normal crystallization from a more or less homogeneous melt. Published data indicate that the members of the batholith comprise a fairly well-defined calc-alkaline series, the geochemistry of which parallels well documented calc-alkaline intrusive sequences elsewhere. Differentiation index versus oxide plots by Flinter[9] also suggest that most members of the batholith are genetically related. It is evident, however, that further petrographic, mineralogical and geochemical data are needed to clarify some of the problems involved in determining the genesis of the batholith.

The predominant granitoid type in the batholith is *adamellite*, less commonly *granodiorite*, with modal quartz generally greater than 20%. *Granites* are less frequent and occur mostly in the Mole Tableland, as well as in the northern and eastern part of the batholith.

Many of the more than one hundred intrusions have been investigated in detail previously for their petrology and geochemistry. In the southern part of the batholith, Chappell[10] and Binns[11] worked in the Moonbi–Moore Creek–Bendemeer area. Flood[12] described the Walcha Road and Kilburnie adamellites. In the Uralla area Vernon[13] studied the granodiorite–adamellite intrusions, and Boesen[14] the Duval and Sunnyside adamellites. The petrology of several intrusions of adamellite–porphyrite around Dundee and Tent Hill has been evaluated by Wilkinson, Vernon and Shaw.[15] Joyce[16] demonstrated the chronological sequence of dyke intrusions in the Yarrowyck area. The western margin of the New England batholith has been studied by Lusk,[17] and Cotton[18] recognized the 'Tingha granite'.

In the eastern part of the batholith Binns[19] described the Round Mountain leucoadamellite and rocks in the Gibraltar Range; Gutsche[20] provided information on rocks in the north-eastern and western part of the batholith. The northern portion of the batholith was described by Shaw[21] and by Phillips,[22] and Lawrence[23] reported on igneous rocks of the Emmaville–Torrington district.

CSIRO investigations

It was clear from the work carried out so far that more information on the geochemistry of granitoid rocks was needed, and that relatively few of the studies mentioned above had been concerned with trace-element concentrations and mineralization in the area. In 1969, therefore, a broad programme of investigations was initiated in CSIRO with the aim of obtaining further geochemical data which might be useful in future mineral exploration. More than 130 samples have been collected and analysed for major- and trace-element contents. The modal composition of the rocks and their content of

*Table 1 List of outcrops sampled**

1.	Ruby Creek granite (1)	18.	Oban River leucoadamellite (1)
2.	Stanthorpe adamellite (2)	19.	Kookabookra adamellite (2)
3.	Bungulla porphyritic adamellite (2)	20.	Llangothlin adamellite (1)
4.	Dundee adamellite–porphyrite from the Tenterfield mass (1)	21.	Day's Creek gabbro (1)
5.	Sandy Flat adamellite (2)	22.	Mt Duval adamellite (2)
6.	Mt Jonblee leucoadamellite (1)	23.	Round Mountain leucoadamellite (2)
7.	Bolivia Range leucoadamellite (1)	24.	Hillgrove adamellite (2)
8.	Mole granite (9)	25.	Gwydir River adamellite (2)
9.	Porphyrite from Emmaville (4)	26.	Yarrowyck granodiorite (1)
10.	Ashford Granite (2)	27.	Uralla granodiorite (2)
11.	Dundee adamellite–porphyrite (1)	28.	?Wilhelmshohe tonalite (1)
12.	Kingsgate granite (2)	29.	Unnamed granitoid (Watson's Creek) (1)
13.	Mt Mitchell adamellite (2)	30.	Attunga Creek adamellite (1)
14.	Shannonvale granodiorite (4)	31.	Bendemeer adamellite (1)
15.	Tingha granite (6)	32.	Walcha Road adamellite (2)
16.	Unnamed granitoid (Gilgai) (2)	33.	Moonbi adamellite (2)
17.	Ward's Mistake adamellite (5)	34.	Balala granodiorite (1)

*Outcrops run from north to south. Number of samples collected from each outcrop is given in brackets.

accessory minerals were determined. The first results of this work have been reported by Flinter.[24]

The present paper gives the results of the statistical analysis of geochemical data available so far on 72 of the above samples, from 34 outcrops (Table 1), the majority of the samples comprising granites, adamellites and granodiorites.* The suite of rocks also contained porphyrites, as well as a tonalite and a gabbro (Table 1). The location of the samples is shown in Fig. 1.

Owing to the complexity of the batholith, the naming of certain types of rocks and the delineation of some outcrops presented problems in the past and it was hoped that an objective comparison by use of the technique of 'cluster analysis' would not only help in rock classification but would also yield information on inter-element associations and on the likelihood of mineralization.

Statistical method

Sample classification (*Q*-mode analysis) was carried out by use of a CDC 3600 computer. Program 'MULTCLAS' was chosen for this analysis, as described by Lance and Williams.[25] The criterion for similarity was based on the 'Euclidean distance', standardized as defined by Burr.[26] In calculating the 'distance' between two samples the differences between standardized variables were squared and summed and the square root of the sum was divided by the number of variables. From the 'distance' matrix obtained for all samples a hierarchical dendrogram was constructed by use of the 'group average' technique,[27] which was based on the method of Sokal and Michener.[28]

The use of the Euclidean distance in *Q*-mode analysis requires that all the variables be orthogonal (i.e. not related). To satisfy this requirement an *R*-mode analysis of the variables was carried out prior to the *Q*-mode analysis to obtain information on correlation between variables. A matrix of correlation coefficients was obtained and a dendrogram was constructed by use of the sorting technique. After consideration of the statistical significance of the correlation coefficient, a limiting value of 0·7 was accepted to distinguish between orthogonal and closely related variables. Only the orthogonal variables were used for the subsequent *Q*-mode analysis of samples.

The construction of dendrograms (both in *Q*- and *R*-mode analysis) tends to introduce some distortion into the initial inter-element measures. A quantitative estimate of the degree of distortion was obtained by calculating the correlation between original data in the similarity matrix and those read from the dendrogram (the higher the correlation, the less the degree of distortion).

The sample groups established by *Q*-mode analysis were then statistically characterized by calculating standard deviations, mean values, coefficients of variations and inter-element correlations on a CDC 3200 computer.

*The Geological Survey of New South Wales nomenclature has been used in this paper.

Results and discussion

Ranges, mean values and coefficients of variation for different elements

In order to obtain a geochemical characterization of the samples the range of concentration, mean value ($\bar{x}$), standard deviation (σ) and coefficient of variation ($CV = \frac{\sigma}{\bar{x}} \cdot 100, \%$) were calculated for each major-element oxide and trace element (Table 2), the results obtained being illustrated in Fig. 2. It is seen that the range of concentration was generally wide with respect to mean values, which was caused by the presence of a gabbro and a tonalite in the samples. These, however, were useful in providing extreme values when the trends of trace-element distributions were established. Fig. 2 also indicates that the mean values were obtained with a reasonable degree of accuracy at the 95% confidence level. Mo and Sn (which represented the main interest in the study from the point of view of mineralization) gave mean values of 1·6 and 4·6 ppm, respectively. From Fig. 3 the number of samples required for obtaining a mean value with a relative accuracy of +10% can be derived at two confidence levels, 95 and 99%.[29] The values of $t/\sqrt{n}$* were calculated from corresponding mean values and standard deviations for each major-element oxide

*Table 2 Summary of statistical data**

	All samples (72)			Group 1 (32)			Group 2 (46)			Group (3) 24		
	Mean	σ	*CV*	Mean	σ	*CV*	Mean	σ	*CV*	Mean	σ	*CV*
SiO_2	69·4	5·1	7·3	67·7	1·9	2·8	67·2	2·4	3·6	75·0	1·4	1·9
Al_2O_3	14·4	1·2	8·5	14·7	0·5	3·4	15·0	0·7	4·7	13·0	0·6	4·6
FeO	3·0	1·4	46	3·3	0·7	21·2	3·5	0·8	22·9	1·6	0·5	31·2
MgO	1·2	1·4	114	1·4	0·5	35·7	1·4	0·5	35·7	0·17	0·16	94·0
CaO	2·3	1·7	74	2·6	0·5	19·2	2·7	0·7	25·9	0·8	0·5	62·6
Na_2O	3·3	0·4	11	3·2	0·2	6·3	3·3	0·3	9·1	3·4	0·3	8·8
K_2O	4·1	0·7	18	4·1	0·4	9·8	4·0	0·4	10·0	4·6	0·3	6·5
H_2O	0·8	0·3	30	0·9	0·2	22·0	0·9	0·2	22·0	0·6	0·15	25
TiO_2	0·38	0·2	58	0·5	0·1	20·0	0·5	0·1	20·0	0·14	0·06	42·9
P_2O_5	0·13	0·08	59	0·17	0·05	29·4	0·17	0·06	35·3	0·06	0·04	66·6
MnO	0·08	0·03	37	0·08	0·01	12·5	0·09	0·02	22·0	0·05	0·02	40·0
Ba	740	400	54	677	244	36·0	793	416	52·5	500	173	34·6
Co	12	5·2	43	11	3·8	34·5	11	4	36·3	—	—	—
Cr†	206	82	40	213	62	29·1	205	65	31·7	188	69	36·7
Cu	11	12	103	13	8·3	63·9	12	7·8	65·0	5·2	4·4	84·7
Ga	20	6·2	31	20	5·3	26·5	19	5·4	28·4	21	7·8	37·1
Li	59	87	147	34	13·6	40·0	49	89	182	81·7	82	100
Mo	1·6	1·0	62	1·4	0·7	50·0	1·4	0·7	50·0	1·9	1·4	73·7
Ni	18	12	70	19	5·9	31·0	18	6·2	34·4	12·1	3·6	29·7
Pb	21	7·3	34	22	6·7	30·5	20	7·0	35·0	23	7·5	32·6
Sc	11	3·6	32	10	1·3	13·0	10	1·5	15·0	10	—	—
Sn	4·6	4·5	98	5	5·7	114·0	4·6	5·2	113	4·6	3·0	65·3
Sr	450	154	34	500	181	36·0	491	169	34·4	372	83	22·3
V	53	38	72	57	21	36·9	66	31	46·9	16	6·9	43·2
Y	27	21	79	19	6·0	31·6	19	6·1	32·1	42	31	73·8
Zr	151	63	42	178	64·9	36·5	170	63	37·1	126	46	36·5

*Mean values for major-element oxides: per cent by weight; for trace elements, ppm; σ, standard deviation; *CV*, coefficient of variation: $\frac{\sigma}{\bar{x}} \cdot 100, \%$.

Group 1: Adamellites+granodiorites; Group 2: adamellites+granodiorites+porphyrites; Group 3: granites +leucoadamellites. Numbers in brackets refer to samples used in the calculations.

†Owing to contamination during crushing, Cr values are not reliable.

*For definition see Fig. 3.

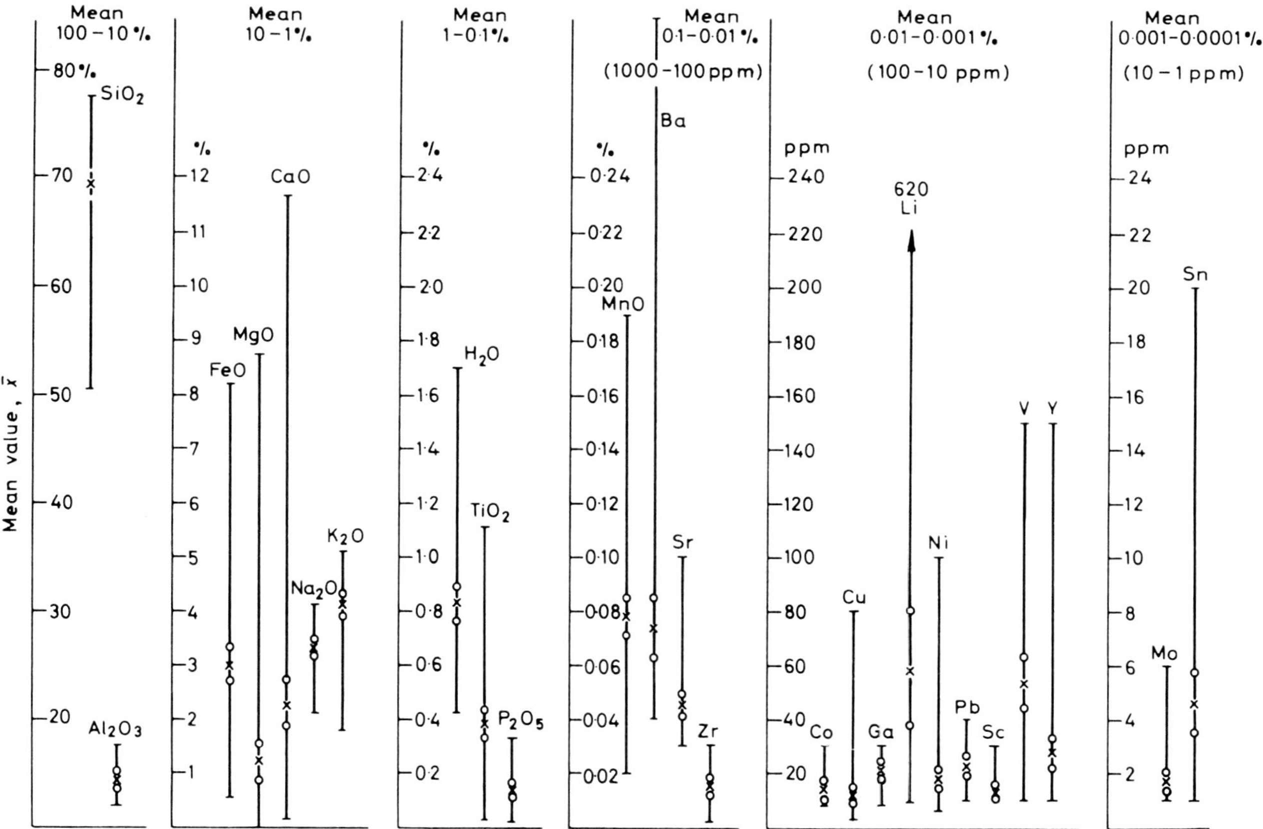

Fig. 2 Composition of rocks (72 samples from New England igneous complex, New South Wales): ×, *mean content*; |, *range*; o|o *95% confidence limit for mean value*, $(v=(t_{95}/\sqrt{n}).CV)$; *v, variation around mean, relative per cent*; t_{95}, *Student's t value at 95% confidence level*; $CV=(\sigma/\bar{x})\cdot 100$, *coefficient of variation*, %; σ, *standard deviation*; $\bar{x}$, *arithmetic mean value*

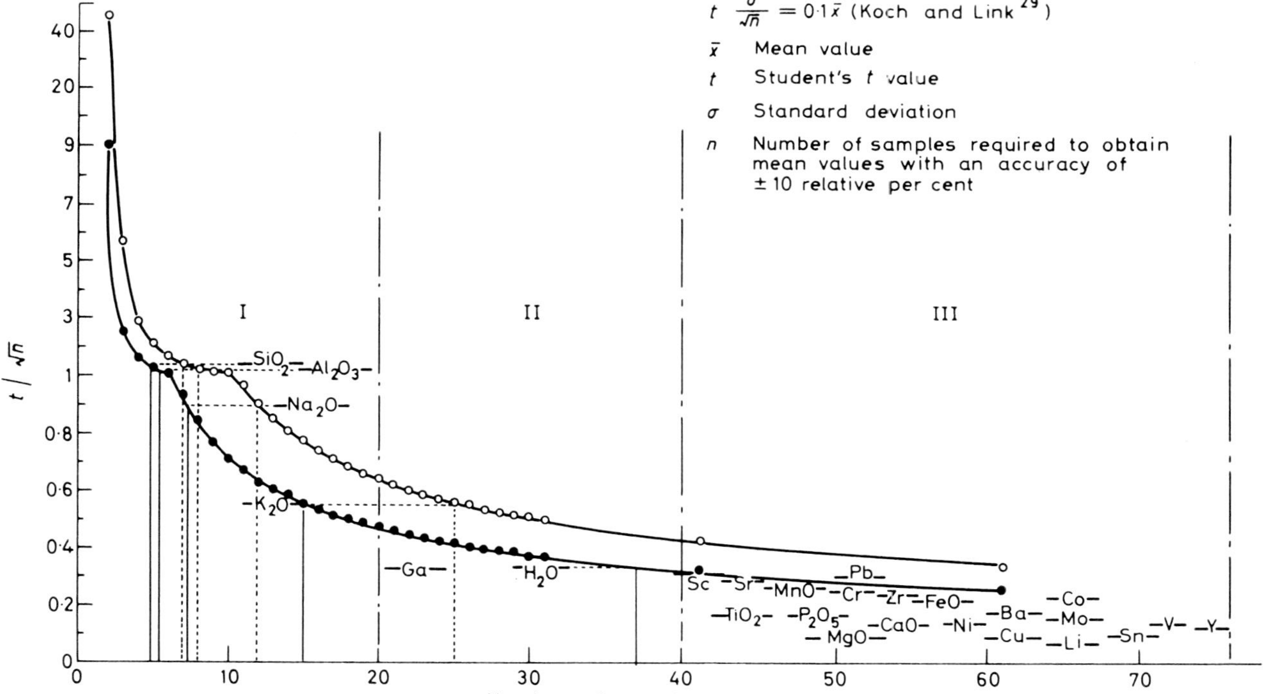

Fig. 3 Number of samples required to calculate mean values with a relative accuracy of ±10% at two confidence levels. Confidence levels: ● 95%; ○ 99%

and trace element, and these are shown as horizontal lines. The intersection of these lines with the two curves in Fig. 3 (for 95 and 99% confidence levels) projected to the abscissa gives the required number of samples. For a number of variables (e.g. SiO_2, Al_2O_3, K_2O, Na_2O, MnO, FeO, H_2O, Ga, Sc, Sr, Pb, Cr, Zr and Co) the 72 samples are sufficient to achieve the above accuracy in the mean value, but for the other variables more samples are required. In Fig. 4 the coefficient of variation values are shown. These are characteristic of the 'spread' (i.e. degree of scatter) of the elements, and they are low (< 20%) for the main constituents of the rocks (SiO_2, Al_2O_3, K_2O and Na_2O), but high for the 'mafic' oxides (FeO, TiO_2, CaO, MgO) and for the trace elements (62% for Mo and 98% for Sn). For most elements the values calculated were well within the coefficients of variation derived by Hazen and Meyer[30] from more than 50 000 observations on 484 ore deposits.

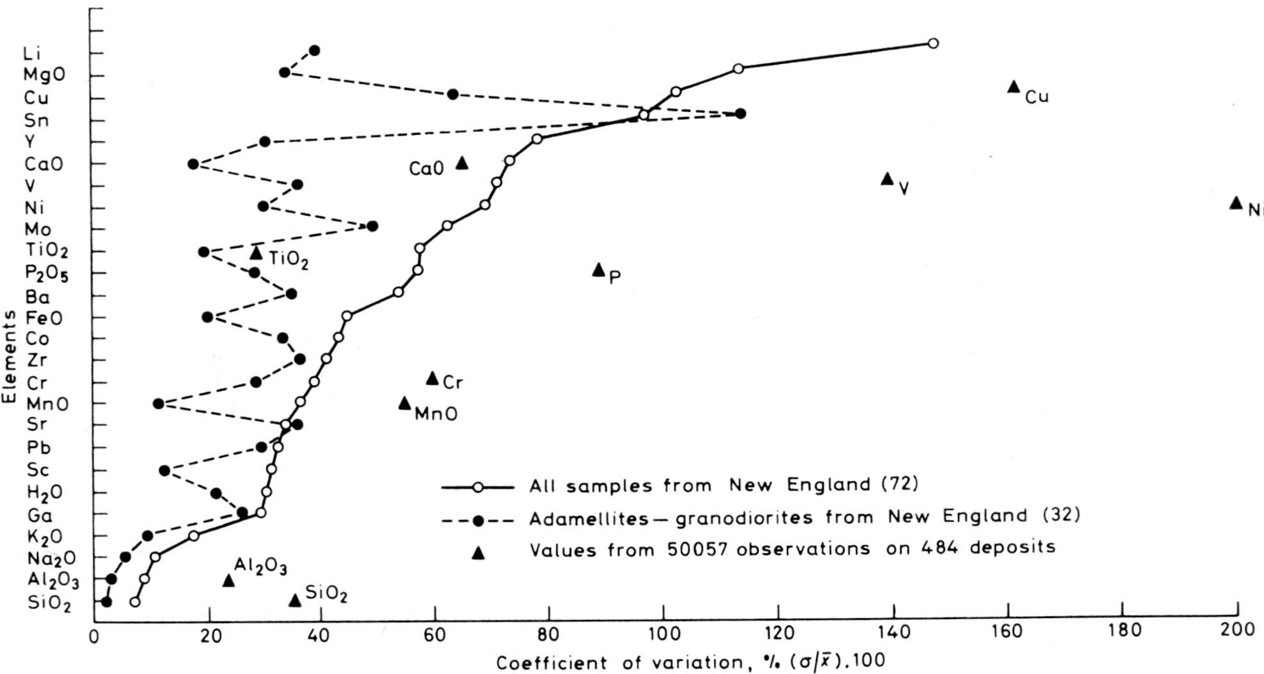

Fig. 4 Dispersion of major-element oxides and trace elements

Cluster analysis of samples based on major-element oxide content

As a first step in the cluster analysis of samples an *R*-mode analysis was carried out to obtain information on inter-element correlations and to select those variables for the *Q*-mode analysis which are orthogonal. A general idea about these correlations can be obtained from the relations of SiO_2 content with the rest of the major-element oxides and with trace elements, as shown in Fig. 5. Only a few positive (direct) correlations were found, namely (in decreasing order) K_2O and Na_2O for the major-element oxides, and Y, Li, Mo, Pb and Ga for the trace elements. The remaining major-element oxides and trace elements are inversely related in amount to the SiO_2 content of the rocks.

The complete dendrogram of inter-element relations is shown in Fig. 6. It should be noted that this relationship concerns only the amounts present *in the rock* and does not necessarily relate to the element correlations in minerals. The elements not related at the 0·38 and 0·70 correlation coefficient levels can be read. The elements not related at the $r = 0{\cdot}70$ level were chosen as being statistically very highly significant (level of significance 99·9%) for the purpose of *Q*-mode analysis.

When the four major-element oxides (SiO_2, Al_2O_3, Na_2O and TiO_2) were considered the cluster pattern shown in Fig. 7 was obtained. The correlation coefficient between values in the 'distance' matrix and those read from the dendrogram was 0·82. The high correlation coefficient indicates that the degree of 'distortion' in constructing the dendrogram was not excessive. The dendrogram (Fig. 7) shows that three distinctly different types of rocks are present. In the first group are the adamellites–granodiorites–porphyrites; in the second the granites and leucoadamellites; and in the third a gabbro and a tonalite

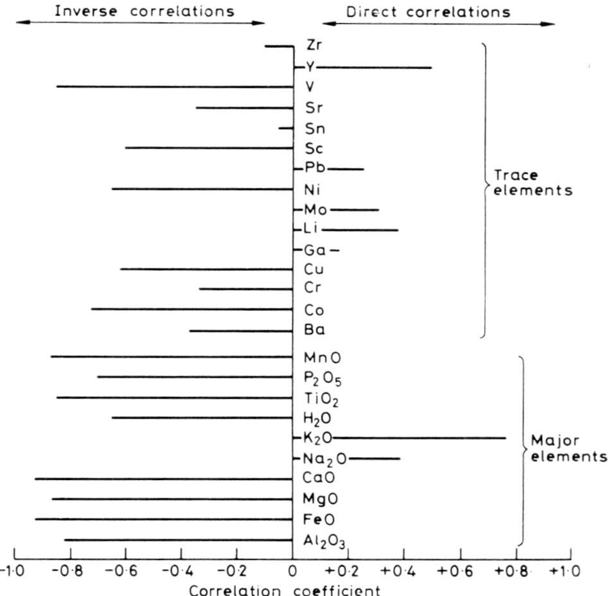

Fig. 5 Correlation of major-element oxides and trace elements with silica content (72 samples from New England igneous complex, New South Wales)

which were very different from the rest of the samples. In the first two groups Euclidean distance values were <0·4 between the 'similar' types. Each group is characterized by the number and type of samples it contains and by the mean SiO_2 content. The symbols defined in Fig. 7 are used in Fig. 8 to show the distribution of the different types of rocks in the outcrops sampled.

The results in some cases revealed inconsistencies in the nomenclature of intrusions, as shown, on the one hand, by the Tingha granite, where all samples are geochemically similar to adamellites in other outcrops, and, on the other hand, by samples in the Stanthorpe adamellite, which had granitic composition. In addition, they indicated the necessity of taking further samples and possibly modifying the contact boundaries within several outcrops. For example, of four samples collected from the Ward's Mistake adamellite one showed a granodioritic composition and another resembled the porphyrites in composition. Similarly, the Shannonvale granodiorite had samples of granodioritic and adamellitic composition, and in the Mole granite two granodioritic samples were collected. The rest of the outcrops appeared to be fairly homogeneous (based on the number of samples analysed) as far as major-element geochemistry is concerned. The majority of the samples—about 66%—fell in the category of adamellites–granodiorites–porphyrites; the minority were true granites and leucoadamellites.

When the whole composition of rocks was taken into consideration, i.e. the *Q*-mode analysis

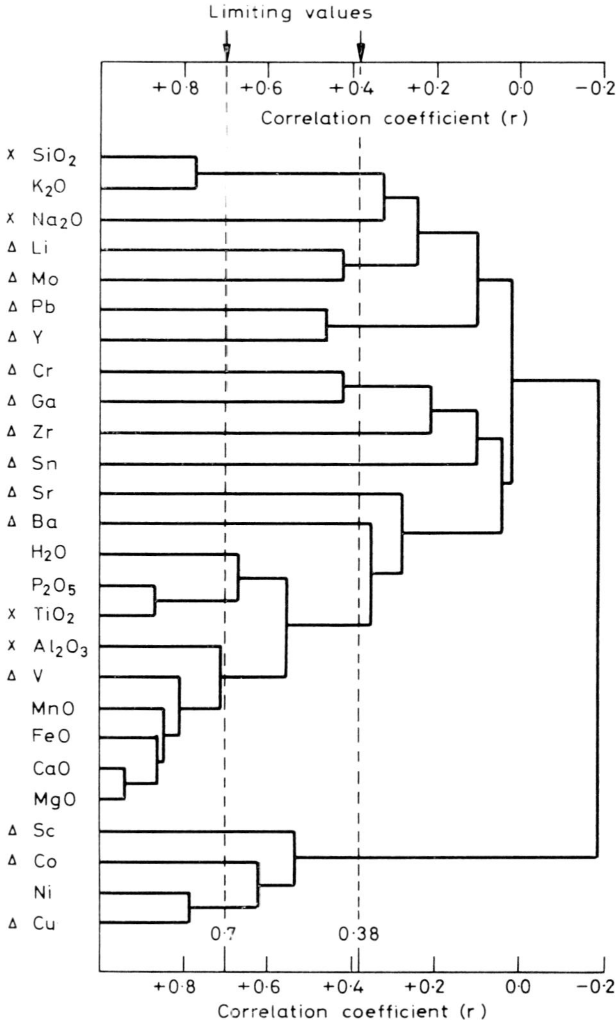

Fig. 6 Dendrogram showing element associations (R-mode analysis). ×, major elements and △, trace elements selected for Q-mode analysis at r = 0·7 level

was carried out with major *and* trace elements, a different pattern of clustering emerged (see Fig. 9). Most of the granites are interspersed with the adamellites–granodiorites–porphyrites. There are only five granites which are very

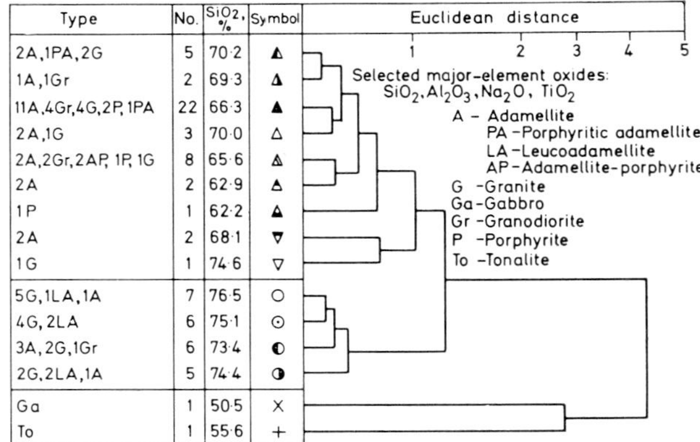

Fig. 7 Q-mode analysis of selected major-element oxides (r = 0·7) (72 samples)

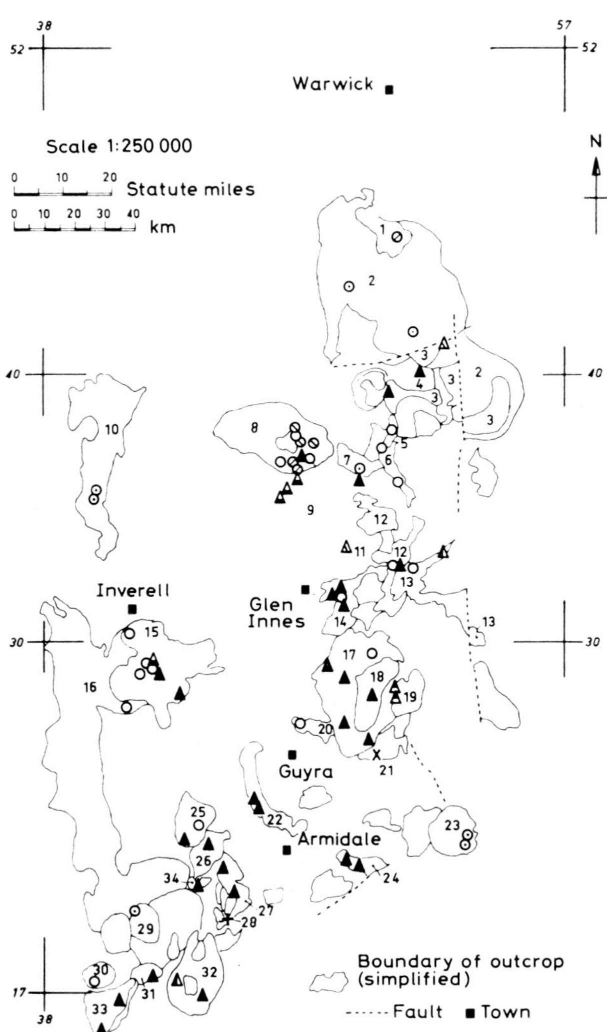

Fig. 8 Clustering of samples based on major-element oxide content. Numbers refer to list of outcrops (see Table 1) (for symbols see Fig. 7)

different from those in the first group, and these were all collected from the Mole batholith. The samples of gabbro and tonalite again showed a distinct difference from the rest of the samples.

Type	No.	SiO_2, %	Symbol
6G,5A,1LA,1Gr	13	70·0	○
3G	3	76·5	○
13A,6Gr,3G,1PA, 1AP, 1P	25	67·0	▲
1P, 1AP	2	64·2	△
1A,1PA	2	68·5	△
2A	2	70·8	△
4LA,4G,3A,1Gr	12	74·7	⊙
2A,1G	3	63·8	△
2P,1G	3	66·1	△
1G	1	76·5	⊘
1G	1	75·7	⊘
1G	1	74·6	⦸
2G	2	76·6	⦸
1To	1	55·6	+
1Ga	1	50·5	×

Euclidean distance: 1 2 3 4 5 6

Selected major-element oxides: SiO_2, Al_2O_3, Na_2O, TiO_2

Selected trace elements: Ba, Co, Cr, Cu, Ga, Li, Mo, Pb, Sc, Sn, Sr, Y, Zr

Fig. 9 Q-mode analysis of selected major-element oxides and trace elements (r=0·7) (72 samples) (for abbreviations see Fig. 7)

Distribution of trace elements

Since the cluster patterns obtained with major plus trace elements suggested that the factors responsible for the distribution of trace elements were distinctly different from those which determined the major-element composition of the rocks, a *Q*-mode analysis of the samples was carried out by use of selected trace elements only. Orthogonal trace elements were used as selected from the *R*-mode analysis (see Fig. 6). They were Ba, Co, Cr, Cu, Ga, Li, Mo, Pb, Sc, Sn, Sr, V, Y and Zr. Fig. 10 shows the cluster patterns obtained; they are essentially similar to those given in Fig. 9. The majority of granites was similar to the adamellites–granodiorites–porphyrites, although the degree of similarity was less ('distance' about 0·6–0·9) than between similar groups based on major-element oxide content. Five granites from the Mole batholith and one sample of Ruby Creek granite showed distinctly different trace-element composition, and the same applied to the samples of gabbro and tonalite.

There are three possible explanations of the similarity of trace-element composition in the different types of rocks: (*a*) the intrusions are coeval, and, thus, all 'magmas' contain similar trace-element contents; (*b*) the trace-element content of rocks was not determined by the composition of the magmas but by post-magmatic alteration processes, which have affected most

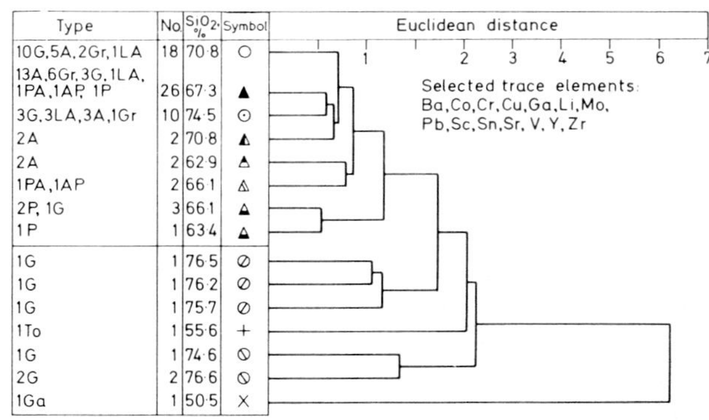

Type	No.	SiO_2, %	Symbol
10G,5A,2Gr,1LA	18	70·8	○
13A,6Gr,3G,1LA, 1PA,1AP, 1P	26	67·3	▲
3G,3LA,3A,1Gr	10	74·5	⊙
2A	2	70·8	△
2A	2	62·9	△
1PA,1AP	2	66·1	△
2P, 1G	3	66·1	△
1P	1	63·4	△
1G	1	76·5	⊘
1G	1	76·2	⊘
1G	1	75·7	⊘
1To	1	55·6	+
1G	1	74·6	⦸
2G	2	76·6	⦸
1Ga	1	50·5	×

Fig. 10 Q-mode analysis of selected trace elements (r=0·7) (72 samples) (for abbreviations see Fig. 7)

of the region; (*c*) the magmas intruded at different times carried varying contents of trace-elements, but the partitioning of these during crystallization reduced the differences of trace-element concentrations in the end-products.

Of these possibilities, the first is the least likely because, even if the magmas originally contained similar amounts of trace elements, their final content would have been strongly affected by partitioning during crystallization into the different rock types. Similarly, it is unlikely that crystallization of magmas carrying widely different trace-element contents (as in (*c*)) would result in rocks with essentially similar trace-element distributions. The most likely explanation is the second, which involves regional metasomatic processes affecting in a similar way the trace-element distributions in most of the rocks, with the exception of the Mole batholith.

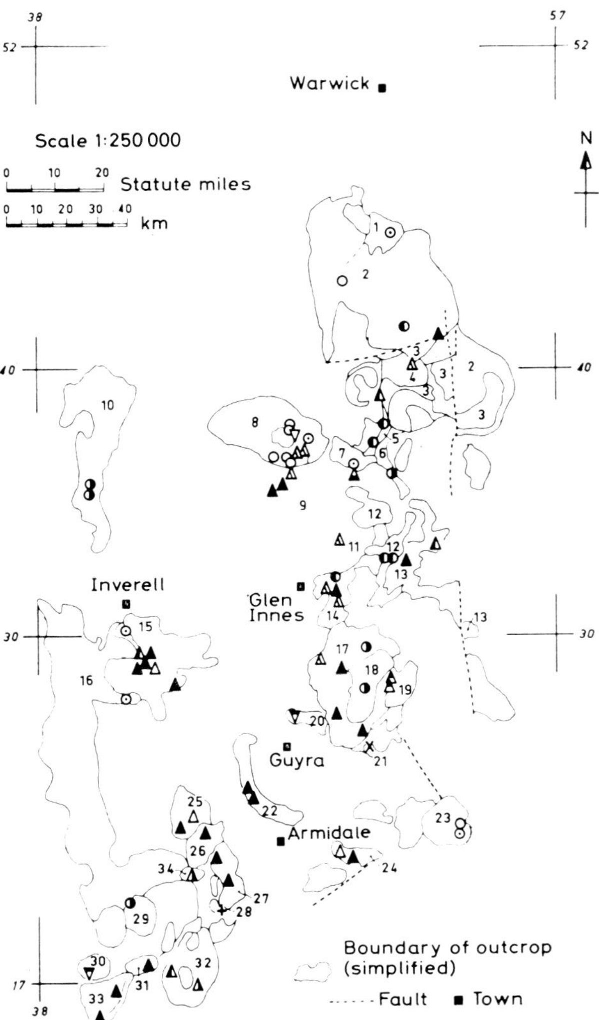

Fig. 11 Clustering of samples based on trace-element content. Numbers refer to list of outcrops (see Table 1) (for symbols see Fig. 10)

In Fig. 11 the similarity of samples with regard to trace-element content can be seen (symbols used are defined in Fig. 10). As was mentioned earlier, the main feature is the general similarity of trace-element distribution in the different types of rocks. In certain outcrops, however, where one would expect the trace-element concentrations to be the same, there are differences, which are thought to be indicative of local alteration events. One example of this is the Tingha granite, where the major-element composition is that of adamellites, but the trace-element distributions in some of the samples are similar to those found in the neighbouring unnamed granite (at Gilgai). Differences were also noted in the trace-element patterns of samples within the Ward's Mistake adamellite, Shannonvale granodiorite, Kingsgate granite and Mole granite. It is noteworthy that 64 of the 72 samples are in the main group where the degree of dissimilarity is less than that between granites–leucoadamellites and adamellites–granodiorites based on major-element oxide content.

Inter-element correlations

To obtain information on the preference of various elements for the two main rock types two groups of samples were chosen, based on the cluster patterns shown in Fig. 8: (*a*) 32 adamellites plus granodiorites (mean SiO_2 content, 67·7% —the 'core' of the first group); and (*b*) 24 granites plus leucoadamellites (mean SiO_2 content, 75·0%). The mean values calculated for the above groups, together with those obtained for all 72 samples, are shown in Table 2. The ratios of

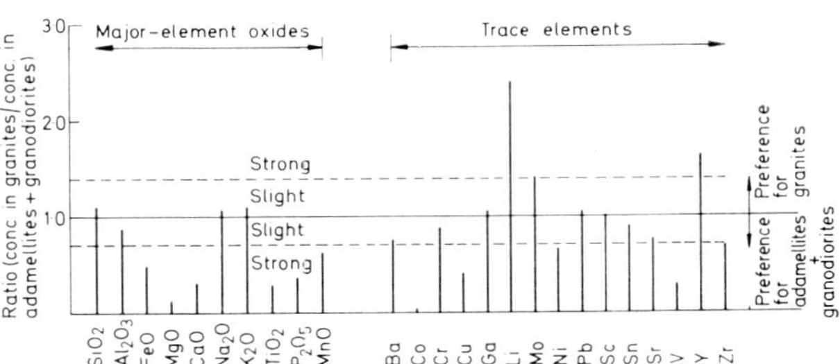

Fig. 12 Element ratios in the two main groups of granitoid rocks

the mean concentration of each element in granites plus leucoadamellites to that in adamellites plus granodiorites are shown in Fig. 12.

Preferences were shown as indicated below.

Major elements

For *granites*—slight preference shown by SiO_2, Na_2O and K_2O

For *adamellites*—slight preference shown by Al_2O_3; strong preference by FeO, MgO, CaO, TiO_2, P_2O_5 and MnO

Trace elements

For *granites*—strong preference shown by Li, Mo and Y; slight preference by Ga and Pb; no preference by Sc

For *adamellites*—slight preference shown by Ba, Cr, Sn, Sr and Zr; strong preference shown by Co, Cu, Ni and V

These preferences conform, in general, with observations made on many rocks from other districts. The preference of Sn for adamellites, however, is unexpected, and that of Ba and Zr for adamellites also appears to be anomalous.

After the main characteristic groups in the samples had been defined, inter-element correlations in the rocks were re-examined to determine whether such correlations can be more meaningful for a homogeneous than for a heterogeneous group of samples. The more homogeneous nature of the adamellite–granodiorite group, for example, is indicated by the lower values of the coefficient of variation for various elements (Fig. 4).

Of the complete *linear* correlation coefficient matrices obtained, those which indicate definite correlations are listed in Table 3 ((*a*) and (*b*)), which also give the correlation coefficients for all

*Table 3 Correlations within groups**

(a) Correlations with major-element oxides

	Group 1	Group 2	Group 3	All samples
SiO_2–FeO	−0·84 (32)	−0·91 (46)	0·09 (24)	−0·93 (72)
SiO_2–MgO	−0·80 (32)	−0·73 (46)	−0·69 (24)	−0·87 (72)
SiO_2–CaO	−0·81 (32)	−0·90 (46)	−0·63 (24)	−0·94 (72)
SiO_2–K_2O	0·07 (32)	0·38 (46)	−0·25 (24)	0·77 (72)
SiO_2–TiO_2	−0·77 (32)	−0·85 (46)	−0·68 (24)	−0·85 (72)
SiO_2–MnO	−0·56 (32)	−0·73 (46)	−0·21 (24)	−0·87 (72)
SiO_2–Mo	0·00 (29)	−0·01 (40)	0·53 (21)	0·32 (62)
SiO_2–Pb	−0·35 (18)	0·03 (25)	0·37 (20)	0·25 (45)
SiO_2–Sn	−0·34 (32)	−0·21 (45)	0·14 (24)	−0·05 (70)
SiO_2–V	−0·60 (32)	−0·70 (45)	−0·70 (21)	−0·86 (68)
SiO_2–Y	−0·15 (32)	−0·13 (46)	0·47 (24)	0·50 (71)
FeO–MgO	0·59 (32)	0·70 (46)	0·18 (24)	0·86 (72)
FeO–CaO	0·57 (32)	0·78 (46)	−0·21 (24)	0·87 (72)
FeO–TiO_2	0·83 (32)	0·87 (46)	0·14 (24)	0·86 (72)
FeO–Sn	0·55 (32)	0·36 (45)	−0·33 (24)	0·12 (70)
FeO–V	0·54 (32)	0·68 (45)	0·22 (21)	0·86 (68)
MgO–CaO	0·67 (32)	0·75 (46)	0·57 (24)	0·94 (72)
MgO–TiO_2	0·62 (32)	0·76 (46)	0·77 (24)	0·65 (72)
MgO–Co	0·48 (23)	0·52 (33)	—	0·73 (35)
MgO–Ni	0·55 (32)	0·50 (46)	0·08 (22)	0·86 (70)
MgO–V	0·34 (32)	0·49 (46)	0·75 (21)	0·73 (68)
MgO–Zr	0·27 (32)	0·13 (46)	0·11 (24)	−0·06 (71)
CaO–TiO_2	0·62 (32)	0·85 (46)	0·55 (24)	0·70 (72)
CaO–P_2O_5	0·72 (32)	0·75 (46)	0·38 (24)	0·58 (72)
CaO–MnO	0·63 (32)	0·76 (46)	0·65 (24)	0·86 (72)
CaO–Pb	0·57 (18)	0·10 (25)	−0·28 (20)	−0·22 (45)
CaO–V	0·68 (32)	0·76 (45)	0·50 (21)	0·81 (68)
CaO–Zr	0·32 (32)	0·08 (46)	0·17 (24)	0·00 (71)
Na_2O–K_2O	−0·59 (32)	−0·44 (46)	−0·37 (24)	0·26 (72)
Na_2O–Mo	0·03 (29)	0·08 (40)	0·43 (21)	0·22 (62)
Na_2O–Y	−0·39 (32)	−0·32 (46)	0·24 (24)	0·14 (71)
Na_2O–Zr	−0·37 (32)	−0·37 (45)	−0·20 (24)	−0·13 (71)
K_2O–Ba	0·40 (30)	0·01 (42)	−0·53 (7)	−0·07 (50)
K_2O–Sc	−0·29 (28)	−0·30 (39)	—	−0·69 (44)
K_2O–Sn	+0·01 (32)	+0·02 (45)	−0·15 (24)	0·02 (70)
MnO–Pb	0·46 (18)	0·09 (25)	−0·42 (20)	−0·26 (45)
MnO–V	0·47 (32)	0·66 (45)	0·03 (21)	0·80 (68)

Table 3—continued

(b) Correlations between trace elements

	Group 1	Group 2	Group 3	All samples
Ba–Ga	0·09 (30)	−0·07 (42)	0·64 (7)	−0·14 (50)
Ba–Pb	0·30 (16)	−0·20 (23)	0·51 (6)	−0·10 (29)
Ba–Sc	−0·37 (26)	−0·14 (37)	—	−0·21 (38)
Ba–Sn	0·49 (30)	0·12 (41)	−0·46 (7)	0·15 (49)
Ba–Sr	0·47 (26)	0·19 (34)	0·64 (7)	0·28 (42)
Ba–V	0·27 (30)	0·45 (42)	0·54 (7)	0·43 (50)
Co–Cu	0·10 (23)	0·03 (33)	—	0·59 (35)
Co–Ga	−0·33 (23)	−0·45 (33)	—	−0·48 (35)
Co–Ni	0·23 (23)	0·16 (33)	—	0·65 (35)
Co–Pb	−0·48 (11)	−0·22 (16)	—	−0·22 (16)
Cu–Mo	0·29 (29)	0·28 (40)	0·04 (21)	−0·02 (18)
Cu–Ni	0·24 (32)	0·28 (46)	−0·37 (22)	0·78 (70)
Cu–Sc	0·49 (28)	0·32 (39)	—	0·64 (44)
Ga–Li	−0·54 (32)	−0·07 (46)	−0·32 (24)	−0·14 (72)
Ga–Ni	0·38 (32)	0·18 (46)	0·01 (22)	−0·07 (70)
Ga–Y	0·37 (32)	0·48 (46)	0·38 (24)	0·36 (71)
Ga–Zr	0·48 (32)	0·45 (45)	0·38 (24)	0·35 (71)
Li–Mo	−0·07 (29)	−0·08 (40)	0·47 (21)	0·42 (62)
Li–Sn	0·00 (32)	−0·07 (45)	−0·02 (24)	−0·06 (70)
Li–Sr	−0·16 (26)	−0·14 (35)	0·57 (18)	−0·08 (54)
Li–V	−0·36 (32)	−0·34 (45)	−0·41 (21)	−0·39 (68)
Mo–Sn	0·42 (29)	0·34 (40)	0·12 (21)	0·18 (62)
Ni–Pb	0·19 (18)	0·42 (25)	−0·26 (18)	0·06 (43)
Ni–Sc	0·12 (28)	−0·11 (39)	—	0·58 (44)
Ni–V	0·04 (32)	−0·11 (45)	−0·07 (19)	0·45 (66)
Ni–Y	0·24 (32)	0·11 (46)	−0·14 (22)	−0·26 (69)
Ni–Zr	0·58 (32)	0·43 (45)	0·22 (22)	−0·04 (69)
Pb–Sn	−0·42 (18)	−0·37 (24)	0·26 (20)	−0·12 (44)
Pb–Sr	0·38 (14)	0·25 (19)	−0·07 (16)	0·02 (35)
Pb–V	0·20 (18)	−0·15 (25)	−0·21 (17)	−0·23 (42)
Pb–Y	−0·29 (18)	−0·45 (25)	0·72 (20)	0·46 (45)
Pb–Zr	0·38 (18)	0·28 (25)	0·17 (20)	0·15 (45)
Sc–Y	0·24 (28)	0·29 (39)	—	−0·06 (43)
Sn–Y	0·31 (32)	0·28 (45)	0·19 (24)	0·12 (70)
Sr–V	0·40 (26)	0·33 (34)	0·28 (15)	0·44 (50)
Y–Zr	0·51 (32)	0·47 (45)	0·13 (24)	−0·03 (70)

*Correlations are characterized by correlation coefficient. Negative values indicate inverse correlation.

Group 1: adamellites–granodiorites; Group 2: adamellites–granodiorites–porphyrites; Group 3: granites–leucoadamellites.

Numbers in brackets refer to samples used for correlation calculations.

For significance of correlation coefficients see Appendix.

samples. (For the significance of correlation coefficients see Appendix.)

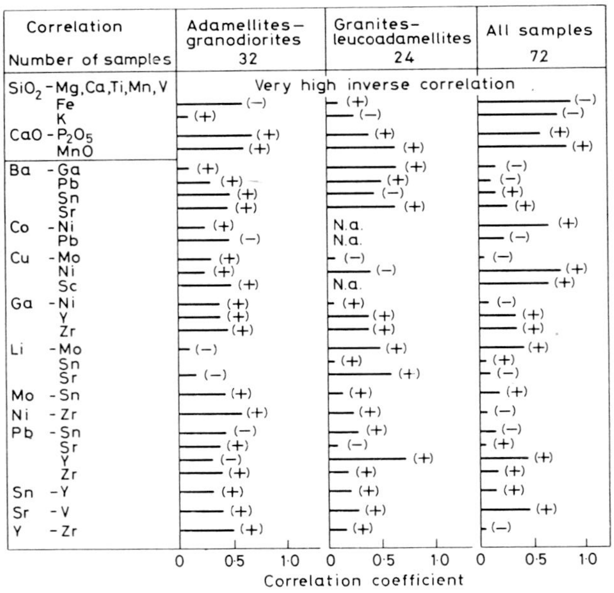

Fig. 13 Inter-element correlations within groups: (+) direct relation; (−) inverse relation; N.a., correlation coefficient is not available

The more important correlations are shown in Fig. 13. SiO_2 showed a very high inverse correlation with Mg, Ca, Ti, Mn and V for both rock types. The SiO_2–Fe correlation was high and inverse for adamellites and for all the samples, but not for granites. The only high positive correlations found for major-element oxides were between CaO and P_2O_5 and CaO and MnO, and these were valid for both rock types. Statistically significant positive *linear* correlations between trace elements (in alphabetical order) were found as indicated below.

In *granites–leucoadamellites*—Ba with Ga, Pb and Sr; Ga with Y and Zr; Li with Mo and Sr; and Pb with Y (and to a smaller extent with Sn)

In *adamellites–granodiorites*—Ba with Pb, Sn, Sr; Co with Ni; Cu with Mo, Ni, Sc; Ga with Ni, Y, Zr; Mo with Sn; Ni with Zr; Pb with Sr, Zr; Sn with Y; Sr with V; Y with Zr.

These correlations are important because they may prove helpful in finding association of certain minerals. In assessing their practical significance, however, it must be remembered that the value of the correlation coefficient depends on the range of concentrations of the elements in question and *sensu stricto* is valid only for the range represented by a particular group of samples. As was stated previously, the correlation coefficients given for 'homogeneous' groups of samples can be regarded as more meaningful than those calculated for 'heterogeneous' groups. So far, only linear correlation coefficients have been calculated and the absence of linear relations does not preclude the existence of more complicated curvilinear correlations, which will be investigated in further detail later.

As was mentioned earlier, the main interest in the general study of these granitoid rocks was in the occurrence of tin–molybdenum mineralization. In the granites and leucoadamellites, where cassiterite would be expected to occur, no close correlation between tin and other elements was found, although a positive correlation of Sn with SiO_2 and Li was expected. On the other hand, Mo showed a good correlation with Li (correlation coefficient, 0·47; level of significance, 97%).

In the adamellites–granodiorites a statistically highly significant relationship was found between Sn and Mo (significance level, 98%) and definite correlations were also established between Cu and Mo and Sn and Y (coefficient of correlation, ~0·3; level of significance, 90%).

The finding that interrelations between certain trace elements vary with rock type is important and should be considered in exploration for minerals containing the elements in question.

Conclusions

The statistical analysis of geochemical data on 72 samples from the New England batholith, New South Wales, led to the following conclusions.

(1) The samples consisted of a wide variety of granitoid rocks, comprising granites, leucoadamellites, adamellites, granodiorites, as well as porphyrites. One gabbro and one tonalite were also present. The SiO_2 content of the samples varied between 50·5 and 77·5%, with a mean value at 69·0%.

(2) The coefficient of variation, indicating the degree of scatter, was low (<20%) for some of the major-element oxides (SiO_2, Al_2O, Na_2O and K_2O), but was higher for others and for the trace elements. Most of the values, however, were well within those which have been calculated for a large number of ore deposits by other authors.

(3) The number of samples appeared to be sufficient for obtaining mean values for the majority of the elements with a relative accuracy of ±10% at the 95% confidence level. For some elements, however (in particular trace elements with a high coefficient of variation), more than 72 samples would be needed to obtain mean values with a reasonable degree of confidence.

(4) The *R*-mode analysis of variables showed interrelations between certain elements. The following independent (orthogonal) variables were selected at the 0·7 correlation coefficient level for *Q*-mode analysis: SiO_2, Al_2O_3, Na_2O, TiO_2

(major-element oxides) and Ba, Co, Cr, Cu, Ga, Li, Mo, Pb, Sc, Sn, Sr, V, Y and Zr (trace elements).

(5) The *Q*-mode analysis of samples based on major-element oxide content showed the presence of two distinctly different groups: (1) adamellites–granodiorites–porphyrites (46 samples); and (2) granites and leucoadamellites (24 samples). Two samples, a gabbro and a tonalite, were very different from all other rocks. Rocks of the first group seem to predominate in the southern and central part of the batholith, and granites and leucoadamellites occur mostly in the north.

(6) The results of *Q*-mode analysis were in good general agreement with previous mapping. In some cases, however, similarity between various outcrops (at present classified as different types of rocks) was found. On the other hand, inhomogeneities within certain outcrops (where the composition was expected to be uniform throughout) were revealed. Further sampling and the reconsideration of nomenclature and the boundaries of rock types within the outcrops in question would appear to be necessary.

(7) The *Q*-mode analysis based on trace-element content showed that the distribution of the latter was more uniform throughout the batholith than that of the major elements. This is thought to be the result of post-magmatic processes of a regional nature. On the other hand, differences in trace-element distribution were noted in certain outcrops where uniform concentrations were expected—suggesting that localized alteration processes were operative in these areas.

(8) Major- and trace-element distributions in the two main types of rocks (adamellites–granodiorites and granites–leucoadamellites) conform, in general, with observations from other districts. Of the trace elements, Li, Y, Mo and Ga show preference for the granites, but Sn (with a mean value of 4·6 ppm for the whole suite of rocks) is slightly more concentrated in the adamellites. The preference of Ba and Zr for adamellites also seems to be anomalous.

(9) High linear correlations between the elements were found in some cases. In the granites–leucoadamellites the correlations of Pb with Y (and to a smaller extent with Sn), Li with Mo and Sr, Ba with Ga, Pb and Sr, as well as Ga with Y and Zr, were the most significant. In adamellites and granodiorites, among the significant correlations obtained were those of Mo with Sn and Cu, Sn with Y, Co with Ni, and Cu with Ni and Sc. The study of curvilinear correlations of a more complicated nature is in progress.

(10) In general, the technique of cluster analysis appears to be a useful tool in improving the mapping and geochemical characterization of rocks.

Acknowledgment

The authors are indebted to Mr. B. H. Flinter, whose samples and data were used for the statistical analysis; to Messrs. J. C. Corbett, N. Morgan, S. Goadby and R. Cosstick for analytical data; and to Mr. J. Harris for assistance in data preparation for computing. The advice of Miss J. Hayhurst and Mr. P. Milne of the CSIRO Division of Computing Research, and Mr. N. I. Fisher of the CSIRO Division of Mathematical Statistics, is also gratefully acknowledged.

References

1. Wilkinson, J. F. G. Intrusive rocks. The New England batholith. *J. geol. Soc. Aust.*, **16**, 1969, 271–8.
2. Voisey, A. H. Tectonic evolution of north-eastern New South Wales, Australia. *J. Proc. R. Soc. N.S.W.*, **92**, 1958, 191–203.
3. Voisey, A. H. Geological structure of the Eastern Highlands in New South Wales. In *Geology of Australian ore deposits* (Melbourne: Congress and Australasian IMM, 1953), 850–62. (*5th Empire Min. Metall. Congr., 1953, vol. 1*)
4. Lawrence, L. J. and Markham, N. L. The petrology and mineralogy of the pegmatite complex at Bismuth, Torrington, N.S.W. *J. geol. Soc. Aust.*, **10**, 1963, 343–64.
5. Spry, A. Flow structure and laminar flow in bostonite dykes at Armidale, New South Wales. *Geol. Mag.*, **90**, 1953, 248–56.
6. Binns, R. A. and Richards, J. R. Regional metamorphic rocks of Permian age from the New England district of New South Wales. *Aust. J. Sci.*, **27**, 1965, 233.
7. Andrews, E. C. The geology of the New England Plateau, with special reference to the granites of Northern New England. Pt. II, general geology. *Rec. geol. Surv. N.S.W.*, **8**, 1905, 108–52.
8. Wilkinson, J. F. G. Some feldspars, nephelines and analcimes from the Square Top intrusion, Nundle, N.S.W. *J. Petrol.*, **6**, 1965, 420–44.
9. Flinter, B. H. The differentiation index applied to the New England igneous complex. Unpublished work.
10. Chappell, B. W. Granitic intrusions from the southern end of the New England batholith. *J. geol. Soc. Aust.*, **16**, 1969, 278–82.
11. Binns, R. A. Hornblendes from some basic hornfelses in the New England region, New South Wales. *Mineralog. Mag.*, **34**, 1965, 52–65.
12. Flood, R. The geology of the Woolbrook–Walcha Road area, N.S.W. B.Sc. thesis, University of New England, 1964.
13. Vernon, R. H. The geology and petrology of the Uralla area, N.S.W. *J. Proc. R. Soc. N.S.W.*, **95**, 1961, 23–33.
14. Boesen, R. S. The Duval and Sunnyside adamellites. *J. geol. Soc. Aust.*, **16**, 1969, 283–4.
15. Wilkinson, J. F. G. Vernon, R. H. and Shaw, S. E. The petrology of an adamellite–porphyrite from

the New England batholith (New South Wales). *J. Petrol.*, **5**, 1964, 461–88.
16. Joyce, A. S. Geology of the Yarrowyck area. B.Sc. thesis, University of New England, 1964.
17. Lusk, J. Geological features and mineralization in the Gulf Creek area, N.S.W. B.Sc. thesis, University of New England, 1961.
18. Cotton, L. A. The tin deposits of New England, N.S.W. Part I. The Elsmore–Tingha district. *Proc. Linn. Soc. N.S.W.*, **34**, 1909, 733–81.
19. Binns, R. A. Granitic intrusions and regional metamorphic rocks of Permian age from the Wongwibinda district, north-eastern New South Wales. *J. Proc. R. Soc. N.S.W.*, **99**, 1966, 5–35.
20. Gutsche, H. W. Granitic rocks west of the Clarence basin. *J. geol. Soc. Aust.*, **16**, 1969, 297–9.
21. Shaw, S. E. Granitic rocks from the northern portion of the New England batholith. *J. geol. Soc. Aust.*, **16**, 1969, 285–90.
22. Phillips, E. R. The adamellites of the Liston area. *J. geol. Soc. Aust.*, **16**, 1969, 290–4.
23. Lawrence, L. J. Igneous rocks of the Emmaville–Torrington district. *J. geol. Soc. Aust.*, **16**, 1969, 294–7.
24. Flinter, B. H. Mineralogy of the heavy fractions of New England granitoids. Paper presented to Symposium on mineralization in acid igneous rocks, ANZAAS Conference, Brisbane, May, 1971.
25. Lance, G. N. and Williams, W. T. Mixed-data classificatory programs. I. Agglomerative systems. *Aust. Comp. J.*, **1**, 1967–68, 15–20.
26. Burr, E. J. Cluster sorting with mixed character types. I. Standardization of character values. *Aust. Comp. J.*, **1**, 1967–68, 97–9.
27. Lance, G. N. and Williams, W. T. A general theory of classificatory sorting strategies. 1. Hierarchical systems. *Comp. J.*, **9**, 1966/67, 373–80.
28. Sokal, R. R. and Michener, C. D. A statistical method for evaluating systematic relationships. *Sci. Bull. Kansas Univ.*, **38**, 1958, 1409–38.
29. Koch, G. S. Jr. and Link, R. F. The coefficient of variation—a guide to the sampling of ore deposits. *Econ. Geol.*, **66**, 1971, 293–301.
30. Hazen, S. W. Jr. and Meyer, W. L. Using probability models as a basis for making decisions during mineral deposit exploration. *Rep. Invest. U.S. Bur. Mines* 6778, 1966, 83 p.

Appendix

*Significance of correlation coefficient**

Degrees of freedom, $n-1$, where n is number of samples	Level of significance, %				
	90	95	98	99	99·9
1	·988	·997	·999	1·000	1·000
2	·900	·950	·980	·990	·999
3	·805	·878	·934	·959	·992
4	·729	·811	·882	·917	·974
5	·669	·754	·833	·874	·951
6	·621	·707	·789	·834	·925
7	·582	·666	·750	·798	·898
8	·549	·632	·716	·765	·872
9	·521	·602	·685	·735	·847
10	·497	·576	·658	·708	·823
11	·476	·553	·634	·684	·801
12	·457	·532	·612	·661	·780
13	·441	·514	·592	·641	·760
14	·426	·497	·574	·623	·742
15	·412	·482	·558	·606	·725
16	·400	·468	·543	·590	·708
17	·389	·456	·528	·575	·693
18	·378	·444	·516	·561	·679
19	·369	·433	·503	·549	·665
20	·360	·423	·492	·537	·652
25	·323	·381	·445	·487	·597
30	·296	·349	·409	·449	·554
35	·275	·325	·381	·418	·519
40	·257	·304	·358	·393	·490
45	·243	·287	·338	·372	·465
50	·231	·273	·322	·354	·443
60	·211	·250	·295	·325	·408
70	·195	·232	·274	·302	·380
80	·183	·217	·256	·283	·357
90	·173	·205	·242	·267	·337
100	·164	·195	·230	·254	·321

*From FISHER, R. A. and YATES, F. *Statistical tables for biological, agricultural and medical research, 6th edn* (Edinburgh: Oliver and Boyd, 1963), Table VI.

519.272:550.84(425.1)

Use of cluster analysis in geochemical prospecting, with particular reference to southern Derbyshire, England

R. C. Obial, M.SC., PH.D., D.M.T.

Endeavour Minerals, N.L., Manila, Philippines; formerly Department of Geology, University of Leicester, England

C. H. James, A.R.S.M., PH.D., D.I.C., F.I.M.M.

Department of Geology, University of Leicester, England

Synopsis

Cluster analysis was applied to a multi-element geochemical stream-sediment survey carried out south of the Derbyshire Limestone Dome. The –80-mesh fractions of 170 samples were analysed for Al, Ca, Mg, K, Fe, Ti, Mn, Ba, Co, Cr, Cu, Ga, Li, Ni, Pb, Sr, Zn and Zr by a direct-reading spectrometer. Cold-extractable Zn, analysed colorimetrically, was also included in the data matrix. The samples and the results of the multi-element analysis were taken to represent objects and variables, respectively, for cluster analysis, and both *R*- and *Q*-mode analyses were performed on the data matrix. The *R*-mode analysis showed three main groups of elements, from which the following associations were deduced: (1) elements associated with clay minerals and resistates; (2) elements associated with Mn and Fe oxides; and (3) elements associated with the dominant type of mineralization in the area.

When all 20 elements were considered in a *Q*-mode cluster analysis, an areal grouping of similar samples was shown which outlined both the lithology of the area and the most intense anomalies. Various combinations of elements were also tried for *Q*-mode analysis, each elaborating on the areal similarities and dissimilarities of the samples on the basis of the elements considered.

By the selection of suitable combinations of elements, areas of mineralization could be distinguished. It appears that the technique may also be capable of distinguishing anomalies related to virgin mineralization from those associated with old mine workings or other contaminating influences.

The development of rapid instrumental techniques of analysing geological materials for a wide range of elements has broadened the field of regional geochemical surveys. By use of multi-element analysis of regional stream-sediment samples, it is possible to scan a vast area for a wide range of mineral deposits, to aid in geological mapping where thick overburden obscures bedrock, and to outline possible geochemical provinces. Apart from these potential geological applications, Webb and co-workers[1] have demonstrated its multi-purpose potential in the fields of agriculture, land-use surveys and public health.

Multi-element analysis was applied to a regional stream-sediment survey south of the Derbyshire Limestone Dome. The potential mineralization in the area consists of lead–zinc vein extensions present in the Carboniferous Limestone, but buried by a thick sequence of younger sediments. Codner[2] attempted to detect possible leakage anomalies in the overlying Edale Shale by analysing stream sediments collected over the area for copper, lead and zinc. Also of importance is the mineralization associated with Triassic unconformities. Much of the mineralization in the dolomitized limestone (typified by Golconda mine[3]) could be regarded as a mineralization of this type, and similar metallic accumulations are known to occur elsewhere (e.g. in Leicestershire), even when other rock types are in unconformable relationship with the Trias. A considerable part of the study area is unconformably overlain by the basal sediments of the Trias, serving as a potential search area for this type of metallic accumulation.

The present study was, therefore, conceived to amplify Codner's copper–lead–zinc survey in terms of a wider range of elements, and to focus attention on the mineralization at the base of the Trias. As in any reconnaissance programme, this survey must be regarded as a preliminary scan designed to determine areas of interest for further exploration work.

Multi-element regional surveys of the type undertaken here inevitably involve the accumulation of a large bulk of analytical data, and the geochemist is often confronted with the problem of data-handling and interpretation. Under such circumstances visual, subjective approaches to interpretation would be tedious, time-consuming and often inadequate. Although in some cases such methods are still essential, they could be greatly improved by computer-orientated techniques. The object of this study, therefore, is to develop methods of increasing the predictive efficiency of geochemical prospecting by examining the applicability of certain multivariate methods as aids in the interpretation of multi-element geochemical data.

Cluster analysis is a multivariate technique extensively used by numerical taxonomists.[4] It relies on an object or sample being defined by a number of attributes or variables. The attributes are quantified or numerically coded, if they are multi-state characters. A data matrix is therefore obtained, consisting of a number of samples with their corresponding set of suitably coded or quantified characters. Similarly, coefficients are calculated between each pair of samples or variables, depending on whether samples (*Q*-mode) or variables (*R*-mode) are being clustered. Most clustering methods consist of grouping the samples or variables on the basis of the computed similarity coefficient, the nucleus of clusters being formed by joining the samples with highest similarity coefficient and gradually admitting more as the similarity coefficient is lowered. Other clusters are eventually initiated until, finally, all the samples are linked. Various clustering methods exist, depending on the criterion of entry of a sample into a cluster, e.g. weighted and unweighted pair-group average linkage, and single-linkage methods.[4] The end-product of the clustering procedure can be represented as two-dimensional hierarchical diagrams called *dendrograms*.

The computer program used is contained in a series of ALGOL procedures written by Sackin.[5] Dendrograms may be drawn by hand, but they are usually produced on a line printer or digital plotter. An example of a *Q*-mode dendrogram is given in Fig. 6.

Contemporaneously with the survey by Codner[2] a large-scale geochemical mapping survey was conducted by Nichol and co-workers[6] on a nationwide basis. This survey covered the peripheral area of the Carboniferous Limestone Dome.

General geology of area

The study area was bounded by eastings SK/416000–435000 and northings SK/345000–355000 of the Ordnance Survey Grid of Great Britain (Fig. 1). Four well-defined streams drain the area: the Bradbourne, Henmore, Ecclesbourne and Cutlers Brooks. The first three streams mainly dissect the shale, and the fourth is in the overlying Triassic and Drift formations.

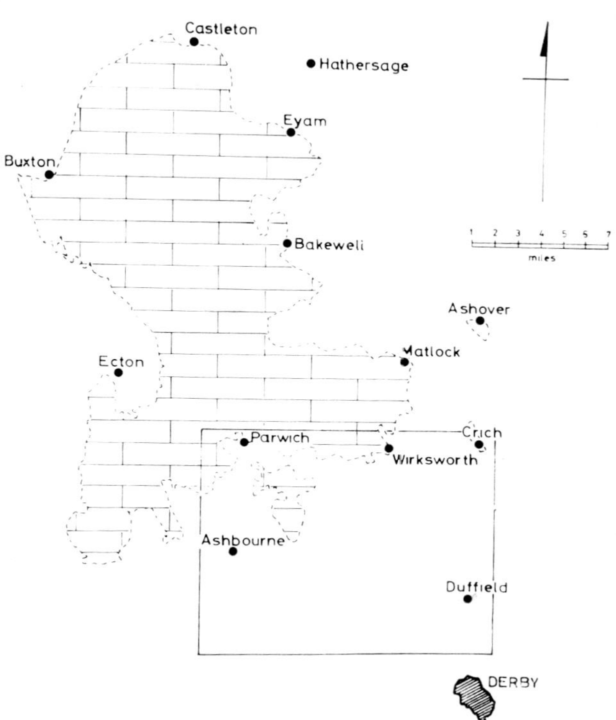

Fig. 1 Locality map showing field area in relation to Derbyshire lead–zinc mining field and area of outcrop of Carboniferous Limestone

The geology of the study area is depicted on the series of geochemical maps. The southern fringe of the main Carboniferous Limestone mass crops out in the extreme north of the study area, and sizable inliers occur at Bradbourne and Kniveton, together with an isolated one at Turnditch. The upper facies of the limestone and younger Carboniferous rocks cover the northern half of the area and disappear southward beneath a cover of Triassic and Drift formations. A detailed description of the stratigraphy and structure of the Carboniferous rocks has recently been given.[7,8]

The limestone exposed on the northern fringe of the study area is dolomitized, although the cause of dolomitization is still in dispute.[7] Such dolomitized limestone acts as the host to the Golconda type of mineralization[3] and could prove to be an important environment of mineralization in the area.

The Triassic formation overlies the Carboniferous rocks unconformably in the study area, the unconformity being defined by the absence of the Permian formations. Farther east, near Mansfield, where the stratigraphic sequence is continuous, the Permian is represented by the Magnesian Limestone. It passes on to the base of the Trias, which consists of the Mottled Sandstones and the Pebble Beds. These strata make up the Bunter and dominate the Triassic rocks in the study area. A detailed stratigraphy of the Trias has been given by Taylor.[9]

Glacial Drift deposits, mainly till and lacustrine, are of minor occurrence in the area, but they are widespread southwards.

Mineralization

The mineralization of the Carboniferous Limestone mass has been described in some detail by Ford.[10] It is relevant here, however, to focus attention on the limestone mineralization south of the main Derbyshire dome, since the proximity of this type of deposit to the study area indicates the likelihood of possible extensions southwards.

The Golconda type of mineralization[3] occurs in the dolomitized parts of the Carboniferous Limestone. The type locality is the once-productive Golconda mine near Brassington. The ore minerals are chiefly galena and baryte, with a few traces of fluorite. Sphalerite, chalcopyrite and chalcocite are also present in residual traces.

A lesser-known type of deposit, which may be of some economic potential, occurs along unconformities of the Trias with other rock types. A number of occurrences of this type have been recorded by King[11] and Ford and King,[12] and a geochemical survey that traced certain anomalies to the base of the Trias has been described by Taylor and co-workers.[13] Although most of these occurrences lie in northern Leicestershire, the similarity in the geological environment suggests that ore accumulations of this type may also be present in the study area.

The ore minerals consist of galena, cerussite, baryte, chalcocite, copper carbonates, vanadium compounds and some uraniferous hydrocarbons. Baryte is ubiquitous in the Permo–Triassic, often constituting the cement of the Triassic detritus.

Techniques

Stream-sediment sampling

The samples employed in this study were originally collected by Codner[2] in 1967. The average sample density for the survey was 1·7 samples per square mile over a total area of some 100 square miles.

The samples were oven-dried at a temperature of 110°C, and sieved through 80-mesh (nominal B.S.S.) nylon bolting cloth; −80-mesh material was used in all subsequent analysis.

Analytical methods

With the exception of cold-extractable zinc (cxZn), all analyses were carried out on the University of Leicester A.R.L. 29000B direct-reading optical emission spectrometer by use of the method described by Davenport.[14]

Analytical control was obtained by replicate analysis of routine samples.[15] With the exception of two elements (Cu and Ga), precisions better than 20% at the 95% confidence level were obtained (see results in Table 1), and these were

Table 1 Analytical precision of spectrometric determinations of stream sediments, soils and Triassic sands

Element	Precision, %*	Range
Al_2O_3	7·15	1·05–15·00%
CaO	8·06	0·05–10·00%
Fe_2O_3	2·05	1·05–8·30%
K_2O	9·41	0·85–4·35%
MgO	11·37	0·70–4·00%
TiO_2	3·40	0·31–1·1%
Mn	3·86	185–5000 ppm
Ba	8·05	183–1000 ppm
Co	17·98	4–28 ppm
Cr	14·35	20–275 ppm
Cu	33·73	4–296 ppm
Ga	31·58	6–19 ppm
Ni	9·34	25–1000 ppm
Pb	19·99	5–1000 ppm
Sr	10·56	18–237 ppm
Zn	11·18	40–820 ppm
Zr	10·39	80–1000 ppm

*Precision at 95% confidence level.

considered to be satisfactory for the present project. Some idea of the accuracy of the technique can be obtained from the analysis of U.S. Geological Survey standard rocks and the comparison with Flanagan's[16] compiled data (Table 2). Once again, mostly satisfactory results were obtained.

Cluster analysis

Cluster analysis as an aid in the interpretation of geochemical stream-sediment data has already been introduced on the basis of the present work.[17,18] A more detailed account is provided here to illustrate the technique more fully and to discuss possible variations and the type of information that could be deduced. Various forms of data input have been tried in order to determine both the quality of the information extracted and how well it correlates with the known geological parameters. The use of cluster analysis used in combination with principal-components analysis is also examined.

Table 2 Comparison of results of analysis of U.S. Geological Survey standard rocks, values expressed in ppm

Element	*G–2* *A**	*B*†	*GSP–1* *A**	*B*†	*AGV–1* *A**	*B*†	*PCC–1* *A**	*B*†
Mn	265	204	326	280	728	631	889	1008
Ba	1950	1212	1360	1083	1410	1000	6·9	
Co	4·9	5·3	7·5	7·3	15·5	14·3	112	106
Cr	9	4	13·2	8	12·9	4	3090	1629
Cu	10·7	5	35·2	25·3	63·7	50·3	10·40	6
Ga	20·2	20·6	18·8	20	18·4	17·6	12·4	1·5
Li	42·7	20	36·2	19·3	12·1	6	‡	‡
Ni	6·4	1	10·7	6	17·8	17·3	2430	2638
Pb	28·7	32	52·4	46	35·4	32·3	13·3	13·3
Sr	463	376	247	243	657	531	0·3	‡
Zn	74·9	80·6	143	91	112	91	53	8
Zr	316	294	544	501	227	198	‡	13

*Flanagan.[16]
†Spectrometric analysis (means of three determinations undertaken periodically during routine analysis).
‡Below limit of detection.

Choice of clustering method

Three methods of cluster analysis were compared during the course of this work: (1) weighted pair-group average method (WPGM); (2) unweighted pair-group average method (UPGM); and (3) single-linkage method.

WPGM and UPGM fall into the group methods devised by R. R. Sokal and C. D. Michener (reference 4, pp. 182–5) and are generally similar to C. Spearman's variable-group methods. The mathematical details of these clustering methods were described by Sokal and Sneath.[4]

Dendrograms *A*, *B* and *D* (Fig. 2) show an *R*-mode analysis (clustering of variables) in which the three methods are compared. Both WPGM and UPGM show very similar dendrogram structures, and similar elemental associations are exhibited by all three methods. Single linkage differs from the other two in dendrogram structure, since more than 50% of the variables are tightly nested in comparison with the few remaining variables. The strongly hierarchical structure it imposes on the dendrogram appears to be an inherent feature of single linkage, which is especially apparent in *Q*-mode cluster analysis. The pair-group methods are, on the other hand, agglomerative in nature, building up clusters from individuals: they thus show better groupings than single linkage.

The cophenetic correlation coefficient was devised by R. R. Sokal and F. J. Rohlf (reference 4, pp. 189–94) as a measure of degree of fit to a set of data and as an index of comparison of the clustering methods. It is, therefore, a measure of how well the relationships, as depicted by the correlation matrix, are preserved by the clustering methods. It has been found that the pair-group methods UPGM and WPGM have the highest cophenetic correlation coefficients, and that the single linkage has the lowest. The single-linkage method, therefore, is least used by numerical taxonomists. Gower[19] recommended WPGM for general purpose classification, whereas Farris,[20] in examining the mathematical background of the cophenetic correlation coefficients, found that it is maximized by UPGM. UPGM has found wide acceptance among numerical taxonomists, other classification workers and the few geologists who have applied the method—notably Parks[21] and McCammon.[22,23] UPGM was, therefore, adopted in this work.

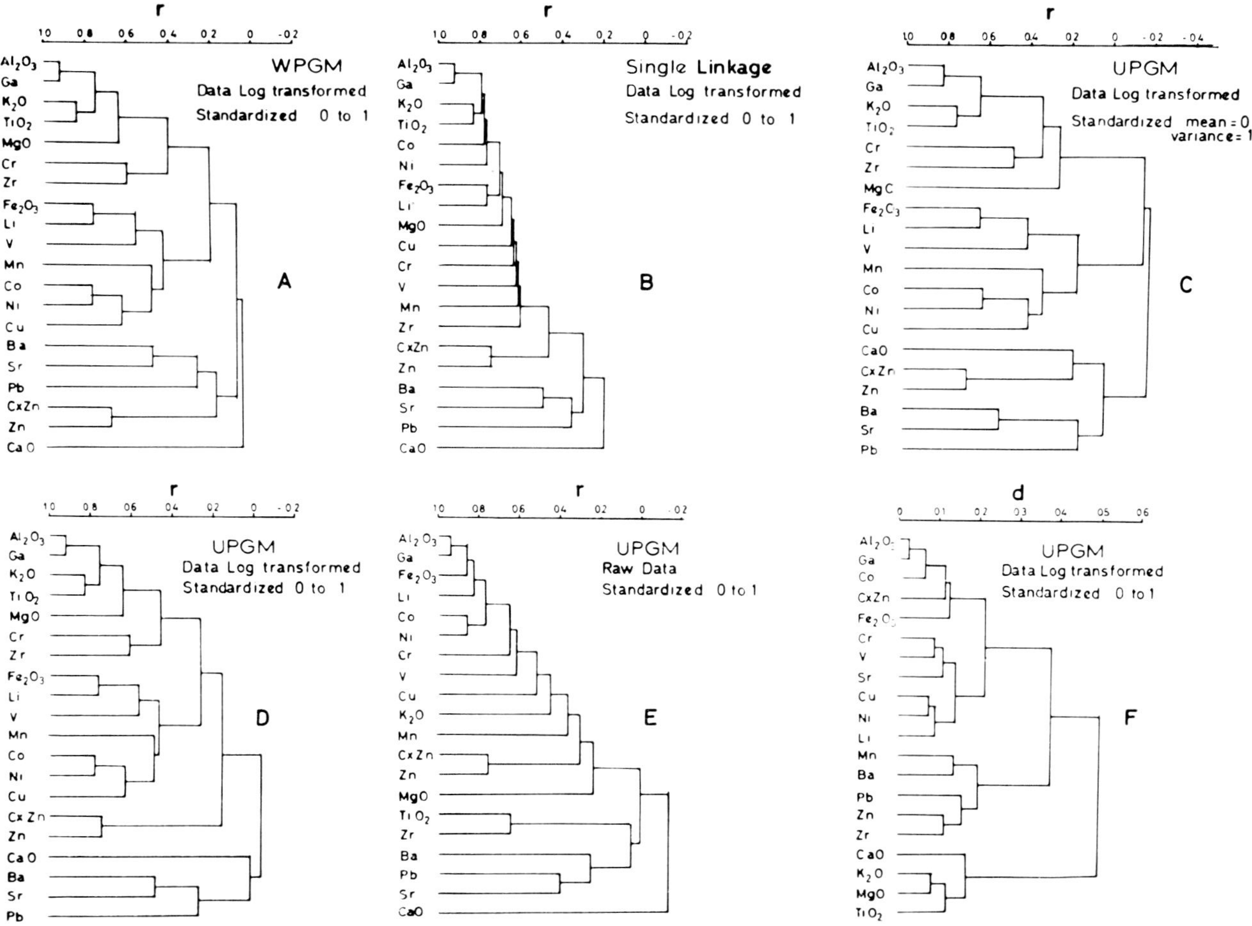

Fig. 2 Comparison of various methods of cluster analysis (r, correlation coefficient; d, distance coefficient)

The choice between raw and logarithmically transformed data was also investigated. Fig. 2 (*D* and *E*) shows the dendrograms produced in each case. It is clear that considerable differences in the dendrogram structures exist. The log-transformed data show discernible groupings, whereas in the raw data major groups are obscured. When compared with the results of single-element interpretation, the *R*-mode log-transformed data depict the characteristic association of the elements better. For *Q*-mode analysis the log-transformed data gave a better sorting result than that of the raw data. Figs. 3 and 4 show *Q*-mode analyses for 20 elements and 170 samples by use of raw and log-transformed data, respectively, the dendrograms obtained being shown in Figs. 5(*A*) and 6, respectively. The log-transformed data show a simpler classification, and the results are readily interpreted in terms of the geology of the area. The use of log-transformed data is also consistent with the apparent lognormality of most geochemical data shown by the cumulative frequency plots. Another advantage of log-transformed data is that they minimize the effect of the few, high, spurious results, which tend to exaggerate the dissimilarities beyond proportion.

A wide variety of similarity coefficients exists, but they mostly cater for either coded, multi-state characters or mixed-mode data. Two similarity coefficients were tried—correlation (r) and distance coefficients. Dendrograms *D* and *F* (Fig. 2) show a marked dissimilarity in the elemental associations obtained when two coefficients are used on the same data. The distance coefficient, although producing a better-structured dendrogram, is inferior to that produced by correlation coefficients in terms of interpretable geochemical associations. Owing to the quantitative nature of the data, the correlation coefficient is deemed better than taxonomic distance as a measure of similarity of geochemical entities. Taxonomic distance as found by other workers is best suited for mixed-mode data.

As the data are in different ranges of concentrations, they were standardized after being logarithmically transformed—but prior to the calculation of correlation coefficients. This step was necessary in order to equalize the influence of the variables or samples during *R*- and *Q*-mode cluster analyses. Two options of standardization were investigated—that which ranged the values from zero to one and that which expressed the data in terms of variances from the data mean, the mean being expressed as zero. Dendrograms *C* and *F* (Fig. 2) show a very close similarity in the clustering of elements, except for a slight reversal between the cxZn–Zn and CaO groups for the variance-type standardization. Standardization by taking variances is, however, probably less desirable because most of the elements have multi-modal distributions. Moreover, a higher correlation is shown by ranging from zero to one. It was therefore used in this work.

An essential task in cluster analysis is to take groups at a certain level of similarity. It is, therefore, of interest to compare the clusters produced by random sets of numbers with those obtained from cluster analysis of data. Rohlf and Fisher[24] have synthesized random data sets for cluster analysis. By use of UPGM they found that the branching of the dendrograms is restricted to a narrow region, i.e. 0·05–0·31 correlation coefficient. Another feature of random data sets is the numerous clusters developed with very few operational taxonomic units.

To illustrate the dendrograms produced by random data sets and their log-transformed equivalents, a 50 × 20 matrix of random numbers was clustered by use of UPGM. The results obtained (Fig. 7) show basically the same features as were found by Rohlf and Fisher.

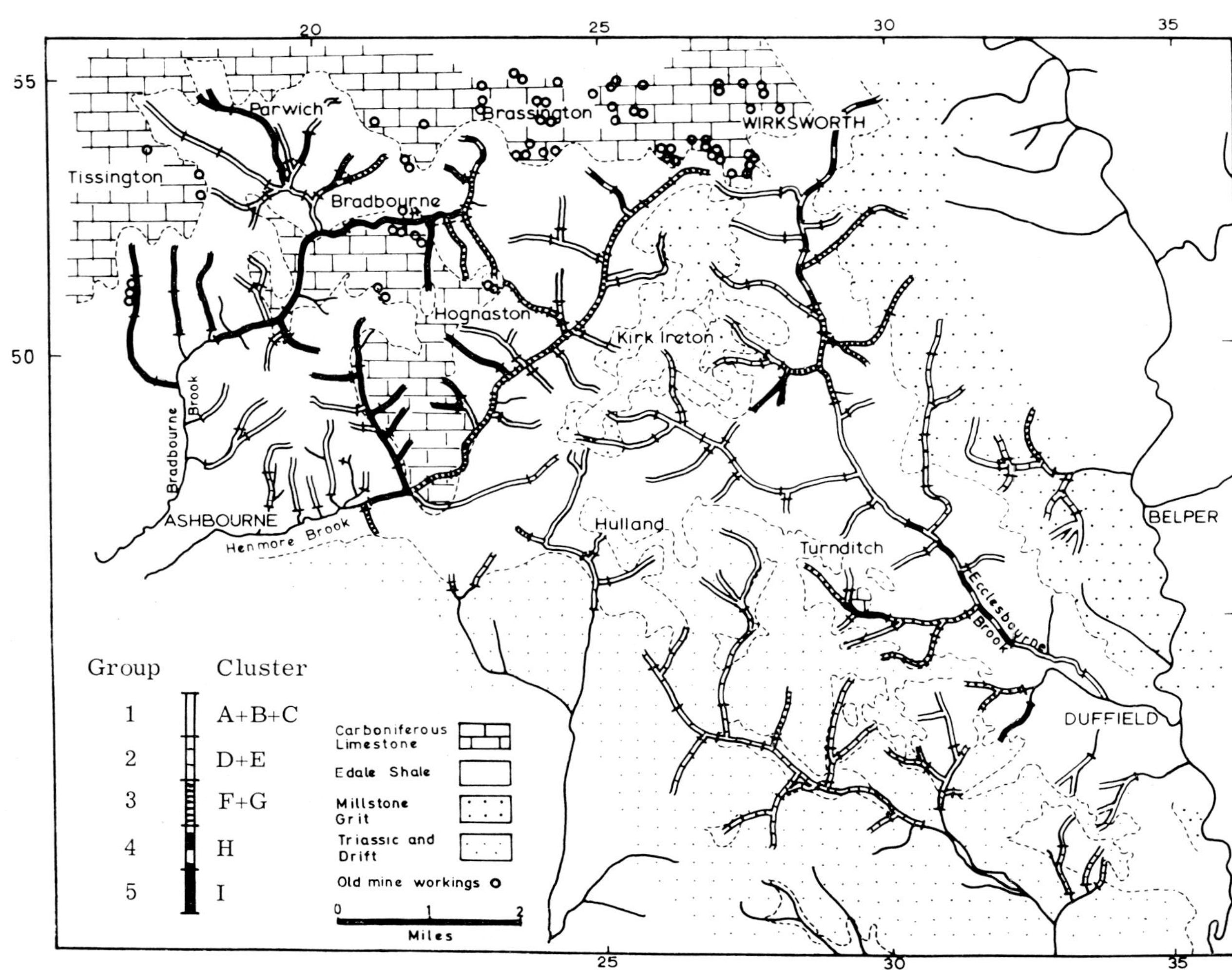

Fig. 3 Q-mode cluster analysis of 170 samples: 20 elements (data in raw form)

R-mode cluster analysis

Nineteen elements, together with cxZn, were considered for *R*-mode cluster analysis. Dendrogram *D* (Fig. 2) summarizes the elements considered and their similarities. At 0·45 *r* six distinct clusters are apparent: (1) Al_2O_3, Ga, K_2O, TiO_2, MgO, Cr and Zr; (2) Fe_2O_3, Li, V, Mn, Co, Ni and Cu; (3) cxZn and Zn; (4) CaO; (5) Ba and Sr; and (6) Pb.

The first cluster implies an association of elements of the clay mineral fraction and resistate

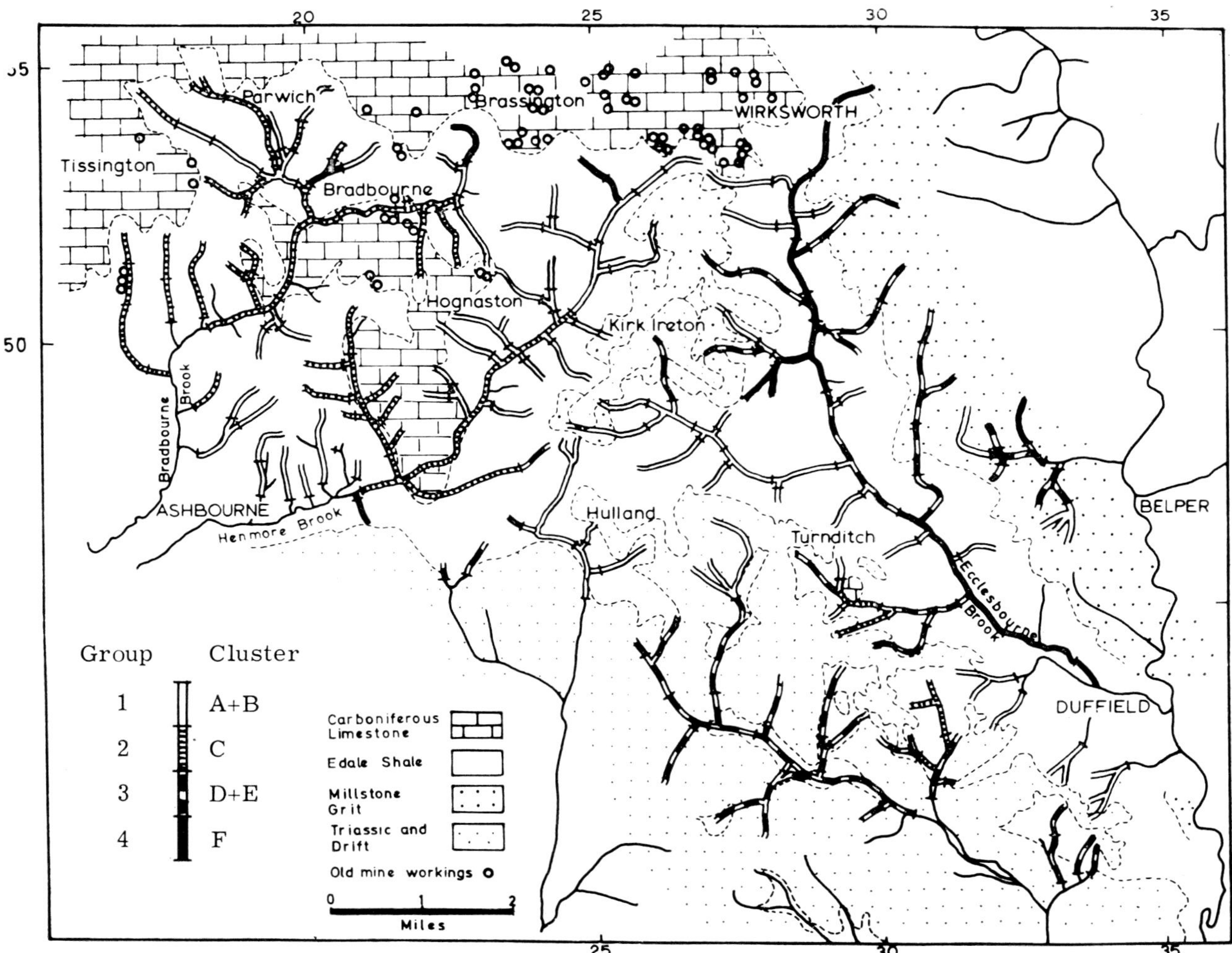

Fig. 4 Q-mode cluster analysis of 170 samples: 20 elements (data log-transformed)

minerals; the second appears to be a suite of elements that is associated with Mn and Fe oxidates; the third is an obvious correlation and mainly related to mineralization. CaO and Pb were shown as single clusters because of their low level of similarity with Ba–Sr. The 'loose' fifth cluster exemplifies the dominant elements associated with the prominent mineralization.

An inherent characteristic of a dendrogram is the distortion resulting from the two-dimensional representation of the elemental relationships. As could be seen from the *R*-mode analysis for 20 elements (Fig. 2*(D)*), cxZn and Zn formed a distinct cluster quite independently of the others. The close similarity between these two in the areal patterns is quite obvious. It is of interest, however, to know to which major group Zn belongs. A repeat *R*-mode analysis was, therefore, undertaken with the exclusion of cxZn (see Fig. 7*(B)*). Zn shows an association with the Mn and Fe cluster—suggesting a possibility of its being partly enriched with these oxidate constituents.

CaO, being derived mainly from the limestone, exhibits a dissimilarity from Ba, Sr and Pb, which are invariably present in the lithologies in which Ca is comparatively deficient. Pb, Ba and Sr are also the most important elements associated with the mineralization. Although Pb belongs to the same cluster as Ba and Sr, it shows a low level of correlation to the cluster, and is, therefore, considered separately. This seems to be a reflection of the subtle dissimilarity in the areal patterns of Pb compared with those of Ba and Sr.

The distribution patterns of the elements in the Mn and Fe cluster show a close similarity. The most notable localities are those in which anomalies have been defined, especially near Tissington (SK/185535) and south of Kirk Ireton (SK/274477). Mn has been reported to be associated with the dolomitized limestone and is known to have been mined in the past (T. D. Ford, personal communication). It is, therefore, of prime importance to check whether the anomalies outlined are related to the mineralization or are due to unrelated local enrichments. Although Pb and Ba could be enriched in Mn oxides, the abundance and extensive areal dispersions of both elements may have masked the scavenging effect of Mn and Fe on these elements.

The known close association of Pb and Zn in the Derbyshire ore field does not appear to have been preserved in the secondary environment, Zn being more closely associated with other elements than Pb (Fig. 7*(B)*). This is readily apparent when the dispersion patterns of the individual elements are compared. The marked difference is thought to be mainly due to the difference in geochemical character of the two elements, since Zn is much more mobile than Pb in the weathering environment. It may also be due to causes other than the difference in mobilities, such as the enrichment in Mn oxides implied by the *R*-mode analysis. Although Zn might be enriched in Mn oxides, it could still be indicative of 'virgin' anomalies. This aspect should, however, be studied further.

An important feature of the *R*-mode dendrogram is that the elements group according to the complexity of the populations recognized in their respective distributions. Thus, the first cluster consists of elements that are essentially unimodal in their distribution. Cr is a notable exception in this group. The second group, with the exception of Mn, represents elements which are largely bimodal in their distribution. The cxZn and Zn clusters belong more to the second major group than to the third, which consists of more complex multi-modal populations.

Q-mode cluster analysis

Q-mode analysis endeavours to compare and classify samples on the basis of their different element contents. A larger correlation matrix is, therefore, computed initially (in this case 170 × 170). Because of the present limitation of the computer (memory 65K), the program was designed to cluster up to 180 samples with 50 variables.

A fundamental characteristic of dendrograms is that the number of clusters defined increases as the criterion of similarity is increased. One could,

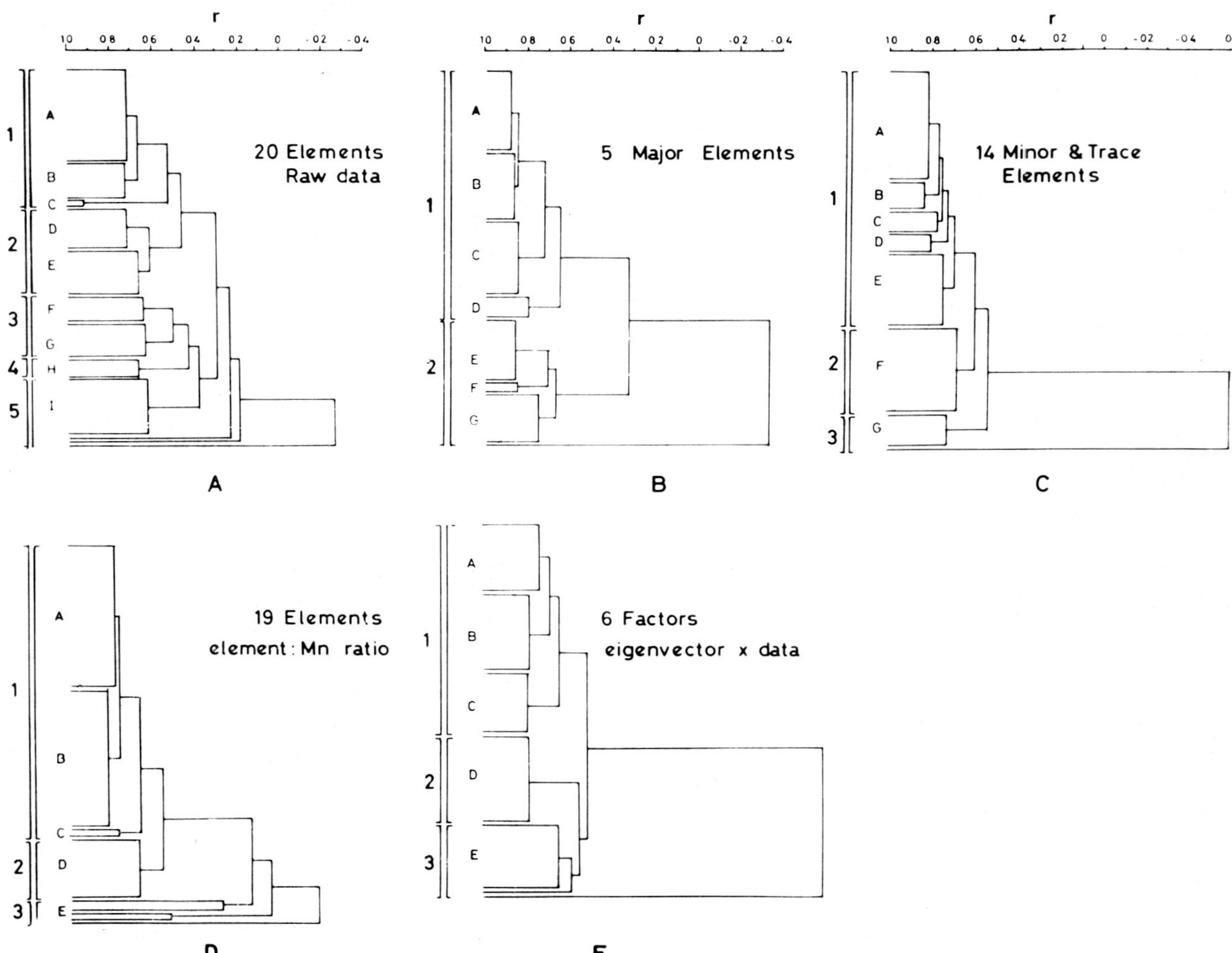

Fig. 5 Dendrograms for various Q-mode cluster analyses (UPGM) of 170 samples (various combinations of elements)

therefore, take groups at any level of similarity, depending on the purpose of the classification. This is analogous to the methods of factor analysis,

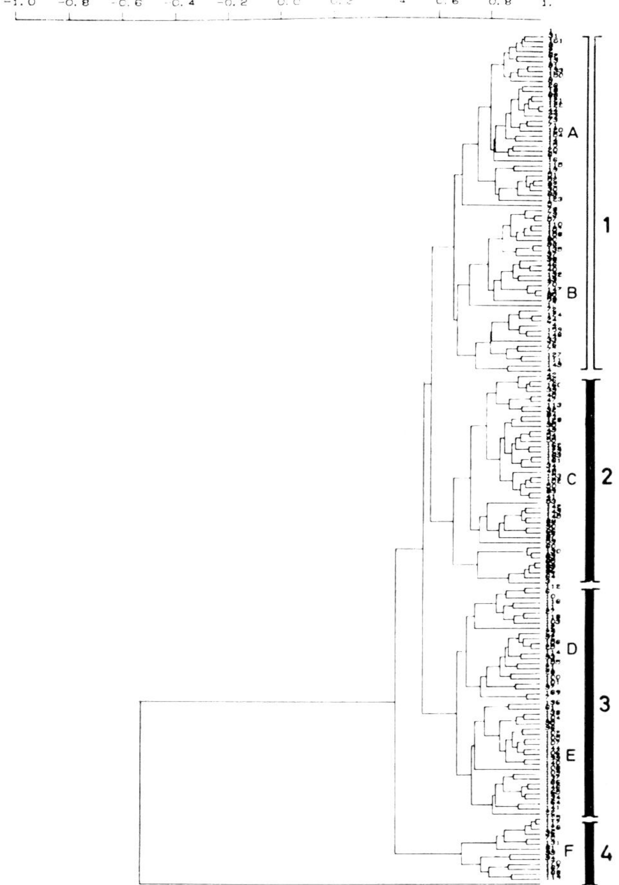

Fig. 6 Dendrogram for 170 samples and 20 elements (UPGM)

particularly principal-components analysis, where a subjective choice of the number of factors is made depending on the object of the study. Thurstone[25] has pointed out two main restrictions, common to statistical analysis, that do not apply to these types of multivariate analysis: (1) that the experimental sample should be representative of a general population or be carefully randomized and (2) that the data should have a normal or known frequency distribution.

Parks[26] reiterated that conventional statistical tests of significance are difficult to apply to cluster and factor analyses because of these restrictions. The level of similarity at which clusters, or groups of clusters or number of factors in principal-components analysis are taken to be significant, therefore, remains a subjective aspect of these methods. It must be borne in mind, however, that the prime purpose of multivariate analysis is simplification—that is, expressing the large number of variates in terms of a few interpretable ones.

Sokal and Sneath[4] introduced the term *phenon-line* to indicate the level of similarity at which groups are distinguished during *Q*-mode cluster analysis. Because taxonomists usually require a relatively rigid classification, both in grouping and position of the samples in the hierarchic structure, it was suggested that groups produced by cluster analysis be defined by the *phenon* at which they were taken. Since the hierarchical structure is of lesser importance in the present application of cluster analysis, however, more emphasis was laid on the number of groups or clusters that were formed at the different similarity levels. Preliminary work on a correlation matrix with a size of 170×170 indicated that a reasonable number of interpretable groups or clusters are produced at a similarity level of from 0·3 to 0·6 *r*. A rough guide to the lower limit of similarity for significant clusters is placed at 0·35, corresponding to the level of correlation of random sets as described by Rohlf and Fisher.[24]

In the following *Q*-mode analyses the dendrograms in Figs. 6, 8, 9 and 10 were drawn by a digital plotter; others were summarized and are shown in Fig. 5. The length of the bars opposite the dendrograms is proportional to the number of samples in each group.

Q-mode analysis for 20 elements

Fig. 4 illustrates *Q*-mode analysis by use of UPGM for all 20 elements. The dendrogram in Fig. 6 summarizes the groups of clusters taken. At a similarity of 0·65 *r* four main groups of clusters are shown (a group may consist of one or more clusters of samples). The important feature revealed by this analysis is the areal grouping of samples, outlining the basic lithology of the area. Superimposed on this is the indication of the most intense anomalies depicted by cluster *F*—the least correlated cluster in the group. Thus, the shale is indicated mainly by clusters $A+B$, the limestone area by cluster *C* and the Millstone Grit, Trias and Drift by clusters $D+E$. The Millstone Grit and the Trias and Drift show as one group—implying an inherent similarity in their bulk mineralogical compositions since both lithologies are sandy in character. Most of the intense anomalies depicted in cluster *F* are related to known old mine workings. A few samples in the southern half of the area, however, i.e. near Hulland, Bradley and Belper, cannot be explained in terms of contamination or other causes, and could be related to possible mineralization. They therefore merit some attention.

Q-mode analysis for major elements

When the five major elements Al, Ca, Fe, K and

Mg were considered for *Q*-mode analysis, two distinct groups emerged at 0·6 *r* (Figs. 5*(B)* and 10). The groups were subdivided into their respective clusters and plotted as such. The first group consists of clusters *A, B, C* and *D;* the second consists of clusters *E, F* and *G.* When plotted on the map, the first group coincides with the non-calcareous lithology, and the second outlines the limestone areas. Several group 2 samples occur east and southeast of the area—indicating a possible source of calcareous materials. were not shown as a distinct cluster—implying that there is no dissimilarity of anomalous samples when compared to background samples in terms of bulk chemical composition.

Q-mode analysis of minor and trace elements

When the minor and trace elements (excluding cxZn) are considered for *Q*-mode analysis, three main groups emerge at 0·65 *r* (Figs. 5*(C)* and 11). Many clusters make up the first group. The

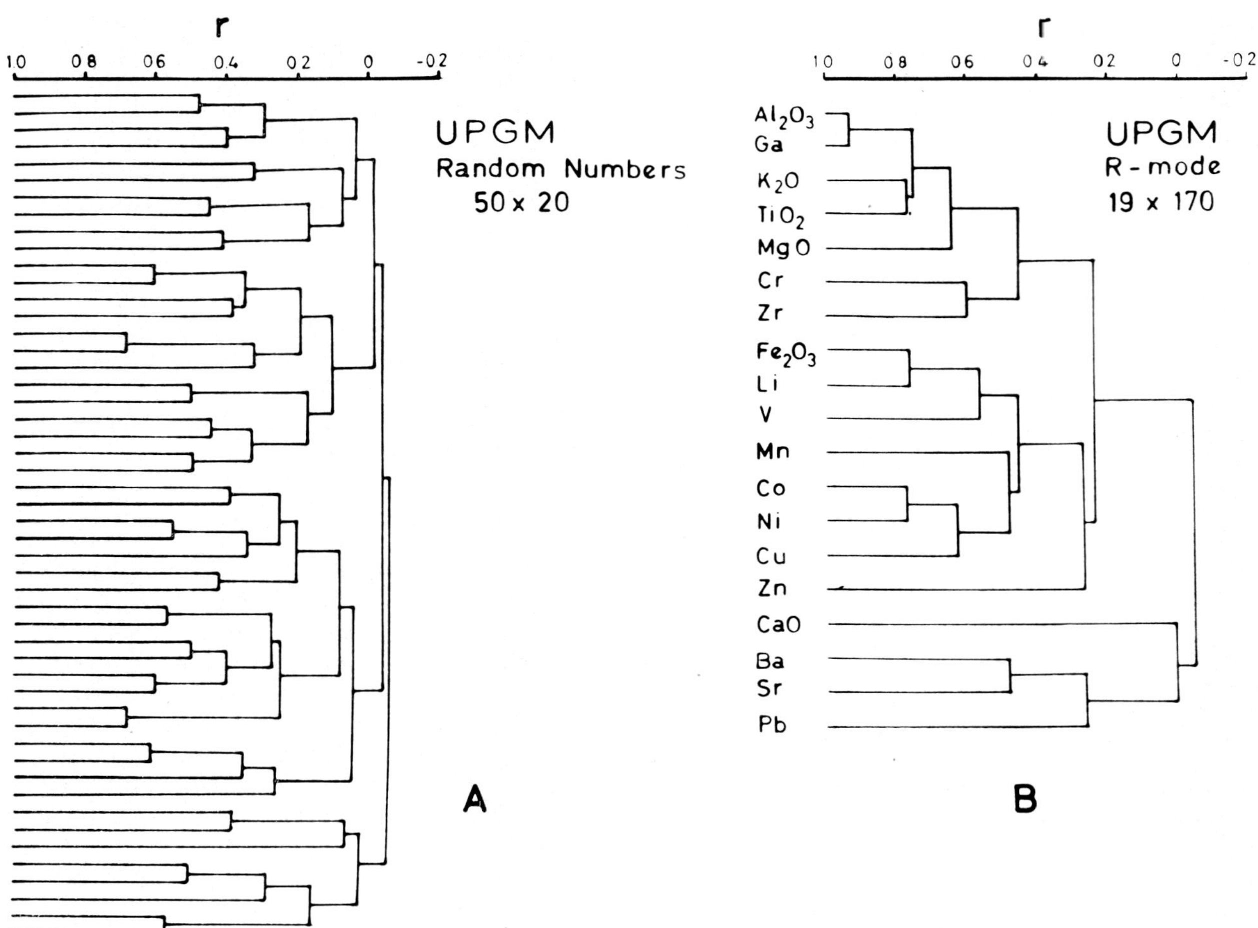

Fig. 7 R-mode cluster analysis of random numbers and 19 elements

The Turnditch limestone inlier was well represented by group 2 samples. The clusters in group 1 subdivide the lithology into shale, Trias and Drift. It is interesting to note the differences shown by the Millstone Grit and by the Trias and Drift—in contrast to the similarity they showed when all 20 elements were considered. This may imply certain differences in major-element composition. The main elemental influences in this classification are CaO and MgO. The contrast of the lithologies with respect to the limestone is strongly expressed by these two major elements. It is also important to note in this respect that the anomalous samples individual clusters of group 1, plotted separately, show no distinct, cohesive areal pattern, but as a group it dominates the shale and limestone areas, although a number of samples also appear in the area of Trias. Group 2, consisting of one cluster of samples, is dominant in the Trias, Drift and Millstone Grit; Group 3 indicates the most intense anomalies, which are known to be mostly due to mining contamination. The other anomalies shown by the 20-element *Q*-mode analysis (Fig. 4) are indicated in group 2, in which these samples are merged with the other group 2 samples. The anomaly near Tissington is not distinguished in this instance. The most intense anomalies, however, form a group which is least correlated with the rest of the samples.

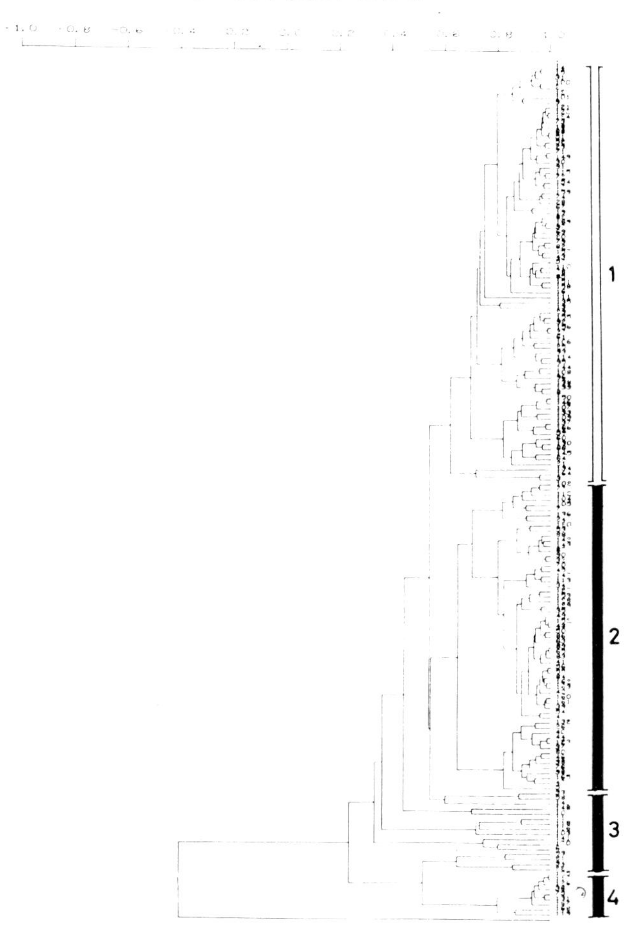

Fig. 8 Dendrogram for 170 samples and 'centroid' elements

Q-mode analysis for 'centroid' elements

In the classification of large data matrices certain variables may prove to be redundant: their effect in the sorting procedure is, therefore, to obscure recognition of significant groups. Thus, it is desirable to remove 'background noise' in order that only those variables which effectively add to classification are considered. It is also desirable that the relatively few anomalous samples be discernible on the basis of a large variance in the data of a few elements. This approach is analogous to principal-components analysis, in which the factors that contribute most of the data variability are preferentially taken. Rohlf and Sokal[27] found a good agreement between the taxonomic results derived from factor and cluster analyses, as well as concluding that all the factors considered were represented by the main groups of clusters. An orthogonal (i.e. uncorrelated) relationship is, therefore, implied by each group.

The following *Q*-mode analysis attempts such an approach by removing Ga, K_2O, MgO, Zr, Fe_2O_3, V, Co, cxZn, Ba and Pb from the list of variables. From the *R*-mode analysis (Fig. 2*(D)*) it is clear that Al_2O_3 and Ga, K_2O and TiO_2, Cr and Zr, etc., are similarly dispersed elements, and one could thus substitute one for any other, or for the whole group. Thus, the variables were reduced from 20 to 10 elements—Al_2O_3, TiO_2, Cr, Li, Mn, Ni, Cu, Zn, CaO and Sr. The resulting *Q*-mode areal distribution and dendrogram are shown in Figs. 12 and 8, respectively. At 0·55 *r* four major groups are defined, the first and second groups constituting a large percentage of the samples and the remaining groups clusters which contain only a few samples. Group 3 is composed of clusters that should, strictly, be considered as individual groups. For presentation purposes, however, they have been considered as one. Group 1 samples are dominant in the shale, although some are also present in the Trias and Millstone Grit. Group 3 represents various samples that are high in certain elements—mostly those

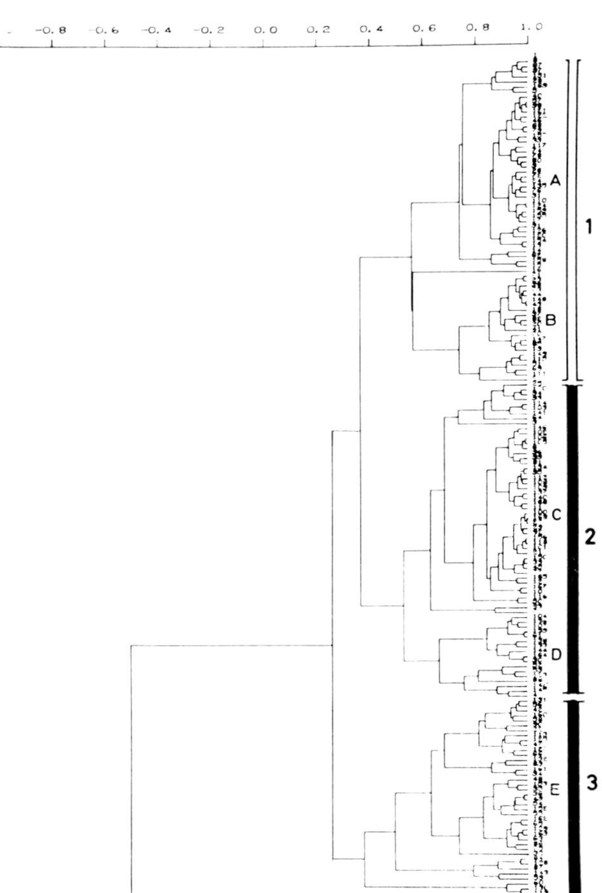

Fig. 9 Dendrogram for 170 samples and elements Al_2O_3, Ga, K_2O, TiO_2, MgO, Cr and Zr

considered to be significant. They are located in the various lithologies and thus show the individual dissimilarities of the samples in this group. Group 4 shows some of the most intense anomalies related to former mining activity. The important anomaly near Tissington is not shown in either group 3 or 4 (the 'anomalous' groups), although it does stand out as being a group 1 entity in a predominantly group 2 area. The importance of this scheme is that it shows the last two groups to be composed largely of anomalous samples and that anomalies due to contamination may possibly be distinguished from those tentatively regarded as due to 'virgin' mineralization.

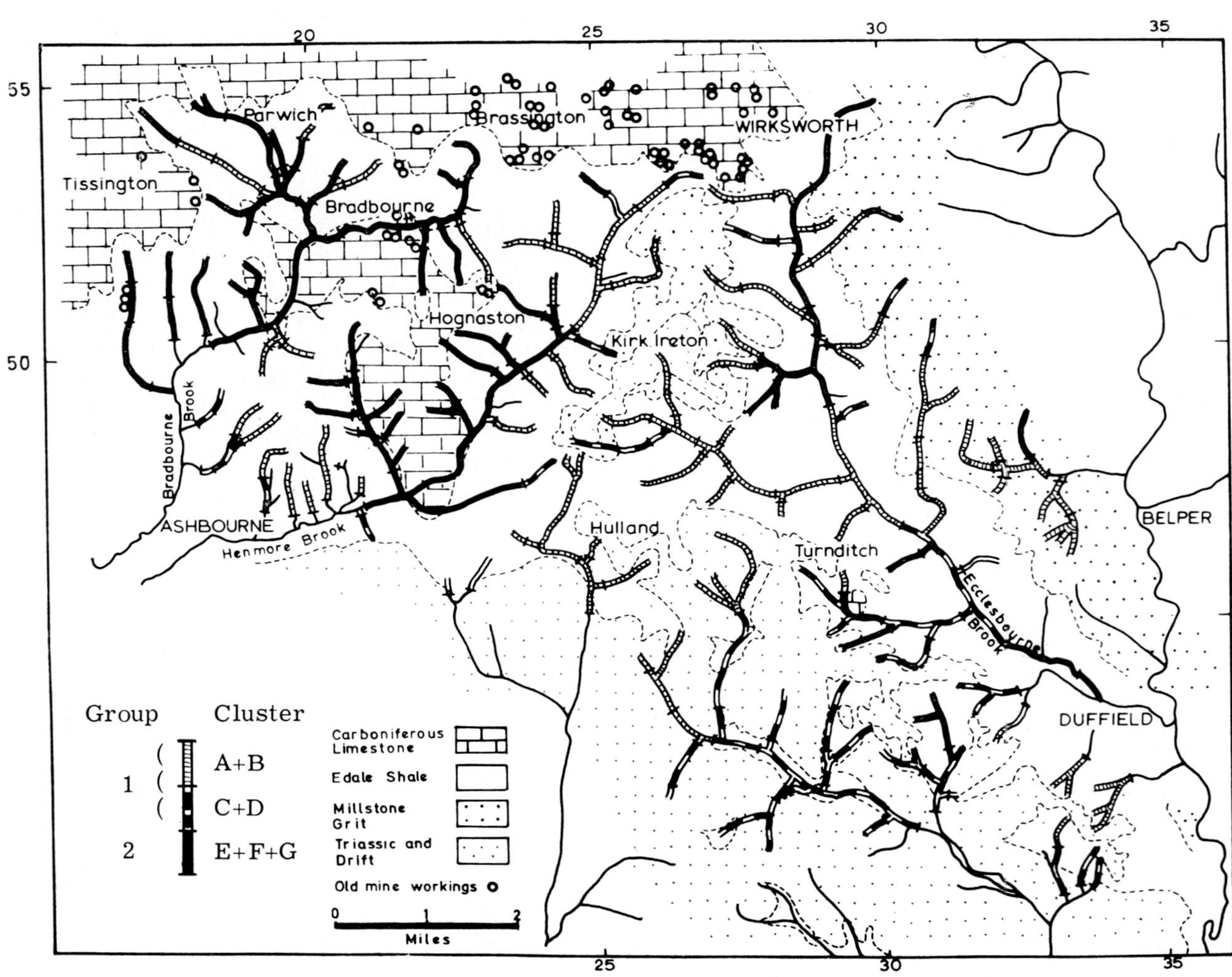

Fig. 10 Q-mode cluster analysis of 170 samples: major elements

Q-mode analysis of group 1 elements

The elements placed in group 1 by *R*-mode cluster analysis were next considered for *Q*-mode analysis in order to see the areal variations of samples with respect to the elemental associations outlined. Figs. 9 and 13, respectively, show the dendrogram and similarity map of the result of this *Q*-mode analysis, considering Al_2O_3, Ga, K_2O, TiO_2, MgO, Cr and Zr as variables. This group was interpreted to be an association of elements with clay minerals and resistates. The main groups were taken at 0·37 *r* level of similarity (shown in Fig. 9). Sizable clusters form at a relatively lower similarity level. The main feature of the similarity map is the delineation of the main lithologies. Group 1 samples show a domination of the shale areas; group 2 samples characterize the Millstone Grit and Trias and Drift and group 3 samples the limestone–shale area. The main classifying elements are Al_2O_3 and MgO, which show an inverse relationship for the calcareous and non-calcareous lithologies. The hierarchical ranking of the groups appears to imply a changing composition of stream sediments in terms of element association. The predominance of CaO and MgO as major constituents in the limestone area possibly reflects this change in the composition.

Q-mode analysis for elements to Mn ratio

To examine the effect of Mn as a major scavenging element the 19 elements (excluding Mn) were considered for *Q*-mode analysis. In this treatment the ratio of the element content to that of the corresponding Mn result was calculated and substituted in the data matrix. Thus, a reflection of the sample similarity was obtained in terms of its Mn content. The dendrogram and similarity map are shown in Figs. 5(*D*) and 14, respectively. The dendrogram shows a high level of similarity related to former mining activity and also that near Tissington. Minor cluster *C*, which is least correlated with group 1, indicated three anomalous samples.

Group 2 samples (cluster *D*) are largely in the Trias and Millstone Grit, whereas those of group 3 (cluster *E*) are scattered. Cluster *E* indicates some of the significant anomalies, e.g. near Bradley, Hulland and near Belper.

An attempt was made to amplify the association of elements and similarities of samples based on the *Q*-mode classification of element to Mn ratios.

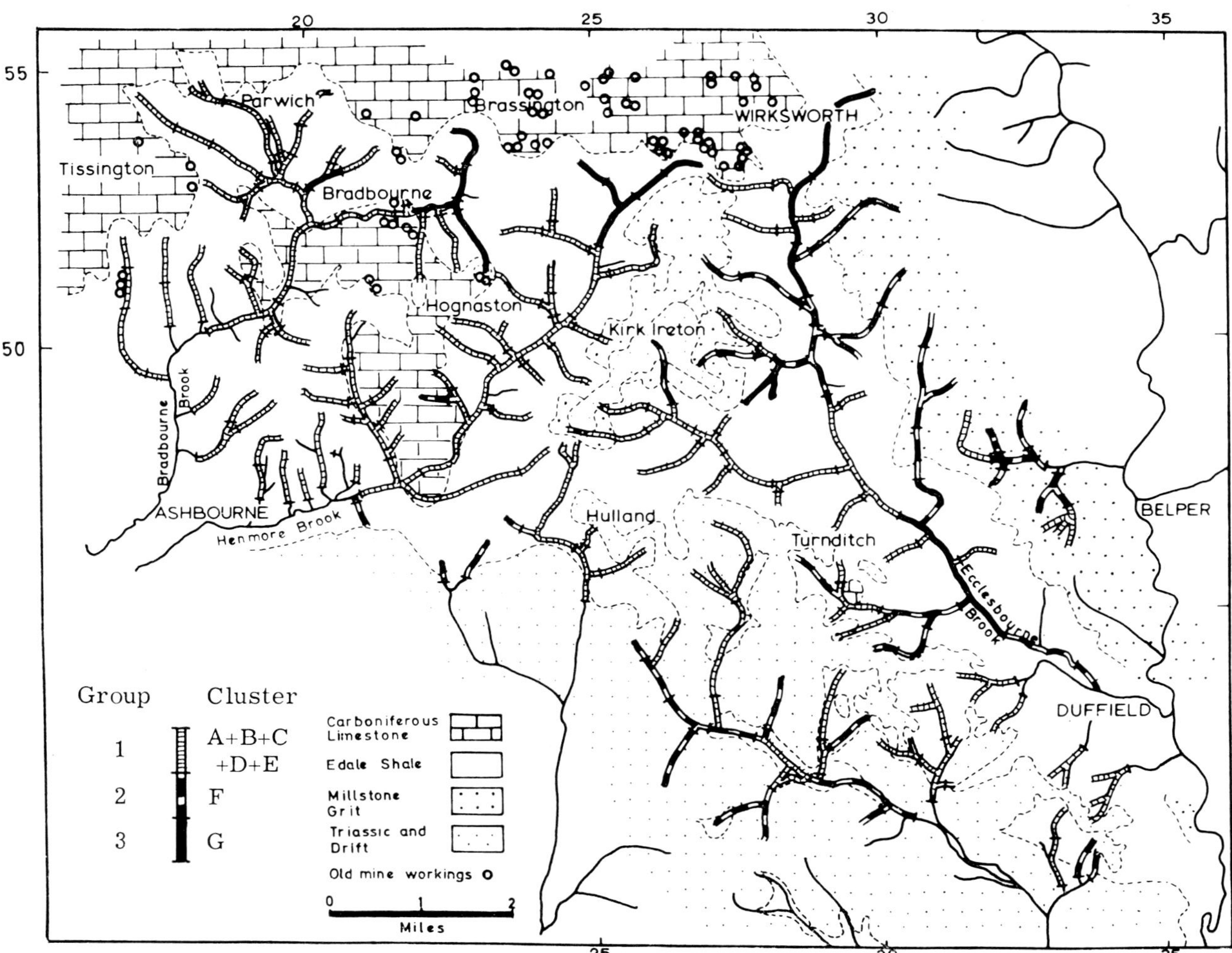

Fig. 11 Q-mode cluster analysis of 170 samples: minor and trace elements

for most of the samples. At 0·65 *r* three groups are depicted, the first being composed of three clusters and the third consisting of individual samples highly uncorrelated with the rest. Areally, clusters *A* and *B* have highly correlated samples and outline mainly the shale and limestone areas. Cluster *A* defines the shale, but also includes some samples from the Trias and a few in the Millstone Grit, and cluster *B* is more definitive of the limestone area. Included in cluster *B* are anomalies mainly Clusters *A* and *B* from the *Q*-mode dendrogram (Fig. 5(*D*)) were individually re-subjected to *R*- and *Q*-modes of cluster analyses. The results of the *R*-mode analyses are shown in Fig. 15, and the *Q*-mode analysis of cluster *B* is shown in Fig. 17. The different association of elements based on Mn contents is shown by the samples from the shale (cluster *A*) and limestone (cluster *B*) (Fig. 15).

The *R*-mode analysis of cluster *B* shows the

distinct grouping of the elements associated with the dominant mineralization (group 2); the major-element oxides, CaO and MgO, formed a separate cluster—accounting for the major influence of the limestone lithology.

For cluster *A* the elements associated with the dominant mineralization are not defined in the same fashion as in cluster *B*. Moreover, the dendrogram for cluster *A* shows a large number of elements with high levels of correlation (groups 1 and 2), the elements of groups 3 and 4 being the least correlated.

The difference in elemental association between the shale and limestone lithologies based on Mn contents further implies a lithological influence in the distribution of Mn. In addition, it is believed that the limestone lithology has influenced the pH of the proximate streams and, consequently, has affected the distribution of Mn and the other elements.

Q-mode analyses were undertaken for both clusters *A* and *B*; however, only cluster *B* showed significant similarities of samples. The dendrogram and areal similarity map are shown in Figs. 16 and 17. Four main groups are revealed at 0·4 correlation, the least correlated group joining the rest at a lower correlation coefficient. The prominent feature of this classification is the grouping of the most intense anomalies related to mining contamination, depicted as the least correlated group (group 4). The anomalies near Tissington and Parwich, which are probably less affected by contamination of this kind, comprise group 2. The other groups show most of the samples that are high in Mn, but with relatively lower concentrations of a number of trace elements.

The results of these *Q*-mode analyses imply that the concentrations of the elements considered change proportionally with the Mn contents and vary with the different lithologies. A further application of this method, therefore, is that of possibly distinguishing trace-element anomalies related to local Mn enrichments from a general increase caused by lithology. In this area the

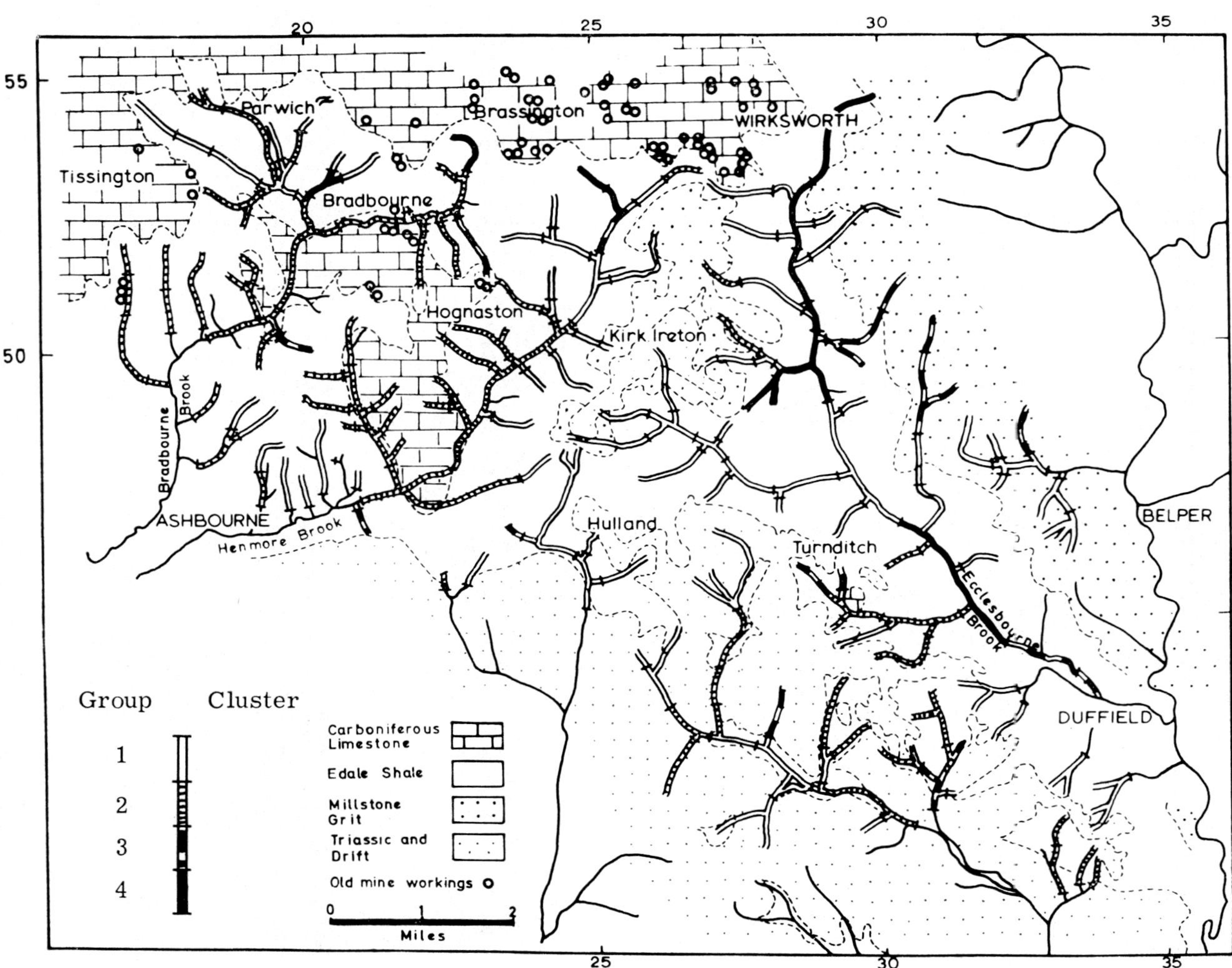

Fig. 12 Q-mode cluster analysis of 170 samples: 'centroid' elements

anomalous samples with low Mn contents were distinguished as being the least correlated members of the group. In the area studied it was also possible to amplify the similarities of samples in the cluster defined by the high Mn limestone lithology—resulting in the classification of samples related to mining contamination, possible anomalies and background.

Comparison of clustering with principal-components analysis

R-mode cluster and principal-components analyses (PCA) show highly compatible results (Fig. 2*(D)* and Table 3). The major elemental groups depicted by both methods may be broadly categorized into: (1) elements associated with clay minerals and resistate minerals; (2) elements associated with Mn and Fe; and (3) elements associated with the dominant mineralization.

The first group can be regarded as being influenced largely by the main lithologies present. This was well shown by the *Q*-mode cluster analysis for this elemental group in which the major rock types were outlined (Fig. 13).

The Mn and Fe elemental suite is also lithologically controlled, but to a lesser extent than the first group. The enrichment of Mn in the dolomitized limestone relative to the 'normal' Derbyshire limestone has been demonstrated in the wallrocks of the Masson Hill deposits.[18] Areas of localized enrichment in the stream sediments were, however, apparent throughout the area; consequently, the lithological control of Mn distribution is not universally true: indeed, in certain secondary environments (e.g. peat bog areas) the processes of secondary dispersion may influence the distribution pattern more than bedrock.[28]

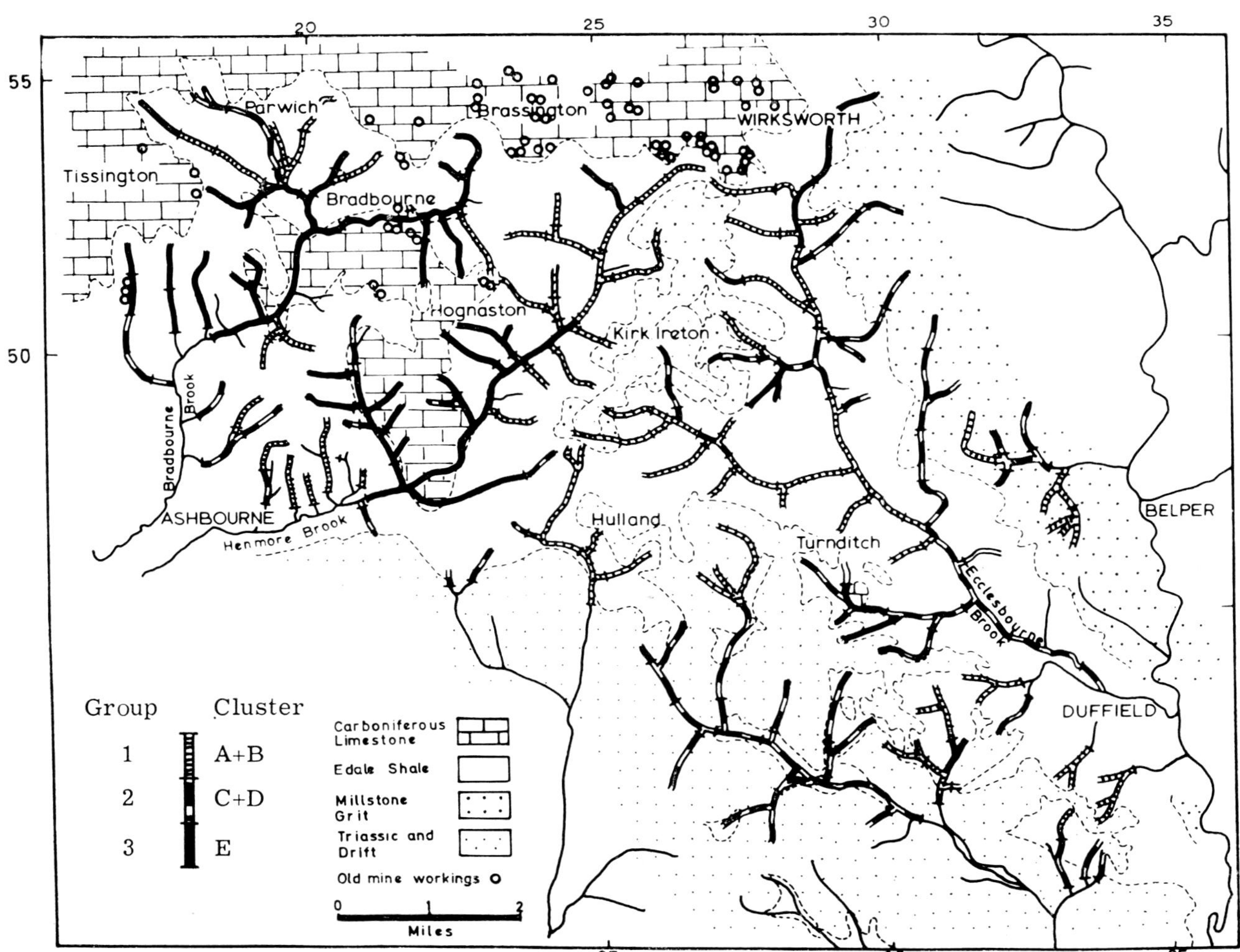

Fig. 13 Q-mode cluster analysis of 170 samples: elements interpreted as associated with clay minerals

The elements associated with the dominant type of mineralization, surprisingly, did not show high correlations with one another. This feature was well shown by both techniques: in cluster analysis they were shown as the least correlated group and individually clustered at a low level of correlation;

Table 3 Varimax rotated principal components of a six-factor model

Principal components	Variance accounted for, %	Variables contributing significantly* to principal components
P1	25·21	(1) Al_2O_3, K_2O, MgO, TiO_2, Co, Ga (2) Fe_2O_3, Cr, Ni, Zr
P2	21·05	(1) Fe_2O_3, Mn, Cr, Cu, Li, Ni, V (2) Co, Ga
P3	11·70	(1) Ba, Sr, Mn (2) Co, Ni, Pb, Zr
P4	10·52	(1) cxZn, Zn (2) Cu
P5	7·53	(1) CaO (2) MgO
P6	7·16	(1) Pb, Cr, Zr (2) Cu
Total	83·19	

*(1) Greater than or equal to 0·500 loading; (2) greater than or equal to 0·30 loading, but not exceeding 0·500.

in PCA, whether a six- or a ten-factor model is taken, the elements appear individually as factors and not collectively as one principal component. The probable reason for this is the differing geo-

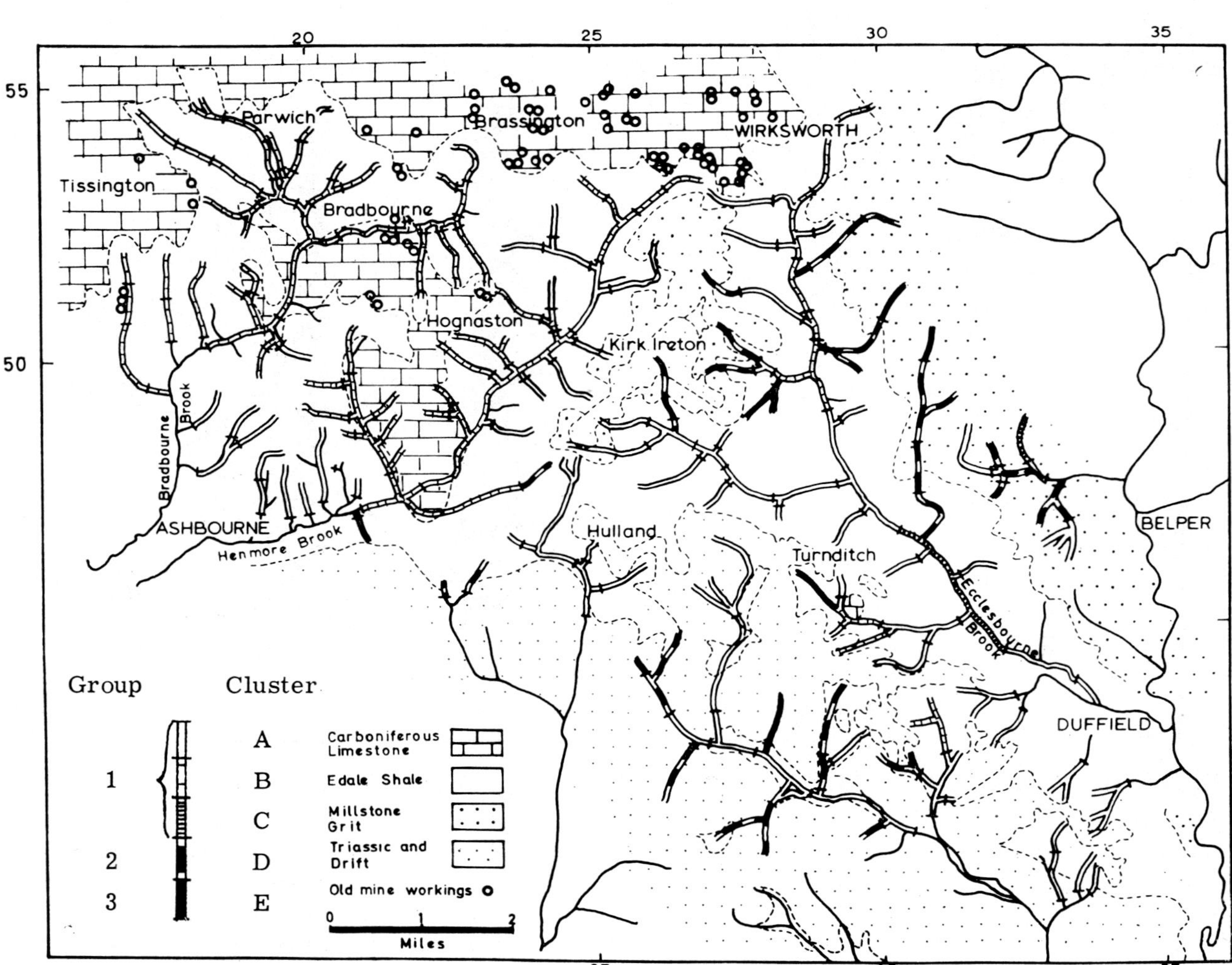

Fig. 14 Q-mode cluster analysis of 170 samples: element/Mn ratios

chemical character of the ore elements in the weathering environment. Another contributing factor is that Ba and Sr are ubiquitous in other lithologies, whereas Pb is largely contributed by the mineralization in the limestone mass. The apparent low values of Zn in the Trias emphasize their dispersion dissimilarities still further.

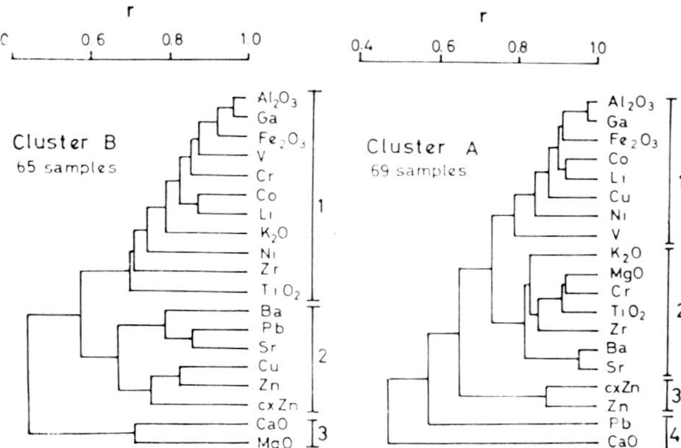

Fig. 15 R-mode cluster analysis (UPGM) of samples comprising clusters A and B of dendrogram for element/Mn ratio. Variables expressed as element/Mn ratios

Some discrepancies were shown by the two methods, however. The association of CaO and MgO is well known in the area, these elements being contributed mainly by the dolomitized limestone. In *R*-mode cluster analysis the two elements are shown to belong to different groups of clusters that are widely separated (Fig. 2(*D*)), whereas in PCA (Table 3) the two elements are grouped to form a factor. MgO also showed as a primary variable in the first factor—suggesting a subsidiary contribution of MgO by rock types other than limestone.

Similarly, Co and Ni, which by *R*-mode cluster analysis appear to prefer a Mn and Fe association, are shown to have significant loadings with different factors in PCA.

These discrepancies are thought to be the result of the strongly two-dimensional implications of dendrograms when attempts are made to simplify multi-dimensional relationships. This method, however, has the advantage of summarizing in one diagram the important elemental associations and the degree to which they are correlated.

It is not possible at present to assess the advantages and disadvantages of both methods with regard to the classification of samples, since *Q*-mode cluster analysis is not comparable with the method of generating factor scores from *R*-mode PCA: they are fundamentally different approaches. Although it is possible to undertake a *Q*-mode PCA, the limitations of the I.C.L.4130 computer storage is a prohibitive consideration. In *Q*-mode cluster analysis the same limitation applies, but it is less severe. It is possible, however, to compare certain *Q*-mode cluster analysis results with their equivalent factor score maps. Fig. 13 represents the *Q*-mode cluster analysis of elements associated with clay minerals, as does Fig. 18, which shows the factor scores for the first principal component for a six-factor model. The *Q*-mode cluster analysis procedure is far superior in indicating the various lithological units, the corresponding factor score lacking coherent geographical grouping. The same could be said when the results of the *Q*-mode analysis of element/Mn ratio are compared with the factor score map of component 2. The grouping is far better with the *Q*-mode analysis than with the corresponding factor score map.

The above comparisons involve a considerable number of elements in both cases. When a few variables or single-element factors are considered, however, an advantage clearly lies with the factor score maps. The use of a few variables, especially when correlated, is not particularly suitable for cluster analysis. Factor score maps based on a few significantly loaded elements are more readily interpretable. Single-element factor score maps, however, contribute very little extra information to that given by the corresponding single-element map.

The closed-array problem for geochemical data, originally discussed by Chayes[29] and amplified by Miesch and co-workers[30] and Miesch,[31] is one of

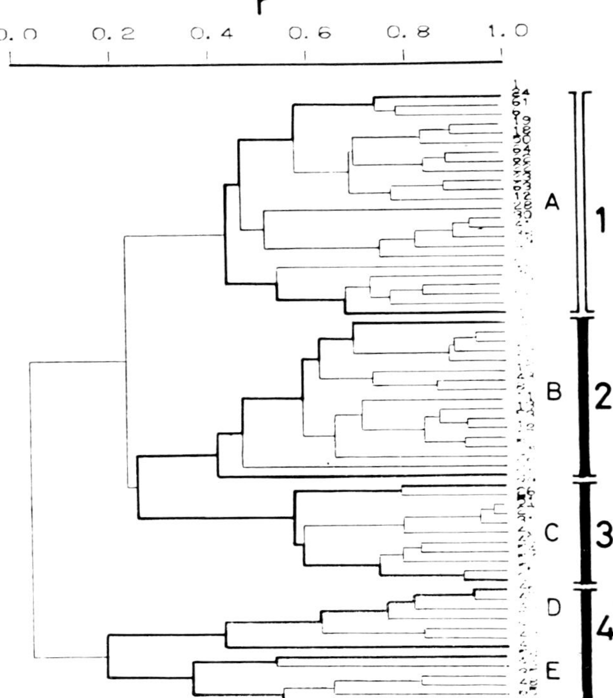

Fig. 16 Q-mode cluster analysis of cluster B: element/Mn ratios

the drawbacks in correlation coefficient analysis. This relates more specifically to major-element analysis, where the sum of the determined constituents must equal 100%. Therefore, an increase in some constituents must necessarily mean a decrease in others, and, hence, a negative difference both from transformed (i.e. taking ratios) and untransformed data. They considered the latter for interpretation. The effect of the constraint due to constant sum is less severe when a considerable number of variables are taken for analysis. In *Q*-mode analysis, where

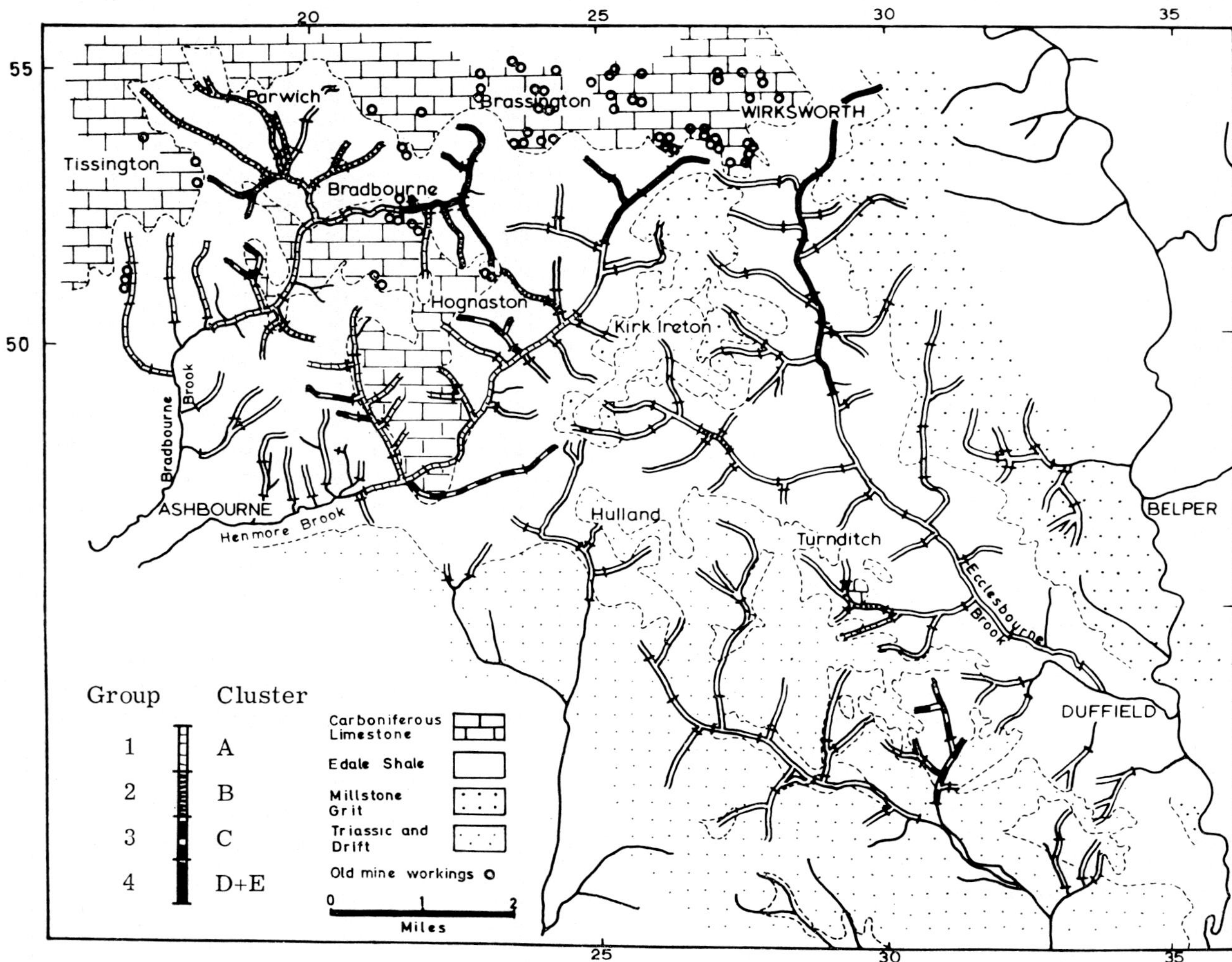

Fig. 17 Q-mode cluster analysis of samples from cluster B (Fig. 15): element/Mn ratios

correlation is built in, even though the relationship is not essentially genuine. This effect is more pronounced in *R*-mode analysis where sums, by rows, of the variables add up to a constant. The overall effect of this constant-sum problem is to mask to a varying degree inherent geochemical correlations, which could render interpretation difficult.

Several methods were tried, notably by Webb and Briggs[32] and Miesch and co-workers,[30] to minimize this constraint. Miesch and co-workers, who employed factor analysis for a set of analytical data on tektites, adopted the method of taking ratios to silica to transform the original data matrix. Webb and Briggs, conscious of this problem of constant sum, applied *R*-mode PCA to nine-element chemical data and found little samples are classified, this problem does not apply.

In the present statistical analysis this problem was minimized by not including the results for SiO_2, which is a major diluent in stream sediments. Moreover, the analytical precision of the method employed for silica is beyond the limits of precision considered acceptable. The data matrix was, therefore, kept as an open array. Clearly, further research should be directed towards this end to evaluate the degree to which the constraint imposed by constant sum affects geochemical correlations.

Conclusions

The multi-element geochemical survey conducted south of the Derbyshire Limestone Dome disclosed complex dispersion patterns of elements mainly influenced by lithology, the secondary environment and mineralization. The patterns related to mineralization are complicated by former mining and smelting activities. It was possible, however, to categorize broad elemental associations and to ascribe such relationships to various geological factors. These associations are: (1) elements associated with clay and resistate minerals; (2) elements associated with Mn and Fe; and (3) elements associated with mineralization.

The first elemental group consists of Al, Ga, K, Ti, Cr, Zr, Mg and Ca. This group depicts the lithological variation and clearly defines the calcareous, arenaceous and argillaceous rock types. The second group (Fe, Mn, Ni, Co, V, Li and Cu) also exhibits a pronounced association with lithology, although local enrichment of Mn and some of these elements is also present. The apparent general relationship of Mn patterns to lithology in this area may not be true elsewhere, such as in peat bog areas, where the secondary environment dominantly influences the distribution of Mn and other associated elements.[28] The elements associated with the prominent type of mineralization (i.e. mineralization in the limestone) are widely dispersed, but are especially predominant near the limestone masses. These elements are Ba, Sr, Pb and Zn. A low degree of correlation is depicted between these ore elements, and this is manifested in their individual dispersion patterns.

Thus, cluster analysis effectively highlighted certain elemental and sample associations that were difficult to deduce by means of the more conventional methods. Cluster analysis involves making fewer assumptions than most methods of factor analysis; nevertheless, the compatibility of the results of *R*-mode cluster and principal-component analyses attest to their effectiveness and to the consistency of the associations revealed. For the classification of samples, *Q*-mode cluster analysis, in particular, seems to be a more attractive method than calculating factor scores from *R*-mode principal-component analysis.

The versatility of cluster analysis was further exploited by subjecting various combinations of elements to *Q*-mode analysis; in each case new

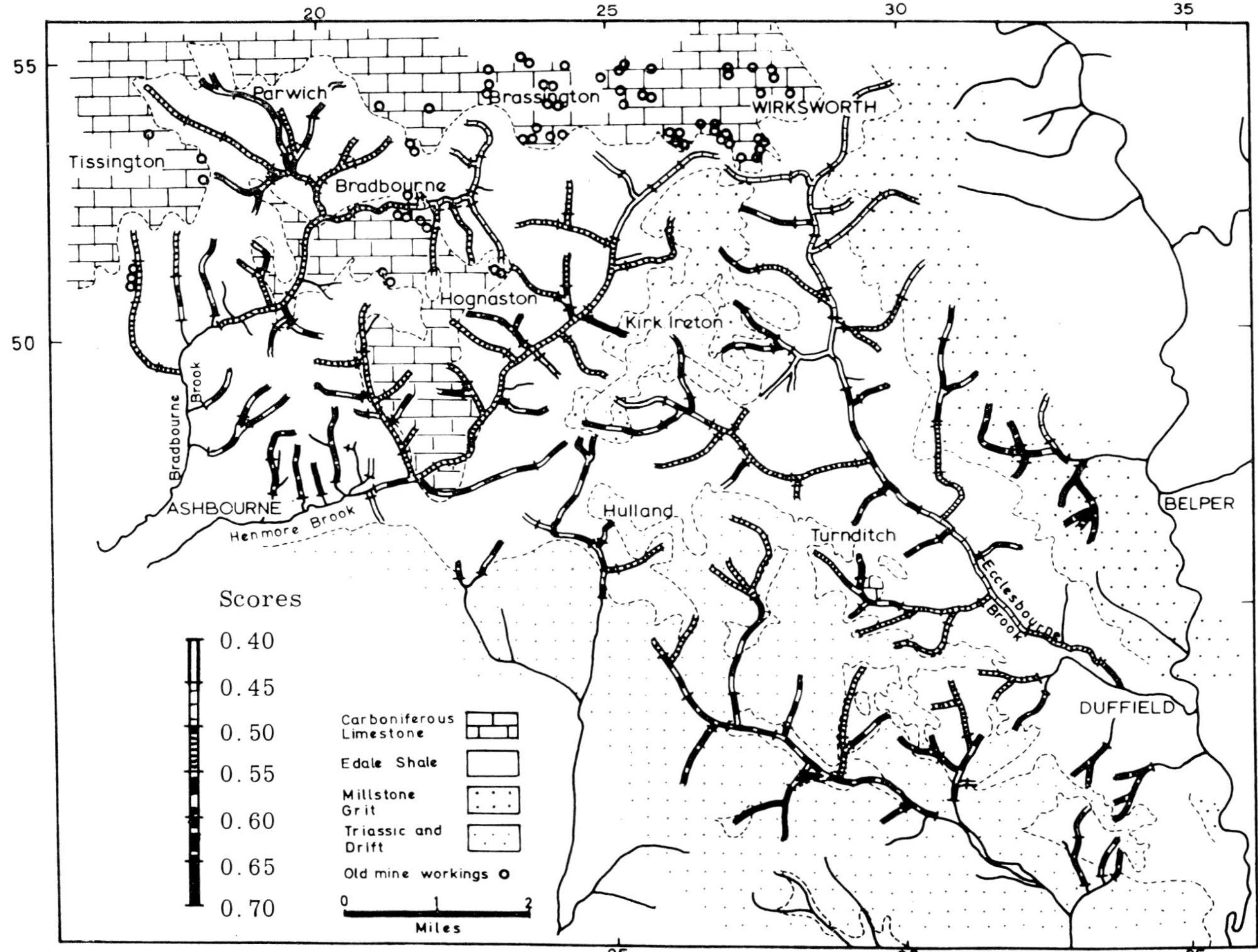

Fig. 18 Varimax scores for factor 1

information was revealed on the similarity of samples. Anomalous samples frequently appeared in these cases as the least correlated members. In certain cases they were also shown as the least correlated samples in a group. Their uniqueness and their small numbers make them dissimilar to the rest of the samples and, hence, they are reflected at low levels of correlation. If too few elements are considered for *Q*-mode cluster analysis, a poor classification of samples is produced; on the other hand, too many variables produce a 'noisy' classification.

Tests on various data input for the multivariate methods revealed that standardized, logarithmically transformed data produced more significant and consistent results. Standardization by ranging from zero to one is preferable to taking means and variances, since the latter assumes a single, normal distribution. The correlation coefficient as a measure of similarity seems particularly suitable to purely quantitative data rather than taxonomic distance.

The limitation on the size of the data matrix imposed by computer storage capacity may be dealt with by (1) reducing sample density, (2) splitting samples according to the stream orders they represent and (3) splitting the area and making a composite map.

Factor score maps are not subject to this limitation. A further development along this line is the multiple-component analysis of McCammon,[22] where an unlimited number of samples could be treated.

It is not proposed that statistical methods should replace the more usual methods of geochemical interpretation: rather they are aimed at extracting and highlighting features of geochemical data that cannot readily be deduced by conventional methods. They offer an obvious advantage in that a large bulk of data can be treated with less subjectivity. They also provide facilities for mixed-mode data treatment, whereby qualitative data could be included in the multivariate analysis, and thus enlarge the scope of examination of the various factors that affect geochemical dispersion.

Acknowledgment

This work was originally carried out as part of a Ph.D. research course by one of the authors (R.C.O.) supervised by the other (C.H.J.). Grateful thanks are due to the Colombo Plan Authority, The Philippine Government and the British Council for providing financial support for the project.

Particular thanks are due to Professor P. H. A. Sneath and M. J. Sackin, both of the M.R.C. Unit, Leicester University, for the assistance and encouragement they provided in discussing the application of cluster analysis to geochemical data, and also for the use of certain of their computer programs. Sincere gratitude is also extended to the staff of the Leicester University Computer Laboratory, who have given great assistance during the course of this work.

Dr. T. D. Ford and R. J. King were most helpful in providing data on the geology and mineralization of Derbyshire. The help of Mrs. N. Farquharson, Mrs. V. Rutherford and Miss K. Langley in the preparation of diagrams is gratefully acknowledged. Thanks are also due to C. C. Codner for permission to use his Derbyshire reconnaissance stream-sediment samples.

References

1. Webb, J. S. Nichol, I. and Thornton, I. The broadening scope of regional geochemical reconnaissance. In *Rep. 23rd Int. geol. Congr.* (Prague: Academia, 1968), vol. 6, 131–48.

2. Codner, C. C. The trace distribution of lead and zinc over shales south of the Derbyshire orefield. M.Sc. dissertation, University of Leicester, 1967.

3. Ford, T. D. and King, R. J. Layered epigenetic galena–barite deposits in the Golconda mine, Brassington, Derbyshire, England. *Econ. Geol.*, **60**, 1965, 1686–701.

4. Sokal, R. R. and Sneath, P. H. A. *Principles of numerical taxonomy* (London and San Francisco: Freeman, 1963), 359 p.

5. Sackin, M. J. Taxonomy package program, 1969. Private distribution.

6. Nichol, I. *et al.* Regional geochemical reconnaissance of the Derbyshire area. *Rep. Inst. geol. Sci.* no. 70/2, 1970, 37 p.

7. Ford, T. D. and Kent, P. E. The Carboniferous Limestone. In *Geology of the East Midlands* Sylvester-Bradley, P. C. and Ford, T. D. eds (Leicester: Leicester University Press, 1968), 59–82.

8. Ford, T. D. The Millstone Grit. In *Geology of the East Midlands* Sylvester-Bradley, P. C. and Ford, T. D. eds (Leicester: Leicester University Press, 1968), 83–94.

9. Taylor, F. M. Permian and Triassic Formations. In *Geology of the East Midlands* Sylvester-Bradley, P. C. and Ford, T. D. eds (Leicester: Leicester University Press, 1968), 149–73.

10. Ford, T. D. The stratiform ore deposits of Derbyshire. In *Sedimentary ores: ancient & modern (revised)* James, C. H. ed. (Leicester: Geology Department, University of Leicester, 1969), 73–96.

11. King, R. J. Mineralization. In *Geology of the East Midlands* Sylvester-Bradley, P. C. and Ford, T. D. (Leicester: Leicester University Press, 1968), 112–37.

12. Ford, T. D. and King, R. J. Mineralization in the Triassic rocks of south Derbyshire. *Trans. Instn*

Min. Metall. (Sect. B: Appl. earth sci.), **77**, 1968, B42–3.

13. Taylor, D. Nichol, I. and Webb, J. S. Enrichment of base metals in the Triassic sandstones of south Derbyshire. *Trans. Instn Min. Metall. (Sect. B: Appl. earth sci.)*, **76**, 1967, B214–5.

14. Davenport, T. G. Geochemical studies of the bauxite deposits of the McKenzie region, Guyana. Ph.D. thesis, University of Leicester, 1970.

15. Garrett, R. G. The determination of sampling and analytical errors in exploration geochemistry. *Econ. Geol.*, **64**, 1969, 568–79.

16. Flanagan, F. G. U.S. Geological Survey standards II: first compilation of data for the new U.S.G.S. rocks. *Geochim. cosmochim. Acta*, **33**, 1969, 81–120.

17. Obial, R. C. Cluster analysis as an aid in the interpretation of multi-element geochemical data. *Trans. Instn Min. Metall. (Sect. B: Appl. earth sci.)*, **79**, 1970, B175–80.

18. Obial, R. C. Multi-element geochemical studies in the southern Pennine orefield and north Leicestershire, England. Ph.D. thesis, University of Leicester, 1970.

19. Gower, J. C. A comparison of some methods of cluster analysis. *Biometrics*, **23**, 1967, 623–37.

20. Farris, J. S. On the cophenetic correlation coefficient. *Syst. Zool.*, **18**, 1969, 279–85.

21. Parks, J. M. Fortran IV program for Q-mode cluster analysis on distance function with printed dendrogram. *Kansas Univ. Comp. Contr.* 46, 1970, 32 p.

22. McCammon, R. B. Multiple component analysis and its application in classification of environments. *Bull. Am. Ass. Petrol. Geol.*, **52**, 1968, 2178–96.

23. McCammon, R. B. The dendrograph: a new tool for correlation. *Bull. geol. Soc. Am.*, **79**, 1968, 1663–70.

24. Rohlf, F. J. and Fisher, D. R. Tests for hierarchical structure in random data sets. *Syst. Zool.*, **17**, 1968, 407–12.

25. Thurstone, L. L. *Multiple-factor analysis: a development and expansion of the vectors of mind* (Chicago: Chicago University Press, 1947), 535 p.

26. Parks, J. M. Cluster analysis applied to multivariate geologic problems. *J. Geol.*, **74**, 1966, 703–15.

27. Rohlf, F. J. and Sokal, R. R. The description of taxonomic relationships by factor analysis. *Syst. Zool.*, **11**, 1962, 1–16.

28. Nichol, I. Horsnail, R. F. and Webb, J. S. Geochemical patterns in stream sediment related to the precipitation of manganese oxides. *Trans. Instn Min. Metall. (Sect. B: Appl. earth sci.)*, **76**, 1967, B112–5.

29. Chayes, F. On correlation between variables of constant sum. *J. geophys. Res.*, **65**, 1960, 4185–93.

30. Miesch, A. T. Chao, E. C. T. and Cuttitta, F. Multivariate analysis of geochemical data on tektites. *J. Geol.*, **74**, 1966, 673–91.

31. Miesch, A. T. The constant sum problem in geochemistry. In *Symposium: computer applications in the earth sciences* Merriam, D. F. ed. (New York: Plenum Press, 1969), 161–76.

32. Webb, W. M. and Briggs, L. I. The use of principal component analysis to screen mineralogical data. *J. Geol.*, **74**, 1966, 716–20.

519.272:550.84

The pattern recognition problem in applied geochemistry

R. J. Howarth, B.SC., PH.D., M.I.M.M.

Applied Geochemistry Research Group, Imperial College, London, England

Synopsis

Available pattern recognition techniques, drawn mainly from the field of electrical engineering, are reviewed, and the application of feature selection, pattern classification (discriminant analysis) and pattern analysis (cluster-seeking techniques) methods in exploration geochemistry are discussed. Examples are drawn from regional stream-sediment multi-element survey data.

The collection of data in exploration geochemistry frequently results in the measurement of several variables, normally chemical analyses, for hundreds or even thousands of samples. It could be said that the aim of an applied geochemical survey is, basically, to interpret multi-element data with a view to the location either of specific exploration targets, such as areas of favourable mineral potential, or patterns of element distribution indicative of particular geological or environmental phenomena in the region mapped.

We are concerned here with aspects of multivariate data treatment when we have *a priori* knowledge of the classes to be distinguished and when we do not have this information. The problem of choosing the best subset of measurements from all those available to the investigator is fundamental to both techniques. Attention is drawn to work carried out by electrical engineers in the pattern recognition field which may not be familiar to geochemists investigating the solution of similar problems.

It will be helpful to define some of the terms used in this discussion (following Ball[2]) in order to prevent confusion. By *measurement* we mean one of several numerical values (e.g. element concentrations) which form a component of the pattern. By *pattern* we mean the collection of measurements characterizing the sample on which they were determined and considered as an entity for the purpose of subsequent analysis and classification. A *cluster* may be visualized as a set of points (corresponding to the patterns) contained in a high-dimensional (measurement) space where the point density is high compared to the density in surrounding volumes. This is a 'natural' cluster in the sense of Forgy.[11]

In the discussion which follows we will be concerned with two aspects of the pattern recognition problem: first, techniques for *pattern classification* when we have *a priori* knowledge regarding the nature of the classes, and, secondly, *pattern analysis* when *a priori* knowledge may not be available and is not used in the cluster-seeking. In either case the aim is to minimize the probability of misclassification over all classes. Ball[2] and Wishart,[41] among others, discussed the difficulties of developing cluster-seeking methods, which detect 'natural' clusters, in contrast to minimum-variance and hierarchical methods, which often cause a spurious 'chopping up' of such groups.

Feature selection

The initial data are a series of measurements collected for a variety of reasons both in the field and the analytical laboratory. Frequently, field observations will include both non-quantitative and semiquantitative information—in contrast to laboratory measurements, which (with the increasing use of fully automatic methods) yield values on a continuous-measurement scale. Field data often include coded information regarding sample location, type and lithology, and the nature of the associated rocks. In stream-sediment surveys the presence of manganese–iron precipitation and pH observations are often included, as well as information regarding the type of environment from which the sample was taken. Laboratory analyses will contribute a large number of measurements by a variety of analytical techniques, some of which will necessarily be more time-consuming and costly than others. The decision as to what to measure is usually made subjectively *prior* to the commencement of the exploration project. *A priori* knowledge is helpful in the identification of particularly useful measurements, e.g. the use of 'pathfinder' elements.

The classification is then based on a subset of measurements selected from all the available observations. These selected 'features' are supposed to be invariant or less sensitive with respect to the commonly encountered variations and distortions, and also to contain less redundancies than the original measurement set.[12] The process of feature selection should be undertaken, even subjectively, with regard to the nature of the goal, cost, reliability and associated 'noise', and redundancy among the measurements.

Most published geochemical studies do not include evaluation of measurement reliability (either from the point of view of subjectivity in data recording by the field team or sampling and laboratory error). Similar feature selection criteria could obviously be applied to geological as distinct from geochemical attributes in mineral exploration strategies.

Much work has been done by electrical engineers and others in the pattern recognition field on the problems of selecting the best feature set which has, so far, had little application by geologists. A brief summary of current techniques follows, based mainly on the work of Fu and coworkers.[12–15]

Coordinate transformation

A given n-dimensional measurement space is transformed into a space of lower dimensions such that a good approximation of the pattern characteristics can be achieved in the new space. Most results, so far, have been obtained with linear transformations of the principal-components type. It has been shown[39] that the first k coordinates corresponding to the eigenvectors for the k largest eigenvalues of the covariance matrix will minimize the error (in the mean-square sense) with respect to any other ordering of k out of the n measurements. Alternatively, if we want to choose a set of features which will minimize the mean-square distances between patterns belonging to the same class, then the eigenvectors corresponding to the k smallest eigenvalues must be chosen.[13]

Estimation of the covariance matrix is subject to the difficulties discussed by Ball:[2] (*a*) different data structures can give rise to the same covariance matrix; (*b*) the effect of erroneous data can be considerable in a particular covariance matrix, affecting the statistics of the entire data set; (*c*) polymodality in measurement space will cause the covariance matrix to be an inadequate description of the data structure; (*d*) predominant subsets within the data can overwhelm subsets that occur less frequently; and (*e*) the number of samples needed to adequately estimate the covariance matrix should not be less than roughly ten times the number of measurements. Thus, the optimality of the technique is defined over *all* the classes, and it makes no explicit provision for discrimination between them.[15]

Divergence and separability criteria

These criteria are statistics which reflect either the 'goodness' of each measurement at conveying information about the category to which the pattern belongs, or maximizing the separability of the classes in terms of the (Euclidean) distance between the group means. In general, the improvement in separation between classes on addition of a new measurement depends not only on the magnitude of the correlation coefficients but also on the magnitude of the differences between group means.[14] These criteria depend both on the assumption of a multivariate Gaussian frequency distribution and also reliable estimation of the covariance matrix. Hence, with real-world data the addition of a measurement might not always improve the discrimination between classes. This has been analytically proved on the basis of the mean accuracy of the pattern recognition system by Hughes.[23]

Sequential forward and backward selection

In this case the improvement in classification success rate or class separability is determined as measurements are either successively added to, or deleted from, the initial set. This stepwise procedure is commonly based on either the F-statistic[10] or Mahalanobis' D^2 distance[9] as measures of the amount of additional group separation which would result if that measurement were to be included in the feature set. The assumptions of multivariate Gaussian distributions and homogeneous covariance matrix again apply. Reyment[34] concluded that the Mahalanobis D^2 is relatively robust for moderate amounts of heterogeneity of the covariance matrix. These stepwise selection techniques are sub-optimal, since different solutions may arise between step-forward and step-backward strategies, and the order in which the measurements are considered will also affect the solution.

Fu and co-workers[14, 15] discussed the application of a non-parametric sequential technique and concluded that, in general, the sequential backward selection procedure gave better classification results (better feature sets) than the sequential forward procedure. Comparison of this method with parametric procedures, similar to those described above, showed that since no assumptions are made regarding the nature of the class distributions and the independence between measurements, the non-parametric method is a much more effective classification tool.

Empirical approach

By *empirical approach* is meant the evaluation of the effectiveness of all possible combinations of k measurements out of the n measurements taken, some criterion such as the overall recognition accuracy being used as a measurement of the utility of the feature set. The main disadvantage here is that the number of possible combinations increases very rapidly with n,* as can be seen from inspection of any table of binomial coefficients. Fu and Min[14] concluded, however, that 'to find an absolute optimum feature set, it is necessary to evaluate the effectiveness of *all* the possible subsets of the given features'.

We will briefly consider the practical considerations in implementing these feature selection methods. The difficulty with the empirical method is the large amount of computer time which may be involved when the number of measurements to be evaluated is large, although it is the only method to guarantee an absolute optimum solution. When the number of measurements is large, sequential selection methods are to be preferred in order to save computation time, although their solutions are sub-optimal. The coordinate transformation approach is also fast computationally, but it is defined over the ensemble of classes and makes no explicit provision for discrimination between them. Comparison with the empirical and sequential selection techniques shows that the coordinate transform generally gives less good classification results for the same number of features ($k < n$), but that this difference decreases as k approaches n. This technique may, however, be useful as a decorrelation pre-processor prior to clustering based on a (Euclidean) distance metric. Aspects of feature selection are illustrated later.

* Total number of combinations to be evaluated for n measurements is given by

$$\sum_{k=1}^{n} \binom{n}{k} = \sum_{k=1}^{n} \left(\frac{n!}{k!\,(n-k)!}\right)$$

Pattern classification

Techniques

In pattern classification we have *a priori* knowledge of the classes which we are interested in distinguishing and we aim to find a decision rule which discriminates well among measurements from patterns belonging to these classes. Application of this decision rule to an unknown data set will assign each sample to one of the categories purely on the basis of the feature set. These techniques are commonly known as 'discriminant analysis'.

A variety of different methods for accomplishing discriminant analysis exist, many of which have not yet seen application in geology.

Maximum likelihood decision rule

The application of the maximum likelihood principle to pattern classification makes use of Bayes' rule for conditional probabilities to determine which class was most likely to produce the observed pattern. For a given pattern $\mathbf{X} = (x_1, x_2, \ldots x_n)$ let $p(\mathbf{X}/i)$ be the probability that $\mathbf{X}$ occurs, given that it is a pattern belonging to category i. Let $p(i)$ be the *a priori* probability of occurrence of category i, and $p(\mathbf{X})$ be the probability that $\mathbf{X}$ occurs (regardless of its category). Then Bayes' rule states that

$$p(i/\mathbf{X}) = p(\mathbf{X}/i)p(i)/p(\mathbf{X})$$

It may be shown[12] that if the penalty for misclassification of a sample belonging to the ith class (L) is equal for each class ($L = 0$, $i = j$; $L = 1$, $i \neq j$), then the decision boundary between classes i and j is given by

$$p(i)p(\mathbf{X}/i) - p(j)p(\mathbf{X}/j) = 0$$

the maximum likelihood decision rule. An example of this applied to geochemical data assuming normal distributions with identical covariance matrices was given by Wignall.[40]

Sequential classification

If the observed measurements are sequential in nature, or the cost of making the next observation must be taken into account (e.g. if the use of costly equipment or time is required to take the next measurement), sequential classification methods can be used. A 'trade-off' between the probable misclassification error and the number of measurements to be made can be obtained by taking the measurements sequentially and terminating the process by making a decision as soon as a desirable classification accuracy has been achieved. Parametric and non-parametric techniques for this have been fully described by Fu.[12]

Discriminant functions

In general, the discriminant function $D_i(\mathbf{X})$ associated with the ith category is defined such that the value of $D_i(\mathbf{X})$ is largest for all samples belonging to this class, i.e. $D_i(\mathbf{X}) > D_j(\mathbf{X})$, where $i \neq j$. In measurement space the decision boundary between the ith and jth class is given by

$$D_i(\mathbf{X}) - D_j(\mathbf{X}) = 0$$

Many different forms of discriminant function exist and include the linear discriminant, piecewise linear and polynomial discriminant functions.[32, 38]

Of these, the linear discriminant has appeared most frequently in geochemical applications (e.g. Chayes and Velde,[5] Link and Koch,[28] Rhodes,[35] Wignall,[40] Koch and Link,[26] Kepezhinskas,[25] Cameron, Siddeley and Durham[4] and Govett[16]). Bugayets *et al.*,[3] Connor[6] and Connor and Gerrild[7] have implemented quadratic discriminants.

Potential function

The potential function is a smooth continuous function which reflects the decreasing influence of a sample point in measurement space as the distance between it and surrounding points decreases. Examples of the types of function in use have been noted by Lemke and Fu[27] and Meisel.[30] The empirical discriminant function applied to geochemical data by Howarth[21, 22] could be regarded as a special type of potential function.

Nearest-neighbour classification rule

Under this rule a sample is assigned to the same class as its nearest neighbour in measurement space, or to the class most frequently represented among its r nearest neighbours—in other words, taking a majority vote.

Testing the classifier

Many of the examples of pattern classification in the geochemical literature have derived classification rules based on a set of data for which the classes to which each of the samples belong is known (the *training* set). In few cases, however, has the probable success rate of the classifier been evaluated on a second *testing* set of data for which the classes are known but which are presented to the classifier as unknown samples. This may be a consequence of the fact that standard discriminant function 'packages' often make no provision for the inclusion of separate testing sets, and it has been common practice to use the training set as a testing set following calculation of the discriminant functions by resubstitution. As was shown by Hills,[19] this results in a biased test of the effectiveness of the classifier.

To illustrate this, a set of post-1920 analyses of sedimentary and igneous rocks analysed for the 11 major oxides[21] were divided at random into two groups. The overall percentage of successful classifications (over 7 classes) was then estimated for all 2047 possible element combinations—first, for the training set used as a testing set, and, second, with the independent testing set. The majority of results (Fig. 1) showed an (overoptimistic) bias of up to 30% in the success rates estimated from the training set.

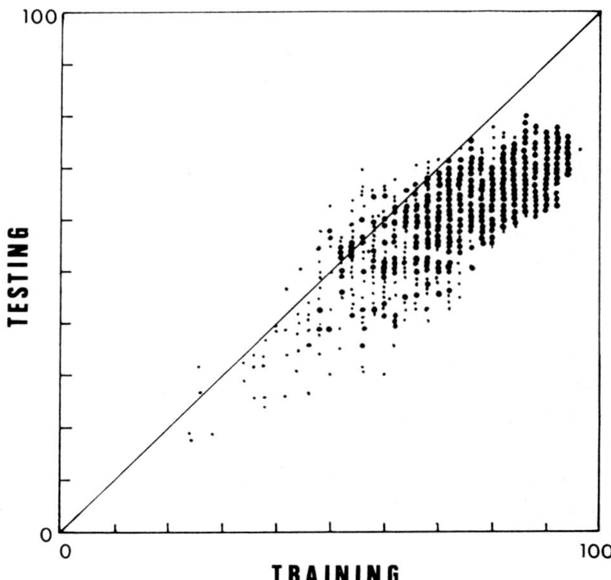

Fig. 1 Comparison of percentage of correct classifications over all classes based on training and independent testing sets for 2047 combinations. Larger dots represent more than one data point

Independent test sets have been used in geochemical studies to evaluate the overall classifier performance,[6, 21] but the small size of the total available data set may make the creation of separate training and testing sets inconvenient. This will be particularly true when the representativity of the data sets is affected by the small sample size.

The optimum solution would seem to be a technique elhich we may call 'training with successive wimination', originally proposed by Kanal and Chandrasekaran.[24] They recommended that in order to partition a limited data set into training and testing sets without overlap *all* the data should be used. Before testing on a given individual, however, it is deleted from the training set, and no item is ever tested on its own training data. With this method a data set of M items yields $(M-1)$ training and M testing items.

Feature selection in pattern classification

The effect of feature selection on pattern classification will be illustrated with a set of regional stream-sediment data in an area of mixed bedrock geology in Devon. We are concerned here with the training set of 153 samples used in an earlier study.[22] Visual estimates from univariate frequency distributions suggested that to distinguish between 5 categories (shales with sandstone, sandstone with shales, granite, basic igneous and areas of Cu and Pb–Zn mining activity) of drainage sediments the elements Pb, Sn, V, Cu, Zn, Ti, Ni, Co, Mn, Cr and As (in order of the measurements as recorded during analysis) would be 'best'. All possible (2047) combinations of these elements were evaluated by use of the empirical discriminant function (*EDF*) algorithm[38] and training with successive elimination.

The 'best' elements considered on their own (Table 1 (*a*)) were vanadium, cobalt and chromium, with overall success rates of 71·9, 67·3 and 67·3%, respectively. From this one might reasonably expect that the optimum combination of two measurements might be V–Co or V–Cr. Table 1 (*b*) shows that this is not so, however, the optimum combination being Ni–V (86·9%)—a great improvement on V–Co (74·5%), V–Cr (78·4%) or Co–Cr (75·2%). It can, therefore, be expected that the geochemist's 'intuitive' estimations of the 'best' combinations of three or more elements may become worse. The effect of adding another feature will, in general, depend not only on the nature of the correlation with features chosen earlier but also on the separation of the means for each category.[14]

Table 1 Comparison of EDF overall success rates for two-feature combinations with results based on single measurements (153 samples, 5 categories) for Devon stream-sediment data

(*a*) *Single measurements*

Measurement no.	1	2	3	4	5	6	7	8	9	10	11
Element	Pb	Sn	V	Cu	Zn	Ti	Ni	Co	Mn	Cr	As
Success rate (%)	52·9	49·7	71·9	60·8	61·4	54·9	65·4	67·3	47·7	67·3	53·6

(*b*) *Combinations* $\binom{11}{2}$ *of measurements i and j*

i \ *j*	2	3	4	5	6	7	8	9	10	11
1	60·1	86·3	67·3	73·9	71·2	69·9	74·5	58·2	79·7	64·1
2		84·3	71·9	77·8	71·2	67·3	70·6	64·1	81·7	72·5
3			81·7	84·3	77·1	86·9	74·5	73·2	78·4	74·5
4				68·6	80·4	81·0	75·8	68·0	86·9	67·3
5					70·6	83·0	75·2	77·1	83·0	66·0
6						75·8	69·3	65·4	77·8	68·0
7							71·9	71·9	76·5	79·7
8								66·0	75·2	77·1
9									68·0	67·3
10										75·2

The frequency distributions of overall success rates for all possible combinations of $\binom{n}{k}$ features (Fig. 2) show that the success rate increases rapidly with k to a certain point and then becomes relatively stable, the addition of a new

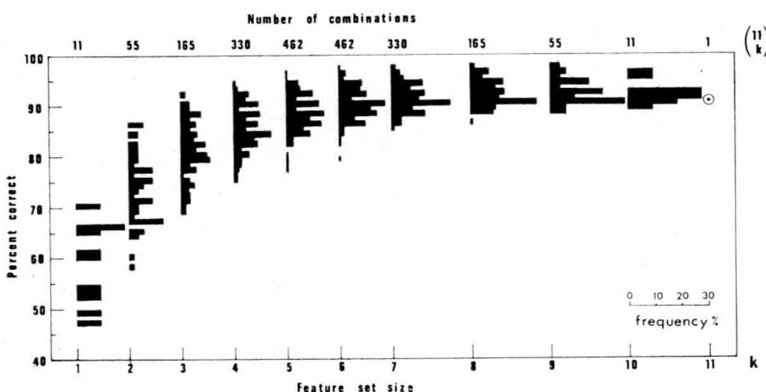

Fig. 2 Frequency distributions of percentage of correct classifications as a function of feature set size (k) for all (2047) combinations of 11 elements using training with successive elimination

feature making no further improvement of the situation. The best result is obtained with a combination of 8 features (98·0%, Table 2), representing a significant improvement over the 11 measurements originally chosen, which gave an overall success rate of 91·5%.

current feature set) show that the step-backward procedure is to be preferred (Fig. 3). This result agrees with the conclusion of Fu, Min and Li[15] based on a different stepwise feature selection technique.

Connor[6] used backward stepwise selection with a linear discriminant function. Testing on the training set, he too found that this led to biased estimates of the overall success rate compared with an independent testing set. Shurygin[37] and Cameron, Siddeley and Durham[4] selected optimum feature sets by use of measures of class separability based on the training set.

Pattern classification of stream-sediment geochemistry

Two examples of the application of different classification techniques to regional stream-sediment reconnaissance data will be briefly discussed to illustrate the implementation of these methods in practice.

Table 2 EDF overall success rates for best element combinations in Devon stream-sediment data

No. of variables k	No. of combinations $\binom{n}{k}$	Overall success rate, %	Features*
1	11	71·9	3
2	55	86·9	3, 7
3	165	92·2	1, 3, 6
4	330	94·1	3, 4, 6, 7
5	462	96·1	3, 4, 6, 8, 10 3, 4, 6, 7, 8
6	462	97·4	1, 3, 4, 6, 7, 8 1, 3, 4, 5, 6, 7, 9
7	330	97·4	1, 3, 4, 6, 7, 8, 9 1, 3, 4, 6, 8, 9, 10 1, 3, 5, 6, 8, 9, 10
8	165	98·0	1, 3, 4, 5, 6, 7, 8, 9
9	55	97·4	1, 3, 4, 5, 6, 7, 8, 9, 10
10	11	96·1	1, 3, 4, 5, 6, 7, 8, 9, 10, 11
11	1	91·5	1, 2, 3, 4, 5, 6, 7, 8, 9, 10, 11
	2047		

*1=Pb; 2=Sn; 3=V; 4=Cu; 5=Zn; 6=Ti; 7=Ni; 8=Co; 9=Mn; 10=Cr; 11=As.

When the number of measurements to be evaluated is large, it is no longer practical to investigate all possible combinations. The alternative is to use a step-forward or step-backward procedure. Results obtained on this same data set by use of the empirical discriminant function with stepwise feature selection (in which the variable causing the greatest improvement in the overall success rate is added to, or deleted from, the

The first example[40] used linear discriminant functions with the Bayes rule to distinguish between mean trace-element data for 460 'drainage cells' (defined on the basis of local stream configurations[8]) in southeastern Pennsylvania and based on 2800 stream-sediment samples. Cells in which mines of known type occurred or for which a 'highly anomalous' mean value occurred were included in the training set (T. K. Wignall,

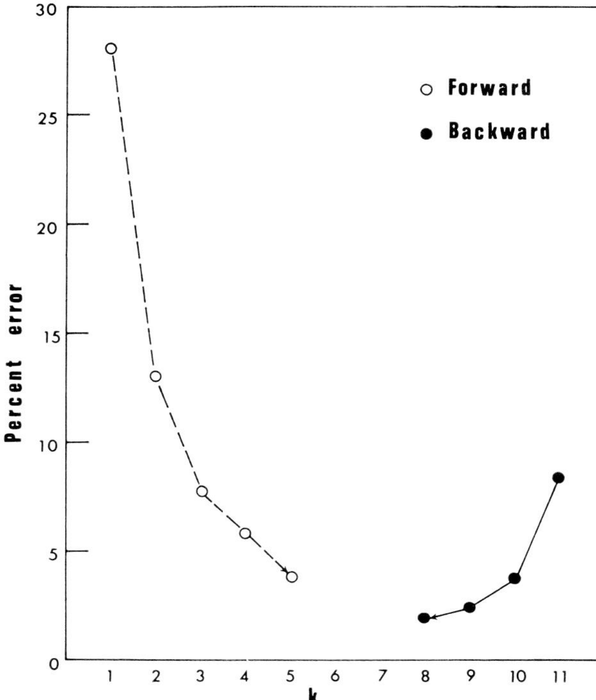

Fig. 3 Percentage of misclassification errors as a function of feature set size (k) for stepwise feature selection methods. 'Best' feature set obtained after 45 and 31 combinations by step-forward and step-backward strategies, respectively, compared with 2047 combinations examined by the empirical feature selection technique (Fig. 2)

personal communication). The results for the training set used as a testing set for mineralization classes—copper, iron, lead–zinc, chromium, copper–iron, silver and gold based on Zn, Cu, Ni, Co, Cr, V, Mn and Fe values—are given in Table 3. It should be noted that Dahlberg[8] pointed out that the 'classification groups entitled "gold" and "silver" should be largely disregarded'

Table 3 Classification success rates for stream-sediment drainage cells in southeastern Pennsylvania containing mineralization. Based on Wignall,[40] Fig. 3

True class	Class assigned Cu	Fe	Pb–Zn	Cr	Cu–Fe	Ag	Au	?
Cu	9	3	3	2	3	2	0	2
Fe	1	4	0	0	0	1	2	1
Pb–Zn	2	1	1	0	1	1	0	0
Cr	1	0	0	3	1	2	1	0
Cu–Fe	1	1	1	1	4	0	1	0
Ag*	0	0	0	0	0	3	0	0
Au†	0	0	0	0	0	0	2	0

Overall successful classifications = 26/61 or 43%.

*Preferably included with Pb–Zn[8].

†Preferably included with Cu–Fe[8].

since they occur as by-products of the lead and copper–iron deposits, respectively. Wignall (personal communication) suggested that the apparently poor classification rate of 43% is improved to 95% (58/61 correct) if the problem is regarded as a two-category one of distinguishing mineralized or unmineralized drainage cells. This illustrates the difficulty of correctly defining the problem to be solved and, hence, adequately defining the data for the training set.

The second example is an application of the empirical discriminant function with the Bayes rule to the classification of approximately 1200 stream-sediment samples from Devon.[22] The area comprises Devonian shales with intercalated sandstones, a Carboniferous sequence consisting of shale or slates with minor sandstones, basic igneous lavas and tuffs in the Dinantian, followed by Namurian shales with thin turbiditic sandstones, and massive sandstones with

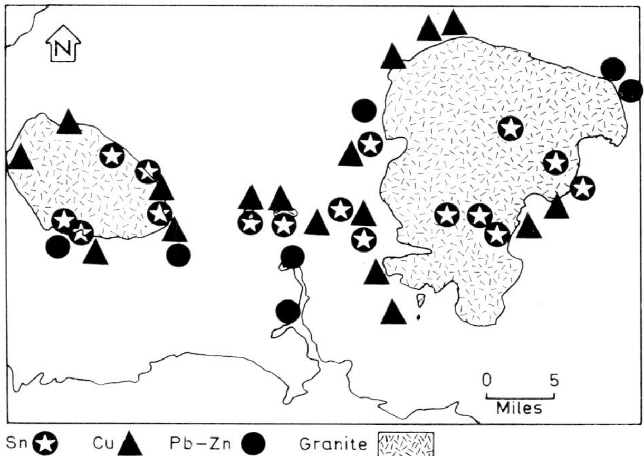

Fig. 4 Principal mineralized districts in Devon stream-sediment survey area

minor shales in the Westphalian; the Permian consists mainly of a mixture of coarse clastics and sandstones. The granite bosses of Bodmin and Dartmoor are associated with tin, copper and lead–zinc mineralization (Fig. 4). The classification results based on *training areas* for seven categories based on $\log_{10}$-transformed Ag, As, Cu, Mo, Ni, Pb, Sn, Ti, V and Zn conformed extremely well with the known geology, and an area of geochemical difference caused by secondary environment changes which enhanced Mn and Fe in the streams over part of Dartmoor was recognized as 'unknown' (Fig. 5). The 'basic igneous' category training set was based on an area of mixed lavas and tuffs associated with Dinantian sediments, and the majority of the samples assigned to this category drain either basic igneous or Dinantian rocks. This again illustrates the importance of selecting representative training data.

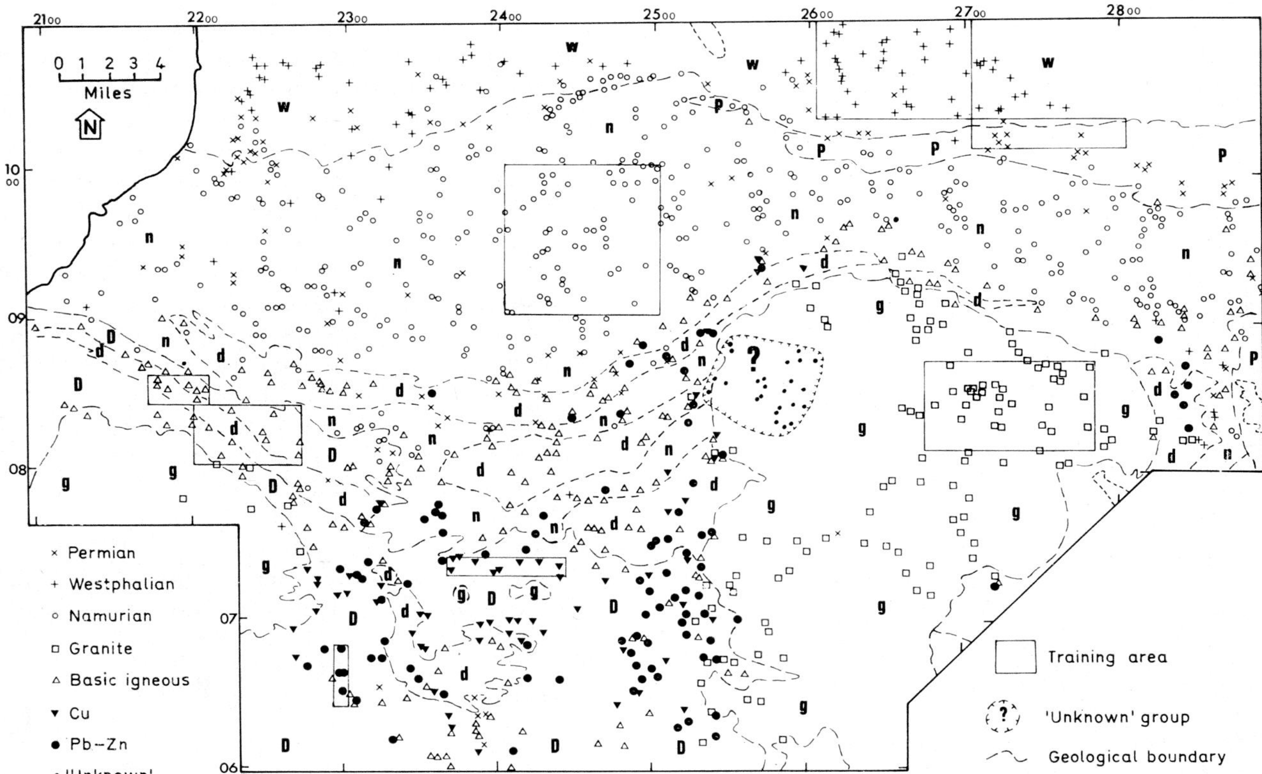

Fig. 5 Results of empirical discriminant function classification of stream sediments in Devon area. Principal geological units are indicated thus: Devonian, D; Carboniferous, Dinantian, d, Namurian, n, Westphalian, w; Permian, p; granite, g. Geology after Institute of Geological Sciences

It may be expected that a certain number of misclassifications arise from the sampling and analytical errors inherent in the data. A possible approach to this is currently being investigated by the writer and Mr. R. Castillo-Muñoz by use of a moving majority vote classification method (analogous to moving average). The loss incurred for an incorrect classification may also be adjusted by use of Bayes' rule (see Howarth[21]) if it is important to detect all samples probably belonging to an important, but 'rare', category such as gold mineralization.

Provided that the training data are adequate, these techniques can be extremely effective, and once the decision rules have been formulated they are extremely fast to implement computationally.

Pattern analysis

In the pattern analysis problem we have a set of multidimensional features representing patterns about which we have no *a priori* knowledge. We must group the samples on the basis of the features alone into clusters which (hopefully) represent an inherent 'natural' category of the observed environment. There are many criteria of similarity on the basis of which the grouping proceeds, but since there is no universally agreed optimal similarity criterion the methods (see, for example, Ball,[2] Wishart[41] and Anderson[1]) tend to be *ad hoc*, results varying with the techniques and parameters used.

The human eye is remarkable in its ability to detect groups of 'similar' points in two-dimensional scatter plots, or fields of similar colour and intensity. These are frequently used to define 'natural' groups of samples on the basis of two measurements. Techniques such as the three-component colour-mapping[29] may be used to detect spatial patterns of three elements visually with considerable success. In higher dimensions, however, the visualization problems defeat the human and we have to rely on procedures which produce some form of classification automatically. A survey of the literature shows that there is a wide diversity of 'clustering' techniques, each with its own difficulties of interpretation, caused both by differences in the algorithms (e.g. divisive and clumping methods) and in visual representation of the results (e.g. as a two-dimensional dendrogram).

Literature on the application of 'clustering' techniques to the interpretation of multi-element geochemical results (e.g. Obial[33] and Rhodes[35]) appears to have concentrated on agglomerative methods. Gower[17] remarked that these have the disadvantages of (1) building clusters by searching for hyperspheres, (2) forcing entries into a hierarchical classification, regardless of the 'natural' groupings, and (3) may have considerable distortion inherent in the two-dimensional tree representation.

Sammon[36] developed a method based on the non-linear mapping of sample points from n-dimensional space to two dimensions such that the inherent 'natural' structure of the data is preserved under the mapping. This is achieved by fitting all the data points in the two-dimensional space such that their inter-point distances approximate the corresponding distances in the original n-space. He demonstrated that the method is highly efficient in identifying hyperspherical, hyperellipsoidal and other complex data structures. The resultant mapping (a two-dimensional scatter plot) is easily appreciated by the human eye and is suited to visual detection of clusters based on point density. This technique is intuitively appealing for use as a tool by the geochemist attempting to analyse complex multi-element survey data with little or no *a priori* information.

Comparison with various sets of geological and geochemical data previously analysed by agglomerative methods suggest that the non-linear mapping is a much better representation of the data structure than the comparable dendrogram (Howarth, in press).

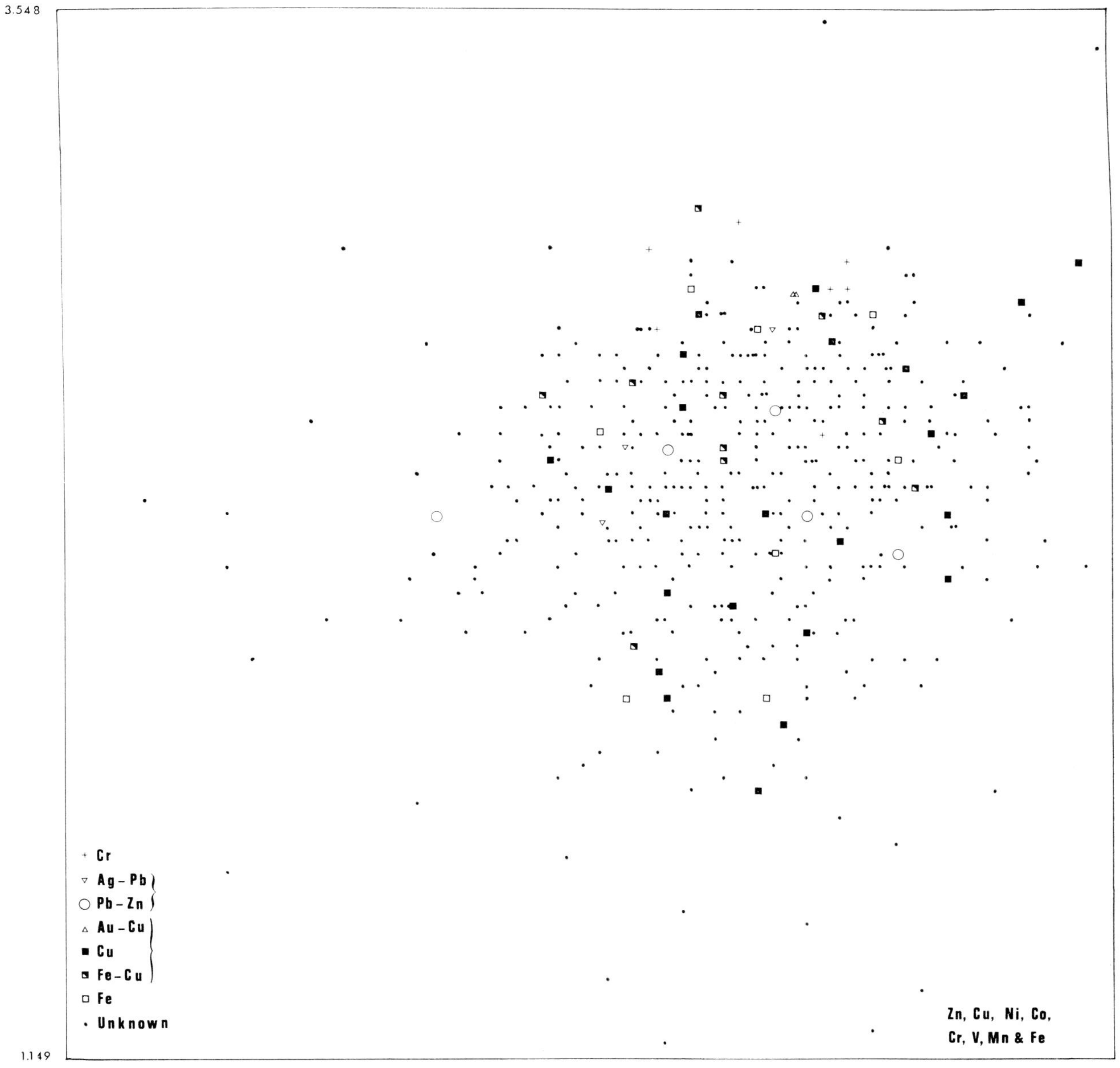

Fig. 6 Non-linear mapping (8-dimensional $\log_{10}$-transformed data) of mineralized training set and 'unknown' data for stream-sediment drainage cell mean compositions from southeastern Pennsylvania

The non-linear mapping algorithm would seem to be ideally suited for interactive computer implementation as an experimental tool in conjunction with pattern classification and factor analysis techniques for visualizing data structure, and it is hoped to begin work to implement such a system shortly.

As an illustration of this potential, consider the non-linear mappings of the two data sets discussed earlier in the pattern classification context. Fig. 6 shows the non-linear mapping (NLM) of the 8-dimensional drainage cell data used by Wignall.[40] Only those cells in the training set known to contain mineralization (Wignall, personal communication) have been shown in Fig. 6 in addition to the samples of 'unknown' type. The mapping would suggest immediately that the separability between the various mineralized cell categories is poor, and the results of Table 3 could have been expected from such an analysis.

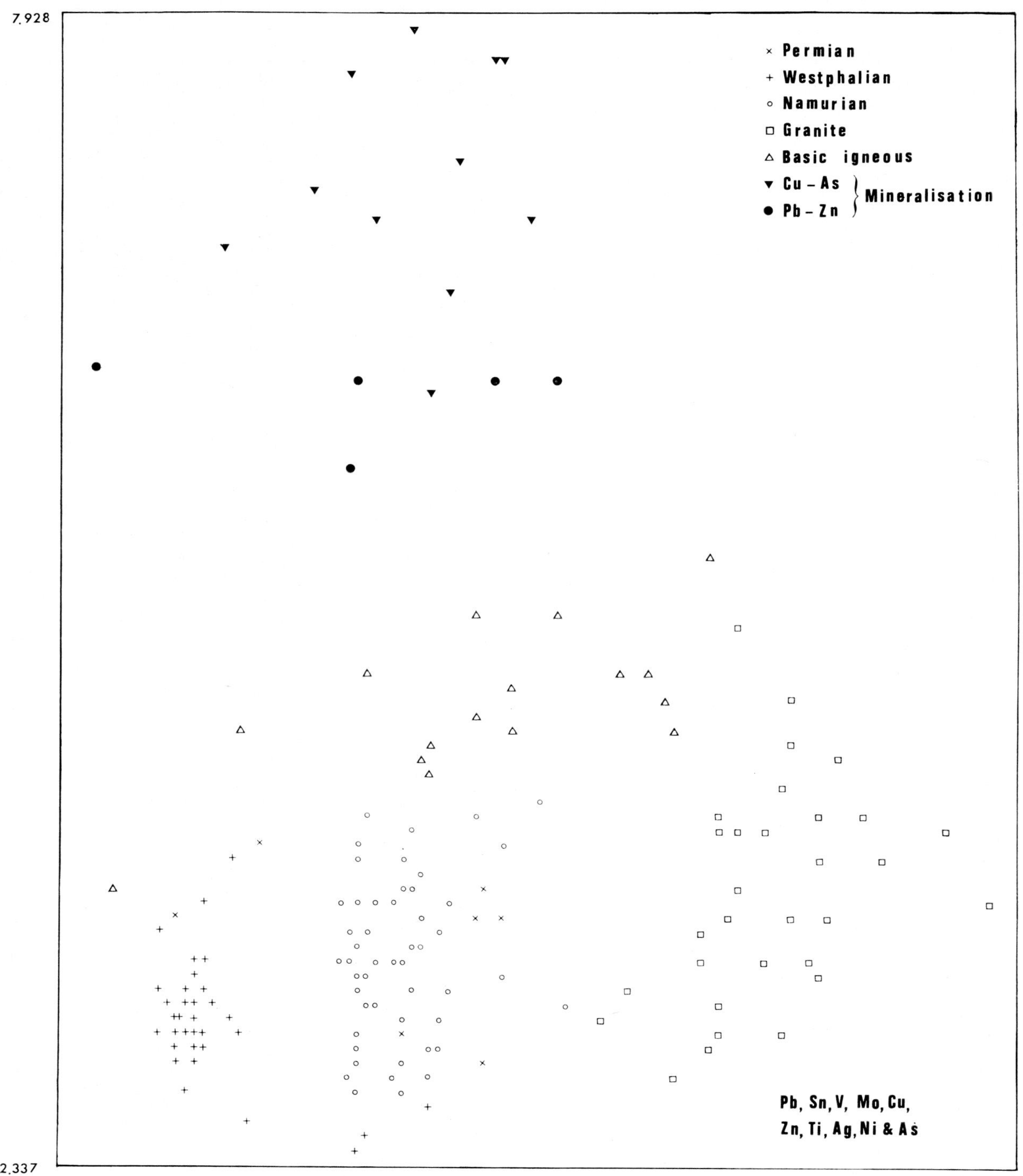

Fig. 7 Non-linear mapping (10-dimensional $\log_{10}$-transformed data) for training set of Devon stream sediments (Fig. 5)

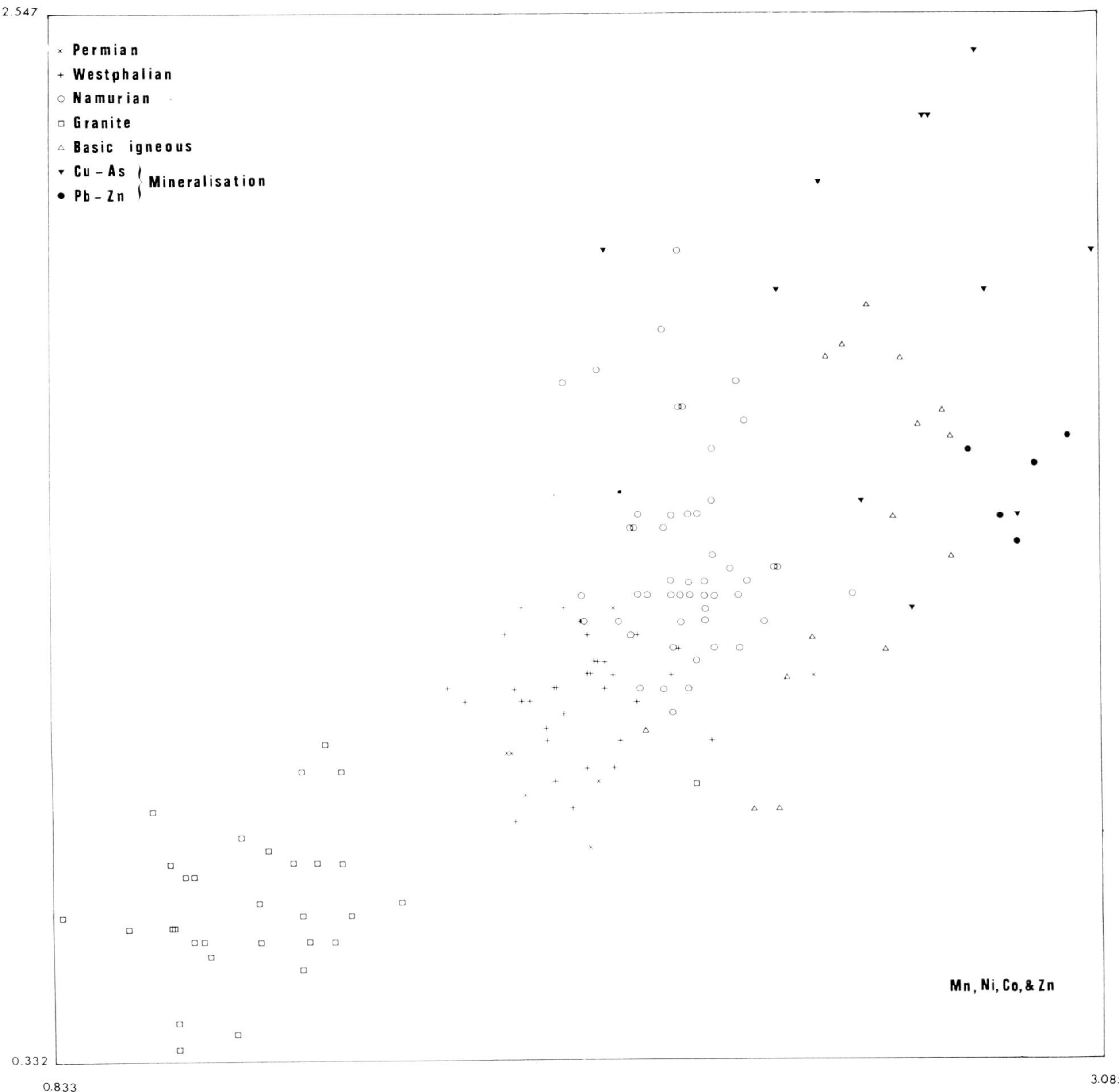

Fig. 8 Non-linear mapping (log_{10}-transformed data) for training set of Devon stream-sediments based on Co, Ni, Mn and Zn only

In contrast, Fig. 7 shows the NLM results for the training samples used for the southwest England study.[22] The separability between the classes is good, with the exception of samples from the Permian, which overlap both the Westphalian and Namurian to some extent and probably account for some of the apparent Permian misclassifications in Fig. 5. J. W. Sammon (personal communication) says that the NLM algorithm is particularly suited to the detection of 'outlying' samples which could then be easily removed from the data set in an interactive environment, and here we have an example where such a technique could have been used with advantage. Fig. 8 shows the linearity of the log_{10}-transformed Mn, Ni, Co and Zn values for the same data. Such a visualization could be an extremely useful check on the probable error to be incurred when applying a linear regression or coordinate transformation to such data, e.g. prior to a factor analysis.

The next example is taken from the postgraduate research of Mr. Castillo-Muñoz, and illustrates the preliminary results of the pattern analysis of exploration reconnaissance stream-sediment geochemical data by use of the NLM method. A regional survey of approximately 800 samples taken at a density of one per square mile over the Denbighshire area of North Wales covered an area of mixed geology.[31] Lower

Palaeozoic shales and mudstones with some interbedded acid volcanics in the southeast of the area are succeeded by Carboniferous shales and sandstones with the Carboniferous Limestone Series at the junction with the Lower Palaeozoic sediments. The sandstones and pebble beds of the Permo–Trias lie unconformably over these rocks in the Vale of Clwyd in the centre of the area and at the eastern edge. Lead–zinc and some copper mineralization is associated with the Lower Palaeozoic rocks and the Carboniferous Limestone Series. Extensive contamination from mine workings and smelters is associated with the ore fields.

peats, peaty gley and podzol soils over the moorland in the western part of the area, correlated with changes in the secondary environment.[20] The second group is related to the mine and smelter contamination of the ore fields in the east. It is hoped to sort samples related to the 'mineralization' around the western ore fields from group 5 samples at a later stage. Thus, NLM analysis with no prior information has satisfactorily sorted out the major features of the geochemical environment.

Moving majority vote techniques have also been successfully applied to the results of this type of analysis to improve the spatial patterns. The NLM

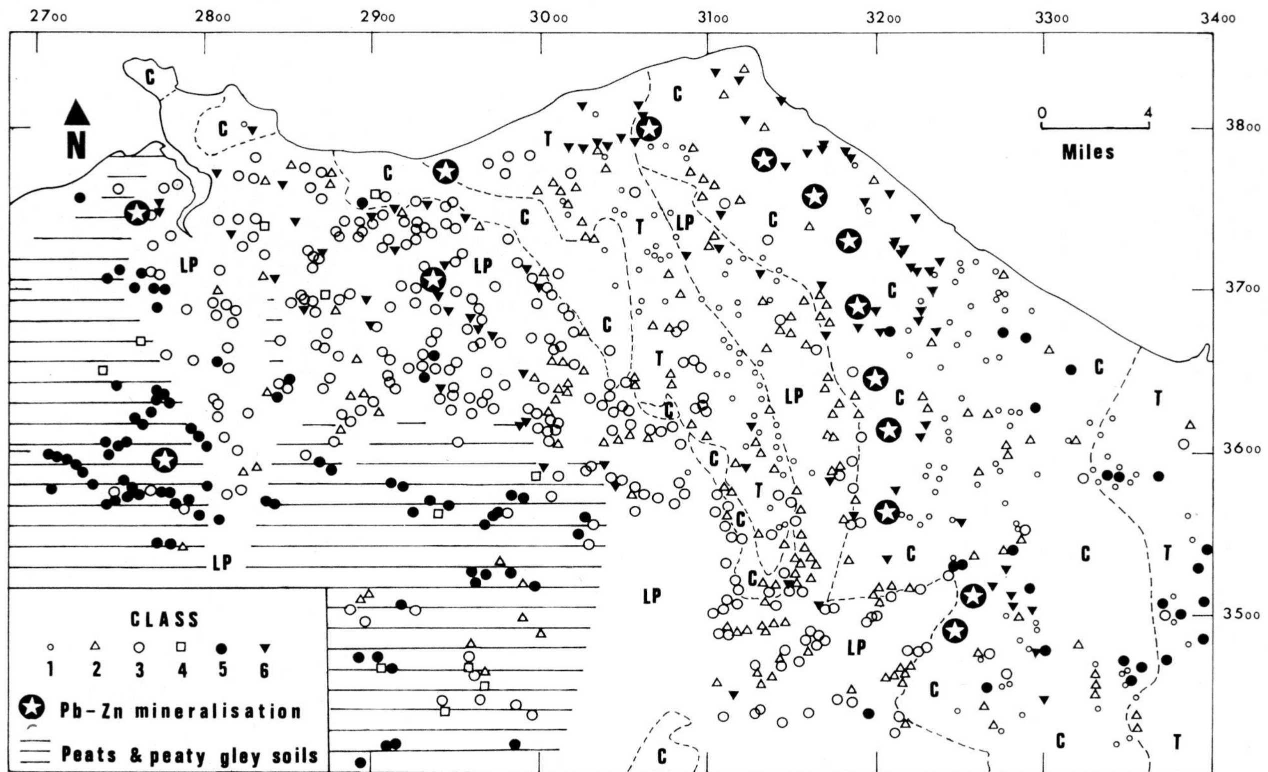

Fig. 9 Clusters obtained by non-linear mapping of stream-sediment compositions in regional survey of Denbighshire, based on 12 elements. Principal geological units are indicated thus: Lower Palaeozoic, LP; Carboniferous, C; Permo–Triassic, T. Geology after Institute of Geological Sciences

Fig. 9 shows the sample distribution resulting from a simplification of the NLM plot into six categories based on Pb, Ga, V, Mo, Cu, Zn, Ti, Ni, Co, Mn, Cr and Fe. Class 1 is associated with the sandstones and shales of the Carboniferous and Permo–Triassic. Classes 2 and 3 are mainly associated with the Lower Palaeozoic: although there appears to be a marked geochemical difference between the western and eastern rocks related to lower Co, Fe, Ni, Ti, V, Cr and Mn contents to the east,[31] no marked lithofacies changes appears to accompany this feature. Two main groups of 'anomalous samples' (classes 5 and 6) exist. The first of these can be related to enhanced trace elements associated with manganese enrichment, particularly in the streams on poorly drained

algorithm also appears to be ideally suited for observing class separability, given *a priori* sample identifications (which are not taken into account when the mapping is calculated) which appear on the plot.

The inclusion of spatial coordinate information has proved difficult to implement by using the (x, y) coordinates as additional variables with methods such as the EDF and NLM algorithms. By the use of majority vote techniques which take into account the adjacency properties of the samples in a sequential (spatial) scan of the data, however, we come some way towards resolving these difficulties. The results of such a technique applied to digitized 10 mile : 1 in maps of Northern Ireland based on K, Ca, Mg and Sr in stream

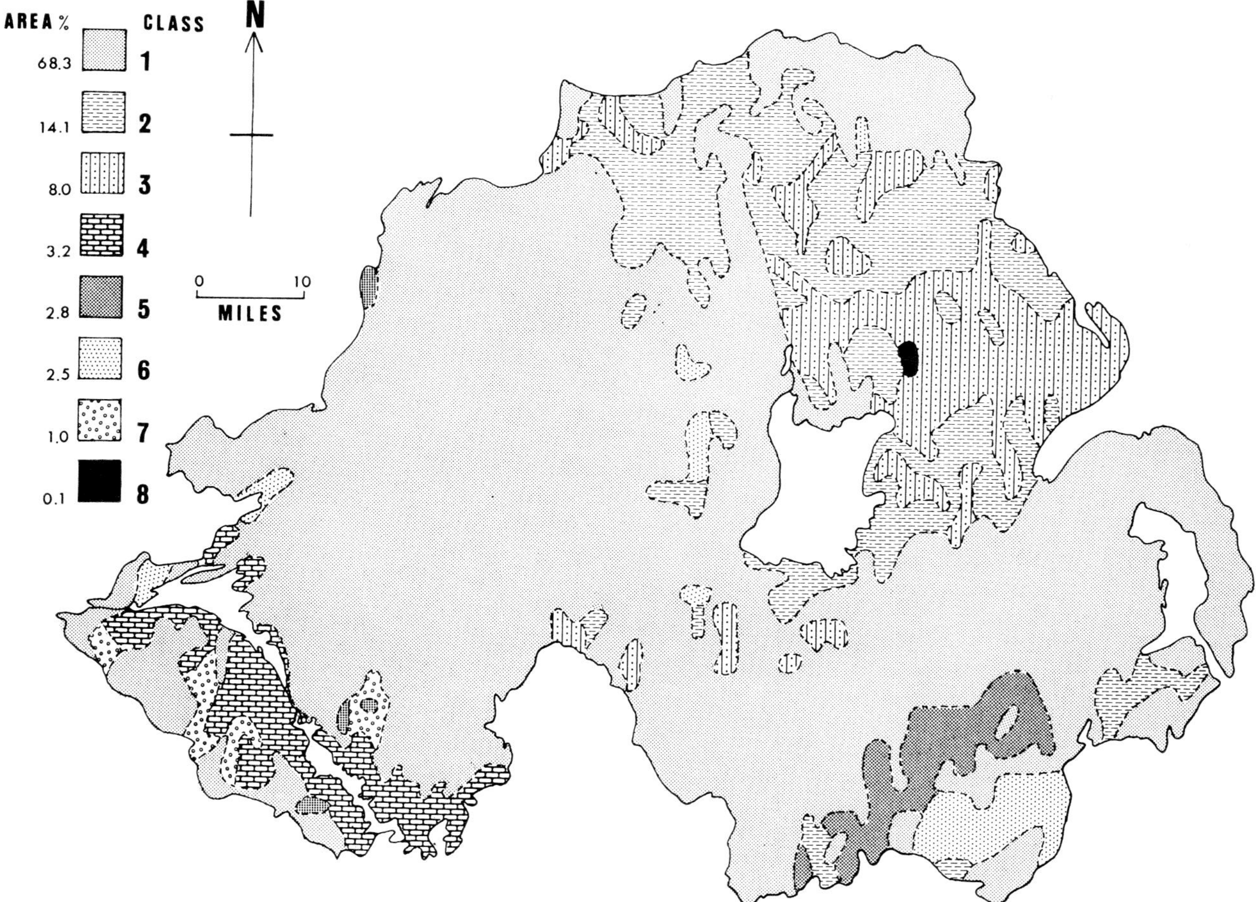

Fig. 10 Results of clustering taking spatial position into account based on regional maps of stream-sediment Ca, K, Mg and Sr distribution over Northern Ireland

sediments are shown in Fig. 10. The area of approximately 5200 square miles was represented by about 3200 map cells. Eight classes account for 100% of the surveyed area (Fig. 10), and are related to the geology as follows: class 1—major sedimentary cover, mainly shales and sandstones; classes 2 and 3—plateau basalts; classes 4 and 7—Carboniferous limestones and associated clastics; class 5—Newry granodiorite; class 6—Mourne Mountains granite; and class 8—rhyolite intrusion. It should be possible to use methods of this type to cluster very large quantities of data based on digitized maps (or moving-average results) where the data are organized into a regular grid of cells.

Once large quantities of regional exploration data have been separated into major 'natural' groups, it should be possible to apply techniques such as factor analysis to each group in turn to clarify the geochemical relationships, without the risk of error inherent in a 'global' approach to the total data set.

Conclusions

The pattern recognition approach to exploration geochemistry embraces both data classification and analysis. The problems of the selection of adequate feature sets and the evaluation of probable classifier performance accuracy seem to have been given insufficient attention in the literature. It would appear that an integrated approach to the data analysis problem centred on interactive methods of computer analysis is desirable. We must seek methods to make the best use of the spatial relationships inherent in geochemical survey data if we are to fully understand the data structure. Techniques for data analysis which do not rely on assumptions regarding the nature of the frequency distributions underlying the data are being developed in other disciplines, and they offer powerful tools for our use. Cluster-seeking and factor analysis methods should be regarded as complementary tools for data analysis.

Acknowledgment

The project of which this paper forms a part has been supported by a Natural Environment Research Council grant for an investigation, under the direction of Professor J. S. Webb, into the applicability of computer methods to the analysis of regional geochemical data. The writer is grateful to Mr. Castillo-Muñoz for permission to include the results of a part of his current research. T. K. Wignall is thanked for supplying the writer with a copy of the data from southeastern Pennsylvania. The non-linear mapping algorithm was originally developed at the Rome Air Development Center, New York: permission to use it is gratefully acknowledged. Computer time has been provided by the Imperial College Computer Centre.

References

1. Anderson, A. J. B. Numeric examination of multivariate soil samples. *Math. Geol.*, **3**, 1971, 1–14.
2. Ball, G. H. Data analysis in the social sciences: what about the details? *AFIPS Conf. Proc.*, **27**, pt 1, 1965, 533–59.
3. Bugayets, A. N. *et al.* Some questions of the application of statistical methods in taking decisions in pegmatite exploration and estimation. Paper presented at 2nd Siberian symposium on the application of mathematical methods and computers in geology and geophysics, Novosibirsk, 1967. Abstract in *Geocom. Bull.*, **3**, no. 12, 1970, 262.
4. Cameron, E. M. Siddeley, G. and Durham, C. C. Distribution of ore elements in rocks for evaluating ore potential: nickel, copper, cobalt and sulphur in ultramafic rocks of the Canadian Shield. In *Geochemical exploration* (Montreal: CIM, 1971), 298–314. (*CIM Spec. vol. 11*)
5. Chayes, F. and Velde, D. On distinguishing basaltic lavas of circumoceanic and oceanic-island type by means of discriminant functions. *Am. J. Sci.*, **263**, 1965, 206–22.
6. Connor, J. J. A geochemical discriminant for sandstones of Mississippian and Pennsylvanian age in Kentucky. *Colo. Sch. Mines Q.*, **64**, no. 3 1969 17–34.
7. Connor, J. J. and Gerrild, P. M. Geochemical differentiation of crude oils from six Pliocene sandstone units, Elk Hills U.S. Naval Petroleum Reserve No. 1, California. *Bull. Am. Ass. Petrol. Geol.*, **55**, 1971, 1802–13.
8. Dahlberg, E. C. Contribution to discussion of reference 40. *Econ. Geol.*, **65**, 1970, 220–2.
9. Davis, J. C. and Sampson, R. J. Fortran II program for multivariate discriminant analysis using an IBM 1620 computer. *Comp. Contr. Kansas St. geol. Surv.* 4, 1966, 8 p.
10. Dixon, W. J. ed. *BMD biomedical computer programs* (Berkeley: University of California Publications, 1968), 600 p.
11. Forgy, E. W. *Detecting 'natural' clusters of individuals* (Santa Monica, Calif.: Western Psychological Association, 1963), 10 p.
12. Fu, K. S. *Sequential methods in pattern recognition and machine learning* (London: Academic Press, 1968), 227 p.
13. Fu, K. S. On the application of pattern recognition techniques to remote sensing problems. *Tech. Rep. Sch. electr. Engng, Purdue Univ.* TR-EE 71–13, 1971, 91 p.
14. Fu, K. S. and Min, P. J. On feature selection in multiclass pattern recognition. *Tech. Rep. Sch. elect. Engng Purdue Univ.* TR-EE 68–17, 1968, 109 p.
15. Fu, K. S. Min, P. J. and Li, T. J. Feature selection in pattern recognition. *IEEE Trans.*, **SSC–6**, 1970, 33–9.
16. Govett, G. J. S. Interpretation of a rock geochemical exploration survey in Cyprus—statistical and graphical techniques. Unpublished work.
17. Gower, J. C. A comparison of some methods of cluster analysis. *Biometrics*, **23**, 1967, 623–37.
18. Griffiths, J. C. Application of discriminant functions as a classification tool in the geosciences. *Comp. Contr. Kansas St. geol. Surv.* 7, 1966, 48–52.
19. Hills, M. The use of the sample in estimating errors in discriminant analysis. *Biometrics*, **21**, 1965, 263. (abstract).
20. Horsnail, R. F. Nichol, I. and Webb, J. S. Influence of variations in secondary environment on the metal content of drainage sediments. *Colo. Sch. Mines Q.*, **64**, no. 1 1969, 307–22.
21. Howarth, R. J. An empirical discriminant method applied to sedimentary-rock classification from major-element geochemistry. *Math. Geol.*, **3**, 1971, 51–60.
22. Howarth, R. J. Empirical discriminant classification of regional stream-sediment geochemistry in Devon and east Cornwall. *Trans. Instn Min. Metall.* (*Sect. B: Appl. earth sci.*), **80**, 1971, B142–9.
23. Hughes, G. F. On the mean accuracy of statistical pattern recognizers. *IEEE Trans.*, **IT-14**, 1968, 55–63.
24. Kanal, L. and Chandrasekaran, B. On dimensionality and sample size in statistical pattern classification. *Proc. natn. Electron. Conf.*, **24**, 1968, 2–7.
25. Kepezhinskas, V. V. Distinguishing between basalt and andesite of the Cenozoic Kurile Kamchataka volcanic province by means of the discriminant function. *Dokl. Akad. Nauk SSSR*, **193**, 1970, 1147–50; *Dokl. Acad. Sci. USSR*, **193**, 1970, 173–5.
26. Koch, G. S. and Link, R. F. *Statistical analysis of geological data, volume 2* (New York: Wiley, 1971), 438 p.
27. Lemke, R. R. and Fu, K. S. Application of the potential function method to pattern classification. *Proc. natn. Electron. Conf.*, **24**, 1968, 107–12.
28. Link, R. F. and Koch, G. S. Linear discriminant analysis of multivariate assay and other mineral data. *Rep. Invest. U.S. Bur. Mines* 6898, 1967, 25 p.
29. Lowenstein, P. L. and Howarth, R. J. Automated colour-mapping of three-component systems and its application to regional geochemical reconnaissance. In *Geochemical exploration 1972* (London: Institution of Mining and Metallurgy, 1973), 297–304. (*Proc. 4th Int. geochem. Expl. Symp. 1972*)
30. Meisel, S. W. Potential functions in mathematical pattern recognition. *IEEE Trans.*, **C-18**, 1969, 911–8.
31. Nichol, I. *et al.* Regional chemical reconnaissance of the Denbighshire area. *Rep. Inst. geol. Sci.* 70/8, 1970, 40 p.
32. Nilsson, N. J. *Learning machines* (New York: McGraw-Hill, 1965), 137 p.
33. Obial, R. C. Cluster analysis as an aid in the interpretation of multi-element geochemical data.

Trans. Instn Min. Metall. (*Sect. B: Appl. earth sci*), **79**, 1970, B175–80.

34. Reyment, R. A. Observations on homogeneity of covariance matrices in paleontologic biometry. *Biometrics*, **18**, 1962, 1–11.

35. Rhodes, J. M. The application of cluster and discriminatory analysis in mapping granite intrusions. *Lithos*, **2**, 1969, 223–37.

36. Sammon, J. W. Jr. A nonlinear mapping for data structure analysis. *IEEE Trans.*, **EC-18**, 1969, 401–9.

37. Shurygin, A. M. Finding parameters which carry information for purposes of classification by means of a linear discriminant function. *Dokl. Akad. Nauk SSSR*, **193**, 1970, 1371–3; *Dokl. Acad. Sci. USSR, Earth Sci. Sect.*, **193**, 1970, 113–5.

38. Specht, D. F. Generation of polynomial discriminant functions for pattern recognition. *IEEE Trans.*, **EC-16**, 1967, 308–19.

39. Watanabe, S. Karhunen–Loève expansion and factor analysis, theoretical remarks and applications. In *Trans. 4th Prague Conf. on information theory . . . 1965* (Prague: Academia, 1967), 635–60.

40. Wignall, T. K. Generalized Bayesian classification functions: *k* classes. *Econ. Geol.*, **64**, 1969, 571–4.

41. Wishart, D. Mode analysis: a generalization of nearest neighbour which reduces chaining effects. In *Numerical taxonomy* Cole, A. J. ed. (London: Academic Press, 1969), 282–311.

519.211:550.84:338.58

A statistical approach to the classification of geochemical anomalies

R. I. Dubov, DR.

Institute of Geochemistry, Academy of Sciences, Irkutsk, U.S.S.R.

Synopsis

A theory is developed for the classification of geochemical anomalies with a view to the estimation of the profit to be expected from exploitation of any associated orebodies. In particular, an algorithm is devised for the solution of the inverse problem—the theory of the quantitative characterization of geological targets, and of the processes that affect them, based on observed data, account being taken of the physical nature of these processes.

Many anomalies discovered by geochemical exploration are of no commercial interest. Their classification and display, being of practical interest, is usually based on experience, which geologists do not always have, and is especially difficult when anomalies have contradictory features. The correct conclusion may then only be obtained by quantitative comparison of these features, with the help of mathematics.

The most widely used mathematical methods, however, determine the class to which the anomaly belongs, with an appraisal of the error risk of this classification, but without calculation of the commercial potential. It is evident that an absolutely correct, unique classification makes it possible to take a decision which is also economically correct. But the solution is not absolutely correct and unique because the probability of the anomaly belonging to various classes is different from zero. The minimizing of the classification error risk is not yet the minimization of the commercial risk.

Let us suppose that the probability estimates for assignment to commercial or non-commercial targets for anomalies of a certain type are, respectively, 0·1 and 0·9. If we take all these anomalies to be non-commercial, then we are right in 90% of the cases. But the consequent refusal to undertake further prospecting of these anomalies may be commercially unjustified if the necessary prospecting is cheap and the profits of finding a commercial deposit in one case out of ten cover all the costs.

It is apparent that the estimate of the profit expectation is to be calculated—such a profit possibly to be obtained by prospecting with a definite method, the probability of anomalies belonging to various classes being taken into account.

The present paper is devoted to the theory of the calculation of such an estimate and to the related theory of the solution of the so-called inverse problem, i.e. the theory of the quantitative characterization of geological targets and of the processes that affect them based on observed data, consideration being taken of the physical nature of these processes.

Initial assumptions

Target classification and profit

Let us introduce the following symbols: q is the anomaly under consideration or the vector of its observed features (concentrations of chemical elements, geological aspects of the anomaly location, etc.); r is the target related to the anomaly or the vector of unknown characteristics of the anomaly (size of the orebody related to it, etc.); A_J is the profit expectation from exploration of the anomaly with method J (a trench or borehole, etc.); $a_J(R)$ is the profit to be actually received following exploration with method J and establishment of the fact that the target r belongs to class R.

If R is the class of non-commercial targets, then $a_J(R)$ is a negative quantity; its positive value is equal to the cost of carrying out work associated with method J.

In the simplest case r can belong to one of the classes of the finite population $\{R_j\}$, $j = \overline{1,N}$* (for instance, if $N = 2$, classes R_1 and R_2 of ore-bearing and barren targets are possible).

Let $P(R_j|q)$ be the conditional probability of $r \in R_j$ if the target r is related to the anomaly q. Then, by definition

$$A_J|q = \sum_{j=1}^{N} a_J(R_j) \cdot P(R_j|q). \qquad (1)$$

The use of a continuous space v of classes R is usually more desirable (for example, in anomaly classification on the basis of the quantities of expected reserves of the related ore). In this case, by definition

$$A_J|q = \int_v a_J(R) \cdot p_\psi(R|q) \cdot dv \qquad (2)$$

where dv is the element of the space v with point R; $p_\psi(R|q)$ is the probability density of the case $r \in R$ by q.

The quantities $P(R_j|q)$ and $p_\psi(R|q)$ can be estimated on the basis of previous experience in prospecting for anomalies of the type q, i.e. on the basis of the known data (training material). If we now calculate the quantity $A_J|q$ by equations 1 or 2, we shall find out for which method J this quantity is maximized, i.e. we shall make the optimum decision about the subsequent exploration of the anomaly. In particular, this may be a decision not to prospect the anomaly further.

The calculation of these estimates poses a serious problem. Even the comparatively simple formula 1 is not always available because of the lack of training data for estimating each of the quantities $P(R_j|q)$, taken separately. Therefore, the special use of approximation is necessary for the appraisal of a variable $P(R|q)$ as the function of q and R. Such an approximation is all the more necessary for the quantity $p_\psi(R|q)$ contained in formula 2. It is possible to do this uniformly on the basis of the following assumptions.

Let V be the product of spaces of points q and r. Let us first consider the obvious case when these points are one-dimensional. Then V is a two-dimensional space with the axes q (abscissa) and r (ordinate). Let us suppose that the points q and r of the axes can belong to only one of the classes of the populations, respectively, $\{Q_i\}(i = \overline{1,M})$ and $\{R_j\}(j = \overline{1,N})$, and let us denote each of these classes by an interval of arbitrary fixed length ΔQ and ΔR on the associated axis with centres corresponding to points Q_i and R_j. Then the event $q \cap r$ by $q \in Q_i$, $r \in R_j$ belongs to the class $Q_i \cap R_j$ represented in the space V by a square with its centre at the point (Q_i, R_j) and with sides of length ΔQ and ΔR. We shall call this event $Q_i \cap R_j$ or the point (Q_i, R_j).

In accordance with well-known assumptions of the probability theory

$$P(Q_i \cap R_j) = P(Q_i) \cdot P(R_j|Q_i) \qquad (3)$$

Here

$$P(Q_i) = \sum_{l=1}^{N} P(Q_i \cap R_l). \qquad (4)$$

Therefore, from equation 3

$$P(R_j|Q_i) = \frac{P(Q_i \cap R_j)}{P(Q_i)} = \frac{P(Q_i \cap R_j)}{\Sigma_{l=1}^{N} P(Q_i \cap R_l)} \qquad (5)$$

The result of calculations by use of formula 5 will not change if, instead of the discrete distributions of q, r and $q \cap r$ on the points Q_i, R_j and (Q_i, R_j), we use the uniform continuous distributions of the same quantities on associated intervals of length ΔQ, ΔR and on a square of area $\Delta Q \times \Delta R$, with the same probabilities but with the conventional probability densities being equal. Hence

$$p_\psi(Q_i) = P(Q_i)/\Delta Q; \qquad p_\psi(R_j) = P(R_j)/\Delta R;$$
$$p_\psi(Q_i \cap R_j) = P(Q_i \cap R_j)/(\Delta Q \cdot \Delta R) \qquad (6)$$

All the points of the region, representing one class, are identified in these expressions.

Substituting expression 6 into 5 we have

$$p_\psi(R_j|Q_i) = P(R_j|Q_i)/\Delta R \qquad (7)$$

* This notation means $j = 1, 2, 3, \ldots, N$.

Then we obtain in a similar manner

$$p_\psi(R_j|Q_i)\cdot\Delta R = \frac{p_\psi(Q_i\cap R_j)\cdot\Delta Q\cdot\Delta R}{\Sigma_{l=1}^{N} p_\psi(Q_i\cap R_l)\cdot\Delta Q\cdot\Delta R} = \frac{p_\psi(Q_i\cap R_j)}{\Sigma_{l=1}^{N} p_\psi(Q_i\cap R_l)} \quad (8)$$

Or, after division of the left- and right-hand sides of expression 8 by ΔR

$$p_\psi(R_j|Q_i) = \frac{p_\psi(Q_i\cap R_j)}{\Sigma_{l=1}^{N} p_\psi(Q_i\cap R_l)\cdot\Delta R}. \quad (9)$$

It is now possible to write, because of continuity of the distribution

$$\sum_{l=1}^{N} p_\psi(Q_i\cap R_l)\cdot\Delta R = \int_{V_{Q_i}} p_\psi(Q_i\cap R)\cdot \mathrm{d}R \quad (10)$$

where V_{Q_i} is the section Q_i of space V, i.e. the population of points having an abscissa Q_i in the space V; R are the ordinates of these points (possessing the values associated with classes R_l, $l = \overline{1,N}$). If we substitute expression 10 into 9, we obtain

$$p_\psi(R_j|Q_i) = \frac{p_\psi(Q_i\cap R_j)}{\int_{V_{Q_i}} p_\psi(Q_i\cap R)\cdot \mathrm{d}R}. \quad (11)$$

Now, for the case in which $q\in Q_i$, corresponding to equation 7, let us substitute the quantity $P(R_j|Q_i) = p_\psi(R_j|Q_i)\cdot\Delta R$ in expression 1 and replace the sum by the integral, done in a similar way in expression 10. We then obtain for $q\in Q$, taking into consideration equation 11 and omitting i and j:

$$A_J|Q = \int_{V_Q} a_J(R)\cdot p_\psi(R|Q)\cdot \mathrm{d}R = \int_{V_Q} a_J(R)\cdot p_\psi(Q\cap R)\cdot \mathrm{d}R \Big/ \int_{V_Q} p_\psi(Q\cap R)\cdot \mathrm{d}R \quad (12)$$

These expressions and equation 1 are equivalent, the first of the two equations in equation 12 is also analogous to equation 2 (space V_Q has the same meaning here as space v in equation 2). Consequently, both expressions 12 can be used for both discrete and continuous distributions of r and q. It is convenient to take the associated interval ΔR and/or ΔQ equal to 1, in the case of the discrete distributions of r and/or q; we then obtain the simplest ratio of actual probabilities and introduced probability densities (if both r and q are discrete, then it is possible to take $P(Q_i\cap R_j)$ numerically equal to $p_\psi(Q_i\cap R_j)$).

Let us use the designation of the event $Q\cap R$ by point (Q, R) of the space V and write the second of the two equations in expression 12 in the form

$$A_J|Q = \int_{V_Q} a_J(R)\cdot p_\psi(Q, R)\cdot \mathrm{d}R \Big/ \int_{V_Q} p_\psi(Q, R)\cdot \mathrm{d}R \quad (13)$$

The quantity $a_J(R)$ is supposed to be defined here on the basis of economic considerations. It is therefore sufficient to approximate the function $p_\psi(Q, R)$ for the appraisal of the quantity $A_J|Q$.

Let us note that this operation is usually considerably simpler than the direct approximation of functions of the type $P(R|Q)$ and $p_\psi(R|Q)$, since such an approximation requires the availability of sufficient training data for each of the different fixed classes Q. But the points (Q, R) associated with a variety of known targets can be randomly distributed in the space V, corresponding to the law under consideration.

Principles for an approximation of the functions given for points in the space under consideration have been worked out earlier. These can also be used for the approximation of the function $p_\psi(Q, R)$. They are considered later in this paper. But it is also necessary to take into account the fact that the function $p^*(Q, R)$, approximating the function $p_\psi(Q, R)$, should be subject to limitations related to specific features of the probability density: this function must be non-negative over the entire space V and must also satisfy the equation

$$\int_V p^*(Q, R)\cdot \mathrm{d}V = 1 \quad (14)$$

It is evident that formula 13, and all that was mentioned in connexion with it, is also correct in cases when the features r and q are multidimensional. The symbols Q and R in designation of the point (Q, R) then become sets of its coordinates in associated spaces (the product of these spaces is the space V).

Information, convolution and inverse problems

Every geochemical anomaly is usually characterized by different ratios of concentrations of a series of elements in some number of samples (the number is not defined exactly because anomaly boundaries are ill-defined). The simultaneous use of all information about the training data anomalies would, therefore, require the creation of a function $p^*(Q, R)$ which would be too complicated even for modern computers.

In many techniques the anomaly classification is made by use of ratios of concentrations of several chemical elements observed in only one sample (for instance, in the sample taken from the anomaly maximum), and a considerable part of potentially useful information is not utilized by this simplification.

In order to utilize useful information about the anomalies more completely, it is necessary to 'convolute' it, i.e. to obtain a relatively few characteristics which include the useful information in the large volume of original data. One of the first of such characteristics was the halo productivity approximately proportional to the productivity of the orebody outcrop under a cover of eluvium talus.[4] This characteristic is defined as the integral of anomaly concentration of the element of interest in the halo.

The possibility of relating the residual halo and the most probable sizes of the target connected with it to the mean concentration of the element of interest in this target and the parameter of dispersion characterizing the halo formation processes was found later.[1] Such a convolution of information in the form of characteristics of the exploration target and related processes is the process of the interpretation of an anomaly. This question has an independent importance and it is, therefore, considered at greater length.

It is evident that the more complete is the consideration of the physical nature of the anomaly formation processes, the greater is the accuracy of the interpretation, other things being equal. Much attention has always been paid to similar interpretational problems in geophysics, where these problems were termed 'inverse' problems.

In geochemistry the theory of inverse problems began to develop later—this is connected partly with the complication of geochemical phenomena. But it is now possible to propose very general principles giving rise to the possibility of overcoming these difficulties. Let us formulate the assumptions which form the basis of the theory of the solution of inverse problems.

In connexion with relatively widespread statistical methods, the words 'concentration distribution' usually imply the probability or density probability function of the concentration. But the objective of investigations solving the inverse geochemical problems is the spatial distribution, i.e. the concentration function of coordinates. Further, we shall sometimes omit mention of the statistical or spatial character of a studied distribution if it is obvious from the context or formulas.

Unless the geochemical observations were complicated by noise (errors, fluctuations of concentrations in samples, etc.), we should observe the spatial distribution of concentrations on points (x, y, z), which may be written down in a general form for each element as

$$C_\psi(x, y, z) = F\{C_0(\xi, \eta, \zeta)\} \quad (15)$$

where $\{C_0(\xi, \eta, \zeta)\}$ is the set of function values of the spatial concentration distribution in points (ξ, η, ζ); F is an operator representing the influence of the processes under study. For instance, the function $C_\psi(x, y, z)$ can represent the element distribution in the sampled upper eluvium talus horizon by lithogeochemical surveys on loose rocks. The function $C_0(\xi, \eta, \zeta)$ may then be an unknown distribution of the element in bedrocks. Operator F then represents the redistribution of the substance by exogenetic processes.

If the right-hand side of expression 15 had a known form and a definite number of parameters determining one-to-one the values $C_\psi(x, y, z)$, in the presence of equal number of these values, we should substitute values of $C_\psi(x, y, z)$ in expression 15 and obtain a set of the same number of equations. Solving such a set relative to the unknown parameters, we should then solve the inverse problem.

But in reality the observations $C(x, y, z)$ are complicated by noise. In addition, the information about the specific form of the right-hand side of expression 15 is usually imperfect. Therefore, we have to make use not of equation 15 but the equation

$$C^*(x, y, z) = F^*\{C_0^*(\xi, \eta, \zeta)\} \quad (16)$$

where $C_0^*(\xi, \eta, \zeta)$ and F^* are the function and operator approximating, respectively, the function and operator $C_0(\xi, \eta, \zeta)$ and F; $C^*(x, y, z)$ is the function calculated by use of the observed values $C(x, y, z)$ as an approximation of function $C_\psi(x, y, z)$. If we choose F^* and C_0^* in expression 16 so that the resulting function C^* satisfies the approximation demands in the best way, then we solve an inverse problem. Thus, the solving of an inverse problem becomes an approximation problem. We therefore pass on to the consideration of the propositions that proved to be necessary for such an approximation.

Basic assumption of approximation

Geochemical problems are very varied, and it is necessary to use an approximation for dealing with conditions which are not foreseen by well-known methods. Considerable difficulties are caused by a lack of *a priori* limitations. It is necessary to take into account the following propositions in developing methods of solution for these problems.

There is not a unique correspondence between the observed characteristics of geochemical anomalies and associated geological bodies and processes, because of the complexity of natural phenomena. Therefore, values of $p_{\psi}(R|q)$ are not equal to zero for different classes of R for the fixed feature q. But the expectation of the difference for any targets r is obviously reduced the smaller the difference of features q associated with them, using arbitrary conventional definitions of these differences. Just because of this, we initially explore anomalies of the type observed mainly at already known large deposits and omit more numerous anomalies of the kind found mainly at less important targets (other things being equal, when there is lack of other information or uncertainty about the roles of other observed features). Such an approach leads deliberately to better luck. We can be wrong in individual cases, but the average quantity of the deposits found by this is increased (i.e. the expectation of their difference from large deposits is then reduced) in comparison with exploration of anomalies such as those which occurred mainly at small deposits.

In choosing methods for problem-solving with complicated conditions, and shortage of information, we are also forced to create algorithms, which only, on average, reach optimal solutions, and the expectation of distinguishing the obtained solution from an unknown exact solution is less than that obtainable by other available algorithms.

It is possible to express the corresponding general assumption in the following form by use of the terminology of functional analysis. Let w_0, w_1, w_2 be any objects studied (bodies, processes, functions, algorithms and the like); $u_0 = u(w_0)$, $u_1 = u(w_1)$, $u_2 = u(w_2)$ are features of the objects; L_w and L_u are operators of distance in object space W and feature space U, respectively; $\mathbf{E}$ is the expectation operator. Then, other things being equal, if

$$L_u(u_1, u_0) < L_u(u_2, u_0)$$
$$\mathbf{E}(L_w(w_1, w_0)) < \mathbf{E}(L_w(w_2, w_0)) \qquad (17)$$

then, i.e. the smaller the distance between points, representing the features of any object in space U, other things being equal, the smaller is the expectation of the distance between the points representing these objects in space W; distances in these spaces can be defined in different ways.

In an earlier paper[3] the proof of proposition 17 was produced on the basis of stricter reasoning. Suppose that w_0 is an unknown function; w_1, $w_2, \ldots, w_n$ are available (known) functions and it is necessary to select one of them to approximate function w_0. Suppose also that there are values $u_1, u_2, \ldots, u_n$ of some quantity (feature) u, corresponding to these functions, and the value u_0 is the known feature of function w_0. Then, in accordance with assumption 17, such a function must be chosen as the optimum, for which the quantity $L_u(u_i, u_0)$ is minimum, other things being equal. This means that the quantity L_u or any other quantity identically defining the quantity L_u is in this case the criterion of approximation.

We shall take advantage of this possibility of using diverse features for the choice of function p^* and C^*.

Let us note that the features u can embrace arbitrary information about objects w, including those which identically determine the object w_0. In the last case the symbol $\mathbf{E}$ in expression 17 can be omitted, for the expectation of a fixed quantity is equal to this quantity. Hence, the expression 17 includes, in particular, the case when the available information (contained in the feature u) is sufficient to choose an algorithm which is optimum in each individual case (not only on average).

Assumption 17, together with the conclusion about the choice of an approximating function and a criterion of approximation, can be called the basic assumption of approximation.

Algorithms of calculations

Rank approximation

Suppose that we have events $Q_k \cap R_k$, where $k = \overline{1,N}$, forms the training material. The events may be, for example, the results of exploration of N haloes chosen at random in the study area. It should be remembered that the points (Q_k, R_k) in space V are associated with these events.

Suppose, initially, that we have already chosen the form of the function $p^*(Q, R)$ and it contains a definite number L of free (not fixed) parameters and is subject to evaluation on the basis of the training data. We imply here and below that $L << N$.

The values of free parameters in the function $p^*(Q, R)$ can be found on the basis of the well-known principle of maximum likelihood and the above specification of the probability density approximation, i.e. from the condition

$$\prod_{k=1}^{N} p^*(Q_k, R_k) = \max \qquad (18)$$

by the additional condition (equation 14).

Suppose, for instance, $p^*(Q, R)$ is an exponential parabola of the form

$$p^*(Q, R) = \exp(A_{00} + A_{10}Q + A_{01}R + A_{11}QR + A_{20}Q^2 + A_{02}R^2) \qquad (19)$$

where Q and R are one-dimensional quantities,

A_{ij} are free parameters. If $Q = Q_k$ and $R = R_k$, $k = \overline{1,N}$, we shall obtain a set of N expressions for $p^*(Q_k, R_k)$ from equation 19. Substituting them in equation 18 we obtain an associated condition of the form

$$\exp\left(A_{00}+A_{10}\sum_{k=1}^{N}Q_k+A_{01}\sum_{k=1}^{N}R_k+A_{11}\sum_{k=1}^{N}Q_kR_k + A_{20}\sum_{k=1}^{N}Q_k^2+A_{02}\sum_{k=1}^{N}R_k^2\right) = \max \quad (20)$$

Equation 14 in this case has the form

$$\int_a^b\int_c^d \exp(A_{00}+A_{10}Q+A_{01}R+A_{11}QR+A_{20}Q^2 + A_{02}R^2)\cdot dQ\cdot dR = 1 \quad (21)$$

where a, b and c, d are boundaries of intervals of possible values Q and R. The calculation of the coefficients A_{ij} satisfying conditions 20 and 21 by a given Q_k, R_k is a routine task that can be solved by well-known algorithms (by variation and numerical integration).

The problems with other forms of function $p^*(Q, R)$ are solved in a similar manner.

The algorithm for the selection of the approximating function, when the form of this function is known and it is only necessary to evaluate its free parameters, we shall call the approximation algorithm of the first rank. Often, we do not exactly know the form of the optimum function p^* and ought to choose it by events $Q_k \cap R_k$ from a certain class $\{p_L^*\}$ of functions differing in the number L of free parameters. In particular, the functions p_L^* may be parabolas of various orders or a composite of general normal distributions, etc.

Considering each function p_L^* as a possible optimum one, we ought to apply the approximation algorithm of the first rank to this function. Therefore, we shall further suppose that the free parameters in the functions p_L^* are already fixed by this algorithm. The subsequent choice of the optimum function will be made with the help of the above basic assumption of approximation. For this it is necessary to evaluate a characteristic playing the role of the quantity u in expression 17. Under the conditions of the problem under consideration this characteristic must be calculated for each function p_L^* by utilizing the information contained in the training material, i.e. in the position of the point set $\{(Q_k, R_k)\}$ in space V.

The expectation of the square of distances of the arbitrary set of points $\{(Q, R)\}$ from a point (Q_l, R_l) may be considered as a characteristic of the position of that set relative to the point (Q_l, R_l) (the square of a distance is in some cases more easily calculated than the distance itself). For the observed distribution $p(Q, R)$ this quantity is, by definition

$$\rho_l = \int_V (L((Q, R), (Q_l, R_l)))^2\cdot p(Q, R)\cdot dV \quad (22)$$

where $L((Q, R), (Q_l, R_l))$ is the distance between points (Q, R) and (Q_l, R_l). In particular, if Q and R are one-dimensional quantities, the following definition could be used:

$$L((Q, R), (Q_l, R_l)) = \sqrt{(Q-Q_l)^2+(R-R_l)^2}$$

It is obvious that the calculation of the square of L is preferable since it is not necessary to extract the root.

If the distribution $p(Q, R)$ is defined by a position of a point set $\{(Q_k, R_k)\}$, then the quantity ρ_l determined from expression 22 can be calculated by a simpler formula, corresponding to known expressions:

$$\rho_l = \frac{1}{N}\sum_{k=1}^{N}(L((Q_k, R_k), (Q_l, R_l)))^2 \quad (23)$$

ρ_l characterizes the observed distribution of events $Q \cap R$. Analogous quantities can be calculated for the distribution of events $Q \cap R$ described by functions $p_L^*(Q, R)$ and $p_\psi(Q, R)$ (with the corresponding substitution of the sum by the integral):

$$\rho_{Ll}^* = \int_V (L((Q, R), (Q_l, R_l)))^2\cdot p_L^*(Q, R)\cdot dV \quad (24)$$

$$\rho_{\psi l} = \int_V (L((Q, R), (Q_l, R_l)))^2\cdot p_\psi(Q, R)\cdot dV \quad (25)$$

Quantities $\Delta\rho_{\psi l} = \rho_{\psi l}-\rho_l$ and $\Delta\rho_{Ll}^* = \rho_{Ll}^*-\rho_l$ characterize differences of the distributions of events $Q \cap R$, described by functions $p_\psi(Q, R)$ and $p_L^*(Q, R)$, from that distribution $p(Q, R)$ which is represented by the set of events $Q_k \cap R_k$. Hence, $\Delta\rho_{\psi l}$ and $\Delta\rho_{Ll}^*$ can be used for the creation of an approximation criterion. One of the possible ways of doing this is as follows.

For each point in the set $\{(Q_k, R_k)\}$ we find the nearest one and make pairs $((Q_i, R_i), (Q_j, R_j))$ of them. One point (Q_k, R_k) can be included in several different pairs. We represent each point of each pair as a point (Q_l, R_l) and calculate the associated quantities $\Delta\rho_{Li}^*$ and $\Delta\rho_{Lj}^*$.

Let $n_c(p_L^*)$ and $n_d(p_L^*)$ be the numbers of cases when the quantities $\Delta\rho_{Li}^*$ and $\Delta\rho_{Lj}^*$ for points of one pair have equal and opposite signs, respectively. In a similar way, the numbers $n_c(p)$ and $n_d(p)$ are defined by the ratios of the signs of the quantities $\Delta\rho_{\psi i}$ and $\Delta\rho_{\psi j}$. Let us designate

$$Z_L^* = \frac{n_c(p_L^*) - n_d(p_L^*)}{n_c(p_L^*) + n_d(p_L^*)}; \qquad Z_\psi = \frac{n_c(p_\psi) - n_d(p_\psi)}{n_c(p_\psi) + n_d(p_\psi)}. \tag{26}$$

Let us show that the quantity $\mathbf{E}(Z)$ can be used as a feature for approximation. For this it is necessary to prove that the value $\mathbf{E}(Z_\psi)$ is known, in spite of the uncertainty of the function p_ψ, and that values $\mathbf{E}(Z(p_L^*))$ are changing with the change of number L.

Let us limit ourselves by the case when it is supposed that the observed distribution p differs from that described by the function p_ψ only in connexion with the display of uncorrelated noise. The signs of $\Delta\rho_{\psi i}$ and $\Delta\rho_{\psi j}$ for each point pair $((Q_i, R_i), (Q_j, R_j))$ are also uncorrelated. Hence, the parts of these expressions have equal expectations when these signs are equal or opposite:

$$\mathbf{E}\left(\frac{n_c(p_\psi)}{n_c(p_\psi) + n_d(p_\psi)}\right) = \mathbf{E}\left(\frac{n_d(p_\psi)}{n_c(p_\psi) + n_d(p_\psi)}\right). \tag{27}$$

Corresponding to the second of the two expressions in equation 26, this means

$$\mathbf{E}(Z_\psi) = \mathbf{E}\left(\frac{n_c(p)}{n_c(p) + n_d(p)}\right) - \mathbf{E}\left(\frac{n_d(p)}{n_c(p) + n_d(p)}\right) = 0 \tag{28}$$

Hence, it is proved that the value $\mathbf{E}(Z_\psi)$ is known.

The values of free parameters of each function $p_L^*(Q, R)$ are found by the principle of maximum likelihood. Hence, if the number L increases, then the values of $p_L^*(Q, R)$ tend to the values $p(Q, R)$ of the observed distribution at all points (Q, R). But the functions $p_L^*(Q, R)$ and $p(Q, R)$ define quantities ρ_l and ρ_{Ll}^* corresponding to formulae 22 and 24. Therefore, ρ_{Ll}^* tends to ρ_l (i.e. $\rho_{Ll}^* \to \rho_l$ for all indexes l) as L increases, but this means the decrease to zero of the part of the expressions when both values ρ_{Li}^* and ρ_{Lj}^* are on the same side of the function ρ_l, i.e. the part of the expressions when quantities $\Delta\rho_{Li}^*$ and $\Delta\rho_{Lj}^*$ have equal signs. That means, from the definition of n_c and n_d, that $n_c(p_L^*) \to 0$ and, corresponding to expressions 26, $Z_L^* \to -1$. The expectation of a fixed quantity is equal to this quantity, i.e. it is possible to write in this case $\mathbf{E}(Z_L^*) \to -1$.

If the number of parameters in the function p_L^* decreases, then the distinction of this function from the function p increases, and, hence, corresponding to equations 22 and 24, the expectation of the distinction of the function ρ_{Ll}^* from ρ_l in points (Q_i, R_i) and (Q_j, R_j) increases. Thus, the expectation of the part of the expression when values ρ_{Li}^* and ρ_{Lj}^* are on the same side of the function ρ_l also increases. Then, corresponding to the definitions of $\Delta\rho_{Li}^*$, $\Delta\rho_{Lj}^*$, $n_c(p_L^*)$ and $n_d(p_L^*)$, the expectation of $n_c(p_L^*)/(n_c(p_L^*) + n_d(p_L^*))$ increases, the expectation of $n_d(p_L^*)/(n_c(p_L^*) + n_d(p_L^*))$ decreases, and $\mathbf{E}(Z_L^*)$ increases.

If the function p_L^* is identical to the function p_ψ (and has the same number of parameters), then, obviously, $\mathbf{E}(Z_L^*) = \mathbf{E}(Z_\psi) = 0$. If this identity is absent and the number of parameters in the function p_L^* is less than in p_ψ, then the quantity $\mathbf{E}(Z_L^*)$ certainly increases and becomes positive.

Thus, the change of $\mathbf{E}(Z_L^*)$ with the change of the number L is proved, and this can serve as a feature u. Simultaneously, we established the special feature of the quantity $\mathbf{E}(Z_L^*)$: it has a zero crossing.

The quantity Z is one-dimensional and it is therefore possible to take the simplest definition for the distance between values $\mathbf{E}(Z_L^*)$ and $\mathbf{E}(Z_\psi)$ in feature space. This definition is the absolute value of the difference between these quantities, i.e.

$$L_u(\mathbf{E}(Z_L^*), \mathbf{E}(Z_\psi)) = |\mathbf{E}(Z_L^*) - \mathbf{E}(Z_\psi))|. \tag{29}$$

The quantity Z_L^* is fixed: it is calculated directly from the training data for the function p_L^*. Therefore, corresponding to the property of the expectation of a fixed quantity, $\mathbf{E}(Z_L^*) = Z_L^*$. $\mathbf{E}(Z_\psi)$ corresponding to equation 27 is equal to zero. Therefore, expression 29, defining the approximation criterion, takes the form

$$L_u(\mathbf{E}(Z_L^*), \mathbf{E}(Z_\psi)) = |Z_L^*| = \frac{|n_c(p_L^*) - n_d(p_L^*)|}{n_c(p_L^*) + n_d(p_L^*)}. \tag{30}$$

This takes a minimum value when

$$Z_L^* = \frac{n_c(p_L^*) - n_d(p_L^*)}{n_c(p_L^*) + n_d(p_L^*)} \tag{31}$$

is the nearest to zero. Therefore, Z_L^* can also serve as the approximation criterion (the proximity of this quantity to zero is evaluated by many simpler calculations than is the proximity of quantity 30 to the minimum).

Thus, the optimum approximation function of the given class is a function which, after use of the approximation algorithm of the first rank, has a value Z_L^* nearest to zero. Let us note that it is unnecessary to do calculations for all functions of the class considered. If a zero crossing is discovered by any change of the number L, it is sufficient to make a choice from any functions with adjacent values of L in which this crossing takes place.

Let us call this algorithm for the optimum approximation function search the approximation algorithm of the second rank.

There are cases when it is necessary to decide the question: 'From which class is it better to choose the approximation function?' This question is not considered in detail here as it is solved in a similar manner to that already published for functions of other types.[2] The algorithm of second rank is then used for each of the classes of functions compared; the population of function-representatives is obtained as a result: one function for each class. From these functions one is to be chosen that has the minimum number of parameters. This is the approximation algorithm of the third rank. Cases occur, when two or more functions simultaneously satisfy the criteria mentioned, that require additional information.

Inverse problem solution

For the solution of the inverse problem for each element considered it is necessary to find a function $C^*(x, y, z)$, which is optimum for the approximation of the function $C_\psi(x, y, z)$, corresponding to equations 15 and 16.

By analogy with the previous reasoning we initially suppose that the form of the function $C^*(x, y, z)$ is given and it is necessary only to find the values of the L free parameters.

The set $\{C(x_k, y_k, z_k)\}, (k = \overline{1,N})$ of observed concentration values can be directly used here as the feature of the function $C_\psi(x, y, z)$, and the set $\{C^*(x_k, y_k, z_k)\}$ of values of function $C^*(x, y, z)$ at the identical points can be used as the feature of that function. The distance between such features may be defined differently, depending on the aims and conditions of the investigations. In many cases it is possible to consider that noise in the values $C(x_k, y_k, z_k)$ is conditioned by errors having a symmetrical statistical distribution. It is then possible to take the mean square of the difference of values $C_k = C(x_k, y_k, z_k)$ and $C_k^* = C^*(x_k, y_k, z_k)$ as such a distance, i.e. a quantity

$$D^* = \frac{1}{N} \sum_{k=1}^{N} (C_k - C_k^*)^2 \qquad (32)$$

corresponding to the least-squares theory.

If the values of some function $g(C)$ of the quantity C have a symmetrical distribution, but not values C themselves, then the corresponding transformation must be introduced:

$$D^* = \frac{1}{N} \sum_{k=1}^{N} (g(C_k) - g(C_k^*))^2 \qquad (33)$$

In particular, in geochemical exploration we often deal with rapid spectrographic analyses, in which errors are approximately lognormally distributed, then $g(C) = \log C$.

Values of the free parameters are found in this case by the method of least squares, minimizing the quantity D^*.

The rule for determination of the approximation function $C^*(x, y, z)$ is called the approximation algorithm of the first rank with regard to the above discussion.

If the function $C^*(x, y, z)$ must be chosen from a definite class $\{C_L^*(x, y, z)\}$ of functions differing by the number L of free parameters, then, as before, we suppose in the beginning that the approximation algorithm of the first rank is applied to each function. In that case, all parameters in the functions are fixed, and a possible minimum quantity $D^* = D_L^*$ is associated with each function. This quantity decreases with the increase of the number L of parameters: corresponding to the theory of least squares, $D_L^* \to 0$ if $L \to N$.

By control of the sampling data it is possible to calculate the error estimate, and from that the estimate of quantity D_ψ, which is the mean square deviation of the values C_k (or $g(C_k)$) from corresponding real values.

In this case the values D_L^* and D_ψ can be considered as features of functions C_L^* and C_ψ, and, according to the basic assumption of approximation, such an approximation function C_L^* must be chosen as an optimum one, for which the quantity

$$L_u(D_L^*, D_\psi) = |D_L^* - D_\psi| \qquad (34)$$

is minimum.

Quantity 34 takes the minimum value when the quantity

$$\Delta D_L^* = D_L^* - D_\psi \qquad (35)$$

is the nearest to zero. Hence, expression 35 can be applied as the criterion of approximation. It is not difficult to show that this quantity goes over zero when number L changes from too small to too large values, by analogy with the quantity $Z(p_L^*)$.

Thus, the search for the optimum function C_L^* with the help of the criterion ΔD_L^* is absolutely analogous to the search for the optimum function p_L^* with the help of the criterion $Z(p_L^*)$.

Corresponding to the term introduced above, the described rule of search for the optimum function C_L^* from a given class is the approximation algorithm of the second rank. The variety of classes, from which the approximation functions are chosen, is determined by geochemical phenomena. It is sometimes not clear, *a priori*, which class is optimum; the approximation algorithm of the third rank can then be applied.[1]

The algorithms described above for the cases of search of optimum approximation functions $p^*(Q, R)$ and $C^*(x, y, z)$ differ only by virtue of the approximation criterion. This distinction is immaterial because the choice of approximation criteria was determined only by available information.

There is sometimes a need to solve an inverse problem without information about errors. It is then expedient to use a criterion based on very common statistical conformities with a law, analogous to quantity 31. This has been described in more detail elsewhere.[2]

Thus, we deal with the very general approximation theory, where diverse criteria can be used in accordance with conditions of problems to be solved.

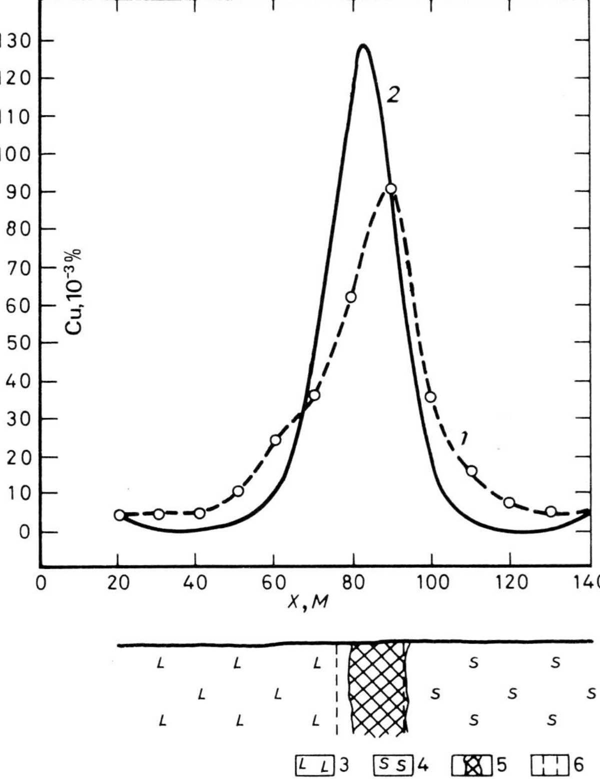

Fig. 1 Results of residual copper halo interpretation (calculations were made by L. S. Vil by use of the author's algorithm), based on a metallometric survey of Kasakh geophysical trust. 1, Cu, concentration values obtained by sampling of loose rocks; 2, curve C_{0L}^* of copper concentration in bedrocks calculated by use of inverse problem solution method ($\sigma^* = 12{\cdot}5$ m); 3, diabases; 4, spilites; 5, extent of orebody as shown by mining exploration; 6, orebody boundaries indicated by the algorithm

One-dimensional criteria were considered above, but in a general case criteria may be multi-dimensional. Limiting ourselves to the bounds of the paper, we cannot consider details of the theory of this aspect. We shall only point out that multi-dimensional criteria may be represented as a set of one-dimensional criteria so the choice of the necessary function is subsequently made for each one of them with the application of the above-mentioned assumptions.

Example

The results of the interpretation of a residual copper halo are shown in Fig. 1. Haloes of this type are encountered fairly frequently. Solovov[4] initiated their investigation. It was shown by Dubov[1] that the spatial element concentration distribution in such haloes at the sampling horizon (near the eluvium talus surface) is approximately expressed, after elimination of the downslope displacement, by the equation

$$C_{\psi}(x, y) = \frac{K}{2\pi \cdot \sigma^2} \cdot \int_{-\infty}^{\infty} \int_{-\infty}^{\infty} C_0(\xi, \eta) \cdot \exp\left(-\frac{(x-\xi)^2+(y-\eta)^2}{2\sigma^2}\right) \cdot d\xi \cdot d\eta \tag{36}$$

where K is a coefficient depending on the distribution of the chemical element by fractions of loose rocks (this coefficient is calculated beforehand) and σ is a parameter of the operator representing exogenetic transformation of the bedrock material (this parameter is proportional to the mean distance passed by particles of the element in the time of the halo formation). $C_0(\xi, \eta)$ is the argument of the operator; this is an unknown function representing the spatial distribution of the element in the upper horizon of the bedrocks; parameters of this function are also parameters of expression 36. In accordance with this, the solution of the problem by the set of values $C_k = C(x_k, y_k)$ is expressed by an approximating function of the kind

$$C^*(x, y) = \frac{K}{2\pi \cdot \sigma^{*2}} \int_{-\infty}^{\infty} \int_{-\infty}^{\infty} C_{0L}^*(\xi, \eta) \cdot \exp\left(-\frac{(x-\xi)^2+(y-\eta)^2}{2\sigma^{*2}}\right) \cdot d\xi \cdot d\eta \tag{37}$$

This function is defined by a parameter σ^* and by the function $C_{0L}^*(\xi, \eta)$. In the case shown in Fig. 1 the function $C_{0L}^*(\xi, \eta)$ is chosen from the class of parabolas. It is possible to show that there are different values of parameters of function $C_{0L}^*(\xi, \eta)$ corresponding to different values of σ^*, when function 37 satisfies criterion 34. Thus, criterion 34 does not give a unique solution. However, it is characteristic that in functions $C_{0L}^*(\xi, \eta)$, determined by this criterion, the minimum values of concentrations between maxima monotonically fall and maxima rise when the value σ^* increases. It is, therefore, sufficient to know the real value $C_0(\xi', \eta')$ at any point (ξ', η') in the region of the anomaly, in order to obtain a unique solution: then such a combination of the value σ^* and function C_{OL}^* has been chosen, where the value $C_{OL}^*(\xi', \eta')$ is nearest to the value $C_0(\xi', \eta')$, together with the satisfaction of the approximation algorithms by criterion 35.

Thus, we use here the approximation algorithm of the second rank with a two-dimensional criterion $(\Delta D_L^*, \Delta C_{OL}^*)$, where

$$\Delta C_{OL}^* = C_{OL}^*(\xi', \eta') - C_0(\xi', \eta')$$

In the given example the calculations were made by use of criterion 34 with several values of σ^*. Minimum values of C_{OL}^* and associated values ΔC_{OL}^* were calculated for functions C^* obtained with this criterion and with the assumption that the real minimum values of the concentrations are equal to the background ones known for the rocks encountered here. Subsequently, the precise values of σ^* were calculated, by which criterion ΔC_{OL}^* is the nearest to zero.

If orebodies have a sufficiently large thickness (in comparison with the quantity σ) and distinct boundaries, then these boundaries can be determined from the curve C_{OL}^* by the bend points (maximum gradient). This is shown in Fig. 1. It is then possible to insert the values of an orebody thickness and its productivity (integral of concentration between the mentioned boundaries) obtained for each element studied into a feature q of the residual haloes for calculations of value A_J. It is often more convenient to use the commercial boundaries determined by definite concentrations C_{OL}^*.

On the whole, it is possible to say that the described principles of inverse problem solution give the possibility of representing the influence of processes inverse to those processes represented by the operator F^* in expression 16.

Additional remarks

The principles of approximation described can be applied not only to those aims to which this paper is devoted directly. In particular, the principles can be utilized in other areas which may benefit from the inverse representation of some process, when the results are expressed not in concentration distribution but in a characteristic of other quantities.

For instance, it is often necessary to evaluate the positions of abyssal tectonic breaks, to which known deposits are related, and it may be necessary to solve this problem mainly from the positions of these deposits. These deposits are not found directly in the breaks, but they are distinct because of numerous imperfectly known factors. The influence of these factors can be represented as an operator of the mutual repulsion of forces. Therefore, if we introduce an inverse operator representing this mutual attraction by a definite law, we can then give more exact information about positions of the mentioned breaks. Such a method was tested by the author and S. I. Kostrovitsky for the specification of the positions of tectonic belts through which the material of diamond-bearing kimberlites had come.

The assumptions considered can be made more precise. In particular, by use of the approximation criterion, it is desirable in some cases to take into account the influence of ties put on the calculated coefficients of the approximating function. Then the optimum number of parameters L may be a little more.

The investigation of the statistical distribution of the approximation criteria values is of interest. We utilized above only those that are expectations or modes out of the various possibilities. Taking into account their statistical distributions, we shall be able to get a value of the accuracy of the conclusions about quantity A_J. In addition, there is the possibility of obtaining the values of that quantity by the solution of some inverse problems without using sizable training material.

References

1. Dubov, R. I. Concentrations of chemical elements in diffusion halos. *Geologiya Geofiz. Novosibirsk*, no. 12 1964, 44–55. (Russian text)

2. Dubov, R. I. On generalized regression and its application to quantitative interpretation of data. *Yb. 1969, Siberian Inst. Geochem.*, 1970, 289–96. (Russian text)

3. Dubov, R. I. On the theory of anomalies classification and inverse problems solutions by geochemical explorations. *Yb. 1971, Siberian Inst. Geochem.*, 1972, 305–9. (Russian text)

4. Solovov, A. P. *Bases of the theory and practice of metallometric surveys* (Alma-Ata: Akademiya Nauk Kazakstan SSR, 1959), 266 p. (Russian text)

519.24:620.113:550.84

Total error and other criteria in the interpretation of stream-sediment data

B. Bølviken, SIV. ING.

R. Sinding-Larsen, SIV. ING.

Both of the Geological Survey of Norway, Trondheim, Norway

Synopsis

As part of a lead exploration project in southern Norway, duplicate stream-sediment samples (*A* and *B*) were collected from some 3000 sample sites near roads over an area of approximately 8000 km^2. The $-0 \cdot 18$-mm material of each of the 6000 samples was analysed for HNO_3-soluble Ag, Co, Cu, Fe, Mn, Ni, Pb, V and Zn by use of atomic absorption spectrometry, *A* and *B* samples being analysed in separate groups at different times. Duplicate sampling enabled the metal variability to be estimated within each sample site (total error).

The following problems were studied on various sub-groups of the data: (1) metal variability within sample sites versus metal dispersion between sample sites (analysis of variance); (2) metal variability within sample sites in relation to metal content and Fe–Mn content at sample sites; and (3) metal content relative to stream order.

For all elements in the sub-groups investigated the dispersion between sample sites was significantly greater than the variability within sample sites. In about 20% of the cases studied Mn showed a greater variability within sample sites than the other elements. An increase of the total error at low concentrations may be due to poor analytical sensitivity (Ag, V). Mn has the largest total error at mean concentrations. A type of plot is introduced where the total error is displayed as a belt on both sides of the empirical frequency distribution curve. Mean metal content is highest in second-order streams, and metal dispersion usually increases with decreasing stream order. The general experience of the investigation is that serious errors may easily be introduced in geochemical data: rigid control is necessary to avoid such errors.

During the Caledonian orogeny the Precambrian rocks which covered large areas of Norway and Sweden were overthrust by an Eocambrian rock suite.[17] Cambrian schists which lie on top of the Precambrian crystalline rocks form the thrust plane. Along the thrust zone a number of lead deposits are known, two of which are economic (Fig. 1). The mineralization is mainly galena, disseminated in autochthonous and allochthonous quartzites.[2,4]

In 1969 the Geological Survey of Norway began a lead exploration project along this zone in southern Norway, geochemical methods being used extensively after an orientation survey carried out in 1967–68 had proved that the known lead deposits caused significant anomalies in the stream sediments.

A regional stream-sediment survey has been undertaken, stream intersections with roads being sampled. To date, some 3000 stream intersections, covering an area of approximately 8000 km^2, have been sampled in this way. At each sample site two sub-samples, *A* and *B*, were collected, the analytical results of each site appearing as duplicates. Since the main purpose of the stream-sediment survey was to locate individual mineral deposits, a reliable estimate of the true metal content at each sample site was required. A complete randomization of the sub-samples was impracticable, and the group of the *A* samples and that of the *B* samples were therefore analysed at different times to avoid the introduction of the

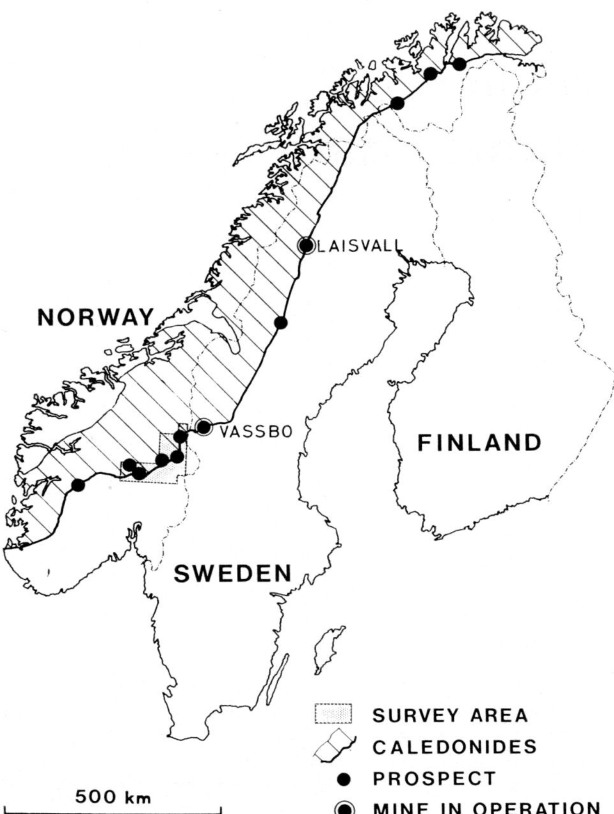

Fig. 1 Lead mineralization along Eocambrian border of the Caledonides

same bias in both sub-samples. The duplicate sampling also enabled this study in data variability to be carried out.

The concepts 'within-site' variability and 'between-site' dispersion can assist in determining whether differences in concentration between sample sites reflect a real trend or merely result from the sum of sampling and laboratory errors.

The following aspects of stream-sediment data have been considered: (1) metal variability within sample sites due to sampling and analytical errors versus metal dispersion between sample sites (analysis of variance); (2) metal variability within sample sites in relation to metal content and Fe–Mn content at the sample site; and (3) metal content in relation to stream order. The aim of the paper is to show the importance of some environmental factors in stream-sediment surveys, and to focus attention on the errors which may normally occur in routine geochemical prospecting data, rather than to devise an experimental design that will reduce the effect of such errors.

Description of survey area

The area is situated on both sides of the border of the Caledonian overthrust in southern Norway (Fig. 1). The topography in the survey area varies from rather smooth in the southern and eastern parts to steep and hilly in the northern and western parts. Altitude ranges between about 100 and 1000 m above sea level. Drainage is generally good, apart from some elevated parts with large bog areas. Overburden consists mainly of glacial till, on average more than 70 cm thick, but varying from zero to several metres.[10] When the overburden is thick, its lithic composition often varies in the vertical direction.[3, 9] The loose material in river valleys is to some extent fluvial or fluvioglacial.

Outcrops are frequent, especially in stream beds. Soils are mostly of the podzol type; both humus and bleached horizon thickness are normally less than 6 cm.[10] Spruce and pine are common up to about 800 m, and birch above this altitude. The lowest part of the area is agricultural land. Climate varies with altitude, the mean temperature being 3–4°C and precipitation ranging from 300 to 700 mm per year.

Streams with severe metal contamination are exceptional and were not considered in the investigation.

Methods

Stream sediments were collected at stream intersections with roads: sample density therefore varies with the density of roads, high density giving a fairly uniform cover of approximately one sample site per 2 km^2 (see Fig. 2).

At each sample site two separate samples, *A* and *B*, were taken 10 m apart and at least 30 m upstream from the road (Fig. 3). The samples were wet-sieved in the field to $-0{\cdot}18$ mm (*ca* 80 mesh). One gramme of the dried sample was digested in hot HNO_3, 1:1, for 3 h, followed by dilution to 20 ml. Ag, Co, Cu, Fe, Mn, Ni, Pb, V and Zn were determined by atomic absorption spectrometry in the diluted solution. The *A* and *B* samples were analysed in separate groups at different times.

Analysis of variance

The theory of error in geochemical data and the general assumptions required for an analysis of variance have been thoroughly reviewed by Miesch.[11] Howarth and Lowenstein[7] have adapted his ideas for the stream-sediment regime and have given a comprehensive summary of the major sources of geochemical sampling errors and their consequences for an analysis of variance.

Raw data relating to the present study are approximately lognormal, the frequency distributions forming more or less straight lines when plotted on logarithmic probability paper (Fig. 4).[8, 18] The data were therefore log-

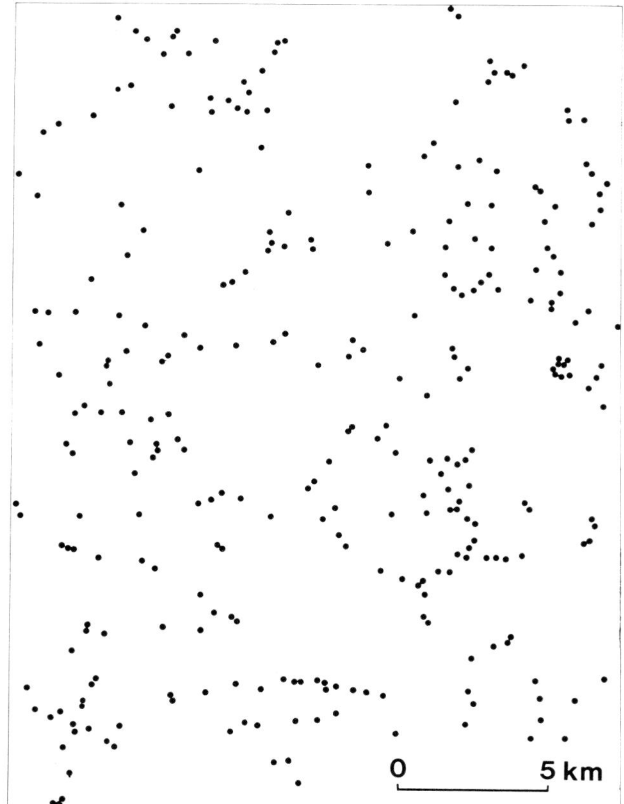

Fig. 2 Stream-sediment sampling at stream–road intersections: example of sample-site distribution. Average density, one sample site per 2 km²

transformed before use in the analysis of variance.

Since the intention, in part, was to test for a 'treatment' effect between the group consisting of

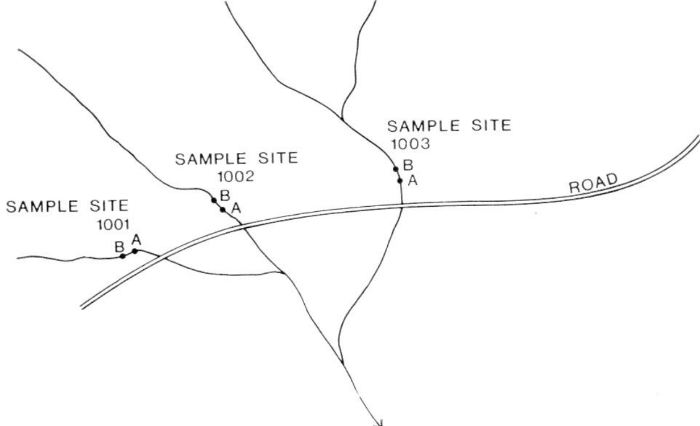

Fig. 3 Stream-sediment sampling model: sample A taken at least 30 m above the road, with a 10-m distance between samples A and B

all the A samples and that consisting of all the B samples in addition to the estimation of the variance components within and between sample sites, the following mixed model was chosen in a two-way parametric analysis of variance:[14]

$$X_{ij} = \mu + a_j + \beta_i + \epsilon_{ij}$$

where i is 1, . . . N and j is 1, 2, X_{ij} is the logarithm of the metal content in the jth sample from the ith locality, μ is the grand mean of the log concentrations of all possible samples in the sample population, a is a fixed effect resulting from the difference between the group of the A samples and that of the B samples due to sampling or analysis (a is the overall bias), β is the difference between the true log concentration at a sample site and the grand mean (β is independent of all ϵ and normally distributed, with mean $\bar{\beta}=0$ and variance σ_{β}^2), and ϵ is the total sample error, including errors in sampling, sample preparation and analysis (ϵ is normally distributed with mean $\bar{\epsilon}=0$ and variance σ_{ε}^2).

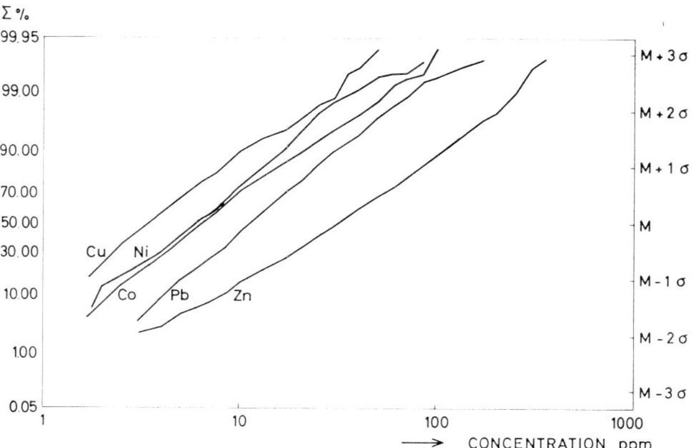

Fig. 4 Cumulative frequency distributions of metal contents in stream sediments. Abscissa gives arithmetic mean of duplicates, left-hand ordinate cumulative per cent and right-hand ordinate scale for estimation of median (M) and standard deviation (σ). 1095 sample sites

It is presumed that neither the variable bias nor the variable precision due to uncontrollable factors is of such magnitude that the necessary conditions for the analysis of variance cannot be fulfilled. It is further assumed that there is no interaction between the two groups of samples (A, B) and sample site. In the two-way analysis of variance the column representing the group of A samples and that representing the group of B samples were considered to be fixed, and the rows representing the sample sites were considered to be random. For a given metal the following two null hypotheses were tested on sub-groups of data:

H_0^1—Mean metal contents are equal at each sample site

H_0^2—Mean of the group of A samples is equal to the mean of the group of B samples

When H_0^1 was rejected, the added variance component among individual sample sites was estimated.

Table 1 shows examples of results from the analysis of variance. For all the elements investigated the dispersion between sample sites proved to be significantly larger than the within-site variability (column 4). This is also true for Ag and V, which showed large variability within sample

Table 1 F ratios from F tests for null hypothesis H_0^1 ($\frac{A_i+B_i}{2}$=constant), H_0^2 ($\Sigma A_i=\Sigma B_i$) and estimated relative magnitudes of variance components in a mixed model of variance

Column 1		2	3	4	5	6	7
						Variance components	
		Group of				Among sites	Within sites
Row	Element	data	N	F_1	F_2	σ_β^2%	σ_ε^2%
1	Pb	1	630	7·0‡	0·6 ns	75·2	24·8
2	Zn	1	630	12·8‡	7·0†	85·5	14·5
3	Ni	1	630	19·3‡	1·8 ns	90·4	9·6
4	Ni	2	1156	4·8‡	97·5‡	56·7	43·3
5	Co	1	630	17·6‡	0·0 ns	88·6	11·4
6	Co	2	1156	9·0‡	0·0 ns	77·8	22·2
7	Cu	1	630	12·4‡	10·7†	85·0	15·0
8	Mn	1	630	12·6‡	4·0*	87·1	12·9
9	Fe	1	630	11·0‡	3·7 ns	83·4	16·7
10	Fe	2	1156	27·1‡	53·8‡	92·3	7·7
11	Ag	1	630	1·9‡	27·9‡	32·5	67·5
12	V	1	630	3·3‡	27·4‡	52·9	47·1

Degrees of freedom for F_1 are $(N-1, N-1)$; for F_2 $(1, N-1)$.
Probability ranges for accepting the null hypothesis: ns=not significant=$P>0{\cdot}05$; *=$0{\cdot}05 \geqslant P>0{\cdot}01$; †=$0{\cdot}01 \geqslant P>0{\cdot}001$; ‡=$P \leqslant 0{\cdot}001$; N, number of sample sites.

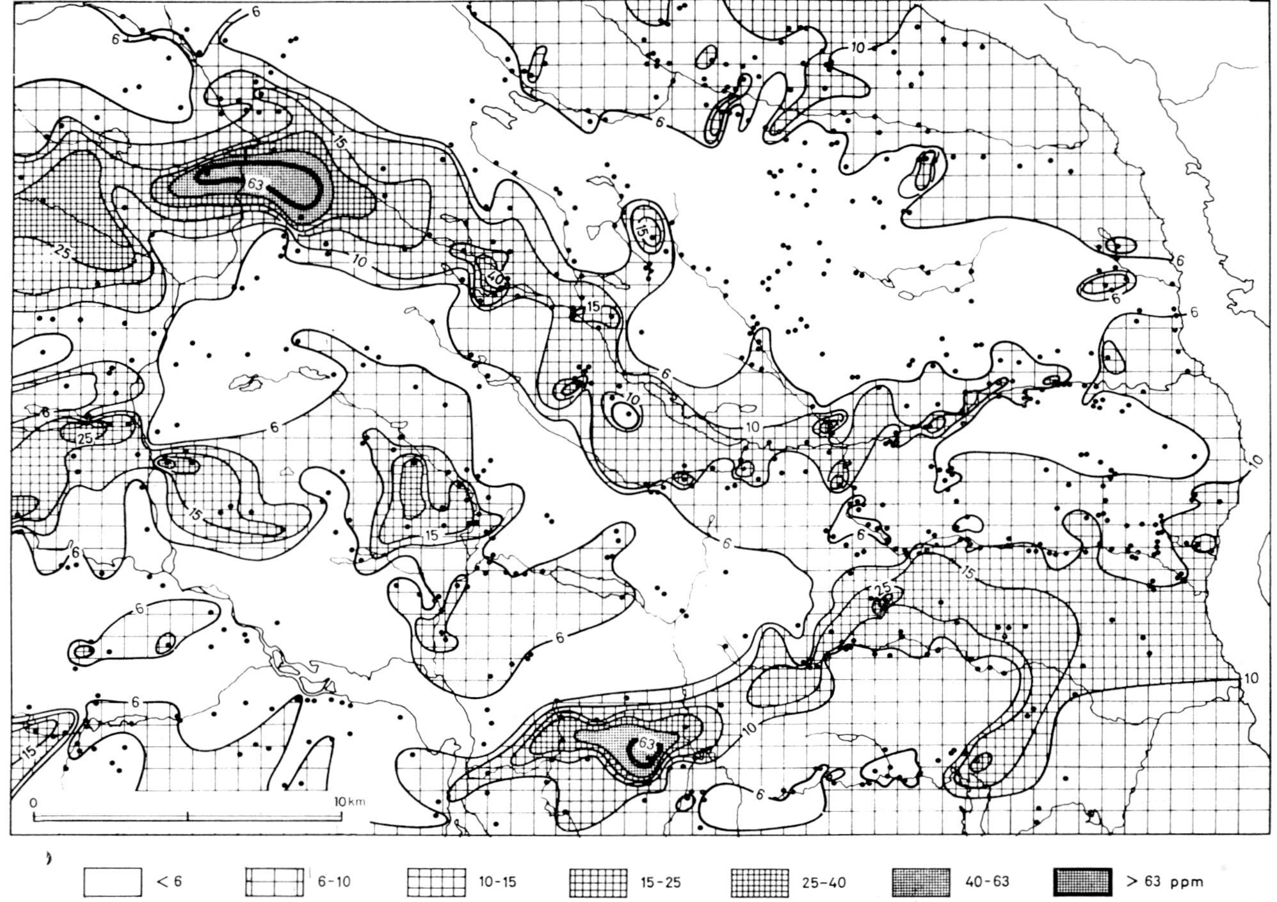

Fig. 5 Contour map of copper concentrations in stream sediments (sample sites indicated by dots)

sites because of poor analytical sensitivity. From the F ratios given in column 5 it is seen that for Ag, Cu, Fe, Mn, Ni, V and Zn there are significant differences between the groups of A and B samples. Since the A and B samples were processed independently, a significant difference between the groups would indicate an overall bias in the sampling or analysis.

When there is a significant difference between the A and B samples, and the within-site variance component is large, the data may be of little value in prospecting. Nickel (row 4) reveals a significant difference between the A and B samples and a relative variance component within sample sites of 43%. Inspection of the data showed a large deviation (400%) between the A and B samples at some 5% of the sample sites. Re-analysis proved that this deviation was caused by a laboratory bias in the A samples.

When there is a significant difference between the A and B samples, and the within-site variance component is relatively small, there can be no large deviation between *many* duplicates. Two possibilities exist: (1) there is a minor deviation between duplicates at many sample sites—this is not very serious in prospecting (see, for example, iron, row 10, which shows a significant difference between the A and B samples and a relative variance component within sample sites of 8%; (2) there is a major deviation between duplicates at a few sample sites—this can be serious if it is not detected by inspection of the data or by computer treatment.

With no systematic sampling bias, and an adequate sample treatment, the difference between the means of the group A samples and the group B samples should be negligible. Table 1 shows that in such cases the variation within sites is about 10–20%, and the variation between sample sites about 80–90%. These values are considered to be typical for the low-density stream-sediment survey described here.

Fig. 5 is an example of a map obtained when the relative variance component within sample

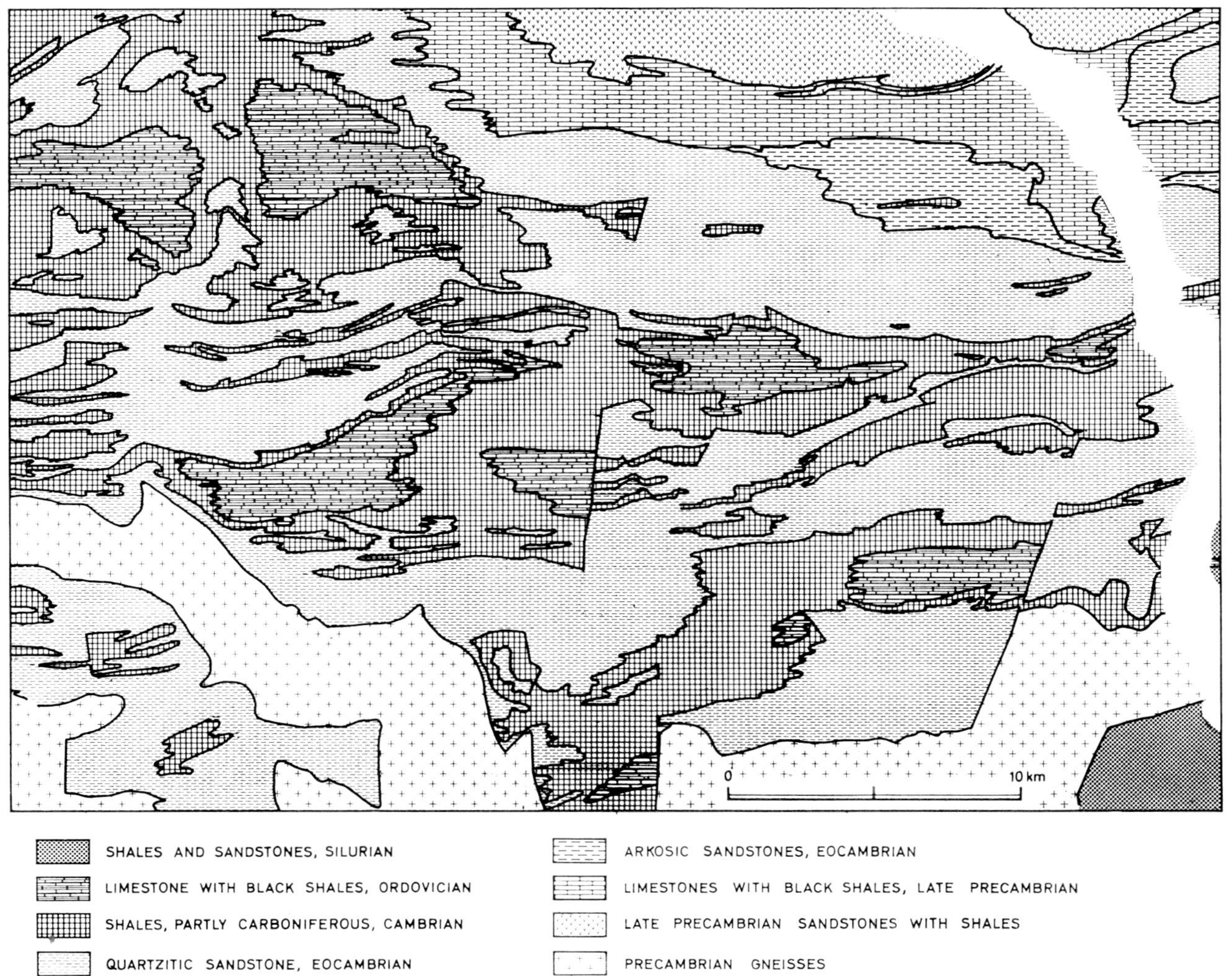

Fig. 6 Geological map of same area as geochemical map shown in Fig. 5. After Bjørlykke[1]

sites was 15%, which means that 85% of the total variability comes from a regional factor. The copper values in the stream sediments give a fairly good picture of the geological trends (Fig. 6), being little influenced by topography (see upper left corner of map (Fig. 5) where the geochemical trend crosses the main valleys). The high Cu trend follows rather closely the pattern of the Ordovician limestones and the Cambrian shales; moreover, the curved pattern of the copper values in the lower right-hand corner is seen to coincide with a fault (Figs. 5 and 6).

Metal variability within sample site

To describe the metal variability within the sample site a quantity log f has been chosen, which is termed the *deviation ratio*. The deviation ratio is defined as the logarithm of the ratio of the concentration in duplicate A to the concentration in duplicate B (log f=log (con A/con B)). The distribution of log f should be symmetrical around the median value $0(f=1)$ if there are no systematic errors in the data (see Fig. 7).

fairly similar for all elements, but the 95 percentiles for f^x differ, and the metals were therefore classified into three groups. The first group, Co,

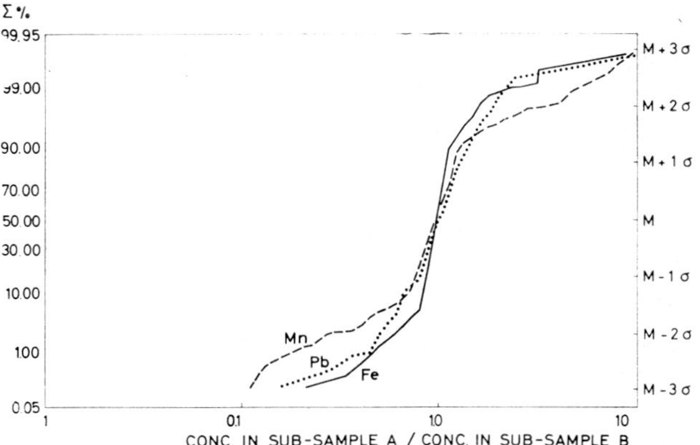

Fig. 7 Cumulative frequency distributions of deviation ratio (log f=log(con A/con B)) for Fe, Mn and Pb (right-hand ordinate gives scale for estimation of median (M) and standard deviation (σ))

Fe, Ni and Zn, which has the lowest 95 percentile for f^x, is represented by Fe in Fig. 7. The second fairly similar group is represented by Pb, and

Table 2 Median and 95 percentile for f^x, ratio of highest to lowest concentrations of duplicates (corresponding to 75 and 97·5 percentiles for f, respectively; see Fig. 7 and text)

	Co	Cu	Fe	Mn	Ni	Pb	Zn
Median	1·13	1·15	1·08	1·19	1·09	1·15	1·10
95%	1·68	1·82	1·63	3·60	1·65	1·91	1·71

In some cases the absolute value of log f ($|\log f|$) is a convenient quantity. This value is equivalent to the logarithm of the ratio of the concentration in the highest duplicate to the concentration in the lowest duplicate. Any percentile of $|\log f|$ can be found from the upper half of the distribution of log f (correspondingly, any percentile of $1-|\log f|$ can be found from the lower half of the distribution of log f).

Cumulative per cent of log $f=100-\frac{1}{2}(100-$cumulative per cent of $|\log f|)$

For example: the cumulative per cent of log f corresponding to the cumulative per cent 50 of $|\log f|=100-\frac{1}{2}(100-50)=75$

The deviation ratio, log f, and its absolute value, $|\log f|$, have both been used in the computations, but for convenience the antilogarithms, f and f^x, respectively, are indicated in the tables and diagrams.

In Table 2 the median values and 95 percentiles of f^x for different metals are listed. Examples of cumulative frequency distribution plots for log f are given in Fig. 7. The median values for f^x are consists of Cu and Pb. The third group, Mn only, has a larger deviation between the duplicates than the other metals at *ca* 20% of the sample sites (Fig. 7). It is believed that this large deviation for Mn is due to precipitation caused by the pH/Eh conditions, as was suggested by Nichol and coworkers[12] (see also later).

Metal variability within sample site in relation to Fe and Mn

The correlation between the deviation ratio of trace metals within the sample sites and the corresponding deviation ratio of Fe and Mn, respectively, within the sites has been studied. Two examples of scatter diagrams (Co versus Fe and Co versus Mn) are shown in Fig. 8. The number of sample sites within each cell in the plot is indicated by means of symbols, and antilogarithms of the deviation ratio are written along the axes. Correlation coefficients for log f are listed in Table 3.

The correlation coefficients for Co, Cu, Ni, Pb

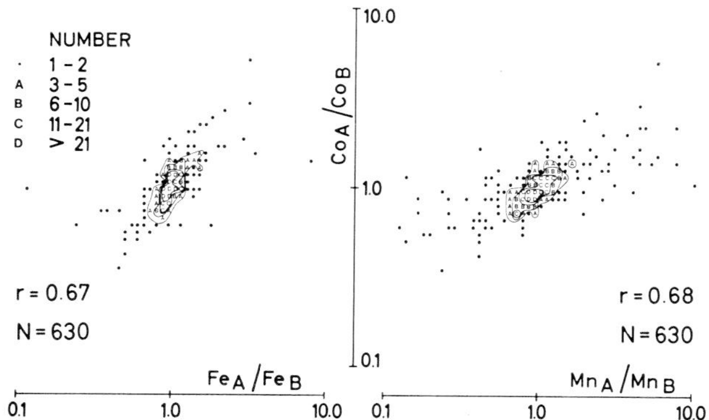

Fig. 8 Scatter diagrams: deviation ratios (log f= log(con A/con B)) for Co versus Fe and Mn, respectively. Number of sample sites within each cell indicated by dots or letters; r, correlation coefficient; N, total number of sample sites

and Zn versus Fe are all positive and similar to those versus Mn; this does not necessarily mean, however, that the relations are similar in detail—see, for example, the difference between the Co versus Fe and Mn patterns in Fig. 8. The correlation coefficients decrease in the order Co > Zn ≥ Ni > Pb > Cu (Table 3).

Table 3 Correlation coefficients for deviation ratio (log f=log (con A/con B)) of trace elements versus deviation ratio of Fe and Mn, respectively

	Co	Cu	Ni	Pb	Zn
Metal versus Fe	0·67	0·37	0·54	0·42	0·55
Metal versus Mn	0·68	0·35	0·47	0·43	0·53

Metal variability within sample site in relation to metal content

Two examples of the relation between mean metal concentration at each sample site and the absolute value of the deviation ratio are given in Fig. 9. |Log f| for V increases with decreasing mean V concentration. This pattern, which indicates poor analytical sensitivity at low concentrations, is more pronounced for Ag and V than for Cu. Co, Ni, Pb and Zn show a |log f| more or less independent of the corresponding metal concentration; |log f| for Fe increases with increasing Fe concentration.

Manganese shows a pattern of large deviations between the duplicates in the middle of the concentration range (Fig. 9). This pattern would not be due to analytical error, since the analytical sensitivity for Mn is adequate at all concentration levels. The following explanation is suggested. Low Mn in the sediment indicates low pH/Eh in the stream water, Mn being kept in solution as bivalent ions. A high manganese content in the stream sediments indicates high pH/Eh in the stream waters, practically all available manganese being precipitated as Mn^{4+}. In both cases the manganese content of the sediments within each sample site should be fairly uniform, and sample error relatively small. Mean manganese concentrations may, however, indicate that pH/Eh conditions are such that minor variations in the stream water (temperature, CO_2 and O_2 content, etc.) could cause local precipitation or dissolution of Mn and a consequent large variability of metal contents within short distances in the stream.

A more illustrative way of displaying the metal variability than that available from the use of scatter diagrams is given below.

Consider N duplicates with averages within a fixed concentration interval. The standard deviation of the average of duplicates can be expressed

$$SD=\sqrt{\Sigma\, di^2/4N}$$

where $di=X_{1i}-X_{2i}$=difference between the log

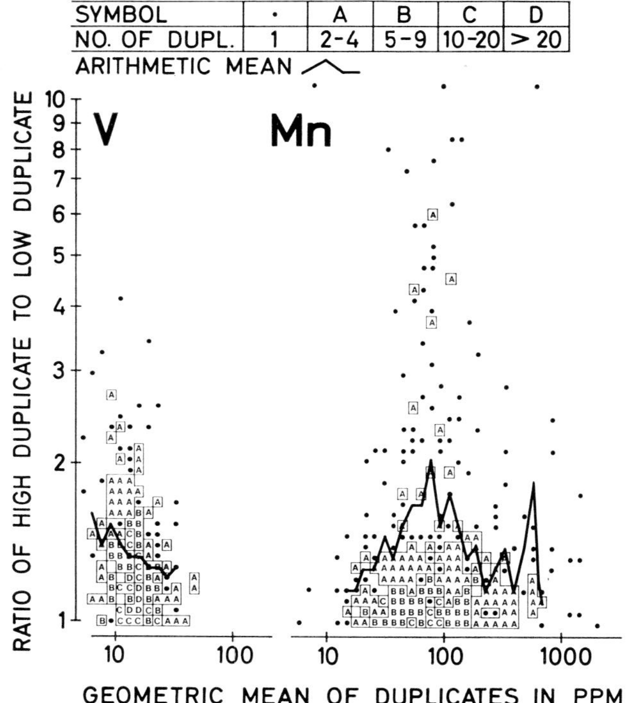

Fig. 9 Scatter diagrams: absolute value of deviation ratio (|log f|=|log (con A/con B)|) for Mn versus concentration of Mn, and absolute value of deviation ratio for V versus concentration of V

concentrations of the duplicate at the ith sample site.

The mean value, M, at a fixed concentration level will be

$$M=\sum_{1}^{N} (X_{1i}+X_{2i})/2N$$

At each concentration level one can calculate an interval which contains the total error with a certain probability. By choosing a probability of 0·95, and assuming lognormality at each level, the interval limits for the duplicates, expressed in ppm, can be written for $N \geqslant 25$

$$L_1=\text{antilog}\ (M-2\sqrt{\Sigma di^2/4N}$$
$$L_2=\text{antilog}\ (M+2\sqrt{\Sigma di^2/4N}$$

These interval limits define how large the total error is at each concentration level. The limits can be marked on each side of the frequency distribution curve, and will be symmetrical around this curve when plotted on logarithmic paper. If the interval limits are connected, an uncertainty belt caused by the total error is obtained. If known at

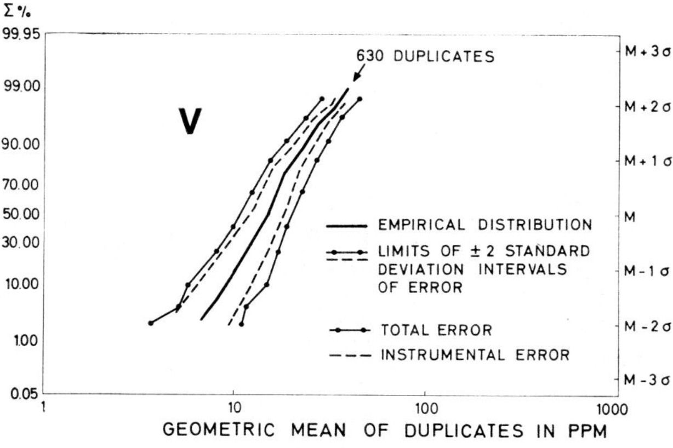

Fig. 10 Error belts around empirical cumulative frequency distribution curve for vanadium: 630 sample sites (right-hand ordinate gives scale for estimation of median (M) and standard deviation (σ))

various concentration levels, the laboratory error can also be plotted in a similar way (Figs. 10 and 11). From this type of plot an estimation of the error at a fixed probability level (here 0·95) can be made. At a given concentration a horizontal line is drawn through the empirical frequency distribution curve. The abscissae corresponding to the intersections of the horizontal line with the error curves indicate the error intervals at the given concentration. Suppose the analytical result for Mn is 100 ppm (Fig. 11): this value would in 95% of cases lie between 90 and 110 ppm if only re-analyses were carried out, and lie between 65 and 150 ppm if re-sampling were carried out

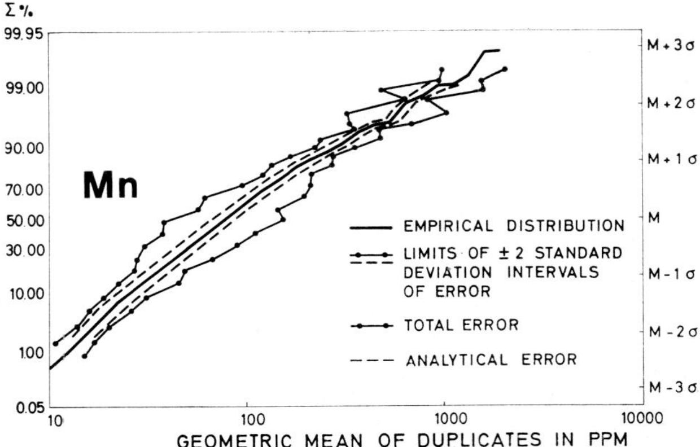

Fig. 11 Error belts around empirical cumulative frequency distribution curve for manganese: 630 sample sites (right-hand ordinate gives scale for estimation of median (M) and standard deviation (σ))

The total error belts shown in Figs. 10 and 11 contain the same data as those given in Fig. 9. The laboratory errors are calculated from results of replicate analyses of standards. Vanadium and manganese are chosen because they are examples of the two extremes: (*a*) a relatively large total error, when compared to the between-site dispersion, which may mask the real dispersion between sample sites and tend to cause random geochemical patterns (Fig. 10); (*b*) a relatively small total error, when compared to the between-sample-site dispersion, which will cause more reliable geochemical patterns (Fig. 11).

It is suggested that plots such as those presented in Figs. 10 and 11 could advantageously accompany any geochemical map for general use. The plots may be valuable aids, for example, in any judgment of the quality of analytical results, in deciding if two seemingly different concentrations are, in fact, different and in determining how narrowly contour intervals could be set without being meaningless.

Metal content in relation to stream order

Several systems exist for the setting up of hierarchy of streams,[5, 6, 13, 15, 16, 19] Fig. 12 showing that suggested by Strahler.[16] Streams in the area investigated were classified according to this system, modern photogrammetric maps (scale, 1:50 000) being used as a base. Metal contents and the deviation ratios were calculated for each stream order. Results from the two map sheets DOKKA and GJØVIK are presented in Tables 4 and 5 and Figs. 13 and 14. Only values for stream orders 1, 2 and 3 are given, order 4, which was the highest order recorded, having been

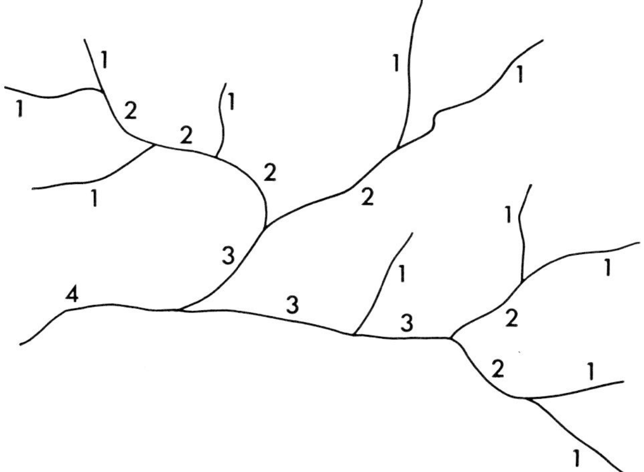

Fig. 12 Stream-ordering system. After Strahler[16]

omitted because of an insufficient number of observations.

Four features are apparent: (1) the metal dispersion between sample sites (expressed by the standard deviation of the log concentrations) decreases with increasing stream order from 1 to 3 for all metals investigated except Cu (Table 4 and Figs. 13 and 14); (2) the mean metal concentration

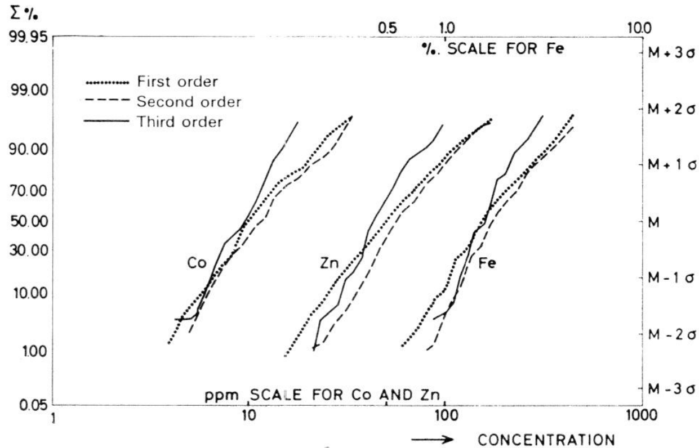

Fig. 13 Cumulative frequency distribution curves for Co, Fe and Zn contents in sediments from streams of different orders: number of samples from first-, second- and third-order streams are 840, 306 and 98, respectively (right-hand ordinate gives scale for estimation of median (M) and standard deviation (σ))

Table 4 Metal contents and dispersion between sample sites in sediments from first-, second- and third-order streams (840, 306 and 98 samples, respectively)

	Order 1			Order 2			Order 3		
	G, ppm	95 %, ppm	*SD*, log ppm	*G*, ppm	95 %, ppm	*SD*, log ppm	*G*, ppm	95 %, ppm	*SD*, log ppm
Co	11	28	0·25	12	30	0·23	9	17	0·17
Cu	7·3	20	0·28	7·7	20	0·23	7·4	15	0·25
Fe	16 000	39 000	0·22	18 000	44 000	0·19	15 000	28 000	0·15
Mn	76	430	0·43	103	530	0·40	67	290	0·34
Ni	14	50	0·28	16	42	0·25	15	38	0·21
Pb	13	30	0·22	13	28	0·19	11	24	0·18
Zn	51	145	0·26	59	145	0·23	44	92	0·16

G, geometric mean of metal concentrations.
95 %, 95 percentile of metal concentrations.
SD, standard deviation of log metal concentrations.

Table 5 Metal variability within sample sites in first-, second- and third-order streams (840, 306 and 98 samples, respectively): ratio of concentration in high sample/low sample

	Order 1		Order 2		Order 3	
	M	95 %	*M*	95 %	*M*	95 %
Co	1·1	1·8	1·1	1·6	1·1	1·7
Cu	1·1	1·8	1·1	1·8	1·1	1·8
Fe	1·1	1·7	1·1	1·5	1·1	1·6
Mn	1·2	4·2	1·2	2·4	1·1	2·0
Ni	1·1	1·8	1·1	1·5	1·1	1·4
Pb	1·2	1·8	1·2	1·8	1·2	1·8
Zn	1·1	1·7	1·1	1·6	1·1	1·7

M, median value of concentration ratio (f^x).
95 %, 95 percentile of concentration ratio(f^x).

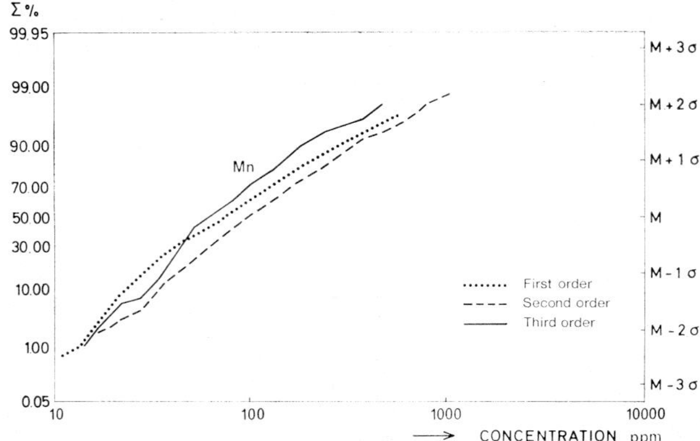

Fig. 14 Cumulative frequency distribution curves for Mn contents in sediments from streams of different orders: number of samples from first-, second- and third-order streams are 840, 306 and 98, respectively (right-hand ordinate gives scale for estimation of median (M) and standard deviation (σ))

is highest in the second-order streams for all the metals investigated (Table 4 and Figs. 13 and 14); (3) at high concentrations (95 percentile) metal contents are similar in first- and second-order streams and lower in third-order streams (Table 4 and Figs. 13 and 14); (4) at high metal concentrations (95 percentile) the within-site variability (expressed as f^x) is larger for Mn than for the other metals investigated (Table 5). For Mn the 95 percentile of f^x decreases markedly with increasing stream order, whereas for the other elements the corresponding f^x is more or less independent of stream order.

Deviation between metal contents in streams of a different order is supposedly due, for the most part, to dilution—this is confirmed by the decreasing metal dispersion between sample sites with increasing stream order (Table 4 and Figs. 13 and 14). Dilution alone, however, is not the only factor, since the mean metal content does not always decrease with increasing stream order.

The low mean metal content in the first-order streams in comparison with the second-order streams can, to some extent, be explained by the geochemistry of Mn. In Norway many streams originate in bogs, and first-order streams would, in many cases, have low pH and Eh values in water from their upper parts. The content of Mn and any heavy metal controlled by Mn will be relatively low in the sediments from this part of the streams. Downstream from a bog area Mn will precipitate due to increases in pH and Eh and the heavy metal contents of the sediments will become higher.[12] Table 5 indicates that precipitation of Mn predominates in the first-order streams (high 95 percentile for the deviation ratio). High metal contents in stream sediments due to the scavenging effect of precipitating Mn would therefore be most pronounced in the lower part of first-order streams or in second-order streams. The mechanism may favour a high mean metal content in the second-order streams and also a high metal dispersion in first-order streams. For stream orders higher than 2 the effect will probably result in relatively low values for both mean metal content and between sample site dispersion.

Another factor is the effect of erosion, which will probably tend to increase heavy metal content with increasing stream order. The topography in the survey area is immature, and the depth of stream erosion increases with increasing stream order. Low-order streams will tend to run on the shallow parts of the overburden, which often may have been transported over long distances and impoverished in soft minerals. Low-order streams may not always be in contact with groundwater. On the other hand, high-order streams will tend to run on the bedrock or on the deeper parts of the overburden, which is often of local origin and to some extent reflects the bedrock. The water of high-order streams will normally be in contact with the groundwater, which generally contains more solutes than the surface water.

As the effects of dilution and erosion counteract, a maximum trace-metal content in the sediments from streams of an order higher than 1 may well result.

Two consequences of the data presented in Tables 4 and 5 and Figs. 13 and 14 may be of interest in prospecting: (1) the threshold for Co, Cu, Fe, Mn, Ni, Pb and Zn should, under appropriate local conditions, be set lower for third-order streams than for first- and second-order streams; and (2) if uniform precision is required for Mn (or metals controlled by Mn), more sub-samples should be taken at each site in first-order streams than in second- and third-order streams.

General conclusions

The general experience from this introductory study points to four significant aspects.

(1) In routine geochemical surveys in which large amounts of data are being considered, serious errors may easily be introduced in the field, in the laboratory or during treatment of the results. If such errors cannot be tolerated, rigid control is required, and some kind of routine replicate sampling procedure may be necessary.

(2) A knowledge of the total error is a basic requirement for decisions on threshold values, the significance of anomalies, contour intervals on maps, etc. The type of error plot introduced in this paper facilitates these decisions. For stream-sediment surveys the dispersion between sample

sites is nearly always significantly higher than the variability within sample sites, regardless of the types of error, because the critical value of the F ratio in the F test approaches one at the degrees of freedom normally encountered in geochemical surveys. A two-way analysis of variance by use of a mixed model with fixed and random effects may help in detecting the different factors which cause unwanted variability in stream-sediment data. Of particular value were the estimated variance components and their relative magnitudes.

(3) Metal dispersion within sample sites depends on such factors as the concentration of the metal under investigation, the concentration of other metals and stream size. Manganese variability within sample sites is, in some cases, exceptionally high, especially in the middle of the manganese concentration range. If a constant precision is required for Mn and trace metals controlled by Mn, a sampling design which varies with the Mn variability at each sample site is needed. Since Eh/pH are basic parameters controlling the Mn precipitation, the sampling design could probably be guided by field estimation of pH and Eh in stream water.

(4) Metal content of stream sediments varies with the size of the catchment area (stream order), but the relation is not exclusively a result of a simple dilution effect. For geochemical prospecting it may be advantageous to set threshold values at different levels for different stream orders. The influence of morphometric properties on the metal content in stream sediment should be investigated further.

References

1. Bjørlykke, A. Geological map from the Gjøvik and Dokka area. *Unpubl. Rep. Norw. geol. Surv.* 1038, 1972, 20p.

2. Bjørlykke, A. En oversikt over stratigrafi, tektonikk og blyforekomster langs den sydlige del av fjellranden. *Unpubl. Rep. Norw. geol. Surv.* 893, Bilag II, 1970, 42 p.

3. Follestad, B. A. Personal communication of State Geologist of Geological Survey of Norway.

4. Grip, E. On the genesis of the lead ores of the eastern border of the Caledonides in Scandinavia. In *Genesis of stratiform lead–zinc–barite–fluorite deposits* Brown, J. S. ed. (Lancaster, Penna.: The Economic Geology Publishing Co., 1967), 208–18.

5. Horton, R. E. Drainage basin characteristics. *Trans. Am. geophys. Un.*, **13**, 1932, 350–61.

6. Horton, R. E. Erosional development of streams and their drainage basins, hydrophysical approach to quantitative morphology. *Bull. geol. Soc. Am.*, **56**, 1945, 275–370.

7. Howarth, R. J. and Lowenstein, P. L. Sampling variability of stream sediments in broad-scale regional geochemical reconnaissance. *Trans. Instn Min. Metall. (Sect. B: Appl. earth sci.)*, **80**, 1971, B363–72.

8. Lepeltier, C. A simplified statistical treatment of geochemical data by graphical representation. *Econ. Geol.*, **64**, 1969, 538–50.

9. Låg, J. Undersøkelser over opphavsmaterialet for Østlandets morenedekker. *Meddr norske SkogforsVes.*, **10**, no. 35, 1948, 223 p.

10. Låg, J. Registrering av hovedtyper av jordsmonn i skogene i Norge. In *Taksering av Norges skoger. Landskogtakseringen 50 år, 1919–1969* (Oslo: Landskogtakseringen, 1970), 143–9.

11. Miesch, A. T. Theory of error in geochemical data. *Prof. Pap. U.S. geol. Surv.* 574–A, 1967, 17 p.

12. Nichol, I. Horsnail, R. F. and Webb, J. S. Geochemical patterns in stream sediment related to precipitation of manganese oxides. *Trans. Instn Min. Metall. (Sect. B: Appl. earth sci.)*, **76**, 1967, B113–5.

13. Scheidegger, A. E. The algebra of stream-order numbers. *Prof. Pap. U.S. geol. Surv.* 525–B, 1965, 187–9.

14. Searle, S. R. *Linear models* (New York: Wiley, 1971), 532 p.

15. Shreve, R. L. Infinite topologically random channel networks. *J. Geol.*, **75**, 1967, 178–86.

16. Strahler, A. N. Dynamic basis of geomorphology. *Bull. geol. Soc. Am.*, **63**, 1952, 923–38.

17. Strand, T. The pre-Devonian rocks and structures in the region of Caledonian deformation. *Norg. geol. Unders.* no. 208, 1960, 170–284.

18. Tennant, C. B. and White, M. L. Study of the distribution of some geochemical data. *Econ. Geol.*, **54**, 1959, 1281–90.

19. Woldenberg, M. J. Horton's laws justified in terms of allometric growth and steady state in open systems. *Bull. geol. Soc. Am.*, **77**, 1966, 431–4.

526.8:535.6:550.84

Automated colour-mapping of three-component systems and its application to regional geochemical reconnaissance

P. L. Lowenstein, B.SC., PH.D., M.I.M.M.

R. J. Howarth, B.SC., PH.D., M.I.M.M.

Both of the Applied Geochemistry Research Group, Imperial College, London, England

Synopsis

The mapping of simultaneous concentration variations of three elements is achieved by use of colour mixing. Examples are presented of subtractive colour maps obtained directly by the superimposition of magenta, cyan and yellow diazo transparencies derived from single-element grey-tone maps produced automatically by use of an LGP 2703 Lasergraphic plotter system. The use of more than three concentration levels per colour component produces a wide range of colour fields which reveal broad-scale geochemical patterns. A scheme containing 27 visually distinguishable colours has been derived for maps in which three elements are each present at any of three concentration levels. These enable unique identification of the element concentrations giving rise to any particular colour field.

Mapping of multi-element stream-sediment reconnaissance data by these techniques has shown a considerable information gain over conventional single-element maps.

The exploration geochemist is often faced with the problem of the interpretation of multi-element survey data, which are commonly represented as single-element maps. The observations may be shown on the map either at discrete sample locations or smoothed in some way to show local and regional 'trends'. To interpret the co-variation of several elements one must either attempt a mental integration of the visual patterns shown by the separate maps or construct a multi-component map by suitable combination of the single-component maps.

Multi-component maps have, until now, tended to use symbols in the form of bar-diagrams, pie-diagrams, wind-rose figures and so on. Although such symbols are capable of representing a large number of components on one map, and may be useful at a single sample site, they are frequently difficult to integrate by eye in order to extract regional patterns of variation. Perception tests with three-component wind-rose representation of regional geochemical data[4] have shown that such symbols 'were very difficult to use where precise results are required from visual [concentration] estimation'.

The method described in this paper is an attempt to represent three-component systems in such a way that the overall patterns of variation are shown to their maximum extent and, at the same time, the composition of each of the three components may be retrieved from any point on the map. The human eye has a great capacity for the integration of discrete colours located at a large number of point sources into visually meaningful patterns—for example, the many separate coloured dots which make up the picture on a colour television screen. It can also distinguish between objects differing only slightly in colour (hue and saturation) and brightness (luminance), although the eye is less sensitive in some colour ranges. It is also known that the information contained in three separate coloured images can be conveyed by a single image resulting from their combination, and is the principle underlying colour television, photography and printing. It is for these reasons that an analogous colour-mapping technique has been developed.

There are two systems of colour representation, one additive and the other subtractive. *Additive colour mixing* occurs when an image is produced by superimposing red, blue and green light, and a wide range of new colours (Table 1) can be produced by varying the proportions of each of these primary colours (e.g. in a colour television display). When all three colours are present at maximum intensity the result is 'white' light. *Subtractive colour mixing* occurs when the complementary colours yellow, cyan (a blue-green) and magenta (a red-blue) are combined, the results being most commonly observed in colour prints and in paintings. When all three colours are present at maximum intensity, the result is 'black' (Table 2). It is easier to visually estimate the relative proportions of the components present in a colour formed by subtractive mixing (e.g. yellow + cyan = green, Table 2) than one formed by additive mixing (e.g. red + green = yellow, Table 1). The dark colours which result in subtractive colour mixing when each colour is present in near-maximum amounts are very suitable for portraying areas on the map containing high concentrations of the three components. The entire range of colours produced by subtractive mixing is also obtainable by overprinting suitable non-opaque yellow, cyan and magenta inks (e.g. in the production of colour illustrations in magazines), and makes the system particularly attractive for the production of maps.

If colour mixing is used for map production, the number of resultant colours will depend on the number of intensity levels present for each component, and there is an upper limit beyond which the eye cannot distinguish between the colours produced. From the point of view of geochemical map interpretation, however, it is extremely important not only to see the composite colour patterns clearly but to be able to relate any observed colour back to the element concentrations

Table 1 *Additive three-component colour mixing*

Components present	*Colour mixture**	*Resultant colour*
None	None	Black
X	Red	Red
Y	Green	Green
Z	Blue	Blue
$X+Y$	Red+green	Yellow
$X+Z$	Red+blue	Magenta
$Y+Z$	Green+blue	Cyan
$X+Y+Z$	Red+green+blue	White

*Individual components represented by maximum brightness.

Table 2 *Subtractive three-component colour mixing*

Components present	*Colour mixture**	*Resultant colour*
None	None	White
X	Magenta	Magenta
Y	Yellow	Yellow
Z	Cyan	Cyan
$X+Y$	Magenta+yellow	Red
$X+Z$	Magenta+cyan	Blue
$Y+Z$	Yellow+cyan	Green
$X+Y+Z$	Magenta+yellow+cyan	Black

*Individual components represented by maximum brightness.

of the three original components. The latter can be achieved by use of a system in which the three components are each present at one of three concentration levels (corresponding, for example, to background, near-anomalous and anomalous concentrations) represented in the single-component images by light, mid and dark grey, respectively. These three grey tones should be such that the 27 possible combinations yield distinct, visually identifiable, colours.

Method

The multi-element survey data are presented in the first instance as single-element grey-tone maps obtained by overprinting characters on a computer line-printer.

The basic computer program used[3] plots a symbol corresponding to the concentration level of the geometric mean of all data points falling within a 'map cell', i.e. the space occupied by a line-printer character at the map scale chosen for display. This program has now been extended to produce smoothed grey-tone maps by use of a moving-average technique. Digital versions of the (smoothed) maps are obtained on punched cards, which record the grey-tone level of each map cell, and are used for further processing. It is obvious that the quality of the grey tones presented by a line-printer map leaves much to be desired, and that an image recording the grey tones as photographic density levels is a prerequisite for any colour-combination technique.

The production of such images from the digitized line-printer maps has been made possible by use of a Dresser LGP 2703 Lasergraphic plotter system. This consists of an LGP 2000 plotter with an on-line Raytheon 703 computer. Plotting instructions are generated by a computer program written for an off-line IBM 360 computer, and are stored on magnetic tape for transmission to the plotter via the Raytheon computer. The LGP plotter is a high-speed computer-controlled device which uses dimensionally stable Kodak 2496RAR film as a plotting medium, and can produce maps up to 40 in wide and 100 ft long. A laser optics system converts the plotting instructions into a photographic quality positive image with up to 16 grey levels. These instructions create variable-intensity raster scans across the area of film on which the image is being plotted (analogous to a television scan system). Light from a helium–neon laser is polarized and passes through an analyser, whose position relative to the polarizer is controlled by the program, thus modulating the beam intensity before it is scanned across the film by a rotating mirror. Subsequent scans follow, and the whole cycle is repeated until the image is complete. The 0·005-in spot size of the laser beam is satisfactory for all but the most stringent cartographic requirements. At the end of plotting the exposed film is cut off and developed.

The positive grey-tone prints are of excellent quality, and the plotter has been extensively used in geophysical data processing.[1] We have used from three to seven grey levels per element, as visual distinction between the grey tones in excess of this becomes insufficiently clear in the plotter film images for successful colour-combination work.

The grey-tone plotter images are converted into images in the complementary colours by positive to positive contact printing on to ammonia-developed colour dyeline (diazo) films. Owing to the differing characteristics of the photographic and diazo films, the use of a contact screen between the original and the copy when printing produces an accurate reproduction of the grey levels in colour. In practice, the production of the coloured three-component maps is a two-stage process. First, a contact screened positive print is made on to sepia diazo film, at which stage any additional artwork or titling information may be added. A different screen (angled at 15°, 45° and 75°) must be used for each colour in order to avoid the production of interference patterns on combining the coloured images. The appropriate master is then used to produce either a yellow, cyan or magenta diazo film transparency. These printing operations are carried out by use of a vacuum-bed multiple-proofing unit to ensure maximum master to print contact, and the use of dimensionally stable polyester base diazo film ensures high-quality prints. Film registration is achieved by use of standard register punch equipment. Both printing and subsequent developing stages are extremely rapid, finished prints being obtained in a few minutes, and once the screened masters have been made, numerous colour prints can be rapidly obtained. It should also be possible to make plates for offset lithography by this technique.

The production of triplets of colour prints with more than three levels for each element in the first instance facilitates the rapid investigation of all possible three-component colour combinations. These reveal the maximum detail in the data and are helpful for the identification of broad-scale patterns. For the practical retrieval of the element concentration values at any point on the map, three concentration classes per element are used. It is then possible to produce a 27-colour key on which each colour field can be uniquely identified in terms of the three components (see Fig. 4).

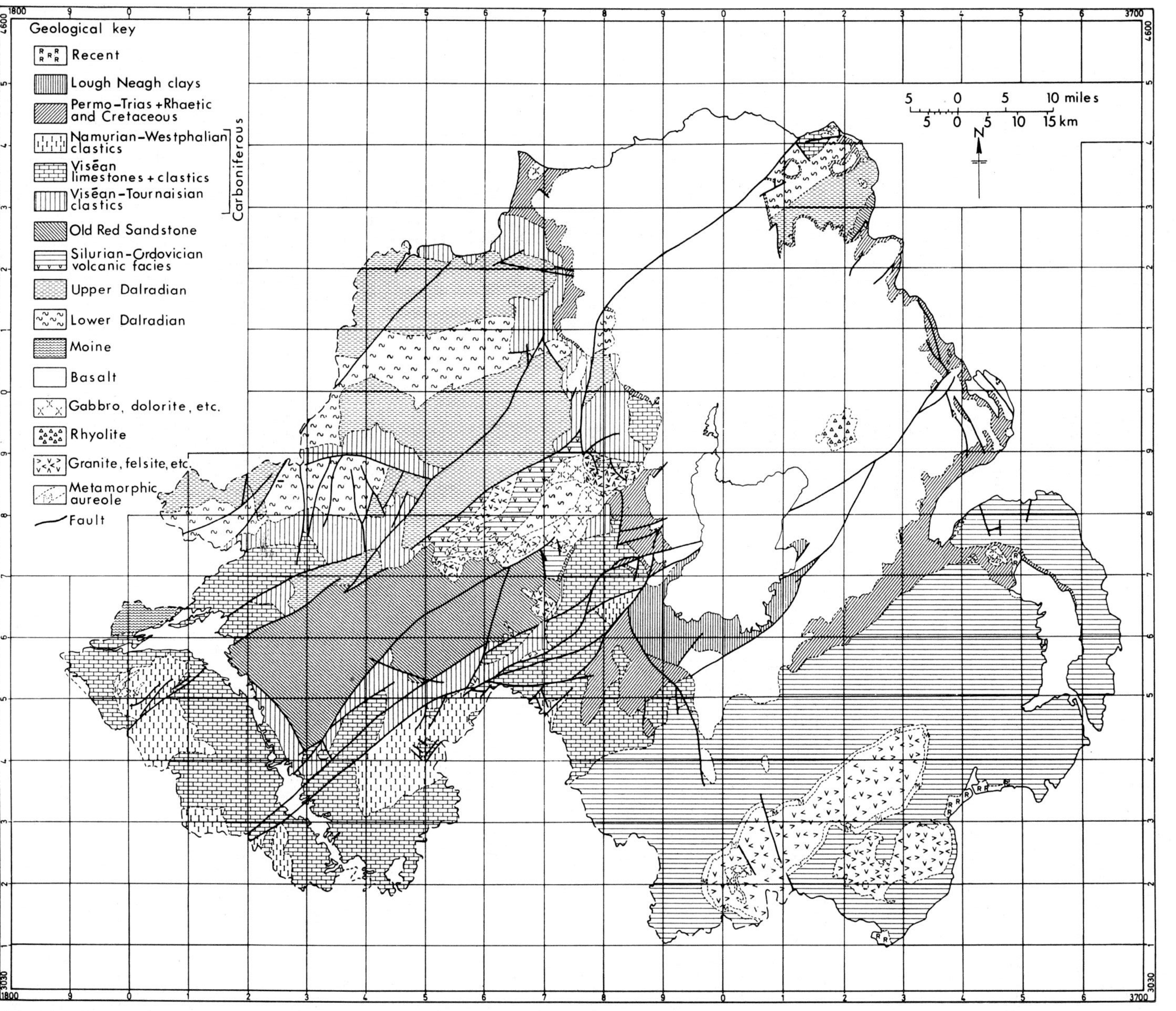

Fig. 1 Geological map of Northern Ireland. After Geological Survey of Northern Ireland[2]

Examples

The maps used to illustrate this technique are the result of an Applied Geochemistry Research Group reconnaissance survey of Northern Ireland, which involved the collection of nearly 5000 stream-sediment samples taken at a density of one sample per square mile at intersections of roads and tracks with the drainage network. Analyses of the unground sieved −80-mesh (*ca* 200-μm) fractions were made by use of an A.R.L. 29000B 'Quantometer' automated emission spectrograph. A comprehensive account of the instrumental operating conditions and a regional interpretation of the geochemical results based on factor analysis has been given by Young.[5] The original maps were produced at a scale of ten miles to the inch.

The first example shows the simultaneous distribution of lead, copper and zinc and has been produced by combining maps which contain five classes (i.e. intensity levels) for each element:

Lead (red) 0, 15, 30, 70, ⩾ 150 ppm
Copper (blue) 0, 14, 28, 56, ⩾ 112 ppm
Zinc (yellow) 0, 60, 120, 240, ⩾ 480 ppm

The resultant map (Fig. 2) shows a large number of colour fields, which both correlate with the geology (Fig. 1) and uniquely define areas of mineral exploration potential.

The old mining district near Keady (28453340) is represented by red, orange and yellow hues, and the tendency for yellow to predominate indicates that zinc is the dominant base metal. This is borne out by the ubiquitous sphalerite in the carbonate gangue in the old mine dumps in the area.

The Mourne Mountains granite and the Silurian rocks in its metamorphic aureole are represented by magenta and red colours—suggesting that here lead is dominant. Stringers containing galena are known to occur within the granite, and in the Silurian rocks at Moneylane (33913365), Wateresk (33873367) and Bryansford (33383339).[5] The most important lead–zinc mine in Northern Ireland, at Conlig (34963756), shows up on the colour map as a

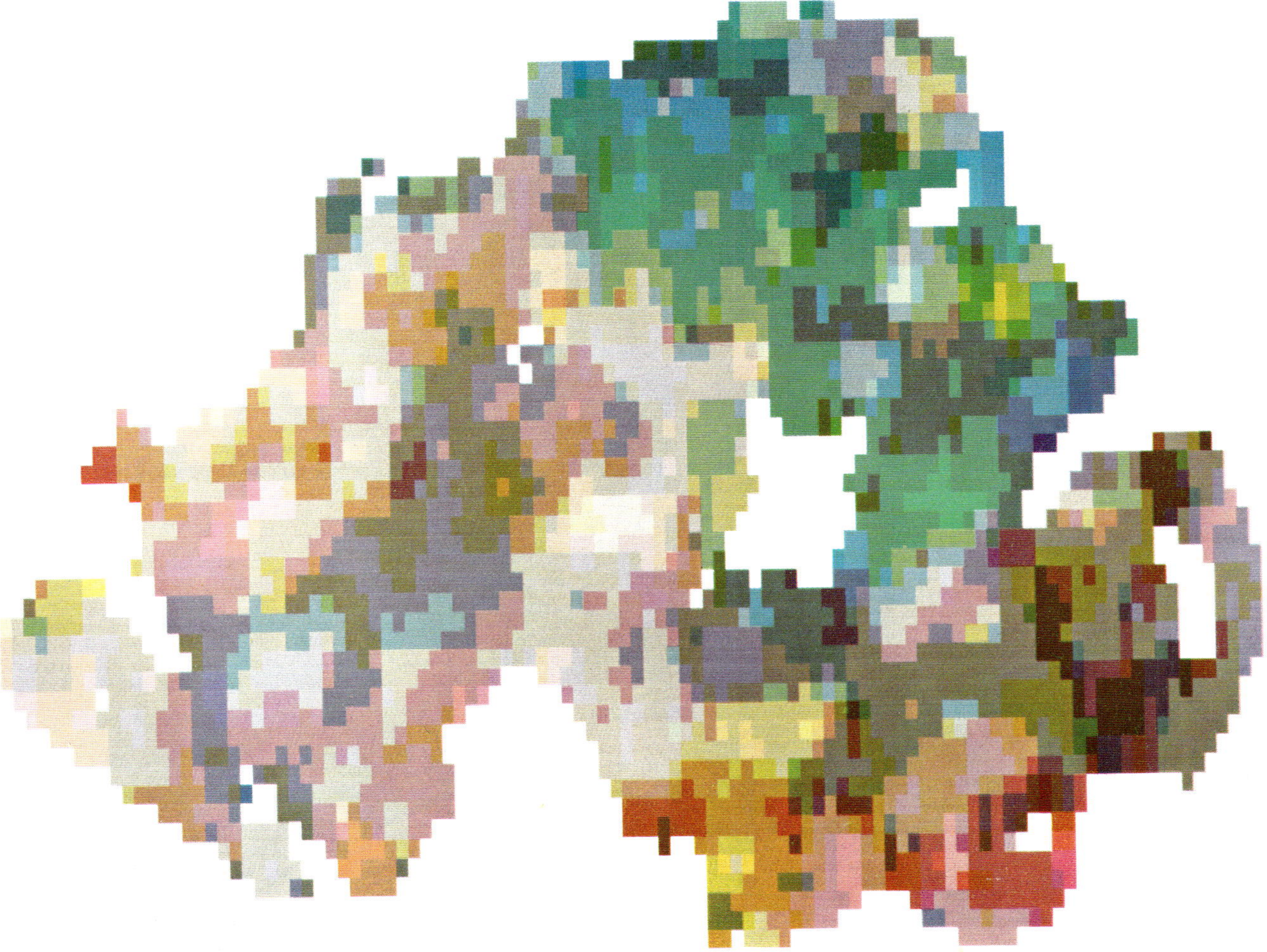

Fig. 2 Multi-level three-component geochemical map of Northern Ireland for lead (red), copper (blue) and zinc (yellow). For explanation see text

dark brown patch to the north of Lough Erne, and is represented by high concentrations of lead, zinc *and* copper. This mineralization is associated with a thick Tertiary dolerite sill. An elongate feature with similar colouring on the map trends NNW–SSE across Co. Down from near Belfast to Downpatrick (34863447), and is in an area with no recorded mineralization. It runs approximately parallel to a linear feature shown on a recent aeromagnetic map,[2] and may indicate the presence of new mineralization which could be associated with the Killough (35393357)–Ardglass (35583373) dyke swarm.

Finally, high concentrations of copper characterize the Antrim Basalts and show up as a blue-green area on the map. Native copper is known to occur in the basalts.

A three-level map for the same elements (Fig. 3) shows similar, but more clearly defined, colour fields, and the element concentrations at any point can be retrieved by use of the corresponding 27-colour key (Fig. 4).

The final example (Fig. 5) shows the simultaneous distribution of potassium (6 levels),

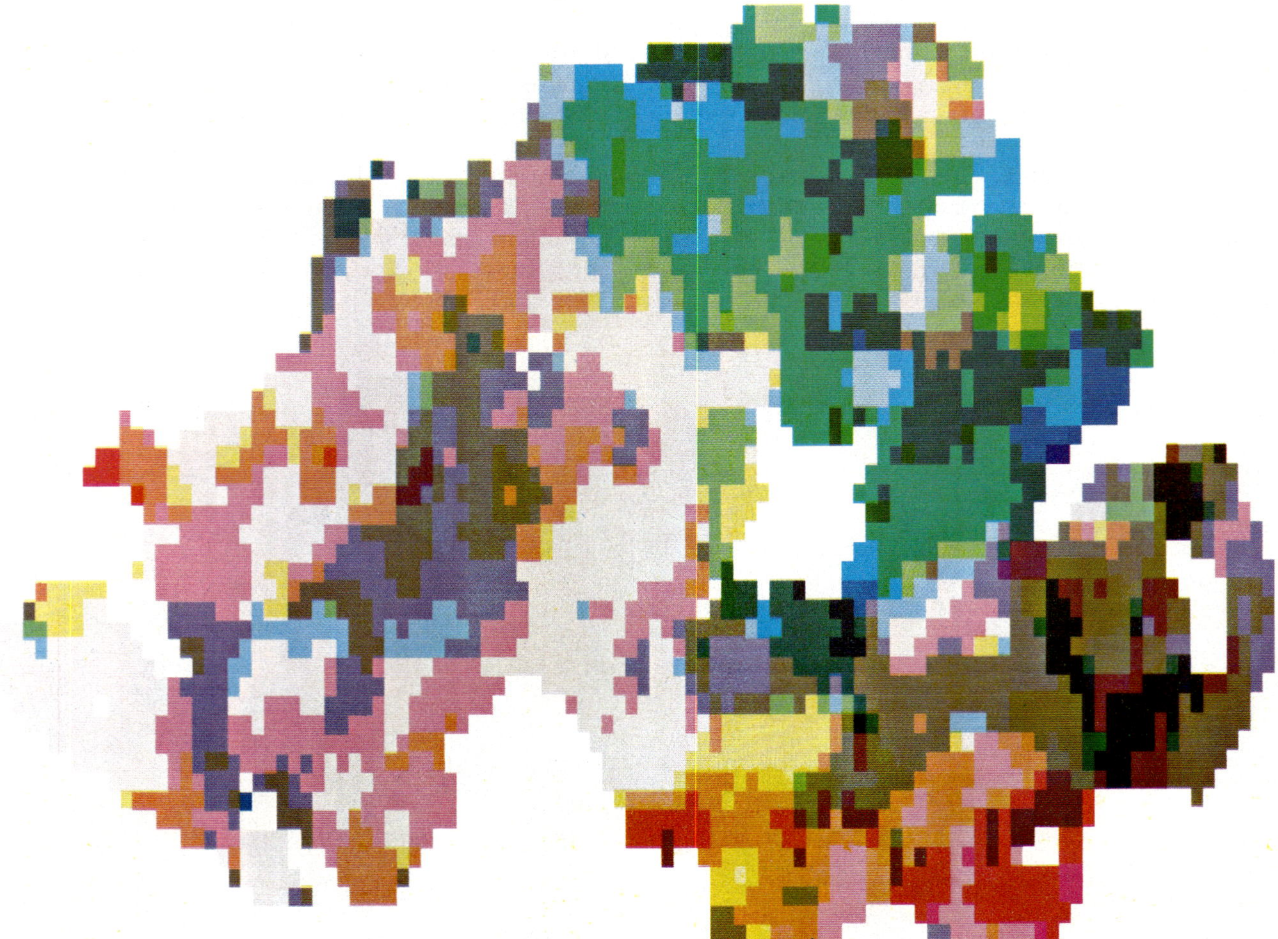

Fig. 3 Three-level three-component geochemical map of Northern Ireland for lead (red), copper (blue) and zinc (yellow). For explanation see text

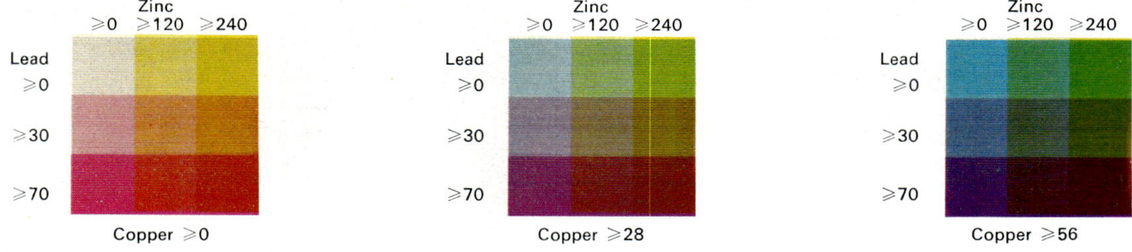

Fig. 4 Three-level three-component key for Fig. 3

Fig. 5 Multi-level three-component geochemical map of Northern Ireland for potassium (red), strontium (blue) and chromium (yellow)

strontium (5 levels) and chromium (6 levels) plotted at the following class intervals:
Potassium (red) 0·0, 0·6, 1·2, 1·8, 2·4, ⩾3·0%
Strontium (blue) 0, 50, 100, 200, 400, ⩾800 ppm
Chromium (yellow) 0, 22·5, 45, 90, ⩾180 ppm
This element combination produces colour fields which closely reflect the geology (Fig. 1) and reveal a number of interesting features. The Mourne Mountains granite (magenta), the Newry granodiorite (violet and blue), the Antrim basalts (green to yellow-green), the Tardree rhyolite (purple), and the Carboniferous limestones of Co. Fermanagh (cyan) stand out clearly. More subtle colours characterize the Silurian rocks of Co. Down and Co. Armagh. Differences in colour fields over single geological units serve to distinguish arenaceous horizons (pink colours) within the Carboniferous, and show that there is a clear geochemical difference between the Upper Dalradian rocks in Co. Londonderry (pinks and greys) and those in Co. Tyrone (magentas), and that both can be distinguished from the Lower Dalradian (lilacs). The 'Central Inlier' of Co. Tyrone (lilac and grey) can be differentiated from the surrounding rocks. Colour changes within the Old Red Sandstone correlate with known facies changes. The plot cell size in this map is slightly different from that used in Figs. 2 and 3.

Conclusions

The technique described here could be of use in the mapping of any three-component system. Production of high-quality single- and multiple-element maps can be achieved rapidly by use of the LGP 2703 Lasergraphic plotter system. Mapping of multi-element stream-sediment reconnaissance data by these techniques has shown that there is a considerable information gain over conventional single-element maps. It has been highly successful in distinguishing between bedrock lithologies, as well as delineating areas of mineral exploration potential and regions of agricultural trace-element deficiency, and would appear to be a valuable aid to the interpretation of complex geochemical systems.

Acknowledgment

The project of which this forms a part has been supported by a Natural Environment Research Council grant for an investigation, under the direction of Professor J. S. Webb, into the applicability of computer methods to the analysis of regional geochemical data. The writers would particularly like to thank Mr. P. S. Ferrer of Seiscom, Ltd., for his considerable help with implementing this technique on the Lasergraphic plotter system and Mr. E. R. Snell of Ozalid, Ltd., for technical advice concerning diazo reproduction methods.

References

1. Dobrin, M. B. Computer processing of seismic reflections in petroleum exploration. In *Computer applications in the earth sciences* Merriam, D. F. ed. (New York and London: Plenum Press, 1969), 41–60.
2. Geological Survey of Northern Ireland. *1:253,440 magnetic anomaly map of Northern Ireland* (Belfast and London: H.M.S.O., 1971).
3. Howarth, R. J. Fortran IV program for grey-level mapping of spatial data. *Math. Geol.*, **3**, 1971, 95–121.
4. Rhind, D. W. Shaw, M. A. and Howarth, R. J. Experimental geochemical maps—a case study of cartographic techniques for scientific research. Paper presented to 6th Technical Conference of the International Cartographic Association, Ottawa, Aug. 1972.
5. Young, R. D. The interpretation of regional geochemical patterns in Northern Ireland. Ph.D. thesis, University of London, 1971.

550.846:546.226:546.3/.8

Hydrogeochemical exploration for sulphide deposits: correlation between sulphate and other constituents

Mario Dall'Aglio

Geochemical Group, CNEN, Rome, Italy

Franco Tonani

Resources and Transport Division, United Nations, New York, U.S.A.

Synopsis

In a hydrogeochemical reconnaissance survey for uranium, carried out in a 20 000-km^2 area of Tuscany, 949 samples of stream waters were collected, on a regular grid. All the samples have been analysed for the major constituents, that is Ca, Mg, Na, K, HCO_3, SO_4, Cl and SiO_2. Automatic processing of the data included simple and multiple correlation, as well as regression analysis. The results obtained suggest that it is possible to discriminate between sulphate from oxidizing sulphide and sulphate of marine origin. The hydrogeochemical aureoles are broad enough to be detected by low-density sampling. Thermal springs, which frequently occur in the same region, cause sulphate aureoles of the same kind as weathering sulphides. Limited data on boron distribution in thermal springs and in waters leaching pyrite ores indicate that boron could be effective in enabling discrimination between haloes formed by oxidizing sulphide and those which originate from thermal springs.

It is proposed to demonstrate that hydrogeochemical exploration for sulphide deposits can be carried out by detecting the sulphate aureoles formed in the supergene environment in the vicinity of oxidizing sulphides. According to Boyle and Garrett,[1] 'sulphate and other anionic constituents in natural waters have received some attention in recent years, but in general, correlation of the results both on the basis of individual constituents and on ratios of two or more constituents have been somewhat disappointing as regards geochemical prospecting. It would seem, however, that further research is advisable, especially as regards the sulphate content of waters since oxidizing sulphide deposits yield vast quantities of sulphate to underground and surface waters'.

Soviet workers have published most of the studies on the characteristics of subsurface and stream waters in the vicinity of oxidizing sulphide ores,[8-11] their papers referring almost exclusively to very detailed geochemical studies, which, moreover, covered very limited areas.

The present paper deals with general criteria applicable to preliminary hydrogeochemical reconnaissance for sulphides with low-density sampling.

The data utilized to critically evaluate the efficiency of this method were obtained in a hydrogeochemical survey for uranium, which covered the whole Tuscany region. Nearly 1000 samples of stream waters were collected, following a regular grid, over a surface area of about 20 000 km^2. All the samples were analysed for major constituents, automatic data processing being widely used to assist in the interpretation.

Characteristics of hydrogeochemical dispersion aureoles originating from oxidizing sulphide deposits

General remarks

The most abundant sulphides, such as FeS, FeS_2, $CuFeS_2$, ZnS, PbS and HgS, have extremely low solubilities in water under supergene conditions. Entry of their component elements into natural waters demands the previous oxidation of the sulphides. Free sulphuric acid is released on oxidation of sulphide both by hydrolysis and as a consequence of sulphur being in stoichiometric excess in comparison with the metals, e.g. as in pyrite. The low pH value observed in waters in the vicinity of sulphide bodies is a product of the above processes. Free sulphuric acid is neutralized by displacement of the weak carbonic acid in rocks and waters, as well as by the leaching of cations from rock silicates. Sulphuric acid is neutralized, in general, within a short time and/or distance. Thus, in large-scale surveys the determination of pH in stream waters is of no practical use in exploration for sulphide ores.

Alteration of metal sulphides produces sulphate and metal ions. The sulphate ion is both stable and mobile in the supergene environment. Metal ions, such as Zn^{2+}, Cu^{2+}, Pb^{2+}, Hg^{2+} and Fe^{3+}, have low geochemical mobilities under normal supergene conditions. Geochemical mobility decreases from zinc to iron in the above list.

At neutral pH values, and in the presence of surface waters of normal composition, lead, copper and zinc precipitate as insoluble secondary minerals, such as malachite, cerussite, smithsonite, etc. All these cations can, moreover, be removed from water by ion-exchange processes, as well as by adsorption by the fine fraction of stream sediments or soils with which they come into contact. Mercury, in particular, is quickly adsorbed by stream sediments.[2] Even living matter and its decay products can remove these elements from natural waters. Therefore, the hydrogeochemical haloes of all these metals, with the partial exception of zinc, have generally limited dimensions. The sulphate ion, on the other hand, having entered into natural waters, is not generally adsorbed or precipitated, and its dispersion aureoles extend over tens of kilometres.[10]

Other favourable factors in the exploration for sulphide deposits by detecting the sulphate aureole, in addition to the high mobility of sulphate, are that the amount of sulphur associated with or broadly related to sulphide deposits of economic importance is, in general, large enough to cause widespread hydrogeochemical dispersion aureoles, and the contrast value, i.e. the ratio between the sulphate content of the waters in contact with orebodies and the sulphate content of normal waters, is generally high. Waters in the vicinity of sulphide ores can attain a very high sulphate content if in contact with the atmosphere. The sulphate content is limited only by the solubility of gypsum, i.e. 2·4 g/l, but there is no significant limit to the solubility of the sulphuric acid. The maximum content of sulphate formed from sulphides is limited by the available amount of oxygen in waters ($<0{\cdot}3$ mM/l) if there is no contact with the atmosphere. This limiting condition, however, does not apply to the supergene environment, where exchange with the atmosphere is possible. Waters leaching crystalline rocks normally have a very low sulphate content—of the order of a few ppm.

Besides these advantages, there are, however, drawbacks to the use of sulphate in exploration. (1) Environmental conditions, especially climatic ones, can drastically limit the applicability of hydrogeochemical methods.[4] (2) The regional variability of the sulphate distribution in the waters in practice hinders the detection of the anomalous samples. The sulphate content of stream waters, unaffected by oxidizing sulphide deposits, varies widely—for example, from the few ppm in waters which leach crystalline rocks to thousands of ppm for waters leaching gypsum. There are, therefore, many processes which introduce sulphate into natural waters—a fact which makes it difficult to identify the sulphate that originates from oxidizing sulphides. (3) Even if one manages to detect the sulphate aureoles originating from sulphide, this exploration method is not a specific technique. In fact, the most widespread sulphide is pyrite, and even its dispersed sub-economic mineralization can cause more marked sulphate haloes than that which originates from ores of sulphides other than pyrite.

Processes introducing sulphate ion into fresh waters

The sulphate found in fresh waters may be introduced by numerous processes. (1) Sulphate is present in rain water, even in the form of free sulphuric acid. Its content is quite variable and concentrations can be as much as tens of ppm. There are three main sources of sulphate in rain: *(a)* air pollution from smoke and combustion gases; *(b)* marine salts, including calcium sulphate, from the sea; and *(c)* atmospheric H_2S, which readily oxidizes to sulphate, from the decomposition of organic matter by anaerobic bacteria.[6,7] (2) Sulphate can be introduced by the leaching of sedimentary rocks containing calcium sulphate by both stream and underground waters. (3) In addition, sulphate is brought to the surface directly, by thermal waters rich in sulphate, and indirectly, by H_2S and other volatile sulphur compounds which reach the surface through fumaroles, thermal springs and large volcanic explosions. These sulphur compounds oxidize to sulphate in the supergene environment. (4) Finally, various artificial origins exist—industrial dumping introduces directly into stream waters vast amounts of sulphate, and fertilizers bring, directly or indirectly, large quantities of sulphate to natural waters.

From a geochemical viewpoint, the origin of sulphate in waters, besides oxidizing sulphide ores, can be reduced to three main groups of processes: *(a) artificial* or *marine origin; (b) marine origin*, which is related to the sulphur already contained as sulphate in sea water; and *(c) circulation of sulphur in the form of H_2S or its products.*

Associations which characterize sulphate in different geochemical environments

The sulphate contained in rain water is not, in general, a relevant complicating factor in hydrogeochemical exploration for sulphide ores, since its content is low in comparison with that normally found in natural waters. Sulphate in rain water is associated with other compounds of marine origin and/or with compounds resulting from industrial pollution.

Marine-origin sulphate is associated both with a stoichiometric amount of calcium—in the case of dissolution of calcium sulphate—and with other elements found in evaporitic formations, e.g. magnesium, boron, chlorine and bromine. These elements and sulphate are geochemically mobile, and anion concentration ratios vary widely—for example, the Cl/SO_4 ratio ranges from zero to infinity; the former occurs in the presence of pure calcium sulphate and the latter corresponds to pure chloride. The sulphate originating from the oxidation of both sulphide deposits, and hydrogen sulphide from hydrothermal manifestations, causes, in natural waters, one or more of the following three types of elemental association: *(a)* association of sulphate with cations essentially in the same proportion as in the country rocks—this occurs when the release of cations is essentially due to depletion processes; *(b)* association of sulphate with the cations already dissolved in the waters—this is caused by the displacement of bicarbonates, or even of chlorides, by sulphate; and *(c)* association with the minor and trace elements contained in sulphide ores and/or in other hydrothermal products.

In all of these associations the sulphate content is generally higher than the calcium content, due to the fact that a part of the sulphuric acid produced by the oxidation of sulphide is neutralized by cations other than calcium. Identification of that part of the sulphate which originates from oxidizing sulphides is necessary to the effective use of this method of exploration. This discrimination procedure can be carried out by studying the correlations between sulphate and other constituents. In this connexion regression analysis has been the technique principally used in the present work.

Hydrogeochemical reconnaissance survey, Tuscany

General

In order to test and critically evaluate the results that can be achieved by this hydrogeochemical prospecting method, from the hydrogeochemical data collected by CNEN in numerous areas in Italy[3] those from Tuscany collected in 1962 were selected—for the following reasons. *(a)* Sulphide deposits of various types exist in Tuscany: they range from the hypothermal pyrite deposits of Gavorrano, Niccioleta and Boccheggiano to the polymetallic sulphide orebodies in the Val di Cecina and in the Monti Romani zone and to the cinnabar deposits in the Monte Amianta area. *(b)* In this region many hot spring systems occur—among them the Larderello geothermal field: these phenomena interfere with the hydrogeochemical search for sulphide deposits. *(c)* Another complicating factor, widely found in Tuscany, is that of outcropping gypsum-bearing sediments.

In 1962 a preliminary geochemical reconnaissance survey for uranium was carried out over the whole Tuscany region. From a 20 000-km^2 area 949 stream water samples were collected on a regular sampling grid. In each water sample all the major constituents were analysed, according to the procedure adopted by the CNEN Geochemical Group, to obtain a complete geochemical classification of each water sample and to be able to make a selection of anomalous uranium samples on the basis of objective criteria and automatic data processing.[5]

In Table 1 the mean and the standard deviation values of the physical parameters (T, pH, electrical conductivity) and major constituents dissolved in the waters from Tuscany are shown, the values of the correlation coefficients for all the element pairs being shown in Table 2. The procedure followed was discussed by Dall'Aglio and Tonani.[5]

Fig. 1 shows a graph relating the concentration of the major constituents to the electrical conduc-

Table 1 Mean and standard deviation values of major constituents of stream water samples from Tuscany

Variables	Number of determinations	Mean value	Standard deviation	Coefficient of variation, %
Temperature, °C	946	18·085	4·549	25
pH	949	7·807	0·494	6
Conductance, ohm^{-1} $cm^{-1} \times 10^{-4}$	949	8·296	6·592	79
Calcium, meq/l	949	5·460	4·173	76
Magnesium, meq/l	949	2·115	2·990	141
Sodium, meq/l	949	1·763	3·495	198
Potassium, meq/l	949	0·100	0·136	137
Bicarbonate, meq/l	949	4·392	2·603	59
Sulphate, meq/l	949	3·557	6·367	179
Chloride, meq/l	949	1·409	3·609	256
Silica, mM/l	936	0·184	0·191	103

Table 2 Linear correlation coefficients for all element pairs: first figure shows value of coefficient; second the number of samples analysed for both parameters taken into account. Stream water samples from Tuscany

	I	pH	Conductance	Ca	Mg	Na	K	HCO_3	SO_4	Cl
SiO_2	−0·122	0·117	0·021	0·023	0·026	−0·005	0·352	0·067	0·017	−0·024
	933	936	936	936	936	936	936	936	936	936
Cl	−0·023	−0·076	0·709	0·024	0·348	0·915	0·417	0·139	0·205	
	946	949	949	949	949	949	949	949	949	
SO_4	0·126	−0·150	0·759	0·869	0·708	0·343	0·236	0·027		
	946	949	949	949	949	949	949	949		
HCO_3	0·210	−0·109	0·369	0·136	0·552	0·275	0·214			
	946	949	949	949	949	949	949			
K	0·020	−0·077	0·466	0·222	0·289	0·474				
	946	949	949	949	949	949				
Na	0·043	−0·088	0·808	0·238	0·542					
	946	949	949	949	949					
Mg	0·156	−0·101	0·786	0·514						
	946	949	949	949						
Ca	0·163	−0·217	0·720							
	946	949	949							
Conductance	0·146	−0·188								
	946	949								
pH	−0·199									
	946									

tance values, i.e. to the total salinity.

All the samples were first arranged in increasing order of electrical conductance and divided into groups of 50 samples each. Within each group the average, as well as the standard deviation values, of all the constituents were computed. In Fig. 1 each point therefore represents the average value of 50 samples.

The waters with the lowest salinity, corresponding to the range of electrical conductance values up to 8×10^{-4} ohm^{-1} cm^{-1}, show the predominance of calcium bicarbonate. These waters mostly correspond to normal leaching of rock. In the conductance range from 0 to 16×10^{-4} ohm^{-1} cm^{-1} the waters acquire, for the most part, a calcium sulphate chemistry, whereas for a higher salinity a sodium chloride chemistry becomes predominant.

Selection of anomalous samples

To select anomalous water samples, i.e. samples from oxidizing sulphide orebodies, the correlations between the sulphate and the other major

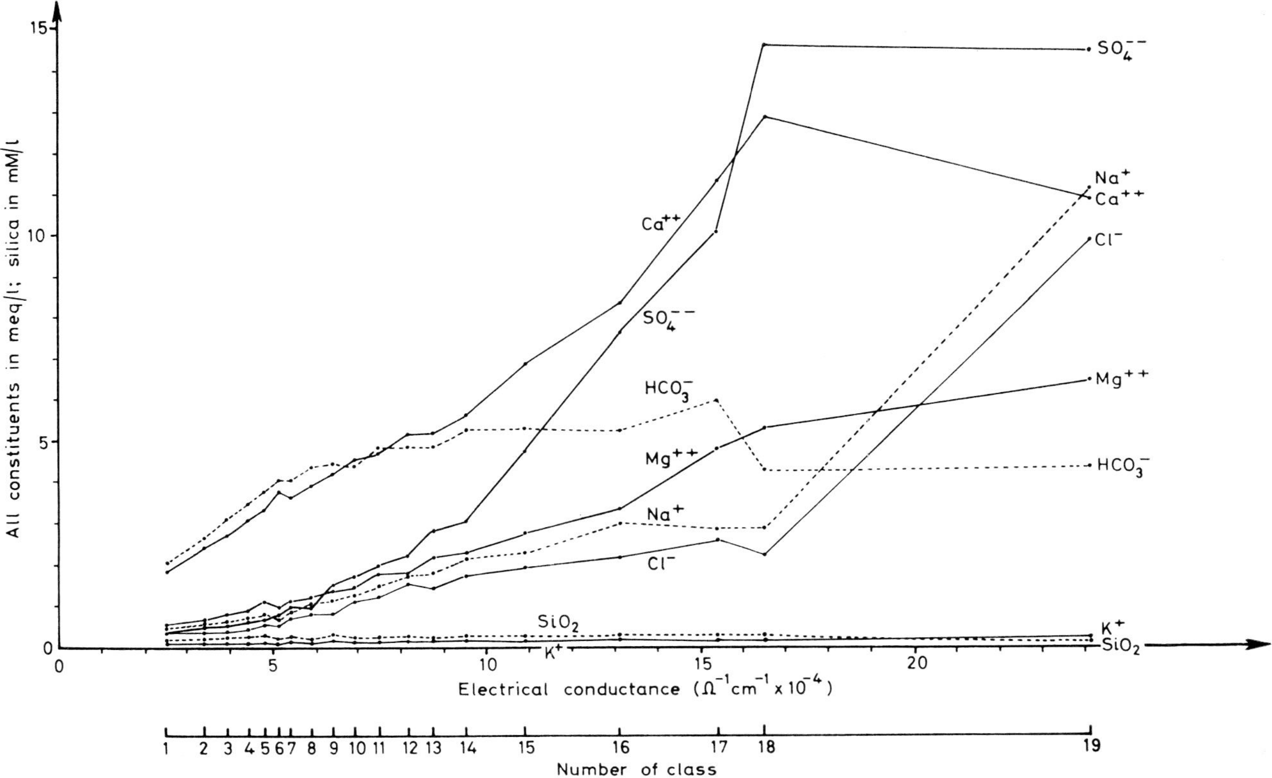

Fig. 1 Pattern of composition of stream waters from Tuscany as a function of electrical conductance values

constituents were first studied in detail. Fig. 1 and the correlation coefficient values presented in Table 1 give an idea of the associations found between the constituents dissolved in the Tuscany waters. The correlations between sulphate and calcium, and between sulphate and chlorine, are of importance in detecting water samples originating from oxidizing sulphide ores. Chlorine and calcium are, in fact, associated with sulphate of marine origin, even if their ratios vary widely. Particular attention must be paid to these associations, since most of the sulphate which circulates in fresh waters is, directly or indirectly, connected with marine sulphate.

In Fig. 2 the regression curves are given for sulphate on chloride and for sulphate on calcium, each point shown representing the average value of 50 samples. (The computing procedure followed to draw these diagrams is the same as that for Fig. 1.) The regression curves shown in Fig. 2 give the best-fit relationship, according to the least-squares method, between SO_4 and Cl and SO_4 and Ca. These curves represent the geochemical associations occurring in the area studied, and they correspond to 'what is usually taking place' in the main water bodies in Tuscany.

Samples which deviate significantly from the best-fit curves are therefore assumed to be anomalous, i.e. to be the product of some unusual process.

In Fig. 3 all the water samples are presented in a SO_4–Ca diagram. Curves of 'anomaly' limits are also given, together with the best-fit curve. These

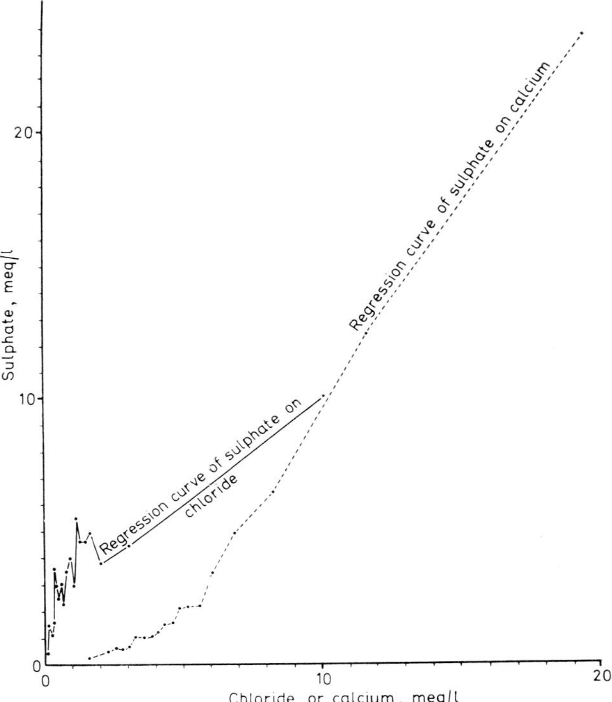

Fig. 2 Regression curves of sulphate content on chloride and of sulphate on calcium: 949 stream water samples from Tuscany

limits were chosen by taking into account the regression curve of Fig. 2 and the value of the standard deviation from the regression (see Table 2). The areal distribution of the anomalous samples so selected is shown in Fig. 5.

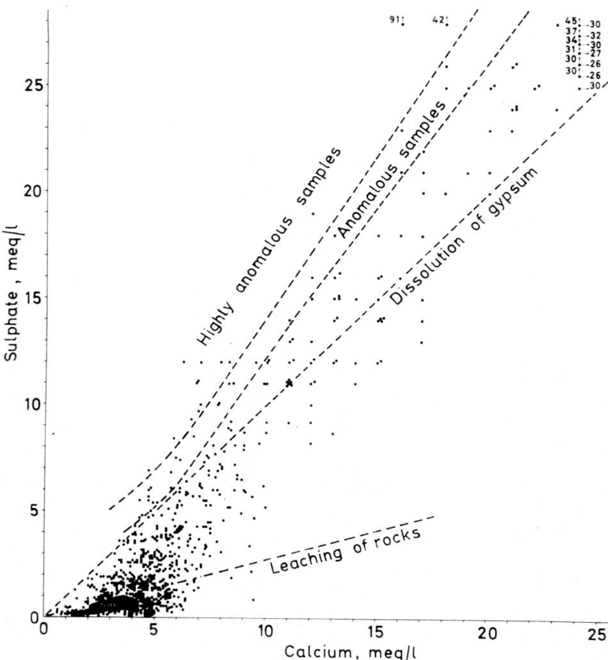

Fig. 3 Sulphate versus calcium content: 949 stream water samples from Tuscany

Simultaneous consideration of the SO_4–Cl and SO_4–Ca correlations would be helpful to the selection of the 'anomalous' samples. A term for the interaction between calcium and chloride would also be included. This appears to be a reasonably complete procedure, which takes into account the most important geochemical factors.

Studies of this kind can be carried out numerically by use of a computer. Our first objective was to set up the best-fit multivariate polynomial by regression analysis. Initially, we studied the univariate regression of sulphate on *(a)* calcium, *(b)* chloride and *(c)* calcium times chloride. A procedure based on proposing polynomials of increasing degree for analysis has been utilized (program IBM POLRG, 360A–CM–03X) to degree 3. The results obtained are shown in Table 3. Computations for the sulphate–calcium couple show that from a strictly statistical viewpoint (accounting for the degrees of freedom by the F test) the best result is achieved by the degree 1 polynomial.

For the sulphate–chloride couple also the best statistical result is obtained by degree 1, and for sulphate–(product of calcium by chloride) degree 3 gives the best fit.

In these conditions, however (non-linear polynomial, non-normal distributions), selection of the 'best' polynomial cannot be made on the sole basis of the level of significance associated with a F value. For example, the intercept of the least-squares straight line with the sulphate axis in a sulphate–calcium graph is $-3{\cdot}7$ meq/l. The level of significance of the degree 3 polynomial does not express its capability to adjust the obtained curve to the foot of the regression line, i.e. below 6 meq/l. Inspection of this part of the curve (Fig. 2) shows that it is significant, in spite of its low weight in the sum of squares. Thus, the degree 3 polynomial was chosen. Similarly, the degree 2 polynomial was chosen to represent the relationship between sulphate and chloride. The degree 3 polynomial for sulphate on product 'calcium times chloride' (interaction of calcium and chloride) has been utilized on the basis of the F test.

Once the most convenient polynomial has been determined, multiple-regression analysis can be carried out with the aim of selecting the anomalous samples. The polynomial used in the present analysis was $SO_4 = a_0 + a_1\,\mathrm{Ca} + a_2(\mathrm{Ca})^2 + a_3(\mathrm{Ca})^3 + a_4\,\mathrm{Cl} + a_5(\mathrm{Cl})^2 + a_6(\mathrm{Ca}\times\mathrm{Cl}) + a_7(\mathrm{Ca}\times\mathrm{Cl})^2 + a_8(\mathrm{Ca}\times\mathrm{Cl})^3$.

Table 4 shows the values of the most significant statistical parameters obtained: these results are very significant, as the values of the multiple correlation coefficient and of F indicate.

For each water sample the sulphate content was then computed by use of the intercept and regression coefficients values given in Table 4. The sulphate contents determined are plotted in Fig. 4 versus the sulphate contents computed

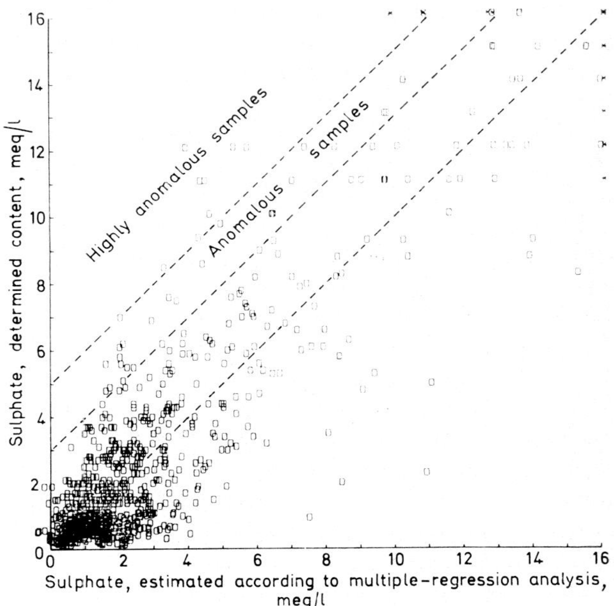

Fig. 4 Actual sulphate content versus sulphate content estimated according to multiple-regression analysis shown in Table 4

Table 3 Polynomial regression analysis of SO_4 on Ca, Cl and Ca × Cl (949 observations)

	Source of variation	Degrees of freedom	Sum of squares	Mean square	*F* value	Improvement in terms of sum of squares
SO_4 on Ca						
Polynomial regression of degree 1						
Intercept–0·3683204*E*–02	Due to regression	1	0·2901990*E*–01	0·2901990*E*–01	0·29216*E* 04	0·2901990*E*–01
Regression coefficients	Deviation about regression	947	0·9406555*E*–02	0·9933005*E*–05		
0·13261*E* 01	Total	948	0·3842645*E*–01			
Polynomial regression of degree 2						
Intercept–0·3070595*E*–02	Due to regression	2	0·2909293*E*–01	0·1454647*E*–01	0·14744*E* 04	0·7303432*E*–04
Regression coefficients	Deviation about regression	946	0·9333521*E*–02	0·9866301*E*–05		
0·11440*E* 01 0·80868*E* 01	Total	948	0·3842645*E*–01			
Polynomial regression of degree 3						
Intercept–0·5647226*E*–03	Due to regression	3	0·2968686*E*–01	0·9895619*E*–02	0·10700*E* 04	0·5939305*E*–03
Regression coefficients	Deviation about regression	945	0·8739591*E*–02	0·9248243*E*–05		
0·91114*E*–01 0·11271*E* 03–0·26114*E* 04	Total	948	0·3842645*E*–01			
SO_4 on Cl						
Polynomial regression of degree 1						
Intercept 0·3048283*E*–02	Due to regression	1	0·1609165*E*–02	0·1609185*E*–02	0·41391*E* 02	0·1609185*E*–02
Regression coefficients	Deviation about regression	947	0·3681726*E*–01	0·3887778*E*–04		
0·36106*E* 00	Total	948	0·3842645*E*–01			
Polynomial regression of degree 2						
Intercept 0·2376167*E*–02	Due to regression	2	0·2491931*E*–02	0·1245965*E*–02	0·32801*E* 02	0·8827467*E*–03
Regression coefficients	Deviation about regression	946	0·3593452*E*–01	0·3798574*E*–04		
0·97702*E* 00–0·13046*E* 02	Total	948	0·3842645*E*–01			
Polynomial regression of degree 3						
Intercept 0·2165546*E*–02	Due to regression	3	0·2598272*E*–02	0·8660906*E*–03	0·22844*E* 02	0·1063407*E*–03
Regression coefficients	Deviation about regression	945	0·3582818*E*–01	0·3791341*E*–04		
0·12176*E* 01–0·33187*E* 02 0·27345*E* 03	Total	948	0·3842645*E*–01			
SO_4 on (Ca × Cl)						
Polynomial regression of degree 1						
Intercept 0·3125646*E*–02	Due to regression	1	0·3097056*E*–02	0·3097056*E*–02	0·83016*E* 02	0·3097056*E*–02
Regression coefficients	Deviation about regression	947	0·3532951*E*–01	0·3730676*E*–04		
0·40040*E* 02	Total	948	0·3842657*E*–01			
Polynomial regression of degree 2						
Intercept 0·2566772*E*–02	Due to regression	2	0·5335044*E*–02	0·2667522*E*–02	0·76257*E* 02	0·2237988*E*–02
Regression coefficients	Deviation about regression	946	0·3309153*E*–01	0·3498046*E*–04		
0·11377*E* 03–0·10935*E* 06	Total	948	0·3842657*E*–01			
Polynomial regression of degree 3						
Intercept 0·1363178*E*–02	Due to regression	3	0·1026067*E*–01	0·3420224*E*–02	0·11475*E* 03	0·4925627*E*–02
Regression coefficients	Deviation about regression	945	0·2816590*E*–01	0·2980517*E*–04		
0·30297*E* 03–0·92682*E* 06 0·66255*E* 09	Total	948	0·3842657*E*–01			

Table 4 Multiple-regression analysis according to the equation

$$SO_4 = a_0 + a_1\ Ca + a_2(Ca)^2 + a_3(Ca)^3 + a_4\ Cl + a_5(Cl)^2 + a_6(Ca \times Cl) + a_7(Ca \times Cl)^2 + a_8(Ca \times Cl)^3$$

Stream water samples from Tuscany region. Number of samples = 949

Variable	Mean	Standard deviation	Correlation *X* versus *Y*	Regression coefficient	Standard error Regression coefficient	Computed *T* value
Ca	54·59577*E*–04	0·417233*E*–02	0·869030*E* 00	−0·314366*E*–01	0·152422*E* 00	−0·206246*E* 00
Cl	14·08689*E*–04	0·360852*E*–02	0·204611*E* 00	−0·750099*E* 00	0·202346*E* 00	−0·370702*E* 01
Ca^2	47·15082*E*–06	0·998912*E*–04	0·832045*E* 00	0·104735*E* 03	0·132928*E* 02	0·787912*E* 01
Ca × Cl	10·77047*E*–06	0·451421*E*–04	0·283907*E* 00	0·146748*E* 03	0·266232*E* 02	0·551204*E* 01
Cl^2	14·94638*E*–06	0·184844*E*–03	0·127530*E* 00	0·126924*E* 02	0·271039*E* 01	0·468286*E* 01
Ca^3	65·09640*E*–08	0·252771*E*–05	0·727541*E* 00	−0·227857*E* 04	0·314100*E* 03	−0·725428*E* 01
$(Ca \times Cl)^2$	21·47449*E*–10	0·334812*E*–07	0·156628*E* 00	−0·424374*E* 06	0·690091*E* 05	−0·614953*E* 01
$(Ca \times Cl)^3$	14·00414*E*–13	0·305783*E*–10	0·133128*E* 00	0·289758*E* 09	0·475969*E* 08	0·608774*E* 01
Dependent						
SO_4	35·56898*E*–04	0·636677*E*–02				
Intercept		0·65395*E*–04				
Multiple correlation		0·88449*E* 00				
Standard error of estimate		0·29831*E*–02				

Analysis of variance for the regression

Source of variation	Degrees of freedom	Sum of squares	Mean squares	*F* value
Attributable to regression	8	0·30063*E*–01	0·37579*E*–02	0·42229*E* 03
Deviation from regression	940	0·83649*E*–02	0·88989*E*–05	
Total	948	0·38428*E*–01		

E–04 = $\times 10^{-4}$, *E*–06 = $\times 10^{-6}$, etc.

according to the regression law. The straight line, with slope equal to 1 and intercept zero, represents, for practical purposes, the regression surface. Hence, anomalous samples can be selected inasmuch as they deviate significantly from the straight line.

The areal distribution of the anomalous samples so selected is shown in Fig. 5.

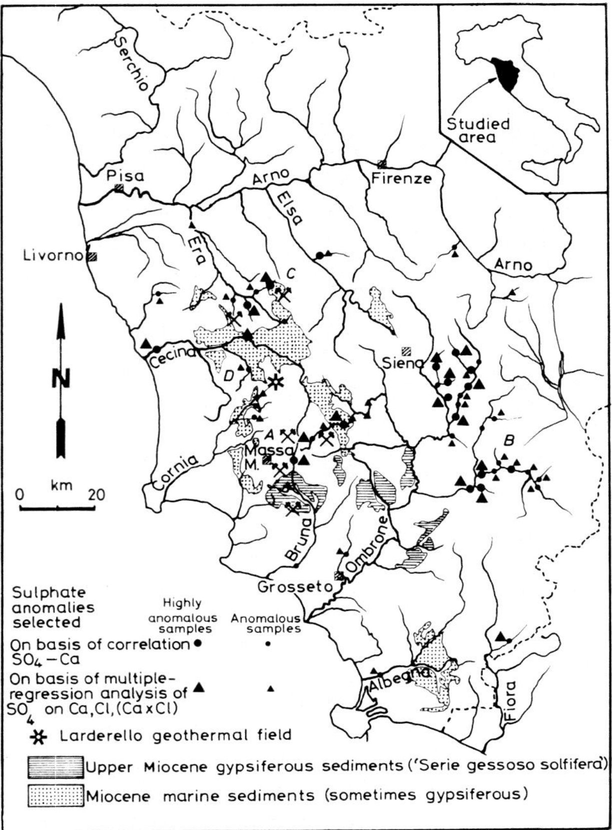

Fig. 5 Map of sulphate anomalies detected in Tuscany: (a) on basis of correlation between sulphate and calcium (circles) and (b) on basis of multiple-regression analysis of sulphate on calcium, chloride and on product Ca × Cl (triangles). Improvement obtained by multiple-regression analysis is shown, e.g. in stressing anomaly A, which originates from very important pyrite mine

Discussion

The geographic distribution of anomalous samples in Fig. 5 shows four main sulphate aureoles. *Anomaly A* comes from the area surrounding the town of Massa Marittima, where important pyrite deposits are located, some of which are mined at the present time (Boccheggiano and Niccioleta). *Anomaly B* originates from an area which includes known systems of thermal springs. The springs are located along the faults that border the Rapolano–Mt. Cetona horst. *Anomaly C* outlines a wide zone where numerous sulphide orebodies are known; in addition, a system of thermal springs is located in Val d'Era. The hydrological situation is similar to that outlined for anomaly *B*. *Anomaly D* appears to originate from the Larderello geothermal field. Sulphide orebodies are also known in this region.

A few additional anomalous samples are scattered throughout, but they are not clustered to form individual dispersion aureoles.

The anomalies found point out that hydrogeochemical prospecting methods, based on the accurate and complete study of the distribution of the major constituents and their associations, define the geochemical background of sulphate in the stream waters as related to such regional factors as the leaching of evaporite formations. Thus, sulphate from other sources can be detected even in areas contaminated by evaporite rocks. The method, however, seems not to be effective in discriminating sulphate aureoles caused by sulphide deposits on the one hand and thermal springs on the other. In both cases the anomalies partly originate from the production of excess sulphuric acid. Another process is important for the formation of sulphate aureoles from thermal springs. The waters of many springs become supersaturated with respect to calcium carbonate once they are equilibrated with CO_2 pressure at the surface. The precipitation of calcium carbonate will substantially change the chemistry of these waters—from a calcium–magnesium carbonate type to a magnesium sulphate type.

Table 5 gives an example of such variations in connexion with the area investigated. The first water sample was collected at the outlet of an important thermal spring in the Rapolano zone (St. Giovanni spring), and the second was collected 1 km downstream along a creek fed exclusively by this spring. Considerable amounts of travertine have been precipitated along the creek.

The changes observed in the chemical composition are self-explanatory. The second sample will be anomalous according to the procedure proposed here, and it will tend to become more and more anomalous as the precipitation of calcium carbonate goes on.

Generally speaking, waters related to thermal springs can be distinguished from those related to oxidizing sulphide orebodies, as thermal springs originate from specific water bodies. The boron content of waters can be effective in enabling one to discriminate between the two origins. In fact, the boron content of many thermal waters in Tuscany is very high—unlike its content in waters leaching sulphide deposits, which is very low. Moreover, the increase in the boron content can be specifically related to hydrothermal circulation itself as a consequence of disruption of stratification of deeper waters, stream leakages, and so on.

Table 5 Changes in water composition due to precipitation of travertine from thermal waters: (1), water sample collected at outlet of spring; (2), water sample collected 1 km downstream

Sample	Ca	Mg	Na	K	HCO_3	SO_4	Cl	SiO_2	B
(1)	40	19	16	1·1	43	25	9·4	0·62	0·9
(2)	21	17	16	1·1	25	25	9·4	0·46	1·0

All concentrations in meq/l, except silica and boron, which are in mM/l.

Sulphate and boron of a few thermal water samples from the Rapolano area, and of waters leaching pyrite deposits (mine waters from Boccheggiano mine) have been plotted in Fig. 6, the difference between these two types being clearly shown. Unfortunately, there is a lack of information about the distribution of boron in stream waters of Tuscany, since this element had not been analysed at the time when this survey was carried out.

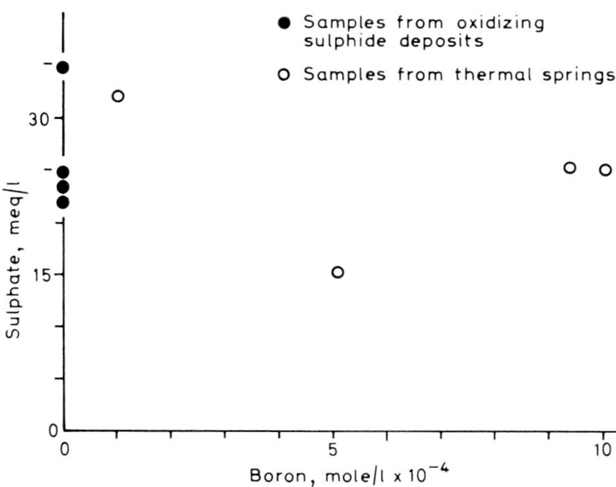

Fig. 6 Sulphate–boron diagram. Water samples from thermal springs and oxidizing sulphide deposits

Conclusions

The feasibility of hydrogeochemical reconnaissance surveys for sulphide deposits, on the basis of the distribution of sulphate in stream waters, has been suggested. The method is effective, and seems to be generally applicable, provided that it is implemented by the collection of fairly complete hydrogeochemical information. Geochemical processes which cause important changes in the sulphate content in stream waters, independently of sulphide deposits, take place, in fact, throughout the area investigated. Correlation studies afford effective discrimination between sulphate of different origins. This is also possible by use of simple and cheap data processing, such as simple correlation.

The results of the study performed in the Tuscany region can be summarized as follows.

(1) This exploration method can be used successfully, especially in preliminary large-scale geochemical surveys. It is advisable that the information collected be implemented by studying the distribution of some other indicator elements, e.g. boron. The information obtained is, moreover, very useful to an understanding of some of the tectonic, hydrologic and metallogenetic aspects of the region under study.

(2) This method is effective in discriminating between the sulphate aureoles originating from the leaching of marine formations and the aureoles originating from the oxidizing of sulphide deposits and/or of hydrogen sulphide of hydrothermal origin.

(3) Geothermal fields, as well as systems of thermal springs, give rise to dispersion aureoles of sulphate which display the same phenomena as those caused by oxidizing sulphide deposits.

(4) This prospecting method must be classified as a method suitable for preliminary reconnaissance rather than a specific technique. Discrimination between oxidizing sulphide deposits and hydrothermal activity is less precise than identification of those types from sulphate of marine origin.

(5) Other geochemical parameters, such as the boron content in waters, can implement this discrimination. These criteria must, however, be selected according to the specific geochemical feature of the region investigated. These criteria are only special cases of the general philosophy outlined above, more effective methods resulting as more geochemical information is utilized.

(6) The sulphate aureoles detected in the survey of the Tuscany region were spread over a large area, which indicates that sulphate circulates steadily in the supergene environment. Sulphate is not removed from surface waters, except by gypsum precipitation, which occurs only under extreme evaporation.

Acknowledgment

The authors are indebted to Dr. G. P. Giannotti for geological assistance, and to C. Gigli for the automatic processing of the data.

References

1. Boyle, R. W. and Garrett, R. G. Geochemical prospecting—a review of its status and future. *Earth Sci. Rev.*, **6**, 1970, 51–75.
2. Dall'Aglio, M. The abundance of mercury in 300 natural water samples from Tuscany and Latium (Central Italy). In *Origin and distribution of the elements* Ahrens, L. H. ed. (Oxford, etc.: Pergamon, 1968), 1065–81.
3. Dall'Aglio, M. Distribution and circulation of the major elements and of some trace elements in surface waters of Italy. *Publ. intn. Ass. scient. Hydrol.* no. 78, 1968, 79–91.
4. Dall'Aglio, M. Planning and interpretation criteria in hydrogeochemical prospecting for uranium. In *Uranium prospecting handbook* Bowie, S. H. U. Davis, M. and Ostle, D. eds. (London: IMM, 1972), 121–34.
5. Dall'Aglio, M. and Tonani, F. Storage and retrieval of geochemical data. *Italy, RT/GEO CNEN*, 1965, 21 p.
6. Eriksson, E. On the geochemistry of chloride and sulfate. In *Atmospheric chemistry of chlorine and sulfur compounds* Lodge, J. P. Jr. ed. (Washington, D.C.: American Geophysical Union, 1959), 124–7. *(Publ. Am. geophys. Union* 652)
7. Gorham, E. The influence and importance of daily weather conditions . . . *Phil. Trans. R. Soc.*, **B241**, 1958, 147–78.
8. Kovalev, V. F. *et al.* Hydrochemical methods of prospecting for copper sulfide ores in the southern Urals. *Geokhimiya,* 1961, 596–603; *Geochemistry*, 1961, 638–47.
9. Kraynov, S. R. Use of surface flow of spring water for hydrochemical prospecting for ore deposits. *Razv. Okhr. Nedr.*, **24**, no. 3 1958, 42–5; *Int. Geol. Rev.*, **2**, 1960, 259–62.
10. Polikarpochkin, V. V. *et al.* Geochemical prospecting for polymetallic ore deposits . . . *Int. Geol. Rev.*, **2**, 1960, 236–53.
11. Stremyakov, A. Ya. Prospection hydrochimiques des gîtes minéraux situés dans les pergelisols. *Razv. Okhr. Nedr.*, **24**, no. 3 1958, 46–7; *Prosp. Prot. Sous-Sol*, **3**, 1958, 43–5.

543.42.662:546.49

Zeeman spectrometer for measurement of atmospheric mercury vapour

J. C. Robbins

Scintrex Ltd., Concord, Ontario, Canada

Synopsis

There has been considerable recent interest in the use of mercury vapour as a tracer for various classes of mineral deposits; as a very volatile toxic metal, mercury and its components are also receiving attention as an environmental pollutant. In the past the accurate determination of the vapour by atomic absorption has been notoriously unreliable—due mainly to the many possible interferences. A dual-beam dual-wavelength spectrometer has been developed by use of the 2536-Å resonance line. The dual or reference wavelength is generated by Zeeman-splitting the emission line by repetitively pulsing a magnetic field around the discharge lamp. Differential measurements of the absorption during the periods of 'field' and 'no field' yield a direct value of mercury vapour concentration.

A more advanced technique makes use of the different planes of polarization of the centre and split-emission components. With an electro-optical modulator and analyser the two components may be sequentially routed through the spectrometer. The extension of the technique to other elements and its application to the chemical analysis of airborne particulates are discussed, and mention is made of the possibility of isotopic measurements.

The mercury spectrometer has been extensively field-tested, both in geological and pollution applications. Data from atmospheric and soil-gas tests are presented, and the applicability of the instrument to geochemical exploration is considered.

The normal or background abundance of mercury in the lower atmosphere[1] is probably of the order of 1×10^{-9} g/m^3—equivalent to about 1 part in 10^{13} by volume. Although the ecologist would probably regard this as an acceptably low figure, its direct and quantitative determination is difficult, particularly in view of the much higher concentrations of potentially interfering species in the atmosphere. In this context the pollutants sulphur dioxide and nitrogen dioxide are rarely found to be less than 5 ppb (5×10^{-9}), ozone has an average worldwide concentration of 2–3 ppm, and even the least abundant of the rare gases, xenon, although not an interference, has a mean level of nearly 90 ppb.

Although chemical separation and colorimetric determinations of mercury have been made in the past, most high-sensitivity measurements now are carried out by atomic absorption techniques by use of the 2537-Å mercury resonance line. Where the optical techniques developed to date differ is in the means chosen to reduce the interference effects of the other atmospheric constituents that absorb strongly in the ultraviolet. The attempts to solve these problems can be categorized into three groups—amalgamation, twin-beam and dual-wavelength methods.

In the amalgamation approach[2] mercury from the sample is absorbed on to a noble metal surface and later volatilized into a simple spectrometer. The main advantage of this method is the ability to integrate very low concentration levels over a period of time. It is, however, difficult to adapt this technique for the fast response required for mobile surveys. In the twin-beam approach[3,4] the absorption due to an incoming sample stream is compared to the absorption of the same sample after passage through a chemical filter that selectively removes the mercury content.

The dual-wavelength methods depend on the measurement of the absorption by the sample vapour at two different wavelengths, at one of which mercury absorbs strongly and a second where the absorption is very much less. If these 'signal' and 'reference' wavelengths are closely spaced, the absorption due to substances other than mercury can be considered constant and a differential measurement will isolate the presence of mercury.

The signal wavelength radiation has conventionally been provided by a low-pressure mercury lamp, but the techniques for generation of the reference wavelength have been quite varied. Continuum sources, such as deuterium or xenon arcs, in conjunction with suitable monochrometers to limit the spectral band pass, have been used. Similarly, a suitably spaced spectral line from another element, such as platinum, has served as the reference source. Ling[5] developed a technique by which the 2537-Å emission line of a heated mercury resonance lamp was sufficiently broad that it showed little attenuation by mercury vapour, and, hence, was useful as the reference generator.

Barringer[6] has described the use of a heated mercury lamp where part of the radiation was split off and passed through a cell containing saturated mercury vapour. After passage through the cell, the centre of the thermally broadened line was effectively absorbed, leaving the 'wings' only of the line to act as the reference. Mechanical means were used to recombine the beams for passage through the sample cell to the detector.

The Zeeman spectrometer described here also uses a single mercury lamp as a source, but it runs at normal temperatures.

A magnetic field is applied to the lamp so that a significant proportion of the radiation is shifted to a slightly different wavelength, due to the Zeeman effect. If a sufficiently strong field is used, the radiation may be displaced beyond the absorption profile, and, hence, may be used as the reference wavelength.

Zeeman spectrometry

For optimum rejection of interferences and for minimum Zeeman magnet power consumption the spectral separation of the signal and reference frequencies should be minimized. The limit is set by the finite width of the emission and absorption line profiles.

The emission profile of the 2537-Å line of a low-pressure mercury discharge lamp is shown in Fig. 1. This may be resolved into a number of components, identified with the various isotope constituents of natural mercury and, in the case of the isotopes with odd mass number, their associated hyperfine structure.

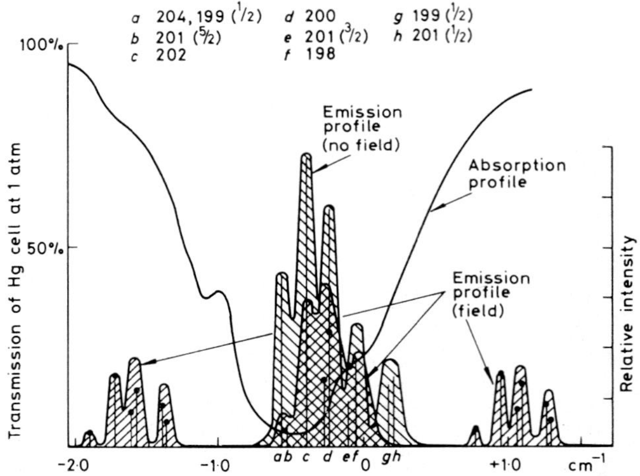

Fig. 1 Emission and absorption profiles of natural mercury

The width of each component is set by Doppler broadening. For this transition and a typical lamp temperature of 300–350°K the width (at half the peak height) can be estimated as 0·05 cm^{-1} approximately. Since the spacing between the outermost components of the emission envelope is 0·74 cm^{-1}, we can expect an overall emission line width of approximately 0·79 cm^{-1}.

The absorption profile of mercury vapour under normal atmospheric conditions is considerably broader and displaced slightly towards lower frequencies in comparison with the emission profile. The absorption profile shown in Fig. 1 is taken from Hadeishi and McLaughlin,[7] who have independently developed a rather different approach to a Zeeman spectrometer for mercury determinations. From this profile we can estimate a width of about 1·5 cm^{-1}.

The dual-wavelength or frequency concept requires that the absorption by the atomic species in question should be relatively small at the reference frequency. Comparison of the emission and absorption profiles given in Fig. 1 shows that a displacement of the emission radiation by 1·2–1·3 cm^{-1} is sufficient to reduce the attenuation to about 15% from a peak value of 90%. The magnetic fields required for such a shift are now considered.

The Zeeman effect is the modification by a

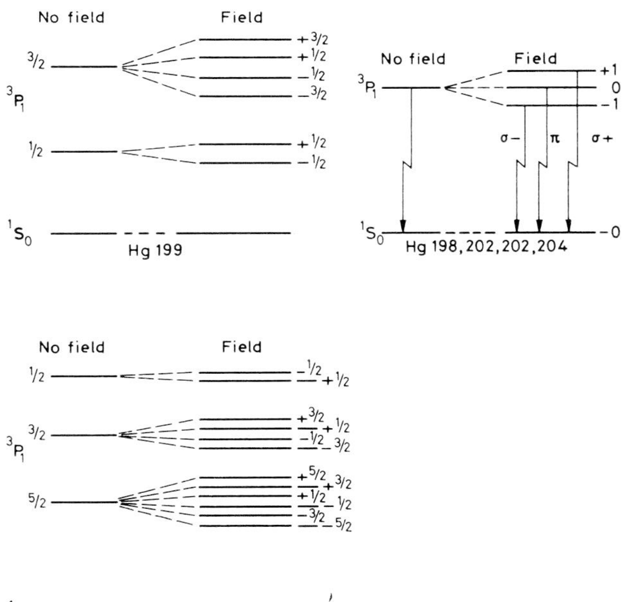

Fig. 2 Zeeman sub-levels of natural mercury

magnetic field on the energy levels involved in a particular transition (Fig. 2). In the case of the $^3P_0$–$^1S_0$ mercury line for the isotopes 198, 200, 202 and 204, the energy increment in the upper level is given by

$$\delta\nu = (g_T \beta / L) H$$

where g_T is the Landé g factor, β is the Bohr magneton, L is Planck's constant and H is the magnetic field. Numerically, $\delta\nu$ here is approximately 2 GHz/kgauss (equivalent to 0·066 cm^{-1}/kgauss). Viewed transversely to the magnetic field, all three components are seen, the displaced components σ^+ and σ^- being polarized perpendicularly to the field, the centre π component having parallel polarization.

For the isotopes with odd mass numbers (199 and 201), with nuclear spins of $\frac{1}{2}$ and $\frac{3}{2}$, respectively, the individual hyperfine components are split in a weak magnetic field to yield a multiplicity of lines (Fig. 2). In the strong field used where the splitting is more than comparable to the separation of the hyperfine components, however, the coupling between the atomic angular momentum and the nuclear spin breaks down and the Zeeman pattern tends to revert to the simple triplet pattern. The centre (π) component is located approximately at the centroid of the original hyperfine components.

The spectral redistribution of energy during the application of a 20 kgauss magnetic field is also shown in Fig. 1. The σ^+ and σ^- now lie at the wings of the absorption profile, leaving the π components essentially undisturbed. Thus, if we periodically pulse the magnetic field and make differential absorption measurements on a mercury sample, a maximum modulation efficiency of about 40% can be expected. In practice, efficiencies closer to 50% are achieved, which indicates that the absorption profile may be rather narrower than that indicated by Hadeishi and McLaughlin.

A higher modulation efficiency could be achieved by the elimination of the π component with a suitable polarizer; however, the transmission of sheet polarizers in the ultraviolet is sufficiently poor that any gain in efficiency is nullified by the loss in light level.

In the present instruments the Zeeman magnet consists of a laminated core with tapered mild steel pole pieces. The current required for 20 kgauss is about 3 A for the 2·5-mm gap now used. The average power consumption of the magnet system pulsed on a 25% duty cycle is less than 20 W.

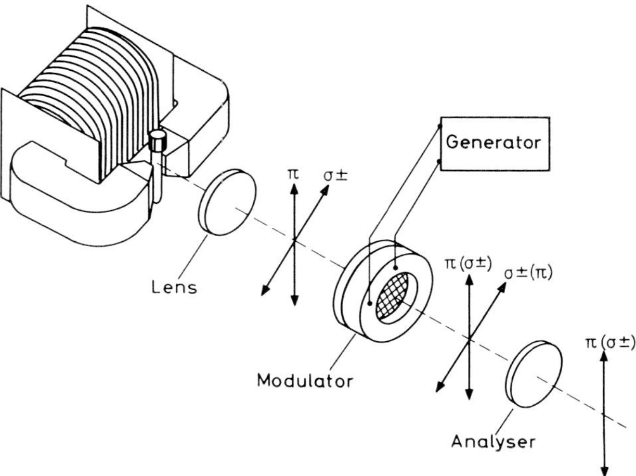

Fig. 3 Zeeman system with electro-optical modulation

A second approach uses the existence of the orthogonally polarized σ and π components. The radiation is directed through an electro-optical modulator and a suitable analyser orientated so that the π component only is transmitted (this is shown schematically in Fig. 3). When an appropriate voltage is applied to the modulator, the planes of polarization of the components of light incident on the analyser are rotated through 90°, thereby allowing it to pass the σ components. The fundamentals of this technique have been tested satisfactorily in the laboratory.

The advantages of this method include the use of a permanent magnet, requiring, of course, no power. Also, in a system limited by the ever-present $1/f$ lamp noise, substantial signal to noise gains can be made by use of the much higher modulation frequencies feasible with electro-optical modulation.

Successful Zeeman-modulated experiments involving the absorption and fluorescence of lead, zinc and cadmium have recently been carried out. The aim of these experiments is the development of rugged, but sensitive, spectrometers for field camp use.

Another development is the application of magneto-optical scanning to lead isotope assays to provide a precise, but low-cost, alternative to mass-spectroscopic methods. In principle, all the elements that show sufficient isotope shift in their hyperfine spectral structure, i.e. those towards either end of the periodic table, should be capable of analysis by spectrochemical techniques.

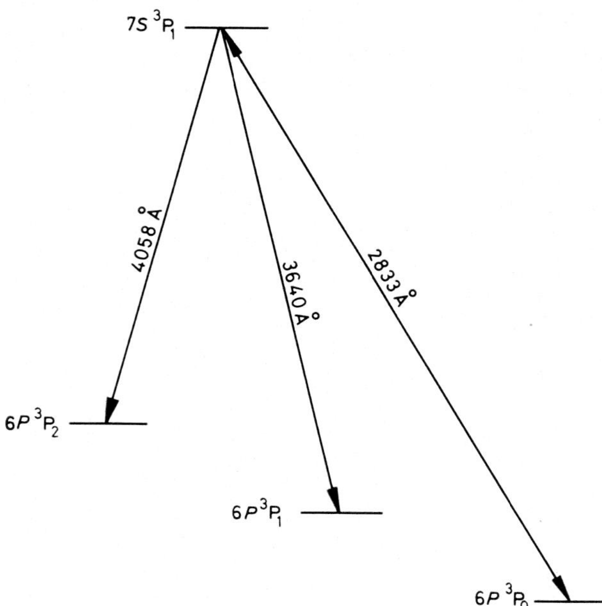

Fig. 4 Energy levels involved in the resonance and line fluorescence of lead

The present concept is that of selective excitation of line fluorescence of the various spectrally shifted isotopic components. The absorption takes place at 2833 Å, but the fluorescence is observed at 4058 Å (the energy levels involved are shown in Fig. 4).

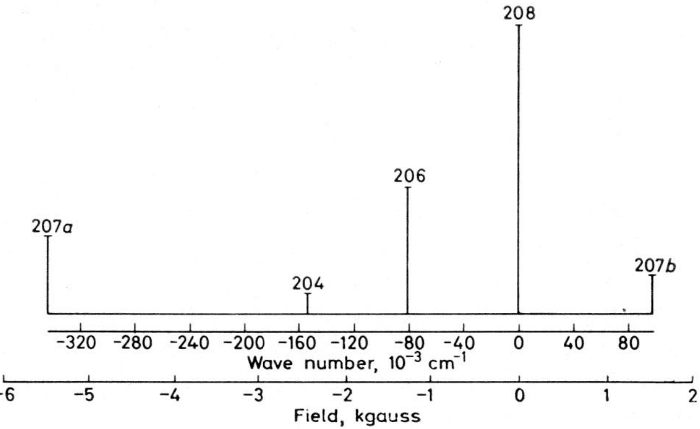

Fig. 5 Hyperfine structure of the Pb 2833-Å line

The light source is a ^{208}Pb lamp situated in a magnetic field that can be varied. As can be seen from Fig. 5, the component viewed longitudinally to the field can be made to sweep sequentially across the absorption frequencies of the isotopes 208, 206, 204 and 207. The resultant fluorescence is measured as a function of the magnetic field and the counts for the various isotopes are stored in separate registers.

Development of mercury instrumentation

Atmospheric sensor

Two basic types of mercury spectrometers have been developed, the earlier type for the measurement of trace concentrations in the atmosphere being shown diagramatically in Fig. 6. In this application, where an almost unlimited sample is available, the greatest sensitivity is achieved by the largest practical sample volume and absorption path length.

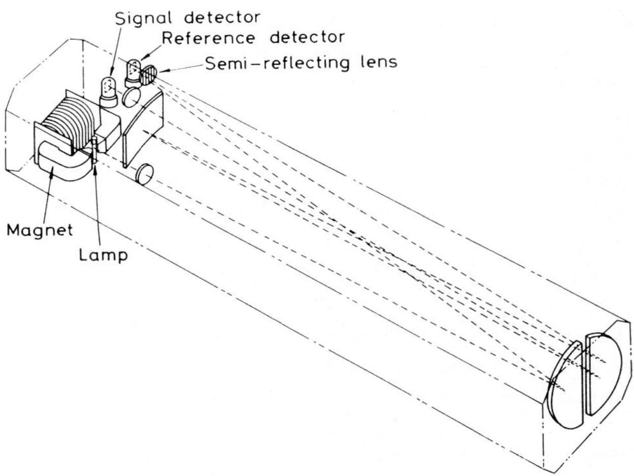

Fig. 6 Schematic diagram of atmospheric sensor

The prototype instrument employed a multipass arrangement to efficiently fold a 20-m path length into a 1·25-m sample cell by use of high-reflectivity (>98%) mirrors. A reference detector was positioned to receive a fraction of the lamp radiation after the first traverse of the cell, the output from this detector being subtracted from that of the signal detector to cancel to a large extent the effects of lamp noise and drift. Noise levels were about 1 ng/m^3 (peak–peak).

Later versions of this unit had a 12-m path length with improved magnet drive and signal processing electronics. Noise limits of these instruments are 2–3 ng/m^3, with response times of about 5–10 sec.

Soil-gas sensor

A more recent development has been a much smaller device capable of back-pack operation for field measurements of mercury in soil gases or in many other sample media. With a sample cell of only 250 ml, this spectrometer has the same *weight* sensitivity as the atmospheric sensor, but a proportionately lower *concentration* sensitivity. Typical noise levels are 30×10^{-12} g Hg. The layout of the spectrometer is shown schematically in Fig. 7, and a general view of this unit is given in Fig. 8. The weight, including batteries and pack frame, is 43 lb.

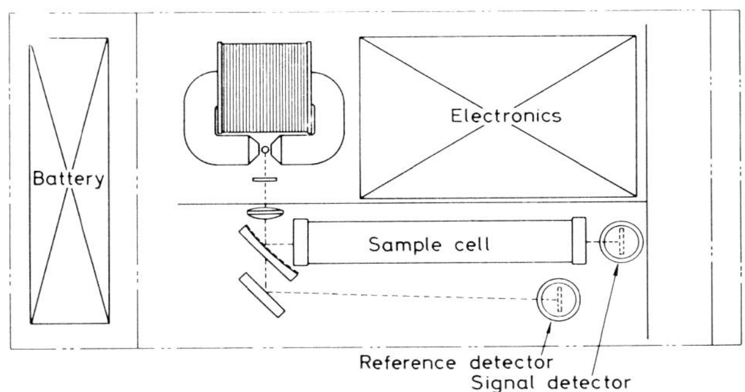

Fig. 7 Schematic plan of back-pack spectrometer

Its use in soil-gas measurements depends on the observation that only a very limited amount of air can be extracted from the pores and crevices of a soil volume before the occurrence of dilution by surface air. The sample is drawn into the spectrometer by means of a hand pump connected to the sample cell exhaust. Repeatedly, it has been found that the first sample (approximately 0·25 l) withdrawn contains the highest levels of mercury, if present at all, subsequent pumping almost invariably showing much lower levels.

Fig. 8 Soil-gas measurements with back-pack spectrometer

As a battery-operated unit in the field, the complete system is run intermittently to conserve battery power. The lamp is, however, run continuously, but the rest of the electronics is switched on only some 10–15 sec before a measurement is made. In normal field use the batteries provided yield a full day's operation and can be recharged in about 6 h.

A battery eliminator can be clipped to the spectrometer in place of the battery pack when line power is available and the unit is to be used continuously. This system has been used successfully in a field camp for the routine determination of mercury in soil samples, power being supplied by a small gasoline generator.

Field methods

Mercury in the atmosphere

Measurements of mercury concentration in the atmosphere have mainly been made by use of truck installations. Over suitable terrain, profiles made at traverse speeds of 5–10 miles/h have been obtained; but, more often, spot measurements along a line have been made with the vehicle at rest.

For work in the U.S.A. and Canada the intake height was about 5 ft above the ground, but for the Australian results described later the intake was mounted on the front bumper to reduce the height to about 1 ft. Air is drawn in continuously, and, to obtain a zero level, the flow is diverted through a filter treated with palladium chloride before entry to the spectrometer.

For airborne pollution studies, carried out with a Bell Jet Ranger helicopter, the spectrometer was slung in the baggage compartment; the filter and control unit were positioned in the cabin, the air intake being through one of the cabin windows.

Mercury in soil gases

Small-diameter holes are spiked in the ground to a uniform depth, typically ranging from 18 to 36 in. A probe is inserted to seal the top of the hole and the soil gas is withdrawn, through the spectrometer, by a hand pump.

Soil-gas surveys are economically carried out by two-man crews, one operator carrying the spike, hammer and probe to prepare the sample holes, and the second carrying and operating the spectrometer. The station spacing and the number of readings taken at each station will obviously vary with the terrain and the type of survey undertaken. We have generally used 50-ft stations, three measurements with separate holes being taken at each station.

The sample rate depends on the terrain and the ease with which the sample holes can be made. The actual measurement time is less than 15 sec, and, with practice, the three measurements at a station should not take longer than 1½–2 min.

Care has to be taken to avoid contamination of sample lines—particularly if the anomalous mercury levels are large. For this reason they are kept as simple and short as possible, spares being carried in case contamination is suspected.

Mercury concentrations below the detection limit of the instrument used directly may be measured by accumulating on a silver collector the vapour from separate holes. (A small amount of silver wool, packed firmly into the centre of a 7-mm quartz tube, some 12 cm long, has been employed. The collector is heated by a small propane or butane torch and the integrated mercury load is drawn into the spectrometer.)

Although this technique is slower than direct measurement, it may be more convenient where the ground is damp or waterlogged and also when the terrain is so rough that back-packing the spectrometer would be difficult.

The use of silver collectors with solar-pumped plexiglass tents[8] may prove of great interest where direct measurements are inadequate—particularly in those regions where sunlight is relatively predictable. Similarly, small cylinders of silver foil can be suspended in holes drilled in the ground. Recovered a few hours later for analysis, this type of collector may prove to be a very inexpensive exploration method.

Mercury in various media

The advantages of being able to run economical and reliable geochemical measurements in the field need hardly be stressed. The portable spectrometer described has been used in field camps for measurements of mercury in soils and rocks. With a basic full-scale meter sensitivity of $1{\cdot}5 \times 10^{-9}$ g Hg, the gain is usually attenuated by a factor of 10 or 100 to allow reasonable sample weights to be used.

Direct pyrolysis, by heating the sample in a test tube with a propane torch, is a fast and quantitative method for most types of samples, including sulphides and air-dried *B* horizon soils. With the instrument's inherently high SO_2 rejection, mercury levels in pyrite, chalcopyrite, galena and sphalerite have been measured successfully.

A field version of the Hatch and Ott method[9] for the extraction of mercury vapour has also been used successfully over a period of time, but it is slower and less sensitive than the direct pyrolysis methods.

Field results

Atmospheric

The modified prototype spectrometer has been used extensively in *Western Australia*, a strong correlation being noted between atmospheric mercury levels and Cu–Zn sulphide mineralization in the volcanic rocks of West Pilbara. J. G. Baird (personal communication) has found that all of

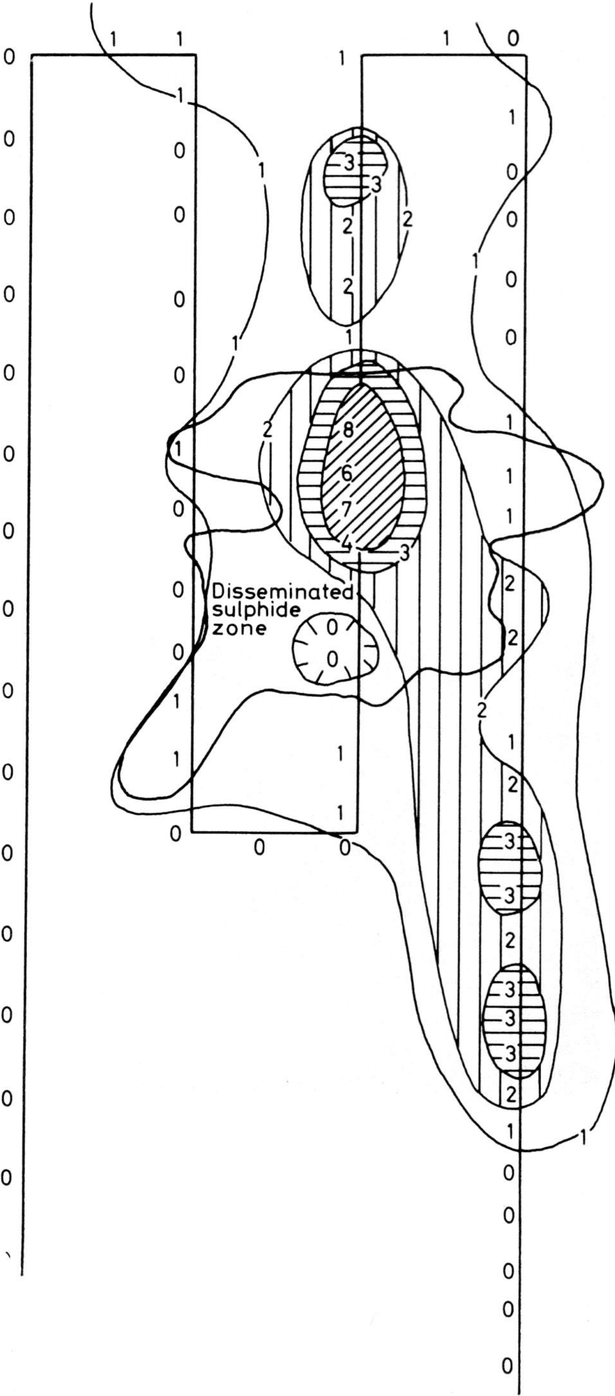

Fig. 9 Sherlock Station (B): scale, 1 in=400 ft; mercury concentration along path of vehicle given in ng/m³

the known cases of sulphide mineralization visited in this district have registered varying, but anomalous, levels; conversely, anomalous levels have not been found to occur over areas known to be unmineralized. The noise level of the equipment was 1 ng/m^3, concentration levels greater than 2–4 ng/m^3 being considered anomalous; the maximum recorded was 11 ng/m^3. Typical results are shown in Figs. 9 and 10.

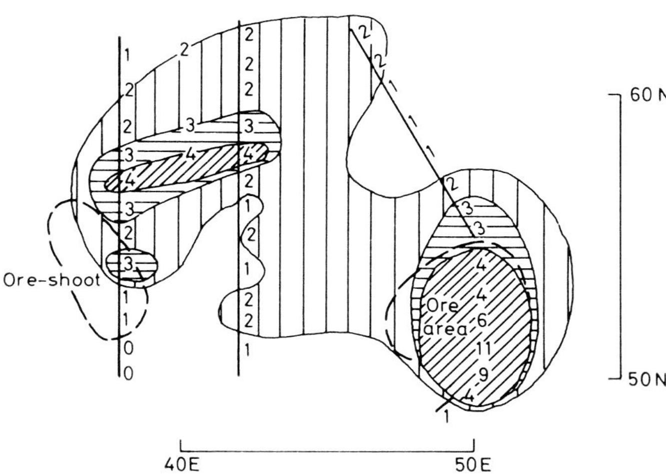

Fig. 10 Whundo: scale, 1 in=500 ft; mercury concentration along path of vehicle given in ng/m^3

Quantitative replication of data, both in concentration levels and exact location of peaks, has depended on the meteorological factors of temperature, wind and precipitation. Also, earlier work over nickel sulphides in ultramafic rocks had proved negative. It is possible (J. G. Baird, personal communication) that any results at that time may have been masked by contamination in the truck used.

Work in *North America* preceded the Australian studies and was carried out with a prototype spectrometer, which had a detection limit of about 5 ng/m^3. This sensitivity was assumed to be adequate for the 30–50 ng/m^3 anomalous levels found earlier[10] over various types of mineralization in the southwestern United States.

Seven major open-pit copper producers and some 20 lead–zinc–silver and gold mine sites were visited in Utah and Arizona. In no case were mercury levels in excess of 5–6 ng/m^3 recorded. The same applied to the Sudbury, Kirkland Lake and Noranda areas in Canada. Slightly higher levels (10–12 ng/m^3) were found at Cobalt, Ontario.

High levels were obtained over the Ord and Sunflower mercury mines in the Matzatzal Mountains, northeast of Phoenix. Fig. 11 shows the levels obtained over a prospect pit at Ord: the great variation with time of the concentration,

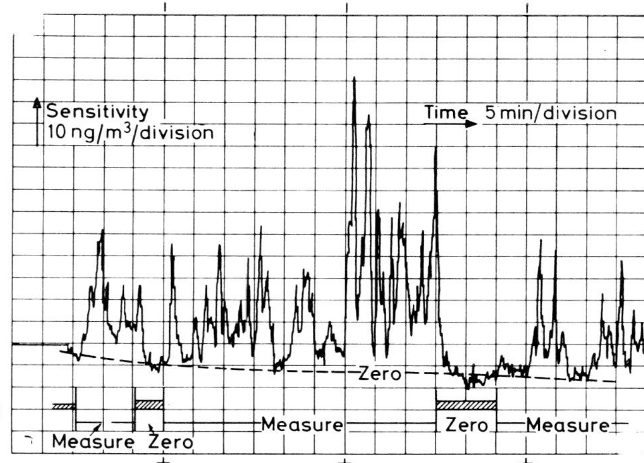

Fig. 11 Ord mine, Arizona: variation of mercury concentration with time (vehicle stationary)

even when measured comparatively close to the ground, is readily apparent. This order of meteorological noise must diminish the prospect of fast and detailed airborne geochemical mapping.

In view of the later Australian results, it can only be said of the work in North America that upper bounds have been set, but that lower, but still equally anomalous, levels may exist over some types of mineralization.

Soil gas

The prototype back-pack instrument, which had a detection limit of better than 50×10^{-12} g, was completed late in 1971 and was briefly field-tested over the Clyde Forks mercury prospect described by Jonasson and Allan.[11] Previously, in the summer, a short test programme had been conducted with the Geological Survey of Canada with the much larger HGP spectrometer; as expected, the soil gas over the mineralized zone was strongly anomalous.

A soil-gas profile obtained with the small spectrometer is shown in Fig. 12. Weather

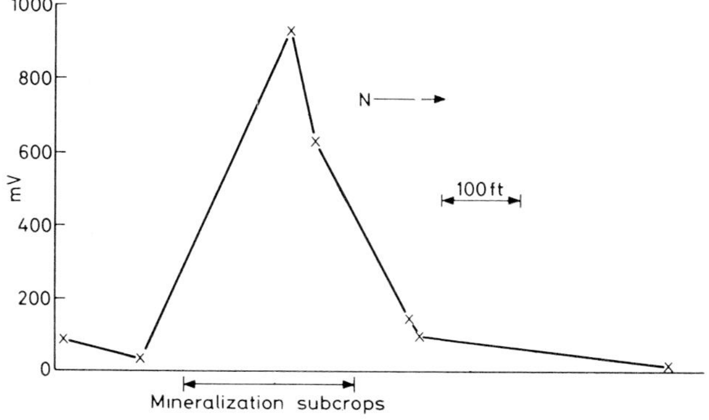

Fig. 12 Clyde Forks, Ontario: soil-gas profile (north–south)

conditions were very poor, it having rained a few hours earlier, and snow had started to fall before the measurements were completed, but the anomaly–background ratio was found to be similar to the earlier results.

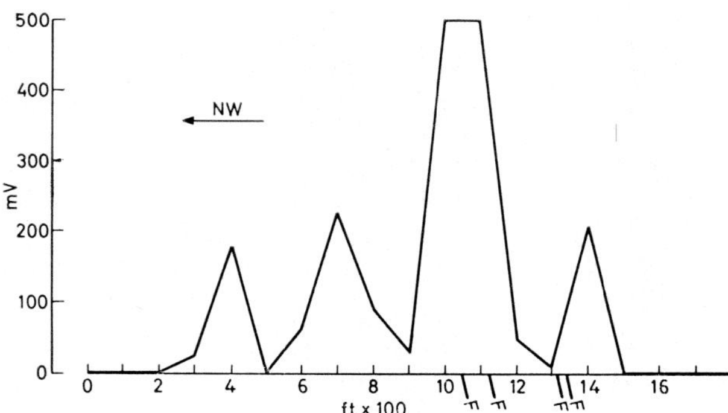

Fig. 13 Keel deposit, Ireland: soil-gas profile (northwest–southwest)

More recently, work has been carried out in Ireland, and a soil-gas profile (D. Burns, personal communication) across the zinc–cadmium deposits at Keel is shown in Fig. 13.

On the negative side, no soil-gas anomalies were detected in the course of a short field trip to lead and zinc prospects in the United Kingdom and France that were believed to be anomalous in mercury. In one case, although an anomaly background ratio of at least 20:1 in field measurements of the mercury content of soil samples from the *B* horizon was obtained, there was no detectable soil-gas equivalent.

Summary

The design and operation of sensitive mercury spectrometers that utilize Zeeman modulation to provide dual-wavelength discrimination against potentially interfering species have been described. This technique, which avoids the need for delicate or moving optical components, is ideal for robust equipment capable of withstanding the abuse inevitable in the field.

The geochemistry of mercury and its application to exploration have been studied for many years, and the subject has been recently and comprehensively reviewed by Jonasson and Boyle.[12] There still, however, appears to remain a great lack of theoretical treatment or experimental data, such as exists for radon, concerning the actual processes responsible for the migration and diffusion of mercury vapour into the atmosphere.

Measurements of atmospheric concentration in North America were disappointing, but the more recent work in Australia is encouraging. The soil-gas technique is certainly less affected by meteorological conditions and, in some ways, is more akin to conventional geochemical exploration methods. Whether the measurements are made directly or via some type of collector, soil-gas surveys represent a rapid and low-cost technique for reconnaissance or detailed work.

In areas such as many parts of the Canadian Shield, where there is extensive outcrop and soil-gas methods are inapplicable, the back-pack spectrometer can be better used for reliable field measurements of mercury in rock, soil or stream-sediment samples.

Acknowledgment

It is a pleasure to acknowledge the support and encouragement of Dr. H. O. Seigel, particularly during the more difficult phases of the development. The authors also wish to thank George Alexander, without whose electronic capability the project would not have been completed successfully.

References

1. Fleischer, M. Summary of the literature on the inorganic geochemistry of mercury. *Prof. Pap. U.S. geol. Surv.* 713, 1970, 6–13.

2. Vaughn, W. W. A simple mercury vapor detector for geochemical prospecting. *Circ. U.S. geol. Surv.* 540, 1967, 8 p.

3. Williston, S. H. Mercury in the atmosphere. *J. geophys. Res.*, **73**, 1968, 7051–5.

4. James, C. H. and Webb, J. S. Sensitive mercury vapour meter for use in geochemical prospecting. *Trans. Instn Min. Metall.*, **73**, June 1964, 633–41.

5. Ling, C. Portable atomic absorption photometer for determining nanogram quantities of mercury in the presence of interfering substances. *Analyt. Chem.*, **40**, 1958, 1876–8.

6. Barringer, A. R. Interference-free spectrometer for high-sensitivity mercury analyses of soils, rocks and air. *Trans. Instn Min. Metall. (Sect. B: Appl. earth sci.)*, **75**, 1966, B120–4.

7. Hadeishi, T. and McLaughlin, R. D. Hyperfine Zeeman effect atomic absorption spectrometer for mercury. *Science, N.Y.*, **174**, 1966, 404–7.

8. McMarthy, J. H. Jr. *et al.* Mercury in soil gas and air—a potential tool in mineral exploration. *Circ. U.S. geol. Surv.* 609, 1969, 16 p.

9. Hatch, W. R. and Ott, W. L. Determination of sub-microgram quantities of mercury by atomic absorption spectrophotometry. *Analyt. Chem.*, **40**, 1968, 2085–7.

10. McCarthy, J. H. Jr. Gott, G. B. and Vaughn, W. W. Distribution and abundance of mercury and other trace elements in several base- and precious-metal mining districts. *Circ. New Mex. St. Bur. Mines Miner. Resour.* 101, 1068, 99–108.

11. Jonasson, I. R. and Allan, R. J. Snow: a

sampling medium in hydrogeochemical prospecting in temperate and permafrost regions. In *Geochemical exploration 1972* (London: IMM, 1973), 159–74. *(Proc. 4th intn. geochem. Explor. Symp., London, 1972)*
12. JONASSON, I. R. and BOYLE, R. W. Geochemistry of mercury and origins of natural contamination of the environment. *CIM Trans.*, **75**, 1972, 8–15.

533.9:546:550.84

Microwave-induced argon plasma emission system for geochemical trace analysis

W. T. Meyer, B.SC., M.SC.

K. C. Y. Lam Shang Leen, B.SC., M.SC.

Both of the Applied Geochemistry Research Group, Imperial College, London, England

Synopsis

A microwave-induced argon plasma emission system was adapted for the quantitative analysis of vapour evolved on the rapid heating of soils, rocks and solution residues in an induction furnace. Detection limits for cadmium, mercury, lead, zinc, copper and iodine were determined in the range 10^{-9}–10^{-11}. Thermal evolution profiles derived from controlled heating of soils and rocks provided information on the forms in which metals were present by comparison with synthetic and natural standards, and trace quantities of galena were identified in soil from South Wales by use of this technique. Fractional analysis of metal vapour evolved on heating soils to 500°C gave improved anomaly–background contrast in areas of transported soil in Labrador and Ireland. Cadmium evolved from soils at less than 500°C was shown to be related to the presence of cadmium-rich sphalerite mineralization, at the same time discriminating against cadmium derived from a zone of sub-economic oxides at Keel, Eire.

A plasma may be defined as a mass of ionized gas in which the concentrations of electrons and positive ions are in equilibrium. This state may be produced in gases at atmospheric pressure by dissociating and ionizing gas molecules into electrons and positive ions by use of the energy of a high-frequency electromagnetic field. High-frequency discharge plasmas at reduced pressure have been studied as a means of excitation of atomic vapours since Gatterer[1] demonstrated the intense spectral emission of halogens, sulphur and selenium in 1948, but the plasma torch[2,3,4] generated at atmospheric pressure has proved to be of greater analytical significance. Investigation of the plasma torch as a source of atomic vapour for absorption and emission spectrometry has been based on the improved atomization efficiency provided by the high electronic temperatures (5000–10 000°K) of the plasma, coupled with the reduction of compound formation through the use of an inert carrier gas. Analytical studies by the use of radio-frequency discharges have been made by a number of workers,[5-20] frequencies from 3·4 to 40 MHz, at powers from 400 to 5000 W, being employed. Veillon and Margoshes[17] found that a plasma torch operated at 4·8 MHz gave improved emission signals for elements such as B, Ta and Ti which are normally difficult to atomize and excite in chemical flames, but that many of the elements studied were subject to unknown interferences. Greenfield *et al.*,[8,9] Hoare and Mostyn,[10] Wendt and Fassel[18,19] and Dickinson and Fassel,[5] however, have demonstrated the relative absence of matrix effects and chemical interferences on spectral absorbance or emission from plasmas generated at a variety of radio frequencies.

Microwave (2450 MHz) plasmas can be induced at lower wattages than radio-frequency discharges by use of equipment commonly available in laboratories for operating the electrodeless discharge lamps[21, 22] used increasingly as sources in atomic fluorescence analysis. Initial studies with the application of microwave plasmas as emission sources were concentrated on the use of the sensitivity of the system as a detector of organic components in gas chromatography,[23-27] both band and atomic emission spectra being monitored to record the presence of a variety of elements, including sulphur, phosphorus and the halogens. The investigation of these relatively low-power plasmas has since been extended to the analysis of a wide range of elements with encouraging results. Runnels and Gibson [28] determined several metals by use of a 2450-MHz argon plasma at 25 W with detection limits in the range 10^{-11}–10^{-12} g on a sample size of 10^{-5}–10^{-6} g, evaporating samples of volatile metal compounds into the gas stream with a filament heater. Further studies by Murayama and co-workers,[29] Aldous *et al.*[30] and Fallgatter and co-workers[31] have examined the trace-analysis potential of microwave-induced plasmas for elements in solution.

The major difficulty in the application of high-frequency discharge plasmas in analytical chemistry lies in the method of sample introduction. Plasma shape and emission characteristics tend to change as the inert gas is diluted by the sample, usually introduced as a vapour or aerosol, and extinction of the discharge often results. Microwave plasmas induced by use of low-wattage generators cannot tolerate the introduction of samples which make up an appreciable percentage of the argon flow.[28] The problem of sample introduction has received considerable attention, which has resulted in the development of a number of techniques for the insertion of solutions and solids that usually involve preliminary vaporization of the sample through heating, or by a combination of nebulization and desolvation. Typical of the methods used for introduction of the sample into microwave plasmas were the volatilization of thin films of material from a heated filament by Runnels and Gibson[28] and of solution aliquots (0·12 μl) by Aldous *et al.*[30] Direct introduction of a sample mist produced by nebulization was achieved by Murayama and co-workers,[29] who used a plasma operated at 400 W, and Fallgatter and co-workers[31] and Hingle and co-workers[32] have used nebulization of solutions followed by low-wattage plasma excitation. The analysis of solutions by use of a radio-frequency argon plasma at atmospheric pressure has been carried out by Kleinmann and Svoboda[11] through evaporation of a 30-μl sample from a graphite disc heated by ohmic current, and ultrasonic atomization and powder injection have been developed by Wendt and Fassel,[18] West and Hume[20] and Hoare and Mostyn.[10] Methods of sample introduction in related spectroscopic analysis which may be applicable in plasma emission and absorption are laser vaporization,[33] or heating of a suitable crucible, similar to the 'Delves' cup,[34, 35] in the gas stream.

The aim of the research described in this paper was to develop a system for the detection and analysis of metal vapours, evolved from heating geochemical samples to temperatures in the region of 1000°C, to investigate observations by Meyer[36] that this volatile fraction provided enhancement of base-metal anomalies in areas of transported overburden. A microwave-induced argon plasma emission system similar to that developed by Aldous *et al.*[30] was adopted as the excitation and detection unit, and vapours were evolved by induction heating of 10-mg samples of soil or rock powder in a graphite micro-crucible, solutions being introduced into the plasma by vaporization of the residue from 10-μl sample aliquots dried on a steel micro-crucible.

Apparatus

The sample introduction, plasma excitation and signal detection systems of the experimental apparatus are illustrated in Fig. 1. Factors considered in the design of this system were the ability

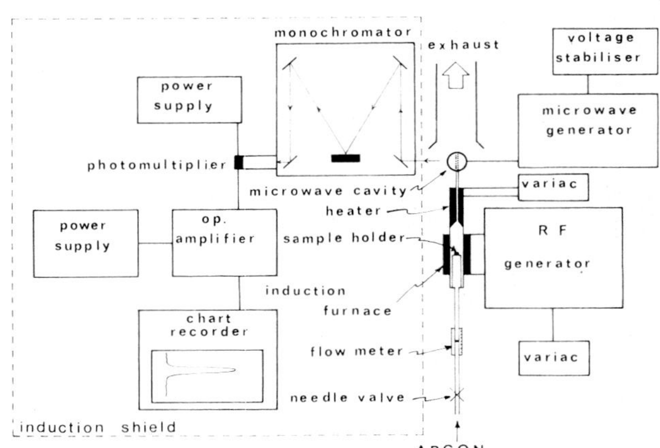

Fig. 1 Block diagram illustrating microwave-induced argon plasma emission system

to detect volatile elements evolved from rapid heating of dry, sieved soil and stream-sediment samples, pulverized rocks and solution residues at a rate of not less than 30 samples per hour; and determination of emission intensity profiles due to vapour evolved during slow, controlled heating of these samples as a means of identifying the forms in which metals occur at trace levels in geochemical environments.

Sample introduction

A Leco induction furnace (Laboratory Equipment Corporation, Michigan, U.S.A.), operating at a frequency of 14·6 MHz with an output of 1500 W, was used to heat micro-crucibles of graphite or steel within the quartz combustion tube (35 mm in internal diameter), which had been modified by the addition of a 5-cm length of transparent quartz tubing with a 2-mm internal diameter. The induction furnace was fitted with a gas flow system, and had a convenient mechanical sample changer, which enabled crucibles to be removed and replaced in less than 20 sec. Argon flow through the combustion cell was monitored by an accurate flow meter (Meterate (1GPE/1), Glass Precision Engineering, Hemel Hempstead, Hertfordshire, England) and controlled through a needle valve and pressure regulator. The graphite micro-crucibles (4 mm × 6 mm in diameter) were prepared from spectrographic electrode rods with a conical cavity capable of holding about 20 mg of powdered material. Metal micro-crucibles were prepared from steel discs (0·1 mm × 7 mm in diameter) with a central indentation capable of holding a 10-μl sample. Crucible temperature was regulated by use of a variable transformer to control voltage to the induction coil. Maximum temperatures achieved at full power were found to be 1100°C for the graphite and 840°C for the steel crucible, as determined by a thermocouple and verified by optical pyrometer measurements. Linear voltage increase from 0 to 100% over a period of 500 sec provided a heating rate of approximately 2°C/sec for the evolved vapour experiments with the graphite crucible, rapid increase to maximum voltage giving a temperature of 850°C in 10 sec. The quartz tube leading from the induction furnace to the microwave cavity was wound with heating wire and controlled through a variable transformer to provide temperatures up to 550°C so that condensation on the walls of the tube could be reduced.

A further aspect of this system was that a radio-frequency discharge could be maintained within and above the cavity of the graphite crucible at maximum power. The heat generated by this plasma produces vapour which would not normally have been evolved by induction heating of the graphite crucible alone. Another effect of the plasma was to produce an aerosol of microscopic particles by surface erosion of solid samples, in a manner similar to that described by Jones and co-workers,[37] which could lead to the use of the plasma aerosol generator as a source of particles for microwave or radio-frequency excitation.[38] Studies on the analytical potential of the plasma as an aerosol generator are continuing.

Vapour excitation

Excitation of the vapour evolved on sample heating in the argon stream was achieved by microwave induction at 2450 MHz produced by a Microton 200 microwave generator (Electro-Medical Supplies, Wantage, Berkshire, England) operating on a stabilized line voltage, with coupling through an Evanson one-quarter wavelength resonant cavity (214L). The plasma, about 1 mm in diameter, was formed largely above the 2-mm quartz tube making up the outlet of the combustion furnace, but approximately one-third of the total length of about 20 mm was located within the tube. Effluent gases were removed through overhead exhaust ducting.

Detection system

Radiation from the plasma was dispersed by a plane grating, symmetrical Czerny–Turner 0·3-m monochromator (Hilger and Watts D 330 single), with bilaterally adjustable slits. The vertical entrance slit was aligned with the plasma by locating the point of maximum emission from a suitable argon line with the slit approximately 5 cm from the source. A photomultiplier (EMI 9601 B) connected to a negative EHT supply of 1·0 kV (EDT Supplies, Ltd., London) was used to detect the light emitted from the monochromator, and the signals were amplified and then displayed on a Servoscribe recorder. The weak and transient signals of the emission system required the development of a rapid electronic amplifier/integrator, such as that proposed by Aldous *et al.*,[30] with two operational amplifiers, and the circuit diagram for this amplifier formed the basis of the unit constructed for the system by Dr. P. L. Lowenstein of the Applied Geochemistry Research Group at Imperial College, London. Improved stability of the amplifier/integrator was obtained by substituting chopper-stabilized operational amplifiers (Analog Devices 183L and 232J) for the Philbrick PF 85 AU units in the original design. Power for the amplifier/integrator was provided by a Coutant OA 10 stabilized supply unit. The system was ready for operation after a warm-up time of 30 min. An expanded aluminium cage provided induction shielding for the electronic detection system to reduce the effects of radio- and microwave-frequency interference.[39]

Operating procedure

The spectral line selected for study was located

on the monochromator–detector system by adjusting to the maximum intensity of light emitted by the appropriate electrodeless discharge lamp fixed within the microwave cavity. Dry sample powders to be analysed were weighed or transferred to the cavity of the graphite crucible by use of a volumetric estimate of the optimum 10 mg provided by a perspex scoop; and 10-μl aliquots of sample solutions were transferred to the steel crucibles by micro-pipettes (Drummond Microcaps). The solution aliquots were then slowly evaporated under an infrared lamp (50°C), leaving a residue for emission analysis. All crucibles were preheated and tested for contaminants before being used for analysis.

The crucible containing the sample was then placed on the asbestos block support and raised into position at the centre of the induction coil. Argon flow was adjusted (1·85 l/min was found to provide a stable plasma and gave reproducible results for most applications), and, after allowing 20–30 sec to flush air from the system, the plasma was initiated by the spark from a 'Tesla' vacuum tester. The plasma was tuned to give maximum stability at an operating power between 60 and 80 W, and would remain stable for long periods without readjustment.

For analysis of the volatile fraction of a solid sample or a solution residue the crucible was rapidly heated to the desired temperature by use of the induction furnace, the resulting peak of emission intensity being observed on the chart recorder. Typical peaks derived from the rapid heating of solid powders, aqueous solutions and organic solvent residues are shown in Fig. 2.

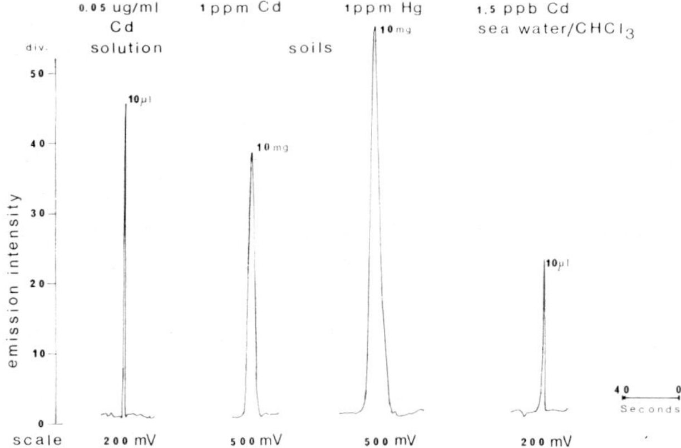

Fig. 2 Typical emission signals from rapid heating of geochemical samples. Broad evolution peaks of soil samples compared with solution residues show that absolute detection limits decrease with increasing sample size when peak heights are used

Production of emission intensity curves was achieved by slowly increasing the power to the induction coil at a fixed rate, marking the chart at regular intervals corresponding to the calibrated temperature.

Problems

Few problems were found in the operation of the system provided that the plasma was tuned correctly and the quartz tubing was regularly cleaned. Cleaning of the 2-mm tube was accomplished by passing a pipe cleaner through the jet at regular intervals, stubborn films being removed by dipping the pipe cleaner in a volatile organic solvent. Soils with a normal content of organic carbon could be analysed without extinguishing the plasma, but some difficulty was experienced with organic-rich soils and damp material. An increase in the microwave power of 5–10 W above that required for a stable plasma and a faster rate of argon flow, however, permitted the analysis of organic-rich samples. Flooding of the system by vapour produced during heating of the sample was also controlled by use of a heating rate which matched the rate of evolution of volatile products from the particular set of samples and standards.

The hourly rate of sample analysis (30) was governed by the time required for air to be flushed from the argon flow system after changing the sample crucible, and an attempt was made to minimize the problem by incorporating a bypass argon system around the combustion tube. The bypass system was discontinued as only a slight decrease in the time taken between samples was achieved, and the tubing introduced a turbulence which tended to lower the stability of the plasma.

Results

Analytical characteristics

Investigation of the analytical properties of the system for the detection of trace elements in geochemical samples was concentrated on cadmium and mercury, but emission from a variety of metals and non-metals was observed and applied in controlled heating experiments. Typical signals derived from rapid heating of rock base standards and solution residues are reproduced in Fig. 2, and the calibration curve for cadmium in solution residues from 10-μl aliquots is shown (Fig. 3) for the interval from 0 to $1{\cdot}0\times10^{-9}$ g (0–100 ppb). Solid standards used in these experiments were prepared by adding 10 ml of the appropriate aqueous solution of the element to 10 g of Johnson Matthey Spec-pure rock base (No. 1) ground to pass through a 200-mesh

(B.S.S.) sieve. Absolute alcohol was added to form a slurry, which was thoroughly mixed before drying under an infrared lamp. The dry standards were further ground and mixed, absolute alcohol was added to form a slurry, which was stirred and dried as before, and the resulting powder was homogenized for 12 h on a roller mill. Aqueous standards were prepared by dissolving soluble metal salts in deionized water.

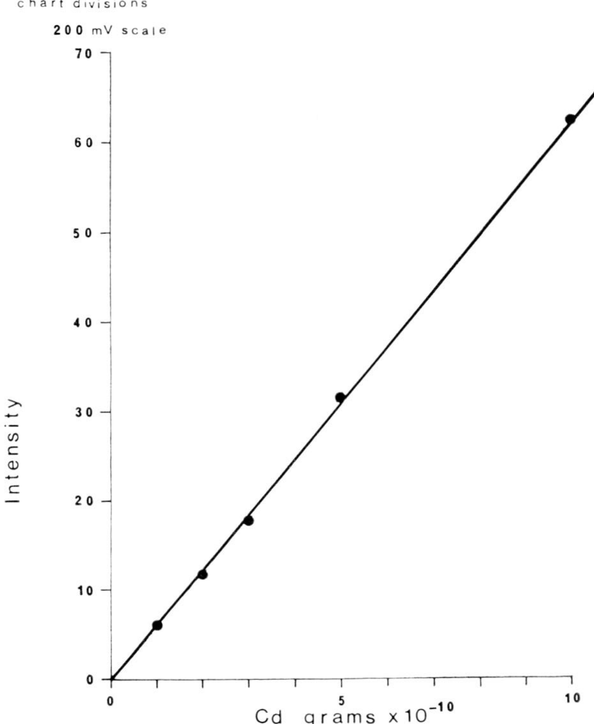

Fig. 3 Calibration curve for cadmium in solution residues (each point is average of six replicate analyses)

Limits of detection for cadmium and mercury evolved from heating standard powders and solution residues to 800°C are given in Table 1, together with the detection limits for lead, zinc, copper and iodine from the analysis of solution residues. A signal to noise ratio of 2 was used in determining the limits of detection. Lower limits of detection in terms of grammes were found for the analysis of solution residues due to the rapid evolution of metal vapour on heating the small samples (Fig. 2). A mean precision of 15·4% at the 95% confidence level was obtained on 10-mg samples from a standard series of Irish soils containing mercury over the range 200–1200 ppb (10^{-9} g/g) by use of Craven's[40] procedure.

The effects of chemical and spectral interferences are presently being investigated, but it should be noted that the method of fractional separation of the elements on heating tends to isolate the required metal from interfering species. Similar studies by Aldous *et al.*[30] on the spectral effects of arsenic, cobalt and antimony when present in a one thousand fold molar amount compared with cadmium showed that only arsenic interfered, and the arsenic emission peak could be identified.

Geochemical applications

Microwave-induced plasma emission systems have a number of geochemical applications in addition to the quantitative analysis of readily volatile metal vapours derived from rapid heating of soils, sediments and rocks. Identification of the forms in which trace elements occur in geochemical samples can be accomplished by the examination of emission profiles derived from slow heating of samples and comparison with standards of known composition. The sensitivity of the method for the analysis of small samples also facilitates the direct analysis of natural waters for a wide range of elements. Elements which have been determined by use of microwave-induced plasmas, in addition to those listed in this study, are antimony, selenium, arsenic, iron, boron and beryllium;[30] aluminium, silicon, calcium, vanadium, manganese, nickel, gallium, germanium, zirconium,

Table 1 Detection limits and equivalent sample concentrations for plasma emission analysis of 10-mg solid and 10-μl solution samples

Element	Wavelength, nm	Detection limits, g		Concentrations, ppm	
		Solid	Solution	Solid	Solution
Cadmium	228·8	$5{\cdot}0\times10^{-8}$	$0{\cdot}9\times10^{-11}$	0·005	0·0009
Mercury	253·7	$4{\cdot}0\times10^{-8}$	$5{\cdot}0\times10^{-11}$	0·004	0·005
Lead	282·3	*	$1{\cdot}7\times10^{-11}$	*	0·0017
Zinc	213·9	*	$1{\cdot}7\times10^{-11}$	*	0·0017
Copper	217·9	*	$6{\cdot}8\times10^{-11}$	*	0·0068
Iodine	206·2	*	$3{\cdot}9\times10^{-9}$	*	0·39

*Not determined.

molybdenum, ruthenium, palladium, indium, tellurium, barium, tungsten, platinum, gold and bismuth;[29] and silver and chromium.[28] Combination of a number of these elements in a multi-element detection system could find widespread application in exploration geochemistry. The following sections deal with some geochemical applications of the single-element plasma emission system.

Fractional analysis of metal vapours

The forms in which metals occur and their associations with other elements in soils and stream sediments have been the subject of numerous studies, notably by Ellis *et al.*,[41] Nichol and co-workers,[42] and Dvornikov and co-workers.[46] Identification and separation of the metal species in soils and sediments that were derived from the weathering of sulphide ore rather than the barren country rocks lead to improvement in the contrast between sulphide-related anomalies and background. Partial chemical attacks,[43] rather than total analysis, have been shown to provide the required increase in contrast in some cases,[41, 44] and they have contributed to the understanding of the forms in which metals occur. Under conditions of active oxidation of a sulphide ore, a proportionally larger concentration of metals would be expected to exist in low-energy bonding sites and colloidal precipitates than in the case of metals liberated from rock-forming minerals.

A parallel approach was applied in a study[36] of the secondary dispersion of silver from the Adeline Cu–Ag prospect[45] in the drift-covered Seal Lake district of Labrador. The –80-mesh (B.S.S.) fractions of soil samples, collected at a depth of 12–16 in on a traverse across the axis of the sulphide mineralization, were heated in an induction furnace and the evolved gas was mixed with air and acetylene. Continuous silver evolution profiles were monitored by flame excitation with a Techtron AA4 absorption spectrophotometer equipped with a chart recorder. Two distinct peaks were observed—a low-temperature absorption maximum of readily volatile metal and a high-temperature peak probably due to silver associated with the major rock-forming minerals. Similar evolution profiles were noted in studies of soils and rocks by use of arsenic, zinc, lead and mercury.

Analysis of the Adeline soils for the first peak, or more volatile fraction, gave an improved contrast between the anomaly and background when compared with a perchloric acid attack (Fig. 4). These results were sufficiently encouraging to justify further investigation of the method as a means of detecting mineralization in areas of deep overburden, where the fractional volatile anomaly would be submerged in the background variations of total analysis, and known methods of analysis of weak acid extracts lacked the required sensitivity.

An examination of the thermal evolution profiles for cadmium derived from heating soils from the area of the Keel zinc–cadmium prospect

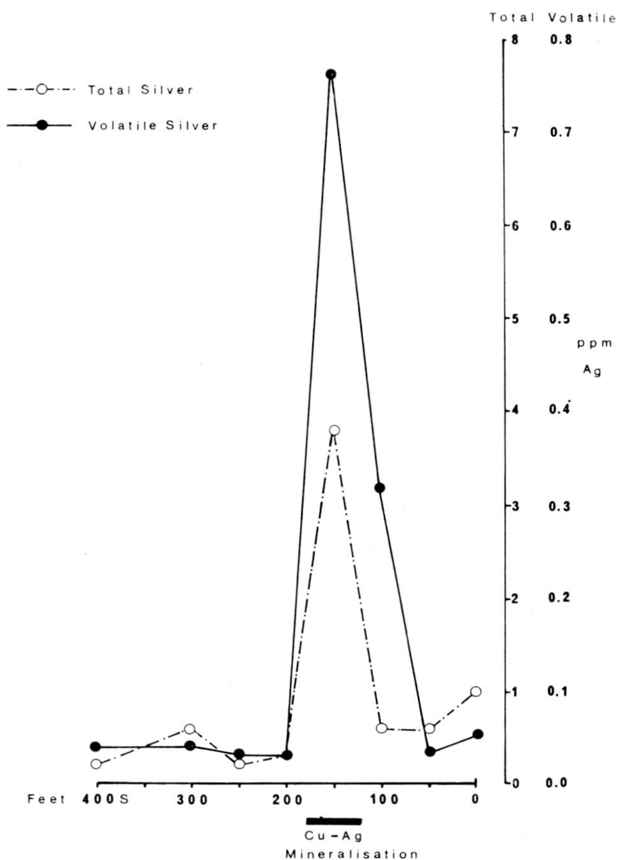

Fig. 4 Volatile and 'total' silver from atomic absorption analysis of soils collected on traverse across drift-covered copper–silver mineralization on Adeline Island, Seal Lake area, Labrador ('total' refers to perchloric acid digestion). After Meyer[36]

was made in an attempt to assess the significance of the analysis of a readily volatile metal fraction in geochemical exploration of a till-covered region in Ireland. The secondary dispersion of trace metals at Keel has been studied in considerable detail by Evans,[47] who concluded that the major proportion of the metal making up the soil anomaly was derived from an oxidized and leached 'decomposed zone' representing the weathered product of sub-economic disseminated sulphides rather than the potential orebodies of massive sphalerite associated with high-angle east–west faults. Different methods of partial chemical extraction failed to reveal a metal, or metal fraction, that could be correlated with the sulphide veins as opposed to the 'decomposed zone'. Thermal evolution profiles for cadmium in soil overlying sulphide mineralization, sulphide-bearing material from the vein wallrock, soil overlying the 'decomposed zone' and a sample of the 'decomposed zone' are illustrated in Fig. 5. A distinct peak at a temperature slightly below 200°C was detected on heating the soil overlying

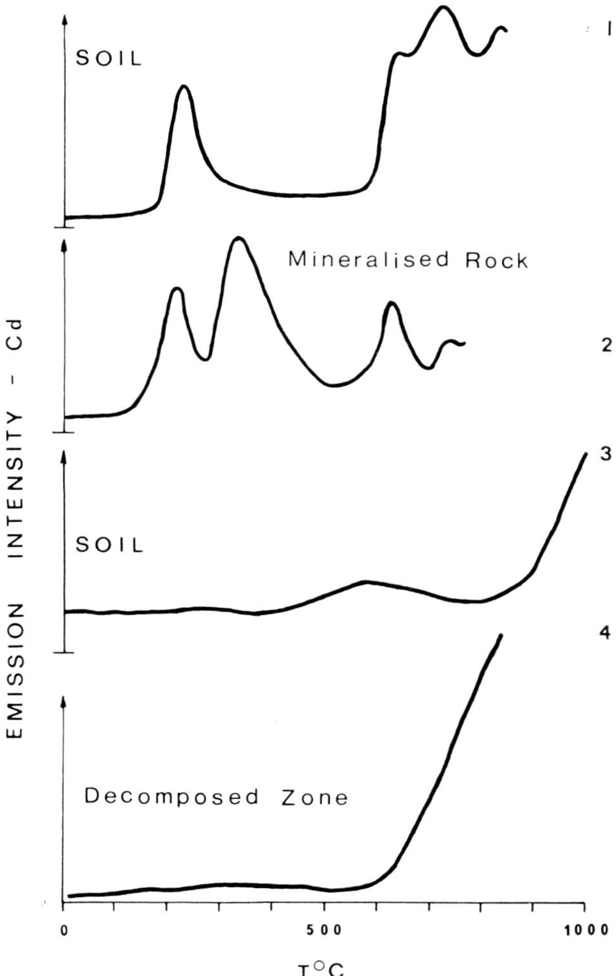

Fig. 5 Thermal evolution profiles for cadmium from soil (1) overlying mineralized rock (2), and from soil (3) overlying the 'decomposed zone' (4) at Keel, Eire

sulphides, and further cadmium evolution took place above 600°C. The mineralized sample displayed a more complex profile—probably indicating the presence of cadmium in a variety of sulphide, chloride and oxide forms analogous with those determined for mercury.[48] Soil overlying the 'decomposed zone' and the sample from the 'decomposed zone' exhibited less complex profiles with cadmium evolution at higher temperatures—probably indicating the presence of an oxide, or cadmium in association with iron and manganese oxides or hydroxides. Insufficient 'fingerprint' profiles from synthetic forms of cadmium were available for a positive identification of the cadmium species.

Recognition of a low-temperature emission peak in the soil overlying sulphide mineralization suggested that analysis of the readily volatile fraction (below 500°C) of cadmium would detect the presence of sulphides, but discriminate against metal derived from the 'decomposed zone'. Samples collected by Evans[47] on a traverse crossing both the 'decomposed zone' and the projected surface intersection with faults that were mineralized at depth were analysed by the plasma emission system with rapid heating to 500°C. Results obtained showed that the volatile cadmium fraction reached a maximum concentration over the area of the projected sulphides, whereas the total cadmium determined by atomic absorption on a hot nitric–perchloric extraction gave equivalent peaks for both sources and did not exhibit the high contrast between anomaly and background of the volatile method (Fig. 6).

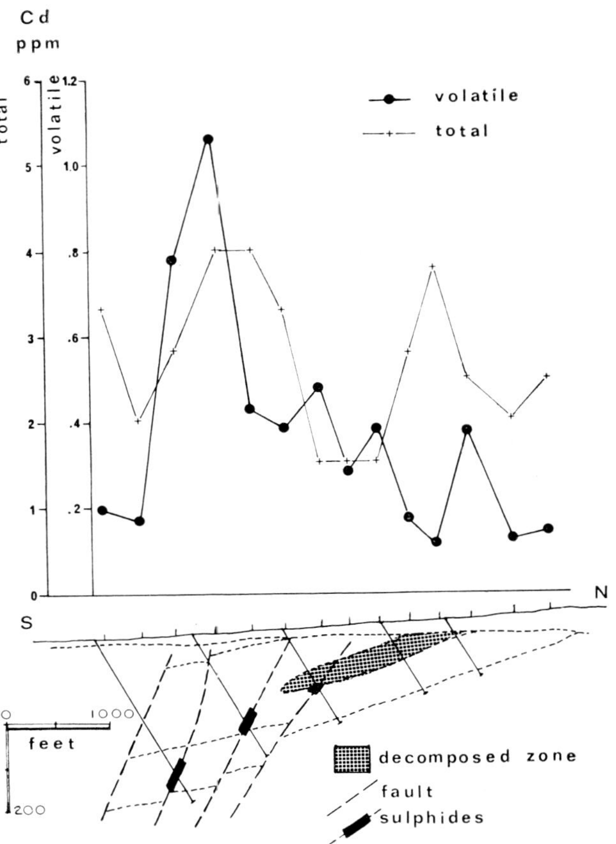

Fig. 6 Volatile cadmium (<500°C) in soils from Keel Zn–Cd deposit determined by plasma emission compared with 'total' cadmium analysed by atomic absorption. Atomic absorption results, after Evans,[47] are based on a nitric–perchloric digestion

Geochemical interpretation by identification of trace-metal forms

Gas evolution profiles have proved to be useful in thermal analysis by pyrolysis[48] and by combination with simultaneous DTA,[49–52] generally with the use of a thermal conductivity detector for measurement of the gas phase. Mass spectrometers have also been applied as the detection system.[54, 55] The plasma emission system provides a means of measuring evolution profiles for the wide variety of metals and non-metals that can be vaporized in the inert gas stream, with the possibility of correlation of evolution profiles for a number of elements.

Thermal evolution profiles for identification of metal forms in the interpretation of geochemical anomalies were applied in a study of concentrations of lead in soil from South Wales. During an investigation of mineral exploration problems revealed by the multi-element stream sediment survey of England and Wales being completed by the Applied Geochemistry Research Group at Imperial College, a linear lead anomaly was detected in soils from Carmarthenshire at some distance from the nearest known mineralization. An unusual feature of the anomaly was the absence of associated metals (Fig. 7) such as zinc, cadmium or silver. At this point, thermal evolution profiles for PbS, $PbSO_4$ and PbO in rock base were compared with the profile derived from soil collected on the axis of the anomaly (Fig. 8). The close correlation between the soil and galena profiles suggests that the anomaly was probably due to the persistence of PbS in soil derived from the weathering of an unexposed vein of galena.

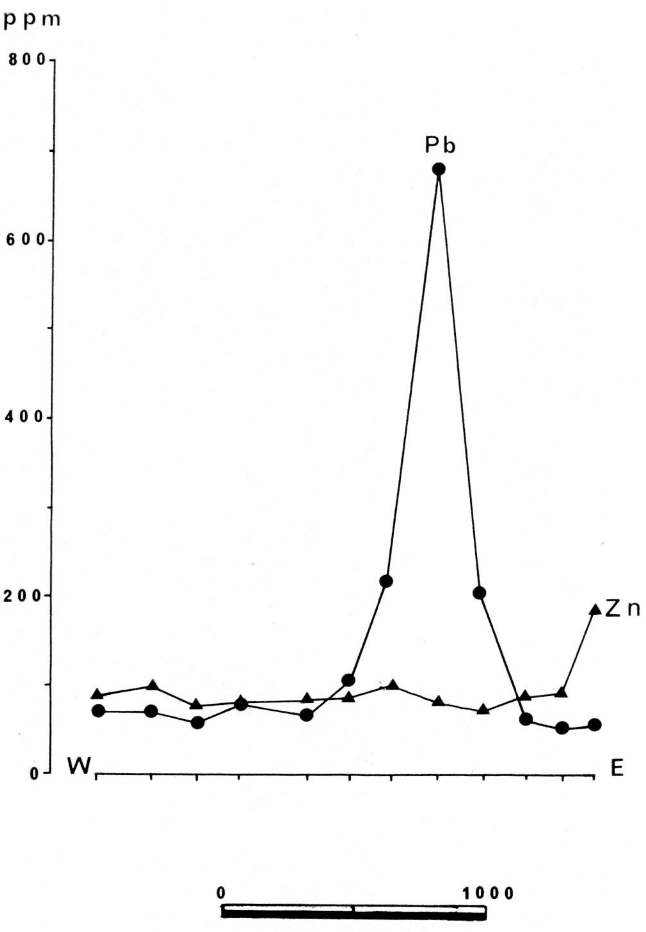

Fig. 7 Spectrographic results for lead and zinc in soils collected across axis of a linear anomaly in Carmarthenshire, South Wales

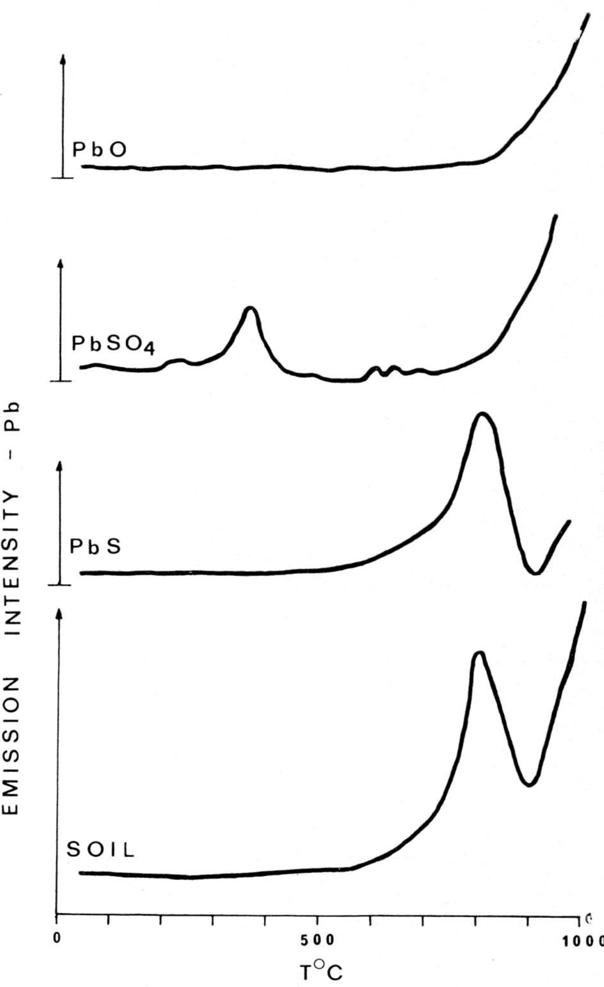

Fig. 8 Thermal evolution profiles for lead from the controlled heating of PbO, $PbSO_4$ and PbS in rock base, and soil from axis of the anomaly illustrated in Fig. 7

Direct analysis of trace metals in natural waters

The low detection limits obtained on small samples provide a simple technique for hydrogeochemical surveys by reducing the amount of sample required from each survey site. Preliminary studies on the direct analysis of sea water have shown that the interference due to the scattering effect of smoke derived from the major salts is not as severe as that reported for direct atomic absorption analysis with a heated graphite atomizer reported by Segar and Gonzalez.[57] Direct analysis of mercury in sea water with the argon plasma system has been accomplished by modifying the argon flow system to flush a reaction vessel for the reduction of saline forms to metallic mercury. A detection limit of 0·2 ppb has been obtained for mercury in a 5-ml sample of sea water by this technique, which has the advantage of speed for routine determinations as the plasma is not extinguished between analyses. Related techniques for the determination of arsenic, selenium and tellurium are being investigated.

Summary

The development of a microwave-induced argon plasma emission system for the analysis of geochemical samples has led to an improved understanding of the formation of trace-metal patterns through the identification of thermal evolution profiles characteristic of the forms in which the metals occur. Anomaly contrast was enhanced through the quantitative analysis of volatile metal fractions related to the weathering of sulphide ores—a factor which should be of importance in exploration for deposits covered by transported overburden. The sensitivity of the system for a wide variety of elements suggests that the technique may be useful in the direct analysis of natural water samples.

Productivity and discrimination of metal species could eventually be improved by conversion to multi-element analysis with a multi-channel detection system. Microwave-induced plasma emission systems of the type described here could be constructed for less than £3000, and as many laboratories are equipped with components such as chart recorders, flow meters and microwave generators the overall cost could be reduced in many cases.

Acknowledgment

Financial support for research into the application of fractional volatile analysis in geochemical exploration was provided by British Newfoundland Exploration, Ltd., Montreal, Canada. The authors were directed towards the investigation of plasma emission by discussions with Visiting Professor A. R. Barringer. Advice on construction of the microwave-excitation and detection device was freely given by Dr. R. M. Dagnall, Dr. K. M. Aldous, B. L. Sharp, and colleagues in the Department of Chemistry, Imperial College, London. Fernando Urquidí contributed samples and analytical assistance in the study of lead in soil from South Wales.

References

1. GATTERER, A. Zur Spektrochemie der Metalloide F, Cl, Br, J, S, Se. *Spectrochim. Acta*, **3**, 1948, 214–32.
2. COBINE, J. D. and WILBUR, D. A. The electronic torch and related high frequency phenomena. *J. appl. Phys.*, **22**, 1951, 835–41.
3. RODDY, C. and GREEN, B. The radio-frequency plasma torch. *Electronics World*, **65**, Feb. 1961, 29–31; 117.
4. WATSON, M. D. FERGUSON, H. I. S. and NICHOLLS, R. W. Electrical and optical studies of the argon plasma jet. *Can. J. Phys.*, **41**, 1963, 1405–19.
5. DICKINSON, G. W. and FASSEL, V. A. Emission spectrometric detection of the elements at the nanogram per milliliter level using induction-coupled plasma excitation. *Analyt. Chem.*, **41**, 1969, 1021–4.
6. FAGAN, A. W. and KLEIN, H. M. Spectrochemical analysis of semiconductor grade boron tribromide with a plasma jet. *Analyt. Chem.*, **40**, 1968, 2041–2.
7. FASSEL, V. A. and DICKINSON, G. W. Continuous ultrasonic nebulization and spectrographic analysis of molten metals. *Analyt. Chem.*, **40**, 1968, 247–9.
8. GREENFIELD, S. *et al.* Atomic absorption with an electrodeless high-frequency plasma torch. *Analytica chim. Acta*, **41**, 1968, 385–7.
9. GREENFIELD, S. JONES, I. L. and BERRY, C. T. High-pressure plasmas as spectroscopic emission sources. *Analyst, Lond.*, **89**, 1964, 713–20.
10. HOARE, H. C. and MOSTYN, R. A. Emission spectrometry of solutions and powders with a high frequency plasma source. *Analyt. Chem.*, **39**, 1967, 1153–5.
11. KLEINMANN, I. and SVOBODA, V. High frequency excitation of independently vaporized samples in emission spectrometry. *Analyt. Chem.*, **41**, 1969, 1029–33.
12. KOROLEV, F. A. and KVARATSKHELI, IU. K. The plasmatron (plasma jet) as a light source for spectroscopy. *Optika Spektrosk.*, **10**, 1961, 398–402; *Optics Spectrosc.*, **10**, 1961, 200–2.
13. MARGOSHES, M. and SCRIBNER, B. F. The plasma jet as a spectroscopic source. *Spectrochim. Acta*, **15**, 1959, 138–45.
14. MAVRODINEANU, R. and HUGHES, R. C. Excitation in radio-frequency discharges. *Spectrochim. Acta*, **19**, 1963, 1309–17.
15. OWEN, L. E. Stable plasma jet for excitation of solutions. *Appl. Spectrosc.*, **15**, 1961, 150–2.
16. TRUITT, D. and ROBINSON, J. W. Spectroscopic studies of radio frequency induced plasma. Part 1—development and characterization of equipment. *Analytica chim. Acta*, **49**, 1970, 401–15.
17. VEILLON, C. and MARGOSHES, M. An evaluation of the induction-coupled, radio-frequency plasma torch for atomic emission and atomic absorption spectrometry. *Spectrochim., Acta* **23B**, 1968, 503–12.

18. WENDT, R. H. and FASSEL, V. A. Induction-coupled spectrometric excitation source. *Analyt. Chem.*, **37**, 1965, 920–2.
19. WENDT, R. H. and FASSEL, V. A. Atomic absorption spectroscopy with induction-coupled plasmas. *Analyt. Chem.*, **38**, 1966, 337–8.
20. WEST, C. D. and HUME, D. N. Radiofrequency plasma emission spectrophotometer. *Analyt. Chem.*, **36**, 1964, 412–5.
21. DAGNALL, R. M. PŘIBIL, J. R. and WEST, T. S. Preparation of microwave-excited electrodeless discharge tubes for titanium, vanadium, and zirconium for use as spectral-line sources. *Analyst, Lond.*, **93**, 1968, 281–5.
22. DAGNALL, R. M. SILVESTER, M. D. and WEST, T. S. Electronic modulation of microwave-excited electrodeless discharge lamps for use in atomic-fluorescence spectrometry. *Talanta*, **18**, 1971, 1103–9.
23. MCCORMACK, A. J. TONG, S. C. and COOKE, W. D. Sensitive selective gas chromatography detector based on emission spectrometry of organic compounds. *Analyt. Chem.*, **37**, 1965, 1470–6.
24. BACHE, C. A. and LISK, D. J. Determination of organophosphorus insecticide residues using the emission spectrometric detector. *Analyt. Chem.*, **37**, 1965, 1477–80.
25. BACHE, C. A. and LISK, D. J. Selective emission spectrometric determination of nanogram quantities of organic bromine, chlorine, iodine, phosphorus, and sulfur compounds in a helium plasma. *Analyt. Chem.*, **39**, 1967, 786–9.
26. BACHE, C. A. and LISK, D. J. Determination of iodinated herbicide residues and metabolites by gas chromatography using the emission spectrometric detector. *Analyt. Chem.*, **38**, 1966, 783–4.
27. MOYE, H. A. An improved microwave emission gas chromatography detector for pesticide residue analysis. *Analyt. Chem.*, **39**, 1967, 1441–5.
28. RUNNELS, J. H. and GIBSON, J. H. Characteristics of low wattage microwave induced argon plasmas in metals excitation. *Analyt. Chem.*, **39**, 1967, 1398–405.
29. MURAYAMA, S. MATSUNO, H. and YAMAMOTO, M. Excitation of solutions in a 2450 MHz discharge. *Spectrochim. Acta*, **23B**, 1968, 513–20.
30. ALDOUS, K. M. *et al.* A microwave-induced argon plasma system suitable for trace analysis. *Analytica chim. Acta*, **54**, 1971, 233–43.
31. FALLGATTER, K. SVOBODA, V. and WINEFORDNER, J. S. Physical and analytical aspects of a microwave excited plasma. *Appl. Spectrosc.*, **25**, 1971, 347–52.
32. HINGLE, D. N. KIRKBRIGHT, G. F. and BAILEY, R. M. A simple low-power reduced-pressure microwave plasma source for emission spectroscopy. *Talanta*, **16**, 1969, 1223–5.
33. WHITEHEAD, A. B. and HEADY, H. H. Laser-spark excitation of homogeneous powdered materials. *Appl. Spectrosc.*, **22**, 1968, 7–12.
34. DELVES, H. T. A micro-sampling method for the rapid determination of lead in blood by atomic-absorption spectrophotometry. *Analyst, Lond.*, **95**, 1970, 431–8.
35. CLARK, D. DAGNALL, R. M. and WEST, T. S. The determination of mercury by atomic-absorption spectrophotometry with the 'Delves sampling cup' technique. *Analytica chim. Acta*, **58**, 1972, 339–46.
36. MEYER, W. T. Geochemical exploration and research at British Newfoundland Exploration Ltd. Unpublished paper presented at RTZ Group Exploration Conference, Toronto, 1969.
37. JONES, J. L. DAHLQUIST, R. L. and HOYT, R. E. A spectroscopic source with improved analytical properties and remote sampling capability. *Appl. Spectrosc.*, **25**, 1971, 628–35.
38. N.R.D.C. Provisional British Patent Specification 190 67/71, 1971.
39. KLINGMAN, G. W. Reduction of interference from radio-frequency heating equipment. *Trans. A.I.E.E.*, **68**, 1949, 718–24.
40. CRAVEN, C. A. U. Statistical estimation of the accuracy of assaying. *Trans. Instn Min. Metall.*, **63**, Sept. 1954, 551–63.
41. ELLIS, A. J. *et al.* Application of solution experiments in geochemical prospecting. *Trans. Instn Min. Metall. (Sect. B: Appl. earth sci.)*, **76**, 1967, B25–39.
42. NICHOL, I. HORSNAIL, R. F. and WEBB, J. S. Geochemical patterns in stream sediment related to the precipitation of manganese oxides. *Trans. Instn Min. Metall. (Sect. B: Appl. earth sci.)*, **76**, 1967, B113–5.
43. HOLMAN, R. H. C. A method for determining readily-soluble copper in soil and alluvium—introducing white spirit as a solvent for dithizone. *Trans. Instn Min. Metall.*, **66**, Oct. 1956, 7–16.
44. BROWN, A. G. Dispersion of copper and associated trace elements in a Kalahari sand environment, Northwest Zambia. Ph.D. thesis, University of London, 1970.
45. BRUMMER, J. J. and MANN, E. L. Geology of the Seal Lake area, Labrador. *Bull. geol. Soc. Am.*, **72**, 1961, 1361–82.
46. DVORNIKOV, A. G. VASILEVSKAYA, A. E. and SHCHERBAKOV, V. P. Mercury dispersion aureoles in the soils of the Nagol'nyi Range. *Geokhimiya*, 1963, 478–83; *Geochemistry*, 1963, 501–9.
47. EVANS, D. S. Secondary dispersion of mercury and associated trace elements at Keel, Eire. Ph.D. thesis, University of London, 1971.
48. WATLING, R. J. Identification of trace mercury compounds in rocks as a guide to sulphide mineralization at Keel, Eire. *Trans. Instn Min. Metall. (Sect. B: Appl. earth sci.)*, **81**, 1972, B47–8.
49. ROGERS, R. N. YASUDA, S. K. and ZINN, V. Pyrolysis as an analytical tool. *Analyt. Chem.*, **32**, 1960, 672–8.
50. MURPHY, C. B. HILL, J. A. and SCHACHER, G. P. Differential thermal analysis and simultaneous gas analysis. *Analyt. Chem.*, **32**, 1960, 1374–5.
51. AYRES, W. M. and BENS, E. M. Differential thermal studies with simultaneous gas evolution profiles. *Analyt. Chem.*, **33**, 1961, 568–72.
52. GARN, P. D. and KESSLER, J. E. Effluence analysis as an aid to thermal analysis. *Analyt. Chem.*, **33**, 1961, 952–4.
53. GARN, P. D. Some problems in the analysis of gaseous decomposition products. *Talanta*, **11**, 1964, 1417–32.

54. Findeis, A. F. *et al.* Gas evolution profiles and differential thermal analysis. *Thermochim. Acta*, **1**, 1970, 383–7.
55. Langer, H. G. and Gohlke, R. S. Mass spectrometric thermal analysis (MTA). *Analyt. Chem.*, **35**, 1963, 1301–2.
56. Hegedüs, A. J. Über die simultane gravimetrisch-thermisch (GTA) und massenspektrometrisch-thermisch (MSTA) Analyse von polykristallinem Zinkoxid. *Mikrochim. Acta*, no. 1, 1971, 40–5.
57. Segar, D. A. and Gonzalez, J. G. Evaluation of atomic absorption with a heated graphite atomizer for the direct determination of trace transition metals in sea water. *Analytica chim. Acta*, **58**, 1972, 7–14.

550.844:629.135.45

Geologic mapping and geochemical sampling by use of the light aircraft

C. L. Sainsbury, B.A., M.S., PH.D.

Formerly U.S. Geological Survey, Denver, Colorado, U.S.A., now AirSamplex Corporation, Indian Hills, Colorado, U.S.A.

Synopsis

Reconnaissance geological mapping and geochemical sampling from a light aircraft have been proved to be feasible. In areas of fair to good exposures a 15-min quadrangle can be mapped in detail sufficiently good for publication at a scale of 1:250 000 in as little as one day. With sufficient time very accurate geologic maps can be made for publication at larger scales. Maps for special use, such as in the search for ore deposits, can be prepared to emphasize gossans, alteration zones and structural features in detail not often shown on published maps.

Devices have been developed by which samples of soils, rock fragments, stream sediments and diverse types of vegetation can be collected while the aircraft remains airborne, although samples must be collected from sites amenable to the sampling methods. Other devices have been developed for use with the helicopter. With them, samples can be taken from beneath tree cover, or along streams where trees have prevented collection of sediments by standard methods.

The methods have been applied in commercial exploration in the western United States and in Alaska, with a considerable reduction in geologic mapping and geochemical sampling costs.

The steadily rising costs of geologic mapping and geochemical exploration have led to the concentrated study of new and cheaper methods of finding orebodies. In recent years the use of all types of airborne techniques has been greatly expanded, and for one reason—costs are lowered and rate of coverage per unit cost is increased. But the use of the airborne techniques has forced departure from the well-established 'rock-in-the-box' techniques upon which much of geologic and geochemical mapping was based. In the final analyses, either a geologic map or a geochemical anomaly must be based upon physical samples.

The methods described in this paper represent an attempt to keep a foot in the past, firmly based upon physical samples, yet to get the other foot into the present by utilizing the speed of the aircraft, with its consequent rapid rate of ground coverage and low unit costs. These methods are not intended as a panacea for all the problems that confront the exploration industry: they do, however, offer some promise in lowering costs, as well as increasing the scope of geochemical sampling.

Other airborne techniques have been called the 'higher and higher' ones, whereas those described here are termed the 'lower and lower' ones—for reasons which will be obvious. It is contended that the human eye, when coupled to a trained and working brain, is highly superior to any combination of cameras and computers. The human eye can recognize subtle variations of diverse types that may give the exploration man a clue to an orebody, and the human brain can interpret these variations much better than any computer.

The mapping methods described were developed over a period of several years while the writer was employed by the U.S. Geological Survey in geologic mapping on the Seward Peninsula, Alaska. During the initial work between 1960 and 1965 the writer used his private aircraft in support of the field operations in the usual ways, such as the moving of camps and supplies, airborne reconnaissance and support of personnel. Initially, two mile to the inch quadrangles were mapped in five summers, in an area of good exposures, but with very complicated structure.[4,5] Major deposits of fluorite, containing notable amounts of beryllium minerals, found during this mapping are currently being developed at an estimated cost of about $80 000 000.[2]

In 1967 the project was reorientated toward the reconnaissance mapping of the entire Seward Peninsula—an area of about 25 000 square miles—at a scale of 1:250 000. This mapping was supported by helicopter, which was used during three of four summers for an average of about 50 hours flying time per summer. During this mapping the aircraft was used in progressively more geologically orientated tasks—principally, the compilation of geologic maps. As pilot skill developed, the aircraft was used much like a helicopter for landing and ground checks of geologic units being mapped from the air. In this work an average day would consist of approximately 4–6 h of flying, supplemented by ground traverses several miles long from the landed aircraft. The mapping rate was so improved that in five summers more than 20 000 square miles of the Seward Peninsula was mapped. This amounts to roughly eighty 1:63 360 scale quadrangles in an area of complex geology involving numerous thrust plates.[5] During this time a second party mapping well-exposed areas of the Seward Peninsula, and, with the use of a helicopter for each of four summers, mapped approximately fourteen 1:63 360 scale quadrangles with standard mapping techniques. The maps of both parties have been published by the Geological Survey, at a scale of 1:250 000, and they can be compared to illustrate the dependability of the two different mapping techniques.[1,6,7,8] Several of the 1:63 360 scale maps prepared by the new methods were sufficiently detailed that they were published* at that scale.[9,10,12,13]

In late 1969 an attempt was made to map an area, completely unknown geologically, by the airborne techniques, and six maps (scale, 1:63 360) were prepared at the rate of one quadrangle per day. No landings were made, and units were identified only from the air. These maps were made available to the companion party who completed the mapping of the area during the following years, and they were included in the 12 quadrangles published.[1] The airborne maps were remarkably similar to the maps finally published after several summers' helicopter mapping. This is convincing proof of the efficiency and dependability of the new methods. In fact, the only thrust fault shown on the map of the 12 quadrangles is one which was mapped by the airborne techniques—a further proof of the dependability of the new methods, for the entire area is characterized by thrusting.

Although the six quadrangles mapped by air were of a reconnaissance nature, many altered zones along faults and larger areas of stained and altered igneous rocks were mapped. It was the frustration of being able to spot such interesting areas and yet be unable to sample them as they were mapped that led the author to consider the possibility of securing samples from the light aircraft. That the interesting gossans and altered areas were indeed worthy of sampling is illustrated by the fact that several hundred claims were staked subsequently to cover many of these altered rocks. Deposits containing lead, silver, molybdenum and fluorite were found in some of the alteration zones that had been plotted from the air (personal observation, 1971).

Equipment used

The aircraft used in developing the methods and in the first commercial applications was a Piper Super Cub with a 135-hp engine. The aircraft was equipped with oversize tyres ('tundra tyres' in the north), and in later field tests with an 82-in propeller. The Super Cub can be flown safely at speeds as low as 50 miles/h, which facilitates landings on short bars and ridgelines when ground traverses are necessary. With the long propeller, the aircraft is practically stall-proof, and can be manoeuvred intricately at speeds as low as 70 miles/h, which enables safe entry into steep and narrow canyons, where the best outcrops are commonly found. In commercial work the aircraft is fitted with special modifications and devices which require openings in the fuselage and modifications to the airframe, necessitating rigid inspection by the Federal Aviation Agency and operation under Restricted Category while sampling.

Although other aircraft may be as suitable as

*These maps may be obtained from the U.S. Geological Survey Branch of Alaskan Economic Geology, 345 Middlefield Road, Menlo Park, California, 94205, U.S.A.

the Super Cub, the writer is familiar only with that aircraft, and it would be difficult to improve upon it at any reasonable cost. It has been said in the north that few pilots ever develop the skills that will allow them to make full use of the capabilities of the Super Cub, with which the writer agrees. In fact, the aircraft will out-perform any pilot, and when troubles arise, the pilot usually is at fault.

Mapping methods

Geologic mapping is done directly upon topographic maps when they are available in suitable scale, or upon aerial photographs where maps are not available. For airborne mapping one mile to the inch maps are generally sufficiently accurate that most contacts can be placed by reference to the contours. On slopes and ridgelines, geologic features such as contacts, attitude of bedding and cleavage, etc., can be placed by flying the aircraft close to the outcrop and reading the altimeter. This enables extremely accurate placement of contacts at the same points where conventional mapping methods usually are most accurate.

In preparing the map the pilot-geologist flies the aircraft along specific traverses chosen to yield the desired information. The map is held upon the knee, or upon a small board fastened to the knee with a rubber band, and the contacts, faults, etc., are drawn lightly in by pencil. As geologic contacts are the most important part of geologic maps, and as faults often are recognized most easily by the displacement of visible contacts, traverses along contacts yield the most information in a unit time. For initial mapping the aircraft is flown approximately 3000 ft above ground, at which height the major features of interest are most easily seen. From this height faults, offsets of bedding, gossans and linear alteration zones can be seen and plotted, and the aircraft does not pass features at bewildering speed. If the feature being plotted is temporarily lost when the pilot-geologist lifts his eyes from the map to look ahead, or to clear for obstructions, the aircraft is merely circled back to the last known point, and the feature being mapped is 'picked up' again. The position of the aircraft is generally marked by a finger that is kept on the map. This is especially valuable, for the map is constantly being turned so that the map direction coincides with the flight direction, and proper orientation is difficult.

After the main features have been plotted, the aircraft is flown closer to the ground in order to identify rock units, as well as to observe minor features, such as cleavage, small stained areas, old prospect pits and so on. On steep slopes with good exposures even minor drag-fold axes can be seen and placed with a considerable degree of accuracy. In field tests in one area two geologists were continuously on the ground making detailed observation of attitudes of cleavage, folds and contacts. Rapid coverage by air of the same areas yielded attitudes that were in most cases within a few degrees, and always within acceptable limits, of those plotted by ground work. Contacts usually were picked at exactly the same place, though at times those plotted by air were more accurately located. Of course, ground work, which is supplemented by observations made on the ground by the pilot-geologist during traverses from the landed aircraft, is invaluable and necessary for detailed geologic mapping. When the pilot-geologist is operating alone, landings on river bars and scrutiny of the gravels are a great aid in helping to identify the lithologic types that outcrop within that drainage basin.

In areas where landings are not possible, a certain number of samples to verify lithologic types can be obtained by the use of devices described later.

Almost of equal value is the information which can be gained with the use of a small airborne scintillometer, for most rock units within any map area are characterized by a specific radioactivity. The variations between such closely related rock types as granite and monzonite can be determined by the scintillometer, which is also a great aid in differentiating between light-coloured orthogneiss versus paragneiss. Orthogneisses normally give a relatively constant and smooth radiometric response, whereas paragneisses usually give a highly varied and erratic radiometric response. Use of the scintillometer at low levels enables one to pick geologic contacts under tundra and other thin superficial cover, as well as on forested slopes. In fact, the 'tricks of the trade' are innumerable, and more are gained with work in each new area.

With practice, the experienced geologist-pilot can make a remarkably accurate geologic map extremely quickly. The newcomer to the method will tend to rush things because of the assumed high cost of the aircraft, and the natural tendency to wander from one interesting feature to another. This tendency to wander, and thus take isolated, disconnected observations, must be avoided—as in ground mapping. Because the aircraft covers ground at a high speed, one is able to fly every foot of visible geologic contacts, which results in a map superior in many ways to one where contacts are drawn between points of observation. Maps prepared by helicopter are most susceptible to great improvement by aircraft mapping techniques. In fact, in one area in Alaska that was mapped by both techniques, major errors were found in the published map which took weeks to map by helicopter, but a little over one day to map by light aircraft.

Costs of geologic mapping

Geologic mapping carried out for the Geological Survey resulted in such an increase of mapping efficiency as to be bewildering and unacceptable to those not versed in the use of the aircraft. In spite of intense desire and several requests, the writer was unable to arrange tests under controlled conditions to obtain dependable information on the costs per map and the rate of mapping by use of the new techniques versus established photogeologic techniques.* As no specific figures were available based upon those costs normally met in private industry, actual costs of the initial geologic mapping were not assessable. The fact that the productivity of the party which employed the aircraft techniques was such that 7 mile to the inch quadrangles could be mapped in seven weeks is proof of a vast increase in mapping efficiency (approximately thirty-fold). In commercial work the use to which the maps are to be put varies greatly, and maps are varied to suit such needs. For use in exploration for mineral deposits, such features as gossans, faults, veins, silicified zones, breccia pipes and vegetation and radiometric variations are immensely valuable and are shown in considerable detail on the geologic map. Most published maps show a minimum of such features. The addition of details to the geologic map slows the rate of mapping, but gives a whole new depth of value to the map, especially when geochemical surveys are to be done in the same area. Hence, costs per quadrangle are highly variable. In country where exposures are good, however, a dependable mile to the inch reconnaissance map can be prepared by one skilled in the art at a rate of approximately one quadrangle per week. Such a map contains a wealth of economic data. In the most favourable circumstances a reconnaissance map can be prepared in one or two days of flying time at a cost of about $1000 per quadrangle. As detail is added, or more ground work is required, the cost rises, but it is always an order of magnitude cheaper than that of the map prepared by standard ground techniques.

* One photogeological specialist summed up the situation with the remark, 'If you think I'm going to preside over the demise of photogeology, you've got another think coming!', which probably was an overstatement of the case.

Geochemical sampling

As was noted earlier, the ability to recognize potentially mineralized ground from the aircraft, but not to be able to sample it at the time it was mapped, led the author to consider sampling from the aircraft. For the initial evaluation of a gossan even a few soil samples are sufficient to give the exploration man an idea as to which gossans are the most likely to be of interest. The early identification of mineralized gossans enables the planning of subsequent work in the most productive manner.

The necessary devices were ultimately built in a crude prototype form, and were used subsequently in initial field tests. Because patent coverage is being sought on new devices and on the general methods for use of the devices from the airborne vehicle, only the prototypes are presented here. More sophisticated equipment, as well as new devices for special uses, is being designed currently, and several have been field-tested successfully.

Devices for use with the light aircraft fall into two main categories, the first being those which are dropped from the aircraft (or helicopter) and then retrieved into the airborne vehicle. The second type of sampling device is firmly affixed to a shock-absorbing nylon rod carried in a steel tube attached to the aircraft. With the first type of device, samples are taken while the aircraft is well above the ground; use of the second type requires that the aircraft be flown as low as 2 ft from the surface or object to be sampled.

Droppable samplers consist of a device valuable for use in sampling soils, such as that shown in Fig. 1, and others for sampling sands, etc.

The device for sampling soils consists of a tubular steel rod milled at the forward end to accept a polyethylene sample holder (not shown in the prototype device), and fitted with fins to

Fig. 1 Prototype sampler used in collecting soil samples from the light aircraft

stabilize it in flight. A hole drilled rearward at an angle to the outside of the tube allows the sample holder to be pushed out with a rod after the sampler is retrieved into the aircraft. The sample remains in the container, which is capped and numbered. A towing ring, which passes through the steel body forward of the aerodynamic centre, allows the device to be recovered smoothly by use of a nylon cord. Retrieval is done in commercial work by use of a rewind winch operated electrically by the aircraft battery. In initial tests in Alaska the device was merely dropped by hand from the side window of the Super Cub and retrieved by hand rewinding of the cord. This prototype sampler proved capable of recovering soil samples from damp Arctic soils, but proved undependable in sampling dry or dusty soils—largely because of wind erosion of the sample as the sampler was retrieved into the aircraft. A modified device, which retains dry soils, has been developed by AirSamplex Corporation.

A second device was used to obtain fragments of rocks where lithologic control was needed for geologic mapping, or to sample gossans covered with small fragments of broken bedrock. This insultingly simple device is pictured in Fig. 2. It

Fig. 2 Prototype 'device' used in collecting rock fragments while aircraft remained airborne

consists of a suitable weight covered with a layer, about $\frac{1}{2}$ in thick, of tenacious adhesive. In use, it is dropped from the aircraft into loose fragments of rocks. On impact, fragments embed themselves in the adhesive and are recovered into the aircraft. In initial tests standard aircraft putty was used as the adhesive, but other more suitable adhesives are being tested.

To collect samples with any of the droppable devices the aircraft is flown as close as is practicable to the surface to be sampled—generally between 10 and 20 ft. At the appropriate moment the device is dropped, contacts the ground to collect the sample and is retrieved into the aircraft or hovering helicopter.

Samplers which are extended from the aircraft on a nylon rod consist of sand and sediment scoops (Fig. 3), scoops for gathering fragments of loose rocks and sharp-edged 'baskets' for sampling the tops of vegetation in geobotanical work. As can be surmised, their use requires a skilled

Fig. 3 Prototype scoop sampler used to collect sand and sediment samples from the light aircraft. Sampler is shown in retracted position, with filled sample bag, beneath belly of Super Cub

pilot-geologist capable of recognizing the suitability of the material to be sampled and of operating the aircraft safely within a few feet of the ground.

To collect samples with the 'scoop' devices the nylon rod with the sampler attached is extended by hand through an opening in the bottom of the aircraft until the sampler extends a few feet below the level of the aircraft wheels (Fig. 4).

In initial tests nylon rods were jointed together and held by hand while they rested against the

Fig. 4 Prototype scoop sampler showing approximate attitude at which sampler encounters material to be sampled

rear of the pilot's seat. In the commercial work the nylon rod was continuous and extended through an opening in the overhead windshield of the Super Cub. Within the aircraft the rod was contained with sliding fit within a steel tube firmly fixed to the overhead aircraft tubing, and to a clamp welded to the rear of the pilot's seat. A ring clamp, which depressed a rubber cushion against the nylon rod through a slot in the tube, was sufficient to hold the rod in the retracted position, or, when extended, to collect the sample. For safety, only the pressure of this rubber cushion held the nylon rod in the tube—if the scoop hung up firmly, the nylon rod would be stripped out of its containing tube with no damage to the aircraft or harm to the pilot. The various types of scoops used in commercial work were detachable by hand by the pilot-geologist after the rod had been brought back into the aircraft.

To secure a sample by the scoop samplers the geologist-pilot selects the area to be sampled and makes practice runs at the right height above ground to test the wind and approach conditions. The nylon rod is then extended by hand to the desired position, the scoop being locked into the proper place by a single twist of the clamp. The aircraft is flown into position as if to land on the area to be sampled, but with excess speed and height. As the aircraft passes the area to be sampled, it is lowered until the scoop scrapes lightly across the surface. In most cases the sample bag attached to the scoop is easily filled. Depending on the size and position of the area to be sampled, one or more passes must be made to secure the required weight of sample. Sand bars, beaches and flat surfaces on ridgelines are sampled with considerable ease. Other areas, especially those on sloping surfaces, are more difficult. These require a greater degree of pilot skill, for the aircraft often must be simultaneously banked and lowered gently toward the surface to be sampled. With the practice and experience gained from the commercial work, the writer has found that even very small surfaces on steep slopes can be sampled with the scoop sampler. This device became the one most used in securing samples for lithologic control during geologic mapping or sampling of altered rocks.

Initial field tests of devices

The prototype devices described for use with the light aircraft were tested by the author. A private company has field-tested a device built by the author for use with the helicopter, and capable of securing samples from soils beneath moderate to heavy tree cover. New devices are being designed, built and tested by AirSamplex Corporation; these are not discussed or illustrated in this paper because of the desire to obtain proper patent coverage prior to the release of such information. This paper will discuss only the results of initial tests obtained with the prototype devices presented in this paper. Some data on mapping costs and rates based on commercial experience are included.

Initial tests of the soil samplers were made on the Seward Peninsula, Alaska, over the large fluorite–beryllium–tin lodes found by the writer and Thomas E. Smith between 1962 and 1964.[3,4] All the soils were damp, and were easily collected with the author alone in the aircraft. In collecting stream sediments with the prototype scoop sampler (Fig. 3) two people were required, and these tests were performed with the aid of Daniel R. Shawe. In the commercial work all devices were installed for operation with the pilot-geologist alone in the aircraft.

Test samples were sent to Kenneth J. Curry, U.S. Geological Survey, who prepared them for analyses. All were analysed by means of the flame spectrometer, checks of many splits being made by John C. Hamilton, who employed a different flame spectrometer in a different laboratory. Because the area sampled had been studied before by detailed sampling of soils, rocks and vegetation,[14] the results obtained from the airborne methods could be compared directly with those obtained by the standard ground methods. Only a part of the results obtained from these tests is given in Table 1. Analytical results for only some of the elements of interest are also shown. Highly anomalous amounts of many different metals were found in the samples collected with the airborne devices, especially tin, beryllium, lead, zinc, silver, copper, arsenic and mercury. Samples collected from the air at or near samples collected previously on the ground gave analytical results directly comparable. The unusually high content of many metals is notable, and it can be concluded that airborne sampling of soils or stream sediments from such an area would certainly alert any exploration man to the possibilities of hidden ore deposits.

The samples collected during the initial field tests in the Lost River area were taken by use of the prototype device, which was dropped by hand from the Super Cub, and retrieved by hand. As this sampler was not milled to accept the sample holders that were incorporated into later models, the sampler had to be cleaned by hand between samples. This was done by use of a toothbrush and a jug of water carried behind the seat. Such cleaning required considerable time between samples. Nevertheless, in the Lost River area 28 samples were taken in a period of about one and one-half hours; these were strung out along a mineralized fault zone for 7 miles. Control samples from unmineralized ground were taken several miles from the ore deposits. Two landings were made to collect samples by hand from the impact points of the air-dropped sampler as checks against these samples. Hence, speed of sampling was considerably reduced over the optimum.

Within the area sampled, topography was steep to gentle, and problems with wind and turbulence were moderate. Target areas varied from very large to very small: for instance, the sediment sample was taken from sediment in a small, dried-up pool alongside the stream in which the muddy sand sampled was but a few feet wide and no more than 8 ft long. The rock fragments were obtained from an old prospect trench on a steep slope. Some soil samples were taken where climb-out after sampling was a problem. All in all, the sampling area could be considered moderately difficult and typical of many areas that might be sampled from the air.

Table 1 Metal content, ppm, of soils, stream sediment and rock fragments collected by light aircraft, Lost River fluorite, beryllium and tin deposits, Alaska

Field no.	Sample type	Ag	As	Be	Cu	Pb	Sn	Zn	Hg	Remarks
70 Asn 236	Soil	1·5	300	300	100	700	150	300	1·0	Rapid River
70 Asn 236*a*	Soil	2·0	200	150	150	500	30	220	0·65	Rapid River
70 Asn 236*c*	Soil	1·0	400	150	100	500	150	360	0·45	Rapid River
70 Asn 236*d*	Soil	2·0	600	700	150	700	150	500	0·60	Rapid River
70 Asn 237	Soil	20·0	80	20	100	500	150	83	1·3	Rapid River
70 Asn 238	Soil	1·5	400	150	150	1000	700	700	0·12	Lost River
70 Asn 239	Soil	1·0	130	30	150	1500	300	340	0·18	Lost River
70 Asn 240	Soil	1·0	300	150	100	2000	700	900	1·0	Lost River
70 Asn 241	Stream sediment	3·0	1500	150	700	1000	+1000	2000	0·15	Lost river
70 Asn 243	Rock fragment	30·0	5000	+1000	150	7000	700	2000	0·3	Trench
70 Asn 244	Rock fragment	2·0	60	30	50	300	150	500	0·3	Trench
70 Asn 245	Soil	0·7	40	15	70	200	70	230	0·16	Lost River
70 Asn 246	Soil	1·0	200	150	100	300	70	300	0·55	Lost River
70 Asn 247	Soil	0·7	200	70	30	150	200	230	0·26	Lost River

Hg determined by mercury detector; all other values are semi-quantitative spectrographic analyses by K. J. Curry and J. C. Hamilton, U.S. Geological Survey.

Costs of geochemical sampling

To date, especially in the commercial work, airborne sampling has been done only as a part of a programme of geologic mapping and geochemical sampling in which samples were taken as needed. These samples tested geochemical anomalies as found, and identified rock units as they were being mapped. In a typical day in commercial work perhaps only 10–20 samples are taken, but these might be collected at places as much as several miles from each other. In such work the costs of sampling are minimal and are integral to the overall mapping programme.

In a programme designed entirely for reconnaissance geochemical sampling by light aircraft the rate of collection of samples, as well as the degree of sample coverage, would depend entirely on such factors as topography, availability of sample sites, and tree and brush cover, which, if thick enough, would essentially prevent sampling by light aircraft. Other types of geochemical sampling jobs can be done with a cost per sample that will be highly variable. In sampling beach sands, river bars in jungle rivers during the dry season or stream waters the rate of sample collection would be very high. In geobotanical surveys, such as the grid sampling of tops of conifers in the heavy rain forests of such areas as British Columbia, or the grid sampling of mesquite in the southwestern U.S.A., samples could be collected at a high rate of speed by use of the light aircraft and the modified scoop sampler. Reconnaissance surveys requiring the collection of alluvium in the dry washes of the western U.S.A., Mexico, Brazil or Western Australia could be carried out at a great rate of speed with the scoop sampler, but the points at which samples were collected would have to be varied to suit the terrain. With the scoop sampler modified to collect the tops of jungle vegetation, reconnaissance geochemical surveys could be carried out in jungle country, which would offer the first practical approach to geochemical prospecting of such a region. Such geobotanical prospecting, of course, would require prior ground studies to identify those trees that store anomalous amounts of metal over mineralized ground.

Problems encountered in sampling

Several problems that were encountered during tests of the sampling devices should be mentioned so that they may be anticipated by anyone who attempts to use these methods. A nylon cord was selected for the retrieval line for the drop samplers because of its elasticity and strength. The cord used had a breaking strength of about 400 lb—sufficient to retrieve the samplers under most conditions. If, however, the sampler hangs up, or buries itself too deeply for retrieval, the nylon cord merely stretches smoothly and breaks, with no detectable effect on the aircraft. If the cord is too short, however, there is danger if the sampler buries deeply, and then is pulled forward by the elasticity of the cord with sufficient added velocity that the sampler may overtake the aircraft. By using at least 100 ft of cord, this possibility is avoided.

If the scoop sampler hangs up (to date this has never happened), the nylon rods used are sufficiently supple that they will merely strip out of the tube which contains them and pass out of the aircraft. In practice, the scoop samplers have encountered large rocks during sampling of scree slopes, but the flexibility of the rods is such that the sampler merely scraped over the obstruction with no ill-effect. In fact, to test the suppleness of the rods the aircraft was landed several times with the scoop sampler fully extended. The rods merely bent to the desired curvature to allow the scoop to scrape along the ground, with no damage to the aircraft or the sampler.

Sampling by helicopter

Samplers for use with the helicopter have been developed by AirSamplex Corporation. Their use allows the collection of soil samples from tree-covered slopes or stream sediments from streams lined by trees spaced so closely that the helicopter cannot descend. Of these, only the soil sampler has been adequately tested. It is scheduled for commercial production in 1973, when patent coverage will be complete. Given time, AirSamplex Corporation expects to have devices capable of a wide range of sampling jobs, some of which will allow sampling of areas previously incapable of being sampled, even by helicopter.

Integrated programmes

With the experience gained to date, a programme designed to achieve complete geologic mapping and geochemical sampling of large areas is suggested. For initial reconnaissance the geologist-pilot, equipped with a suitable light aircraft modified for airborne sampling devices, completes the geologic map. During mapping, gossans are plotted in detail, but only a sufficient number of samples are taken by air to verify map units and to test the accessible gossans that cannot be tested by traverses from the landed aircraft. Reconnaissance geochemical surveys may be done during the mapping. Following completion of a geologic map, a helicopter piloted by a second geologist-pilot is brought in, and all the detailed sampling is done, devices being used where necessary. A helicopter flown at lowest gross weight by a geologist-pilot would be relatively safe because of the reserve performance. Alternatively, a geologist-pilot capable of flying either the Super Cub or the helicopter would work with minimum support, mapping until the need for detailed sampling arose, at which time the helicopter could be used. Such a programme would require a very small number of people and would be highly mobile and productive. Very large areas could be covered in a relatively short time. The more remote the area, the greater would be the cost savings because of the simplified logistics. With proper care and radios, safety of operators and personnel could be assured. The commercial experience accumulated to date shows that an area 100 miles on a side can be covered by reconnaissance geologic maps, and numerous gossans, altered rocks and geologic units sampled, in a field season of approximately 90 days. Total costs for such a programme would be approximately $95 000.

Other suggested uses

In addition to routine geologic mapping and geochemical sampling, other uses for the methods can be visualized, especially in the fields of environmental and health studies—for instance, studies of trace elements or pesticides in farm soils, which could be of regional extent, could be done quickly and at reasonable costs by the use of the light aircraft. Studies of suspended sediments in streams during floods could be made so rapidly that samples from widely spaced areas in the drainage basin could be considered as almost simultaneously collected. Snows could be sampled with the light aircraft rapidly and cheaply—and from altitudes far above the range of most helicopters. Although the airborne mercury detector enjoyed a short period of testing and great promise, it has fallen into disuse because of the inability to identify the source of the airborne mercury. If the airborne mercury detector were used with the sampling devices described here, however, airborne anomalies could be ground-checked immediately by the collection of soil or rock samples taken during the airborne mercury survey. A similar use probably could be developed in obtaining the so-called 'ground truth' samples that would be usable in interpreting many types of geophysical anomalies in many terrains. Numerous other uses have been discussed, and others could certainly be thought up by an imaginative group of scientists, exploration geologists and geochemists.

Conclusions

The methods discussed here are new, but sufficient experience has been gained already to show that they will work. Most of the limitations which have been found can be overcome either by varying the methods, combining them with standard methods or by the development of improved equipment. The writer is especially pleased by the commercial acceptance of the geologic mapping methods, for herein lies the least difficult part of the job. These airborne methods offer accelerated mapping of large, inaccessible and unmapped parts of the world, especially in those so-called 'underdeveloped' nations, which cannot afford either the costs of standard geologic mapping or the delay of the benefits derived from geologic maps.

In conversations, and in the review of this paper, the writer has been asked to comment on the possibility of having geologists not trained as pilots to use the methods. In his opinion the optimum is achieved by a pilot-geologist, for many reasons; but, with practice, a geologist working with the same pilot over a period of time could achieve a great proficiency in geologic mapping. He could increase his rate of mapping by several times at least. On the other hand, the necessity for two people reduces the performance capabilities of the aircraft. The inherent reluctance of any pilot to subject a passenger to the dangers of low flight in an aircraft would, no doubt, seriously curtail the collection of samples by light aircraft. Nevertheless, the author has a waiting list of young, adventurous pilot-geologists who would like to become involved in the methods.

Probably the greatest deterrent to the use of the methods will be the simple reluctance of people to accept a new approach. During the development of the methods and since then the writer has been constantly reminded that the methods are 'too dangerous'. Yet an honest evaluation of this factor must be made. Certainly, the methods are not as dangerous as crop-dusting, yet most crops are now dusted by aircraft. Few exploration managers hesitate to send their crews out in helicopters, which, usually, are dangerous because they are operated at full gross weight or over gross weight in all types of terrain. Never a year passes without someone in exploration with whom the writer is personally acquainted being involved and seriously injured in a helicopter crash—yet exploration by helicopter goes on. Geologists are drowned in boats, killed in crashes of field vehicles, mashed and maimed on freeways *en route* to the office after the field season is over, attacked by bears and bitten by snakes—yet exploration goes on. But the writer knows of only one field man who was injured in a light aircraft, and that when struck by the propeller of a taxiing float plane while he was on the floats.

If the methods described become used to their full potential, most probably it will be done with the help of the exploration industry rather than by Government agencies. In exploration, costs are more critically scrutinized, new methods more often are sought out and developed, and new ideas of doing things are more easily received and acted upon. Too often, the bureaucracies are too compartmented, jealousies between competing units are detrimental to the adoption of new ideas, and increased efficiency is easily cast into doubt by desire to maintain the *status quo*. To supplant a method which has existed essentially unchanged for more than a hundred years requires courageous people who are willing to test new methods and, when proven, to adopt them. When requests for field-checking of maps cannot be honoured, when devices cannot be tested by those responsible for the development and testing of new methods of exploration, and when massive opposition is raised by those unacquainted with the substance of a new method, there is little likelihood of change. Such reactions are the norm in the bureaucracy.

A final remark on why the author asked for, and was granted, by the Commissioner of Patents, the commercial rights to the patents filed to protect his devices and methods: simply, the writer has great faith in the methods, and a certain exuberance about them, similar to that experienced by the geologist who finds a major orebody. Faced with reassignment to a new area and into a new job, principally the compilation of data unrelated in any way to his methods, he could not accept this end for them, and, following an invitation from private industry to put the techniques to work commercially, he took early retirement from the Geological Survey to pursue this course. Whether the methods prove successful or not depends entirely upon the writer and the commercial competitiveness of them: he has great faith that they have much to offer the exploration industry. Initial work completed to date suggests that this may be true.

Acknowledgment

The author wishes to acknowledge specifically the help of his assistants Travis Hudson, Rodney Ewing and William R. Marsh, who participated enthusiastically in the full-scale test of the new methods in 1971, during which seven of the 1:63 360 scale quadrangles referred to in the text were mapped. Mr. Harry Smedes, U.S. Geological Survey, visited the area in 1970 and field-checked the six maps prepared at the rate of one quadrangle per day. He concluded that they were accurate enough for reconnaissance mapping and publication at a scale of 1:250 000, as published subsequently in modified form. Mr. Daniel R. Shawe visited the field camp in 1970 and participated in the initial test of the scoop sampling device. The help of all these people is gratefully acknowledged. Samples collected during tests of sampling devices were analysed in the laboratories of the U.S. Geological Survey by John C. Hamilton and Kenneth J. Curry.

The acknowledgments would not be complete without a 'tip of the hat' to the many unbelievers and dedicated obstructionists within the Geological Survey; they provided an added incentive to prove that the methods had merit.

Since taking early retirement from the Geological Survey to pursue the development and utilization of the methods, the author has had the opportunity to apply them successfully in commercial work through the faith of J. Keith Hayes and M. C. Stacy, United States Steel Corporation; Mr. Joseph Bach and Robert Laverty, Wyoming Minerals Corporation; and Mr. Murray Watts, Pan Central Explorations, Ltd. He is forever in the debt of these individuals, who were willing to take a gamble on such heretical methods of geological mapping and exploration.

References

1. Miller, T. P. *et al.* Preliminary geologic map of the eastern Solomon and southeastern Bendeleben quadrangles eastern Seward Peninsula, Alaska. *Open File Rep. U.S. geol. Surv.*, 1972, 11 p.
2. Big fluorite–tin mine at Lost River, Alaska. *Nth. Miner*, **58**, no. 26, 14 Sept. 1972, pp. 1 and 6.
3. Sainsbury, C. L. Beryllium deposits of the western Seward Peninsula, Alaska. *Circ. U.S. geol. Surv.* 479, 1963, 18 p.
4. Sainsbury, C. L. Geology and ore deposits of the central York Mountains, western Seward Peninsula, Alaska. *Bull. U.S. geol. Surv.* 1287, 1969, 101 p.
5. Sainsbury, C. L. The A. J. Collier thrust belt of the Seward Peninsula, Alaska. *Bull. geol. Soc. Am.*, **80**, 1969, 2595–6.
6. Sainsbury, C. L. Geologic map of the Teller quadrangle, western Seward Peninsula, Alaska. *U.S. geol. Surv. Map* I-685, 1972, 4 p.
7. Sainsbury, C. L. Hummel, C. L. and Hudson, T. Reconnaissance geologic map of the Nome quadrangle, Seward Peninsula, Alaska. *Open File Rep. U.S. geol. Surv.*, 1972, 28 p.
8. Sainsbury, C. L. *et al.* Reconnaissance geologic map of the western half of the Solomon quadrangle, Seward Peninsula, Alaska. *Open File Rep. U.S. geol. Surv.*, 1972, 10 p.
9. Sainsbury, C. L. *et al.* Reconnaissance geologic maps of the Solomon D–5 and C–5 quadrangles, Seward Peninsula, Alaska. *Open File Rep. U.S. geol. Surv.*, 1972, 11 p.
10. Sainsbury, C. L. *et al.* Reconnaissance geologic map of the Nome C–2 quadrangle, Seward Peninsula, Alaska. *Open File Rep. U.S. geol. Surv.*, 1972, 12 p.
11. Sainsbury, C. L. *et al.* Reconnaissance geologic map of the Nome C–3 quadrangle, Seward Peninsula, Alaska. *Open File Rep. U.S. geol. Surv.*, 1972, 6 p.
12. Sainsbury, C. L. Smith, T. E. and Kachadoorian, R. Reconnaissance geologic map of the Nome D–3 quadrangle, Seward Peninsula, Alaska. *Open File Rep. U.S. geol. Surv.*, 1972, 8 p.
13. Sainsbury, C. L. *et al.* Reconnaissance geologic map of the Solomon D–6 quadrangle, Seward Peninsula, Alaska. *Open File Rep. U.S. geol. Surv.*, 1972, 21 p.
14. Sainsbury, C. L. Hamilton, J. C. and Huffman, C. Jr. Geochemical cycle of selected trace elements in the tin–tungsten–beryllium district, western Seward Peninsula, Alaska. *Bull. U.S. geol. Surv.* 1242–F, 1968, 42 p.

543.53:546.79:552.121

Fission-track radiography of uranium and thorium in radioactive minerals

Harold A. Wollenberg, M.S.

Lawrence Berkeley Laboratory, University of California, Berkeley, California, U.S.A.

Synopsis

The fission-track method is a quick, relatively simple and inexpensive technique for the determination of the location and abundance of uranium and, in some cases, thorium in thin or polished sections of rocks. Thermal neutrons induce fission in ^{235}U, and ^{232}Th and ^{238}U fission with fast-neutron bombardment. Therefore, the sections with appropriate track detectors are exposed, first, to thermal neutrons to induce only U to fission, and then to fast neutrons for fission of U plus Th. The detectors are etched to reveal the damaged areas (tracks) caused by passage of massively charged fission fragments. High-quality muscovite mica is the preferable track detector for minerals with U contents greater than 10–15 ppm—mainly because tracks in mica are easy to recognize. Polycarbonate plastic (Lexan or Makrofol) is preferred as a track detector for low contents of U and Th, because this plastic contains essentially no inherent U; therefore, it has no background track density. Thorium is determined successfully if the Th/U ratio in the mineral is sufficiently large. Relative errors (from counting statistics) in Th are less than 25% if Th/U is greater than 3; for ratios less than 3 the errors increase rapidly and exceed 40–50% if Th/U is less than 1.

The method was applied to the study of U and Th in granitic rocks of the Sierra Nevada batholith, California, and in strongly contrasting mineralized terrains in south and east Greenland. The intense U and Th mineralization of south Greenland is associated with lujavrites, peralkaline rocks wherein the mineral steenstrupine contributes the most to whole-rock radioactivity. Mean values of Th and U in steenstrupine are 23 000 and 6200 ppm, respectively. Adjacent grains show less radioelement variation than is seen in the granitic rocks. Lujavrites of lower radioactivity are characterized by eudialyte, whose U and Th contents are at least an order of magnitude less than those of steenstrupine. The distribution and abundance of radioelements were also investigated in monazite, thorite and pigmentary material of the lujavrites.

Fission-track examinations of thin sections of mineralized fault-zone breccia from east Greenland indicated that U is strongly associated with limonitic material in thin fluoritized veinlets, and is occasionally found concentrated near the edges of reddish opaque grains. By combination of the results of fission-track studies and an electron-microprobe analysis of the minerals, the abundance of U and Th may be correlated with that of other elements.

Determination of the location of radioactive minerals in thin sections of rock has been accomplished for many years by alpha-track autoradiography. Briefly described, α-particles radiating from U- and Th-bearing zones near the surface of a thin section are registered on an overlying photographic emulsion, which, when developed, locates sites of radioactivity and permits quantitative determination of their α-particle activity.[2, 3, 9] A drawback of the technique is that even minerals with relatively high radioactivity (several hundred to a few thousand ppm uranium) require exposure times of the order of ten days to yield a statistically significant number of countable α-tracks. Further, in a collection of α-tracks it is very difficult to differentiate between those from the decay series of Th and those from the U-decay series, and it is also possible that there is not secular equilibrium in the U-decay series. Therefore, unless the Th/U ratios of the radioactive minerals are known through independent measurements or are assumed, α-particle autoradiography does not furnish independent determinations of U and Th.

The fission-track method, developed in the mid-1960s by Fleischer, Price and Walker,[7] permits determination of the location and individual abundances of uranium and, under favourable conditions, thorium, in thin or polished rock sections after short exposures (generally a few minutes) to reactor neutrons. Over the past five years several workers have applied the method to study the uranium contents, ranging usually from tens to hundreds of ppm, in thin and polished sections containing accessory and rock-forming minerals—see, for example, Kleeman and Lovering[13] and Wollenberg and Smith.[19] Uranium contents as low as a few tens of ppb have been measured in finely powdered preparations of whole-rock ultramafic samples.[6] Much less has been accomplished in the study of the occurrence of thorium, because of limitations of the method, which are discussed later.

Fission-track techniques and their physical bases are described. Applications of these techniques are illustrated by a discussion of the distribution of uranium and thorium in contrasting intrusive occurrences. It is intended to illustrate the usefulness of the fission-track method to an understanding of the geochemistry of radioelements in mineralizations of possible future economic importance.

The fission-track method

Significant in the application of the fission phenomena to earth sciences are the facts that (1) ^{232}Th and ^{238}U fission readily under irradiation by fast neutrons; (2) ^{235}U fissions most readily under thermal-neutron (energy $<0{\cdot}025$ eV) irradiation; and (3) ^{238}U fissions spontaneously (^{232}Th, ^{235}U and ^{234}U also fission spontaneously, but at much slower rates than ^{238}U). The energy threshold for fission of ^{238}U and ^{232}Th is 0·9 MeV; therefore, in rock samples thermal-neutron irradiation causes only ^{235}U to fission. The spontaneous fission of ^{238}U, though necessary for fission-track age determinations of a mineral, also yields a background of fission tracks in natural materials used as track detectors.

A fission track is formed when a massively charged fragment, emanating from a fissioned nucleus, enters the structure of a detector (generally a good insulating solid such as mica, plastic or glass), causing damage by charge repulsion. The resulting damaged area is enhanced by etching the detector with the appropriate reagent: 40–48% hydrofluoric acid (HF) for mica or glass, a strong base (6 N KOH or NaOH) for plastic. One can then view the tracks or damage pits easily with an ordinary microscope.

The formation of tracks depends upon the specific ionization, dJ/dx,[8] which determines the number of ions formed along the path of the fission fragment. dJ/dx is a function of the effective charge of the fragment, the velocity of the fragment, the speed of light, the ionization energy of the outer electrons of the detecting material, the mass of the electron and a constant which depends on the detecting material.

For our purposes, fragments from the fission of U or Th, with atomic mass numbers ranging approximately from 90 to 140, and with average energies of 85–90 MeV, register rapidly in muscovite mica, plastic or glass. With proper etching conditions, tracks from the spontaneous fission of ^{238}U have been observed in several rock-forming and accessory minerals.

If uranium-free detectors are used, the lower limits of detection depend mainly on the magnitude of the integrated neutron flux (regulated in most cases by the duration of the exposure). For example, Fisher[6] used an integrated flux of 10^{16} n/cm^2 to measure U concentrations as low as a few ppb. The uranium content and geologic age of natural mica detectors combine to furnish a background track density which, essentially, governs their detection limit—1–5 ppm with an integrated flux of $\sim 10^{13}$ n/cm^2. Therefore, such detectors are applicable to neutron-induced fission of U and Th in concentrations from several tens to thousands of ppm.

Computations

The equation which governs the density of fission tracks in a detector placed directly on the flat surface of material containing U, and irradiated with neutrons, is

$$\rho = \varphi \sigma E n \qquad (1)$$

where ρ is the track density, tracks/cm^2, φ is the integrated neutron flux, n/cm^2, σ is the neutron-induced fission cross-section (~ 580 barns for thermal neutron fission of ^{235}U; 1 barn $=10^{-24}$ cm^2), E is the efficiency of detection (0·5 in the present case) and n is the number of ^{235}U nuclei per square centimetre of thin-section surface within the effective range of fission fragments. For fragments from U and Th fission the effective range in most silicate and phosphate minerals is ~ 3 mg/cm^2.

Thus

$$n = (3\ Ca/A) \times 10^{-3} \qquad (2)$$

where C is the concentration of radioelement, g/g, a is Avogadro's number and A is the atomic mass number of the radioelement.*

In practice it is not necessary to use the above equations to determine element concentrations if standards are included in the neutron exposure. The equations are most useful in estimating the neutron flux required for an irradiation.

Procedures

The basic procedure to obtain fission tracks from uranium and/or thorium is quite simple. Solid-state track detectors, usually mica or plastic, are mounted on a series of samples and standards, which are then exposed to reactor neutrons. On completion of the exposure, the detectors are removed and etched in the appropriate reagents. The resulting tracks are then viewed and counted with the aid of a microscope.

Table 1 Comparison of detector materials

Natural muscovite	Polycarbonate plastic
Very durable	Surface scratches easily
Tracks easy to recognize, readily distinguished from scratches	Tracks perpendicular to surface appear as small dots, easily confused with scratches
Etching time, ½–4 h	Etching time, 10–15 min
Track size regulated by varying etching time without danger of appreciable surface loss	Surface removed readily as etching progresses; not possible to enlarge tracks after 10–15 min
Uranium content plus geologic age yield background tracks from spontaneous and induced fission. This governs detection limit (1–5 ppm U)	Essentially no background tracks, detection limit governed by neutron exposure
	Print of rock surface forms after etching[14]

Choice of detectors

In choosing a fission-track detector one must weigh the advantages and disadvantages of different materials. This discussion is limited to the two most commonly used in geologic applications: lexan polycarbonate plastic (Lexan or Makrofol) and muscovite mica; they are compared briefly in Table 1. Tracks in muscovite are long hexagons if etched for a short time, or develop into diamond shapes as etching progresses (Fig. 1(*a*)). Tracks appear in Lexan (Fig. 1(*b*)) as dark points (from fission fragments entering the detector perpendicular to the surface) or as dark lines, the apparent lengths of which depend on the entry angle of the fragment and its starting position in the sample. Lexan's principal advantage is that it has essentially no background uranium content, whereas even the highest-quality mica shows a background of tracks from spontaneous and induced fission of its ^{238}U con-

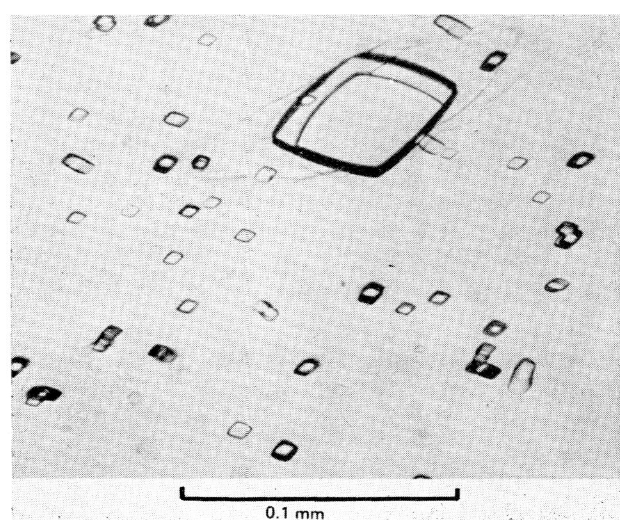

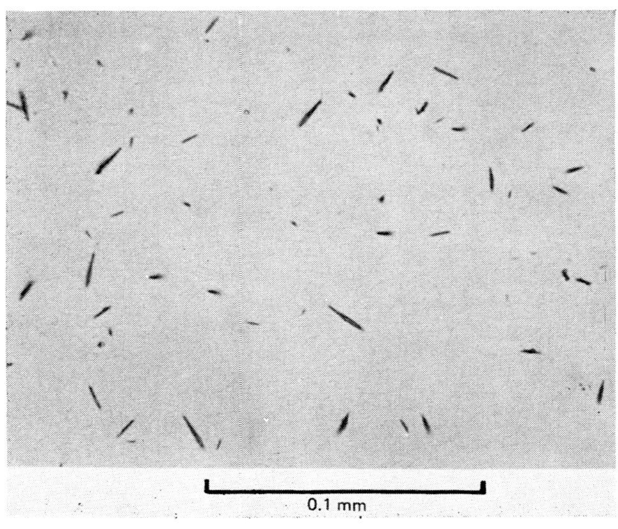

Fig. 1 Fission-track etch pits in muscovite (a) (top) and Lexan polycarbonate (b) (bottom). Track density, $2{\cdot}25 \times 10^5$ t/cm². Mica was etched in 48% HF for ~1 h, the Lexan in 6 N KOH at 60° C for 15 min

*The ratio of ^{238}U to ^{235}U, 139, is essentially constant in nature.

tent. When examining minerals with U contents >10 to 15 ppm, detector background is not a problem because track densities from the sample are considerably greater than background. If a low content of U is expected in the sample, it may be preferred to use a Lexan detector or expose a mica detector without the sample to obtain the effective background value.

Standards

To furnish track-density–element-content calibrations, and to serve as neutron flux monitors, glass standards containing known amounts of U and Th are irradiated with the samples. The neutron flux may be depressed within a stack of thin-section-detector sandwiches; therefore, standard glasses inserted at the centre and ends of the stack yield close approximations of the true calibration factors for each thin section. Soda disilicate ($Na_2O.2SiO_2$) is convenient as the standard glass medium, primarily because of its low melting temperature ($\approx 1100\,°C$). The range of fission fragments in glass is roughly the same as in most minerals (approximately 3×10^{-3} g/cm^2).

Irradiation

Exposure, processing and observation conditions for different uranium concentrations are summarized in Table 2. The choice of irradiation conditions depends on the element to be determined and its expected concentration. Easily observable track densities range from 10^4 to 10^6 tracks/cm^2. To determine an appropriate thermal-neutron flux for a uranium irradiation, desired track density is combined with the estimated U concentrations of the samples in equations 1 and 2. The samples are exposed in the thermal column of the reactor (a position outside the core, surrounded by graphite to exclude fast neutrons), where the thermal/fast neutrons ratio is highest (preferably $\geqslant 1000$).

Conversely, a fast-neutron irradiation is used for thorium. The sample–detector package is placed as close as possible to the core of the reactor, and is encased by at least a 3-mm thickness of cadmium to permit irradiation by as 'pure' a flux of fast neutrons as possible. In the flux estimate for a thorium irradiation one must include the large contribution to the observed track density from ^{238}U.

The estimated integrated neutron flux is then

$$\varphi = \rho / E(\sigma_{Th} n_{Th} + \sigma_U n_U) \tag{3}$$

where ρ is the desired track density, σ_{Th} and σ_U are the fast neutron (neutron energy ≈ 3 MeV) fission cross-sections for ^{232}Th ($\approx 0{\cdot}3$ barn) and ^{238}U ($\approx 1{\cdot}1$ barn), respectively, and n_{Th} and n_U are the estimated concentrations of Th and ^{238}U nucleii.

Etching

Etching time is varied to achieve an optimum track size for a given track density. This is most important if tracks are counted from photographs of the detectors. In this respect mica has an advantage over Lexan in that track size in mica can be controlled by HF etching time; very little of the mica's surface is removed, even after etching for several hours. After 10–15 min the strong base etchant used for Lexan rapidly removes its surface, causing a continuous depletion of tracks with time—first the shallower tracks, then those of greater penetration.

Observation

To determine the location of the sites of radioelements the detector with the developed tracks can be replaced on the thin section, or matching areas of sample and detector can be photographed separately and the pictures compared. The image of the rock surface on Lexan, observed by Kleeman and Lovering[13] after intense neutron irradiations, aids in rematching plastic detectors to thin sections following etching. Otherwise, small fiducial marks on mica or plastic detectors and thin sections may be matched, or obvious track-density patterns matched with their mineral

Table 2 Exposure, processing and observation conditions for various evenly distributed uranium concentrations

U content, ppm	Neutron exposure, n/cm$^2 \times 10^{13}$	Track density, t/cm^2	Etching time, h*	Magnification	Tracks/field
50	2	$1{\cdot}2 \times 10^4$	3–4	$\times 36$	~ 500
500	1	6×10^4	2–3	$\times 80$	~ 450
5000	1	6×10^5	1–2	$\times 270$	~ 400

*Etching of high-quality muscovite mica in 40–48% HF at room temperature.

grain contributors. Associated track densities can be counted in the microscope, from photographs, or, if contrast and spacing permit, by machine. In the case of a thermal-neutron exposure for U, sample and standard track densities are compared directly to give the uranium content. Errors are from track-counting statistics.

Thorium determination

The use of fast neutrons permits a determination of Th content if the Th/U ratio is great enough. Since practically all Th-bearing minerals also contain some U, it is necessary, first, to determine U by a thermal-neutron exposure. Then the sample, with a new detector and U and Th standards, is exposed to fast neutrons (energies >1 MeV, usually 3–4 MeV). The resulting track density is from ^{238}U *and* ^{232}Th, but, knowing the U content, and having simultaneously exposed a U standard, the U contribution is determined. The contribution from Th is then the difference between the total track density and the contribution from U.

The main difficulty with this method is that the relative errors due to counting statistics are rather large; the cross-section ratio for fast neutrons favours U over Th by a factor of ~ 3. The error in the thorium determination comes from the accumulation of counting statistics in measurements of U and Th standards, and from measurements of the samples' track densities from fast- and thermal-neutron exposures. Fig. 2 relates Th/U ratios and relative errors in Th for typical exposure and counting conditions of accessory-mineral samples. Relative errors are $<25\%$ if Th/U is greater than ~ 3; for ratios <3 the errors increase rapidly, and they are considered unacceptably high ($>40\%$) for Th/V <1.

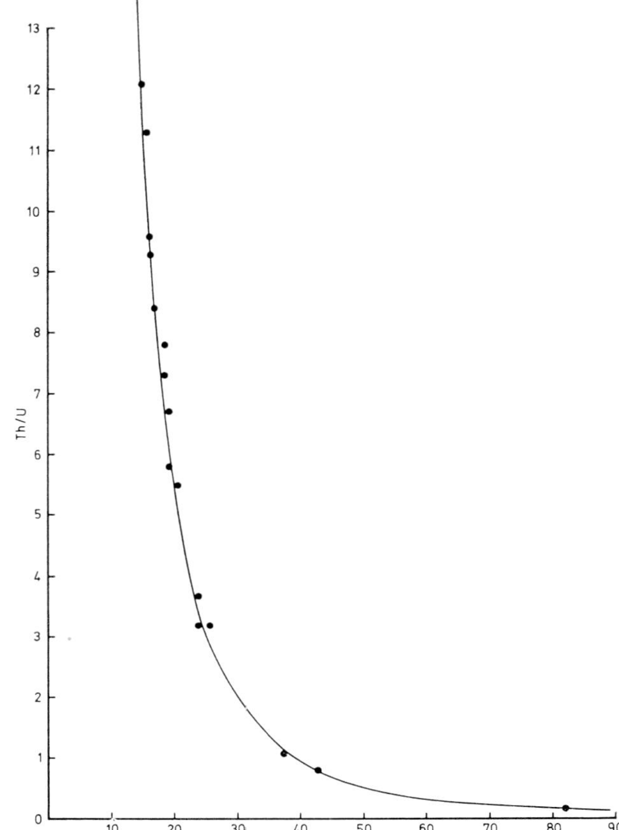

Fig. 2 Th/U ratio versus relative error in Th for typical exposure and counting conditions

Results and discussion

The fission-track method has been applied to the study of the distribution of radioelements in calc-alkaline granitic rocks of the Sierra Nevada batholith, California; in peralkaline rocks of the Ilímaussaq intrusion, south Greenland; and in a mineralized fault zone in east Greenland. A brief review of these studies serves to demonstrate the usefulness of the method in discerning the geochemistry of radioactive mineral assemblages.

Radioactive minerals of the Sierra Nevada batholith

As part of a comprehensive investigation of the distribution of radioelements in the granitic rocks of the Sierra Nevada batholith,[19] the writer determined the locations and abundances of uranium and thorium in rock and grain-mount thin sections. Photomicrographs of fission tracks from uranium registered in mica, superimposed on thin sections, are given in Fig. 3.

In the mica-rich Sierran samples U is associated mainly with biotite, occurring near the edges of grains or along dislocations within the grains. In other Sierran granitics U and Th are found primarily within the non-magnetic accessory minerals. In 11 of 13 samples, representing rocks typical of the batholith's various plutons (whole-rock averages: U, 3·9; Th, 14·0 ppm), sphene, averaging 250 ppm U, contributes from 9 to 38% of the rocks' uranium (F. C. W. Dodge, U.S. Geological Survey, private communication); the abundance of sphene does not exceed 0·6%. (Sphene makes up less than 0·004% of the other two samples: thus, it is not a significant contributor to their whole-rock U.) Zircon, though less prevalent than sphene, is an important contributor to whole-rock uranium.

Only five of the samples contain appreciable thorium in sphene. Although thorium has not been investigated extensively in grain mounts of other Sierran accessory minerals, examination

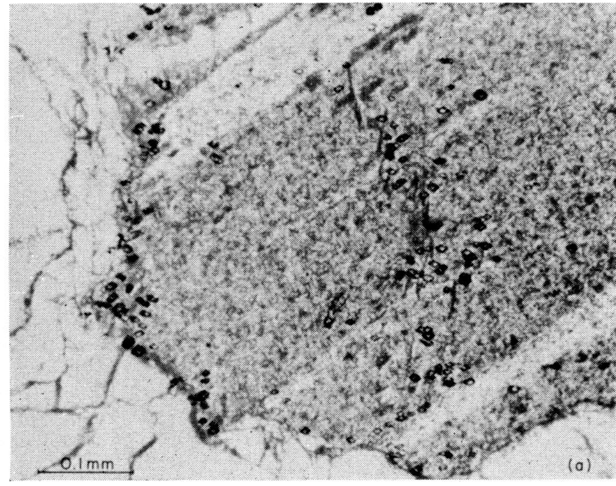

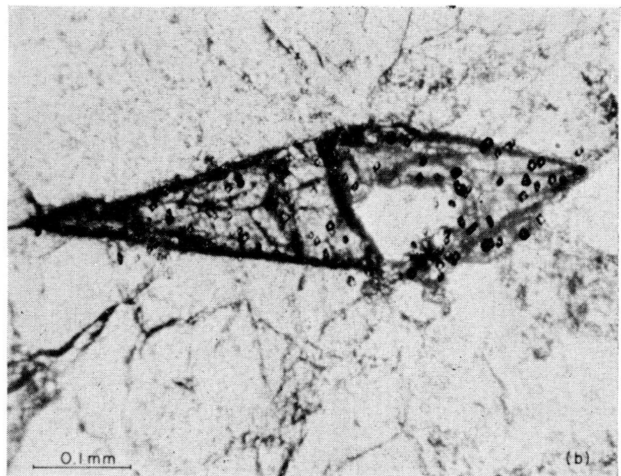

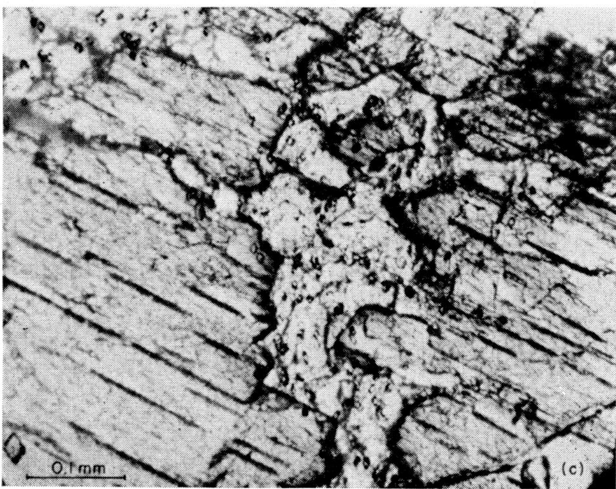

Fig. 3 Fission tracks in detecting mica superimposed on uranium-bearing minerals of Sierra Nevada granodiorites: (a) tracks associated with U, rimming the end of a grain of biotite and at discontinuities within the biotite; (b) tracks associated with U (120 ppm) in a grain of sphene; (c) tracks from U (160 ppm) in sphene, encompassed by hornblende

of thin-section radiographs indicates that it is concentrated in monazite and allanite. Two grains of allanite from a high-radioactivity granodiorite contain 8900 and 7200 ppm Th; U is, respectively, 237 and 217 ppm. In many instances uranium and thorium vary by factors of two to four between grains of the same accessory mineral within a single thin section.

Radioactive minerals of the Ilímaussaq intrusion

The radioelement and rare-element contents of the strongly differentiated Ilímaussaq alkaline intrusion[1, 10, 11, 15, 18, 20] have been intensively studied. The intrusion (age, *ca* 1000 m.y.) consists of an early augite syenite followed by a sequence of peralkaline nephelene syenites.[17] Lujavritic zones contain relatively high concentrations of U, Th, Zr, Nb and rare-earth elements. The highest radioelement concentrations occur in rocks of the Kvanefjeld plateau, near the intrusion's northwest border, where lujavrites are in contact with sheared cap-rock lavas. In this area steenstrupine* is the most abundant radioactive mineral; whole-rock uranium ranges from a few hundred to 800 ppm, and thorium from 200 to 2000 ppm. In the southern part of the intrusion lujavrites, rich in eudialyte† and containing appreciable Zr, have U and Th in the range of 30–60 ppm. Also examined by the fission-track technique were monazite, thorite and uranium-rich pigmentary material in lujavrite of the Kvanefjeld plateau.

Steenstrupine

Steenstrupine is characterized by a broad range in the amount of U and Th occurring, even from grain to grain, in a single thin section. This is exemplified in samples from a borehole drilled in lujavrites of the Kvanefjeld plateau. In a single thin section of rather homogeneous-appearing brown steenstrupine U ranges from 2700 to 7000 ppm and Th from 13 700 to 25 700—approximately factor-of-two differences between proximate grains. There is close agreement, however, between the overall average Th/U (3·1) in steenstrupine of the thin sections of the borehole with a Th/U ratio of 2·7 in corresponding whole-rock samples. This substantiates the conclusion from observation of the thin sections that in this lujavrite the abundance of steenstrupine essentially controls whole-rock uranium and thorium. The ratios of whole-rock and steenstrupine Th and U contents indicate that this mineral makes up to 2–4% of the rock.

*$Na_2Ce(Mn,Ta,Fe)H_2((Si,P)O_4)_3$.
†$(Na,Ca,Fe)_6Zr(OH,Cl)(Si_3O_9)_2$.

The distribution of U and Th within a large steenstrupine grain from a pegmatite vein is illustrated graphically in Fig. 4. The traverse shows a rather uniform distribution of uranium, whereas thorium varies by a factor of two to three. The error bars associated with the thorium points are from the counting statistics.

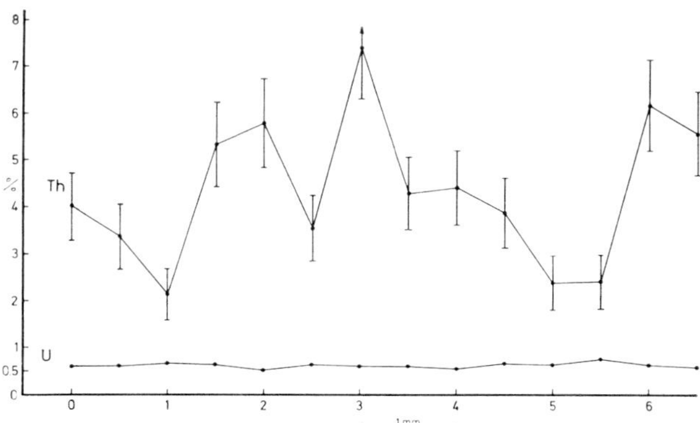

Fig. 4 Distribution of U and Th in profile across a large steenstrupine grain from a pegmatite

Other steenstrupine grains are distinguished by their marked zoning, which is exemplified in reddish steenstrupine (described by Buchwald and Sørensen[3]). Here the centres of the grains appear somewhat isotropic under crossed nicol prisms, whereas the outer areas are anisotropic (Fig. 5). Buchwald and Sørensen noted a marked difference in α-particle activity of these grains: higher activities occurred in the central portions and lower in the anisotropic borders. This difference is reflected in the U and Th contents, which are approximately twice as great in the isotropic centres as in the border areas.

The frequency distribution of radioelements can be plotted from fission-track radiography data. Fig. 6 gives histograms of U, Th and Th/U in steenstrupine, illustrating the broad variations in uranium and thorium in this mineral—especially from samplings of a 160-m portion of the Kvanefjeld borehole core.

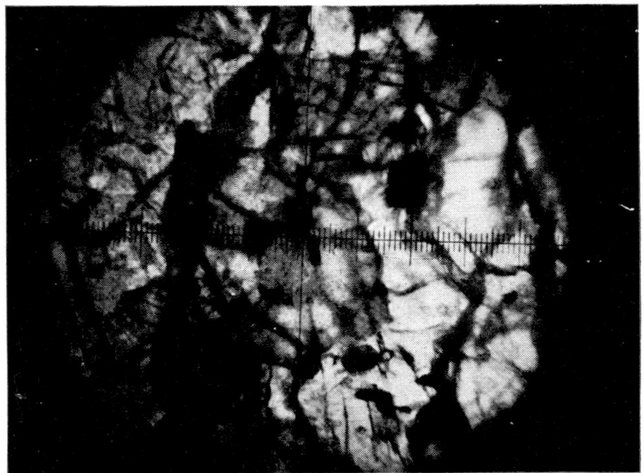

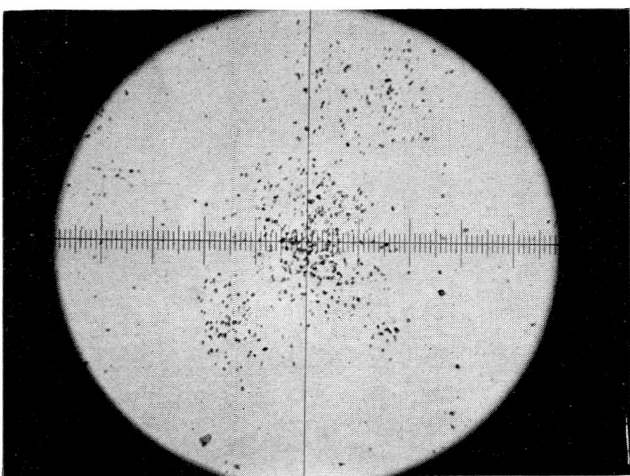

Fig. 5 Zoned reddish steenstrupine showing fission tracks associated with U concentrated in the grain's isotropic centre (4200 ppm), and a lower U content in the rim zone (2000–3000 ppm). Field diameter, 1·2 mm

Eudialyte

There are large differences in the uranium content of eudialyte in different rock types. Uranium, averaging 209 ppm, appears rather evenly distributed in unaltered eudialyte from a borehole (Fig. 7) drilled in 'green' (aegirine–arfvedsonite) lujavrite in the southern portion of the Ilímaussaq intrusion. This contrasts with considerably lower U (50–100 ppm) in eudialyte closely associated with steenstrupine in a borehole on the Kvanefjeld plateau. The latter eudialyte has a range of U similar to the 60 ppm reported by Hamilton[10] in a sample from the northern portion of the intrusion. It appears that essentially all of the radioactivity in lujavrite of the southern part of the Ilímaussaq intrusion is contributed by U and Th in eudialyte. The ratio of average U contents of eudialyte to average U in corresponding whole-rock samples (35 ppm) is approximately 6—in rough agreement with an estimate by Ferguson[5] of 10% eudialyte in green lujavrite.

Pigmentary material

Pigmentary material closely associated with arfvedsonite laths in lujavrite of the Kvanefjeld plateau (Fig. 8) contains up to 2·5% U, with little or no thorium. (This material is similar to that described by Hansen[11] as occurring in radioactive veins near the Ilímaussaq intrusion.) This strong contribution solely from uranium, superimposed on those of Th and U in steenstrupine, may account for the relatively low whole-rock

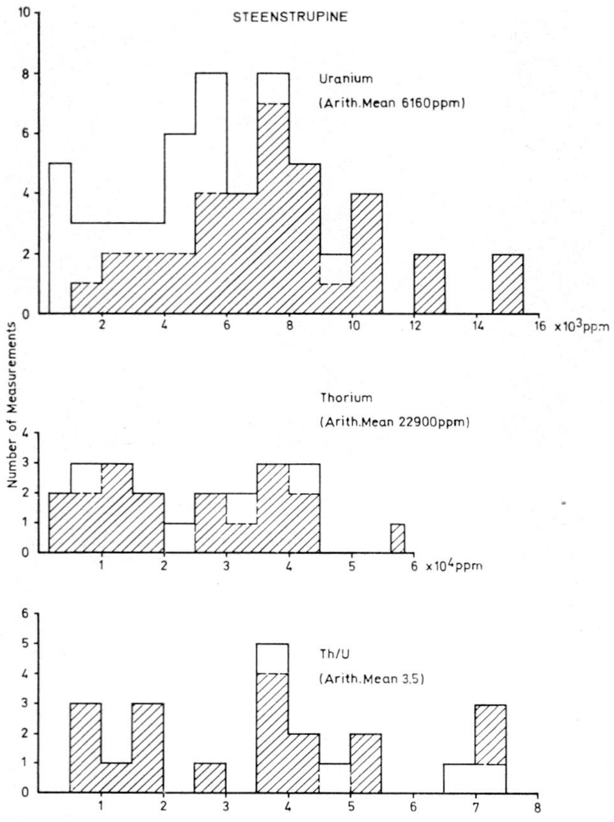

Fig. 6 Frequency distributions of U, Th and Th/U in steenstrupine: lined areas represent steenstrupine from a borehole on the Kvanefjeld plateau; unlined areas represent reddish and pegmatitic steenstrupine

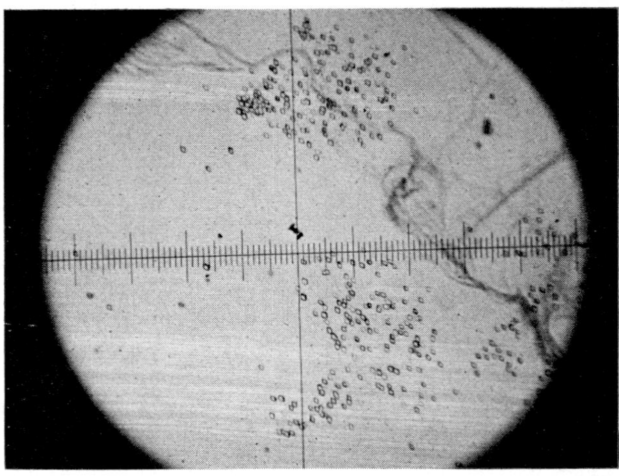

Fig. 7 Unaltered eudialyte in aegirine–arfvedsonite lujavrite showing a rather even distribution of U (~ 200 ppm). Field diameter, 1·2 mm

Th/U ratio (1·54) observed in some arfvedsonite lujavrites of the Kvanefjeld.[15] The apparent close association of pigmentary material with arfvedsonite, a mineral characterized by a predominance of Fe^{2+} over Fe^{3+}, may indicate reducing conditions in favour of the selective localization of uranium.

Table 3 Comparison of α-track data with radioelement contents

Description	α tracks/cm²/h*	U, ppm	Th, ppm
Steenstrupine in lujavrite, Kvanefjeld area	1300–12 000	820–15 000	1720–40 000
Sample 18467b-1, zoned red steenstrupine:			
anisotropic borders	3100–5300	2000–2700	7700–20 000
isotropic centres	6200–8300	4200–7200	33 000 (1 measurement)
Eudialyte in green lujavrite, southern area	100–640	210†	

*From Buchwald and Sørensen.[3]
†Mean value of borehole samples.

Thorite

Material considered likely to be thorite has been observed in veinlets of the Kvanefjeld.[15] In thin section (Fig. 9) it appears light yellowish-brown, and is associated with greenish cloudy monazite. The high thorium (28–55%) and uranium contents (2–3%) determined by fission-track radiography are consistent with those reported for thorite in reference texts.[16]

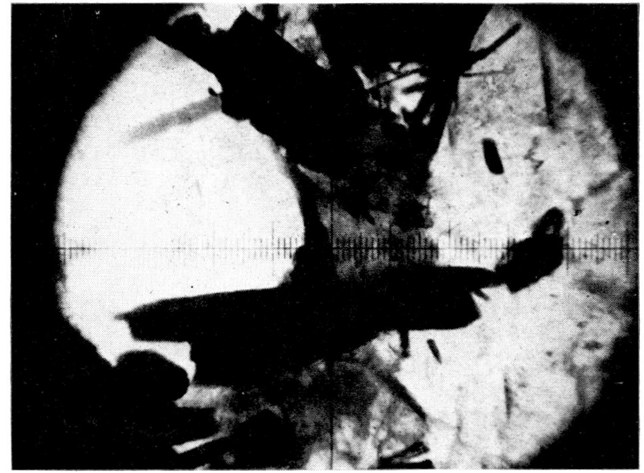

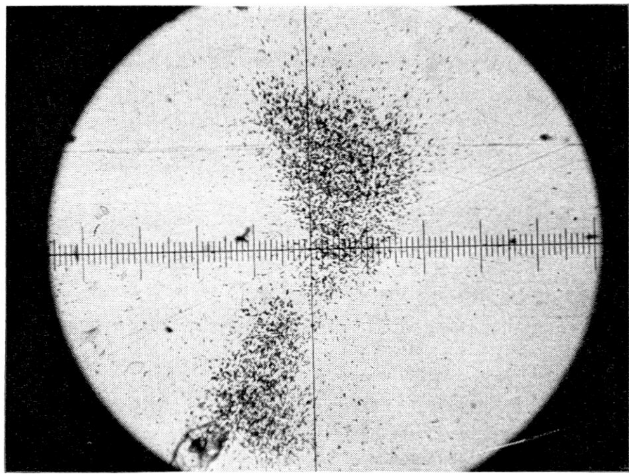

Fig. 8 Pigmentary material (1·4–2·6% U) in and between grains of arfvedsonite in lujavrite. Field diameter, 1·2 mm

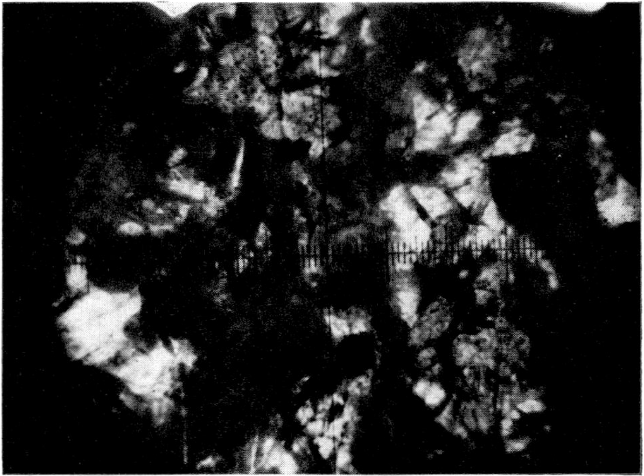

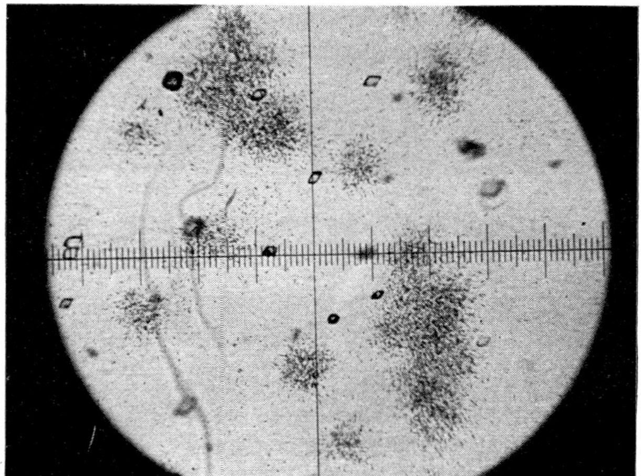

Fig. 9 Thorite (28–55% Th, 2–3% U) associated with greenish cloudy monazite in a veinlet in lujavrite. Large tracks are from a 16-h etch prior to the fast-neutron exposure. Field diameter, 1·2 mm

Monazite

Monazite, which occurs generally in cloudy-appearing aggregates of small elongate crystals (Fig. 10), was examined in some drill cores and surface samples of lujavrite from the Kvanefjeld plateau. Monazite associated with the thorite described above contains approximately 25% Th, and has a high Th/U ratio. Considerably lower Th contents (1–2%) were measured in monazite from medium- to coarse-grained lujavrite of the Kvanefjeld; Th/U ratios in this monazite fall in the range 9–14. These are similar to the ratios in thorite, and contrast with the considerably lower Th/U measured in steenstrupine in the same thin sections.

Comparison with alpha-track autoradiographic data

A comparison of uranium and thorium values with corresponding α-track densities (Table 3), reported by Buchwald and Sørensen,[3] indicates relatively good concurrence. Low α-track densities of eudialyte from the green lujavrite mentioned earlier correspond to U ranging from 160 to 410 ppm. Considerably higher α-track densities, varying over nearly a factor of 10 in steenstrupine of Kvanefjeld lujavrite, match broad ranges in U and Th in thin sections from the boreholes.

There is also good agreement between the U and Th contents and the α-track data for isotropic and anisotropic zones in the red steenstrupine shown in Fig. 5. To directly compare Buchwald and Sørensen's observed α-track densities with those calculated from fission-track data the mean U and Th values for the anisotropic and isotropic zones were entered in the formula of Coppens,[4] which relates radioelement

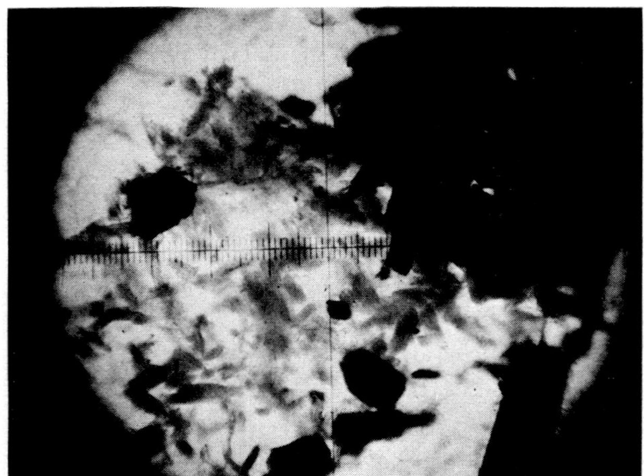

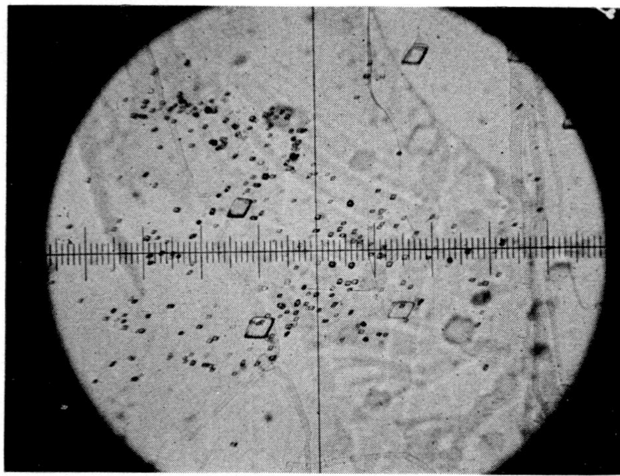

Fig. 10 Fission tracks from U plus Th (U ~ 1200 ppm, Th ~ 12 000 ppm) in monazite (grey) with arfvedsonite (black) in medium to coarse lujavrite. Large tracks are from a 16-h etch prior to the fast-neutron exposure. Field diameter, 1·2 mm

content with α-track density. The resulting values, 8200 and 3500 α/cm²/h for isotropic and anisotropic zones, respectively, fall within the reported α-density ranges.

An east Greenland fault-zone occurrence

The distribution of uranium in thin sections from a mineralized fault zone was investigated by the fission-track method. The fault forms the boundary between crystalline rocks of the Caledonian fold belt to the west and Palaeozoic–Mesozoic sedimentary deposits of Jameson Land to the east. The geology of the area has been described in general terms by Kempter.[12] The samples from which the thin sections were prepared consist of moderately to strongly oxidized fault breccia, mineralized with varying intensity, principally by fluorite and barite. Megascopic examinations, coupled with γ-ray spectrometric determinations, indicate that the concentration of whole-rock uranium generally varies with the intensity of fluoritization.

From fission-track radiography (Fig. 11) it is apparent that U is closely, but not always directly, associated with the fluorite. U also occurs with yellowish to brownish limonite that lines fluorite veinlets, is intergranular between

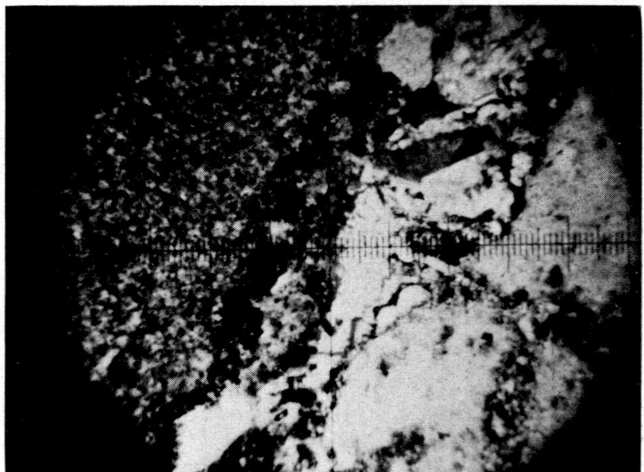

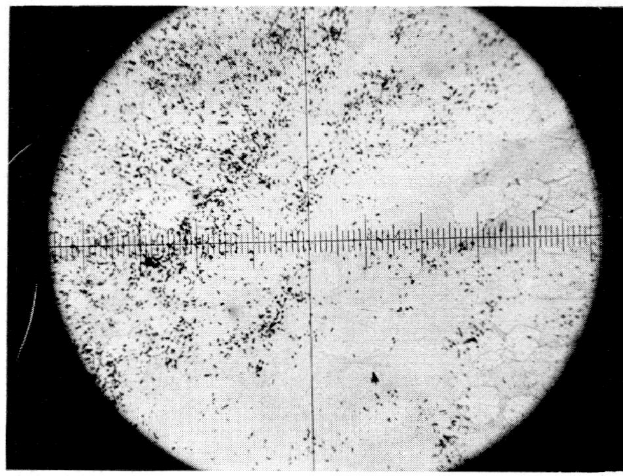

Fig. 11 Strong association of U (up to 4000 ppm) with limonite (light grey) and fluorite (dark and intermediate) in fluoritized fault-zone breccia, east Greenland. Field diameter, 1·2 mm

fluorite grains, or exists separately from the veinlets. Uranium concentrations in this material range from several hundred to more than 4000 ppm. Less common associations of uranium are illustrated by very intense fission-track densities, occurring with yet unidentified translucent grains, and with U enrichment at the margin of reddish semi-opaque grains, possibly hematite (Fig. 12).

There is little consistency between whole-rock uranium contents and those of the mineralized zones in the thin sections. In this environment uranium is confined to thin veinlets, which are very unevenly distributed within the fault-zone breccia.

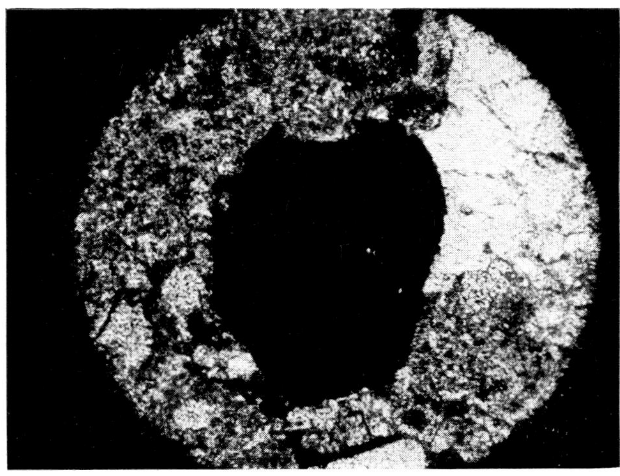

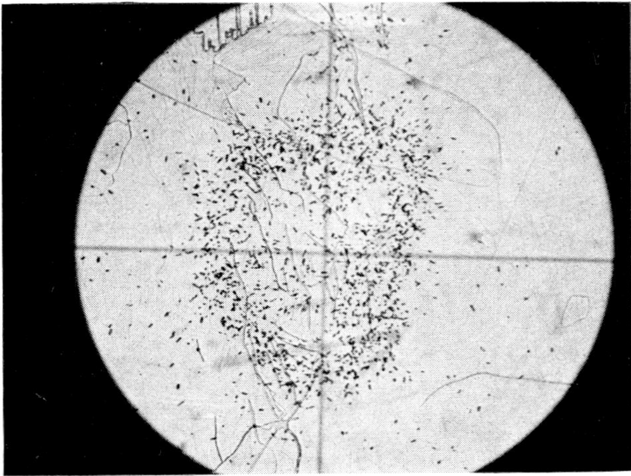

Fig. 12 Enrichment in U (2500 to 3000 ppm) at margin of an opaque grain (hematite?) in fluoritized fault-zone breccia, east Greenland. Field diameter, 0·88 mm

Summary and conclusions

The fission-track method is a quick, simple and inexpensive way to determine the location and abundance of uranium, and, in some cases, thorium, in uncovered thin sections. The method can be used to determine accurately U concentrations ranging from several ppb up to several per cent. The accuracy for the determination of thorium is quite poor when $Th/U < 1$, but improves to about 25% (relative error) when $Th/U > 3$.

Fission-track radiography of granitic-rock thin sections from the Sierra Nevada batholith indicates that uranium associates most strongly with biotite in rocks rich in this mica; otherwise, U occurs most frequently with the accessory mineral sphene, and Th with monazite and allanite. Such associations suggest that the uranium and thorium balance in a rock unit may be obtained by combining data from whole-rock γ-ray spectrometric analyses with data from fission-track measurements on populations of separated mineral grains.

The fission-track results and recent whole-rock γ-ray spectrometric measurements in the Ilímaussaq intrusion[15] indicate that the abundance of steenstrupine governs whole-rock radioactivity in lujavrites of the Kvanefjeld area, and eudialyte controls radioactivity in agpaitic rocks in the remainder of the intrusion. In lujavrite large disparities were observed between radioelement contents of steenstrupine grains only a few millimetres apart. Uranium varies much less in eudialyte of the aegirine–arfvedsonite drill core from the southern portion of the intrusion. Measured uranium and thorium contents of steenstrupine agree well with α-track densities observed in autoradiographs by Buchwald and Sørensen.[3]

In a hydrothermally mineralized fault zone in east Greenland fission-track radiography reveals that uranium, in concentrations up to several thousand ppm, occurs closely associated with limonitic material in strongly fluoritized veinlets. Thus, the intensity of fluoritization generally controls whole-rock radioactivity.

As with almost all geochemical techniques, the fission-track method is most valuable when its results are combined with chemical data obtained by other methods, e.g. electron-microprobe examinations that disclose the distributions of rare-earth elements, Zr and Ti in accessory minerals, and reveal the chemical nature of uranium-rich pigmentary material. Microprobe data would aid in identifying scattered uranium-rich minerals such as those observed in an east Greenland fault zone.

Acknowledgment

The author is grateful to the Danish Atomic Energy Commission, Research Establishment, Risø, for support of fission-track studies while he was a visiting scientist during 1969–71. Much of the Greenland-orientated work reported here was accomplished in cooperation with Leif Løvborg, Risø, and Professor Henning Sørensen and John Rose-Hansen, Institute of Petrology, Copenhagen University.

The study of radioelements in the Sierra Nevada batholith is part of a continuing cooperative effort between the Lawrence Berkeley Laboratory and the United States Geological Survey. Mineral concentration data and uncovered Sierran thin sections were kindly provided by Paul C. Bateman and Franklin C. Dodge of the Survey. The support of H. Wade Patterson and Alan R. Smith, Health Physics Department, Lawrence Berkeley Laboratory, is greatly appreciated. Part of the work reported here was done under the auspices of the U.S. Atomic Energy Commission.

References

1. Bohse, H. Brooks, C. K. and Kunzendorf, H. Field observations on the kakortokites of the Ilímaussaq intrusion, southwest Greenland, including mapping and portable X-ray fluorescence analyses for zirconium and niobium. *Rep. geol. Surv. Grønland*, in press.
2. Bowie, S. H. U. Nuclear emulsion techniques. In *Nuclear geology* Faul, H. ed. (New York: Wiley, 1954), 48–64.
3. Buchwald, V. and Sørensen, H. An autoradiographic examination of rocks and minerals from the Ilímaussaq batholith, South West Greenland. *Meddr Grønland*, **162**, no. 11, 1961, 43 p.
4. Coppens, R. Sur la mesure de la radioactivité des roches par l'émulsion photographique. *Bull. Soc. fr. Minér. Cristallogr.*, **75**, 1952, 57–8.
5. Ferguson, J. Geology of the Ilímaussaq alkaline intrusion, South Greenland. *Meddr Grønland*, **172**, no. 4, 1964, 83 p.
6. Fisher, D. E. Homogenized fission track determination of uranium in whole rock geologic samples. *Analyt. Chem.*, **42**, 1970, 414–6.
7. Fleischer, R. L. Price, P. B. and Walker, R. M. Tracks of charged particles in solids. *Science, N.Y.*, **149**, 1965, 383–93.
8. Fleischer, R. L. *et al.* Criterion for registration in dielectric track detectors. *Phys. Rev.*, **156**, 1967, 353–5.
9. Hamilton, E. The distribution of radioactivity in the major rock forming minerals. *Meddr Grønland*, **162**, no. 8, 1960, 41 p.
10. Hamilton, E. The geochemistry of the northern part of the Ilímaussaq intrusion, S.W. Greenland. *Meddr Grønland*, **162**, no. 10, 1964, 104 p.
11. Hansen, J. A study of radioactive veins containing rare-earth minerals in the area surrounding the Ilímaussaq alkaline intrusion in South Greenland. *Meddr Grønland*, **181**, no. 8, 1968, 51 p.
12. Kempter, E. Die Jungpaläozoischen Sedimente von Süd Scoresby Land. *Meddr Grønland*, **164**, no. 1, 1961, 125 p.
13. Kleeman, J. D. and Lovering, J. F. Uranium distribution in rocks by fission-track registration in Lexan plastic. *Science, N.Y.*, **156**, 1967, 512–3.
14. Kleeman, J. D. and Lovering, J. F. Lexan plastic prints: how are they formed? *Radiat. Eff.*, **5**, 1970, 21–6.
15. Løvborg, L. *et al.* Field determination of uranium and thorium by gamma-ray spectrometry, exemplified by measurements in the Ilímaussaq alkaline intrusion, South Greenland. *Econ. Geol.*, **66**, 1971, 368–84.
16. Nininger, R. D. *Minerals for atomic energy* (New York: Van Nostrand, 1954), 367 p.
17. Sørensen, H. On the magmatic evolution of the alkaline igneous province of South Greenland. *Rep. geol. Surv. Grønland* no. 7, 1966, 20 p.
18. Sørensen, H. Hansen, J. and Bondesen, E. Preliminary account of the geology of the Kvanefjeld area of the Ilímaussaq intrusion, South Greenland. *Rep. geol. Surv. Grønland* no. 18, 1969, 40 p.
19. Wollenberg, H. A. and Smith, A. R. Radiogeologic studies in the central part of the Sierra Nevada batholith, California. *J. geophys. Res.*, **73**, 1968, 1481–95.
20. Wollenberg, H. A. Kunzendorf, H. and Rose-Hansen, J. Isotope-excited X-ray fluorescence analyses for Nb, Zr, and La+Ce on outcrops in the Ilímaussaq intrusion, South Greenland. *Econ. Geol.*, **66**, 1971, 1048–60.

543.53:550.84

Application of fission-track and neutron activation methods to geochemical exploration

S. H. U. Bowie, D.SC., F.R.S.E., F.I.M.M.

P. R. Simpson, M.A., M.SC.

C. M. Rice, B.SC., PH.D.

All of the Institute of Geological Sciences, Geochemical Division, London, England

Synopsis

Fission-track radiography, supported by a delayed neutron method of analysis and field observations, has been used to investigate uranium mineralization controls in parts of Scotland. The areas selected were the Criffel granodiorite, Aberdeenshire granites, Helmsdale granite and Old Red Sandstone sediments of the Orcadian cuvette, including the Ousdale arkose. Special attention was given to the comparison of rocks from mineralized and non-mineralized environments.

The study has indicated that magmatic fractionation is unlikely to have resulted in uranium enrichment associated with the Criffel granodiorite and the Helmsdale granite. The most likely control is proximity to a major fault system, uranium and associated mineralization being developed in tension faults. The Criffel granodiorite mineralization clearly post-dates granitic intrusion, as was previously shown by uranium and lead dating. In the case of the Helmsdale granite and Ousdale arkose, however, it is considered that uranium was introduced in Lower ORS times simultaneously with transcurrent movement on the Helmsdale Fault system which permitted the development of favourable dilatant zones in which faults and fractures were enriched in uranium, fluorite and pyrite.

Fission-track radiography has proved particularly effective in (*a*) defining the background non-mineralized situation for the granitic rocks, (*b*) distinguishing between syngenetic, diagenetic and epigenetic uranium in the sediments and (*c*) documenting the changes in mineralogy and uranium location as enrichment proceeds. In granites not known to have associated uranium mineralization of economic grades, there is some enrichment of uranium in faulted and unfaulted granite in association with martite, hematite, limonite, chlorite and sericite.

The study of ^{235}U fission tracks in a polycarbonate plastic such as Lexan provides precise information on the distribution of uranium in rock sections, particularly at concentrations from 0·05 ppm upwards. In this work the partition of uranium in granites with and without associated uranium mineralization is compared in polished thin sections of rocks from several areas in Britain. Normal and abnormal concentrations of uranium in sediments are also examined, together with field evidence, in an attempt to throw new light on uranium mineralization controls. Whole-rock uranium values to supplement the fission-track data were determined by a delayed neutron method.[1, 2]

Autoradiography[3] has been widely used in the study of the distribution of uranium in rocks, especially where it is concentrated as point sources comprising, for example, discrete grains of uraninite in a quartz-pebble conglomerate matrix. In this type of situation the autoradiographs provide information on the location of thorium- and uranium-rich mineral phases. Time-consuming measurements of alpha-particle track lengths are necessary, however, before a distinction can be made between activity due to uranium, thorium or daughter products such as radium. The method is also relatively insensitive to low concentrations of these elements. The fission-track technique supplements autoradiography in that it is much more sensitive and specific to uranium or uranium together with thorium depending on the experimental conditions.

The present emphasis in the search for uranium in the form of disseminated low-grade mineralization in large volumes of sedimentary rocks requires sensitive and precise methods for the study of uranium distribution, and this is provided by the fission-track method. Furthermore, several geological events or processes may have had a control on uranium distribution, and fission-track studies can be used to assist in recognizing these both on a local and regional scale.

The method was applied to samples of granitic rock from two uranium districts which had been examined in detail on behalf of the United Kingdom Atomic Energy Authority (UKAEA)—Helmsdale and Dalbeattie. In addition, granites from Aberdeenshire not known to have any associated uranium mineralization were studied. Normal and abnormal amounts of uranium in the post-tectonic molasse facies sediments of the Orcadian cuvette were also studied since it is in such an environment that disseminated deposits of uranium might be expected to occur.

Experimental method

The fission-track method has been described elsewhere,[4, 5] and is only briefly summarized here. It is based on the fission of ^{235}U, which occurs in a thermal neutron flux. If the fission fragments so produced are allowed to bombard an overlay of a suitable detector, such as Lexan polycarbonate film, the tracks can subsequently be rendered visible by etching the film in a solution of 6 N NaOH at 90°C for a few minutes. The Lexan print may then be compared directly with petrographic features observable in the sample. The matching precision between Lexan print and sample is generally about 50 μm; but, where necessary, this can be reduced to about 10 μm. Quantitative results can be obtained if glass standards with known amounts of uranium are irradiated in the neutron flux along with the sample. Since the standard and sample receive a similar dose, quantitative comparisons can be made between the track density in the Lexan prints from the standard and sample. Several irradiations may be required, each at different neutron dosage levels, in order to characterize the uranium distribution in a sample.

The technique employed in this study involves the preparation of polished thin sections 1 in × 1·25 in in size. This is convenient for transmitted light and reflected light microscopy and is a commonly used size in electron-microprobe analysis. The sections are overlaid with Lexan film cut to the dimensions of the slide and held in place with a small piece of adhesive tape. The sections are loaded into a pure aluminium cylindrical container with a screw top. To prevent contamination the sections are loaded into the container in pairs between pure aluminium spacers with the Lexan overlays in direct contact back to back for each pair of sections. When the container is full, the cap is screwed down firmly to prevent movement during irradiation. Otherwise, distortion of the fission-track distribution pattern could result and make comparisons difficult. One container will take up to 18 sections, including standards. The technique is well suited to the routine study of uranium distribution in relatively large batches of sections, such as might result from systematic geochemical surveys. The study described here involved 200 sections treated in this way.

Uranium in granites

Introduction

During the UKAEA uranium exploration programme in Scotland, a spatial association between uranium mineralization and certain granites was

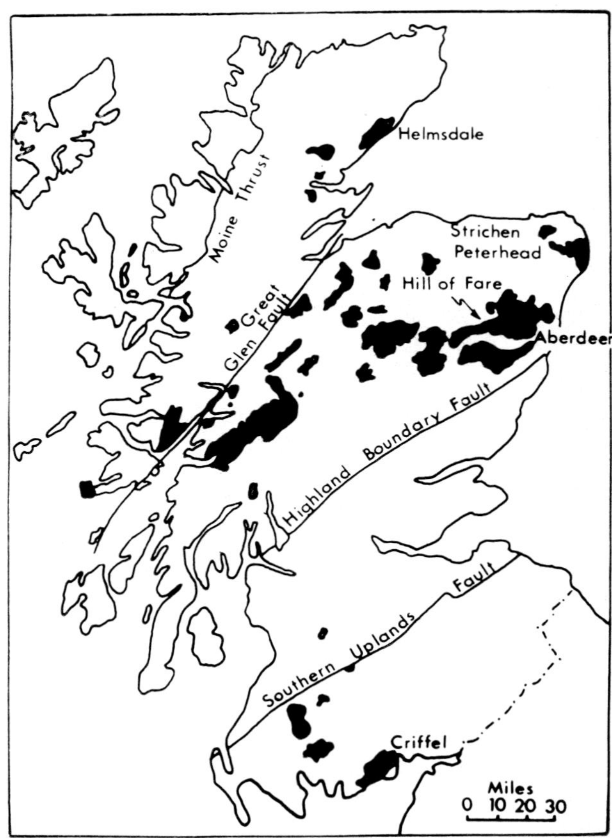

Fig. 1 Caledonian granites; named granites were sampled in this study

noted.[6] These included the Criffel granodiorite and the Helmsdale granite (Fig. 1), which are representatives of a suite of intrusions emplaced at a late stage in the development of the Caledonian orogen—the Newer Caledonian Granites of Read.[7] In the aureole of the Criffel granodiorite uranium is associated with late crosscutting fractures and veins associated with a major coastal fault system, and at Helmsdale a coastal transcurrent fault cuts the granite. The Lexan plastic method was applied to determine whether there are differences in the distribution of uranium in the Helmsdale granite and the Criffel granodiorite as compared with Newer Caledonian granites in Aberdeenshire with no known uranium mineralization. The distribution of uranium in non-mineralized Aberdeenshire granites is described first since these are considered to represent the background level for Newer Caledonian granites. Most of the latter have an annular structure with an early outer phase intruded by a later inner phase. The separate phases of the mineralized intrusions are compared in order to investigate the possibility of enrichment related to magmatic fractionation. The relative importance of this process is compared with the effects of later alteration and faulting.

Aberdeenshire granites

Approximately 40 samples were collected from 15 localities in the Hill of Fare, Peterhead and Strichen granites. Polished thin sections were cut from the majority of these and examined petrographically and for uranium distribution. In addition, total uranium contents were obtained by the delayed neutron method.

Petrography and uranium distribution

The granites are of three main types: (*a*) hornblende–biotite granites (granodiorites); (*b*) biotite granites; and (*c*) muscovite granites. One or more of these granite types occurs at each locality visited: for example, at Craigenlow both hornblende–biotite and biotite granites occur, and in the Hill of Fare granite all three types are present. When the samples were collected, attention was paid to evidence of magmatic differentiation, uranium enrichment at contacts between different granite types, faulting and weathering effects.

In the majority of samples studied uranium is located mainly in the accessory minerals. Very minor quantities, amounting to only a few parts per million, are associated with patches of biotite, particularly where it is chloritized. There is no apparent enrichment of uranium in secondary iron hydroxides at grain boundaries or in shear zones. Neither is it enriched in pegmatite phases nor is there any appreciable partitioning between different granite types.

The uraniferous accessory minerals are sphene, zircon, monazite, allanite and apatite. Monazite and allanite are less common than the other minerals and apatite contains least uranium. Sphene is mainly confined to the hornblende–biotite granites. A further important uranium location is in martite (Fig. 2), and in several samples this is the main host mineral in the rock. Hematitization—which is usually limited to martitization—affects all three rock types and about half of all the samples studied. Martite is not always uraniferous, however. Uranium determination by X-ray fluorescence analysis on separated sphene indicates values of 150–3000 ppm. These data are consistent with the fission-track study, which indicated that sphene contains appreciable amounts of uranium in the Aberdeenshire granites investigated.

The 33 samples analysed for total uranium include representatives of all the rock types collected (Table 1). They show limited variation with a range of 1·4–6·6 ppm and a mean of 3·2 ppm. This is lower than a recently quoted[8] clarke for granite of 4·0 ppm. The mean values for uranium in the different types of granite show a tendency to increase with increasing acidity, namely hornblende–biotite, biotite, muscovite granite (2·45, 2·70, 3·70). The number of analyses performed was, however, inadequate to ensure that this trend is significant. The Peterhead and Strichen granites contain similar quantities of uranium to that of the Hill of Fare granite (3·95 and 3·4 ppm compared with 2·90 ppm). Most of the samples from the Hill of Fare granite are from the margin of the intrusion, but these contain similar amounts of uranium to samples from nearer its centre.

Samples from a pegmatite, a fault zone and weathered granite do not show appreciable deviations from the mean for all samples. The mean values of 3·4 and 2·6 ppm for hematitized and non-hematitized samples, respectively, do not differ significantly. It is, however, probably significant that, with one exception, all samples with uranium contents greater than 4·0 ppm are hematitized, and in the sample with the highest value of 6·6 ppm uranium is mainly located in martite. Hematitization apparently involves the addition of uranium to granite, and in Aberdeenshire only the incipient stage is attained.

The uranium-bearing accessory minerals in non-hematitized granite, with the exception of allanite, are common detrital minerals which are stable under most weathering conditions. Uranium in these minerals is, therefore, unlikely to enter readily into the hydrogeochemical cycle. In

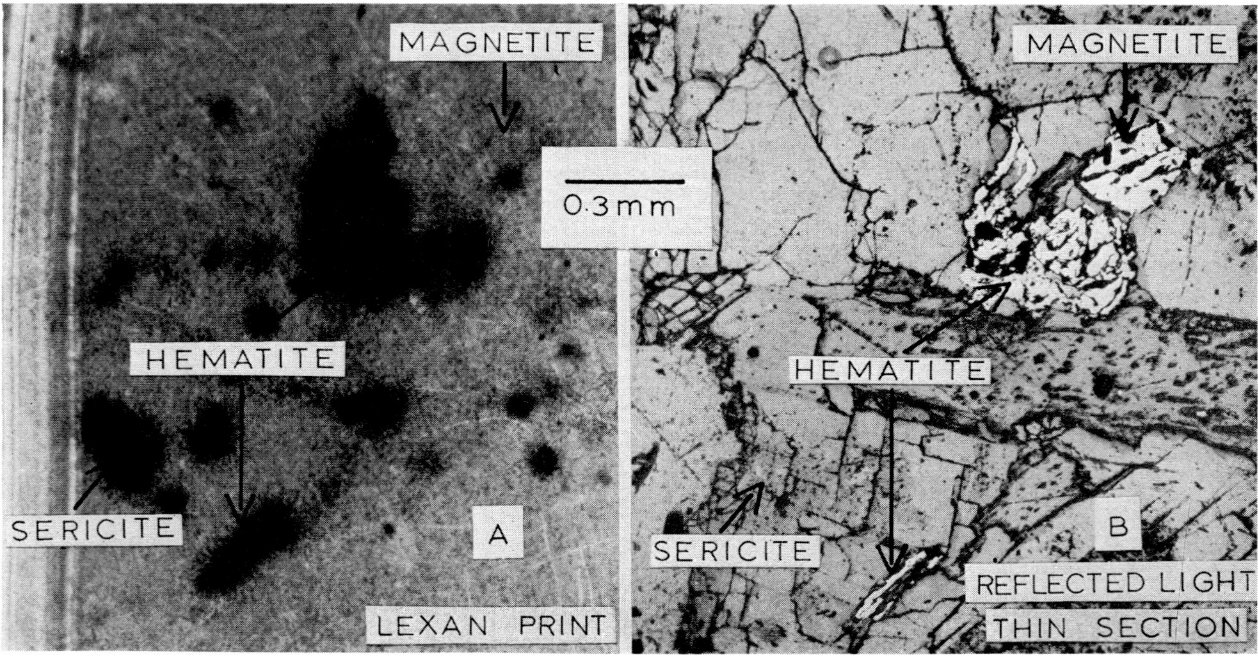

Fig. 2 Hematite, magnetite and sericite are identified in a section of granite in reflected light. Lexan print from this area of the section carries areas of high track density (black), which indicates that the uranium occurs here in association with hematite and sericite but is not associated with magnetite

granites which have been affected by hematitization and the introduction of uranium, additional factors are involved. Hematite readily hydrolyses to limonite in an oxidizing environment and uranium becomes susceptible to removal in solution.

The Criffel granodiorite

The Criffel granodiorite is intruded into deformed Lower Palaeozoic sediments. It has been described by Phillips,[9] who distinguished two main facies: (*a*) an outer main granodiorite and (*b*) an inner porphyritic variety; other minor facies exist but are not relevant to the present study. Accounts of the uranium mineralization in this area have been given by Miller and Taylor[10] and Gallagher *et al.*[6] Miller and Taylor[10] observed that the mineralization is confined to the aureole of the granodiorite and is localized in northwest–southeast feather faults associated with a major northeast–southwest coastal fault. Further uranium mineralization associated with a northwest–southeast vein has recently been recorded in the northern aureole of the granodiorite by Gallagher *et al.*[6] In the present study samples were collected from several localities, including the two main facies of the granodiorite, uraniferous veins near Sandyhills and Beeswing and a uraniferous shatter zone within the granodiorite south of Dalbeattie noted by Gallagher *et al.*[6]

Petrography and uranium distribution

The distribution of uranium in the two main facies was examined, but no difference was observed between them. All samples are affected by partial chloritization, martitization and sericitization; but, with the exception of the most altered sample, the bulk of the uranium is in sphene, the most common accessory, and allanite. In the most altered sample the highest levels of uranium are associated with lozenge-shaped aggregates of rutile crystals. Lower levels of concentration are related to patches of chlorite pseudomorphing biotite. It is possible that the aggregates of rutile crystals represent incipient secondary sphene. The distribution and concentration of uranium in the relatively unaltered Criffel granodiorite is comparable to granodiorite of the Hill of Fare (Table 2). It is, however, worth noting that an aplite vein which cuts the main granodiorite at Craignair quarry contains about three times as much uranium as the granodiorite. The location of the uranium is not known, but this implies some enrichment of uranium during magmatic fractionation. It would seem unlikely that this process has contributed to the main mineralization, which was dated by Miller and Taylor[10] as Triassic–Jurassic in age.

In the shatter zone south of Dalbeattie, uranium is associated with abundant hematite and its alteration products (Fig. 3). A level of 58 ppm was obtained for the strongly hematitized and

Table 1 Uranium levels in samples collected from the Hill of Fare, Strichen and Peterhead granites by DNA (estimated coefficient of variation 2%)

Rock type	Hill of Fare No. of samples	Hill of Fare U ppm mean and range	Strichen No. of samples	Strichen U ppm mean and range	Peterhead No. of samples	Peterhead U ppm mean and range
Hornblende–biotite granite	2	2·45 (2·0–2·9)	—	—	—	—
Biotite granite	10	2·4 (1·6–4·7)	5	3·4 (2·6–4·5)	2	3·95 (3·3–4·6)
Muscovite granite	3	3·7 (2·3–6·6)	—	—	—	—
Mean of above Hill of Fare granites		2·9	—	—	—	—
Others						
Hybrid granite with mafic nodules	1	1·5	—	—	—	—
Weathered hornblende–biotite granite	2	2·1 (1·8–2·4)	—	—	—	—
Sheared and reddened granite from fault zone	1	2·2	—	—	—	—
Late cross cutting basic dykes	3	3·6 (2·6–4·7)	—	—	—	—
Pegmatite	—	—	—	—	1	1·4
Mafic nodule	1	2·0	—	—	—	—

All granites, excluding samples listed under 'Others': mean, 3·0; range, 1·9–6·6.

All samples, including samples listed under 'Others': mean, 3·2; range, 1·4–6·6.

14 samples with hematite: mean, 3·4; range, 1·4–6·6.

19 samples without hematite: mean, 2·6; range, 1·5–4·7.

sheared granodiorite (Table 2). No other uraniferous mineral was observed.

At Beeswing the vein cuts banded quartz–biotite hornfels in the aureole of the granodiorite. The uranium content of the hornfels away from the vein is low (Table 2) and the uranium is evenly distributed and not associated with accessory minerals. Within the vein the main gangue minerals are quartz and carbonate; and the ore minerals hematite and its alteration products and chalcopyrite. The hematite and alteration products are generally uraniferous. Uranium minerals were not observed and a sample of vein gave a low uranium content (Table 2). Sub-spheroidal pitchblende does occur, however, within the hornfels in a zone about 1 cm wide bordering on the vein, and Gallagher *et al.*[6] recorded uraninite coating joints in the hornfels. The Lexan prints also showed that within the above-mentioned zone the general level of non-localized uranium is higher than the normal hornfels.

Similar mineral assemblages were recorded in veins at Sandyhills and Beeswing, although in the former locality pitchblende was observed within the vein. An additional feature at Sandyhills is the presence of significant amounts of uranium in the cleavages of dolomite where the dolomite is associated with pitchblende. The general trend of the uraniferous structures mentioned above is northwest–southeast.

The two main facies in the Criffel granodiorite have similar uranium concentrations and distributions, which indicates that the associated uranium mineralization is unlikely to have been concentrated by magmatic fractionation processes. Furthermore, it is noteworthy that the general level of uranium and its mode of distribution is similar to that of the Aberdeenshire granodiorites.

The Helmsdale granite

The Helmsdale granite cuts rocks of Moinian age. The intrusion has two principal phases—an outer

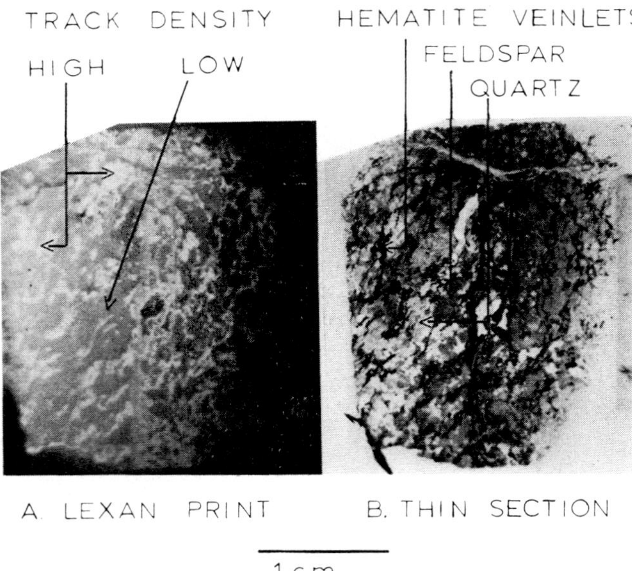

Fig. 3 Sheared and hematitized granodiorite from shatter zone south of Dalbeattie in which uranium is associated with hematite

porphyritic and an inner fine-grained adamellite. The two granite types have a concentric annular structure with gradation from one to the other.[11] The southeast portion of the intrusion is probably faulted out by a major northeast–southwest-trending coastal transcurrent fault described as the Helmsdale fault.[12] It is suggested by analogy with the sub-parallel Great Glen fault[13] that a major movement of the Helmsdale fault system occurred in Lower ORS time. A later movement with a normal component has downthrown the Jurassic. The granite is crosscut by northwest–southeast feather faults, which are part of the Helmsdale fault system. The granite lies close to the basal ORS unconformity of the Orcadian cuvette. It has been suggested[14] that after unroofing of the granite during the late stages of the Caledonian orogeny it was covered by ORS sediments until Tertiary times, when it was unroofed for a second time.

Table 2 Uranium content of samples from Criffel granodiorite

Locality and sample type	U, ppm
Beeswing vein	
(*a*) Hornfels immediately adjacent to and including a barren part of vein (2 samples)	1053 2174
(*b*) Banded hornfels away from vein	2·9
(*c*) Hematite–quartz pod within vein	2·5
Outer main granodiorite	
(*a*) Cowpark quarry	3·8
(*b*) Craignair quarry	5·4
(*c*) Mafic patch (Craignair)	6·7
(*d*) Aplite vein (Craignair)	14·0
Inner porphyritic granodiorite	
(*a*) Kippford quarry (3 samples)	3·0 2·4 2·2
Auchensheen shear zone	
(*a*) Strongly sheared and hematitized sample	58·0
(*b*) Sheared and hematitized sample	7·0
Sandyhills vein	>3000

The porphyritic adamellite consists of orthoclase, plagioclase and quartz, with about 5% of mafic minerals, of which the principal one is biotite. The phenocrysts consist of potash feldspar. The plagioclase is albite or oligoclase and is often enclosed in the potash feldspar. Quartz forms interstitial grains between the feldspars or rounded aggregates. Biotite occurs as thick ragged prisms, often partly chloritized. Apatite, zircon and magnetite, more or less completely converted to hematite, are accessory phases. The zircon is included in the biotite, which is often present in close proximity to the martite. The fine-grained adamellite is modally and compositionally similar. The main difference is the occurrence of potash feldspar forming smaller crystals 1–2 mm across.

Fig. 4 Porphyritic adamellite from Helmsdale in which uranium is associated with chloritized biotite

The fission-track method indicates that the main concentrations of uranium in the granite are associated with areas of chloritized biotite (Fig. 4), partly as a disseminated phase but also as small point sources of high intensity which correlate with the presence of small zircon crystals, mainly in the biotite. Apatite is also enriched in uranium. These relationships apply to porphyritic and fine-grained adamellite alike.

Table 3 Uranium content in relation to granite type and degree of alteration at Helmsdale

Granite type	U, ppm	No. of analyses
Partially altered fine-grained granite	5·7–9·0	3
Partially altered porphyritic granite	5·0–7·5	2
Partially altered intermediate granite	3·5	1
Strongly altered fine-grained granite	13·6	1
Strongly altered porphyritic granite	10·7–18·3	4

The uranium content increases with alteration of the granite (Table 3). Uraniferous veinlets studied by the fission-track method, and containing galena, sphalerite, pyrite and fluorite, cut the granite in a loose boulder of intermediate adamellite type in Caen burn (Fig. 5). Kasolite is found

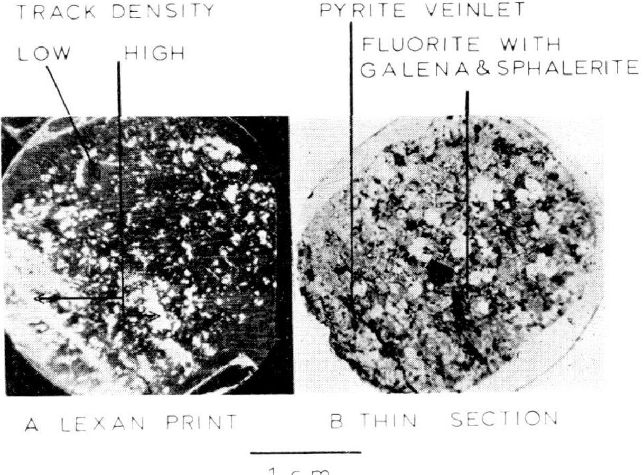

Fig. 5 Loose boulder of intermediate adamellite type from Caen burn, Helmsdale, cut by uraniferous veinlets containing pyrite, galena, sphalerite and fluorite

as a secondary expression of this assemblage on joint coatings. Fluorite is also recognized filling crosscutting shear zones in the granite.[6] Further enrichment of uranium is associated with extensive alteration and the formation of limonite in faults. This process is particularly well developed in a northwest–southeast fault zone which runs along the Allt an Dir to Allt na Muic. Samples from the fault zone comprise finely mylonitized feldspar and strained and fractured quartz in a limonitic matrix. The fission-track method shows clearly that the uranium enrichment in the fault zone is mainly disseminated and not associated with discrete uranium-bearing minerals. Analyses of clay fractions of the drill-core sludges from the fault are also relatively enriched in uranium (100–300 ppm).[6]

Uranium in sediments

The Orcadian province

The sediments in the Orcadian basin (Fig. 6) comprise a molasse facies sequence deposited during uplift and erosion of the Caledonian

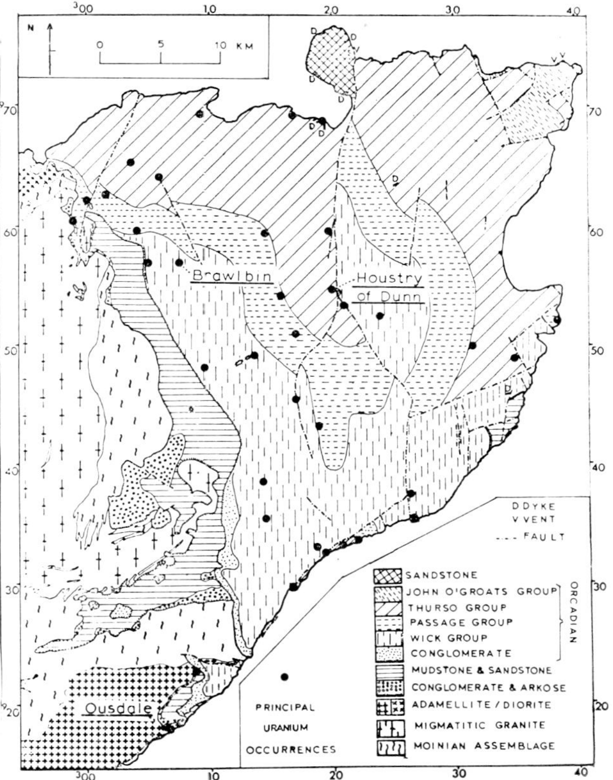

Fig. 6 Old Red Sandstone sediments in Caithness with principal uranium occurrences. After Gallagher et al.[6] *Localities discussed in this work are named*

orogen. Three main groups are recognized. There is a restricted development of Lower ORS at present exposed around the margins of the basin. This is separated by a marked angular unconformity from the overlying Middle ORS, which is apparently the dominant group in the basin. There is limited development of Upper ORS. Significant uranium anomalies are identified only in the two former groups of sediments, and examples taken from these groups are discussed later.

Deposition of the Lower ORS was contemporaneous with the initiation of major fault systems associated with uplift of the orogenic belt. The major northeast–southwest transcurrent faults, such as the Great Glen and Helmsdale fault systems, which involve shifts of the order of sixty-five[13] and fifty-five[12] miles, respectively, were initiated at this time, and sedimentation in the Lower ORS is thought to have occurred in fault-bounded intramontane basins. At the margins of the basins the sediments are comprised of screes and taluses associated with fault scarps. In the centre of these basins are well-developed sequences of mudstones and sandstones. The limited development of the Lower ORS in the Orcadian cuvette may be due either to a restriction in the initial sedimentation or to the subsequent removal of sediment prior to the Middle ORS deposition. In the Midland Valley of Scotland there is widespread andesitic volcanicity associated with this stage in the development of the orogen. Boulders of andesite in the basal Middle ORS are the only evidence for a volcanic episode of Lower ORS age in the Orcadian basin. Mineralization and possible fenitization[15] is, however, associated with the Great Glen fault adjacent to the basal unconformity of the ORS in the extreme southwest of the Orcadian basin.

In contrast with the Lower ORS, the Middle ORS was deposited in a basin stretching from the Moray Firth to Norway and east Greenland. Sedimentation was probably tectonically controlled in a rapidly sinking intermontane basin. There are two main facies—continuous red bed sedimentation around the margin of the basin with a grey facies in the central part consisting of approximately 20 000 ft of calcareous siltstones known as the Caithness Flags. These are thought to represent fluvio-lacustrine deposits laid down at a time when the balance between erosion and deposition was affected by both climatic and tectonic factors. The water level and the pattern of the well-developed sedimentary cycles discussed later are a result of the interaction of these two factors. The Upper ORS comprises red beds which developed across the whole basin and were accompanied by renewed volcanicity. Post Upper ORS faulting was the last major event: the Brough fault, which cuts Middle and Upper ORS sediments in the centre of the basin, is a good example of this.

Most of the rocks on the basin have a carbonate-rich cement and the water presently circulating in the basin is enriched in bicarbonate. In this environment uraninite is unstable and uranium so far seen is present in association with carbonate, phosphate and hydrocarbon and in a dispersed. state Mineral separation of uraniferous phases is, fore, extremely difficult. The examination of rocks in polished thin section in conjunction with the fission-track method, however, enables the relationships of the uranium to the host rock to be studied.

Three areas within the basin have been selected for particular attention in order to examine the relative importance of different geological factors in controlling the distribution of uranium. Knowledge of the controlling factors, particularly at low concentrations, may be helpful in explaining the occurrences and distribution of known uraniferous localities and in predicting where others may occur.

The fission-track method facilitates the evaluation of controls of uranium mineralization, several of which are considered here. Control by lithological features is examined in a series of interbedded mudstones and siltstones of a fluvio-lacustrine Middle ORS type which occur in the centre of the basin. Diagenetic processes involving bioturbation[16] and consolidation of the sediments are studied on the micro-scale and later cross-cutting carbonate veins associated with major basement faults are compared with structures of local origin. Arkoses and mudstones which were formed adjacent to major transcurrent faults active during Lower ORS sedimentation are examined in order to evaluate the relative importance of faulting and sedimentation in controlling the uranium distribution. The effect of recent groundwater may be superimposed on all these processes, and it is also considered here. Areas at Brawlbin, Houstry and Ousdale were selected for detailed examination.

Brawlbin—boreholes 3 and 4

The gamma logs and the variation of uranium as determined by the delayed neutron method on the core are presented in Figs. 7 and 8. The gamma log reflects the distribution of total activity due to potassium, uranium and thorium. In borehole 3 (Fig. 7) the background counts not attributable to uranium amount to approximately 4·5 μR/h, which would be accounted for by 2·8% eK_2O or 27 ppm $eThO_2$. The counts due to uranium are superimposed on this background and there is a tendency for peaks to correspond to intersections composed predominantly of mudstone (10 ppm) and troughs to intersections composed predominantly of siltstone (3 ppm). The upper 25 ft of the core is in disequilibrium and, therefore, probably leached. In borehole 4 (Fig. 8) the higher levels of uranium reached in the mudstone are more obviously indicated. The distribution of uranium in minerals and microstructures cannot be seen in cores and is best studied by the fission-track method.

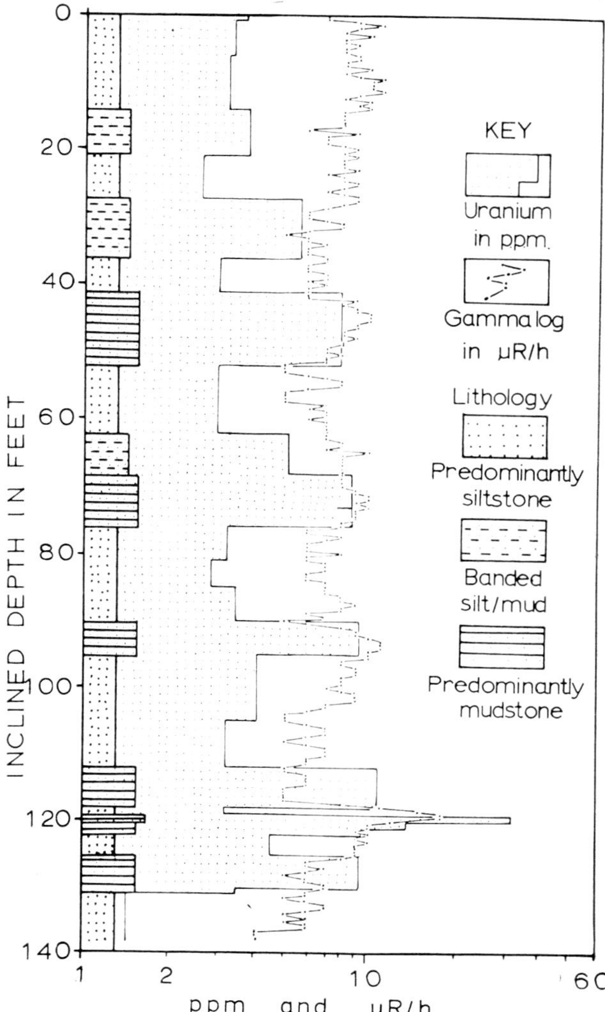

Fig. 7 Brawlbin diamond drill hole 3 in Middle ORS sediments. Uranium assay values obtained by delayed neutron method are presented together with results of gamma scintillometry. Highest uranium values correlate with mudstone horizons, intermediate values with mixed mud and silt, and lowest values with siltstone

The textures and compositions of the rocks in both boreholes are variable, ranging from finely laminated silty mudstone, with poorly sorted coarse and fine fractions, to coarsely banded silty mudstone in which silty lenses up to 3 mm thick occur. The silt is composed of angular fragments of quartz and some feldspar. The calcareous muds are brown in colour and nearly opaque in thin section. White mica occurs as flakes parallel to the banding. Marcasite is present as a fine-grained dissemination throughout the mudstone and siltstone. The crystals frequently occur in banded lenticular clusters with the banding parallel to the bedding. Graded bedding is also present with a

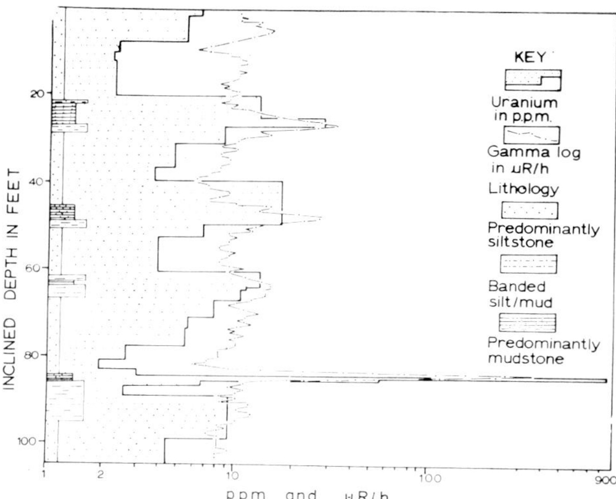

Fig. 8 Brawlbin diamond drill hole 4 in Middle ORS sediments. Uranium assay values obtained by delayed neutron method are presented together with results of gamma scintillometry. Highest uranium values at 85 ft downhole are due to uraniferous collophane and hydrocarbon pellets; elsewhere, generally higher values tend to occur in mudstone rather than siltstone

cycle scale of about 1 cm (Fig. 9).

Burrows are common, but they are distorted by differential compaction of the sediments. In some cases bioturbation is confined to a single bed, leav-

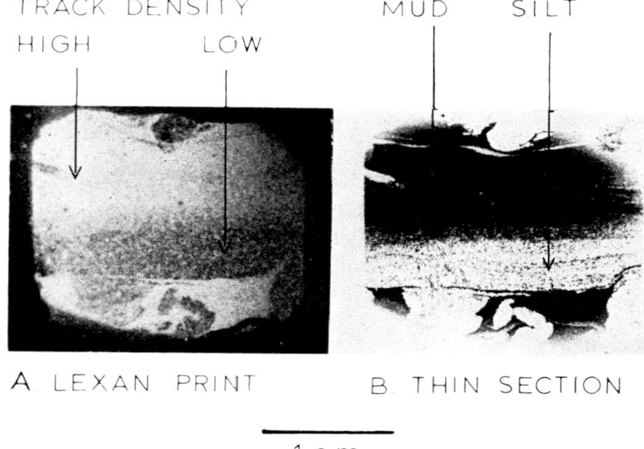

Fig. 9 Laminated and graded silty mudstone from diamond drill hole 3, Brawlbin, at 20 ft downhole. Fission-track distribution indicates that gradational change from silt to mud involves a corresponding gradational increase in uranium content. In disturbed areas at top and bottom of section uranium is similarly enriched in muddy fraction and depleted in silty fraction

ing interbanded mudstones undisturbed, whereas in other cases sedimentary features are obliterated. Calcite veins occur in post-consolidation shears and fractures which cut across the structures described above (Fig. 10). Higher levels of uranium occur in the muddy fraction of borehole 4 and are due to the presence of uraniferous collo-

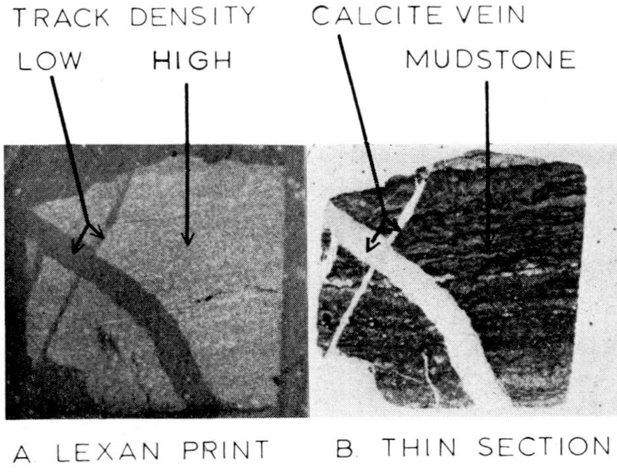

Fig. 10 Laminated mudstone from diamond drill hole 3, Brawlbin, 92 ft downhole, cut by calcite veins which have very low uranium contents compared with disseminated uranium in mudstone (10 ppm)

phane, hydrocarbon pellets and complex hydrocarbon–collophane lenses. Three of the mudstone horizons in borehole 4 contain from 10 to 30 ppm uranium. The fourth horizon at 85 ft contains 900 ppm uranium, which is related to the greater abundance of uraniferous hydrocarbon pellets and a higher level of uranium within the phosphatic matrix. In addition, a contribution from finely dispersed uranium minerals such as coffinite within the hydrocarbon is a possibility. Hydrocarbon pellets were always seen to contain more uranium than the phosphatic matrix.

In the micro-scale textures, for example, the graded bedding described above, uranium is concentrated in the mudstone and depleted in siltstone. The calcite veinlets have less than 0·1 ppm of uranium and have no effect on uranium distribution. All the evidence suggests that uranium was associated preferentially with the fine-grained muddy fraction at the time of deposition. Bioturbation does not alter the preferential association of uranium with the mudstone. This is consistent with the uranium distribution on a larger scale described above (Figs. 7 and 8). The joint application of the fission-track method and uranium analyses on split core therefore enable the time of entry of uranium into these rocks to be defined. The fission-track study indicates that the present distribution is the same as it was at the time of consolidation of the sediments and was not changed by later events.

Houstry of Dunn—boreholes 1 and 2

These drill-holes are sited near the line of the Brough fault in central Caithness, for which the last movement is post Upper ORS in age. The sediments at Houstry of Dunn are of the same general type as those at Brawlbin, the main difference being that the sedimentary features are destroyed in the fault zone by brecciation, which is associated with high uranium values (Fig. 11).

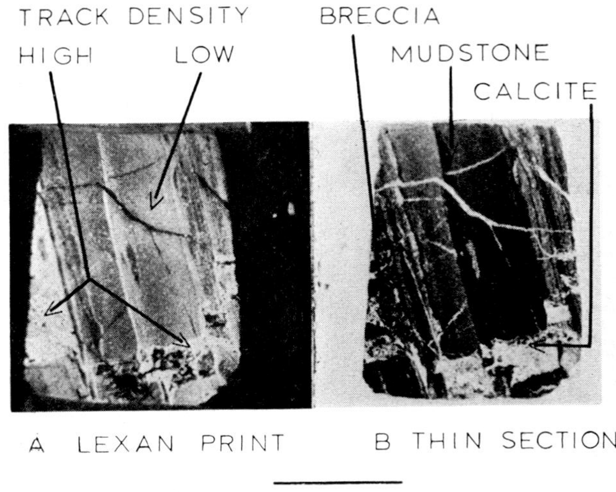

Fig. 11 Brecciated mudstone from diamond drill hole 3, Houstry of Dunn, in region of Brough fault. Undisturbed laminated mudstone in centre has much lower uranium contents than crosscutting calcite veins and brecciated mudstone

Calcite is commonly present in crosscutting veinlets and as the matrix between breccia fragments. Calcitic material is usually strongly uraniferous, in contrast to the background values in the sediments, and differs in this respect from the crosscutting calcite veinlets at Brawlbin. Some fragments in the fault breccia retain their sedimentary lamination undisturbed and in such instances the uranium content is low (5–10 ppm) and apparently related to the lamination. In the zone of maximum brecciation the sedimentary structures are obliterated and uranium is disseminated throughout the breccia at concentrations of up to 740 ppm. It is worth noting that where the Brough fault cuts Upper ORS sediments in which anomalous uranium occurrences have not been recorded no mineralization was detected.[6]

Ousdale

The Lower ORS in southern Caithness consists of the Ousdale arkose, overlain by the Ousdale mudstones. The latter are overlain unconformably and overstepped by the Middle ORS Badbea breccia. The lower reaches of the Ousdale burn follows a northwest–southeast-trending fault associated with the Helmsdale fault[12] discussed earlier. The main transcurrent movement on the Helmsdale fault system was probably in Lower ORS times and contemporaneous with the deposition of the Ousdale arkose. Radiometric anomalies occur in arkose in the Ousdale burn within 500 ft of the present shoreline[6] (Fig. 12). These anomalies are lensoidal in shape. The axis of elongation is in the same direction as the Ousdale burn and the corresponding fault. The anomalies tend to occur in sub-parallel groups which correspond with different horizons in the arkose. Samples from diamond drill core[6] and surface outcrop were examined in this study in order to ascertain the reason for the radiometric anomalies recorded in the arkose. Borehole 2 is 300 ft from the present shoreline and is collared in the burn. Borehole 1 is 50 ft from the shore and is collared in a small knoll at the side of the burn (Fig. 12). Other specimens examined came from the surface outcrop near these and other radiometric anomalies nearby.

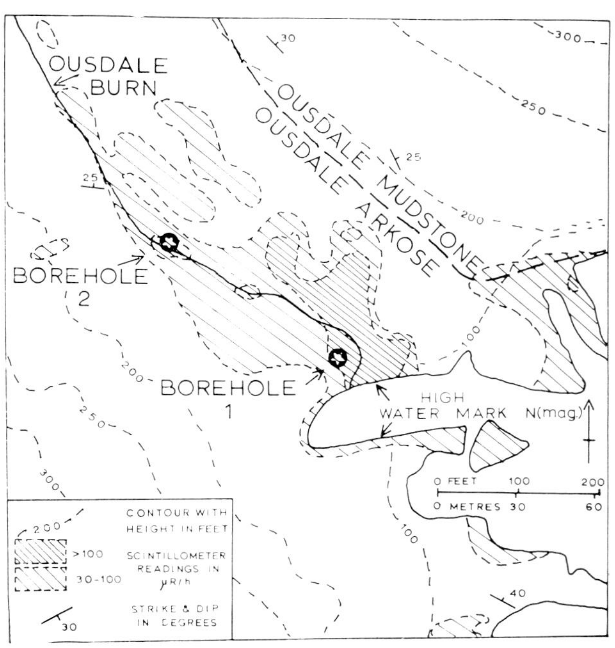

Fig. 12 Radiometric anomalies at Ousdale which have a northwest–southeast trend and intensify towards the coast. Uranium distribution in surface samples from this locality and diamond drill holes 1 and 2 (Fig. 13) is described in text. Map after Gallagher et al.[6]

The Ousdale arkose is composed of poorly sorted angular to rounded clasts of granite. Feldspar crystals exhibit fine-grained iron oxide staining on the periphery of the clasts—suggesting that some weathering occurred prior to the formation of the arkose. The more angular clasts are not appreciably altered and contain less uranium than is present in the fine-grained matrix of the arkose. In one example studied the altered rim of a granitic pebble is depleted in uranium compared with the interior. This suggests that some uranium was removed from the granite by weathering prior to the formation of the arkose. Lithological variations occur within the arkose where abundant small-scale sedimentary features such as graded bedding are present. There is preferential enrichment of uranium in fine-scale sedimentary features and relatively rich discrete bands are concordant with the bedding. Faulting is present in the arkose representing movement on the Helmsdale fault system. The localized mineralization associated with this movement is discussed later.

The overlying Ousdale 'mudstone' is a grey-green massive siltstone comprising, essentially, quartz and micaceous minerals aligned parallel to the bedding. Anomalous concentrations of uranium have not been detected in this formation and the uranium that is present is homogeneously distributed. One lenticle in a thin arkosic horizon within the Ousdale mudstone contains hydrocarbon-bearing uraninite. The uraninite is extremely fine-grained and dispersed through the hydrocarbon. It is associated with chalcopyrite, sphalerite and radian baryte.

In drill-hole 2 (Fig. 13) the base of the hole at 75 ft is in a fine-grained sandstone. This changes abruptly upwards into a siltstone and mudstone succession, 10 ft thick, comprising, essentially, a competent reddish siltstone crosscut by occasional pyritic veinlets. This grades upwards into a greenish carbonate-rich interbanded silty and muddy sequence. The upper portions of the section are greatly disturbed and convoluted. Later veins are also present. At the top of this sequence there

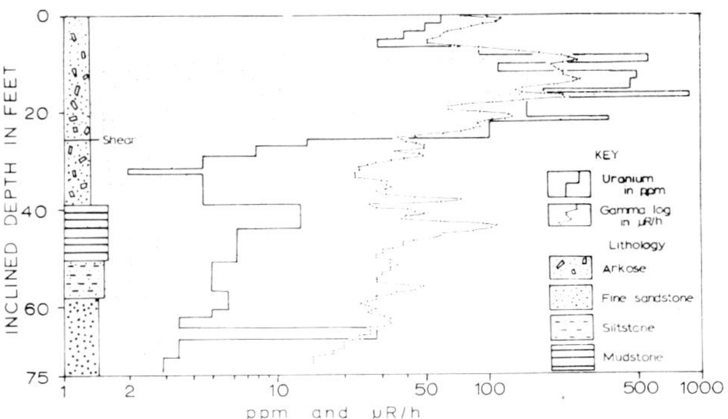

Fig. 13 Ousdale diamond drill hole 2 in Lower ORS sediments. Lower portion contains uranium in amounts comparable with those in Middle ORS sediments at Brawlbin (Figs. 7 and 8), whereas upper portion contains anomalous amounts in fine-grained matrix of the arkose. After Gallagher et al.[6]

is a laminated siltstone crosscut by pyritiferous veinlets. At this point the succession changes from the generally fine-grained and well-sorted sequence already described to one in which the upper 40 ft of the hole is of poorly or unsorted arkose. A minor sheared zone 25 ft from the top of the hole contains marcasite and goethite and cuts across a post-formational quartz veinlet.

The distribution of uranium in this drill core from uranium analyses is compared with the results of the fission-track study. From the base of the hole up to 25 ft below the surface the bulk uranium content varies between 3 and 30 ppm. In this intersection the uranium is closely controlled by sedimentological features, as shown by the fission-track distribution. The higher concentrations are present in the convoluted siltstones and mudstones, particularly their darker and clay-rich fractions. The small-scale crosscutting features do not contain uranium, and, generally, the uranium distribution is comparable with the Middle ORS occurrence at Brawlbin.

is leached. The shearing at 25 ft is associated with introduction of marcasite, goethite and a minor amount of uranium, which is confined to the shear zone. Sheared arkose in this zone has uranium mineralization, but competent arkose is not enriched (Fig. 14).

Drill-hole 1 is entirely in arkose, the uppermost 10 ft of which is leached. The remaining 25 ft of arkose has uranium contents in the region of 300 ppm. Uranium is strongly enriched in the fine-grained matrix, and this is due to some extent to disseminated fluorite and resistate minerals. A more general dissemination of uranium on grain boundaries is particularly well developed in small fine-grained sedimentary structures, as was also noted in borehole 2 and on surface exposure. Calcite and fluorite filled veinlets occasionally cut across the arkose, but most of the fluorite is disseminated.

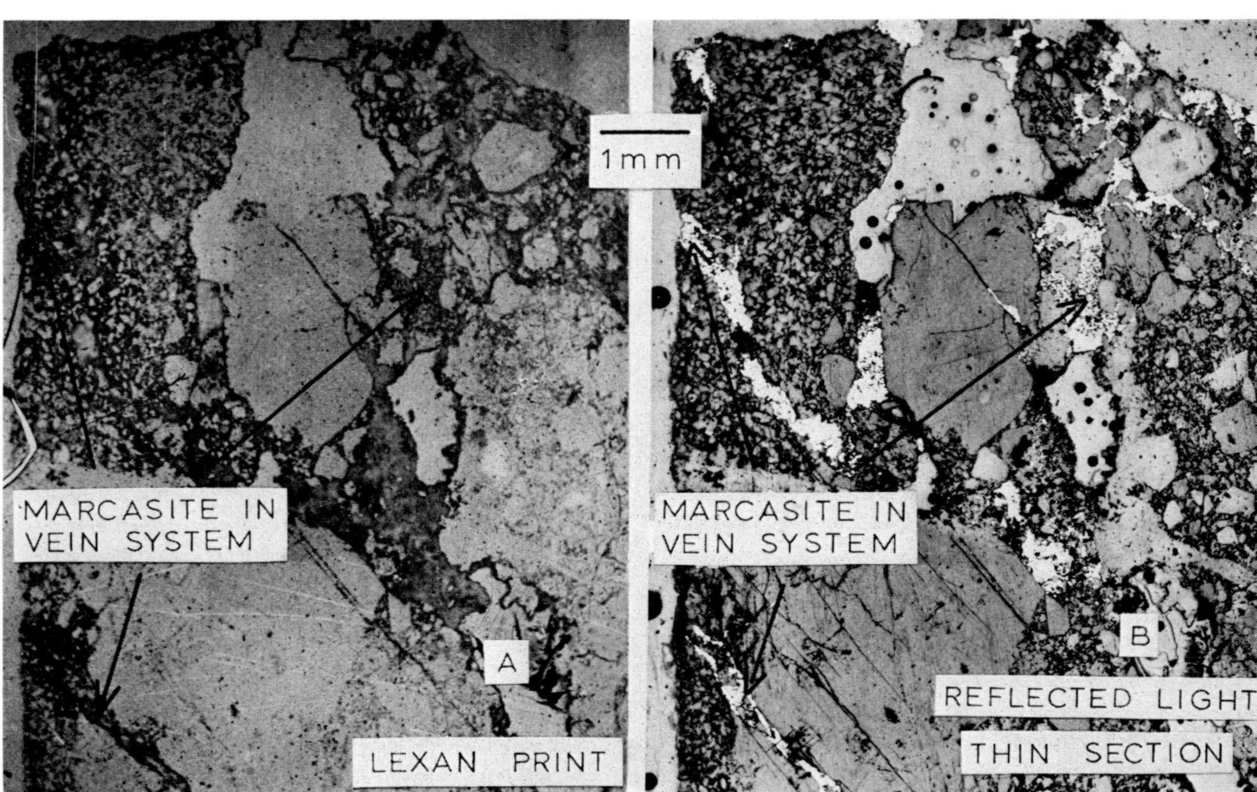

Fig. 14 Sheared arkose from 25 ft in diamond drill hole 2, Ousdale (Fig. 13). Shear zones are uraniferous and contain a system of marcasite veins, but competent arkose contains no uranium or sulphide mineralization

The maximum uranium concentration of 500 ppm is attained between 25 and 15 ft downhole. It is concentrated in the fine-grained groundmass between granitic clasts and pebbles. Disrupted fine-grained sedimentary structures in the arkose are enriched in uranium. Minor amounts of disseminated pyrite, chalcopyrite, covellite and hydrocarbon are present in the intersections with highest uranium content. The uppermost 10 ft of the hole

Oxidized samples from surface anomalies were examined by the fission-track method. In contrast with the arkose from boreholes 1 and 2, which is grey in colour, the surface samples are notable for red staining of the feldspar showing no enrichment in uranium. Cracks are occasionally present which traverse feldspar clasts and groundmass alike. Uranium is enriched in the groundmass compared with the phenocrysts, but it is also concentrated in these post-consolidation cracks in association with kaolinite. By use of the fission-track method detrital crystals of apatite can be located precisely, and in the oxidized zone these are partially or completely altered to torbernite and metatorbernite. The technique is useful, therefore, in assessing the extent to which uranium has been enriched or depleted after consolidation of the arkose.

Comparison of uranium distribution in Helmsdale granite and Ousdale arkose

To help to establish a working hypothesis on controls of mineralization in the Ousdale area it is necessary to evaluate the relative importance of the Helmsdale granite and Ousdale arkose, both of which are shown to be anomalous by surface gamma-scintillometry, regional geochemical reconnaissance and subsurface core sampling as well as in hand specimens collected at surface and in trenches.[6] The uraniferous zones in the arkose are composed mainly of fragments of the Helmsdale granite. It is pertinent to ascertain whether the uranium contents for this granite (Table 3), which are significantly higher than values for Aberdeenshire granites, are due to primary magmatic concentrations of uranium or to post-magmatic processes.

If magmatic enrichment is involved, it is possible that differences would exist between the partially altered outer porphyritic and inner fine-grained types of granite. As is shown in Table 3, however, they are very similar in uranium content in the range 5–9 ppm, but both types differ markedly from the strongly altered equivalents of these rock types which have uranium contents between 11 and 18 ppm. The lower values for partially altered granite of both types correspond with the values for the martitized Aberdeenshire granites.

A positive correlation between uranium content, shearing and alteration is observable both in the field and in thin section. Quartz is often strained, feldspar crystals are bent and biotite is altered to chlorite in the granite samples with the highest uranium content. An extreme case in this type of process is the formation of uraniferous limonite in a fault zone at Allt na Muic from which uranium contents of 100–300 ppm have been recorded in the clay fraction. Hematite is a ubiquitous accessory phase which partially replaces and in some cases pseudomorphs magnetite in most of the samples examined of both types of granite. This feature is also associated with uranium enrichment in the Aberdeenshire granites. An additional factor in the Helmsdale granite is the presence of late crosscutting fine-grained uraniferous vein systems which contain fluorite, pyrite and baryte.

In the Ousdale arkose, on the other hand, there is evidence for sedimentary control of uranium, which reaches anomalous levels in a coarse arkosic sequence where it is preferentially concentrated in finer-grained sedimentary facies. Superimposed on this, however, uranium occurs in narrow crosscutting veins or stockworks which are similar to the pyrite–fluorite vein systems cutting the granite. Disseminated mineralization in the arkose partly corresponds with the vein mineralization and is, therefore, likely to have been introduced in part during deposition of the arkose. The overlying Middle ORS Badbea Breccia is not mineralized, which suggests that introduction of the uranium had ceased in the Ousdale area by that time.[6]

This evidence suggests that uranium mineralization in the Helmsdale granite and Ousdale arkose is associated with fluorite–pyrite vein systems which probably represent mineralization in dilatant zones associated with transcurrent movement on the Helmsdale fault system[4]. Some local redistribution of uranium is related to later weathering processes.

Conclusions

The Lexan plastic fission-track method, combined with neutron activation analysis, provides a valuable means of distinguishing between levels of uranium in normal igneous or sedimentary rocks and those which have subsequently been mineralized. The granites of Aberdeenshire, with no known associated uranium mineralization, have about the clarke for uranium, except where there has been a development of hematite. The uranium distribution and content of the Criffel granodiorite is similar to granites from the Hill of Fare, but there is uranium enrichment associated with northwest–southeast tension faults contemporaneous with a major northeast–southwest coastal fault. This mineralization was introduced into the aureole of the granite and has previously been dated as Jurassic–Triassic.[10]

Uranium mineralization in the Helmsdale granite and Ousdale arkose is also associated with a major coastal fault system—the Helmsdale fault. Stockworks with a development of pyrite and fluorite are also important loci of uranium. There is no evidence of such deep-seated faults in the region of the Aberdeenshire granites studied, though some enrichment of uranium is associated with an oxidation process which also results in the martitization of magnetite and the development of chlorite and sericite. This type of enrichment is best developed in the Helmsdale and Criffel granites and is closely related to the intensity of the faulting.

In the Middle ORS sediments of the Orcadian

cuvette the fission-track and neutron activation methods enable the relative importance of sedimentary, diagenetic and later processes to be differentiated. The variation in syngenetic uranium with lithology is clearly illustrated in Figs. 7 and 9. There is enrichment of uranium—probably also syngenetic or perhaps diagenetic—in association with phosphatic horizons and with hydrocarbon. The main development of uranium, however, up to levels of the order of 1000 ppm is epigenetic and fault-controlled, and is commonly associated with the introduction of calcite.

Similar processes are observed in the Lower ORS of Ousdale, the main controlling feature being tension faults associated with the Helmsdale fault system. The original pattern of uranium distribution in the Ousdale area is masked by later weathering processes, which have resulted in the formation of secondary uranium minerals, and there has been radium enrichment to a depth of 10 or more feet below surface.

Acknowledgment

This paper is published by permission of the Director, Institute of Geological Sciences.

References

1. Ostle, D. Coleman, R. F. and Ball, T. K. Neutron activation analysis as an aid to geochemical prospecting for uranium. In *Uranium prospecting handbook* (London: IMM, 1972), 95–107.
2. Amiel, S. Analytical applications of delayed neutron emission in fissionable elements. *Analyt. Chem.*, **34**, 1962, 1683–92.
3. Bowie, S. H. U. Autoradiography. In *Physical methods in determinative mineralogy* Zussman, J. ed. (London and New York: Academic Press, 1967), 467–73.
4. Price, P. B. and Walker, R. M. A simple method of measuring low uranium concentrations in natural crystals. *Appl. Phys. Lett.*, **2**, 1963, 23–5.
5. Kleeman, J. D. and Lovering, J. F. Uranium distribution studies by fission track registration in Lexan plastic prints. *Atomic Energ. Aust.*, **10**, Oct. 1967, 3–8.
6. Gallagher, M. J. *et al.* New evidence of uranium and other mineralization in Scotland. *Trans. Instn Min. Metall. (Sect. B: Appl. earth sci.)*, **80**, 1971, B150–73.
7. Read, H. H. The geology of central Sutherland (explanation of sheets 108 and 109). *Mem. geol. Surv. Scotland*, 1931, 238 p.
8. Wedepohl, K. H. *Handbook of geochemistry* (Berlin, etc.: Springer, 1969).
9. Phillips, W. J. The Criffel–Dalbeattie granodiorite complex. *Q. Jl geol. Soc. Lond.*, **112**, 1956, 221–38.
10. Miller, J. M. and Taylor, K. Uranium mineralization near Dalbeattie, Kircudbrightshire. *Bull. geol. Surv. Gt Br.* no. 25, 1966, 1–18.
11. Phemister, J. In reference 7, pp. 193–6.
12. Flinn, D. A geological interpretation of the aeromagnetic maps of the continental shelf around Orkney and Shetland. *Geol. J.*, **6**, 1969, 279–92.
13. Kennedy, W. Q. The Great Glen Fault. *Q. Jl geol. Soc. Lond.*, **102**, 1946, 41–76.
14. Godard, A. *Recherches de géomorphologie en Ecosse du nord-ouest* (Strasbourg: Publications de la Faculté des Lettres de l'Université, 1965), 701 p.
15. Deans, T. Garson, M. S. and Coats, J. S. Fenite-type soda metasomatism in the Great Glen, Scotland. *Nature, Lond., phys. Sci.*, **234**, 1971, 145–7.
16. Burollet, P. F. Bryramjee R. and Couppey, C. Contribution à l'étude sédimentologique des terrains Dévoniens du nord-est de l'Ecosse. *Notes Mem. Comp. franc. Petrol.* no. 9, 1969, 83 p.

543.53:553.411.3

Application of neutron activation analysis to the evaluation of placer gold concentrations

Jane Plant, B.SC.

Geochemical Division, Institute of Geological Sciences, London, England

R. F. Coleman, B.SC., A.R.I.C.

Atomic Weapons Research Establishment, Aldermaston, Berkshire

Synopsis

A method of neutron activation analysis for gold in large samples of geologic materials is described in relation to the economic assessment of the valley deposits in the Strath of Kildonan, Sutherland.

The gold content of samples consisting of up to 500 g of unground material is measured by neutron activation analysis by use of the reaction $^{197}Au\ (n, \gamma)\ ^{198}Au$. Samples are irradiated in the thermal column of the Herald Reactor at Aldermaston, in which the thermal flux is 2×10^{10} $n\ sec^{-1}\ cm^{-2}$. Known weights of sample are irradiated for 20 min in sealed polyethylene containers, to which a small gold disc is attached to monitor the neutron flux. After irradiation, the samples are stored for six days. They are then divided into portions not exceeding 130 g in weight and are transferred to 8-cm diameter polyethylene containers. The area under the 0·41 MeV ^{198}Au photo peak on the resulting gamma spectrum is compared with that of a standard sample over a counting time of 10 min. All results are corrected for variations in neutron flux.

The material investigated in the Strath of Kildonan consists of alluvium and fluvioglacial detritus, which infills a glacially overdeepened section of the Helmsdale River Valley. Orientation studies in the tributary drainage of the area indicated that native gold occurs in particles of up to 2 mm in diameter: the concentration of gold in the alluvium is approximately 0·1 ppm; the estimated mean density of the gold is 16·5. Substitution of these values into Gy's sampling formula indicates that the accurate estimation of the gold content in the Strath of Kildonan sediments is highly dependent on analysis of an adequately large sample. The neutron activation method enables up to 500 g of material to be analysed non-destructively at a cost of £2–£3 per sample. Sample preparation is minimal and cost is, therefore, independent of sample size and productivity is high (20 samples per man-day). The combined sampling and analytical reproducibility compares favourably with that attained by atomic absorption spectrophotometric or fire assay methods, which involve the analysis of only 20–100 g of material. No systematic bias occurs in relation to fire assay analysis, and self-absorption of neutrons by gold grains is insignificant for individual particles of less than 40 mg in weight. It is specific, and, hence, fineness corrections essential in rapid fire assay methods are not required.

The limit of detection of the method is 0·01 ppm on 500 g of material, irrespective of small variations in weight. In the Strath of Kildonan investigation determinations were carried out on heavy mineral concentrates obtained from the beneficiation of samples of 360 kg in initial weight. Thus, a limit of detection of better than 0·005 ppm was obtained. The incorporation of the analytical technique into a general procedure, which permitted calculation of the local gold content and classification of the type and source of the unconsolidated sediments, is described.

The application of a neutron activation method of gold analysis to the evaluation of placer gold occurrences is described with particular reference to the Strath of Kildonan, Sutherland, Scotland. The material investigated consists of alluvium and fluvioglacial detritus, which infills a glacially overdeepened section of the Helmsdale River Valley. Tributaries which drain from the north-west of the area investigated—particularly the Kildonan and Suisgill burns—formed the sites of small nineteenth century gold workings.

During the regional geochemical survey of Scotland currently being carried out by the Geochemical Division of the Institute of Geological Sciences the area of anomalous gold values was shown to be greater than had hitherto been known. Follow-up work indicated that concentrations of gold are closely associated with moraine left by the Pleistocene ice sheet and that they are most probably derived from local primary sources. The main objective of the investigation was to make an estimate of the gold distribution and grades in the unconsolidated sediments of the Helmsdale Valley. In addition, the variation of gold content with sediment type and origin was studied, the information obtained being used to predict possible extensions of the known occurrences.

In order to make an accurate estimate of the gold content of different types of sediment it is critically important to obtain a representative sample for analysis; errors inherent in the analytical methods currently available for gold determination are much less significant. In sampling and sub-sampling for gold in placer concentrations the problems encountered are mainly related to the occurrence of gold as particles of native metal in a barren matrix. Native gold has a high specific gravity (13–19) relative to that of the minerals with which it is normally associated in placer deposits. Furthermore, it occurs as flakes, grains and nuggets, and, hence, segregation of gold in the host matrix, occurring naturally as well as during normal bulk-sampling procedures, is highly complex. The malleability of gold limits the application of grinding methods normally used to homogenize samples for chemical analysis. In a discussion on the problems associated with the recovery of a bulk sample Miesch[1] has indicated that serious problems arise if gold grains are larger than 0·015 mm at a mean concentration of 1 ppb. The effect of the size of gold particles on the amount of sample which it is necessary to analyse has also been noted by Pardee,[2] who indicated that the uneven distribution of gold grains at concentrations equivalent to about 2 ppm may make the analysis of less than 30 g of sample material an unreliable indication of tenor.

The neutron activation method described here was devised principally to ensure that a sample of adequate size could be analysed quickly and at relatively low cost. The capability of analysing large unground samples of up to 500 g minimizes preparation costs and reduces the number of samples which must be analysed in order to arrive at a reliable value for the mean gold content. For large samples the method compares favourably with fire assay, and it is superior to rapid atomic absorption methods in which problems of grinding, roasting, extraction and the disposal of toxic reagents increase proportionally with the sample size.

Analytical procedure

The gold content of samples can be measured by neutron activation by use of the reaction ^{197}Au $(n, \gamma)^{198}Au$. Samples are irradiated in the thermal column of the Herald Reactor at Aldermaston, in which the thermal flux is 2×10^{10} n sec^{-1} cm^{-2}. Known weights of sample (up to 500 g) are irradiated for 20 min in sealed 8 cm × 7 cm cylindrical polyethylene containers, and a small gold foil disc is attached to each sample to monitor the neutron flux. After irradiation, the samples are stored for six days. They are then divided into

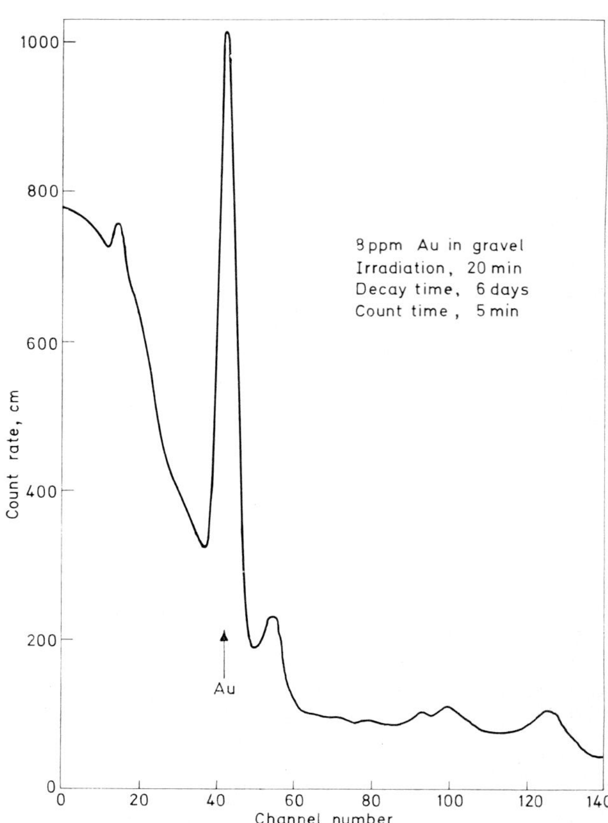

Fig. 1 Gamma spectrum

portions not exceeding 130 g in weight and transferred to 8 cm × 2·5 cm polyethylene containers. The induced activity is determined by gamma spectrometry with the container 8 cm from a sodium iodide detector. With these irradiation and counting conditions the photo peak from the 0·41-MeV gamma ray emitted by ^{198}Au can readily be distinguished in the gamma spectrum (Fig. 1). The activity is compared with that of a standard sample consisting of sand of negligible gold content to which a known quantity of gold solution is added. The sample and standard are irradiated and counted under identical conditions and all results are corrected for variations in the neutron flux.

Interference from other elements in a sand matrix is unlikely to affect the estimation of gold significantly when the conditions outlined above are used. For large samples, however, variations in the neutron flux through the sample and the varying solid angle subtended by different parts of the sample at the detector could give rise to errors. Also, as gold has a high neutron cross-section the effect of the grain size of the gold must be considered.

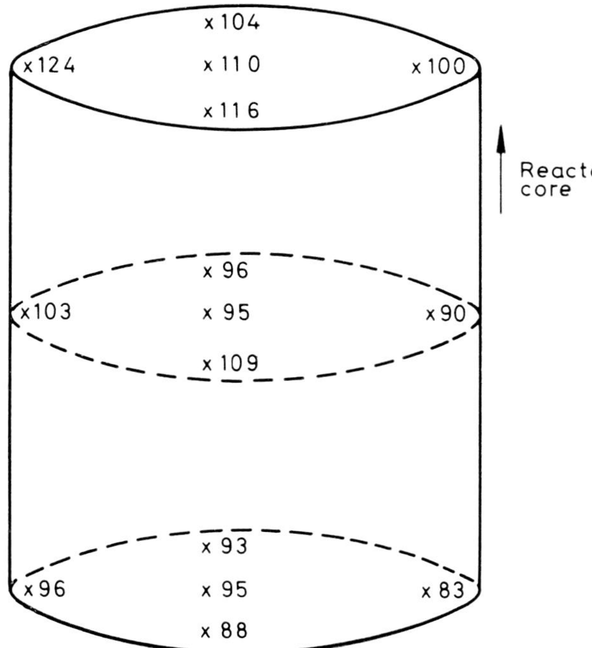

Fig. 2 Relative neutron flux in 450-g gravel sample

The variation in neutron flux through the sample was studied on a sample of 450 g of gravel in which small discs of gold foil were distributed in a regular pattern. After irradiation, the specific activities of the foils were determined and the relative neutron flux of various positions in the sample is shown in Fig. 2. Although the variation in flux is quite large, it has no effect on the analytical method if the sample and standard are homogeneous and have the same volume. In this study the standard is homogeneous, but the sample contains discrete gold particles. The maximum error (25%) would arise if the sample contained only a single particle of gold situated in the region of highest flux during irradiation. As the number of gold particles increases, the error will decrease, tending towards zero as the sample approaches homogeneity. Similarly, the variation in solid angle subtended by different parts of the sample at the detector does not introduce an error when the sample and standard have identical properties. For samples containing particulate gold the error is minimized by reducing the depth of sample by transferring it to a shallower container 2·5 cm high prior to counting. The maximum error under these conditions would amount to 30% if all the gold in the sample occurred in a single particle and were situated close to the wall of the container.

Table 1 Specific activity of gold grains of varying weight

Weight of gold spheres, mg	Specific activity, counts/min/mg
2·3	4720
5·4	4980
6·3	4730
18·4	5030
23·4	4860
35·2	5060

The effect of particle size on induced specific activity was investigated for gold particles up to 2 mm across, which is the size at which, in practice, samples were screened prior to gold analysis. This size limit was selected on the basis of field orientation studies described later. The maximum weight of a single gold grain of −2-mm size was estimated to be less than 40 mg, and determination of induced specific activity was carried out on small spheres of gold ranging from 2 to 35 mg. The results obtained (Table 1) indicate that the specific activity does not decrease with increasing particle size for particles weighing less than 40 mg, and no correction for self-absorption of neutrons is required if a well-thermalized flux is employed.

Precision and sensitivity

The analytical error on a determination at a concentration of 1 ppm is ±20% at the 95% confidence level, due mainly to variations in irradiation and counting geometry. This increases to ±50% near the limit of detection, which is normally 0·1 ppm, but which could be lowered to 0·01 ppm by modification of the counting and irradiation geometry.

The combined error which results from sampling and analysis is a more meaningful parameter than analytical error in assessing the significance of gold analyses, and the former was investigated on three unground 500-g sub-samples from a

Table 2 Effect of sub-sampling weight and analytical technique on measurement of gold in gravel

Technique	Atomic absorption	Fire assay	Activation analysis	Jigging, sorting and activation analysis
Sub-sample weight, g	20	100	500	1800
Gold analysis, ppm	0·015	1·8	6	11
	114	0·26	8	
	0·3	33·9	19	
	0·2	17·9		
	11·2	1·1		
	0·2	1·3		
	3·0	2·5		
	0·9	8·2		
	0·5	7·8		
Range	0·015–114	0·26–33·9	6–19	
Mean	14·5	8·3	11	11

Replicate splits of gravel passing through 0·25-cm aperture screen.

sample of alluvium. The results obtained are compared with analyses by atomic absorption and by fire assay on 9 ground sub-samples of 20 and 100 g, respectively (Table 2). The agreement between the means for the different analytical methods indicates that there is no significant bias in any of the techniques. The results on different sub-samples prepared for analysis by fire assay and atomic absorption vary widely from one another. This variation is related to the size of samples analysed and demonstrates the necessity of performing many replicate determinations when a technique suitable for only small samples is used.

Accuracy

The gold content of eight samples of alluvium was determined by neutron activation and by fire assay (Table 3). There is no systematic bias between the two methods throughout the concentration range examined, and the agreement between the two sets of values is good. A standard of 350 g of material from Cortez mine in the U.S.A., supplied by kind permission of Dr. A. L. Mather of Geochemical and Mineralogical Laboratories Pty., Australia, was also analysed. The preferred value for this standard is 52 ppm by fire assay and 53·5 ppm by atomic absorption. A value of 56 ppm was obtained by neutron activation analysis.

Table 3 Comparison of fire assay and neutron activation analysis results

Gold content, ppm Fire assay	Neutron activation analysis
0·09	0·10
0·26	0·21
0·13	0·19
0·21	0·2
0·83	0·62
0·18	0·2
4·43	3·75
11·5	13·3

Productivity and costs

In this study gold determinations were performed at a rate of 20 samples per man-day at a cost of £2–£3 per sample, independent of sample size. If a sufficiently large number of samples were analysed, automation of the method would result in a considerable reduction in the cost of individual determinations.

Application of the method to Helmsdale Valley alluvium and fluvioglacial material

A preliminary study of the origin of the gold in the Helmsdale Valley area suggests that granite veins associated with a Caledonian injection complex—the Strath Halladale granite of Read[4]—are the primary source of gold. Bedrock in the area is deeply weathered and overlain by ablation moraine and fluvioglacial detritus, and it seems likely that natural beneficiation has occurred as a result of weathering processes during an earlier climatic regime. Further reworking and concentration of the gold is attributable to the

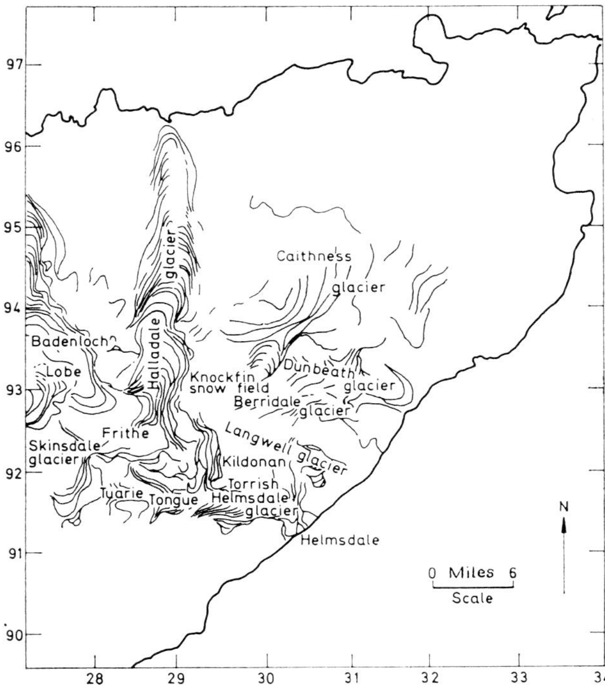

Fig. 3 Successive margins of ice in Helmsdale and Caithness area

melting and retreat of the Pleistocene ice sheet. There is evidence of two ice sheets in the area during the late stages of the Pleistocene glaciation—a local one known as the Knockfin snow field, and the Helmsdale glacier. These joined a major ice sheet to the north and west (Fig. 3).[5]

Orientation and sampling

The errors associated with sample size for a gold occurrence can be estimated[3] if the density, mean concentration and maximum diameter of gold particles are known (see Appendix). Orientation work was therefore carried out to obtain an estimate of these and other parameters characteristic of the occurrences. Several heavy mineral concentrates were collected by panning from tributaries joining the Helmsdale River from the north and northwest. The mean concentration of gold in the concentrates was determined and an estimate of the mean concentration in the original alluvium was also made. A mineralogical examination was performed on each sample. The maximum grain size of gold particles was determined in a composite panning concentrate from the Suisgill area, where the coarsest gold occurs, and electron-microprobe analyses of several grains from the area were carried out to determine fineness* and, hence, mean density. The results of the orientation work are summarized in Table 4. Gold was found to occur only as monomineralic grains, although a few particles contained small inclusions of quartz or feldspar.

*Fineness is a measure of the gold and silver content given by 1000(Au/Au+Ag).

Table 4 Determinations on panning concentrates of 20 g from initial samples of 20 kg

Concentration, a × 10²			
Number of samples			25
Approximate expectation value in panning concentrates assuming approximation to Poisson distribution			100 ppm
Arithmetic mean			115 ppm
Range			0–400 ppm
Estimated concentration in stream alluvium			0·1 ppm
Size analysis, d			
Sieve number, B.S.I.	mm	% retained	
>10	>1·680	—	
−10 +30	−1·680 +0·5	30	
−30 +60	−0·500 +0·252	69	
−60 +85	−0·252 +0·177	1	
−85	−0·177	—	
Fineness of gold			
To obtain approximate density (*r*) determined by electron-microprobe analysis on grain cores.*			
Number of samples		20	
Mean		805	
Standard deviation		73	
Range		672–911	
Approximate density		16·5	

*Analysts, A. Gough and C. M. Rice

If the values obtained are substituted in the formula given by Gy,[3] the minimum sampling error* associated with different sample weights of concentrates and untreated fine alluvium can be ascertained. The approximate minimum sampling error in relation to concentration and sample size is indicated in Table 5. The large sampling error is related to the coarse grain size and the low concentration of the gold in the Strath of Kildonan occurrence; and analysis of a subsample of up to 500 g of this material involves an error of > ±1000% at the 95% confidence level (Table 5). In order to reduce sub-sampling error concentrates obtained in a quantitative way from the initial sample were analysed and the concentration in the initial sample was obtained by

*Gy's formula enables only the error associated with an equiprobable sampling model to be determined. It does not include errors related to bias, which are important in sampling for gold in placer concentrations.

Table 5 Effect of sample weight and gold concentration on sampling error

Sample weight	Minimum sampling error (1 standard deviation as percentage) Au concentration, ppm 100	10	1	0·1
50 g	80	250	800	2,500
100 g	57	180	570	1,800
500 g	25	80	250	800
1 kg	18	57	180	570
10 kg	6	18	57	180
100 kg	2	6	18	57
500 kg	1	3	8	25

Based on Gy's sampling equation[3] (see Appendix). From orientation work d=0·17 cm and r=16·5.

recalculation. The two-stage sampling procedure adopted consisted of size fractionation to remove all material greater than coarse sand in size followed by jigging to obtain a heavy mineral concentrate. Size fractionation is effective in concentrating gold, which, because of its high specific gravity, occurs in smaller particles than most associated mineral grains. It was particularly effective in this investigation because of the marked grading of the morainal sediments and a concentration factor, ranging from ×1 in fine lake sediments to > ×5 in coarse moraine, was obtained by sieving. The sample treatment is summarized in Fig. 4. The analytical results

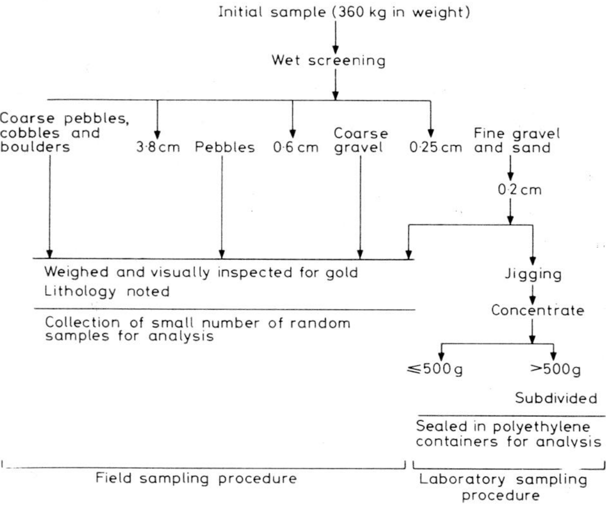

Fig. 4 Sample preparation procedure

obtained for concentrates up to 500 g in weight were used to calculate the concentration of gold in the initial sample, giving a concentration factor > ×720. Thus, a limit of detection of less than 0·005 ppm was obtained with a limit of detection of 0·1 ppm for the concentrates. A corresponding improvement in sampling and analytical precision is also achieved by the beneficiation procedure described since sub-sampling errors are largely eliminated. From Gy's formula it is estimated that the error on a single determination for gold is reduced by a factor of 30 from 1 standard deviation = 800% for 500 g of untreated alluvium to 1 standard deviation = 25% for 500 g of concentrate: this is equivalent to analysing the total initial sample of 360 kg. Recovery was assumed to be complete since neutron activation analyses of the tailings from auriferous samples detected no gold.

Preliminary assessment of gold distribution

The investigation involved test drilling to indicate gold distribution and grades in the vicinity of four cross-traverses of the Helmsdale Valley in a 10-mile section downstream from Kinbrace (Fig. 5). Samples were obtained from 13 boreholes by use of a one-ton Pilcon Wayfarer shell and auger drilling system, which drives casing 0·3–0·25 m in diameter into the ground 0·6 m ahead of a shell baler. The latter is used to recover all material finer than boulders, which are broken by means of a power-driven chisel. The samples obtained by this method consisted of approximately 360 kg of disturbed sediment comprising boulders, cobbles, coarse gravel and sand. In order to determine whether gold was being systematically displaced downwards during drilling, approximately 0·25 m of bedrock was broken with the chisel and recovered together with superficial sediment at the base of each drill-hole.

The material in the drill-holes was classified and logged in the field. The variation of gold content with different types of sediment is shown for the drill-holes on traverse 2 (Fig. 6). Boreholes 6 and 7 intersected sediments interpreted as a lacustrine sequence which are thought to have been derived from ablation of the local ice sheet. Borehole 5 intersected barren moraine deposited by the Helmsdale Valley glacier. In boreholes 6 and 7 five layers can be distinguished: (1) recent alluvium and soil; (2) a succession of varved clay and interbedded sand containing leaves and organic debris (these are interpreted as proglacial lake deposits); (3) a layer of coarse detritus thought to have been flushed into the lake during a period of maximum decay of the ice; (4) a succession comparable to layer (2); and (5) a basal layer of moraine.

The association of gold with ground moraine

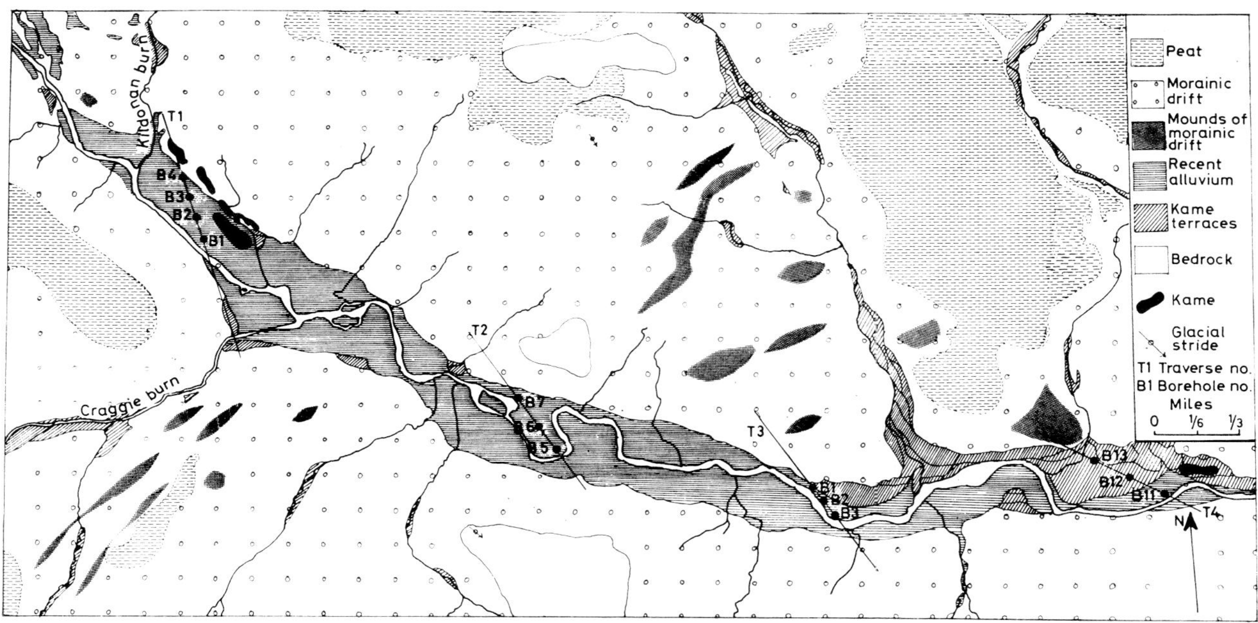

Fig. 5 Distribution of superficial deposits in the Strath of Kildonan

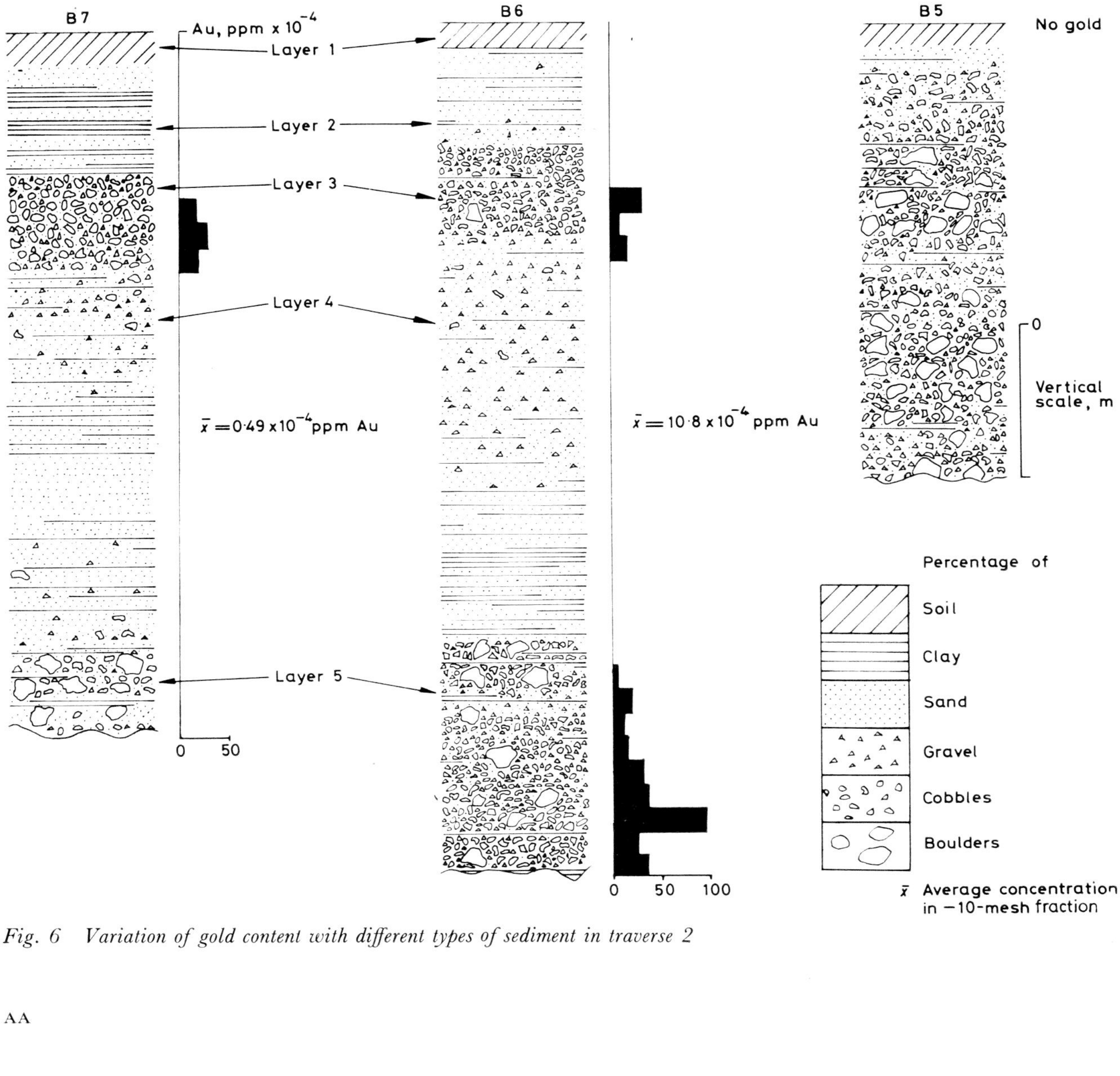

Fig. 6 Variation of gold content with different types of sediment in traverse 2

and coarse fluvioglacial detritus derived from the local Knockfin snow field is readily identified. The good agreement between the concentration of gold and the grain size of sediment in the lacustrine sequence suggests that the gold concentrations are related to hydraulic factors operative during the decay of the Pleistocene ice. The marked agreement also suggests that the downward displacement of gold during drilling was not a significant source of error in this investigation.

On the basis of the investigation described here the sediments were shown to be sub-economic. An analysis of the variation of gold in relation to the late glacial history of the Helmsdale area suggested, however, that higher concentrations of gold might be expected to occur on the north bank of the valley. Preliminary work in this area has confirmed this deduction, although the grade indicated for the morainal material is also sub-economic. It should be noted, however, that the gold in the moraine and the sediments is easily upgraded by size fractionation, and this matter might merit further study.

Conclusions

The principal advantage of the neutron activation method for the evaluation of placer gold concentrations is that it allows large samples to be analysed and, hence, gives more representative results than can be obtained by atomic absorption or fire assay methods, which involve the analysis of only 20–100 g of material.

Other features of the method important in its application to the evaluation of placer gold concentrations are: (1) it is rapid and simple—samples do not have to be ground or roasted prior to analysis and, in general, sample preparation is minimal, so cost is independent of sample size; (2) it shows no systematic bias when compared with fire assay or atomic absorption determinations of gold occurring in particles less than 40 mg in weight; (3) it is specific—fineness corrections essential in rapid fire assay methods are, therefore, not required; (4) it is non-destructive, and since unground samples are analysed, mineralogical studies of gold grains can be performed on selected samples after activation analysis, in addition to such other chemical determinations as may be necessary; (5) with minor modifications to the method it is possible to carry out the entire analysis without breaking the seal on the sample container, thereby eliminating loss or addition of gold in transit; and (6) a limit of detection of 0·1–0·01 ppm can be improved by upgrading the initial sample by size fractionation or concentration of heavy minerals.

Acknowledgment

The authors thank Dr. S. H. U. Bowie, Mr. D. Ostle and Mr. P. J. Moore for helpful discussions and comments on the manuscript and Mr. D. Briggs, Mr. F. Stacey and Mr. A. B. St. John Webb for technical support. The paper is published by permission of the Director, Institute of Geological Sciences, London.

References

1. Miesch, A. T. Chart showing number of particles of metal expected per 5-pound sample for various levels of grade and average particle size. *Open-file Rep. U.S. geol. Surv.* 1967.
2. Pardee, J. T. Beach placers of the Oregon coast. *Circ. U.S. geol. Surv.* 8, 1934, 41 p.
3. Gy, P. Poids à donner à un échantillon. *Revue Ind. minér.*, **38**, 1956, 53–92.
4. Read, H. H. The geology of central Sutherland (Explanation of sheets 108 and 109). *Mem. geol. Surv. Scotland*, 1931, 238 p.
5. Charlesworth, J. K. The late-glacial history of the Highlands and Islands of Scotland. *Trans. R. Soc. Edinb.*, **62**, 1956, 769–928.

Appendix

Gy's basic sampling equation

$$M = \frac{Cd^3}{s^2}$$

M is weight of sample required, g, C is a sampling constant for the particular material to be sampled—this is composed of four parameters characteristic of the material: f, g, l and c (all are dimensionless, except c, which is expressed in grammes per cubic centimetre).

$$C = f \times g \times l \times c$$

General situation	*Value for placer gold concentrations*
f=shape factor	0·2
g=particle-size distribution factor	0·2
l=liberation factor	1
c=mineralogical composition factor	r/a, where r is density of gold, a is assay of the sample, expressed as a decimal, and d is the dimension of the screen which retains about 5% of the grains of gold, cm

s is standard deviation of the results of gold determination as a decimal.

For placer gold the formula is

$$M = \frac{0{\cdot}04 \times r/a + d^3}{s^2}$$

543.422.2:546.682/3:550.42

Atomic absorption determination of thallium and indium in geologic materials

Arthur E. Hubert

Hubert W. Lakin

Both of the U.S. Geological Survey, Denver, Colorado, U.S.A.

Synopsis

Crustal abundance is estimated to be 0·3–1·3 ppm for thallium and 0·1–0·25 ppm for indium. The usual semi-quantitative emission spectrographic detection limit is 50 ppm for thallium and 10 ppm for indium; consequently, these elements are rarely detected by spectrographic analysis and their geochemistry is poorly known.

The atomic absorption method described here provides a sensitivity of 0·2 ppm for thallium and indium in geologic materials. The more detailed geochemical study made possible by this method may prove useful in geochemical exploration. The procedure is indicated below.

An oxidized sample is covered with hydrofluoric acid. The acid is slowly evaporated to dryness. The sample residue is dissolved in a solution of bromine in hydrobromic acid, heated to dispel the excess bromine, and diluted with water. The cool solution is shaken with methyl isobutyl ketone to extract thallium and indium. Interference of iron is minimized by extracting the ketone with 1·5 N hydrobromic acid. Thallium and indium are then measured by aspiration into an air–acetylene flame by use of appropriate cathode lamps and Perkin-Elmer 403 atomic absorption instrument. Fifty samples can be analysed per man-day for the two elements.

Organic-rich samples are ashed in a porcelain dish over a Bunsen flame before the hydrofluoric acid attack.

Sulphides are oxidized with a carbon tetrachloride–bromine–nitric acid mixture, evaporated to dryness, and dissolved in a bromine–hydrobromic acid solution.

The minor elements that occur in sulphide minerals may be more definitive in the study of weathered haloes of orebodies than the major ore metals. It has been observed that iron and manganese oxide precipitates which form cemented conglomerates in streams draining pyritized rocks contained as much copper, zinc and lead as similar precipitates from active mine drainages. Distinguishing mere solution by sulphuric acid of these metals in barren rocks from solution of their sulphides in ore requires another guide. Guides may be provided by such elements as indium and cadmium, which are enriched in sphalerite, and by thallium, silver, bismuth and antimony, which are enriched in galena.

The crustal abundance is estimated to be 0·3–1·3 ppm for thallium and 0·1–0·25 ppm for indium.[3] Semiquantitative emission spectrographic methods commonly used in geochemical exploration studies have a lower detection limit for thallium of 50 ppm and for indium of 10 ppm. Consequently, these elements are rarely detected in nature and their geochemistry is poorly known.

The atomic absorption method described here provides a relatively fast, low-cost and simple procedure capable of detecting thallium and indium to a lower limit of 0·2 ppm. Thus, studies of their geochemistry should be facilitated, and knowledge of their distribution may prove to be a useful tool in geochemical exploration. The U.S. Geological Survey is currently studying the geochemistry of these two elements to determine their usefulness in geochemical exploration.

Reagents and apparatus*

Bromine
Hydrobromic acid, concentrated
Hydrobromic acid, 1·5 N: dilute 172 ml of concentrated HBr to 1 l with water
Methyl isobutyl ketone (MIBK)
Sodium bromate
Hydrobromic acid–1% bromine solution: dissolve 10 ml of liquid bromine in 1 l of concentrated HBr
Carbon tetrachloride
Nitric acid, concentrated
Hydrofluoric acid, 48%
Potassium hydroxide, pellets
Perchloric acid, 72%
Stock indium solution, 1000 γ/ml: dissolve 1·00 g pure indium in 200 ml of hydrobromic acid–1% bromine solution; add concentrated hydrobromic acid to make 1 l
Dilute indium solution, 100 γ/ml: dilute 10 ml of stock (1000 γ) indium solution to 100 ml with concentrated hydrobromic acid
Stock thallium solution, 1000 γ/ml: dissolve 1·00 g pure thallium in 200 ml of hydrobromic acid–1% bromine solution; add concentrated hydrobromic acid to make 1 l
Dilute thallium solution, 100 γ/ml: dilute 10 ml of stock (1000 γ) thallium solution to 100 ml with concentrated hydrobromic acid
Teflon beakers, 100-ml capacity

Procedure

Solution of the sample

A—Digestion of rocks and soils with hydrofluoric, perchloric and nitric acid

Place 2 g of a pulverized rock or soil sample in a 100-ml Teflon beaker. Wet the sample with water and add 15 ml of nitric acid, 5 ml of perchloric acid and 15 ml of hydrofluoric acid. Allow the solution to stand overnight and then evaporate to dryness. Let the Teflon beaker cool, add 10 ml of water, warm the beaker to effect solution, and add two potassium hydroxide pellets. Evaporate the resulting solution to dryness, cool the beaker, and add 25 ml of hydrobromic acid–1% bromine solution. Let the beaker stand for 30 min and then warm to effect solution. Once the sample is in solution add about 0·2 g of sodium bromate. Place the Teflon beaker on a hotplate and boil gently until the volume of the solution is reduced to approximately 15 ml and then allow to cool. The sample is now ready for the extraction step of the procedure.

B—Digestion of rocks and soils with hydrofluoric acid

Place a 2-g pulverized sample in a 100-ml Teflon beaker. Wet the sample with water and add 15 ml of hydrofluoric acid. Allow the solution to stand overnight and then evaporate to dryness. Add 10 ml of water to the residue and warm to effect solution. Add two potassium hydroxide pellets and evaporate the solution to dryness. Then add 25 ml of hydrobromic acid–1% bromine solution to the residue and warm to effect solution. Add a small amount of sodium bromate and boil gently until the volume of the solution is approximately 15 ml and then allow to cool. The sample is now ready for extraction.

C—Digestion of iron and manganese oxides with hydrobromic acid–1% bromine solution

Place 2 g of sample in a 100-ml beaker and add 25 ml of hydrobromic acid–1% bromine solution. Let the mixture stand for 1 h and then warm to effect solution. Add a small amount of sodium bromate and boil the solution gently to reduce the volume of the liquid to 15 ml and cool. The solution is now ready for extraction.

D—Pretreatment of organic materials

(1) Transfer between 40 and 50 g of dry plant material, coal or humus into a porcelain dish and place in a cold muffle furnace. Raise the temperature of the furnace by 50°C each hour up to 550°C, then leave it at 550°C for about 10 h. After cooling, remove the dish and weigh the ash. Use 2 g of this ash.
(2) Place 2 g of organic-rich samples such as soils or black shales in a porcelain crucible and ash with a Meker burner. Cool.

Place ash from above pretreatment in a Teflon beaker and proceed as detailed in digestion procedure *B*.

E—Digestion of sulphides with a solution of bromine in carbon tetrachloride and nitric acid

Place a 2-g sample in a 100-ml beaker with a solution of 5 ml bromine in 5 ml carbon tetrachloride and then add 10 ml of nitric acid. Let this mixture stand overnight before evaporating it to dryness. Then add 25 ml of hydrobromic acid–1% bromine solution to the residue and warm to effect solution. Add 0·2 g of sodium bromate and boil the solution gently until the volume is approximately 15 ml. After cooling, the sulphide sample is ready for the extraction step of the procedure.

*All chemicals should be of reagent grade.

Extraction of the sample solution

Transfer the entire contents (sample and acid) of the beaker to a 25 mm × 200 mm screwcap culture tube with approximately 15 ml of wash water. Add 10 ml of MIBK, replace the screwcap, and shake the tube vigorously for 5 min. Centrifuge the tube until the two layers separate. Transfer all of the MIBK layer to another screwcap culture tube containing 30 ml of 1·5 N hydrobromic acid. Shake the tube for 5 min and then centrifuge to separate the two layers. The thallium and indium remain in the MIBK phase and are ready for determination by atomic absorption spectrometry.

Estimation of thallium and indium

To prepare standard series pipette suitable aliquots of dilute thallium and indium standards into 100 ml of 1·5 N hydrobromic acid. Extract these solutions with 100 ml of MIBK to make a standard series such as 0, 0·1, 1·0, 2·0 and 5·0 γ/ml of MIBK for both thallium and indium.

	Thallium	Indium
Wavelength, Å	2768	3040
Source	Hollow cathode	Hollow cathode
Lamp current, mA	20	20
Slit, Å	20	7
Burner, 4-in titanium flathead		
Air pressure, lb/in²	30	30
Acetylene pressure, lb/in²	8	8
Air flow on instrument gauge	44	44
Acetylene flow on instrument gauge	18	18

In this work we used a Perkin-Elmer 403 atomic absorption spectrophotometer. The operating conditions for this instrument are given in the previous column.

Aspirate a 5·0 γ/ml MIBK standard and adjust the burner placement to give maximum absorption. Change the mode of the instrument to concentration and adjust the digital readout to read 5·00 for the 5 γ/ml standard. The standard series previously made should now give a readout of 00, 0·10, 1·00, 2·00 and 5·00 on the digital readout.

The sample is now aspirated into the flame and the value in γ/ml recorded.

With a 2-g sample extracted into 10 ml, parts per million thallium, or indium, in the sample

$$= \frac{10 \text{ ml} \times \mu\text{g/ml found}}{2 \text{ (wt of sample)}}$$

A digital readout reading of 0·04 γ/ml is considered to be above the noise level of the instrument, permitting a lower limit of sensitivity of 0·2 ppm. If a sample reads higher than the maximum standard value of 5 γ/ml, the MIBK should be diluted to bring the readings within the range of standards. Do not use any standards above 10·00 γ/ml on the digital readout, because of the degression from linearity of the standard calibration curve.

Discussion

A radioactive isotope of thallium was used to determine that thallium bromide can be extracted from a hydrobromic acid solution with MIBK. If the normality of the hydrobromic acid used in the

*Table 1 Determination of thallium and indium by atomic absorption by use of various digestions**

Digestion procedure	*A*	*A* minus HF	*B*	*C*
Thallium determinations				
Mystic mine vein gossan, Empire, Colorado	1·3	<0·2	1·5	<0·2
Gossan, Montezuma, Colorado	5·3	5·2	4·9	5·5
Jasperoid, Drum Mountains, Utah	5·7	5·2	4·1	5·9
Indium determinations				
Mystic mine vein gossan, Empire, Colorado	4·7	3·5	4·5	4·9
Gossan, Montezuma, Colorado	1·4	1·3	1·2	1·4
Jasperoid, Drum Mountains, Utah	1·0	0·8	1·1	0·7

*Values in parts per million; digestion procedures described in text.

extraction is between 0·5 and 6 N, the percentage extraction into MIBK is at least 98. Because there is no suitable radioisotope for indium for study of extraction methods, the amount of indium in both phases of the extraction was determined by atomic absorption. At a concentration of 0·5 N hydrobromic acid approximately 75% of the indium bromide is extracted into MIBK and with a 1 N to 4 N hydrobromic acid solution (and increasing concentrations) the extraction of indium is above 98%.

Table 2 Atomic absorption thallium values (ppm) compared with others on U.S. Geological Survey standard rock samples

	This paper Advocated atomic absorption method	Flanagan[2] Polarograph	Optical spectrograph
Granite, G-2	1·0	0·85	1·3
Granodiorite, GSP-1	1·5	0·71	1·6
Basalt, BCR-1	0·4	0·36	—
Andesite, AGV-1	0·4	0·27	1·6

iron is normally extracted from the MIBK in the 1·5 N hydrobromic acid wash step. If the MIBK layer remains brown in colour from iron, it must be washed a second time with 1·5 N hydrobromic acid.

Thallium and indium values obtained by four different digestions are compared in Table 1. These data show little variation between digestions; however, digestion *A*, the most drastic attack, gives maximum recovery. The hydrofluoric acid digestion *B* is fairly rapid and adequate for geochemical prospecting. The bulk of our data has been based on this digestion.

Precision

Thallium data obtained by our atomic absorption method on some U.S. Geological Survey standard rock samples (Table 2) compare favourably with those given by Flanagan.[2] The indium values given by Flanagan are less than our detection limit.

Thallium and indium values obtained with this atomic absorption method are compared with the values obtained by use of a special emission spectrographic technique in Table 3. The agreement between methods is good.

Reproducibility data for thallium and indium are shown in Tables 4 and 5. The relative standard deviation for thallium ranges from 4·0 to 12·6% and is not apparently related to the quantity present. The relative standard deviation for indium ranges between 2·4 and 27·8%, the poorer

*Table 3 Comparison of atomic absorption and spectrographic values for thallium and indium**

Sample	Thallium Atomic absorption method	Thallium Spectrographic	Indium Atomic absorption method	Indium Spectrographic
USGS glass standard D	7·2	15	42	50
Pierre Shale, Colorado	1·6	<2	<0·2	<1
Pierre Shale, South Dakota	25	20	<0·2	<1
Colloform sphalerite, New Prospect mine, eastern Tennessee	4·4	5	1·8	<1
Galena, sphalerite, pyrite, Ophir, Utah	3·5	<2	32·4	20
Mystic mine vein, gossan, Empire, Colorado	1·5	<2	4·6	5
Gossan, Montezuma, Colorado	1·4	<2	13	15
Gossan, Montezuma, Colorado	5	5	1·2	2
Carbonaceous ore, Carlin, Nevada	8	10	<0·2	<1
Jasperoid, Drum Mountains, Utah	4	3	1	<1
Orpiment and realgar, Getchell, Nevada	400	>200	<0·2	<1

*Spectrographic analyses, in parts per million, by E. L. Mosier, U.S. Geological Survey.

If enough iron is present to colour the MIBK layer brown, it will interfere with the indium determination, possibly by a process called 'light scattering,' as described by Billings.[1] Most of the

*Table 4 Replicate determinations for thallium**

Sample	High, ppm	Low, ppm	Mean	Standard deviation	Relative standard deviation, %
USGS glass standard B	0·5	0·5	0·5	0·02	4·6
USGS glass standard E	14·4	11·7	13·2	1·2	8·8
Mystic mine vein, gossan, Empire, Colorado	1·7	1·4	1·5	0·1	8·5
Gossan, Montezuma, Colorado	5·1	4·4	4·9	0·2	5·0
Jasperoid, Drum Mountains, Utah	4·6	4·0	4·1	0·3	6·5
Carbonaceous ore, Carlin, Nevada	8·7	7·3	7·8	0·5	6·4
Galena, sphalerite, pyrite, Ophir, Utah	3·3	2·8	3·1	0·2	5·6
Colloform sphalerite, New Prospect mine, eastern Tennessee	4·3	3·8	4·1	0·2	4·6
Soil over sphalerite vein, Montezuma, Colorado	5·1	4·7	4·9	0·2	4·0
Humus ash, Montezuma, Colorado	6·2	4·6	5·6	0·5	9·6
Exposed spruce tree root ash, Montezuma, Colorado	4·4	3·0	3·9	0·5	12·6

*Based on five determinations per sample.

*Table 5 Replicate determinations for indium**

Sample	High, ppm	Low, ppm	Mean	Standard deviation	Relative standard deviation, %
USGS glass standard B	0·7	0·6	0·7	0·1	5·6
USGS glass standard E	539	502	520	12·5	2·4
Mystic mine vein, gossan, Empire, Colorado	5	4·2	4·5	0·4	9·2
Gossan, Montezuma, Colorado	1·3	1·1	1·2	0·06	5·2
Jasperoid, Drum Mountains, Utah	1·1	1·0	1·1	0·04	3·7
Carbonaceous ore, Carlin, Nevada	$<0{\cdot}2$	$<0{\cdot}2$	$<0{\cdot}2$	—	—
Galena, sphalerite, pyrite, Ophir, Utah	51	46	47	2·3	4·9
Colloform sphalerite, New Prospect mine, eastern Tennessee	0·5	0·5	0·5	0·02	4·7
Soil over sphalerite vein, Montezuma, Colorado	0·8	0·4	0·5	0·1	27·8
Humus ash, Montezuma, Colorado	0·5	0·3	0·4	0·08	18·2
Exposed spruce tree root ash, Montezuma, Colorado	1·6	1·3	1·4	0·21	15·1

*Based on five determinations per sample.

precision occurring at the lower levels of concentration. The reproducibility is adequate for geochemical prospecting investigations.

Conclusions

The threshold values obtained by this method are near or below the estimated abundances of these elements, which were given by Parker[3] as 0·3–1·3 ppm thallium and 0·1–0·25 ppm indium. Thus, the analytical procedure should produce suitable data to study the usefulness of thallium and indium in geochemical prospecting. By this method, 50 samples can be analysed per man-day for thallium and indium.

Acknowledgment

This paper is published by permission of the Director, U.S. Geological Survey.

References

1. Billings, G. K. Light scattering in trace-element analysis by atomic absorption. *Atom. Absorp. Newsl.*, **4**, 1965, 357–62.
2. Flanagan, F. J. U.S. Geological Survey standards —II. First compilation of data for the new U.S.G.S. rocks. *Geochim. cosmochim. Acta*, **33**, 1969, 81–120.
3. Parker, R. L. Composition of the earth's crust. *Prof. Pap. U.S. geol. Surv.* 440-D, 1967, 17 p.
4. Perkin-Elmer Corporation. *Analytical methods for atomic absorption spectrophotometry* (Norwalk, Conn.: Perkin-Elmer Corporation, 1966).

550.84:543.53:621.386.832

Geochemical mapping and prospecting by use of rapid automatic X-ray fluorescence analysis of panned concentrates

R. C. Leake, B.SC., PH.D.

J. W. Aucott, B.SC., PH.D.

Both of the Institute of Geological Sciences, Geochemical Division, London, England

Synopsis

Geochemical mapping by use of X-ray fluorescence analysis of panned concentrates has been successfully related to regional rock geochemistry in the Cheviot area of northern England and the Scottish Highlands. Multivariate statistical methods have been applied to the absolute data and element ratios and populations have been correlated with the observed geology.

The XRF analysis of panned concentrates is considered to be particularly useful in assisting geological mapping in areas with little contrast between the major rock types. The high precision of the analyses enables interpretation and comparison of sample sites to be based primarily on elemental ratios and interrelationships rather than absolute quantities, so reducing the effect of environmental factors which produce geochemical noise. The influence of secondary precipitation on trace-element distribution is considerably less in panned concentrates than in fine sediment fractions.

Dried concentrates are split into two fractions, one of which is used for analysis and the other for mineralogical examination. The analysis fraction is mixed with an acrylic resin binder, Elvacite 2013, in the ratio 4 parts concentrate to 1 part Elvacite, and ground in a tungsten carbide pot Tema swing mill for 3 min. The resultant powder is pressed into a 32-mm diameter disc under a load of 15 ton for a few seconds. The discs are analysed for Si, Ca, Ti, Mn, Fe, Ni, Cu, Zn, Rb, Sr, Y, Zr, Nb, Ag, Sn, Sb, Ba, Ce, Pb, Th and U and ratioed against a synthetic mineral matrix standard, spiked with 1000 ppm of each of the trace elements, by use of a Philips 1220C automatic X-ray spectrometer. Raw count data are output on paper tape and corrections for mass absorption differences between standard and unknown are completed on an IBM 1130 computer with an 8K store.

In the Cheviot area of the north of England it has been possible to distinguish between different phases of granite, Old Red Sandstone lavas and various sedimentary rocks by use of Zn/Fe, Ni/Fe, Mn/Fe and Nb/Ti together with absolute Cu, Sn, Zr and Th values. With similar ratios in the Scottish Highlands it has been possible to differentiate between basic rocks of Moinian and Lewisian age, Old Red Sandstone sediments, Caledonian granites and different types of siliceous gneiss.

Mineralization has also been recognized either in terms of absolute contents of elements such as Pb or by significant departures of ratios such as Zn/Fe from the norm. In the Cheviots vein Pb, Zn, Ba, Cu and disseminated Cu mineralization has been indicated, and in Wester Ross the presence of disseminated celestite and pyrite and vein pyrite, galena and sphalerite has been demonstrated.

The work described is directed at the development of a technique of regional geochemical prospecting and mapping which utilizes the increased contrast in many trace-element concentrations produced by the removal of light felsic mineral phases, the relatively high precision of X-ray fluorescence spectrometry and the speed and convenience of automatic analysis and computer-orientated interpretative methods. The technique has been applied to an area of relatively simple geology, but widespread mineralization (the Cheviot Hills), and also to an area for which little geological, petrological or geochemical information is available (the extreme north of Sutherland).

Collection of samples

Over both the areas investigated, samples were collected by several different teams over a total of three field seasons. In the Cheviot Hills concentrates were obtained by the classical gold panning method so as to retain a significant proportion of the lighter mineral phases in an approximately constant volume of sediment. In north Sutherland the stream sediments were *dulanged* to remove

Table 1 Instrumental parameters of analysis

Element	Tube, kV × mA	Analysing crystal	Counter	Counting time, sec	Background, degrees, 2θ from peak
Si	60 × 40	P.E.	Flow	4	—
Ca	60 × 40	P.E.	Flow	4	—
Ti	60 × 40	LiF 200	Flow	4	−1·22
Mn	60 × 40	LiF 200	Flow	8	+0·58
Fe	40 × 20	LiF 200	Scint	4	—
Ni	60 × 40	LiF 200	Flow + Scint	10	−0·68
Cu	60 × 40	LiF 200	Flow + Scint	20	−0·37
Zn	60 × 40	LiF 200	Flow + Scint	10	−0·75
Rb	60 × 40	LiF 200	Scint	20	−0·76 and +0·41
Sr	80 × 30	LiF 220	Scint	20	+0·74
Y	80 × 30	LiF 220	Scint	20	−0·29
Zr	80 × 30	LiF 220	Scint	8	−0·38
Nb	80 × 30	LiF 220	Scint	20	−0·55
Ag	80 × 30	LiF 220	Scint	20	−0·31 and +0·53
Sn	80 × 30	LiF 220	Scint	20	−0·32 and +0·52
Sb	80 × 30	LiF 220	Scint	20	−0·30 and +0·50
Ba	80 × 30	LiF 220	Scint	8	−0·44 and +0·21
Ce	80 × 30	LiF 220	Scint	20	−0·25 and +0·50
Pb	60 × 40	LiF 200	Scint	20	−0·31
Th	60 × 40	LiF 200	Scint	20	−0·45 and +0·48
U	60 × 40	LiF 200	Scint	20	−0·32 and +0·85

Analysing peak, $K\alpha$, except for Pb($L\beta_1$), Th($L\alpha$) and U($L\alpha$). Collimator, fine, except for Si and Ca (coarse).

Table 2 Analytical variance (repeated analysis of one concentrate)

Element	Analysis 1	Analysis 2	Analysis 3	Mean	Standard deviation	Coefficient of variance, %
Ca	2·41 %	2·41 %	2·31 %	2·38 %	0·058 %	2·4
Ti	0·68 %	0·69 %	0·72 %	0·70 %	0·021 %	3
Mn	1410 ppm	1350 ppm	1410 ppm	1390 ppm	32 ppm	2·3
Fe	3·68 %	3·68 %	3·70 %	3·69 %	0·01 %	0·3
Ni	46 ppm	46 ppm	45 ppm	45·7 ppm	0·6 ppm	1·3
Cu	50 ppm	50 ppm	49·5 ppm	49·8 ppm	0·2 ppm	0·4
Zn	66 ppm	65 ppm	68 ppm	66 ppm	1·6 ppm	2·4
Rb	185 ppm	195 ppm	187 ppm	189 ppm	5·3 ppm	2·8
Sr	165 ppm	156 ppm	166 ppm	162 ppm	5·5 ppm	3·4
Y	37 ppm	31 ppm	31 ppm	33 ppm	3·4 ppm	10·3
Zr	346 ppm	337 ppm	341 ppm	341 ppm	4·5 ppm	1·3
Nb	19 ppm	16 ppm	13 ppm	16 ppm	2·9 ppm	18
Sn	6 ppm	11 ppm	10 ppm	9 ppm	2 ppm	24
Ce	106 ppm	128 ppm	115 ppm	116 ppm	11 ppm	9·5
Ba	500 ppm	470 ppm	433 ppm	468 ppm	33 ppm	7·1
Pb	23 ppm	23 ppm	21·5 ppm	22·5 ppm	0·9 ppm	4
Th	24 ppm	30 ppm	12 ppm	22 ppm	8·3 ppm	38
U	12 ppm	9 ppm	2 ppm	8 ppm	5·1 ppm	64

virtually all the lighter phases, but this procedure results in the collection of samples of greatly varying weight, which presents analytical difficulties.

Sample preparation

Concentrate samples were dried and split with a riffle sampler to produce fractions of 8 g for analysis, wherever possible. Any material remaining was retained for later mineralogical examination. The 8-g sub-samples were mixed with 2 g of Elvacite 2013 resin and ground in a tungsten carbide Tema swing mill for 3 min as described by Smith.[1] Smaller samples were mixed with Elvacite in the same proportion. The samples were then pressed under a load of 15 ton for a few seconds into discs of 32-mm diameter.

Analysis

The discs were analysed for the elements Si, Ca, Ti, Mn, Fe, Ni, Cu, Zn, Rb, Sr, Y, Zr, Nb, Ag, Sn, Sb, Ba, Ce, Pb, Th and U on a Philips 1220C automatic X-ray spectrometer with a 2·8-kW tungsten anode tube. The instrumental parameters of analysis are shown in Table 1.

The concentrate samples were compared with a mixture of pure minerals spiked with 1000 ppm of each of the trace elements. The raw count data were output on paper tape and the results were calculated on an IBM 1130 computer with an 8K store. The initial step in the calculation of results was to obtain peak minus background counts for each element and to correct for peak overlap, e.g. for the effect of Rb $K\beta$ on Y $K\alpha$. The counts obtained for each element were then ratioed

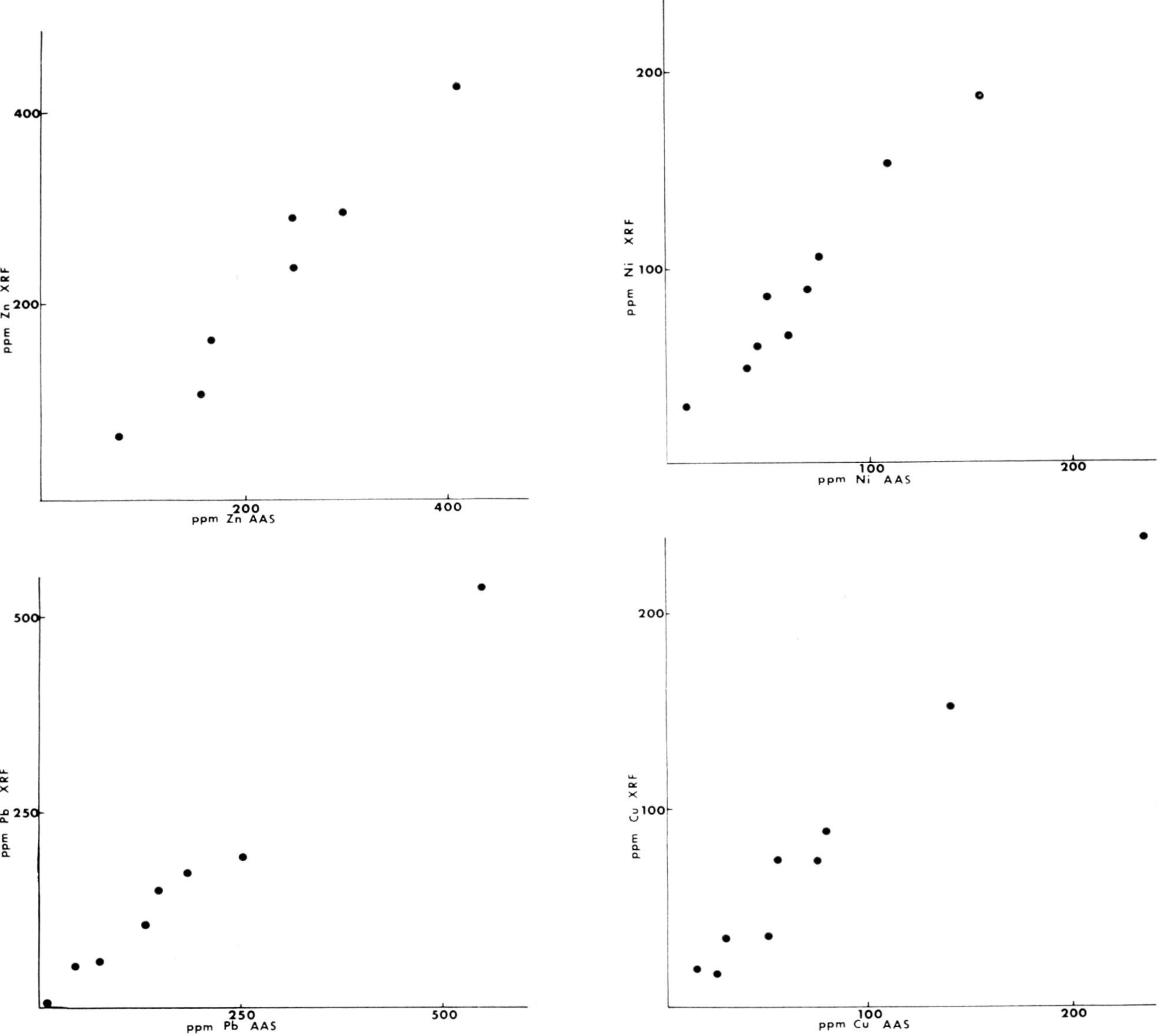

Fig. 1 Comparison of X-ray fluorescence and atomic absorption spectroscopy determinations of Ni, Cu, Zn and Pb in Cheviot concentrates

Table 3 Analysis of concentrates from same site in Cheviot Hills

Element	BB 47	BG 34	BG 37	Mean	Standard deviation	Coefficient of variance, %
Ca	0·21 %	0·45 %	0·51 %	0·39 %	0·16 %	41
Ti	0·29 %	0·31 %	0·31 %	0·30 %	0·011 %	3·7
Mn	590 ppm	600 ppm	630 ppm	606 ppm	21 ppm	3·5
Fe	3·69 %	3·62 %	4·02 %	3·78 %	0·21 %	5·6
Ni	110 ppm	109 ppm	118 ppm	112 ppm	5 ppm	4·5
Cu	30 ppm	23 ppm	37 ppm	30 ppm	7 ppm	23
Zn	195 ppm	160 ppm	180 ppm	178 ppm	17 ppm	9·6
Rb	120 ppm	100 ppm	110 ppm	110 ppm	10 ppm	9·1
Sr	40 ppm	40 ppm	45 ppm	42 ppm	2·8 ppm	6·7
Y	20 ppm	25 ppm	20 ppm	22 ppm	2·9 ppm	13
Zr	200 ppm	190 ppm	200 ppm	197 ppm	5·8 ppm	2·9
Nb	9 ppm	1 ppm	5 ppm	5 ppm	2·7 ppm	54
Sn	5 ppm	6 ppm	2 ppm	4 ppm	2·1 ppm	53
Ba	340 ppm	400 ppm	460 ppm	400 ppm	60 ppm	15
Ce	85 ppm	80 ppm	85 ppm	83 ppm	2·8 ppm	3·4
Pb	19 ppm	10 ppm	35 ppm	21 ppm	12·6 ppm	60
Th	2 ppm	10 ppm	5 ppm	6 ppm	4 ppm	67
U	3 ppm	7 ppm	<1 ppm	3 ppm	3·5 ppm	120

against the counts for the same element in the standard and initial uncorrected concentrations were calculated. A mass absorption factor for elements determined at wavelengths lower than that of the iron absorption edge was calculated from the approximate Si, Ca, Ti, Fe and Ba contents by use of the mass absorption coefficients given by Heinrich.[2] This was then multiplied by the initial elemental values to correct for differences in mass absorption between samples and standard. A further correction was applied to the smallest samples where the thickness was less than

Table 4 Analyses of concentrates from same site in north Sutherland

Element	1	2	3	4	Mean	Standard deviation	Coefficient of variance, %
Ca	4·28 %	3·46 %	3·40 %	8·04 %	4·80 %	2·21 %	46
Ti	0·74 %	0·23 %	0·27 %	1·05 %	0·57 %	0·39 %	68
Mn	2830 ppm	1020 ppm	1080 ppm	2280 ppm	1803 ppm	898 ppm	50
Fe	6·97 %	3·01 %	3·37 %	6·56 %	4·98 %	2·08 %	42
Ni	97 ppm	57 ppm	62 ppm	105 ppm	80 ppm	24 ppm	30
Cu	5 ppm	7 ppm	3 ppm	9 ppm	6 ppm	2·5 ppm	42
Zn	71 ppm	48 ppm	50 ppm	67 ppm	59 ppm	12 ppm	20
Rb	9 ppm	18 ppm	19 ppm	13 ppm	15 ppm	4·6 ppm	31
Sr	137 ppm	267 ppm	264 ppm	315 ppm	246 ppm	76 ppm	31
Y	69 ppm	28 ppm	32 ppm	66 ppm	49 ppm	21 ppm	43
Zr	3240 ppm	610 ppm	805 ppm	5350 ppm	2500 ppm	2245 ppm	90
Nb	8 ppm	<1 ppm	<1 ppm	24 ppm	8 ppm	11·3 ppm	141
Ba	115 ppm	330 ppm	310 ppm	340 ppm	274 ppm	106 ppm	39
Ce	110 ppm	55 ppm	60 ppm	160 ppm	96 ppm	48 ppm	50
Pb	13 ppm	<1 ppm	2 ppm	23 ppm	10 ppm	11 ppm	110
Th	9 ppm	3 ppm	3 ppm	14 ppm	7 ppm	5·3 ppm	76
U	7 ppm	5 ppm	6 ppm	12 ppm	8 ppm	3·2 ppm	40
Ti/Fe $\times 10^3$	106	76	80	160	106	39	37
Mn/Fe	406	339	320	348	353	37	10
Ni/Fe $\times 10$	139	189	184	160	168	23	14
Ni/Zn $\times 10^2$	137	119	124	157	134	17	13

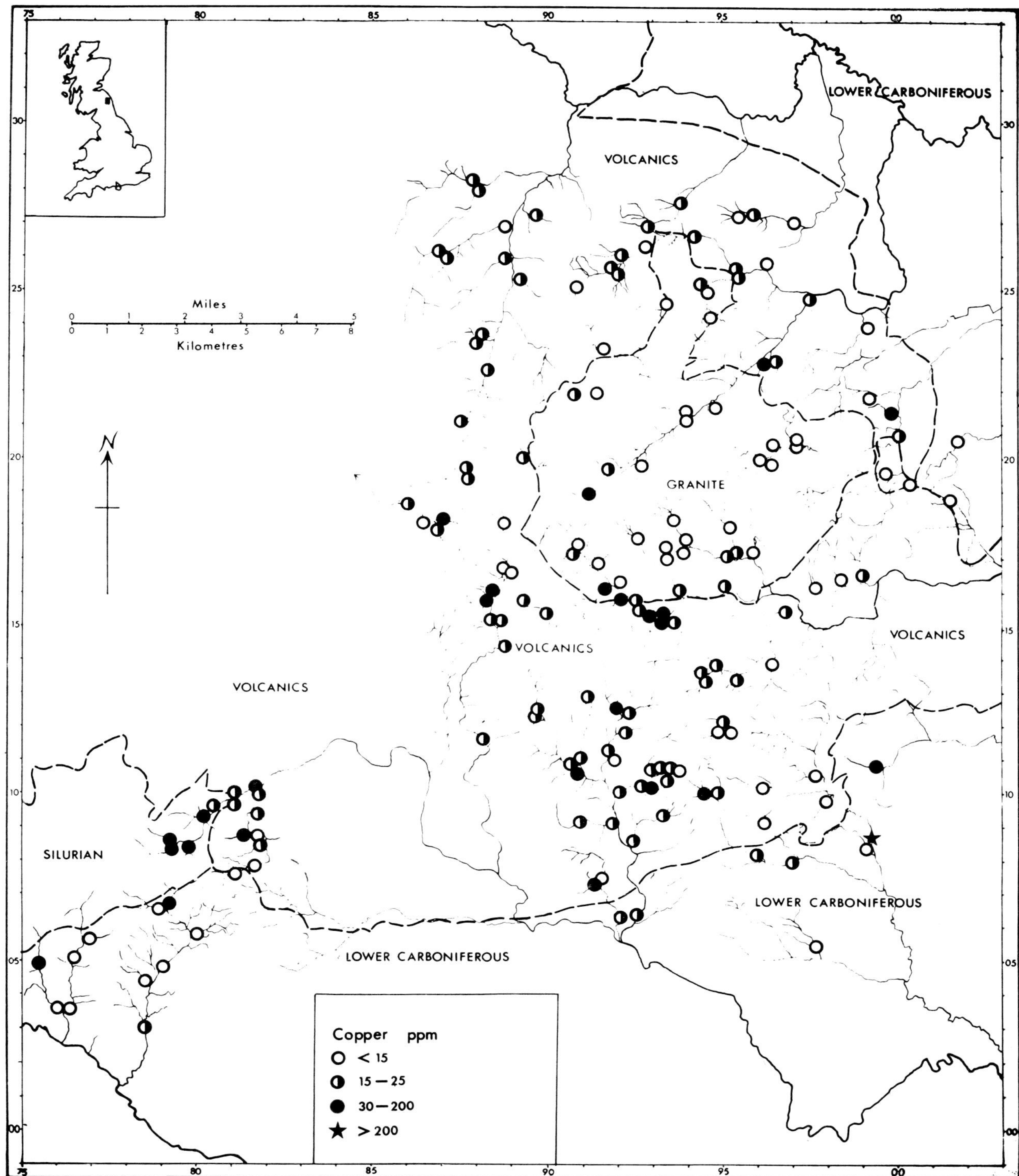

Fig. 2 Panned concentrates: copper, Cheviot

the critical depth of penetration.

The time taken for the determination of the full range of elements with the present system was 27 min per sample. The determination of Cu, Pb and Zn can be achieved in 4 min per sample.

Analytical precision and accuracy

Analytical precision is illustrated by three analyses of a typical concentrate shown in Table 2.

As a check on the accuracy of the XRF analyses the Ni, Cu, Zn and Pb contents of a few panning concentrates were determined by atomic absorption spectroscopy after dissolution in a HF–$HClO_4$ mixture. Graphical comparisons of the results (Fig. 1) reveal relatively close agreement between the two methods.

Sampling precision

Variation of the concentrate obtained during routine sampling of a site by the gold panning technique varies according to catchment geology and mineralogy, but it can be low, particularly in streams draining relatively fine-grained and homogeneous rocks. In the Cheviot Hills the coefficient of variation of three samples due to combined sampling and analytical variance of most elements above the 10 ppm level is less than 10% for streams draining the Cheviot lavas and Silurian sediments. Analyses of three concentrates collected at different times and by different people from a stream draining Silurian sediments are given in Table 3.

The sampling variance obtained on material from the smallest streams draining the Cheviot granite, a coarse-grained and heterogeneous rock, is appreciably greater.

The variance of samples obtained by the *dulang* panning method used in north Sutherland is greater than that for samples obtained by the gold panning technique employed in Cheviot. The variation in composition of four concentrates panned by different operators at different times from a small stream draining the Borgie basic metamorphic complex in north Sutherland is shown in Table 4.

Interpretation of results

Elemental concentrations

The simplest interpretative procedure is to investigate the variation of a single element. This approach is most successful where the sampling precision is relatively high and the element in question exhibits a large range or contrast and where the number of mineral phases containing the element is low. The distribution of copper in

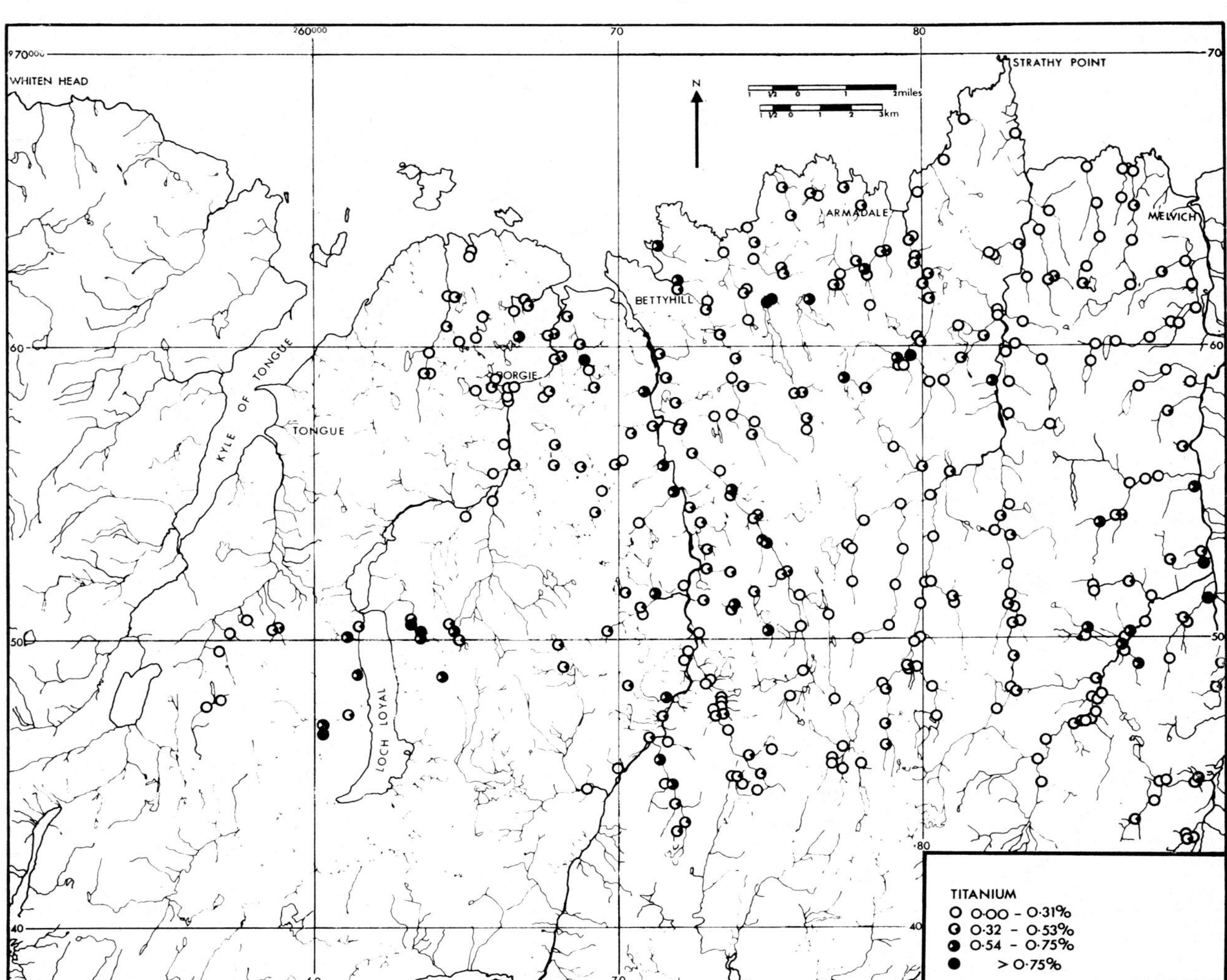

Fig. 3 Stream sediments: titanium, north Sutherland

concentrates from the Cheviot Hills is used to demonstrate this case. A cumulative frequency plot of the copper contents of the 187 concentrates collected revealed the presence of three populations, one with all values less than 15 ppm, a second stretching from 15 to 25 ppm and a third with values greater than 25 ppm. Copper contents within these intervals are shown in Fig. 2. Chalcopyrite and malachite have been identified in several of the concentrates examined. The majority in the lowest class are concentrated over the Cheviot granite and also in streams draining the Carboniferous sediments.

In contrast, the second class predominates over the volcanics. The highest class is interpreted as indicating the presence of vein copper sulphide mineralization. In the majority of these cases either mineralization has been observed upstream of the site sampled or the sample contains other elements also indicative of vein mineralization.

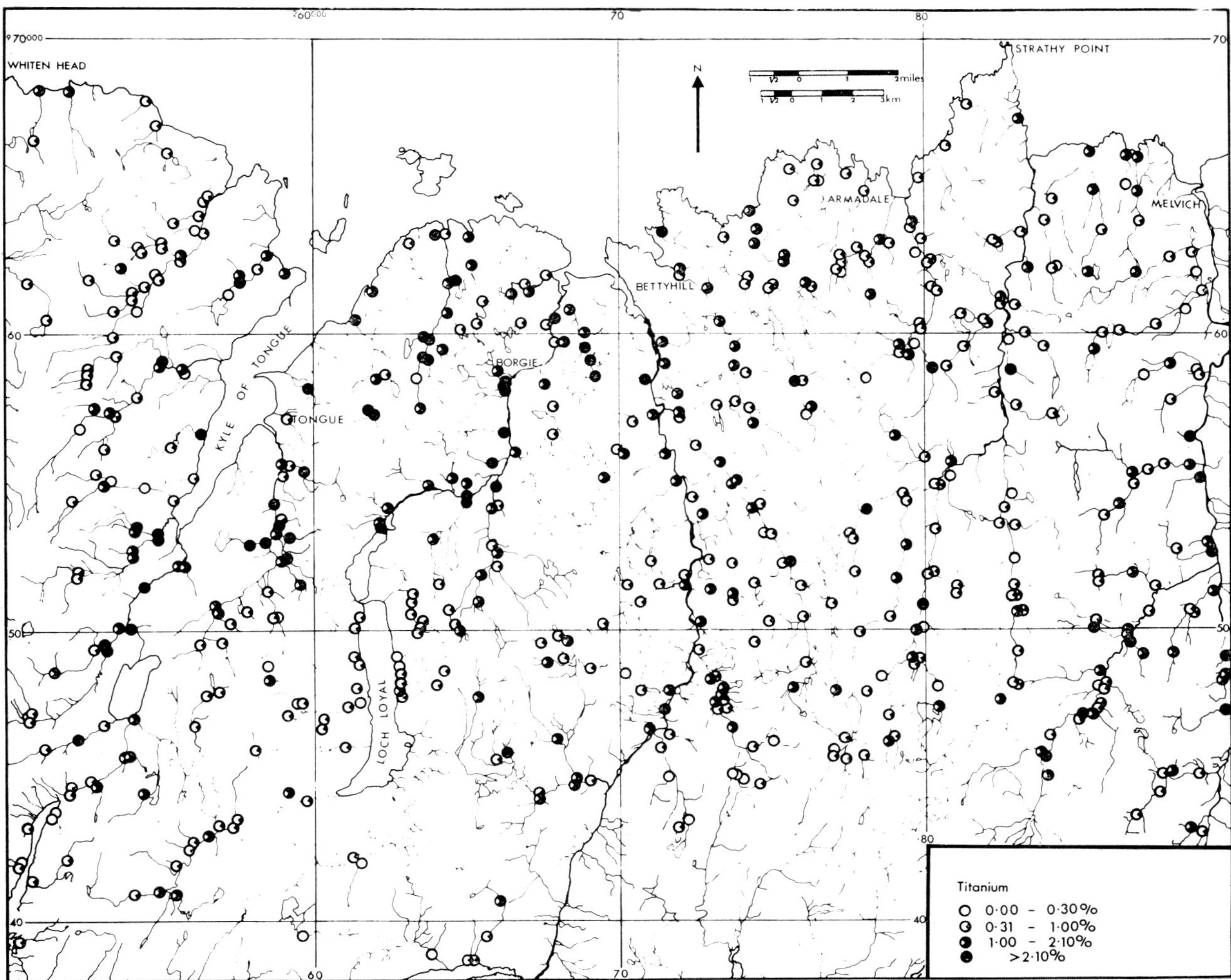

Fig. 4 Panned concentrates: titanium, north Sutherland

Elemental ratios

The investigation of the distribution of elemental ratios is of value in geochemical surveys where the concentrate sampling precision is relatively poor. Correlation coefficients and factor analysis of the concentrate analyses enable pairs of elements showing geochemical association to be identified. High positive correlation coefficients suggest that the two elements occur in the same mineral species and that the ratio is relatively constant in spite of differing proportions of light and heavy minerals collected. Table 4 shows that the coefficients of variation of the elemental ratios are significantly lower than those of the corresponding single elements. Cumulative frequency plots of the ratios are used to distinguish separate populations, maps of which can then be assembled.

The distribution of titanium in north Sutherland, both in the −100-mesh fraction stream sediment and concentrate, illustrates the application of the above procedure. The titanium in −100-mesh fraction stream sediment analysed by optical emission spectrograph (see Fig. 3) shows a considerable scatter of the four classes. Concentrations of titanium are apparent to the east of the Naver Valley where basic bodies within the Moine are common, within the southern part of Strath Halladale, and in the central and northern part of the Ben Loyal syenite complex. The map of the distribution of titanium in the panned concentrate, which is also in terms of four class intervals, shows considerable scatter, but the highest class is concentrated over parts of the Borgie basic metamorphic complex. Other titanium-rich areas are the Tongue basic mass, basic Moine rocks to the west of this, basic-rich Moine in the Naver Valley, and the southern part of Strath Halladale. Fig. 4 therefore serves to define the areas richest in basic rocks.

The corresponding Ti/Fe ratio map, also with four class intervals, is shown in Fig. 5. A narrow area about Armadale on the north coast is delineated by the lowest ratio class. Examination and analysis of rocks from this area indicated a strong correlation between the geochemistry of the concentrates and rocks. The low Ti/Fe ratio characteristic of the area results from the presence of almandine garnet and grunerite amphibole in abundance in the basic rocks and from the scarcity of iron–titanium oxide phases. Other features of the map which can be correlated with the observed rock mineralogy and geochemistry are the distinctions between the basic Lewisanoid and basic Moine complexes. The areas rich in Moine basics are characterized by the occurrence of the third class—for example, the Naver Valley and west of Tongue, whereas the Lewisanoid Tongue mass, a further body west of Melness, and the central parts of the Borgie mass are characterized by the second class. The margins of the Borgie mass contain rocks intermediate in composition between Lewisanoid rocks from elsewhere and the

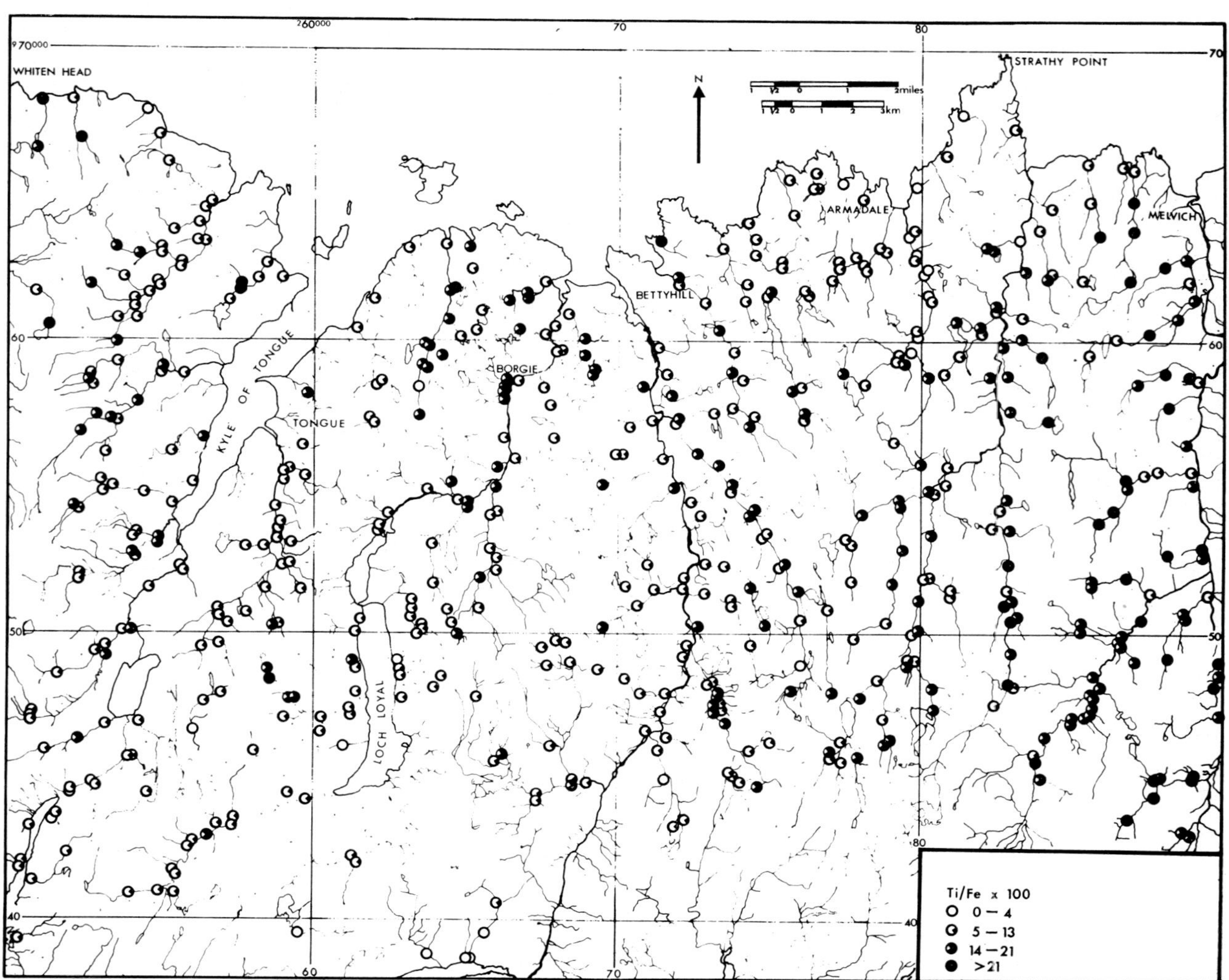

Fig. 5 Panned concentrates: Ti/Fe × 100, north Sutherland

basic Moine. The average Ti/Fe ratio of 17 analysed basic and intermediate Moine rocks is 114×10^{-3} and that of 25 analysed Lewisanoid rocks is 55×10^{-3}. The Lewisanoid basic rocks are characterized by the abundance of hornblende and epidote and the relative scarcity of iron–titanium oxides and the absence of sphene. The Moine basic rocks on the other hand are characterized by the relative abundance of sphene and opaque phases. The highest class is strongly concentrated in Strath Halladale, a region of granite injection. The mean Ti/Fe of four analysed rocks from this region is 187×10^{-3} and can be correlated with abundance of sphene and absence of opaque phases. The patterns revealed by the Ti/Fe concentrate map are thus readily interpreted in terms of the rock mineralogy and geochemistry.

Mineralization by an element such as zinc, which is a minor component of rock-forming minerals, can be recognized by anomalous ratios with an associated element. Indication of small-scale zinc mineralization in the Cheviot Hills is suggested by an anomalously high population of Zn/Fe ratios. This is substantiated by the identification of a few grains of sphalerite in concentrates derived from volcanics south of the Cheviot granite.

Multiple combination of element ratios

A technique of combining concentrate element ratios to produce a multi-element map is illustrated from the distribution of Ti/Fe, Ni/Zn and Mn/Fe in north Sutherland concentrates. The separate ratio maps (Figs. 5–7) have been combined in Fig. 8 by a pattern recognition technique. The main boundaries separating the different populations are sketched in and the three maps have been superimposed upon one another, the two middle Ti/Fe classes having been combined to avoid overcomplication. When the composite

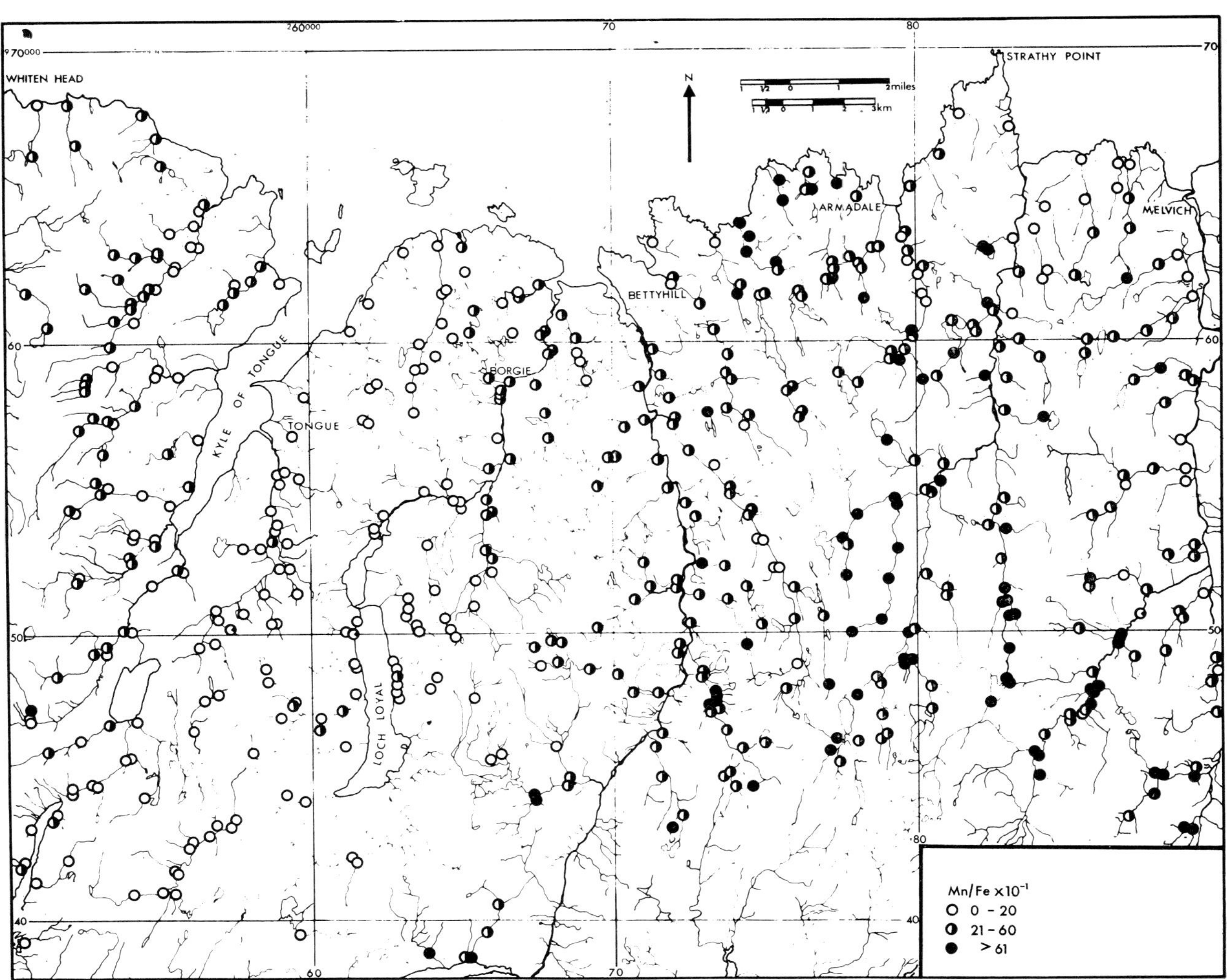

Fig. 6 Panned concentrates: Mn/Fe, north Sutherland

map is compared with the geological map (Fig. 9), obvious correlations between pattern sectors and geological boundaries are apparent. Separate pattern sectors define the rocks west of the Moine thrust, the Ben Loyal syenite complex, the granite east of Loch Loyal, the Lewisanoid rocks in the northern part of the Borgie complex, isolated sector of this pattern to the west of the Kyle can also be correlated with an outcrop of Lewisanoid rocks. The pattern encountered in the Moine rocks immediately west of the Kyle of Tongue reappears to the east of Borgie and can again be correlated with relatively basic-rich Moine gneisses. The pattern which expresses

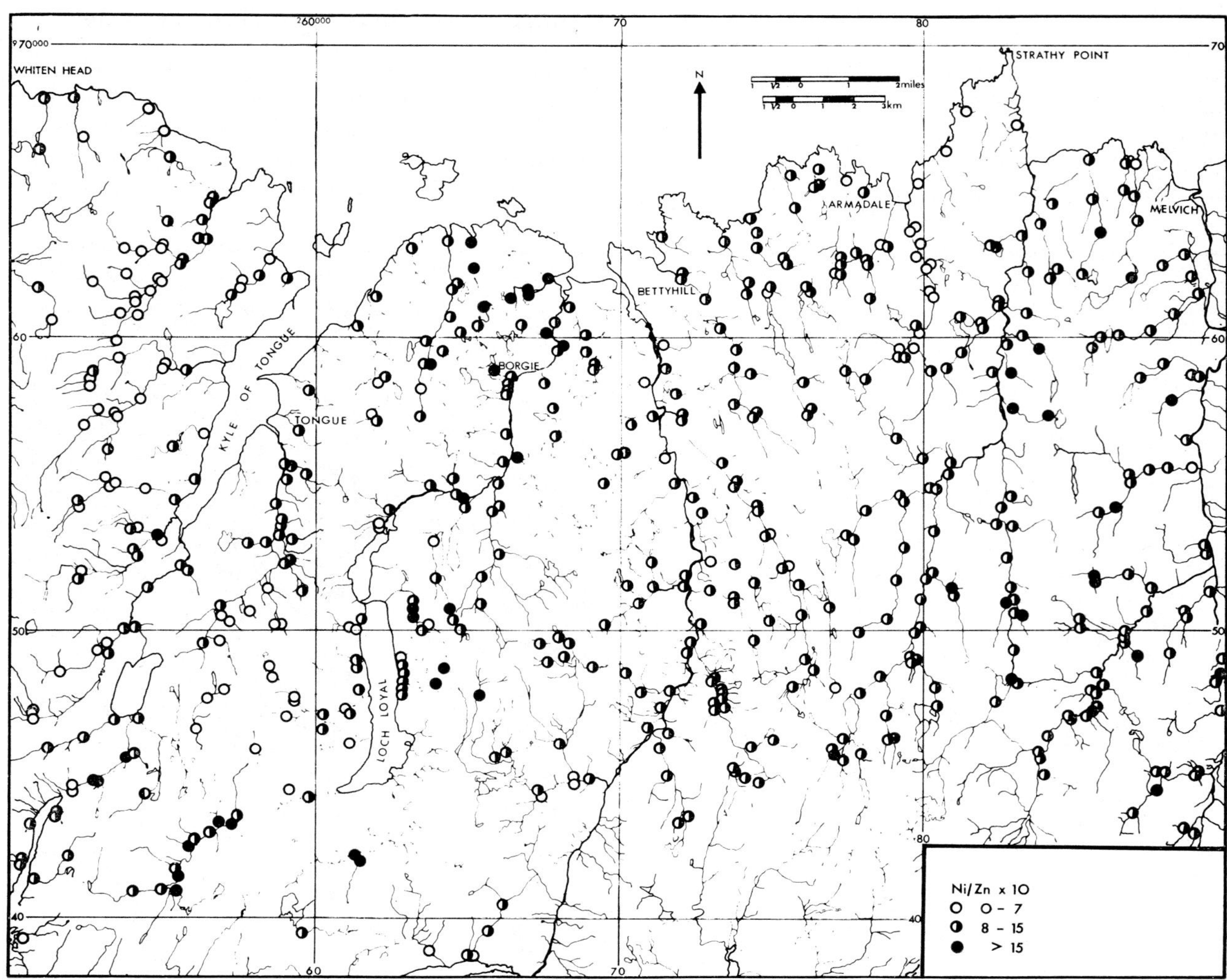

Fig. 7 Panned concentrates: Ni/Zn × 10, north Sutherland

the Lewisanoid rocks southwest of Ben Loyal, the exotic Moine rocks around the Armadale burn and the Old Red Sandstone sediments south of Strathy Point. In addition, six pattern zones parallel to the regional gneissic layering are displayed. East of the Moine thrust is a zone characterized by moderate Ti/Fe and Mn/Fe ratios and by low Ni/Zn ratios. East of this is a zone with higher Ni/Fe ratios, which can be correlated with more basic-rich Moine rocks. East of the Kyle of Tongue is a zone including the Lewisanoid masses, which is characterized by concentrates with moderate Ti/Fe ratios, moderate to high Ni/Zn ratios and low Mn/Fe ratios. An moderate Ti/Fe and Ni/Zn ratios and high Mn/Fe ratios is visible south and west of Armadale and corresponds approximately to the outcrop of the Strathy gneiss. The easternmost zone represents high Ti/Fe ratios, moderate to high Ni/Zn ratios and moderate Mn/Fe ratios and corresponds to the Strath Halladale granite injection complex. This zone is also superimposed upon the extreme east of the Strathy gneiss pattern zone.

Conclusions

Compared with −100-mesh stream-sediment surveys of the same areas covered by concentrate sampling, the method described here has some advantages. These include the ability to study the distribution of elements such as Sb and Th, which are difficult to determine in large numbers of samples by other analytical methods, and the generally better analytical precision of XRF compared with other methods of multi-element analysis. The process of panning serves to remove much of the quartz diluent, thereby concentrating the elements of interest and facilitating their analytical detection. The almost complete removal of secondary element concentrations on clay organic matter and colloids is advantageous in detrital-rich regimes in that false metal anomalies are removed, thereby simplifying the interpretation of the data.

A limitation of the method is the present difficulty in the XRF determination of elements such as Be and B, but it is possible to determine most elements with atomic number greater than 9. In addition to the 21 elements quoted above, K, As, Mo, W and Bi have been determined in selected concentrates, though some of these require the use of an X-ray tube with a Mo target. The analysis time is relatively long compared with direct-reading optical emission spectroscopy, but XRF equipment is amenable to automation and, with a suitable changer, it is possible to run the equipment overnight without attention. The technique is, of course, unsuitable for the investigation of the distribution of an element confined to light mineral phases or of a highly mobile element for which scavenging by clays and organic matter is the main agent for its incorporation into stream sediment.

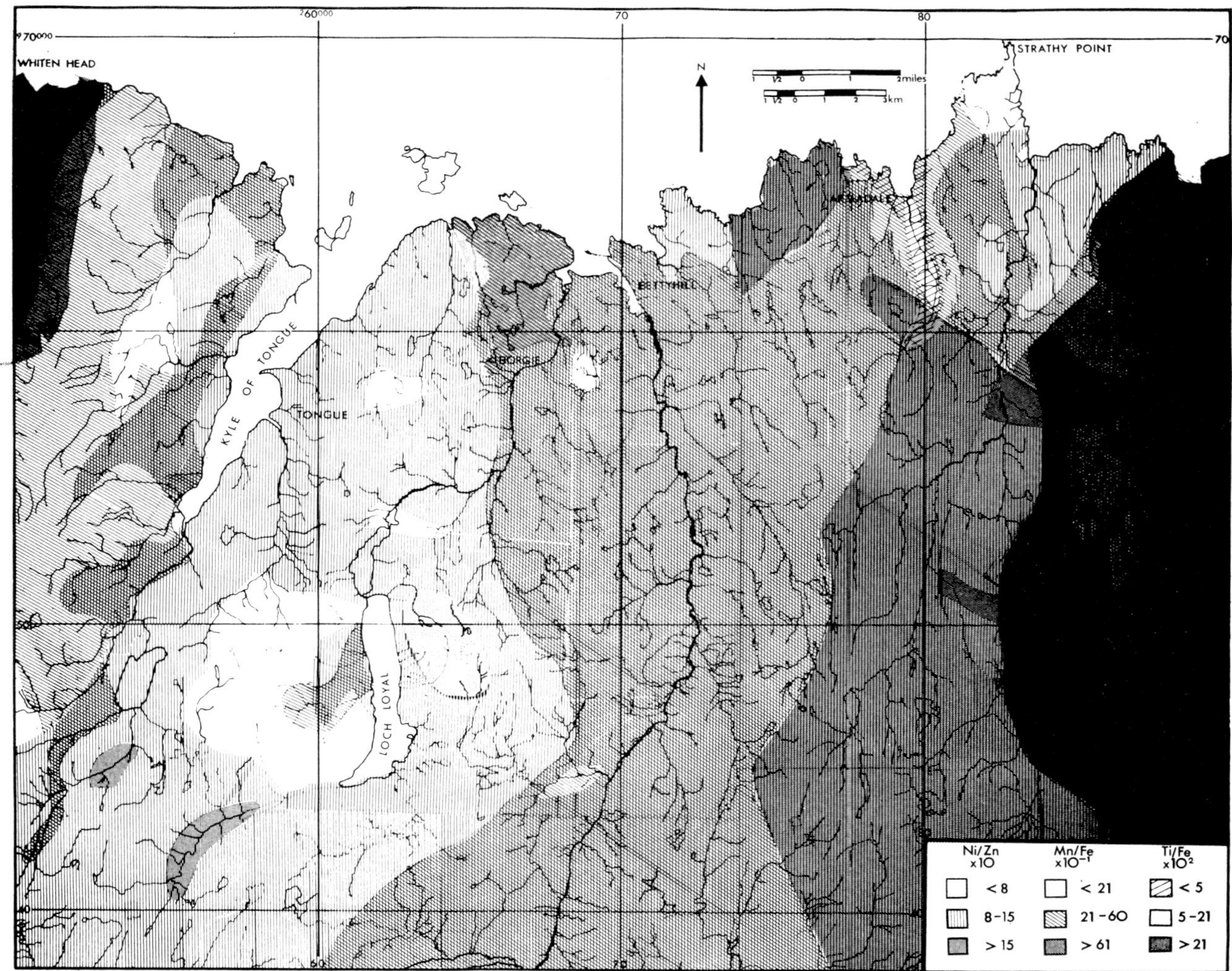

Fig. 8 Panned concentrates: combined Ni/Zn, Mn/Fe and Ti/Fe ratios, north Sutherland

The method of geochemical exploration described above is particularly useful in the survey of regions rich in crystalline rocks of varying mineralogy which have been glaciated. In addition to the detection of mineralization, the

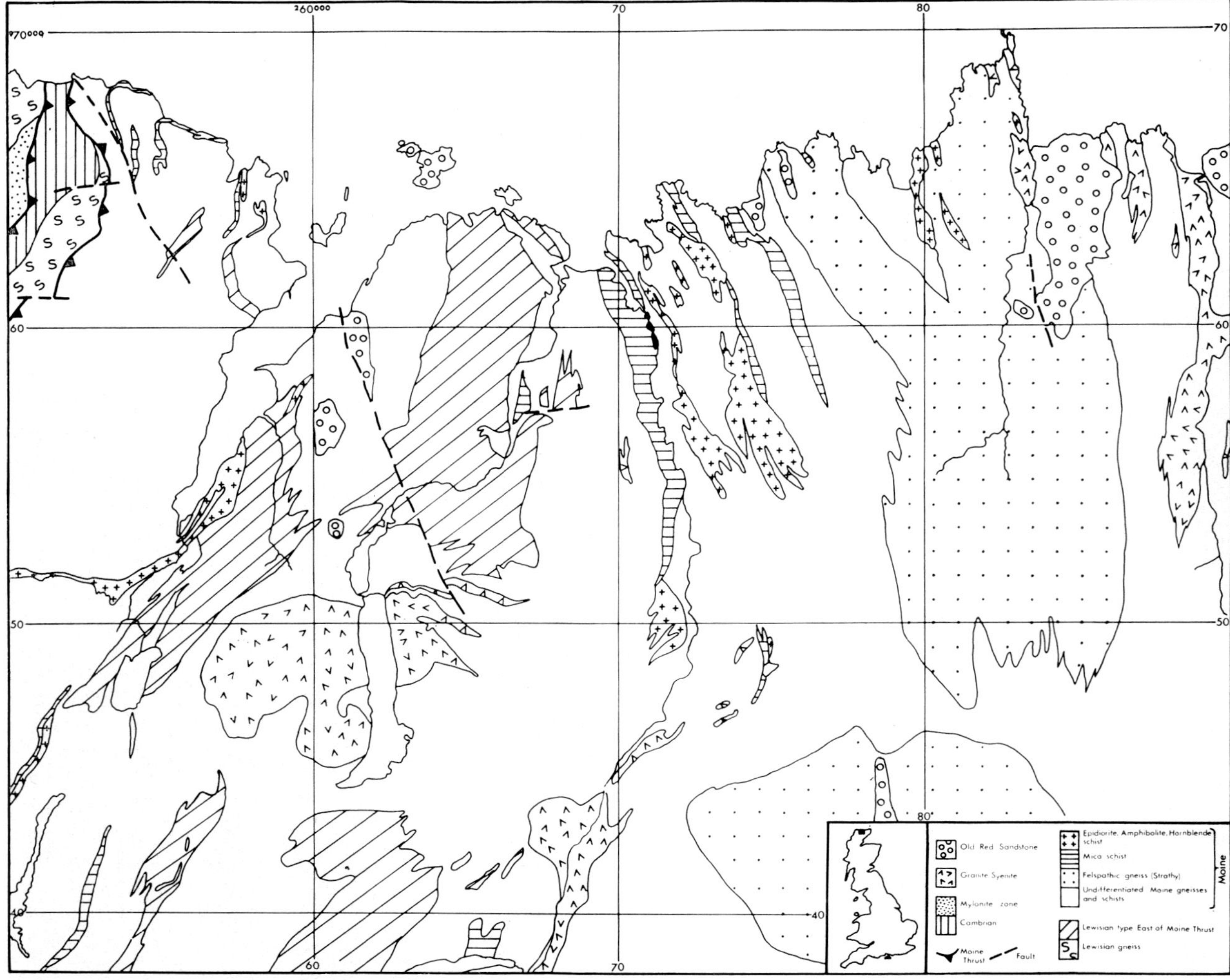

Fig. 9 Geological map, north Sutherland

method enables a great deal of geochemical information to be gathered quickly in little known areas such as north Sutherland, for which little petrological or geochemical information is currently available.

Acknowledgment

The authors wish to thank Dr. J. S. Coats for writing a Fortran program for the calculation of XRF results and Mr. D. Peachey for atomic absorption determinations. Dr. H. Haslam and Mr. P. J. Moore are thanked for helpful discussions and Miss C. Collingborn and student volunteers for help in the preparation of maps and the collection of samples, respectively.

The paper is published by permission of the Director of the Institute of Geological Sciences.

References

1. Smith, T. K. A new method of preparation of heavy mineral concentrate for the determination of trace elements by X-ray fluorescence. *Rep. IGS analyt. Ceram. Unit.* no. 77, 1972.

2. Heinrich, K. F. J. X-ray absorption uncertainty. In *The electron microprobe* McKinley, T. D. Heinrich, K. F. J. and Wittry, D. B. eds. (New York: Wiley, 1966), 296–377.

621.386.832:543.53:553.1

Non-destructive determination of metals in rocks by radioisotope X-ray fluorescence instrumentation*

H. Kunzendorf, M.SC.

Atomic Energy Commission Research Establishment, Risø, DK–4000 Roskilde, Denmark

Synopsis

Radioisotope X-ray fluorescence instrumentation for rapid and non-destructive determinations of metals in rocks is described. Portable apparatuses have been applied in preliminary studies on alkali syenite outcrops of the Ilímaussaq intrusion, south Greenland, and on an outcropping Mo deposit and a heavy mineral deposit in east Greenland. The instrumentation has also been tested in a small stream-sediment survey in east Greenland.

Detection limits found were 0·1% Ti, 0·4% Zr and 0·1% Nb for measurements on cut rock surfaces, and 0·21% Cr, 0·14% Cu, 0·06% Ni, 0·14% Zn, 0·04% Zr, 0·07% Nb, 0·03% Mo, 0·1% Pb and 0·02% La+Ce in pulverized rocks.

The application of portable radioisotope X-ray fluorescence instrumentation is proposed for the investigation of mineralized rock outcrops, especially of the disseminated-ore type, which may lead to the estimation of boundaries and grade of the outcropping ore deposit and its surrounding rocks, for the assay of loose boulders, for the semiquantitative analysis of panned heavy mineral concentrates, and for the assay of drill cores and pulverized rock samples.

Radioisotope X-ray fluorescence spectrometry, which has detection limits of the order of 10 ppm, can be used for the multi-element analysis of stream-sediment samples, drill cores and pulverized rocks in the laboratory. Preliminary work on a drill-core section is presented.

*This paper is based, in part, on a dissertation to be submitted to the Rheinisch-Westfälische Technische Hochschule Aachen, West Germany.

In general, ground geochemical exploration studies are based on trace-element analysis with atomic absorption, spectrographic methods or neutron activation analysis—techniques which have detection limits of the order of 1 ppm metal or less.

In some cases the sampling site and the analytical laboratory are far apart: this may lead to problems concerning further investigations of samples and sites showing anomalous element patterns. Thus, as was mentioned by Salmon,[1] for example, a rapid analytical method used on site may save hours of work and avoid the collection of samples of no importance. In the laboratory, on the other hand, rapid analytical methods are needed with detection limits of the order of 5–10 ppm; these methods should also be of the multi-element type, enabling the simultaneous determination of a series of elements.

Among regions with an increasing exploration activity, as was pointed out recently by Hoy,[2] Greenland presents enormous difficulties for geochemical surveying work. Detailed geological maps necessary for such work do exist (mainly of the west coast of Greenland), and a tectonic/geologic map on the 1 : 2 500 000 scale has been published by the Geological Survey of Greenland (GGU).[3] In planning geochemical exploration in Greenland, however, the following facts have to be considered: (1) the field season is, for this arctic region, relatively short (about three months); (2) transportation is possible only by boat or helicopter—a fact which makes operations by many field teams costly; and (3) the arrival at the analytical laboratory of samples taken in the field is invariably delayed by several months.

To overcome some of these difficulties portable radioisotope X-ray fluorescence (portable RXF), as introduced by Bowie, Darnley and Rhodes,[4] has proved to be one of the few analytical techniques that can be used for *in situ* analysis in Greenland. Although the detection limits of the analytical method employed are normally not better than 200–500 ppm, the speed of analysis (about 30 sec for single-element determinations), low weight (approximately 8 kg) and low price (about £2000) of this type of instrumentation, coupled with the fact that analysis is non-destructive, make this method very appropriate for exploration.

The application of portable instruments to process control and other analytical operations in mining has been described by Cameron and Rhodes,[5] Bowie,[6] Gallagher[7] and Rhodes.[8] As Clayton pointed out,[9, 10] very few applications exist where portable RXF has been used *in situ*, e.g. measurements on rock outcrops, stream sediments or drill cores. This paper includes examples of the successful application of portable RXF instrumentation in the field or field camp in Greenland.

For laboratory use RXF instrumentation is now available which incorporates semiconductor detectors with high-energy resolution to X-rays. These spectrometers are very comparable with the X-ray spectrograph, and have been proved to be superior to it in some cases (e.g. in the determination of heavy metals in rocks). Their introduction has also enabled detailed studies of characteristic features involved in the use of portable instruments (e.g. filter effect, inter-element effects).

Following the early work of Bowman *et al.*,[11] semiconductor X-ray spectrometers incorporating silicon- or germanium-lithium-drifted (Si(Li) and Ge(Li), respectively) detectors have been tested by Burkhalter and Marr[12] in the evaluation of detection limits for the analysis of the heavy elements Ag, W, Pb and U in ores, and by Snyman and Stewart[13] for the determination of U in ores and solutions. A different direction in semiconductor X-ray spectrometry has recently been indicated by Giauque and Jaklevic,[14] who, instead of the radioisotope source, applied an X-ray tube of small size in connexion with a Si(Li) X-ray detection system in their analytical work. The number of fundamental studies of semiconductor RXF spectrometry in the analysis of geological materials is, however, small.

In the present paper the principles of RXF are outlined, and the instruments and techniques used in connexion with geochemical exploration are described. Work with RXF instrumentation in Greenland is presented.

Techniques and instrumentation

Method

The principles of RXF are much the same as those of X-ray spectrography: X-ray fluorescence is used for elemental analysis and the intensities of characteristic X-rays produced in a composite sample are related to the contents of the elements in this sample. In place of the X-ray tube unit and the crystal spectrometer applied in X-ray spectrography, however, a radioisotope and an energy-dispersive detection system are used. A schematic diagram of components of RXF instrumentation is shown in Fig. 1. A radioisotope emits low-energy electromagnetic radiation, which reaches a

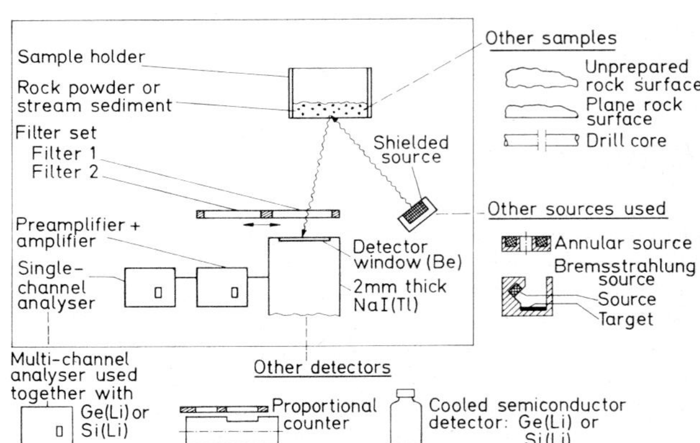

Fig. 1 Schematic diagram of components of RXF instrumentation

sample in a container or the surface of other geological sample material (rock surface, drill core). Characteristic X-rays of the elements in the sample, which are lower in energy than the primary photons, then reach a detector system.

The X-ray detectors used comprise proportional counters, NaI(Tl) scintillation counters and Ge(Li) or Si(Li) semiconductor detectors. Because of the different detectors, RXF may be divided into portable RXF, which generally uses the scintillation counter, and RXF spectrometry, which implies usage of semiconductor detectors. Detector signals are, in both cases, treated by appropriate nuclear electronics. Because of many possible secondary reactions of primary photons from the source and the characteristic X-rays with the atoms of the sample material, the use is recommended of radioisotopes which emit only γ- or X-photons with discrete energies. This simplifies corrections for absorption and scattering in the sample, but the source–target configuration may also be used. The requirements of the radioisotope sources for RXF purposes (low-energy photon emission, long half-life and low price) are met only by a small number of radioisotopes. These sources, generally 1 cm in diameter × 0·5 cm in size, together with some characteristic features and their range of application, are listed in Table 1.

Table 1 Radioisotopes, source strengths, X-ray filters and range of use of portable and semiconductor RXF instrumentation for geochemical exploration purposes

Radio-isotope	Decay mode	Energy and type of radiation emitted	Half-life, years	Portable instrumentation			Semiconductor X-ray spectrometers	
				Source strength, mCi	Element excited	X-ray filter set	Source strength, mCi	Elements excited sufficiently
^{55}Fe	*EC*	5·9 keV Mn *K* α X 6·5 keV Mn *K* β X	2·7	2	Cr V Ti	Ti/V Sc/Ti Sc/Ti	> 5	Cr to P
^{238}Pu	α, γ	13·5 to 20 keV U *L* X	86	10	As Ga Zn Cu Ni Co Fe Mn	Ga/Ge Cu/Zn Ni/Cu Co/Ni Fe/Co Mn/Fe Cr/Mn V/Cr	30	As to Ti
^{109}Cd	*EC*	22·1 keV Ag *K* α X 25 keV Ag *K* β X	1·3	1	Mo Nb Zr	Y/Zr Sr/Y Rb/Sr	> 3	Mo to Rb Fe, if > 1 %
^{241}Am	α, γ	59·6 keV γ 26 keV γ 13·8–21 keV Np *L* X	458	10	La + Ce Ba Sb Sn Cd Ag	Cs/I Sb/I Ag/Cd Pd/Ag Ru/Rh Mo/Rh	> 30	Tm to Rb
^{57}Co	*EC*	136 keV γ 122 keV γ	0·74	1	Pb Hg W Ta	Re/Ir W/Re Er/Tm Tm/Ho	> 3	U, Th Pb to Hf

Generally, for intensity calculations of characteristic X-rays of elements in the rock sample the following contributions to the X-ray intensities have to be considered.

(1) The fluorescent X-ray spectrum recorded by the X-ray detection system is composed of both the characteristic and the scattered X-rays. As geological sample material is mainly made up of light elements (O, Si, Al) for which scattering of low-energy photons is not negligible, scattered photons will be recorded by the detector in quantities which cannot be ignored. This fact is illustrated in Fig. 2, where the characteristic X-ray spectrum of 4975 ppm La is shown in an artificial rock matrix, together with the components of the scattered radiation, obtained with a Si(Li) X-ray spectrometer.

(2) *K* β X-rays of elements with the atomic number Z have nearly the same energy as *K* α X-rays with the atomic numbers Z + 1 (20 < Z < 30), increasing to Z + 4 for the heavy elements (Z > 52). These *K* β X-ray intensities, whose relative intensities compared with the respective *K* α X-ray intensities are well known, have to be subtracted from the *K* α X-ray intensi-

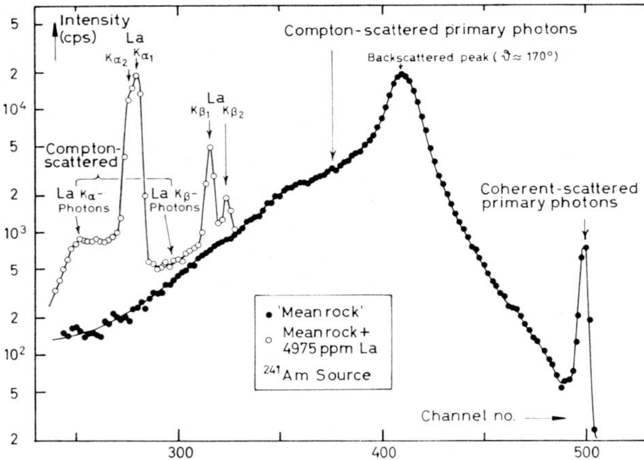

Fig. 2 Radioisotope-excited X-ray spectra of a mean rock sample and 4975 ppm La in this sample material showing both the scattered and the characteristic X-rays

ties used for the analysis.

(3) L X-rays of heavy elements ($Z > 52$) occurring in quantities greater than 1000 ppm in a sample will overlap with energies of K X-rays of elements with $Z < 40$. It is, therefore, necessary to correct also for L X-rays.

(4) Characteristic X-rays are absorbed in the sample material before they reach the X-ray detector. Corrections for this are usually not simple, because the degree of absorption depends on the mean atomic number $\bar{Z}$ of the sample, i.e. the elemental composition. In the same category are absorption effects caused by grain-size variations in the sample material. These effects are reduced by comparing only samples with relatively known grain-size patterns.

The real advantage of RXF lies in its speed—not in the low values of its detection limits, accuracy and precision. Sample preparation should, therefore, be as simple as possible. With RXF spectrometry, for example, quick semi-quantitative checks of samples for a series of trace elements may be more informative than time-consuming whole-rock analyses for trace elements.

Portable instrumentation

Portable RXF instrumentation, according to Rhodes,[8] is available from various manufacturers. The weight of these instruments is normally less than 10 kg, a probe unit, containing source, X-ray filters and detector unit, and a single-channel analyser being coupled together.

Because of the poor energy resolution of the detectors employed, X-ray filters select energy intervals of the composite X-ray spectrum of the sample corresponding to the element to be analysed. In general, for these filters usage is made of the mass-absorption coefficient function $\mu(E)$. The X-ray filter effect is shown in Fig. 3. Two filters belong to one filter set: the filter constructed of Rb, which has an absorption edge at 15·2 keV, and the filter made of a Sr compound (absorption edge, 16·2 kev) make up a pass-band for Zr $K\alpha$ X-rays (15·7 keV).

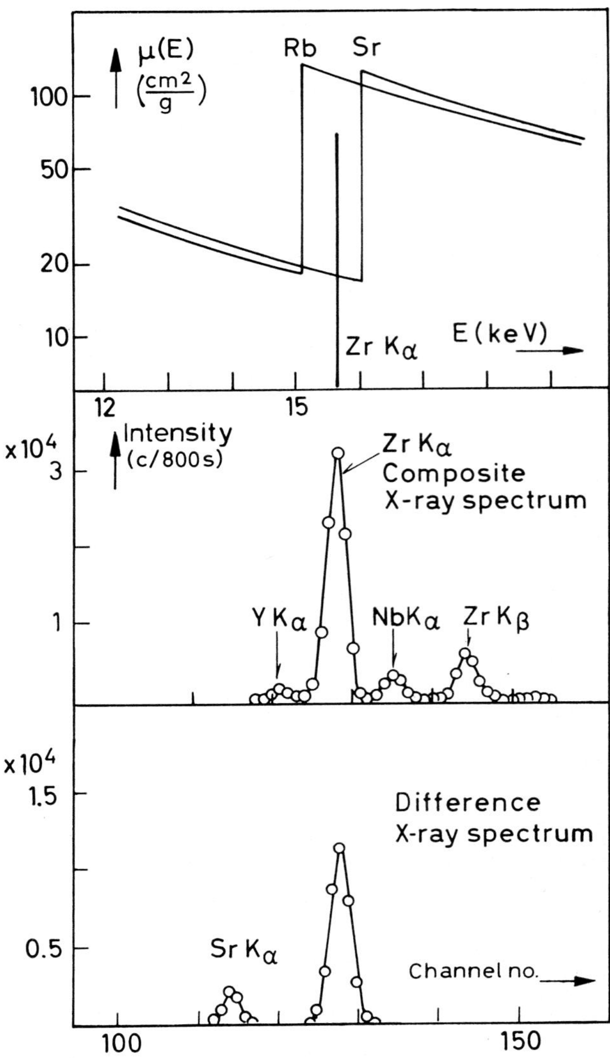

Fig. 3 Mode of operation of X-ray filter set: difference X-ray spectrum has been obtained by subtracting X-ray spectrum through Rb filter from that through Sr filter

The difference count rate (DCR) through both filters (first, count rate through Sr filter) belongs mainly to Zr. This is clearly shown in Fig. 3, where a high-resolution Si(Li) X-ray spectrometer has been used to detect the X-rays of a composite sample containing Zr through both filters. Filter X-rays may, however, be detected if the filters are not properly balanced, i.e. absorption before and after the absorption edges is not made equal for both filter materials. X-ray filter sets for all metals are easy to construct and are available for most elements of economic interest.

The following portable RXF instruments have

been used in this investigation: (1) an early laboratory unit comprising a Xe-filled proportional counter and standard nuclear electronics: it was used to check the feasibility of the RXF method before field-testing (results of this check on cut rock surfaces are given in Table 2); (2) an RXF probe manufactured by Nuclear Enterprises, Ltd., England, used in connexion with a portable gamma spectrometer;[15] and (3) a *Mineral Analyser* supplied by Ekco Electronics, Ltd., England.

Table 2 Detection limits, accuracy and precision of Zr, Nb and Ti measurements of cut alkali syenites by means of laboratory RXF instrumentation incorporating proportional counter (Xe-filled) and X-ray filters

Element	Detection limit, %*	Accuracy, %†	Precision, %‡
Zr	0·4	0·15	< 6 For coarse-grained rocks
Nb	0·1	0·15	< 3 For medium- to fine-grained rocks
Ti	0·1	—	

*Based on the 95% confidence interval for a given count rate in the regression analysis; regression analysis of the linear part of the calibration curve gives the calibration constants a and b, their respective variance, and the variance of the difference count rate y; to any element content $x=(y-a)/b$, 95% confidence limits $x\pm\Delta x(y)$ can be assigned. The detection limit is given by $x_D=\Delta x$.

†Determined as $t_{(n-2)}\cdot\sqrt{\sum(e^2/(n-2))}$, where e is the difference between the contents calculated from the X-ray intensities of the standards by use of the regression line and the chemically determined contents of the standards, n is the number of standards, and $t_{(n-2)}$ the appropriate Student's t distribution value.

‡Determined from replicate analyses.

Steps to be considered in the work with RXF field instrumentation may be exemplified by the procedures involved in the measurements with the *Mineral Analyser*. Equipped with a probe unit and a single-channel analyser with digital readout, the instrument can be used continuously for 5 h. The battery pack, connected to the analyser unit, was then recharged by means of a small motor generator during field work in Greenland. Calibration of the instrumentation was carried out before the field work by successive dilution of ore

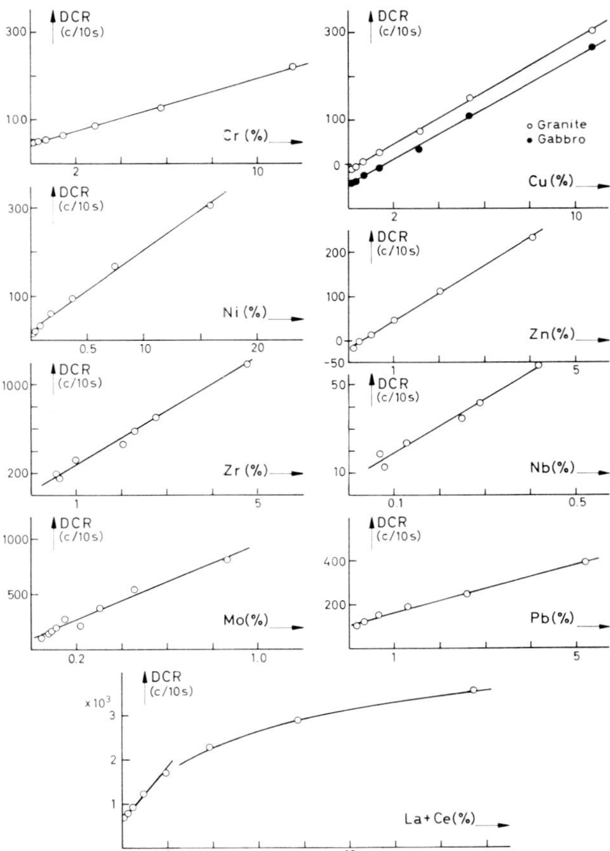

Fig. 4 Calibration curves for chromite (Cr) and rammelsbergite (Ni) in gabbro, bornite (Cu) in gabbro and granite, sphalerite (Zn), monazite (La+Ce) and galena (Pb) in granite, eudialyte (Zr and Nb) in alkali syenite (kakortokite), and molybdenite (Mo) in alkali granite

minerals with appropriate matrix materials (granite or gabbro). Calibration curves for the elements Cr, Ni, Cu, Zn, Zr, Nb, Mo, La+Ce and Pb are given in Fig. 4. Detection limits and accuracy for these measurements are shown in Table 3.[17]

For quantitative measurements on rock outcrops, in principle, only a conversion factor for the curves in Fig. 4 is needed for the calculation of the elemental contents. Certain effects (grain size, surface irregularities, matrix composition), however, necessitate calibration of the field measurements by analysis of chip samples of selected measured areas or measured surfaces of selected hand specimens.

One possibility for the correction of intensities of X-rays effected by absorption in the sample material is the use of scattered radiation, also detectable with the portable instrument. In the analysis of Zr in pulverized alkali syenite (kakortokite) the count rate through the Rb filter may be used to reduce the influence of different mineralogical compositions of the samples on the difference count rates. For this purpose alkali-syenite

Table 3 Detection limits and accuracy for analysis of rock powders by means of portable RXF instrumentation (Ekco Mineral Analyser)

Element	Sample material	Range of elemental content, %	Detection limit, %	Accuracy, %
Cr	Chromite in gabbro	0–12	0·21	0·08
Ni	Rammelsbergite in gabbro	0–1·6	0·06	0·03
Cu	Bornite in granite	0–5·4	0·14	0·17
Zn	Sphalerite in granite	0–4·2	0·14	0·25
Zr	Kakortokite in granite	0–1·0	0·04	0·02
Nb	Kakortokite	0–0·29	0·07	0·02
Mo	Molybdenite in alkali granite	0–0·17	0·03	0·03
La+Ce	Monazite in granite	0–0·49	0·02	0·01
Pb	Galena in granite	0–2·7	0·10	0·14

Detection limits and accuracy as defined in Table 2.

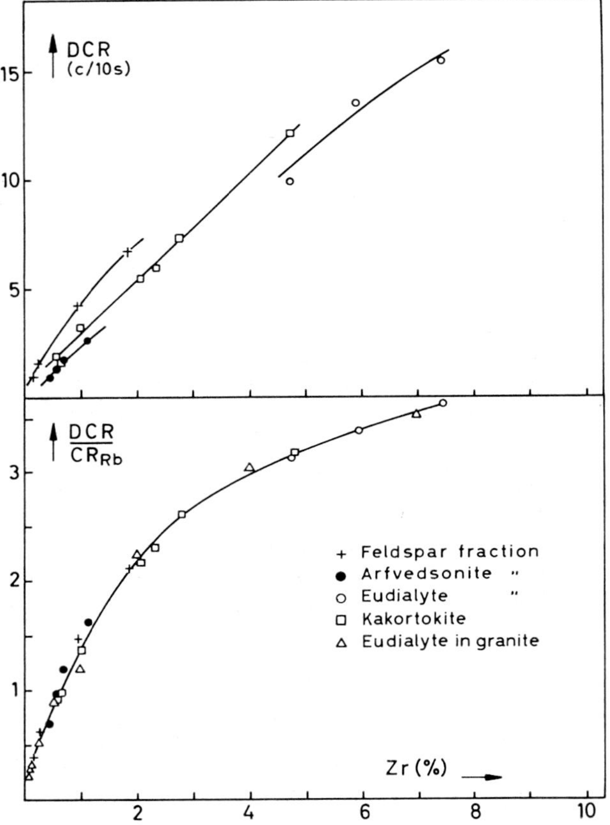

Fig. 5 Suppression of matrix effect in analysis of Zr in alkali syenite (kakortokite) by division of difference count rates (DCR) by count rate through filter with lowermost absorption edge (CR_{Rb}), which mainly contains the scattered radiation

powders were separated into fractions dominated by feldspars, arfvedsonite and the Zr mineral eudialyte. RXF measurements are plotted against colorimetric determinations of Zr in the upper part of Fig. 5. As expected, the fractions dominated by the heavy minerals arfvedsonite and eudialyte yield lower difference count rates than the feldspar fractions and the kakortokite samples due to increased absorption of Zr *K* X-rays in the heavier minerals. By dividing the difference count rates by the count rates through the Rb filter, however, which is, to a first approximation, equal to division by the number of photons incoherently scattered in the sample, all of the RXF measurements are represented by only one calibration curve. Moreover, analyses of samples in which a kakortokite sample with high eudialyte content was successively diluted with a granite powder also fit this curve.

Laboratory instrumentation

Si(Li) and Ge(Li) X-ray detection systems have been marketed during the last three years. They are most efficiently used in connexion with a multi-channel analyser. The energy resolution of these spectrometers is of the order of 200–300 eV for 6-keV X-rays. It is only obtained for systems with relatively small detector volume (<1 cm^3). Hence, source strengths of radioisotopes used in connexion with semiconductor X-ray spectrometers have to be an order of magnitude greater than for portable instruments. The whole X-ray spectrum of a sample is treated for analysis. The need for liquid nitrogen supply and for advanced electronic devices limits the application of these detectors to the laboratory or the very advanced field laboratory.

Detection systems used in this investigation were a Ge(Li) detector with an energy resolution of about 1 keV, and an ORTEC Si(Li) X-ray spectrometer with a detector of 6-mm diameter, 3-mm active depth and a guaranteed system resolution of about 250 eV for Fe *K* α X-rays. The fluorescent X-ray spectra of stream-sediment samples excited by ^{241}Am and recorded with the

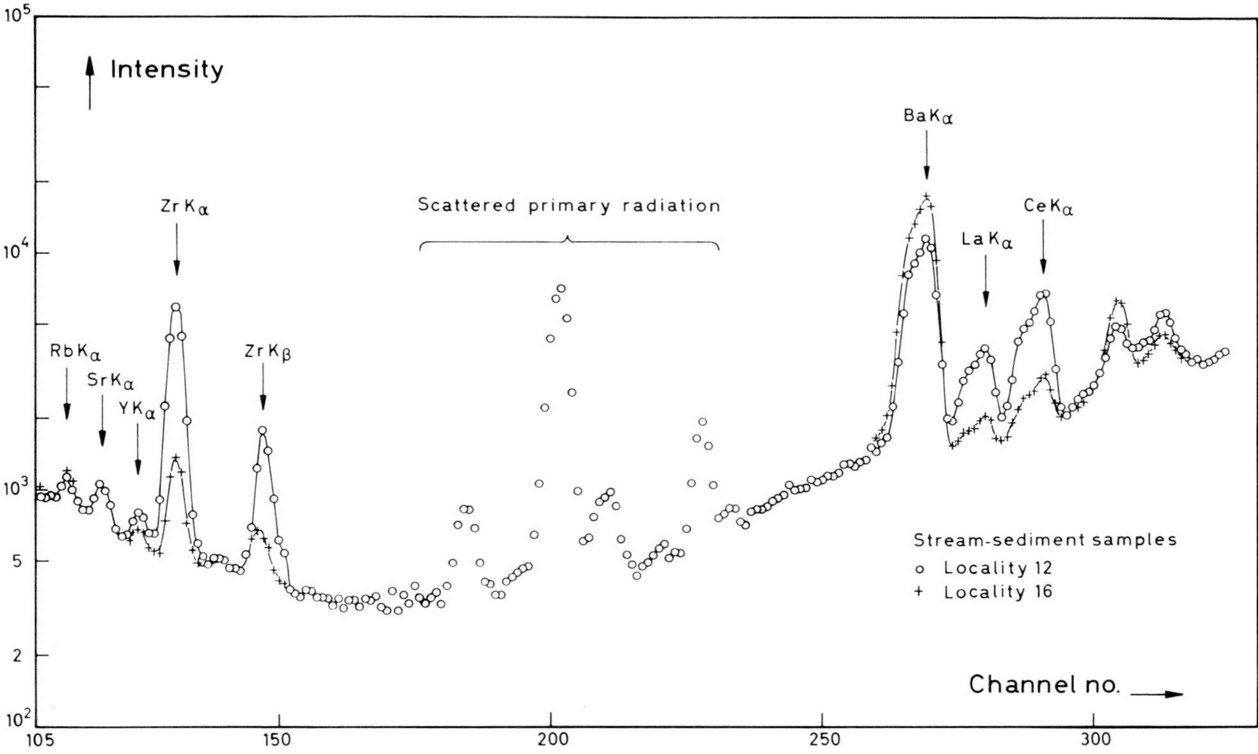

Fig. 6 Characteristic X-ray spectra of elements Rb through Ce in stream-sediment samples from east Greenland. Spectra were obtained with 10-mCi ^{241}Am source, Si(Li) X-ray detection system and 512-channel analyser

Si(Li) X-ray detection system are shown in Fig. 6.

For calibration purposes X-ray intensities have to be related by element concentrations in the sample determined by independent chemical methods. Intensity calculations are usually done with the aid of a computer, which treats the whole X-ray spectrum of a sample. Two methods have been tried: least-squares fitting of the measured X-ray spectrum to a theoretical X-ray spectrum[18] and integration of selected energy intervals, which are mainly composed of the characteristic X-rays of the element to be analysed.[19] Table 4 presents some of the results obtained (detection limits, accuracy and precision). It appears that (cf. Table 2) detection limits are rather pessimistic (dependence on range of elemental contents of standards). Also, accuracy is poor because RXF is based on the accuracy of chemical determinations of elements in these standards. By choosing standards with lower Zr, Nb and Mo contents than those used in Table 3, detection limits an order of magnitude lower can be expected.

Table 4 Typical detection limits, accuracy and precision for RXF by use of semiconductor detectors

Sample material	Elements determined	Detection limit, ppm	Accuracy, ppm	Precision, %	Remarks
Cut rock surfaces	Ce	400	400	2	Ge(Li) X-ray spectrometry; least-squares fitting technique used[18]
	Nd	180	170	2	
Rock powders	Zr	69	66	2	Si(Li) X-ray spectrometry; integration of X-ray peaks used[19]
	Nb	206	296	5	
	Mo	172	159	—	
	Fe	1·35%	1·24%	2	

Detection limits, accuracy and precision as determined in Table 2.

It is worth while to mention that the amount of inelastic scattered radiation obtained by the scattering of primary radiation in the sample may be used to correct X-ray intensities for changes in the mean atomic number $\bar{Z}$ of the sample. In most cases the scattering ratio of inelastic to elastic scattered photons is a function of $\bar{Z}$. Hence, this number mainly influences the magnitude of the mass-absorption coefficient $\mu(E)$ of the sample. X-ray intensities are, however, difficult to correct for interferences.

Results and discussion

Examination of rock outcrops and boulders

RXF measurements on rock outcrops carried out with portable instrumentation are possible under certain conditions: (1) the rock outcrop should be clean and fairly plane; minor surface irregularities do not, however, influence the difference count rate of RXF measurements by more than 10%, because three distance pins on the front side of the probe reduce the sample–source distance effect; (2) the mineralogical composition of the rock must be taken into account; and (3) the texture of the rock determines the number of measurements necessary, since only a few square centimetres are examined in any single analysis.

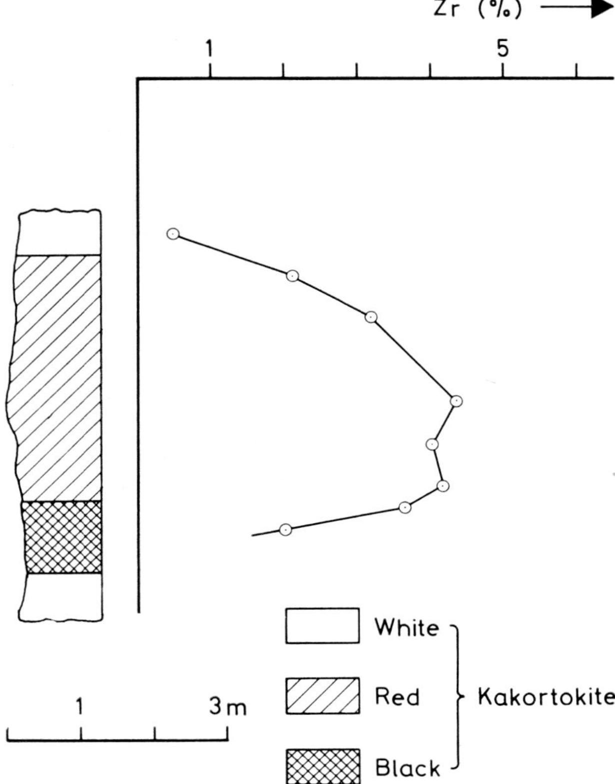

Fig. 7 Distribution of Zr in section of kakortokite unit (vertical section) of Ilímaussaq intrusion as obtained by measurements with portable RXF (note that Zr is highest in red kakortokite layer, where the dominating heavy mineral is eudialyte)

For dense homogeneous rocks the mean abundance of an element in the rock outcrop may be established by only a few RXF measurements, whereas many analyses are needed for coarse-grained rocks and, particularly, for pegmatitic and vein-type mineralization.

Because of the small penetration depth of characteristic X-rays, it is always necessary to sample measured areas by removing the upper few millimetres of these areas and to make check analyses of the sampled material by independent chemical methods.

Four examples relate to the application of portable RXF in Greenland.

(1) The layered kakortokite body (locality 1, Fig. 9) of the Ilímaussaq intrusion, south Greenland[20, 21] contains considerable amounts of the Zr mineral eudialyte.[16] RXF measurements for Zr with portable RXF instrumentation (Nuclear Enterprises, Ltd., probe) along the uppermost kakortokite unit yielded an estimate of the Zr in the outcrop of this unit. RXF measurements of a section of the unit (Fig. 7) show that the Zr content is highest in the layer which also contains most eudialyte (red kakortokite has up to 4·5% Zr). As the layers of the kakortokites have a relatively constant mineral composition and seem to be fairly homogeneous within each unit,[16] systematic RXF measurements on the outcrops of the kakortokites may yield an estimate of the mean Zr concentration in the other kakortokite units and thus, in conjunction with geological information, may enable the approximate size and grade of the orebody to be calculated.

(2) On the basis of 400 field analyses carried out with the *Mineral Analyser* in the Mestersvig region, east Greenland, on the west side of an outcropping low-grade molybdenite deposit (closely spaced RXF measurements, 0·5 m between measurement stations), a first approximation of the Mo contents in the outcrop was established. The instrument was calibrated after return to Denmark by analysing the crushed material of selected ore specimens by atomic absorption spectrography. The spread of the Mo contents of subsequent measurements (Fig. 8) demonstrates the influence of the texture of the rock type on RXF measurements (Mo occurs in molybdenite fissures and veins and the *Mineral Analyser* responds only to Mo if the measuring head is placed on a molybdenite vein). The geometric mean of these field readings gives a fairly good estimate of the Mo content of the rocks investigated.[28]

(3) The same RXF instrumentation was used in the Jurassic sediments of Milneland (locality 2, area A, Fig. 9). The Charcot Bugt Sandstone[22] has been found to be radioactive, mainly in regions where the sediments are in close contact with the

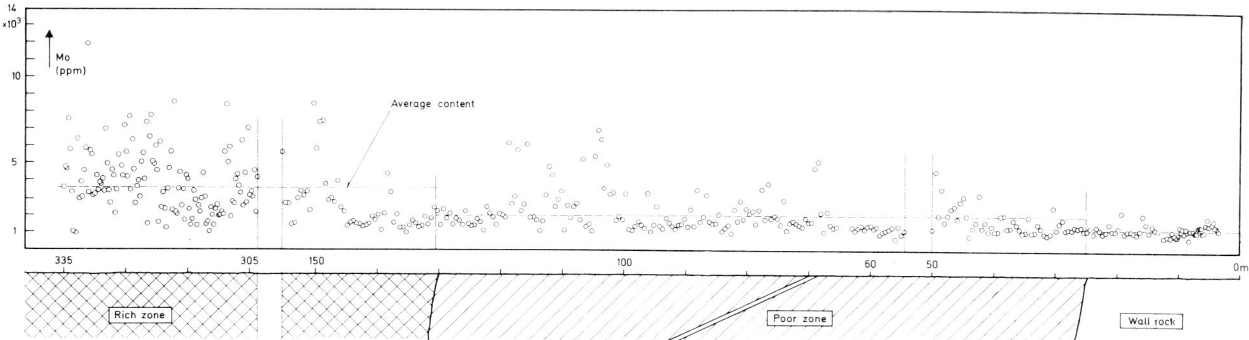

Fig. 8 Mo distribution in molybdenite-bearing alkali granite near Mestersvig, east Greenland, along western part of the outcropping deposit (note spread of measured Mo values—0·5 or 0·25 m between each measurement—in the rich zone, which is due to increased occurrence of MoS_2 veins and fissures in this zone)

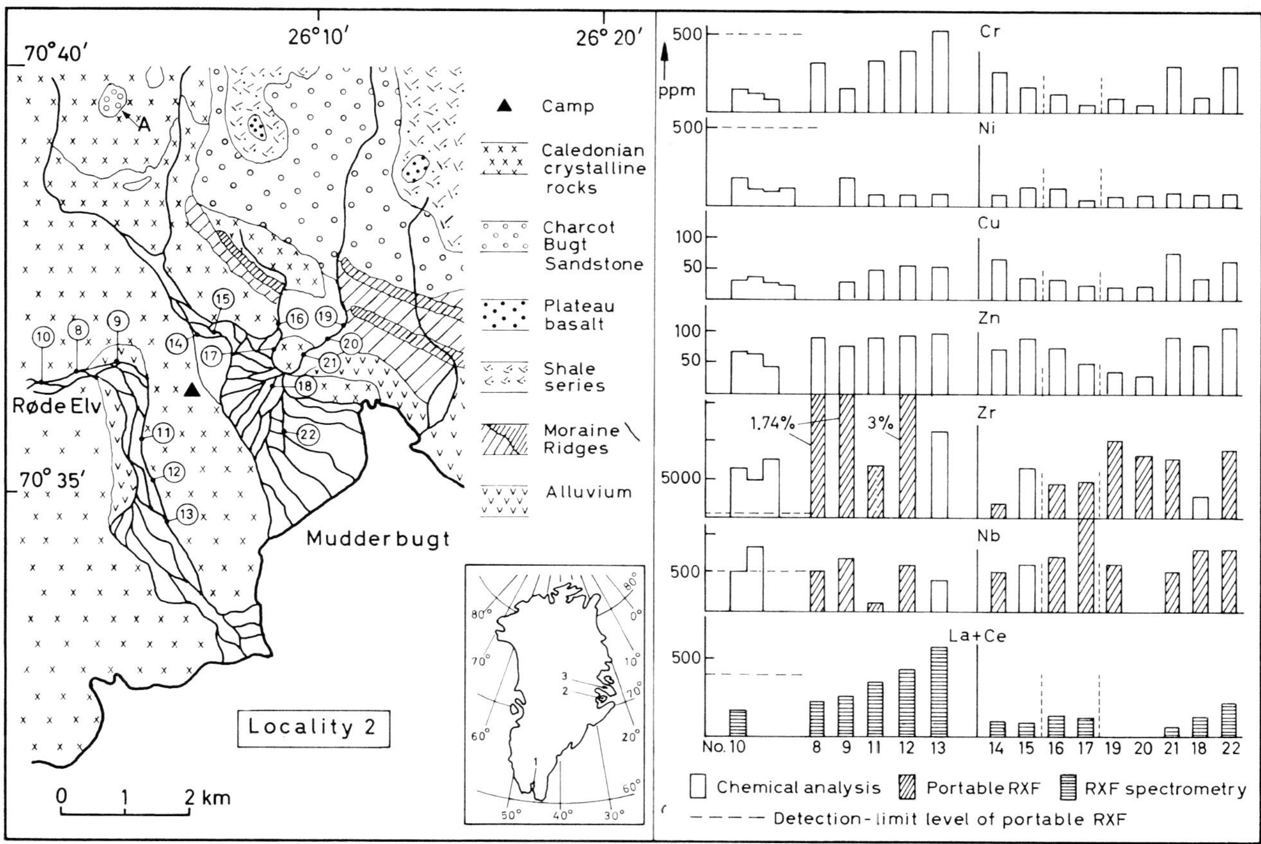

Fig. 9 Regional map of sampling sites from stream sediments, and results of chemical analyses, portable RXF measurements and RXF spectrometry. Detection limit levels of portable RXF, as indicated by dashed lines, are estimates. Areas indicated by 1, 2 and 3 in schematic Greenland map are those where RXF instrumentation has been used in situ

crystalline rocks, representing the bottom of the sandstones.[23] Of the localities with anomalous high values of radioactivity, one was selected for detailed investigation. Zr and La+Ce contents were determined in material from trenches and in surface material from traverses across the trenches. The results of RXF measurements through a trench section are given in Fig. 10. Generally, along the line of measurement Zr is higher than 10%. It seems that there is Zr enrichment towards the bottom of the unconsolidated sediment layer. For La+Ce the same tendency is observed, these elements possibly being associated with the mineral monazite.

(4) In some areas of east Greenland with a particularly difficult terrain hand specimens were collected; in addition to the usual quick petrological examination of the specimens, their fresh surfaces were check-analysed with the portable RXF analyser. Further investigation of the areas was initiated if both petrological examinations and the instrument had detected anomalous

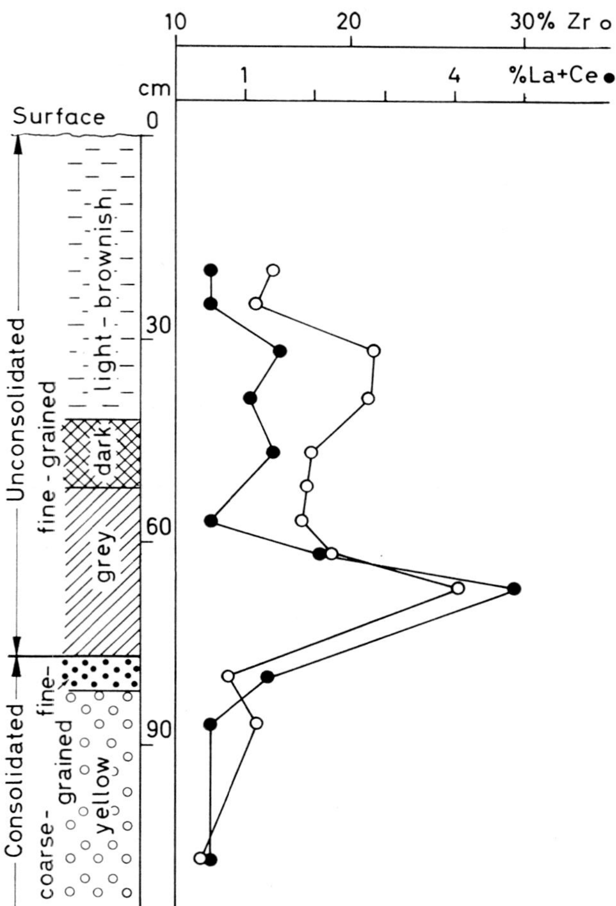

Fig. 10 Distribution of Zr and La+Ce in blasting cave no. 2 (area A, Fig. 9) as obtained by portable RXF measurements (note maximum of both Zr and La+Ce in unconsolidated greyish sediment layer)

amounts of metals in the hand specimen. By this procedure more uncommon ore mineralization was easily recognized on the spot.

Analysis of stream-sediment samples

Examination of heavy minerals in stream-sediment samples for exploration purpose may be carried out by separating the heavy fraction of the stream sediments from the light minerals by means of heavy liquids of a specific gravity greater than 2·6. This fraction is then investigated mineralogically and chemically.[25]

It has been shown by Kalsbeek[26] that most of the minerals, including the heavy minerals of sand samples from the Fiskenæsset area, west Greenland, represent the catchment drainage. These minerals may, therefore, be representative of the rocks that may carry ore mineralization. Thus, rapid analysis of ore and pathfinder metals in the stream-sediment samples, followed by detailed mineralogical examination of the samples, may be a useful tool in geochemical studies in Greenland.

A systematic survey of stream-sediment samples with the *Mineral Analyser* was carried out in the area indicated in Fig. 9 (locality 2). Only the eastern part of the area has been mapped in detail by the Geological Survey of Greenland; the western part of the area is assumed to be mainly composed of crystalline rocks. The general geology[22] is indicated in Fig. 9.

The purpose of the investigation was to test the sensitivity of the portable instrumentation and to determine the procedures which should be applied for a possible stream-sediment survey with this type of instrumentation in east Greenland.

The −0·5-mm fraction of the stream sediments collected along the Røde Elv and the Mudderbugt area (Fig. 9) was treated in the following way: the clay fraction of the sediments was removed by brief washing in a pan; it was then air-dried overnight (stable, dry weather conditions) and examined the next day by means of the portable RXF instrument. All the stream sediments were later analysed for Cr, Ni, Cu, Zn, Mo and Pb (atomic absorption), Zr and Nb (wet chemistry) and Ce (mass spectrometry). A calibration curve for La+Ce for the Si(Li) X-ray spectrometer was established by plotting the La+Ce contents of selected stream-sediment samples versus their respective X-ray intensities. It was then possible to calculate the La+Ce contents in other stream sediments. The results of the stream-sediment survey are given in Fig. 9. Although the anomalous high values for Mo and Pb (>20 ppm Mo and >150 ppm Pb) in the samples are not included, some general conclusions may be drawn.

The low concentrations of Cr, Ni, Cu, Zn (<200 ppm), Mo and Pb in the stream sediments suggest that portable RXF instrumentation has detection limits which are too poor for it to be applicable in further studies for these elements. With regard to Zr, the response to the RXF instrumentation seems to be fairly good—indicating that Zr determinations in stream-sediment samples are possible. Relatively high Nb values in the sediments may be due to the known difficulties of the chemical analysis. The La+Ce contents of the stream sediments from the Røde Elv area are close to the detection limit of the portable instrumentation. Field readings were, however, higher for the stream sediments of the Mudderbugt area—a fact which, most probably, can be ascribed to the higher contents of light minerals in the respective stream sediments, leading to less absorption of characteristic La and Ce X-rays in the samples.

The Zr and La+Ce contents of the stream sediments of the Røde Elv, which originates in the area where crystalline rocks (mainly migmatites) predominate, are higher than those of the sedi-

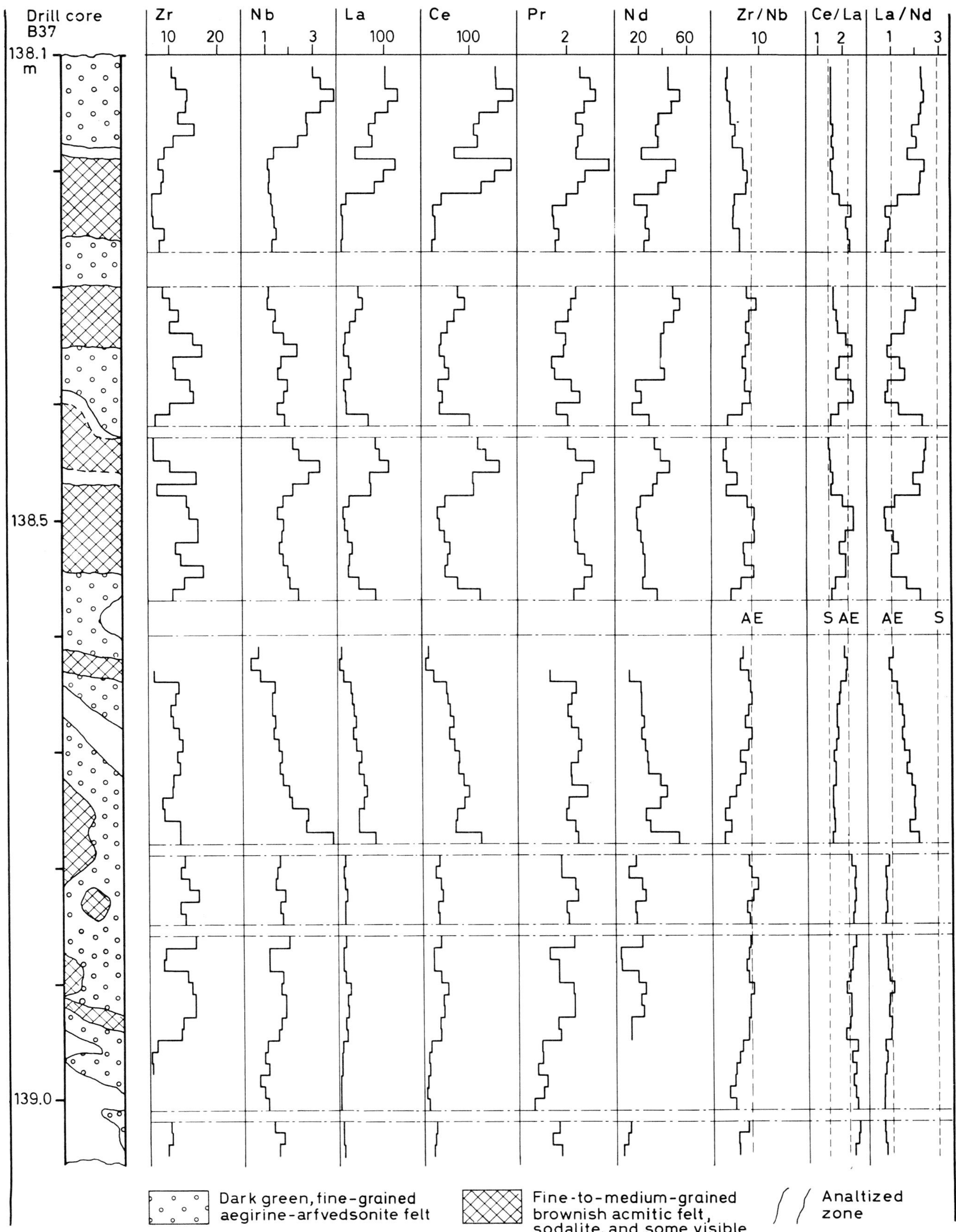

Fig. 11 RXF scanning diagram for Zr, Nb, La, Ce, Pr and Nd in longitudinally cut drill core of Ilímaussaq intrusion. X-ray intensities (counts/sec) were obtained by means of ^{241}*Am source and Si(Li) X-ray detection system. Element ratios Zr/Nb, Ce/La and La/Nd in drill-core section are also shown, and indicated as dashed lines for steenstrupine (S) and aegirine (AE)*

ments of rivers running through areas where Jurassic sediments are present. Although parts of the Jurassic clastic sediments contain high Zr and rare-earth element concentrations, no particular enrichment of these elements is observed in the respective stream sediments. It is not known whether the detrital minerals of the Charcot Bugt Sandstone (zircon and monazite) were derived from migmatites or from post-kinematic granites within the same area.

Portable RXF instrumentation is more useful in the semiquantitative analysis of panned concentrates of stream sediments. Bearing in mind that these concentrates are mainly composed of heavy minerals, which strongly absorb X-rays, field difference count readings obtained with the *Mineral Analyser* (see Fig. 4) strongly indicate the presence of the respective elements in the concentrates in amounts in excess of 1% by weight.

For systematic surveys of stream-sediment samples, however, a laboratory radioisotope X-ray spectrometry system should be employed, this type of instrumentation giving results as good as those of other spectrographic methods. Many elements can directly be determined non-destructively in the samples with only one radioisotope. This is clearly exemplified by the X-ray spectra of a sample from locality 12 and a stream-sediment sample from locality 16 of the area described above; Fig. 6 shows the presence of Rb, Zr, La, Ce and Ba and other elements with different magnitudes in these samples.

Assay of drill cores

Non-destructive multi-element assay of drill cores is one of the most obvious applications of semiconductor RXF spectrometry. Although portable instruments may give good results when semiquantitative drill-core analyses in connexion with drilling operations and guidance to further drilling work are needed, filter and source change operations are time-consuming and make the application of these instruments less useful if the ore is of the multi-element type (e.g. Cu–Ni–Zn or Pb–Zn).

Multi-element assaying of longitudinally cut or unprepared drill cores by means of semiconductor RXF spectrometry in the laboratory is superior to other analytical techniques. Simple scanning apparatus used in connexion with a Si(Li) or Ge(Li) X-ray detection system and operated in the continuous or step-scanning mode may be constructed. The drill core passes a collimator slit, with a preselected width, through which the core is bombarded by the photons of a radioactive source. The whole fluorescent X-ray spectrum of the core is then recorded by a multi-channel analyser and treated by a computer program. Calibration of the multi-element scanning system may be carried out in a way similar to that applicable to plane rock surface—removal of thin layers of selected areas with known X-ray intensities of a series of elements, followed by chemical analysis of the crushed surface material.

The proposed technique may be considered by reference to the preliminary scanning diagrams given in Fig. 11: 1 m of a longitudinally cut drill core from the radioactive drill cores of the Ilímaussaq intrusion[18] has been scanned manually along its plane surface. Instrumentation used for this investigation was the Si(Li) X-ray spectrometer with a 10-mCi ^{241}Am source; the radiation of the source reached the drill core through a 1-cm wide slit, which permitted only 1 cm × 2·5 cm of the drill core to be viewed; backscatter geometry was used. The output of a 512-channel analyser was treated by a computer program with the aid of a 18K Burroughs computer. Net X-ray intensities, corrected for scattered primary and secondary radiation, the backscatter photon intensities and a scattering ratio (inelastic to elastic scattered photons) were calculated.

For the manual scanning of this drill-core section some preliminary conclusions may be drawn: high La, Ce, Pr and Nd contents seem to be associated with the rocks where the Th mineral steenstrupine (S) is visible, and they coincide with the high Th and U values obtained by gamma-spectrometric drill-core scanning of the same areas.[24] These areas are also indicated by low Ce/La ratios, which are characteristic of steenstrupine, whereas La/Nd ratios are high for these areas; Zr and Nb X-ray intensities are slightly influenced by Th *L* X-rays, leading to higher Zr and Nb intensities in the areas with visible steenstrupine than in the areas where the mafic mineral aegirine predominates.

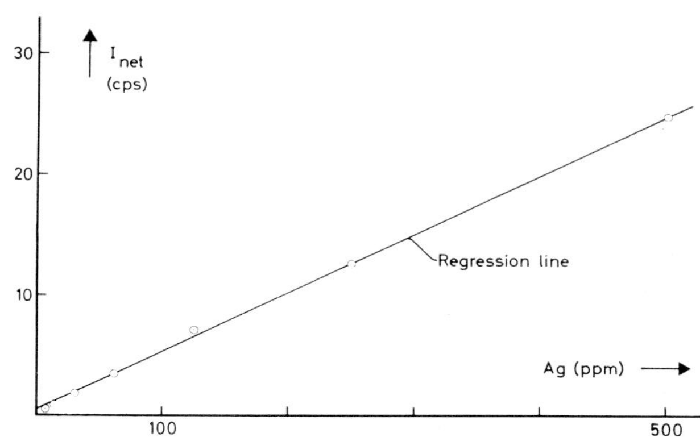

Fig. 12 Calibration curve for Ag in $H_2SiO_3.nH_2O$ powder covering range 0–500 ppm Ag as obtained by Si(Li) X-ray spectrometry

Analysis of rock powders

Although analysis of rock powders is not considered to be non-destructive, portable RXF instrumentation may be applied in the assay of powdered metallic ores if the ore grade is higher than 200 ppm. This type of analysis is particularly useful in the field laboratory which has rock-crushing equipment. Bearing in mind that RXF measurements are carried out within 30 sec, the duration of sample preparation should not exceed 10 min. Most of the known analytical work on rock powders by means of portable instrumentation has, however, been done in the advanced laboratories where other analytical methods might have given better results.

Si(Li) or Ge(Li) X-ray detection systems may be used in the analysis of rock powders containing trace elements of economic importance in quantities greater than 10 ppm. Multi-element analysis of fine-grained rock powders (−200 mesh) requires carefully analysed standard materials; the U.S. Geological Survey standard rock powders W-1, G-2, PCC-1, DTS-1, AGV-1, BCR-1 and GSP-1 may be used for calibration purposes. Many trace elements are, however, present in these standards in quantities lower than the detection limits for the spectrometer systems used. Mixed samples containing chemical compounds of trace elements of varying contents may yield a first approximation of the content of an unknown trace element in a geological sample. Careful sample matrix choice is then recommended. The calibration curve of successively diluted Ag in a $H_2SiO_3.nH_2O$ powder is shown in Fig. 12. The calibration data fit very well with the regression line. A detection limit of 19 ppm Ag and an accuracy of 15 ppm Ag are obtained from the regression analysis. For ore samples, however, different calibration curves are expected, silver usually being found together with other heavy metals, e.g. Pb.

Conclusions

Two techniques of non-destructive radioisotope X-ray fluorescence have been described. The advantages of RXF instrumentation include the possibilities of *in situ* analyses with compact RXF systems, which may be portable apparatuses, the simple operation of these instruments and the speed of analysis, which may be of the multi-element type. The main disadvantage of the method is that only a relatively thin layer of the sample material is sufficiently excited by the radioisotope.

Portable RXF seems to be extremely useful in the following phases of geochemical exploration: (*a*) investigations of mineralized rock outcrops, particularly of the disseminated-ore type; (*b*) assay of loose boulders (e.g. from moraines) with their origin in nearby rocks; (*c*) semiquantitative analysis of heavy mineral concentrates obtained by the panning of stream sediments; (*d*) assay of drill cores to direct drilling operations; and (*e*) analysis of pulverized rock samples.

Generally, detection limits are greater than 1000 ppm for metals in rock outcrops and greater than 200 ppm for metals in rock powders. Values for the accuracy of these analyses are greatly influenced by the accuracy of analysis of standard samples, upon which RXF is based. In most cases values close to the detection limit are obtained. Precision based on repeated measurements is generally better than 10%.

RXF spectrometry with semiconductor detectors is not yet possible on rock outcrops, although drill-hole logging equipment incorporating a semiconductor X-ray detector has recently been reported.[27] By means of the Si(Li) or Ge(Li) X-ray detection systems, multi-element analyses are possible, at the 10 ppm level, of samples of types (*c*) and (*d*) as well as of stream sediments and soil samples. Si(Li) X-ray detectors operated without cooling facilities in a field laboratory and equipped with a multi-channel analyser coupled to a small computer may be introduced in the future. The physical aspects of treating a complex X-ray spectrum superimposed on the scattered radiation from a rock sample composed of many elements have, however, not yet been fully clarified. Detailed studies should lead to a method which corrects the X-ray intensities of the elements in the sample on the basis of the scattered radiation, which is also recorded by the semiconductor X-ray detection system.

At present this type of instrumentation is being tested for multi-element assay of manganese nodules from the Central Pacific. It is intended to use a Si(Li) X-ray spectrometer on board an exploration vessel in the next future.

Acknowledgment

The author wishes to express his gratitude to Professor H. Sørensen, Director K. Ellitsgaard-Rasmussen and H. Hintsteiner for logistic support during the 1969–71 field seasons. Thanks are due to A. Thorboe, E. Sørensen and S. Å. Markland for chemical analysis of samples, and to H. Wollenberg for his contributions to the X-ray work. The invaluable help of S. E. Worm, W. Wedrac, E. Eberharth, H. Bohse and P. Sørensen during the field work and the critical reading of the manuscript by B. L. Nielsen, F. Kalsbeek, J. Rose-Hansen and L. Løvborg are greatly appreciated. Assistance in the laboratory was provided by E. Johannsen. He is indebted to Professor G. H. Friedrich for detailed discussion of the paper and many helpful suggestions and to J. Rasmussen for continuing interest in the X-ray work.

References

1. Salmon, M. L. The role of the instrumental analyst. *Q. Colo. Sch. Mines*, **64**, no. 1 1969, 11–3.
2. Hoy, R. B. Exploration trends for ore minerals—scientists not prospectors. *World Min.*, **24**, no. 8 1971, 26–31; no. 9 1971, 42–51.
3. *Tectonic/geological map of Greenland* (Copenhagen: Geological Survey of Greenland, 1970).
4. Bowie, S. H. U. Darnley, A. G. and Rhodes, J. R. Portable radioisotope X-ray fluorescence analyser. *Trans. Instn Min. Metall.*, **74**, 1965, 361–79.
5. Cameron, J. F. and Rhodes, J. R. X-ray spectrometry with radioactive sources. *Nucleonics*, **19**, no. 6 1961, 53–7.
6. Bowie, S. H. U. Portable X-ray fluorescence analysers in the mining industry. *Min. Mag., Lond.*, **118**, 1968, 230–9.
7. Gallagher, M. J. Portable X-ray spectrometers for rapid ore analysis. In *Mining and petroleum geology* (London: IMM, 1970), 691–729. (*Proc. 9th Commonw. Min. Metall. Congr. 1969, vol. 2*)
8. Rhodes, J. R. Portable radioisotope X-ray analyzers—techniques and applications. *Isotop. Radiat. Technol.*, **8**, 1970–71, 153–61.
9. Clayton, C. G. Applications of radioisotope X-ray fluorescence analysis in geological assay, mining and mineral processing. In *Nuclear techniques and mineral resources* (Vienna: IAEA, 1969), 293–324.
10. Clayton, C. G. Applications of radioisotope X-ray fluorescence analysers in metalliferous mineral exploration and in mining. In *Nuclear techniques for mineral exploration and exploitation* (Vienna: IAEA, 1971), 27–35.
11. Bowman, H. R. *et al.* Applications of high-resolution semiconductor detectors in X-ray emission spectrography. *Science, N.Y.*, **151**, 1966, 562–8.
12. Burkhalter, P. G. and Marr, H. E. Detection limit for gold by radioisotopic X-ray analysis. *Int. J. appl. Radiat. Isotop.*, **21**, 1970, 395–403.
13. Snyman, G. C. and Stewart, G. Uranium determination in ore samples by radioisotope-excited X-ray fluorescence. PEL–203, 1970, 8 p.
14. Giauque, R. D. and Jaklevic, J. M. Rapid quantitative analysis by X-ray spectrometry. LBL–204, 1971, 22 p.
15. Løvborg, L. A portable γ-spectrometer for field use. *Risö Rep.* no. 168, 1967, 36 p.
16. Bohse, H. Brooks, C. K. and Kunzendorf, H. Field observations on the kakortokites of the Ilímaussaq Intrusion, South Greenland, including mapping and analyses by portable X-ray fluorescence equipment for zirconium and niobium. *Rep. Greenland geol. Unders.* no. 38, 1971, 43 p.
17. Kunzendorf, H. Løvborg, L. and Wollenberg, H. Assay of powdered metallic ores by means of a portable X-ray fluorescence analyser. *Risö Rep.* no. 251, 1971, 29 p.
18. Kunzendorf, H. and Wollenberg, H. Determination of rare-earth elements in rocks by isotope-excited X-ray fluorescence spectrometry. *Nucl. Instrum. Meth.*, **87**, 1970, 197–203.
19. Kunzendorf, H. An instrumental procedure to determine Fe, Rb, Sr, Y, Zr, Nb and Mo in rock powders by Si (Li) X-ray spectrometry. *J. radioanalyt. Chem.*, **9**, 1971, 311–24.
20. Sørensen, H. The Ilímaussaq Batholith, a review and discussion. *Bull. geol. Unders. Greenland* no. 19, 1958, 48 p.
21. Ferguson, J. The significance of the kakortokite in the evolution of the Ilímaussaq Intrusion, South Greenland. *Bull. geol. Unders. Greenland* no. 89, 1970, 193 p.
22. Håkansson, E. *et al.* Preliminary results of mapping the Upper Jurassic and Lower Cretaceous sediments of Milne Land, Grønland. *Rep. geol. Surv Greenland* no. 37, 1971, 32–41.
23. Nielsen, B. L. Aeroradiometric survey in the Scoresby Sund region, East Greenland. *Rep. geol. Surv. Greenland* no. 45, 1972, 67 p.
24. Løvborg, L. *et al.* Drill core scanning for radioelements by gamma ray spectroscopy. In press.
25. Boyle, R. W. *et al.* Geochemistry of Pb, Zn, Cu, As, Sb, Mo, Sn, W, Ag, Ni, Co, Cr, Ba, and Mn in the waters and stream sediments of the Bathurst–Jacquet River district, New Brunswick. *Pap. geol. Surv. Can.* 65–42, 1966, 50 p.
26. Kalsbeek, F. The composition of sands from the Fiskenæsset region, south-west Greenland, and its bearing on the bedrock geology of the area. *Rep. geol. Surv. Greenland* no. 40, 1971, 12 p.
27. Langheinrich, A. P. Forster, J. W. and Linn, T. A. Energy dispersion X-ray (EDX) analysis in the nonferrous mining industry. *Isotop. Radiat. Technol.*, **9**, 1971, 173–81.
28. Hintsteiner, E. A/S Nordisk Mineselskab, Copenhagen, Denmark. Personal communication.

541.132:550.822.1:550.84

Instrument for *in situ* measurements of Eh, pH and self-potentials in diamond drill holes

B. Bølviken, SIV. ING.

Ø. Logn, CAND. REAL., LIC. TECHN.

A. Breen, SIV. ING.

O. Uddu, ING.

All of the Geological Survey of Norway, Trondheim, Norway

Synopsis

The instrument described comprises a glass electrode, a platinum electrode and a reference electrode mounted in a perspex support, all held in a 35-mm brass tube which contains an amplifier for the glass electrode circuit. External pressure acting upon the electrodes is compensated by an internal silicon oil phase separated from the exterior by a rubber membrane. Eh and pH are read on volt meters connected to the electrodes by a cable. If an additional reference electrode is placed in the soil at the surface, the natural self-potential (SP) in the drill-hole can be measured. For calibration the electrode housing is filled with standard buffer solution before being lowered into the drill-hole.

The instrument has been tested in drill-holes through a pyrite orebody. pH was successfully measured to a depth of 350 m, readings varying from 6·0 to 9·5. Eh measurements were troublesome, probably due to a lack of suitable redox pairs in the water; values ranged from 0 to +560 mV.

The SP clearly indicated the presence of both massive and disseminated ore types. There appears to be some relationship between Eh, pH and SP. The type of data which the instrument can provide may contribute to a better understanding of chemical processes in groundwater.

Redox potentials (Eh) and acidity (pH) are two parameters fundamental to descriptions of natural chemical systems (see, for example, Garrels and Christ[9]). These parameters are of particular interest in ore prospecting. Sato[18, 19] has discussed the geochemical environment of oxidizing sulphide bodies in terms of Eh and pH, and Hansuld[11] showed how knowledge of Eh and pH can be utilized in geochemical exploration. Habashi[10] suggested that the oxidation of sulphide ores is an electrochemical process, represented by an anodic and a cathodic reaction. One related aspect of this problem is the presence of natural self-potentials (SP) around orebodies. The existence of self-potentials has been known for more than a century,[8] but their origin is not completely understood. A major contribution to the theory of self-potentials is that of Sato and Mooney,[20] who explained them as being due, for the most part, to electrochemical processes caused by Eh differences in the upper part of the lithosphere. An orebody could be considered as a dipole electrode in an electrolyte, the composition of which varies between the poles. As the theory is incomplete, however, more information about natural Eh and pH values is required. Measurements of Eh, pH and SP in groundwaters around ore deposits should, therefore, be of interest both to geochemists and to geophysicists. Such measurements should, preferably, be taken *in situ*, since any water sample collected at such a depth that the temperature and partial pressure of O_2 and CO_2 are not the same as on the earth's surface is likely to change in composition during its transportation to the surface. A procedure for measuring Eh and pH which involves bringing the sample to the instrument would therefore introduce a serious error. At any economic orebody many drill-holes usually exist, and these provide an excellent opportunity for *in situ* measurements of Eh and pH in groundwater.

The technique for measuring SP in drill-holes is well established. The usual procedure is to lower an unpolarizable electrode into the hole and measure its voltage difference to a stationary reference electrode placed in the surface soil near the hole. The unpolarizable electrodes used for this purpose in Norway are Cu–$CuSO_4$ electrodes (Cu wire immersed in a saturated solution of $CuSO_4$ separated from the exterior by a semi-permeable, wooden plug).

Examples of measurements of Eh and pH taken in a similar manner to those of SP in drill-holes are not known.

In situ Eh measurements by use of platinum electrodes in drill-holes should not, however, create any additional or greater problems than those involved in other Eh measurements in natural waters, although, as was pointed out by Sillén,[21] these problems may sometimes be quite serious. In particular, in neutral and high pH waters with a low content of suitable redox pairs, the electrode reactions could be slow and erratic.

In situ pH measurements with a glass electrode at high pressures have been reported by various workers, [1, 7, 14, 15] oceanographic research having featured strongly.[2–6, 17] In principle, analogous measurements in diamond drill holes should be fairly simple, the main practical difficulties being (1) the construction of an instrument of sufficiently small dimensions to allow its passage into the drill-hole; (2) overcoming the high capacity in the glass electrode circuit because of the great distance between the electrode and the volt meter; and (3) elimination of the risk of breaking the glass electrode because of the high water pressure in the hole.

Difficulties (1) and (2) can be tackled by mounting a small transistorized amplifier in the probe, and (3) by use of silicon oil and rubber membranes to compensate for the pressure.

A prototype instrument developed at the Geological Survey of Norway is described here, together with some representative examples of results obtained during the testing of the instrument in a mining area. All definitions are taken from standard textbooks: for Eh and pH, see Krauskopf[13] (pp. 32–5 and 237–55) and for SP see Parasnis[16] (pp. 64–7).

Description of the instrument

Basically, the instrument is constructed as shown in Figs. 1–4.

Three electrodes, a glass electrode, a platinum electrode and a reference electrode, are mounted in a perspex support. The glass electrode used by the authors is a Radiometer G 200 B, the upper insulated part of the electrode having been cut off. The platinum electrode consists of a rectangular piece of platinum sheet, moulded into the perspex, giving a 5-mm by 5-mm exposed area.

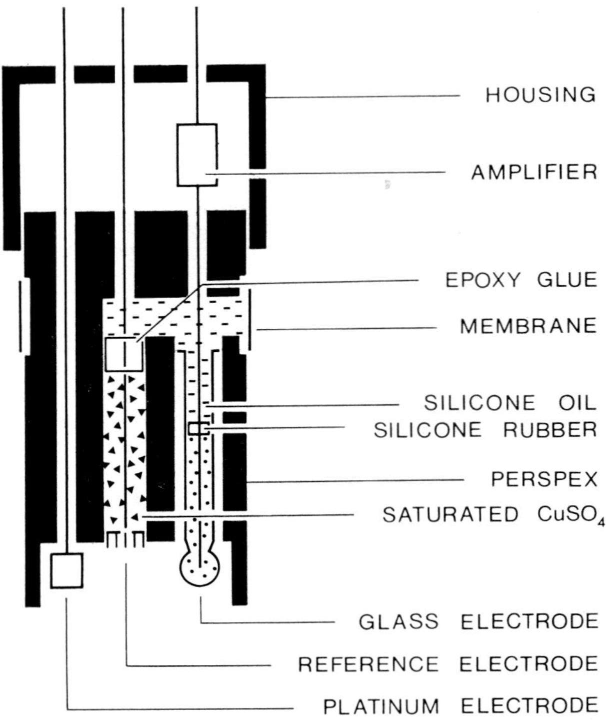

Fig. 1 Electrode assembly for pH, Eh and SP measurements

The reference electrode of the prototype consists of a saturated solution of $CuSO_4$, which is placed in a cylindrical hole in the perspex support. A 2-mm thick copper wire is immersed in the $CuSO_4$ solution, and the lower end is tapped with a screw cap in which a small cotton wool plug acts as a diaphragm. The upper part of the glass and reference electrode chambers has an open passage to a radial hole through the support. A piece of rubber tube is fitted into a groove around the support at this point to cover the radial hole. The hole and the passage to the upper part of the electrode chamber are filled with silicon oil (type M.S 200 Fluid I CTS), which forms a non-conducting phase separated from the water phase within the electrodes by a silicon rubber piston. Any external water pressure acting upon the glass electrode is, therefore, completely compensated.

The electrodes are connected by a cable to volt meters at the surface. The potentials of the glass electrode are amplified by a small amplifier

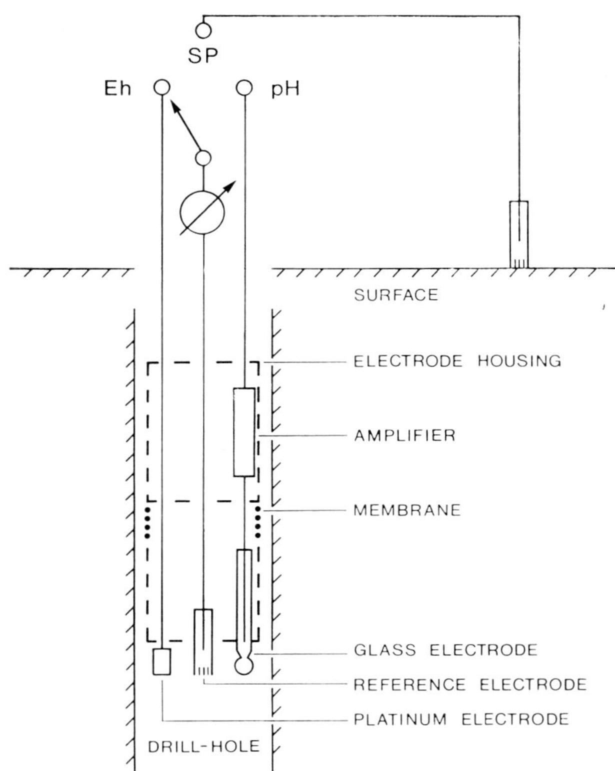

Fig. 2 Principles for in situ measurements of pH, Eh and SP in diamond drill holes

(Analog Devices Model 141*) which, together with four 9-V batteries, is placed above the perspex support in the brass tube that holds it. All connexions are made waterproof with epoxy glue or O-ring couplings.

The following parameters can be recorded: Eh—by measuring the voltage difference between the reference electrode and the Pt electrode; pH—by measuring the amplified voltage difference between the reference electrode and the glass electrode; and SP—by measuring the voltage difference between the reference electrode and a similar stationary electrode.

Calibration

For calibration of the electrode system the lower part of the electrode housing is filled with standard buffer solution and sealed with a plastic tape, which acts as a membrane to compensate the pressure. The instrument is then lowered into the drill-hole, and the calibration is made under the same conditions as the borehole logging. (Calibration should be made in the drill-hole after the logging to avoid any erratic results due to contamination of the hole from the buffer.)

* Analog Devices, 221 Fifth Street, Cambridge, Mass. 02142, U.S.A.

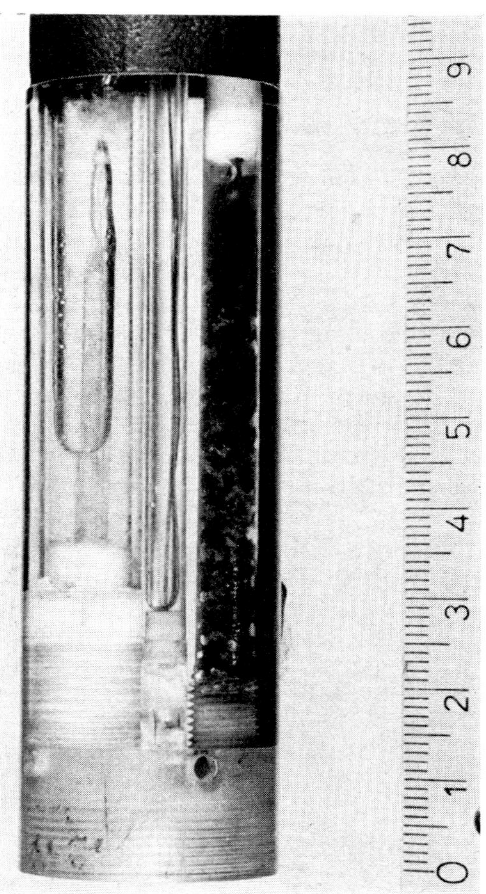

Fig. 3 Probe for pH, Eh and SP measurements in diamond drill holes

When the instrument was tested, the Eh standard used was the ZoBell solution,[22] pH measurements were standardized against pH 4 and 9, and the buffers were made of commercial buffer tablets after National Bureau of Standards formulae. Temperatures of drill-hole water were measured

Fig. 4 Equipment for pH, Eh and SP measurements in diamond drill holes

with thermistors and the pH of buffers was corrected for temperature.

Preliminary results

The prototype instrument has been tested in drill-holes at Joma, a copper- and zinc-bearing massive pyrite deposit in Nord-Trøndelag, Norway. The country rocks of the Joma deposit are greenstones with 5–10% carbonate content (carbonate content in the ore, approximately 2·5%). The mine is at present under development. The holes in which measurements were made were drilled at least five years ago, and were last logged one year before the present Eh and pH measurements were taken. Water levels were 5–10 m below the surface. The holes were 100–350 m deep, and all but one intersected the ore horizon, which varies in thickness from 2 to 35 m.

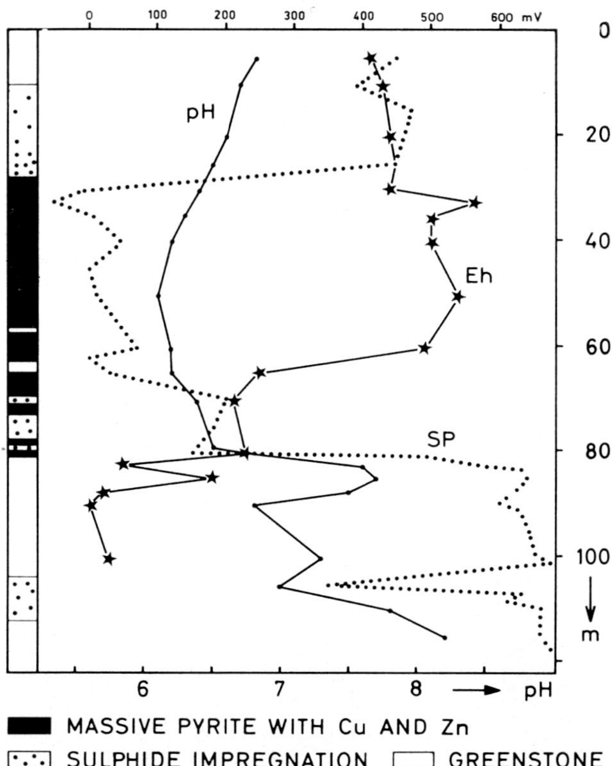

Fig. 5 pH, Eh and SP in drill-hole no. 117, Joma, Nord-Trøndelag, Norway

Some results of, and comments on, these preliminary measurements are given here, and a more complete account of the results will be presented later.

Reproducible Eh results were obtained in only one drill-hole (no. 117), Eh values between 0 and 560 mV being recorded (see Fig. 5). Otherwise, Eh measurements were erratic.

pH was measured down to 350 m (pressure, approximately 35 atm) without damage to the glass electrode. pH of the water in the drill-holes tested varied between 6·0 and 9·5, depending on depth, country rock and degree of mineralization. Examples of results are given from a hole near the outcrop of the ore (no. 117, Fig. 5) and from a hole intersecting the central part of the ore (no. 10M, Fig. 6).

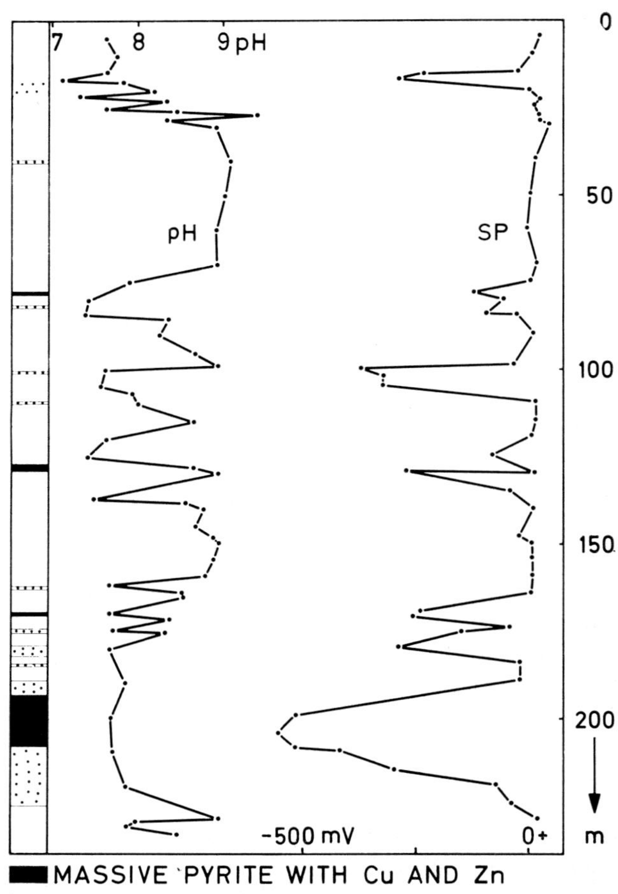

Fig. 6 pH and SP in drill-hole no. 10M, Joma, Nord-Trøndelag, Norway

Groundwater temperatures in the holes varied between 3·5°C at a depth of 15 m and 10°C at a depth of 350 m. The average air temperature during the measurements was about 9°C. After the probe had been lowered into the hole, a waiting time of 5–10 min was allowed to elapse to enable the probe to cool to the temperature of the groundwater. From then on the readings were very stable, and could be taken as rapidly as the hoisting permitted. Reproducibility appeared to be within 0·1 pH unit.

Calibration of the pH system caused no major problems. The readings were stable and the calibration curves are rectilinear.

SP readings taken down the holes clearly

indicate the presence of both the compact and the disseminated types of pyritic ore (Figs. 5 and 6). The results from hole no. 117 (Fig. 5) seem to indicate some correlation between Eh, pH and SP.

Concluding remarks

The difficulty experienced in measuring Eh can be explained by reference to comments on Eh measurements in natural water by Sillén.[21] Because of the high carbonate content of the bedrock in the Joma area, the pH is probably too high to allow a sufficiently high concentration of a suitable redox pair, for example, Fe^{2+}/Fe^{3+}, in the water. Estimation of the redox potentials by measurement of the O_2 content of the water, as proposed by Sillén, would perhaps be a better alternative. O_2 measurements in drill-hole waters could probably be taken by use of an oxygen electrode. It is of interest to note that the only drill-hole in which the Eh measurements were successful (no. 117, Fig. 5) has the lowest pH of all the drill-holes tested: this indicates that measurements of Eh in drill-holes by use of a platinum electrode would probably be generally more reliable in mining districts with a more acid groundwater than that of the Joma area.

The advantage of measuring pH *in situ* as opposed to water sampling is illustrated by the observed pH variation down the drill-holes. In water samples brought to the instrument such local variations as those seen in Figs. 5 and 6 would hardly be detected.

The local variation in the SP values shows that when pH and Eh are being measured *in situ*, the reference electrode should be located near the other electrodes. Measuring systems with any great distance between the reference electrode and the glass/platinum electrodes[15] involve the risk of SP variations being interpreted as changes in pH and Eh.

One reason why the *in situ* pH readings in drill-holes are so stable is probably that the rocks surrounding the drill-hole constitute a perfect shield against foreign fields acting upon the electrical circuit of the glass electrode. This assumption is partly based on the fact that it was very difficult to obtain reproducible pH results with the instrument under laboratory conditions. Consequently, the instrument itself should be improved by replacing the $Cu–CuSO_4$ reference electrode by a Ag–AgCl electrode, which is stable to *circa* 0·01 pH unit and has also been investigated in detail.[2, 3, 4, 12] With this improvement fairly accurate routine pH values in drill-hole waters should be obtained. Future use of the instrument may, therefore, provide data which could contribute to a better understanding of the chemical processes that take place in groundwater—in particular, the electrochemical processes that occur in the vicinity of ore deposits.

Acknowledgment

The authors thank Grong Gruber A/S for permission to work in the Joma area and for allowing them to publish this paper.

References

1. Allgeier, R. J. Hafford B. C. and Juday C. Oxidation–reduction potentials and pH of lake waters and of lake sediments. *Trans. Wis. Acad. Sci. Arts Lett.*, **33**, 1941, 115–33.
2. Ben-Yaakov, S. and Kaplan, I. R. A versatile probe for *in situ* oceanographic measurements. *J. oceanogr. Technol.*, **2**, no. 3 1968, 25–9.
3. Ben-Yaakov, S. and Kaplan, I. R. *p*H–temperature profiles in ocean and lakes using an *in situ* probe. *Limnol. Oceanogr.*, **13**, 1968, 688–93.
4. Ben-Yaakov, S. and Kaplan, I. R. High pressure p_H sensor for oceanographic applications. *Rev. scient. Instrum.*, **39**, 1968, 1133–8.
5. Culberson, C. and Pytkowicz, R. M. Effect of pressure on carbonic acid, boric acid, and the *p*H in seawater. *Limnol. Oceanogr.*, **13**, 1968, 403–17.
6. Distèche, A. and Distèche, S. Effect of pressure on *p*H and dissociation constants from measurements with buffered and unbuffered glass electrode cells . . . *J. electrochem. Soc.*, **112**, 1965, 350–4.
7. Distèche, A. *p*H measurements with a glass electrode withstanding 1500 kg/cm^2 hydrostatic pressure. *Rev. scient. Instrum.*, **30**, 1959, 474–8.
8. Fox, R. W. On the electromagnetic properties of metalliferous veins in the mines of Cornwall. *Phil. Trans. R. Soc.*, **120**, 1830, 399–414.
9. Garrels, R. M. and Christ, C. L. *Solutions, minerals, and equilibria* (New York: Harper and Row, 1965), 463 p.
10. Habashi, F. The mechanism of oxidation of sulfide ores in nature. *Econ. Geol.*, **61**, 1966, 587–91.
11. Hansuld, J. A. Eh and pH in geochemical exploration. *Trans. Can. Inst. Min. Metall.*, **69**, 1966, 77–84.
12. Ives, D. J. G. and Janz, G. J. eds. *Reference electrode* (New York: Academic Press, 1961), 651 p.
13. Krauskopf, K. B. *Introduction to geochemistry* (New York: McGraw-Hill, 1967), 721 p.
14. Le Peintre, M. Mesure du pH sous pression et à température élevée. *Bull. Soc. fr. Electns*, **1**, 1960, 584–91.
15. Manheim, F. In situ measurements of pH and Eh in natural waters and sediments. *Stockholm Contr. Geol.*, **8**, 1961, 27–36.
16. Parasnis, D. S. *Principles of applied geophysics* (London: Methuen, 1962), 176 p.
17. Park, K. Deep-sea pH. *Science, N.Y.*, **154**, 1966, 1540–2.
18. Sato, M. Oxidation of sulfide ore bodies, I.

Geochemical environments in terms of Eh and pH. *Econ. Geol.*, **55**, 1960, 928–61.
19. SATO, M. Oxidation of sulfide ore bodies, II. Oxidation mechanisms of sulfide minerals at 25°C. *Econ. Geol.*, **55**, 1960, 1202–31.
20. SATO, M. and MOONEY, H. M. The electrochemical mechanism of sulfide self-potentials. *Geophysics*, **25**, 1960, 226–49.
21. SILLÉN, L. G. Oxidation state of Earth's ocean and atmosphere. I. *Ark. Kemi.*, **24**, 1965, 431–56.
22. ZOBELL, C. E. Studies on redox potential of marine sediments. *Bull. Am. Ass. Petrol. Geol.*, **30**, 1946, 477–513.

543.062:546.161–2

Determination of total and cold-extractable fluoride in soils and stream sediments with an ion-sensitive fluoride electrode

W. L. Plüger, DR. RER. NAT.

G. H. Friedrich, DR. RER. NAT., F.I.M.M.

Both of Institut für Mineralogie und Lagerstättenlehre, Technische Hochschule, Aachen, West Germany

---

Synopsis

An ion-sensitive fluoride electrode shows a Nernst-type response to fluoride ion activity over more than five orders of magnitude. The usefulness of this type of electrode in geochemical analytical work is demonstrated. Interference by other ions involved in the analysis and the technique by which it is eliminated are discussed.

Depending on time, electrolyte and the nature of the clay mineral, the fluoride adsorption capacity has been experimentally proved on soils in general, and for clay minerals in particular. Based on these studies, methods have been developed for the determination of the cold-extractable fluoride content of soils and stream sediments. These methods have been applied successfully to the determination of the secondary dispersion patterns of various fluorite deposits.

Further methods and applications for the determination of the total fluoride content of soils and stream sediments, by sintering with NaOH and by complexing with $Be(NO_3)_2$ solutions, are also discussed.

The use of fluorspar in the metallurgical and chemical industries is increasing. Because the known reserves of fluorspar may suffice for only 20–30 years, there is an increasing activity in prospecting and exploration for new deposits. Prospecting for fluorspar was generally carried out by use of conventional geological prospecting methods. The applicability of geochemical prospecting methods has been investigated more recently, based on the following concepts: (1) the investigation of dispersion patterns of indicator elements for fluorspar deposits[6] and (2) the investigation of the fluoride dispersion patterns.[1,7]

The application of fluoride-sensitive electrodes to the determination of fluoride in exploration geochemistry has been described by Plüger.[9] His potentiometric method has considerable advantages over other analytical methods for fluoride determination.

In the study described here different methods for the determination of fluoride in stream waters, stream sediments and soils in geochemical exploration surveys are discussed, precision and detection limits being given and adsorption capacities of certain clay minerals and soils investigated. This has been done as a contribution to knowledge of the formation of secondary dispersion patterns.

Instrumentation*

A fluoride-sensitive electrode and a reference electrode are used in conjunction with a direct-reading expanded-scale mV meter.[5] The fluoride sensitivity of the electrode depends on the potential difference between the faces of a LaF_3 single crystal, which is embedded at the end of the

*The electrode manufacture refers to a working range of the fluoride-sensitive electrode from 19 ppb fluoride up to saturated solutions.

electrode (Fig. 1). The outer side of the crystal is in contact with the sample solution; the inner side contacts the internal electrode solution containing 0·1 M NaF and 0·1 M KCl.

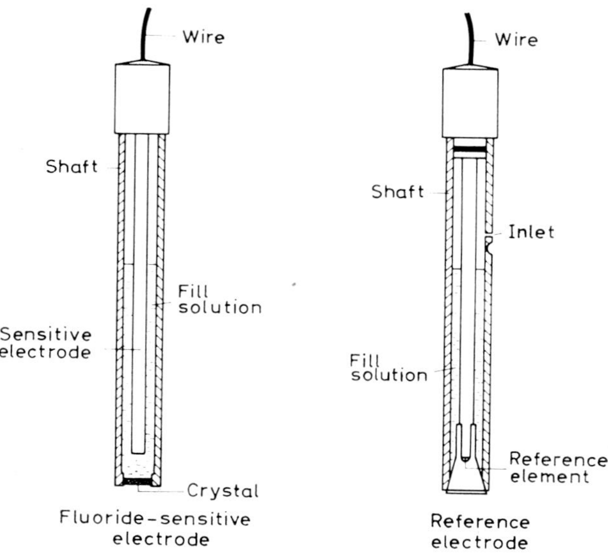

Fig. 1 Cross-section of fluoride-sensitive electrode and reference electrode

The theoretical basis of the potentiometric measurement is given by the Nernst equation:

$$E=E_0+\frac{RT}{nF}\ln a_{ion} \qquad (1)$$

With regard to the valency of the fluorine ion, the substitution of the constants R and F and equation 1, the transformation of the logarithm, can be written

$$E=E_0-59{\cdot}16 \log a_{F^-} \qquad (2)$$

for a 25°C sample solution temperature.

Equation 2 shows that a tenfold change in fluoride activity (a_{F^-}) results in a 59·16-mV change.

The relationships between fluoride activity and fluoride concentration are given by the following equations:

$$a_{F^-}=\gamma\,[F^-] \qquad (3)$$

$$\log \gamma=-k\sqrt{J} \qquad (4)$$

$$J=0{\cdot}5\ \Sigma z_i^2\ c_i \qquad (5)$$

where γ is the activity coefficient, $[F^-]$ is the fluoride concentration, k is a constant, z_i is the charge of each and every ion in the solution and c_i is the concentration of each and every ion in the solution.

The principal interferences in fluoride measurements with the fluoride electrode are noted below.

Hydrogen and hydroxide ions

Hydrogen ions complex the free fluoride to form undissociated HF. This kind of interference, however, appears only when the pH of the solution is less than 4–5. Hydroxide ions have a similar ionic radius to that of fluoride, as well as an equal charge. Interference can be observed if the hydroxide ion amount is greater than one-tenth of the fluoride content.[8]

Polyvalent cations

Substances which are able to complex or precipitate with fluoride (Fe^{3+}, Al^{3+}, Si^{4+}, Be^{2+}, etc.) reduce the free fluoride concentration in the solution. Their masking can be carried out by the addition of sodium citrate to the solution. Fig. 2 shows the effect of various amounts of tri-sodium citrate on electrode potentials of solutions containing various fluoride and ferric ions concentration.

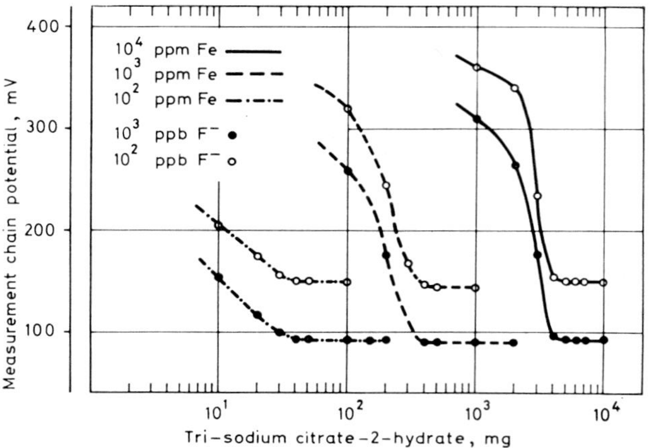

Fig. 2 Effect of various amounts of tri-sodium citrate 2-hydrate on electrode potentials of solutions with various fluoride and Fe(III) concentrations

Anions

Anions which complex or precipitate with the lanthanum of the sensing element increase the lower detection limit of the measurement by also dissolving the fluoride of the crystal.[10]

Fluoride adsorption by clay minerals and soils

The formation of secondary dispersion pattern in soils depends, in part, on the mineralogical composition of the soils. It is well known that high amounts of clay minerals cause an enrichment of elements in soils.

By use of a fluoride-sensitive electrode the investigation of fluoride adsorption by clay minerals and soils can be done without complicated analytical work. It is possible to record continuously the decrease of an initial known fluoride concentration of a solution containing a known amount of solid material. The difference between the initial and the end fluoride concentrations of the solution allows the calculation of the adsorbed fluoride content by the solid. Preliminary studies have shown that the adsorption process takes place in less than one hour.

The fluoride adsorption capacities of the following four different clay minerals were investigated:

Kaolinite	(Zettlitzer Kaolinstandard la)
Fireclay	(Ponholzer Kapsel- u. Masseton 42/44)
Illite	(Engobeton von Eisenberg)
Montmorillonite	(Ca-Montmorillonit von Moosburg)

Studies were made of the adsorption capacities at several initial fluoride concentrations (200, 600 and 1000 ppb fluoride) and at various pH levels (pH 4, 5, 6 and 7). The results obtained are summarized in Table 1 and plotted in Fig. 3. There is a general decrease in the fluoride capacities at increasing pH values of the solutions, but the decreasing trend is different for the four clay minerals. Citric acid and phosphate Sörensen buffers have been used as solutions because they simulate the nature of soil waters.[11]

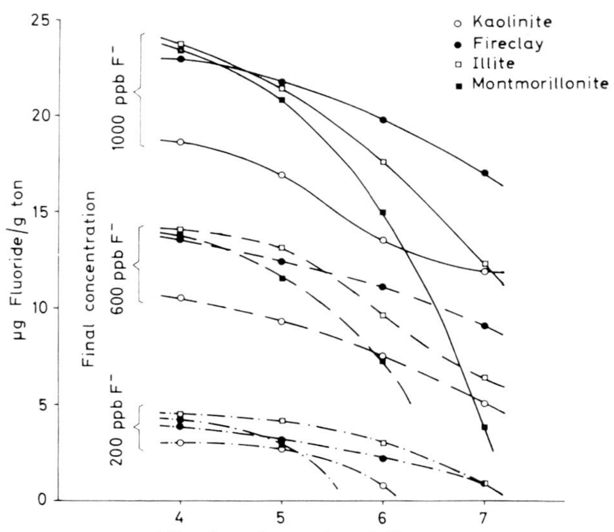

Fig. 3 Effect of solution pH on amount of fluoride adsorbed by different clay minerals at various (initial) concentrations of fluoride in solutions

Table 1 Adsorption capacities ($\mu g\ F^-/g$) of four clay minerals at different initial fluoride concentrations and at various pH levels

pH buffer solution	Initial fluoride concentration, ppb	Adsorption capacity, $\mu g\ F^-/g$ Kaolinite	Fireclay	Illite	Montmorillonite
4·0	200	3·0	3·9	4·2	4·0
	600	10·5	13·5	14·0	13·8
	1000	18·6	22·9	23·7	23·4
5·0	200	2·7	3·1	4·2	2·8
	600	9·3	12·4	13·1	11·6
	1000	16·9	21·7	21·5	20·8
6·0	200	0·8	2·2	3·0	—
	600	7·5	11·1	9·6	7·3
	1000	13·5	19·8	17·6	15·0
7·0	200	—	0·9	0·9	—
	600	5·0	9·1	6·4	—
	1000	11·9	17·0	12·3	3·8

The influence of buffer composition on the adsorption capacities was also shown by use of a mixed citric acid–phosphate buffer (pH 4, 5, 6 and 7).[13] Differences in the adsorption capacities at an initial fluoride concentration of 38 ppm for both buffers are apparent from Fig. 4. The lower adsorption capacities obtained with the McIlvaine buffers are probably due to the phosphate in these buffers, which may partly exchange hydroxide ions from the clay minerals. In the Sörensen buffer phosphate only occurs at pH 7.

The addition of fluoride to a soil sample suspended in both Sörensen and McIlvaine buffers gives rise to a general decrease in the initial fluoride content in the solution. There is a decrease in the adsorption capacities with increasing pH and the adsorption capacities in the Sörensen buffers are higher (Table 2). This is in good agreement with the results obtained with clay minerals.

The adsorption capacities of the soils are generally greater than those of pure clay minerals, which means that fluoride adsorption is due not

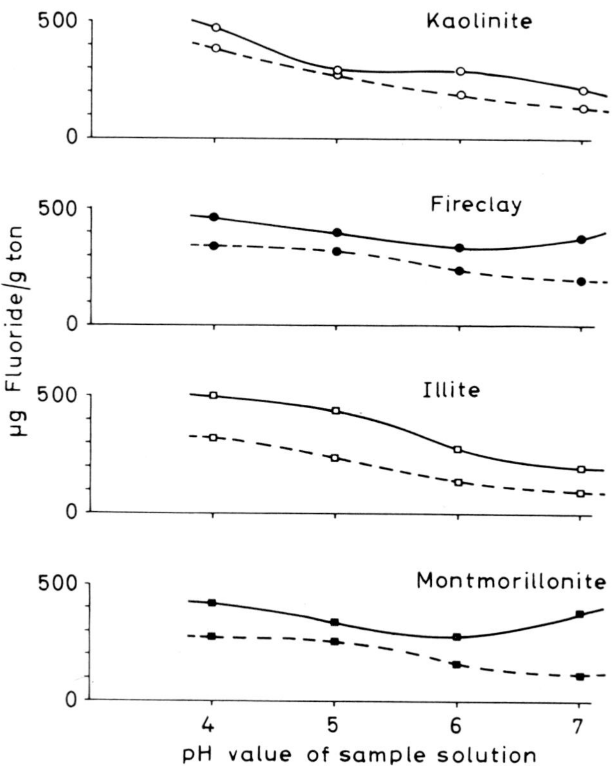

Fig. 4 Effect of Sörensen[11] *(unbroken lines) and McIlvaine*[13] *(broken lines) buffer solutions and solution pH on amount of fluoride adsorbed by different clay minerals*

only to clay minerals but, probably, to adsorption by amphoteric hydroxides and organic compounds. Further investigations into this aspect will be made.

In the case of fluoride adsorption by clay minerals, the sorption takes place by hydroxide–fluoride exchange on the surfaces of the clay minerals. In addition, the decay of clay minerals during weathering results in additional exchangeable hydroxide ions by destroying the Si–O–Si and Al–O–Al groups. Subsequently, the resulting free valencies will be saturated by the dissociation products of the water. The newly formed SiOH and AlOH groups are able to exchange hydroxide ions beneath their isoelectric point.[4]

Free amphoteric hydroxides are also able to exchange hydroxide ions. Since the isoelectric point of these hydroxides varies over a wide range (pH 4·9–9·8 for $Al(OH)_3$ and pH 4·5–8·5 for $Fe(OH)_3$), there may be a notable exchange capacity, depending on the pH of the solutions.

Applications of fluoride-sensitive electrode in analytical techniques

Determination of fluoride in stream water

To determine the fluoride concentration of stream waters a procedure for the determination of fluoride in drinking water[8] can be used. Stream water samples and fluoride standard solutions are diluted 1 : 1 with a total ionic strength adjustment buffer (TISAB) containing 1 M NaCl, 0·25 M acetic acid, 0·75 M sodium acetate and 0·001 M sodium citrate.* The pH of the buffer is between 5·0 and 5·5, and the ionic strength about 1·75. This buffer equals the ionic strength of sample and standard solutions, adjusts the pH and avoids interference by polyvalent cations.

The method has been used in geochemical prospecting surveys by Chrt[1] and Friedrich and Plüger.[7] Replicate measurements on stream water samples show the standard deviation of the technique (Table 3).

Determination of fluoride in soils and stream sediments

Cold-extractable heavy metal and total heavy metal determinations in geochemical prospecting have worldwide application. Determinations of cx fluoride and total fluoride in soils and stream

Table 2 Adsorption capacities of different soil samples from B horizon (granitic bedrock (g), dolomitic bedrock (d)) at initial fluoride concentrations of 1000 ppb F^- at pH 5 and 6

Sample of *B* horizon	Adsorption capacity, μg F^-/g soil Sörensen buffer pH 5	 6	 McIlvaine buffer pH 5	 6
He 226 *(g)*	50	48	50	50
Be 7/7 *(d)*	37	25	35	18
Os 1/4 *(g)*	30	20	20	16
Os VIII/3 *(g)*	25	10	12	9

Sample weight, 1 g; volume of solutions, 50 ml; added fluoride content, 50 μg.

*TISAB solution is commercially available from the electrode manufacturers Orion Research, Inc., Cambridge, Massachusetts, U.S.A.

Summary and conclusions

Experimental studies have shown that the fluoride adsorption capacity of soils depends on time, electrolyte and the nature of the clay minerals. In addition to the amount of different clay minerals, organic compounds and amphoteric hydroxides are responsible for the higher adsorption capacities of soils, and, therefore, they may also influence the formation of secondary dispersion patterns around fluorite deposits.

Geochemical exploration for fluorite can be carried out by the use of different analytical methods for the determination of fluoride in stream waters, stream sediments and soils. The most useful method will be the determination of fluoride in waters. The determination of the cold-extractable fluoride content in stream sediments by treatment with TISAB can be recommended for exploration work in areas in which stream water analysis is not possible. Moreover, the determination of the TISAB soluble fluoride in soils can be used in detailed exploration work. The use of HCl or NaOH solutions instead of TISAB has the advantage of showing higher contrasts between background, threshold and anomaly. The application of NaOH solution is restricted to cold-extractable fluoride analysis of soils.

The determination of total fluoride by NaOH fusion or sintering with KNO_3–Na_2CO_3 gives the best contrast and may be necessary in areas where cold-extractable fluoride values are very low and show no contrast. Fluoride analysis with $Be(NO_3)_2$ solutions as a complexing agent results in lower 'total fluoride'.

References

1. CHRT, J. Methoden der Prospektion und Erkundung auf Flusspat in der ČSSR. *Berg- u. hüttenm. Mh.*, **114**, 1969, 459–64.

2. FEIGL, F. and SCHAEFFER, A. Analytical use of the formation of the berylliumfluoride complex. *Analyt. Chem.*, **23**, 1951, 351–3.

3. FICKLIN, W. H. A rapid method for the determination of fluoride in rocks and soils, using an ion-selective electrode. *Prof. Pap. U.S. geol. Surv.* 700-C, 1970, 186–8.

4. FIEDLER, H. J. and REISSIG, H. *Lehrbuch der Bodenkunde* (Jena: VEB G. Fischer, 1964), 544 p.

5. FRANT, M. S. and ROSS, J. W. Jr. Electrode for sensing fluoride ion activity in solution. *Science, N.Y.*, **154**, 1966, 1553–5.

6. FRIEDRICH, G. Über die Dispersion von Quecksilber in Böden im Bereich einiger Flusspatvorkommem von Nabburg-Wölsendorf. *Erzmetall*, **23**, 1970, 482–6.

7. FRIEDRICH, G. H. and PLÜGER, W. L. Geochemical prospecting for barite and fluorite deposits. In *Geochemical exploration* (Montreal: CIM, 1971), 151–6. (*CIM spec. vol. 11*)

8. *Orion-ionalyzer instruction manual, fluoride electrode model 94–09* (Cambridge, Mass.: Orion Research Inc., 1970), 24 p.

9. PLÜGER, W. L. Die potentiometrische Fluoridbestimmung und ihre Anwendung bei der geochemischen Prospektion von Flusspatlagerstätten. Dissertation, Aachen Technische Hochschule, 1972.

10. ROSS, J. W. Jr. Solid-state and liquid membrane ion-selective electrodes. In *Ion-selectivee lectrodes* DURST, R.A. ed (Washington, D.C.: National Bureau of Standards, 1969), 57–88. (*Spec. Publ.* 314)

11. SCHWIETE, H. E. and POHL, K. D. Struktur und Bindung organischer Substanzen in Tonen. Report of Institut für Gesteinshüttenkunde, Aachen Technische Hochschule, 1968.

12. VAN LOON, J. C. The rapid determination of fluoride in mineral fluorides using a specific ion electrode. *Analyt. Letters*, **1**, 1968, 393–8.

13. VOGEL, A. I. *Text-book of macro and semimicrc qualitative inorganic analysis, 4th edn* (London, etc.: Longmans, Green and Co., 1954), 663 p.

552.1:543.062:546.131.2

Analysis of water-extractable chloride in rocks by use of a selective ion electrode

J. C. Van Loon, B.SC., PH.D.

S. E. Kesler, B.SC., PH.D.

C. M. Moore, B.ENG.

All of the Department of Geology, University of Toronto, Toronto, Ontario, Canada

Synopsis

Previous studies have indicated that chlorine abundances can be useful in distinguishing intrusive rocks genetically associated with hydrothermal mineralization. Because of the difficulty of determining total chlorine in rocks, this potential exploration tool has not been widely utilized. An attempt has been made to render the technique more amenable to inexpensive routine exploration by use of a chloride-selective ion electrode to determine the water-soluble chloride in rock powders such as are commonly used for trace-metal analysis.

There has been a recent upsurge in the development and utilization of selective ion electrodes as analytical tools. A largely overlooked field of application is that of geochemical exploration. In contrast with many traditional chemical techniques, electrodes can be readily used with portable equipment (the Orion 401 expanded scale pH/mV meter was used in this study), which makes mobile field operations possible.

Average granitic rocks contain 100–250 ppm chlorine, of which a significant, but variable, proportion is present as dissolved NaCl in fluid inclusions of *primary* and *secondary* origin. In the case of hydrothermal ore deposits associated with intrusive rocks, this fluid inclusion chlorine should be the most effective indicator of favourable intrusive rocks. Furthermore, when a fluid inclusion is broken during sample preparation, the dissolved chlorine will precipitate on the surface of the fragments, rendering it easily recoverable with cold water leaching.

Analyses made by stirring 1 g of rock powder in distilled water for 2 or 3 min give a range of values of from 10 to 150 ppm chloride (rock) on the suite of granodioritic rocks tested so far. Optimum grain size for the powder is 200 mesh. Further grinding gives no significant increase in chloride values. Replicate analyses of 10 different powder samples from the same rock indicate a repeatability of ±2% at the 95% confidence level. In 90% of the cases solutions could be left in contact with the powders for up to one month without appreciable changes in chloride content.

Selective ion electrodes do *not* respond specifically to the ion of interest. In the case of the chloride electrode, however, only sulphide, bromide and iodide (none of which is important in this application) are likely to interfere in geochemical sample solutions. It must be kept in mind that electrodes measure ion activity, and, hence, corrections in some instances of high salt content solutions may be necessary to give proper concentration values.

Linear electrode response to chloride is obtained over the range 4–40 000 mg/ml, with response times of 10–30 sec, depending on concentration. Chloride in the range 0·1–4 mg/ml can also be determined, but response times are much slower (30 sec–5 min).

The electrode used for this study was constructed at a cost of $3.00 from silver chloride precipitate embedded in dental acrylic. When the membrane surface is coated with a thin layer of silicone oil and the electrode used in conjunction with a sleeve-type double-junction reference electrode, work can be done in highly turbid solutions containing abrasive particles with good reproducibility.

The method was used on a suite of granodioritic intrusive rocks from Puerto Rico. Some of these rocks are from the Utuado batholith and its younger satellites, which have generated two (known) medium-size porphyry copper deposits. Others are from the La Torrecilla and San Lorenzo plutonic complexes, which are not known to be associated with commercial mineralization. The most interesting result of these tests is that whereas total copper in these rocks is highest in samples from the Utuado batholithic complex, the water-soluble chloride is not. Instead, the highest chloride content occurs in unmineralized or weakly mineralized rocks. No significant difference has been detected so far between porphyritic and equigranular rocks. This too is a surprise because it would be expected that porphyritic intrusions would have crystallized in a shallow environment in equilibrium with a more abundant chloride-rich vapour phase. At present the results obtained suggest that within individual pluton complexes high copper values are associated with high chloride values in cases of (presently) marginal or sub-ore grade mineralization. But high copper values are associated with low chloride values in rocks that are host to ore grade mineralization. If this relation is found to be generally applicable, it could be possible to distinguish economic from weak mineralization by using part of the rock powder (used for copper analysis) to obtain a chloride value.

Evaluation of this technique is continuing on other rock suites in both fresh and weathered conditions. Attention is also being given to the possible use of other ion-selective electrodes (e.g. fluoride, cupric and sulphide) for related determinations. The results of these tests are reported.

Previous studies, such as that by Stollery and co-workers,[1] suggested that chlorine abundances can be useful in distinguishing intrusive rocks that are genetically associated with hydrothermal mineralization. Because of the difficulty of determining total chlorine in rocks, this potential exploration tool has not been widely utilized. Average granitic rocks contain 100–250 ppm chlorine, of which a significant, but variable, proportion is present as dissolved NaCl in fluid inclusions of primary and secondary origin. In hydrothermal ore deposits associated with intrusive rocks this fluid inclusion chloride should be the most effective indicator of favourable intrusive rocks. When a fluid inclusion is broken during sample preparation, the dissolved chloride will precipitate on the surface of fragments, rendering it easily recoverable with cold water leaching.

There has been a recent upsurge in the development and utilization of selective ion electrodes as analytical tools, but a field of application which has largely been overlooked is that of geochemical exploration. Chloride selective ion electrodes can be obtained which will respond to chloride in solution linearly from 40 000 to 4 ppm and, with care, will yield acceptable results even as low as 0·1 ppm. The attractive possibility exists that these electrodes could be used to determine the water-soluble chloride concentration of rock powders commonly used for trace-metal analyses. Equipment for selective ion electrode analysis is inexpensive and simple to operate. Electrodes exhibit linear response over a wide concentration range. Many selective ion meters are portable, making mobile field operation possible, should this be necessary.

The majority of modern chloride selective ion electrodes are of two types, one employing a solid membrane ($AgCl$ or $AgCl/Ag_2S$) and the other a liquid ion exchanger incorporating a dimethyl distearylammonium cation. The former is rugged and stable, but is subject to sulphide, bromide and iodide interference. The organic liquid ion-exchange type does not exhibit these interferences, but it is not stable and does not behave suitably in the turbid, gritty rock solutions necessary for rapid geochemical exploration work. The electrode used for the following study was of the solid membrane type.

Experimental

Equipment

An Orion model 401 pH/expanded mV meter was used together with an Orion model 90-02-00 double-junction sleeve-type reference electrode. A single-junction electrode with chloride filling solution should not be used.

Commercially available chloride solid membrane electrodes cost from $150.00 to $250.00. It is possible to construct, from readily available materials and at a cost of about $3.00, an electrode which possesses all the characteristics of commercial models.[2,3] Good membranes can be made by the use of dental acrylics as the inert binder to hold any analytically precipitated silver chloride or mixed silver chloride–silver sulphide precipitate (the active material). The finished membrane can be cemented to the end of a glass or plastic tube and the tube is subsequently filled with 10^{-3} M chloride (NaCl) filling solution. Electrical

Table 1 Comparison of general operating characteristics of typical laboratory constructed and commercial electrodes

Electrode stability, for 40 ppm Cl^-		
	$\bar{x}$, mV	Sx, mV
Clc	34·9 (80 min)	1·2 (n=16)
Cll	28·6 (115 min)	0·5 (n=23)
Cls	79·0 (75 min)	1·2 (n=11)
Electrode response time, for 40 ppm Cl^-, sec		
Clc	10 – 20	
Cll	10 – 20	
Cls	5 – 10	
Average mV/decade concentration change, linear portion		
Clc	52	
Cll	48	
Cls	48	

Working range (linear portion), 4–40 000 ppm, all electrodes.

Clc, commercial liquid filling solution electrode; Cll, laboratory constructed liquid filling solution electrode; Cls, laboratory constructed solid lead connected electrode.

contact is made with the membrane by inserting a reference electrode into the filling solution. The internal Ag–AgCl electrode from a broken glass pH electrode is convenient, if available. A direct electrical connexion can also be achieved by embedding a piece of silver, soldered on the end of the central wire of a shielded cable, in the plastic membrane[2]—this eliminates the need for a plastic or glass tube and the liquid filling solution. Table 1 gives comparative operating characteristic data of commercial and laboratory-constructed electrodes.

Recommended procedure for sample analysis*[2]

Weigh 1·0-g samples of rock powder (—200 mesh) into plastic beakers. Add 5 ml of distilled water to each beaker by use of a transfer pipette. Swirl the beakers for 2–3 min (or use a teflon-coated magnetic stirring-bar and stirrer).

Coat the chloride electrode membrane with a thin layer of silicone oil (e.g. Orion 94-00-03) and rub in well with a paper tissue, removing any excess. Rinse the electrode several times in distilled water. Place the chloride and reference electrodes in the test solution in the beaker. Swirl the beaker gently for 10 sec. Read the mV reading after allowing a response time-interval of an additional 20 sec. Always measure standards (and samples where possible) in the order of most dilute to most concentrated. Rinse the electrodes with a stream of distilled water from a wash bottle between sample solution measurements, but do not wipe the electrodes between readings. Check the standards after every fifth sample.

Behaviour of electrodes in rock solutions

Solutions obtained by the above procedure are turbid and contain very abrasive grains of angular silicate rock fragments. There is a temptation to use the electrodes to stir solutions rather than swirling the beaker. This should be avoided, since gouging and scratching of electrode surfaces may result. Even when a gently swirling motion is used, the question arises as to what effect the solution constituents have on the stability of electrode potentials. Table 2 presents electrode stability data for sample solutions.

As was recommended in the procedure, standards were run after every fifth sample determination. The electrode drift over the entire sample run (24 sample readings) for the 10 ppm standard was 8 and 6 mV, respectively, for the commercial and laboratory-constructed electrodes. This would introduce a maximum error of approximately 15% in the determination if standards were read only at the beginning or the end of the run.

Table 2 Electrode stability in rock solutions (ppm chloride in rock)

	Clc $\bar{x}$	Sx	n	Cll $\bar{x}$	Sx	n
Sa	17	0·3	3	18	0·3	3
Sb	19	0·3	3	17	0·0	3
Sc	19	0·3	3	18	1·2	3

Sa, b and *c* are three cuts of the same sample.

Chloride release study

Grain size

The chloride ion in a sample solution comes from tiny fluid inclusions in the rock, so, obviously, the grain size of the sample will have an effect on the number of fluid inclusions which are fractured. To investigate this effect, a representative rock sample was chosen and ground for various lengths of time in a mortar. Chloride determinations on sample solutions of the resulting powders indicate that —200-mesh material gives maximum chloride recovery. For a typical rock, a sample of —20-mesh grain size gave 14 ppm soluble chloride (24 ppm soluble chloride for a —200-mesh grain size).

Leach time

Rock powders ground to —200 mesh were leached for various lengths of time (up to 2 min) by use of the magnetic stirring and hand swirling methods. In both, a peak was reached in less than 20 sec, which is within the electrode response time. Further contact of the powder with the water (without stirring) for up to one week produced no appreciable change in the chloride content of the sample solution. Samples stirred magnetically and hand swirled sometimes yielded slightly different maximum values, for example, 19 ppm Cl^- when hand stirred against 22 ppm Cl^- when magnetically stirred, but in both cases the maximum was reached within 20 sec.

Weight of sample

Sample weights up to 15 g were leached by the procedure described above, a linear relationship

*Samples which have been stirred magnetically should not be compared with those extracted by swirling. A consistent procedure should be adopted.

Electrode surfaces should not be wiped during a set of measurements but rinsed with distilled water, any drops being gently shaken off.

between the amount of chloride leached and sample weight being obtained.

Sample contamination

Handling

Fist-size rock samples were broken in halves and rinsed in distilled water, one-half being ground after a minimum of handling and the other being subjected to frequent manipulation by perspiring hands, much as might occur under extremely hot field-handling of samples. Both powders were analysed, the results obtained being shown in Table 3.

Table 3 Effect of handling samples

	Clean, Cl^-, ppm	Frequent handling, Cl^-, ppm
Sa	5	5
Sb	6	14
Sc	6	10
Sd	15	16
Se	17	25
Sf	7	12

It is clear that care must be exercised to prevent extraneous contamination. Rinsing of rock specimens with distilled water before crushing will solve this problem.

Contamination by airborne sea salt

Unwashed samples from the Virgin Islands batholith have an average leachable chloride content more than 150 ppm greater than that of other similar plutons in the Greater Antilles, where most of the tests noted here have been made. The Virgin Islands batholith differs from other granodioritic intrusive rocks in the Greater Antilles in being exposed on numerous, very small islands, the best exposures occurring on or near a beach. The apparent explanation for this anomaly is accumulation of salt spray on the rock outcrops and incorporation of this material into the rock powder.

Application of the method

The chloride values obtained by the procedure described above will be largely a function of the volume and concentration of fluid inclusions in the sample.[4]

To test the method as an exploration tool the water-extractable chloride has been determined in samples from the Cala Abajo–Piedra Hueca porphyry copper zone and the nearby unmineralized Utuado and La Torrecilla intrusions in Puerto Rico (Fig. 1). Bradley[5] has pointed out

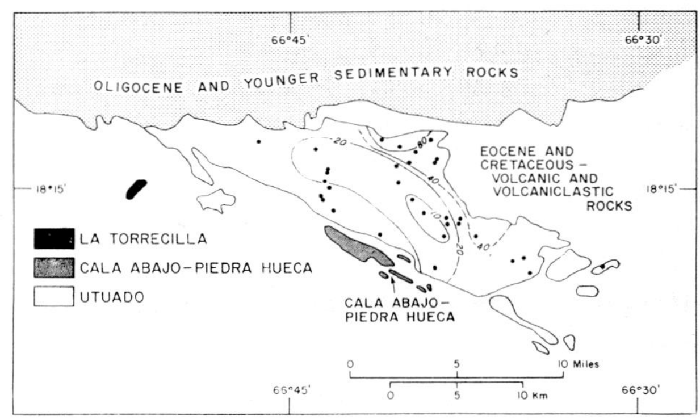

Fig. 1 Generalized geologic map of Utuado area, Puerto Rico, showing location of plutons considered in this study. Small dots in Utuado batholith are control points for contours of water-extractable chloride content. Geology after Briggs,[7] Nelson[9] and Mattson[8]

that the porphyry copper mineralization is found in the Eocene quartz-diorite porphyry stocks that occupy a linear alternation zone south of the Utuado batholith. Barabas[6] has suggested that the Utuado pluton was emplaced at about 60 m.y., and has shown some radiometric evidence for the geologic conclusion of Mattson[8] and Bradley[5] that the mineralized stocks are Eocene in age (40 m.y.). The La Torrecilla pluton has been shown by Barabas[6] to be 40 m.y. old, and, hence, it provides an unmineralized, approximately coeval, pluton for comparison with the mineralized Cala Abajo–Piedra Hueca plutons. No presently known commercial mineralization is associated with the Utuado batholith.

The water-extractable chloride in samples from each of these bodies is shown in Fig. 2. Water-

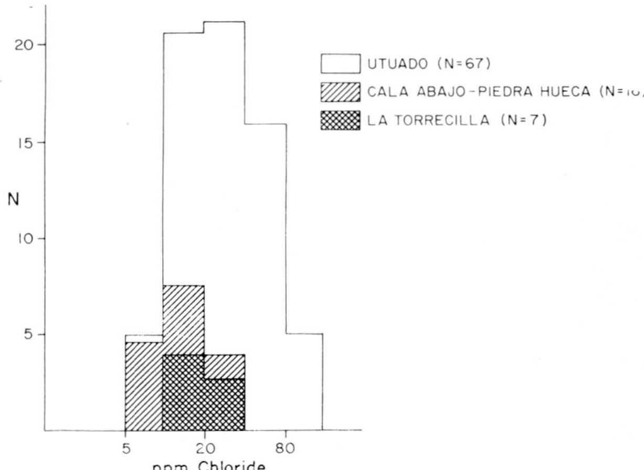

Fig. 2 Water-extractable chloride in intrusive rocks, central Puerto Rico

leachable chloride is roughly exponentially distributed in these rocks and the overlap between the plutons is almost complete. By use of the Wilcoxon–Manning–Whitney test[10] it can be shown that the La Torrecilla and Cala Abajo–Piedra Hueca results do not differ at the 80% confidence limit, although they do differ at the 90% confidence limit. Application of the same test to the Utuado and Cala Abajo–Piedra Hueca results indicates that the values for the two complexes are different at the 99% confidence level.

These results do not support the contention that mineralized zones or even 'fertile' intrusions are anomalously rich in water-leachable chloride. This point is further emphasized by the apparent lack of correlation when copper and zinc analyses of fresh mineralized rock are plotted against leachable chloride for the Cala Abajo–Piedra Hueca complex. A complete scattering of points is obtained.

The most obvious possible explanation for the observed similarity between the La Torrecilla and Cala Abajo–Piedra Hueca complexes, on the one hand, and their differences from the Utuado batholith, on the other, is the differing depths of emplacement of the intrusions. The Utuado consists largely of equigranular to sub-porphyritic rocks with a medium-grained groundmass. It was probably emplaced at depths sufficient to produce a relatively high water pressure during crystallization.[11] In contrast, the Cala Abajo and La Torrecilla plutons are strongly porphyritic with fine-grained, in some cases almost glassy, groundmass. Moreover, the Utuado batholith is more potassium-rich than most of its shallower counterparts, although some of the parts of the Cala Abajo–Piedra Hueca zone are very potassic.[11]

An even more graphic example of the apparent depletion of inclusion-bound chloride in high-level porphyritic stocks is given by comparison of the Minas de Oro and San Francisco stocks in Honduras. The San Francisco stock, which is porphyritic, contains an average of 11 ppm chloride (maximum 18 ppm), whereas the equigranular, deeper-level Minas de Oro stock contains an average of 64 ppm chloride (minimum 30 ppm).

Thus, it seems most likely that the hydrothermal phase, associated with the shallower intrusions, was not retained in the area of the intrusion long enough to form either abundant primary or secondary inclusions. Furthermore, it appears that the intense brecciation and alteration associated with porphyry mineralization, at least in this location, is not associated with an enrichment in local inclusion-bound chloride.

The spatial distribution of chloride values in the Utuado batholith, as shown by contours in Fig. 1, is surprisingly regular. These chloride values do not coincide with known compositional variations in their host rock. The presence of a low chloride core is surprising in the context of the common concept of chloride-rich hydrothermal solutions originating from residual magmas and migrating outward. Thin-section data do not suggest that this pattern reflects an increase in volume of inclusions outward from the batholith core accompanied by constant or even decreasing chloride content of the included fluids. At this stage it appears possible that the distribution of chloride values in intrusive complexes could give information on the configuration and composition of hydrothermal systems during late stages of magmatic crystallization, and in so doing might yield directional information related to the location of genetically associated mineralization.

Conclusion

Selective chloride ion electrodes are useful tools for the determination of water-leachable chloride in rock leach solutions. The required procedure is simple, rapid and reliable. The electrodes operate well even under the adverse conditions presented by the turbid, gritty sample solutions. Inexpensive laboratory-constructed electrodes give results which compare favourably with those obtained by the use of expensive commercial models. Problems which might cause sample contamination (handling, sea spray) can be eliminated by employing a distilled water rinsing of the rocks prior to crushing.

Application of this technique to three intrusive complexes, one of which is host to economic porphyry copper mineralization, indicates that there is no easily observed difference between the average chloride content of mineralized and unmineralized intrusions. High-level stocks, however, seem to retain less water-leachable chloride than do deeper-seated plutons. The distribution of chloride values in the Utuado batholith suggests that these data could yield information on the configuration of hydrothermal systems associated with late-stage magmatic crystallization.

Acknowledgment

J. C. Van Loon would like to acknowledge financial support for this project by the Geological Survey of Canada and by Dr. E. J. Brooker of X-ray Assay Laboratories, Toronto, Canada. S. E. Kesler acknowledges support from the Geological Survey of Canada and helpful assistance from R. A. Bradley, D. P. Cox and T. W. Donnelly in sample collection.

References

1. Stollery, G. Borcsik, M. and Holland, H. D. Chlorine in intrusives: a possible prospecting tool. *Econ. Geol.*, **66**, 1971, 361–7.
2. Van Loon, J. C. Laboratory construction and laboratory and field evaluation of thermoplastic and chloride-selective electrodes with liquid filling solution and solid–solid connections. *Analyt. chim. Acta.*, **54**, 1971, 23–8.
3. Mascini, M. and Liberti, A. An analytical study of a new type of halide-sensitive electrode prepared from silver halides and thermoplastic polymers. *Analyt. chim. Acta*, **47**, 1969, 339–45.
4. Kesler, S. E. Van Loon, J. C. and Moore, C. A. Evaluation of ore potential of granodiorite rocks using water-extractable chloride and fluoride. Paper submitted to *CIM Bull.*
5. Bradley, R. A. The geology of the Rio Vivi porphyry copper deposits, Puerto Rico. *Abstr. geol. Soc. Am.*, **3**, no. 7 1971, 511.
6. Barabas, A. H. K–Ar dating of igneous events and porphyry copper mineralization in west central Puerto Rico. *Abstr. geol. Soc. Am.*, **3**, no. 7 1971, 498.
7. Briggs, R. P. Provisional geologic map of Puerto Rico and adjacent islands. *U.S. geol. Surv. misc. Invest. Map* I-392, scale 1:240,000, 1964.
8. Mattson, P. H. Geologic map of the Adjuntas quadrangle, Puerto Rico. *U.S. geol. Surv. misc. Invest. Map* I-519, scale 1:20,000, 1968; Geologic map of the Jayuya quadrangle, Puerto Rico. *U.S. geol. Surv. misc. Invest. Map* I-520, scale 1:20,000, 1968.
9. Nelson, A. E. Geologic map of the Utuado quadrangle, Puerto Rico. *U.S. geol. Surv. misc. Invest. Map* I-480, scale 1:20,000, 1967.
10. Natrella, M. G. *Experimental statistics* (Washington, D.C.: National Bureau of Standards, 1963). n.p. (*Handbook* no. 91).
11. Donnelly, T. W. *et al.* Chemical evolution of the igneous rocks of the eastern West Indies: an investigation of thorium, uranium, and potassium distributions, and lead and strontium isotopic ratio. *Mem. geol. Soc. Am.* 130, 1971, 181–224.

543.422.2:546.56:550.84:553.43

Flameless atomic absorption and ion-sensitive electrodes as analytical tools in copper exploration

G. H. Friedrich, DR. RER. NAT., F.I.M.M.

W. L. Plüger, DR. RER. NAT.

E. F. Hilmer, DR. RER. NAT.

I. Abu-Abed, DIPL. GEOL.

All of the Institut für Mineralogie und Lagerstättenlehre, Technische Hochschule, Aachen, West Germany

Synopsis

The determination of copper in natural waters can be carried out with an atomic absorption spectrophotometer equipped with a deuterium background corrector and a heated graphite atomizer, as well as with copper-sensitive electrodes (stream water only). Both have definite advantages over methods which involve preconcentration and extraction. The determination of copper in soils, stream sediments, rocks and manganese nodules normally requires an initial preparation step. These materials can be analysed by use of the flameless atomic absorption technique directly as solid or in suspension.

A brief description of the instrumentation of both methods is given. Sensitivity and precision of replicate copper analysis, and the applicability of these techniques to geochemical exploration studies, are discussed.

The increasing activity in exploration for copper in different geological environments demands new, rapid, methods for the determination of copper and possible pathfinder elements in low concentrations. Several methods for the determination of copper in water, soils and rocks have been used with success, the techniques differing principally in the manner in which the sample solution is prepared. The proportion of the total copper extracted from a sample depends on the nature and intensity of the digestion. Exploration based on the use of dithizone or biquinoline for the determination of total copper or easily extractable copper has been widely used in field laboratories and at sample sites, whereas atomic absorption spectroscopy is the most widely used analytical technique in geochemical laboratories. One disadvantage of the method is the necessity for sample digestion in order to spray the sample as solution into a flame, where a portion of the solute is reduced to free atoms of its constituent elements.

Massmann[14] has developed a flameless sampling device by heating a graphite tube (as proposed by L'vov[11]) with electrical current passed through its wall. A modification of the Massmann design, as described by previous workers,[12, 13, 20, 21] installed

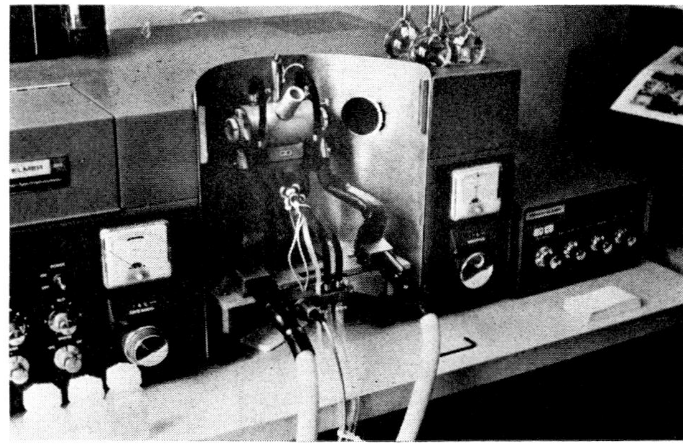

Fig. 1 Heated graphite atomizer HGA–70 installed in Perkin-Elmer 303 atomic absorption spectrophotometer

in a Perkin-Elmer atomic absorption spectrophotometer, was used in the present study.

As a contribution to the application of ion-selective electrodes in geochemical exploration, a cupric ion activity electrode, developed by Orion Research, Inc., has been used to measure the activity of free cupric ion in natural waters. Selectivity, precision and pH effects have been investigated.

Heated graphite atomizer technique

Instrumentation

A Perkin-Elmer Model 303 atomic absorption spectrophotometer, equipped with a deuterium background corrector and a Model HGA–70 heated graphite atomizer, was used for this study. A photograph of the graphite tube furnace is shown in Fig. 1, and a simplified cross-section diagram of the unit in Fig. 2. The removable

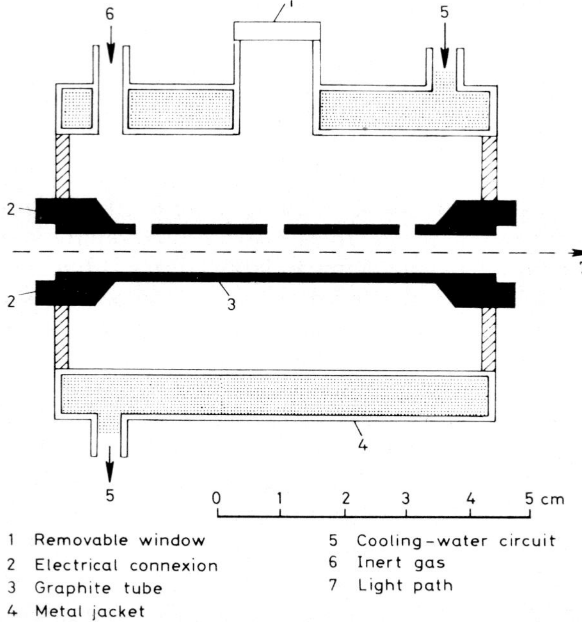

Fig. 2 Cross-section of graphite atomizer. After Massmann,[14] Manning and Fernandez[12] and Fernandez and Manning[13]

graphite tube is 5 cm long and 1 cm in diameter, open at both ends, and installed in a special furnace holder, which replaces the conventional burner equipment. The ends of the graphite tube are supported by two graphite rings and connected to an electrical power supply, which allows rapid heating to temperatures of about 2500–2600°C. The graphite tube is heated by passing high current at low voltage (up to 500 A at 10 V) directly through the tube. Drying, ashing and atomizing of the samples have to be carried out at different temperatures: the power supply therefore includes three separate timers and seven selectable temperature programmes. As inert gas nitrogen was used in our study to prevent burning and oxidation of the sample. A more detailed description of the instrument set-up was given earlier.[12, 13]

Analytical procedure

The sample to be analysed for copper or other elements is introduced into the graphite tube, as a solution or in suspension, with a micro-syringe or a micro-pipette through the removable window (Fig. 2), or as solid material with a special spoon and spoon-holder into one end of the tube. The syringe needle is inserted through a small hole of the upper part of the tube and the sample is sprayed to the bottom of the tube, where it is evaporated and then vaporized. Matrix effects can be eliminated or reduced by choosing definite ashing and atomizing temperatures, and by use of a deuterium background corrector. Normal sample sizes are from 5 to 100 μl, depending on the copper content of the sample.

Sensitivity and reproducibility

The sensitivity of the heated graphite atomizer method is about 1000 times greater than the nebulizer flame technique, depending on the fact that the atoms will be present in the light-path for a longer time. In Table 1 the absorption and absorbance values for different amounts of copper at various wavelengths, and the standard deviations for replicate analysis, are shown. The results of the measurement are given in absolute quantities.

Determination of copper in solid material by flameless atomic absorption

Sample preparation

A small amount (2–100 mg) of solid material, ground to a fine powder, is placed at the bottom of the graphite tube with a special spoon. With the insensitive analytical line at a wavelength of 2441 Å for the determination of copper only 2 mg should be taken for analysis if the expected copper content is in the region of 250 ppm. Samples with higher copper values can be analysed directly only if less than 2 mg of the sample is used.

To avoid the difficulties of weighing and handling such small amounts of solid material (which may not be representative) the following method was used in our study: 50 mg of the solid

Table 1 Standard deviations of copper determinations with heated graphite atomizer technique at various wavelengths

Standard solution used, ppm	Aliquot taken, μl	Absolute Cu content, ng	Wavelength, Å	n	Rec. units, $\bar{X}$, mm	Absorption, %	Absorbance	Standard deviation, mm	Relative standard deviation, %
10	10	100	2492·2	21	103	18·7	0·0899	5·5	5·3
1	50	50	2492·2	21	52	9·8	0·0448	3·8	7·3
1	5	5	3274·0	21	183	34·2	0·1818	13·0	7·1
0·1	50	5	3274·0	21	180	34·0	0·1805	10·0	5·6
0·01	50	0·5	3247·5	21	76	13·8	0·0645	6·6	8·7

material to be analysed was suspended in a 1:1 mixture of demineralized water and ethanol by use of a magnetic stirrer. Aliquots of this suspension were sprayed into the graphite tube with a micro-syringe. This preparative technique has striking advantages in that samples with more than 250 ppm copper or other elements can be analysed and precision is much better because of the homogeneous nature of the suspension, which is stirred continuously during replicate determinations. The standard deviation for copper analysis of a single manganese nodule was 10% (Table 2), with 100 mg of sample and suspension in a mixture of 5 ml demineralized water and 5 ml ethanol. For replicate analysis aliquots of 10 μl were used.

Determination of copper in soils, stream sediments and bog material

Soil samples from Imsbach, Germany (copper mineralization in porphyries), stream sediments from the Osor area, Spain (vein-type fluorite mineralization with base-metal sulphides) and bog material from Sackville, New Brunswick, Canada ('copper swamp') were analysed by use of the graphite atomizer technique (Table 3). Different weights of samples were taken because of the differences in copper content of the chosen material, known from former analysis (see column 9). An initial suspension of 100 or 1000 mg of sample in a mixture of 50 ml demineralized water and 50 ml ethanol was prepared for samples with more than 1% copper. From this suspension 1 ml was pipetted into 5 ml water and 5 ml ethanol, and aliquots of 10 μl were sprayed into the graphite tube.

Nitrogen was used as spoil gas, the temperature of drying was 100°C (for 40 sec), programme 6 was used for ashing (2–5 min at 750°C), and atomizing was carried out for 20–30 sec at 2300°C (8 V). The analytical curve for the determination of copper with the HGA technique at 2492 Å is shown in Fig. 3.

Replicate analysis of soil and bog material can be carried out without difficulty and with sufficient precision, but a decrease of the height of signals obtained for replicate determinations of copper in stream sediments has been observed. This may be due to the presence of large amounts of silicates. The effect can be avoided by baking out the graphite tube at maximum temperature several times after each determination. Following this procedure, all particles of the sample are completely burned and removed, and no 'memory effect' influences the next determination.

The copper values obtained by the HGA technique are in good agreement with previous analytical data.

Table 2 Standard deviation for replicate copper analysis of single manganese nodule by use of a suspension with HGA–70 heated graphite atomizer technique

Sample	Average Cu content, ng	Number of replicate analysis	Standard deviation, ng	Relative standard deviation, %
D3	180	21	18	10

Determination of copper in rocks

Copper was determined in the *G–2* rock standard (Westerly granite) by use of the HGA technique. 1 g of the sample was ground to a fine powder and suspended in 5 ml demineralized water and 5 ml

Table 3 Copper values of different samples analysed by graphite atomizer technique

Sample Locality	No.	Material	Sample weight, mg	Volume of suspension, ml*	Aliquot taken, μl	Replicate determination	Cu, ppm HGA–70	Flame AA
Imsbach	1/0	Soil	100	20	10	5	1040	900
Imsbach	0+5	Soil	100	10	50	5	262	300
Imsbach	0+100	Soil	1000	10	50	5	2·2	5
Osor	110	Stream sediment	1000	10	50	5	12·2	15
Osor	123	Stream sediment	1000	10	50	5	9·6	10
Sackville	1	Bog	100	100/11†	10	5	6·7%	Up to 10%‡
Sackville	2	Bog	100	100/11†	10	5	5·3%	2–5%‡

*1 : 1 mixture of demineralized water : ethanol; †Initial suspension of 50 ml demineralized water and 50 ml ethanol (from this suspension 1 ml was added to 5 ml demineralized water and 5 ml ethanol); ‡Analytical data by R. W. Boyle, Geological Survey of Canada.

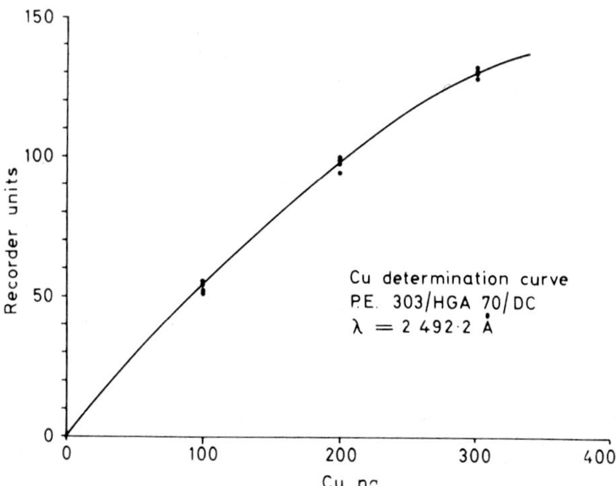

Fig. 3 Analytical curve for determination of copper with heated graphite atomizer at 2492 Å

ethanol. 50- and 100-μl aliquots of the suspension were taken for each replicate analysis. The instrumental set-up and atomizing technique were the same as described before, and special care was taken in baking out the graphite tube at 2600°C after each determination. This procedure was repeated up to five times until no fumes could be observed during the heating period.

Flanagan[6] has carried out spectrochemical determination of copper of the *G–2* rock standard, the data given by him lying in the range 9–14 ppm Cu. The results of our replicate analysis with the HGA technique and procedure as described above are shown in Table 4 (average copper content was 4·5 ppm).

To increase the intensity of the digestion of the rock sample before atomization in the graphite tube, 1 g of the sample was suspended in 10 ml 20% HF. 50-μl aliquots of the suspension were sprayed into the tube and ashed for 5 min at 230°C (programme 3), 50 μl of 20% HF (pure) then being added to the tube. The ashing process was repeated at 230°C, atomization being carried out at 2300°C (8 V). By use of this analytical procedure an average content of 8 ppm copper was obtained for the *G–2* rock standard. This procedure cannot be carried out frequently owing to corrosion of the instrumentation by HF. The direct determination of copper or other elements in silicate-bearing material by use of the HGA technique therefore cannot be recommended at present.

Table 4 Copper values for G–2 rock standard analysed by heated graphite atomizer technique

Sample	Sample weight, mg	Volume of suspension, ml	Aliquot taken, μl	Replicate determination	Cu, ppm HGA–70 *	†	‡
G–2	1000	10*	50	5	4·5		
rock	1000	10*	100	5	4·5		9–14
standard	1000	10†	50	5		8	

*HGA–70 technique: 1 : 1 suspension of demineralized water/ethanol. †HGA–70 technique: sample suspended in 20% HF, with addition of 20% HF (pure) to graphite tube after first ashing process. ‡Analytical data by Flanagan.[6]

Table 5 Copper values for manganese nodules from Central Pacific analysed by graphite atomizer technique (for comparison copper data for a manganese concretion from B horizon of a New Brunswick soil profile are given)

Sample Locality	No.	Sample weight, mg	Volume of suspension, ml	Aliquot taken, µl	Replicate determination	Cu, HGA–70, ppm
Pacific	*D3*	100	10	10	21	1800
Pacific	*D11*	100	10	10	5	1710
Pacific	*D13/B*	100	10	10	5	1500
New Brunswick	1	500	10	10	5	102

Determination of copper in manganese nodules

Manganese nodules from the Central Pacific taken during the 1971 R.V. *Prospector* expedition are being analysed for copper and other elements by use of the HGA technique. Table 5 shows the results of copper analysis of some selected manganese nodules compared with the copper value of manganese concretions from the *B* horizon of a New Brunswick soil profile (see Friedrich and co-workers[7]). The instrument settings are similar to those used for copper determinations in soils.

Analysis of natural waters

Heated graphite atomizer technique

Low concentrations of copper, zinc, cadmium, lead, cobalt and nickel have been determined directly in natural waters with the HGA–70 heated graphite atomizer. The water samples were injected without preconcentration or extraction into the graphite tube with Eppendorf micropipettes (various volumes). Standard solutions were used to test the instrument and to compile calibration curves. The time required for each determination was 100 sec, including sample drying, and the coefficient of variation was calculated as 4·4%. Precision values for several other elements (Zn, Cd, Pb, Co, Ni) ranged from 3·8 to 5·2%.

Table 6 Absorptions obtained from 1 ng (1×10^{-9} g) of several elements by use of heated graphite atomizer technique

Element	Absorption, %
Zn	80
Cd	60
Pb	20
Cu	17
Co	5
Ni	4

Analytical curves for the determination of copper, zinc, cadmium, lead, nickel and cobalt (Fig. 4) show the low detection limits, especially for zinc and cadmium. The percentage absorption obtained from 1-ng samples of the elements is given in Table 6.

For the analysis of stream water samples the instrument settings for the copper determination were the same as described before (model 303 Perkin-Elmer AAS/HGA–70/deuterium background corrector) at 3247 Å and slit 4. Programme 7 (1·5 min at 1100°C) was used for ashing, and atomization was carried out for 20 sec at 2300°C (8 V).

The copper contents of selected stream water samples from the Ramsbeck area, Rheinisches Schiefergebirge, Germany (containing lead–zinc–copper vein-type deposits), were found to be in the range 0·4–6·5 ppb. Stream water samples from the Erzweiler copper district, Germany, contained from 1·3 to 6·8 ppb Cu.

Attempts were made to analyse sea water samples at the same instrumental settings, but the measurements were influenced by the high chloride level of the samples. Owing to the fact that copper will be partly vaporized as $CuCl_2$ at 933°C[22] during the ashing and initial stage of atomization, the copper values of the samples obtained by this method may be too low. To discover the influence of the chloride ion a series of measurements was carried out after the addition of NaCl to copper standard solutions. Fig. 5 shows the effect of sodium chloride on the signals of copper standard solutions. On adding 10 µl of 10% NaCl solution to 10 µl of 100 ppb copper standard solution the absorbance values decreased from about 0·135 to 0·075.

The sea water samples analysed had an apparent copper content between 0 (no Cu detected) and 30 ppb, but the real copper values may be higher (up to 60 ppb, using a factor 2

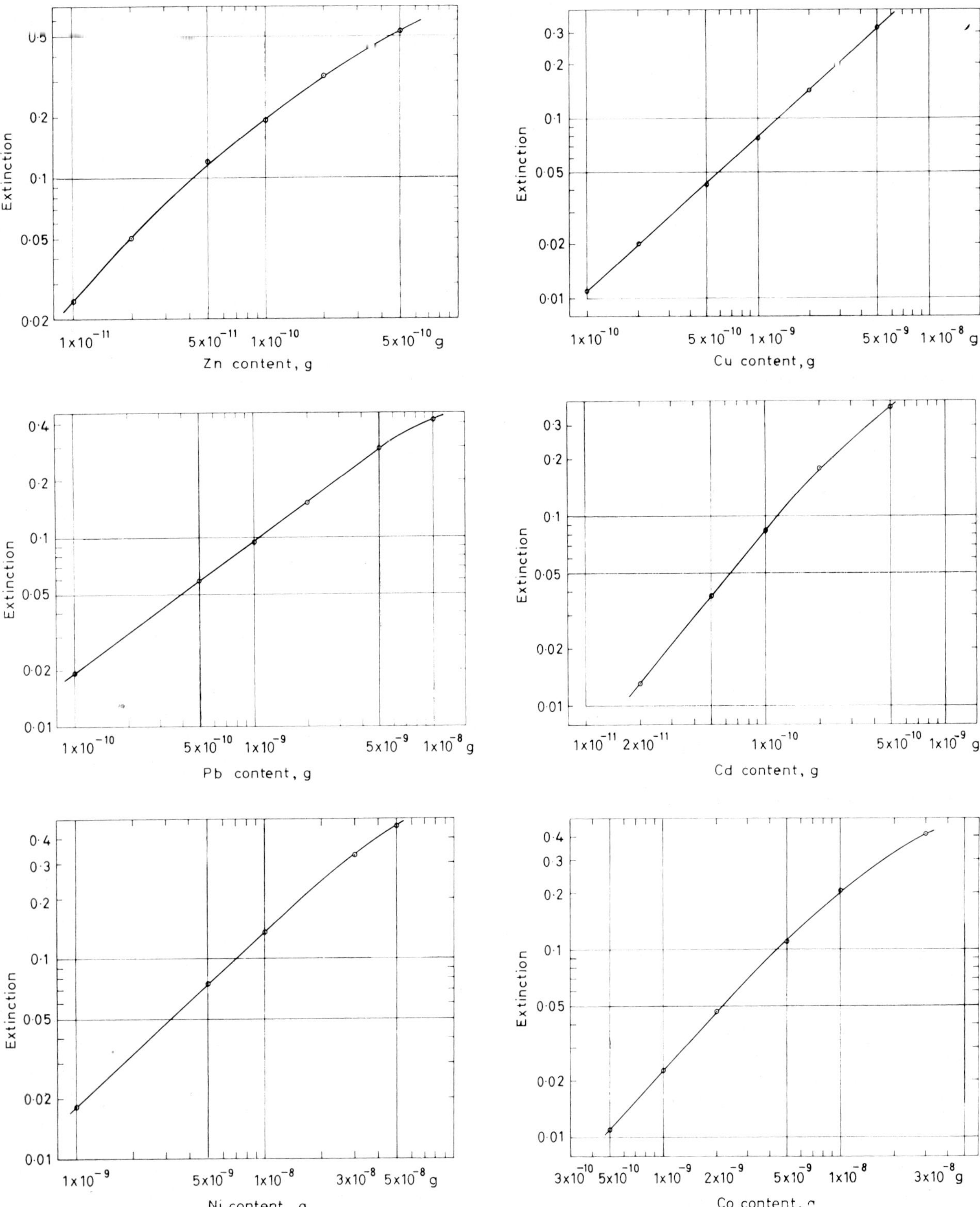

Fig. 4 Analytical curves for determination of copper, zinc, lead, cadmium, nickel and cobalt with graphite atomizer technique (analytical wavelength: Cu, 3247 Å; Zn, 2138 Å; Cd, 2288 Å; Pb, 2833 Å; Co, 2407 Å; Ni, 2320 Å)

according to the results shown in Fig. 5), depending on the NaCl content of the sea water samples and the corresponding loss of copper during analysis.

Cupric sensitive electrode

Various ion-sensitive electrodes have been used in analytical chemistry in the last few years. The theory of the method, the analytical techniques and the applications of 'ion-selective electrodes' have already been discussed.[4] Of all the ion-selective electrodes developed in recent years, the fluoride electrode has probably been the one most used in geochemical exploration studies.[2, 8] The application of ion-sensitive electrodes in explora-

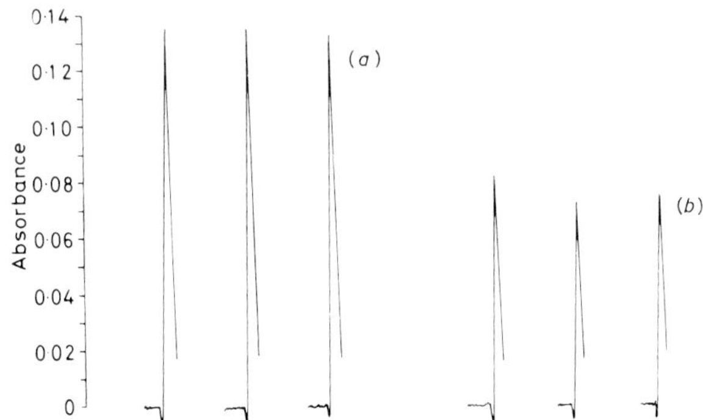

Fig. 5 Effect of sodium chloride on determination of copper by use of heated graphite atomizer technique: (a), 10 μl of 100 ppb copper standard solution (without NaCl); (b) 10 μl of 10% NaCl solution added to 10 μl of 100 ppb copper standard solution

tion geochemistry will certainly increase in the near future, especially in stream water surveys and in marine exploration studies. Studies by Spencer and Brewer (see Durst,[4] p. 393), who used a unique device for measuring continuous F^-/Cl^- profiles in sea water, have shown that special devices are feasible for deep-sea monitoring and that chemical determinations can be carried out directly from submersibles. Measurements of variations of base-metal concentration, in addition to the determination of fluoride/chloride ratios, may help to localize areas of submarine volcanic activity or submarine mineral deposits.

In the present study a cupric ion activity electrode was used to measure the activity of free cupric ion in stream water and in sea water. The Orion 94–29 cupric sensitive electrode is a solid state membrane electrode with a mixed cupric–silver sulphide sensing element. This electrode is used in conjunction with a reference electrode and a direct reading millivoltmeter. The Nernst equation describes the relation between the cupric ion activity in solution and the electrode potential:

$$E = E_0 + 29{\cdot}6 \log \text{activity } Cu^{2+} \quad (\text{mV})$$
$$(25^\circ \text{ solution temperature})$$

Hence, a tenfold change of cupric activity in the sample solution will change the electrode potential by 29·6 mV.

Certain substances—for example, hydroxides, sulphides and acetates—which precipitate or complex cupric ions will decrease the amount of free cupric ions. Furthermore, some cations poison the sensing element and must be absent from the sample solution. Table 7 gives a compilation of the possible interfering materials. The extent of the interference depends (1) on the concentration of the interfering ions, (2) on the pH of the sample solution, (3) on the ionic strength and (4) on the total cupric ion concentration. The lower detection limit of the method with neutral solutions is 10^{-8} M Cu^{2+} (0·6 ppb Cu).[16]

The background copper values of stream waters range from 0·9 to 12 ppb.[19] Therefore, copper contents in the region of 1 ppb should be detectable during geochemical stream water surveys. Preliminary studies with the cupric electrode have shown that small amounts of copper can be detected only in the neutral range of about pH 7. This is in accordance with data given by Orion Research[16] regarding the lower detection limit of the copper electrode (10^{-8} Cu^{++} in neutral solutions). The upper limit for cupric ions in the neutral range is 60 ppb Cu, higher amounts of cupric ions precipitating as $Cu(OH)_2$.

Previous investigations having shown that a constant pH of the solution is necessary for the measurement of cupric ion activities in the range 1–50 ppb, the measurements had to be carried out in buffered solutions. If di-sodium hydrogen phosphate–potassium di-hydrogen phosphate is used as buffer (pH 7),* the cupric ions may be precipitated by the phosphate. The result is a decrease of the cupric ion activity.

The effect of precipitation of cupric ions by phosphate was tested as follows: a 10 ppb copper

Table 7 Effects of interfering materials in potentiometric copper determination using the Orion 94–29 cupric sensitive electrode

Interfering material	Effect
Chloride ions	Decrease cupric ion activity if $[Cu^{2+}]\,[Cl^-]^2 > 1{\cdot}6.10^{-6}$
Bromide ions	Decrease cupric ion activity if $[Cu^{2+}]\,[Cl^-]^2 > 1{\cdot}3.10^{-12}$
Ferric ions	Interfere if $[Fe^{3+}] > 0{\cdot}1\ [Cu^{2+}]$
Mercury and silver ions	Poison electrode sensing element
EDTA, acetate, citrate and ammonia ions and amino acids	Complex cupric ions
Hydroxide, sulphide and phosphate ions	Precipitate cupric ions
Iodide and cyanide ions and ascorbic acid	Reduce cupric ions to cuprous ions

*Titrisol a product manufactured by Merck.

standard solution and a 50 ppb copper standard solution were diluted with buffer solution (1:1), and the cupric ion activity was measured after certain intervals. After 1 h no difference in the potential readings could be observed—indicating that free cupric ion activities were identical in both standard solutions!

To avoid the effect of copper precipitation by phosphate the Titrisol buffer was diluted 1:100 with demineralized water. This buffer was used to prepare the copper standard solutions (1, 10 and 50 ppb), which gave potential readings of 8, 25 and 40 mV. The difference between the 1 and the 10 ppb standard is only 17 mV, i.e. less than the expected ideal Nernst slope of 29·6 mV.

Measurements of the cupric ion content in stream water samples were carried out with the 1:100 diluted buffer, 10- or 20-ml aliquots of the samples being mixed 1:1 with this buffer. The readings of the electrode potentials were unsatisfactory because of a remarkable fluctuation in the instrument needle on the expanded millivolt scale, so no exact measurements could be made and only very rough estimations of the copper content were possible. Standard solutions have to be run after each determination. In Table 8 the range of copper obtained by measurements with the electrode is compared with the results of copper determinations with the heated graphite atomizer technique.

Table 8 Range of copper in different water samples from Ramsbeck area, Germany, analyzed with the electrode and heated graphite atomizer technique

Sample	Copper, ppb Electrode	HGA-70
RW125	0·5–5	2·8
RW135	1·5–3	1·6
RW136	1·5–4	3·6
RW166	5 –12	6·5

The copper content of sea water at 35‰ salinity was determined by Slowey[18] by use of neutron activation analysis, by which he found 0·9 ppb. As already mentioned, the solid state electrode is usable down to total cupric ion concentrations of 10^{-8} M (0·6 ppb), but is interference-free only with respect to the usual divalent cations.[5] Interferences occur[16] if $[Cu^{2+}] \times [Cl^-]^2 > 1 \cdot 6 \times 10^{-6}$ and $[Cu^{2+}] \times [Br^-]^2 > 1 \cdot 3 \times 10^{12}$.

Culkin[3] reported a chloride content of $1 \cdot 94 \times 10^7$ ppb ($5 \cdot 5 \times 10^{-1}$ M Cl^-), and Morris and Riley (see reference 3) reported a bromide content of $6 \cdot 73 \times 10^4$ ppb ($8 \cdot 4 \times 10^{-4}$ M Br^-) in sea water. At these chloride and bromide levels a cupric ion determination with the electrode is possible only if $[Cu^{2+}] < 130$ ppb.

Lewis and Goldberg[9] mentioned an iron level of 3·4 ppb in sea water. Ferric ions affect the membrane electrode surface if the ferric ion level is greater than one-tenth of the cupric ion level.[16] The cupric ion concentration in sea water must, therefore, be at least 38·7 ppb to be measured potentiometrically, and the possible range of measurement is between 38·7 and 130 ppb. None of the sea water samples available to us in the present study has copper contents in this range. The application of the sensitive cupric electrode in marine exploration is thus of doubtful value.

Summary and conclusions

Determinations of copper in stream water, ocean water, stream sediments, soils, rocks and manganese nodules have been carried out by use of the heated graphite atomizer technique. One of the main advantages of this technique is that small amounts of solution (about 25 μl for one determination) are needed. So numerous water samples can be sampled during an exploration survey in addition to stream-sediment sampling in areas where the determination of copper with biquinoline or the copper-sensitive electrode at the sample site is impossible because of difficult terrain. The HGA method may be used with success in marine geochemical exploration surveys. Sea water samples and marine sediment samples (including manganese nodules) can be analysed quickly on board in order to guide the exploration campaign. The possibility of using suspensions has the definite advantage of avoiding methods involving preconcentration and extraction.

The most striking advantages of the heated graphite atomizer technique are (1) rapid analysis (from 2 to 5 min); (2) preconcentration or extraction is not necessary; (3) samples can be analysed in solution, suspension or solid phase; (4) one determination requires a very small amount of sample (from 5 to 100 μl solution or suspension or from 1 to 10 mg solid material; and (5) the absolute analytical sensitivity of the HGA technique is about 1000 times better than the nebulizer flame method.

The use of cupric-sensitive electrodes in stream water surveys can be recommended if rough estimates only of the copper content are necessary. But the application is restricted to waters with low levels of interfering ions. For measurements at the sample site a 1:100 diluted Titrisol buffer may be used.

The application of the electrode in marine geochemical research is very doubtful at present.

As was mentioned by Durst,[5] an ideal electrode should be available which should be 'capable of good response in the parts per billion range, tolerant of massive excess of Na^+, Mg^{++}, Ca^{++}, K^+, Cl^- and Br^-, and should have a fairly low impedance for working in this salty environment'.

Acknowledgment

The authors express their grateful thanks to the Deutsche Forschungsgemeinschaft for their financial help for setting up the instrumentation. They also wish to thank the Metallgesellschaft, AG, Frankfurt/Main, for making available manganese nodules from the 1971 expedition to the Central Pacific, and to Dr. Anil Kumar Ghosh, research associate in their group, for running check analyses.

References

1. Butler, J. N. Thermodynamic studies. In reference 4, 143–89.
2. Chrt, J. Methoden der Prospektion und Erkundung auf Flussspat in der ČSSR. *Berg- u. hüttenm. Mh.*, **114**, 1969, 459–64.
3. Culkin, F. The major constituents of sea water. In *Chemical oceanography* Riley, J. P. and Skirrow, G. eds. (London: Academic Press, 1965), vol. 1, 121–61.
4. Durst, R. A. ed. *Ion-selective electrodes* (Washington, D.C.: National Bureau of Standards, 1969), 452 p. (*Spec. Publ.* 314)
5. Durst, R. A. Analytical techniques and applications of ion-selective electrodes. In reference 4, 375–414.
6. Flanagan, F. J. U.S. Geological Survey silicate rock standards. *Geochim. cosmochim. Acta*, **31**, 1967, 289–308.
7. Friedrich, G. Rosner, B. and Demirsoy, S. Erzmikroskopische und mikroanalytische Untersuchungen an Manganerzkonkretionen aus dem pazifischen Ozean. *Mineral. Deposita*, **4**, 1969, 298–307.
8. Friedrich, G. H. and Pluger, W. L. Geochemical prospecting for barite and fluorite deposits. In *Geochemical exploration* (Montreal: CIM, 1971), 151–6. (*CIM Spec. vol. 11*)
9. Lewis, G. J. and Goldberg, E. D. Iron in marine waters. *J. marine Res.*, **13**, 1954, 183–97.
10. Light, T. S. Industrial analysis and control with ion-selective electrodes. In reference 4, 349–74.
11. L'vov, B. V. The analytical use of atomic absorption spectra. *Spectrochim. Acta*, **17**, 1961, 761–70.
12. Manning, D. C. and Fernandez, F. [J.] Atomization for atomic absorption using a heated graphite tube. *Atom. absorp. Newsletter*, **9**, 1970, 65–70.
13. Fernandez, F. J. and Manning, D. C. Atomic absorption analyses of metal pollutants in water using a heated graphite atomizer. *Atom. absorp. Newsletter*, **10**, 1971, 65–9.
14. Massmann, H. Vergleich von Atomabsorption und Atomfluoreszenz in der Graphitküvette. *Spectrochim. Acta*, **23B**, 1968, 215–26.
15. Oehme, F. Direkt-potentiometrische Konzentrationsbestimmungen mit ionensensitiven Elektroden. Paper presented at ILMAC meeting, Zurich, 1968.
16. *Orion-ionalyzer instruction manual* (Cambridge, Mass.: Orion Research Inc., 1968), 20 p.
17. Paus, P. E. The application of atomic absorption spectroscopy to the analysis of natural waters. *Atom. Absorp. Newsletter*, **10**, 1971, 69–71.
18. Slowey, J. F. The distribution of copper, manganese and zinc in the ocean using neutron activation analysis. *U.S. A.E.C. Rep.* TID 23295, Sect. 2, 1966, 113 p.
19. Turekian, K. K. The oceans, streams, and atmosphere. In *Handbook of geochemistry* Wedepohl, K. H. ed. (Berlin, etc.: Springer, 1969), vol. 1, 297–323.
20. Welz, B. Paper presented at the 2nd Atomic Absorption Spectroscopy Conference, Sheffield, 1969.
21. Welz, B. and Wiedeking, E. Bestimmung von Spurenelementen in Serum und Urin mit flammenloser Atomisierung. *Z. analyt. Chem.*, **252**, 1970, 111–7.
22. Lax, E. ed. *D'Ans Taschenbuch für Chemiker und Physiker 3rd edn* (Berlin: Springer, 1967), vol. 1, 1268 p.

Name index

Subject index

Designed and produced by Barnes Harris Associates and printed in England